T0135092

Lecture Notes in Computer Science 13892

Founding Editors

Gerhard Goos
Juris Hartmanis

The series Lecture Notes in Computer Science (LNCS), including its subseries Lecture Notes in Artificial Intelligence (LNAI) and Lecture Notes in Bioinformatics (LNBI), has established itself as a medium for the publication of new developments in computer science and information technology research, teaching, and education.

LNCS enjoys close cooperation with the computer science R & D community, the series counts many renowned academics among its volume editors and paper authors, and collaborates with prestigious societies. Its mission is to serve this international community by providing an invaluable service, mainly focused on the publication of conference and workshop proceedings and postproceedings. LNCS commenced publication in 1973.

Sergio Rajsbaum · Alkida Balliu ·
Joshua J. Daymude · Dennis Olivetti
Editors

Structural Information and Communication Complexity

30th International Colloquium, SIROCCO 2023
Alcalá de Henares, Spain, June 6–9, 2023
Proceedings

 Springer

Editors
Sergio Rajsbaum 🆔
National Autonomous University of Mexico
Mexico, Mexico

Alkida Balliu
Gran Sasso Science Institute
L'Aquila, Italy

Institut de Recherche en Informatique
Fondamentale
Paris, France

Dennis Olivetti
Gran Sasso Science Institute
L'Aquila, Italy

Joshua J. Daymude
Arizona State University
Tempe, AZ, USA

ISSN 0302-9743   ISSN 1611-3349 (electronic)
Lecture Notes in Computer Science
ISBN 978-3-031-32732-2   ISBN 978-3-031-32733-9 (eBook)
https://doi.org/10.1007/978-3-031-32733-9

This Springer imprint is published by the registered company Springer Nature Switzerland AG
The registered company address is: Gewerbestrasse 11, 6330 Cham, Switzerland

Preface

This volume contains the papers presented at SIROCCO 2023: 30th International Colloquium on Structural Information and Communication Complexity, held on June 6–9, 2023 in Alcalá de Henares, Spain.

SIROCCO is devoted to the study of the interplay between structural knowledge, communication and computing in decentralized systems of multiple communicating entities. Special emphasis is given to innovative approaches leading to better understanding of the relationship between computing and communication.

There were 44 submissions. Each submission was reviewed by 3 program committee members, and a few were reviewed by additional external reviewers. The committee decided to accept 22 papers. The program also included two special events, a session dedicated to Special Models of Computation, and a session dedicated to the 30th Anniversary of SIROCCO. The proceedings include one paper of the former session, and three papers of the latter session, which were reviewed by at least one reviewer.

The program committee selected the following two papers to share the Best Paper Award

- Lila Fontes, Mathieu Laurière, Sophie Laplante and Alexandre Nolin. The communication complexity of functions with large outputs.
- Stefan Schmid, Jakub Svoboda and Michelle Yeo. Weighted Packet Selection for Rechargeable Links in Cryptocurrency Networks: Complexity and Approximation.

and the following two papers to share the Best Student Paper Award

- Abir Islam, graduate student at the University of New Mexico, for the paper: Abir Islam, Jared Saia and Varsha Dani. Boundary Sketching With Asymptotically Optimal Distance and Rotation.
- Sameep Dahal, graduate student at Aalto University, Finland, for the paper: Sameep Dahal and Jukka Suomela. Distributed Half-Integral Matching and Beyond.

The session dedicated to Special Models of Computation was organized by Joshua J. Daymude (Arizona State University, USA), Andréa Richa (Arizona State University, USA) and Christian Scheideler (University of Paderborn, Germany). In this session, researchers discussed the histories and open problems for distributed computing models that connect computer science theory to other interdisciplinary aims. Roger Wattenhofer spoke about neural networks and blockchains, Yuval Emek presented the Stone Age model and other bio-inspired models, Frederik Mallmann-Trenn spoke on models based in neurology, and Joshua J. Daymude discussed the amoebot model of programmable matter.

This year we celebrated the 30th Anniversary of SIROCCO, with a full day of special talks, including personal anecdotes and memories related to SIROCCO and its community, organized by Alkida Balliu and Dennis Olivetti (both of Gran Sasso Science Institute, Italy). The speakers were Pierluigi Crescenzi (Gran Sasso Science

Institute, Italy), Pierre Fraigniaud (Université Paris Cité and CNRS, France), David Peleg (Weizmann Institute of Science, Israel), Michel Raynal (IRISA, University of Rennes, France), Nicola Santoro (Carleton University, Canada), Jukka Suomela (Aalto University, Finland), Sara Tucci (Paris-Saclay University, France) and Rotem Oshman (Tel-Aviv University, Israel).

We would also like to thank the keynote speakers Stefan Schmid (Technical University of Berlin, Germany), Bernadette Charron-Bost (École Normale Supérieure, France), Seth Gilbert (National University of Singapore, Singapore), and Michael Schapira (Hebrew University of Jerusalem, Israel), for their insightful talks, as well as Boaz Patt-Shamir (Tel Aviv University, Israel) for his featured talk as the recipient of the 2023 SIROCCO Innovation in Distributed Computing Prize.

We would like to thank the authors who submitted their work to SIROCCO this year and the PC members and subreviewers for their valuable and insightful reviews and comments. The SIROCCO Steering Committee, chaired by Magnús M. Halldórsson, provided help and guidance throughout the process. The EasyChair system was effectively used to handle the submission of papers and to manage the review process. Without all of these people it would not have been possible to produce these proceedings and the great conference program. We are very greatful also to the organization team, led by Antonio Fernandez Anta and Ernesto Jimenez Merino, who made the conference possible together with their local organization team.

June 2023

Sergio Rajsbaum
Alkida Balliu
Dennis Olivetti
Joshua J. Daymude

Organization

Program Committee

Emmanuelle Anceaume	CNRS/IRISA, France
Alkida Balliu	Gran Sasso Science Institute, Italy
Jérémie Chalopin	LIS, CNRS, Aix-Marseille Université, Université de Toulon, France
Yi-Jun Chang	National University of Singapore, Singapore
Joshua Daymude	Arizona State University, USA
Giuseppe Antonio Di Luna	University of Rome - Sapienza, Italy
Manuela Fischer	ETH Zurich, Switzerland
Luisa Gargano	Università di Salerno, Italy
Olga Goussevskaia	Federal University of Minas Gerais, Brazil
Maurice Herlihy	Brown University, USA
Taisuke Izumi	Osaka University, Japan
Tomasz Jurdzinski	University of Wrocław, Poland
Evangelos Kranakis	Carleton University, Canada
Mikel Larrea	University of the Basque Country UPV/EHU, Spain
Friedhelm Meyer Auf der Heide	University of Paderborn, Germany
Pedro Montealegre	Universidad Adolfo Ibáñez, Chile
Yasamin Nazari	University of Salzburg, Austria
Thomas Nowak	ENS Paris-Scalay, France
Boaz Patt-Shamir	Tel Aviv University, Israel
Andrzej Pelc	Université du Québec en Outaouais, Canada
David Peleg	Weizmann Institute of Science, Israel
Sergio Rajsbaum	Instituto de Matematicas, UNAM, Mexico; IRIF, Paris, France
Andrea Richa	Arizona State University, USA
Cesar Sanchez	IMDEA Software Institute, Spain
Christian Scheideler	University of Paderborn, Germany
Ulrich Schmid	Vienna University of Technology, Austria
Gadi Taubenfeld	Reichman University, Israel
Sébastien Tixeuil	Sorbonne University, France
Jara Uitto	Aalto University, Finland
Nitin Vaidya	Georgetown University, USA

Additional Reviewers

Aarts, Sander
Bamberger, Philipp
Bramas, Quentin
Carnevale, Daniele
Castenow, Jannik
Chrobak, Marek
Cicerone, Serafino
Cohen, Reuven
Cordasco, Gennaro
Das, Shantanu
Daymude, Joshua
De Marco, Gianluca
de Vos, Tijn
Del Pozzo, Antonella
Dereniowski, Dariusz
Devismes, Stéphane
Di Stefano, Gabriele
Emek, Yuval
Felber, Stephan
Feng, Weiming
Garg, Vijay
Ghinea, Diana
Grunau, Christoph
Harbig, Jonas

Hurfin, Michel
Jauregui, Benjamin
Kamei, Sayaka
Knollmann, Till
Koranteng, Ama
Lamani, Anissa
Łukasiewicz, Aleksander
Markou, Euripides
Moreno-Sanchez, Pedro
Naser Pastoriza, Alejandro
Ooshita, Fukuhito
Oshman, Rotem
Pai, Shreyas
Peltonen, Saku
Pourdamghani, Arash
Rapaport, Ivan
Rescigno, Adele
Scheideler, Christian
Schlögl, Thomas
Schmid, Laura
Sudo, Yuichi
Villani, Neven
Wada, Koichi

Contents

30th Anniversary and Special Models of Computations

Degree Realization by Bipartite Multigraphs

Amotz Bar-Noy[1], Toni Böhnlein[3], David Peleg[3(✉)], and Dror Rawitz[2]

[1] City University of New York (CUNY), New York, USA
`amotz@sci.brooklyn.cuny.edu`
[2] Bar Ilan University, Ramat-Gan, Israel
`dror.rawitz@biu.ac.il`
[3] Weizmann Institute of Science, Rehovot, Israel
`{toni.bohnlein,david.peleg}@weizmann.ac.il`

Abstract. The problem of realizing a given degree sequence by a multigraph can be thought of as a relaxation of the classical degree realization problem (where the realizing graph is simple). This paper concerns the case where the realizing multigraph is required to be bipartite.

The problem of characterizing degree sequences that can be realized by a bipartite (simple) graph has two variants. In the simpler one, termed BDR^P, the partition of the degree sequence into two sides is given as part of the input. A complete characterization for realizability in this variant was given by Gale and Ryser over sixty years ago. However, the variant where the partition is not given, termed BDR, is still open.

For bipartite multigraph realizations, there are again two variants. For BDR^P, where the partition is given as part of the input, a complete characterization was known for determining whether the bi-sequence is r-max-bigraphic, namely, if there is a multigraph realization whose underlying graph is bipartite, such that the *maximum* number of copies of an edge is at most r. We present a complete characterization for determining if there is a bipartite multigraph realization such that the *total* number of excess edges is at most t. As for the variant BDR, where the partition is not given, we show that determining whether a given (single) sequence admits a bipartite multigraph realization is NP-hard. On the positive side, we provide an algorithm that computes optimal realizations for the case where the number of balanced partitions is polynomial, and present sufficient conditions for the existence of bipartite multigraph realizations that depend only on the largest degree of the sequence.

1 Introduction

1.1 Background and Motivation

Degree Realization: This paper concerns a classical network design problem known as the *graphic degree realization (GDR)* problem. The number of

This work was supported by US-Israel BSF grant 2018043.

neighbors or connections of a vertex in a graph is called its *degree*, and it provides information on its centrality and importance. For the entire graph, the sequence of vertex-degrees is a significant characteristic which has been studied for over sixty years. The graphic degree realization problem asks if a given sequence of positive integers $d = (d_1, ..., d_n)$ is *graphic*, i.e., if it is the sequence of vertex-degrees of some graph. Erdös and Gallai [6] gave a characterization for graphic sequences, though not a method for finding a *realizing* graph. Havel and Hakimi [9,10] proposed an algorithm that either generates a realizing graph or proves that the sequence is not graphic.

Relaxed Degree Realization by Multigraphs: An interesting direction in the study of realization problems involves *relaxed* (or approximate) realizations (cf. [1]). Such realizations are well-motivated by applications in two wider contexts. In scientific contexts, a given sequence may represent (noisy) data resulting from an experiment, and the goal is to find a model that fits the data. In such situations, it may happen that *no* graph fits the input degree sequence exactly, and consequently it may be necessary to search for the graph "closest" to the given sequence. In an engineering context, a given degree sequence constitutes constraints for the design of a network. It might happen that satisfying all of the desired constraints *simultaneously* is not feasible, or causes other issues, e.g., unreasonably increasing the costs. In such cases, relaxed solutions bypassing the problem may be relevant.

In the current paper we focus on a specific type of relaxed realizations where the graph is allowed to have parallel edges, namely, the realization may be a *multigraph*. It is easy to verify that if *(multiple) self-loops* are allowed, then *every* sequence $d = (d_1, \ldots, d_n)$ whose sum $\sum_i d_i$ is even has a realization by a multigraph. Hence, we focus on the case where self-loops are not allowed.

The problem of degree realization by multigraphs has been studied in the past as well. Owens and Trent [13] gave a condition for the existence of a multigraph realization. Will and Hulett [17] studied the problem of finding a multigraph realization of a given sequence such that the underlying graph of the realization contains as few edges as possible. They proved that such a realization is composed of components, each of which is either a tree or a tree with a single odd cycle. Hulett, Will, and Woeginger [11] showed that this problem is strongly NP-hard.

Degree Realization by Bipartite Graphs: The *bigraphic degree realization (BDR)* problem is a natural variant of the graphic degree realization problem, where the realizing graph is required to be bipartite. The problem has a sub-variant, denoted BDR^P, in which *two* sequences are given as input, representing the vertex-degree sequences of the two sides of a bipartite realizing graph. (In contrast, in the general problem, a *single* sequence is given as input, and the goal is to find a realizing bipartite graph based on some partition of the given sequence.) The BDR^P problem was solved by Gale and Ryser [7,15] even before Erdos and Gallai's characterization of graphic sequences. However, the general problem – mentioned as an open problem over forty years ago [14] – remains unsolved today.

A sequence of integers $d = (d_1, ..., d_n)$ can only be *bigraphic*, i.e., the vertex-degree sequence of a bipartite graph, if it can be partitioned into two subsequences or *blocks* of equal total sum. The later problem is known as the *partition problem* and it is solvable in polynomial time assuming that $d_1 < n$ (which is a necessary condition for d to be bigraphic). Yet, BDP bears two obstacles. First, a sequence may have several partitions of which some are bigraphic and others are not. Second, the number of partitions may be exponentially large in n. Recent attempts on the BDR problem (see [2,3]) try to identify a small set of partitions, which are suitable to decide BDR for the whole sequence. Each partition in the small set is tested using the Gale-Ryser characterization. In case all of them fail the test, it is conjectured that no partition of the sequence is bigraphic. The conjecture was shown to be true in case there exists a special partition that (perfectly) splits the degrees into small and large ones.

Paralleling the above discussion concerning relaxed degree realizations by *general* multigraphs, one may look for relaxed degree realizations by *bipartite multigraphs*. This question is our main interest in the current paper.

1.2   Our Contribution

In this paper, we consider the problem of finding relaxed *bipartite multigraph* realizations for a given degree sequence or a given partition. That is, the relaxed realizations must fulfill the degree constraints exactly but are allowed to have parallel edges. (Self-loops are disallowed.)

To evaluate the quality of a realization by a multigraph, we use two measures. (i) The *total multiplicity* of the multigraph, i.e., the number of parallel edges. (ii) The *maximum multiplicity* of the multigraph, i.e., the maximum number of edges between any two of its vertices. As shown later, these measures are non-equivalent, in the sense that there are examples for sequences where realizations optimizing one measure are suboptimal in the other, and vice-versa.

For relaxed realizations by *general* multigraphs, it follows from the characterizations given, respectively, by Owens and Trent [13] and Chungphaisan [5], how to optimize the two measurements and find the respective multigraph realizations.

For relaxed realization by *bipartite* multigraphs, finding a realization for BDR^P (the given partition variant) that minimizes the *maximum* multiplicity follows from the characterization presented by Berge [12], and the BDR case (single sequence variant) was considered by us in [2]. In the current paper, we review the known literature on the problem of degree realization by general and bipartite multigraphs, strengthen the results of [2] on maximum multiplicity realizations for BDR, and also present additional results on multigraph realizations with bounded *total* multiplicity for BDR^P.

In more detail, Sect. 2 introduces formally the basic notions and measures under study. Section 3 presents known results on multigraph realizations with low *total* multiplicity of parallel edges. The problem was solved for general graphs by Owens and Trent [13]. We provide a characterization for bipartite multigraphs based on a given partition (BDR^P). Both characterizations are translated to Erdös-Gallai and Gale-Ryser conditions, respectively.

One necessary condition for a sequence $d = (d_1, \ldots, d_n)$ to be bigraphic is that it can be *partitioned*. If $d_1 < n$, this problem can be decided in polynomial time. However, for a multigraph realization to exists, the inequality $d_1 < n$ is not a necessary condition, and it follows that BDP^P is NP-hard. We review this matter in greater detail in Sect. 4 and discuss an output sensitive algorithm to generate all partitions of a given sequence which was presented in [2]. In case the number of partitions of a sequence is small, the algorithm allows us to find optimal realizations with respect to both criteria.

In Sect. 5, we discuss sufficient conditions for the existence of approximate bipartite realizations that depend only on the largest degree of the sequence of a given sequence.

2   Preliminaries

Let $d = (d_1, d_2, \ldots, d_n)$ be a sequence of positive integers in non-increasing order[1]. The *volume* of d is $\sum d = \sum_{i=1}^{n} d_i$. For a graph G, denote the sequence of its vertex-degrees by $\deg(G)$. Sequence d is *graphic* if there is a graph G such that $\deg(G) = d$. We say that G is a *realization* of d. Note that every realization of d has $m = \sum d / 2$ edges. Consequently, a graphic sequence must have even volume. In turn, we call a sequence of positive integers with even volume a *degree sequence*. We use the operator $\circ$ to define $d \circ d'$ as the concatenation of two degree sequences d and d' (in non-increasing order).

2.1   Multigraphs as Approximate Realizations

Let $H = (V, E)$ be a multigraph without loops. In this case, E is a multiset. Denote by $E_H(v, u)$ the multiset of edges connecting $v, u \in V$. If $|E_H(v, u)| > 1$, we say that the edge (v, u) has $|E_H(v, u)| - 1$ *excess* copies. Let E' be the set that is obtained by removing excess edges from E. The graph $G = (V, E')$ is called the *underlying graph* of H.

We view multigraphs as *approximate* realizations of sequences that are not graphic. Owens and Trent [13] gave a condition for the existence of a multigraph realization.

Theorem 1 (Owens and Trent [13]). *A degree sequence d can be realized by a multigraph if and only if $d_1 \leq \sum_{i=2}^{n} d_i$.*

To measure the quality of an approximate realization we introduce two metrics. First, the *maximum multiplicity* of a multigraph H is the maximum number of copies of an edge, namely

$$\mathsf{MaxMult}(H) \triangleq \max_{(v,w) \in E} \left(|E_H(v, w)| \right),$$

---

[1] All sequence that we consider are assumed to be of positive integers and in a non-increasing order.

and for a sequence d define

$$\mathsf{MaxMult}(d) \triangleq \min\{\mathsf{MaxMult}(H) : H \text{ realizes } d\} \ .$$

We say that a sequence d is r-*max-graphic* if $\mathsf{MaxMult}(d) \leq r$, for a positive integer r.

Second, the *total multiplicity* of a multigraph H is the total number of excess copies, namely

$$\mathsf{TotMult}(H) \triangleq \sum_{(v,w) \in E} (|E_H(v,w)| - 1) = |E| - |E'| \ ,$$

where E' is the edge set of the underlying graph of H. For a sequence d define

$$\mathsf{TotMult}(d) \triangleq \min\{\mathsf{TotMult}(H) : H \text{ realizes } d\} \ .$$

We say a sequence d is t-*tot-graphic* if $\mathsf{TotMult}(d) \leq t$, for a positive integer t.

2.2 General Multigraphs

Given a degree sequence d, our goal is to compute $\mathsf{MaxMult}(d)$ and $\mathsf{TotMult}(d)$.

First, observe that the best realization in terms of maximum multiplicity is not necessarily the same as the best one in terms of total multiplicity. See example in Fig. 1.

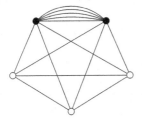   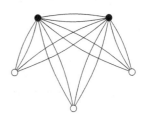

(a) Optimal **TotMult** realization G_1.   (b) Optimal **MaxMult** realization G_2.

Fig. 1. Optimal multigraph realizations for the sequence $d = (8^2, 4^3)$. On the left we have $\mathsf{TotMult}(G_1) = 4$ and $\mathsf{MaxMult}(G_1) = 5$, while on the right we have $\mathsf{TotMult}(G_2) = 7$ and $\mathsf{MaxMult}(G_2) = 2$.

Next, we iterate the characterization of Erdös and Gallai [6] for graphic sequence.

Theorem 2 (Erdös-Gallai [6]). *A degree sequence d is graphic if and only if, for $\ell = 1, \ldots, n$,*

$$\sum_{i=1}^{\ell} d_i \leq \ell(\ell - 1) + \sum_{i=\ell+1}^{n} \min\{\ell, d_i\} \ . \tag{1}$$

Theorem 2 implies an $\mathcal{O}(n)$ algorithm to verify whether a sequence is graphic. Chungphaisan [5] extended the above characterization to multigraphs with bounded maximum multiplicity as follows.

Theorem 3 (Chungphaisan [5]). *Let r be a positive integer. A degree sequence d is r-max-graphic if and only if, for $\ell = 1, \ldots, n$,*

$$\sum_{i=1}^{\ell} d_i \leq r\ell(\ell - 1) + \sum_{i=\ell+1}^{n} \min\{r\ell, d_i\} \ . \tag{2}$$

Notice the similarity to the Erdős-Gallai equations. Moreover, verify that $r \leq d_1$, for any r-max-graphic sequence d. It follows that $\mathsf{MaxMult}(d)$ can be computed in $\mathcal{O}(n \cdot \log(d_1))$.

The problem of finding a multigraph realization with low total multiplicity was solved by Owens and Trent [13]. They showed that the minimum total multiplicity is equal to the minimum number of degree 2 vertices that should be added to make the sequence graphic. We provide a simpler proof of their result.

Theorem 4 (Owens and Trent [13]). *Let d be a degree sequence such that $d_1 \leq \sum_{i=2}^{n} d_i$. Then, d is t-tot-graphic if and only if $d \circ 2^t$ is graphic.*

Proof. Let d be a degree sequence such that $d_1 \leq \sum_{i=2}^{n} d_i$. First, assume that d can be realized by a multigraph H with $\mathsf{TotMult}(H) \leq t$. Let F be the set of excess edges in H. Construct a simple graph G by replacing each edge $f = (x, y) \in F$ with two edges (x, v_f) and (y, v_f), where v_f is a new vertex. Clearly, this does not change the degrees of x and y and adds a vertex v_f of degree 2. Hence the degree sequence of G is $d \circ 2^{|F|}$. Also, G is simple. If $|F| < t$, then one may replace any edge in G with a path containing $t - |F|$ edges, yielding a graph with degree sequence $d \circ 2^t$.

Conversely, suppose the sequence $d \circ 2^t$ is graphic. Let G be a simple graph that realizes the sequence. Pick a degree 2 vertex v with neighbors x and y, replace the edges (v, x) and (v, y) with the edge (x, y), and remove v from G. This transformation eliminates one degree 2 vertex from G without changing the remaining degrees. But it may increase the number of excess edges by one (if the edge (x, y) already exists in G). Performing this operation for t times, we obtain a multigraph H with $\mathsf{TotMult}(H) \leq t$ and degree sequence d. □

The next corollary follows readily with Theorems 2 and 4.

Corollary 1. *Let t be a positive integer, and let $d' = d \circ 2^t$. Sequence d is t-tot-graphic if and only if, for $\ell = 1, \ldots, n + t$,*

$$\sum_{i=1}^{\ell} d_i' \leq \ell(\ell - 1) + \sum_{i=\ell+1}^{n+t} \min\{\ell, d_i'\} \ . \tag{3}$$

Owens and Trent [13] implicitly suggest to compute $\mathsf{TotMult}(d)$ by computing the minimum t such that $d \circ 2^t$ is graphic. Using binary search would lead to a running time of $\mathcal{O}(n \cdot \log(\mathsf{TotMult}(d)))$.

Several authors [16,18] noticed that the equations of Theorem 2 are not minimal. For a degree sequence d where[2] $d_1 > 1$, let $\text{box}(d) = max\{i \mid d_i > i\}$. If Equation (1) holds for the index $\ell = \text{box}(d)$, then it holds for index $\ell + 1$. To see this, consider the equations for the two indices and compare the change in the LHS and RHS. Observe that the RHS increases at least by $(\ell+1)\cdot\ell - \ell\cdot(\ell-1) = 2\ell$ while the LHS only increases by $d_{\ell+1} \leq \ell$. It follows that Equation (1) does not have to be checked for indices $\ell > \text{box}(d)$.

Observation 1 ([16,18]). *A degree sequence d is graphic if and only if, for $\ell = 1, \ldots, box(d)$,*

$$\sum_{i=1}^{\ell} d_i \leq \ell(\ell - 1) + \sum_{i=\ell+1}^{n} \min\{\ell, d_i\} \,. \tag{4}$$

On a side note, it is also known that only up to k many equations have to be checked where k is the number of different degrees of a sequence (cf. [12,16,18]).
Observation 1 helps to simplify Corollary 1.

Corollary 2. *Let t be a positive integer. Degree sequence d is t-tot-graphic if and only if, for $\ell = 1, \ldots, box(d)$,*

$$\sum_{i=1}^{\ell} d_i \leq \ell(\ell - 1) + \sum_{i=\ell+1}^{n} \min\{\ell, d_i\} + t \cdot \min\{\ell, 2\}. \tag{5}$$

Proof. Let d and t be as in the corollary. In case $d_1 = 1$, the sequence d is graphic, i.e., it is t-tot-graphic for any positive integer t.

Hence, assume that $d_1 > 1$. Also, let $d' = d \circ 2^t$. One can verify that Equations (5) are the (reduced) Erdös-Gallai inequalities of Observation 1 for d'. Moreover, $\text{box}(d) = \text{box}(d')$, and the claim follows. □

Corollary 2 implies a simple algorithm to compute $\mathsf{TotMult}(d)$. Let

$$\Delta_\ell(d) = \sum_{i=1}^{\ell} d_i - (\ell(\ell - 1) + \sum_{i=\ell+1}^{n} \min\{\ell, d_i\}),$$

for $\ell = 1, \ldots, n$, be the Erdös-Gallai differences of a degree sequence d. Also, let $\Delta_{\max}(d) = \max_{2\leq\ell\leq\text{box}(d)} \Delta_\ell(d)$. It follows that $t = \max\{\Delta_1, \Delta_{\max}/2\}$ implying a $\mathcal{O}(n)$ algorithm to calculate $\mathsf{TotMult}(d)$.

2.3 Bipartite Multigraphs

In this section, we start investigating whether a degree sequence has a bipartite realization, i.e., if it is *bigraphic* or not. Particularly, we are interested in multigraph realizations where the underlying graph is bipartite.

---

[2] If $d_1 = 1$, we define $\text{box}(d) = 0$. Note that in this case d is realized by a matching graph.

Let d be a degree sequence such that $\sum d = 2m$ for some integer m. A *block* of d is a subsequence a such that $\sum a = m$. Define the set of blocks as $B(d)$. For each $a \in B(d)$ there is a disjoint $b \in B(d)$ such that $d = a \circ b$. We call such a pair $a, b \in B(d)$ a balanced *partition* of d since $\sum a = \sum b$. Denote the set of all partitions of d by $\mathsf{BP}(d) = \{\{a, b\} \mid a, b \in B(d), \ a \circ b = d\}$. We say a partition $(a, b) \in \mathsf{BP}(d)$ is *bigraphic* if there is a bipartite realization $G = (A, B, E)$ of d such that $\deg(A) = a$ and $\deg(B) = b$ are the vertex-degree sequences of A and B, respectively.

Observe that, as in the case of general graphs, the best realization in terms of maximum multiplicity is not necessarily the same as the best one in terms of total multiplicity. See example in Fig. 2.

(a) Optimal TotMult^{bi} realization G_1.   (b) Optimal MaxMult^{bi} realization G_2.

Fig. 2. Optimal multigraph bipartite realizations for the sequence $d = (4^2, 2^2)$. On the left we have TotMult$^{bi}(G_1) = 2$ and MaxMult$^{bi}(G_1) = 3$, while on the right we have TotMult$^{bi}(G_2) = 3$ and MaxMult$^{bi}(G_2) = 2$.

Note that not every graphic sequence has a balanced partition. Yet, if d is bigraphic, then $\mathsf{BP}(d)$ is not empty. The Gale-Ryser theorem characterizes when a partition is bigraphic.

Theorem 5 (Gale-Ryser [7,15]). *Let d be a degree sequence and partition $(a, b) \in \mathsf{BP}(d)$ where $a = (a_1, a_2, \ldots, a_p)$ and $b = (b_1, b_2, \ldots, b_q)$. The partition (a, b) is bigraphic if and only if, for $\ell = 1, \ldots, p$,*

$$\sum_{i=1}^{\ell} a_i \leq \sum_{i=1}^{q} \min\{\ell, b_i\} \ . \tag{6}$$

We point out that Theorem 5 does not characterize bigraphic degree sequences. Indeed, if the partition is not specified, it is not known how to determine whether a graphic sequence is bigraphic or not. There are sequences where some partitions are bigraphic while others are not. Moreover, $|\mathsf{BP}(d)|$ might be exponentially large in the input size n.

We turn back to approximate realizations by bipartite multigraphs. A multigraph is bipartite if its underlying graph is bipartite. Analogue to above, we use the maximum and total multiplicity to measure the quality of a realization. Naturally, let

$$\mathsf{MaxMult}^{bi}(d) \triangleq \min\{\mathsf{MaxMult}(H) : H \text{ is bipartite and realizes } d\} \ .$$

For a partition $(a, b) \in \mathsf{BP}(d)$, we define

$$\mathsf{MaxMult}^{bi}(a, b) \triangleq \min\{\mathsf{MaxMult}(H) : H = (A, B, E)$$
$$\text{s.t. } \deg(A) = a \text{ and } \deg(B) = b\} .$$

Let r be a positive integer. If there is a bipartite multigraph $H = (A, B, E)$ where $\mathsf{MaxMult}(H) \leq r$, we say that d is r-max-bigraphic. Moreover, we say that the partition $(a, b) \in \mathsf{BP}(d)$, where $a = \deg(A)$ and $b = \deg(B)$, is r-max-bigraphic. Miller [12] cites the following result of Berge characterizing r-max-bigraphic partitions.

Theorem 6 (Berge [12]). *Let r be a positive integer. Consider a degree sequence d and a partition $(a, b) \in \mathsf{BP}(d)$, where $a = (a_1, \ldots, a_p)$ and $b = (b_1, \ldots, b_q)$. Then (a, b) is r-max-bigraphic if and only if, for $\ell = 1, \ldots, p$,*

$$\sum_{i=1}^{\ell} a_i \leq \sum_{i=1}^{q} \min\{\ell r, b_i\} . \tag{7}$$

Note the similarity to the Gale-Ryser theorem. Theorem 6 implies that $\mathsf{MaxMult}^{bi}(a, b)$ can be computed in $\mathcal{O}(n \cdot \log(d_1))$ using binary search.

For the second approximation criteria, we bound the total multiplicity of a bipartite multigraph realization. Define

$$\mathsf{TotMult}^{bi}(d) \triangleq \min\{\mathsf{TotMult}(H) : H \text{ is bipartite and realizes } d\} .$$

Additionally, for a partition $(a, b) \in \mathsf{BP}(d)$, we define

$$\mathsf{TotMult}^{bi}(a, b) \triangleq \min\{\mathsf{TotMult}(H) : H = (A, B, E)$$
$$\text{s.t. } \deg(A) = a \text{ and } \deg(B) = b\} .$$

We present our results on determining $\mathsf{TotMult}^{bi}(a, b)$ in the next section. In Sects. 4 and 5, we consider $\mathsf{MaxMult}^{bi}(d)$ and $\mathsf{TotMult}^{bi}(d)$.

3 Multigraph Realizations of Bi-sequences

In this section, we are interested in bipartite multigraph realizations with low total multiplicity, assuming that we are given a sequence and a specific balanced partition. First, we provide a characterization similar to Theorem 4 for bipartite multigraph realizations for a given partition.

Theorem 7. *Let d be a degree sequence and t be a positive integer. Then, d is t-tot-bigraphic if and only if there exists a partition $(a, b) \in \mathsf{BP}(d)$ such that $(a \circ 1^t, b \circ 1^t)$ is bigraphic.*

Proof. Let d, t be as in the theorem. Assume that there is a bipartite multigraph $H = (L, R, E)$ with $\mathsf{TotMult}(H) \leq t$. Hence, there is a partition $(a, b) \in \mathsf{BP}(d)$ where $\deg(L) = a$ and $\deg(R) = b$. Let F be the set of excess edges in H. Construct a bipartite graph G by applying the following transformation. For every excess edge $(x, y) \in F$, add a new vertex x_e to A and a new vertex y_e to B, and replace (x, y) by the two edges (x, y_e) and (x_e, y). Note that x_e and y_e are placed on opposite partitions of G. Since there are t excess edges, G realizes $(a \circ 1^t, b \circ 1^t)$ without excess edges.

For the other direction, assume that there exists a partition $(a, b) \in \mathsf{BP}(d)$ such that $(a \circ 1^t, b \circ 1^t)$ is realized by a bipartite graph $G = (L, R, E)$. Let $x_1, \ldots, x_t$ and $y_1, \ldots, y_t$ be some vertices of degree one in L and R, respectively. Also, for every i, let y_i' (respectively, x_i') be the only neighbor of x_i (resp., y_i). Construct a bipartite multigraph H by replacing the edges (x_i, y_i') and (x_i', y_i) with the edge (x_i', y_i') and discarding the vertices x_i and y_i, for every i. Since this transformation may add up to t excess edges, we have that $\mathsf{TotMult}(H) \leq t$. $\qquad\square$

The above characterization leads to extended Gale-Ryser conditions for total multiplicity.

Theorem 8. *Let d be a degree sequence with partition $(a, b) \in \mathsf{BP}(d)$ where $a = (a_1, \ldots, a_p)$ and $b = (b_1, \ldots, b_q)$, and let t be a positive integer. The partition (a, b) is t-tot-bigraphic if and only if, for all $\ell \in 1, \ldots, p$,*

$$\sum_{i=1}^{\ell} a_i \leq \sum_{i=1}^{q} \min\{\ell, b_i\} + t \ . \tag{8}$$

Proof. Consider (a, b) and t as in the theorem. One can verify that the following equations are the Gale-Ryser conditions of Theorem 5 for the partition (a, b): For all $\ell \in 1, \ldots, p$,

$$\sum_{i=1}^{\ell} a_i \leq \sum_{i=1}^{q} \min\{\ell, b_i\} + t \ , \tag{9}$$

and for all $h \in 1, \ldots, t$,

$$\sum_{i=1}^{p} a_i + h \leq \sum_{i=1}^{q} \min\{p + h, b_i\} + t \ . \tag{10}$$

To finish the proof, we argue that Equation (10) holds for any $h \in \{0, \ldots, t\}$ if Equation (9) holds for $\ell = p$. Recall that $\sum_{i=1}^{q} \min\{p + h, b_i\} = \sum_{i=1}^{q} b_i$ if $p + h \geq b_1$. It follows that Equation (10) holds for indices $h \geq b_1 - p$.

Observe that $\sum_{i=1}^{q} \min\{p+h+1, b_i\} - \sum_{i=1}^{q} \min\{p+h, b_i\} \geq 1$ for $p+h < b_1$, i.e., the RHS of Equation (10) grows by at least 1 when moving from index $p+h$ to index $p + h + 1$. By assumption, Equation (9) holds for $\ell = p$, implying that Equation (10) holds for $h = 0$. Since the LHS of Equation (10) grows by 1 exactly, Equation (10) holds for indices $h < b_1 - p$. $\qquad\square$

Given a degree sequence d with partition $(a, b) \in \mathsf{BP}(d)$, Theorem 8 implies that

$$\mathsf{TotMult}^{bi}(a, b) = \max_{1 \leq \ell \leq p} \left(\sum_{i=1}^{\ell} a_i - \sum_{i=1}^{q} \min\{\ell, b_i\} \right) .$$

It follows that $\mathsf{TotMult}^{bi}(a, b)$ can be computed in time $\mathcal{O}(n)$.

4   Bipartite Realization of a Single Sequence

In this section, we study the following question: given a degree sequence d, can it be realized as a multigraph whose underlying graph is bipartite? Also, if there exists such a realization, we would like to find one which minimizes the maximum or the total multiplicity.

4.1   Hardness Result

Given a sequence and a balanced partition one may construct a bipartite multigraph realization by assigning edges in an arbitrary manner.

Observation 2. *Let d be a sequence and let $(\ell, r) \in \mathsf{BP}(d)$ be a partition of d. Then, there exists a bipartite multigraph realization of (ℓ, r).*

It follows that deciding whether a degree sequence d can it be realized as a multigraph whose underlying graph is bipartite is NP-hard.

Theorem 9. *Deciding if a degree sequence d admits a bipartite multigraph realization is NP-hard.*

Proof. We prove the theorem using a reduction from the PARTITION problem. Recall that PARTITION contains all sequences $(a_1, \ldots, a_n)$ such that there exists an index set $S \subseteq [1, n]$ for which $\sum_{i \in S} a_i = \sum_{i \notin S} a_i$ (see. e.g., [8]). Observation 2 implies that d is a PARTITION instance if and only if d admits a bipartite mulitgraph realization. □

Since PARTITION admits a pseudo-polynomial time algorithm, we have the following.

Theorem 10. *Deciding if a sequence d admits a bipartite multigraph realization can be done in pseudo-polynomial time.*

4.2   Degree Sequences with a Small Number of Balanced Partitions

Bar-Noy et al. [2] describe an *output sensitive* algorithm that given a degree sequence d, computes all balanced partitions of d. The running time of the algorithm is bounded by the number of balanced partitions of d. For instances where this number is small, it allows us to solve BDR in polynomial time. In this section, we extend this approach to finding bipartite multigraphs.

The algorithm relies on the self-reducibility of the PARTITION problem. Let $T_{Part}(n)$ be the time complexity of the best pseudo-polynomial time algorithm for deciding PARTITION.

Lemma 1 ([2]). *The BDR problem admits a polynomial time output sensitive algorithm. More specifically, given an integer sequence $d = (d_1, d_2, \ldots, d_n)$, such that $d_i < n$ for every i, it is possible to find all partitions of d in time $\mathcal{O}(T_{Part}(n) \cdot n \cdot |\mathsf{BP}(d)|)$.*

The above running time becomes polynomial, if $|\mathsf{BP}(d)| = \mathcal{O}(n^c)$ for some constant c. The next result is readily implied with Theorem 6.

Corollary 3 ([2]). *Let d be a degree sequence of length n such that $|\mathsf{BP}(d)| = \mathcal{O}(n^c)$ for some constant c. Then, $\mathsf{MaxMult}^{bi}(d)$ can be computed in polynomial time.*

Similarly, Theorem 8 implies the following.

Corollary 4. *Let d be a degree sequence of length n such that $|\mathsf{BP}(d)| = \mathcal{O}(n^c)$, for some constant c. Then, $\mathsf{TotMult}^{bi}(d)$ can be computed in polynomial time.*

5   Small Maximum Degree Sequences

Towards attacking the realizability problem of general bigraphic sequences, we look at the question of bounding the total deviation of a nonincreasing sequence $d = (d_1, \ldots, d_n)$ as a function of its maximum degree, denoted $\Delta = d_1$.

Burstein and Rubin [4] consider the realization problem for directed graphs with loops, which is equivalent to BDR^P. (Directed edges go from the first partition to the second.) They give the following sufficient condition for a pair of sequences to be the in- and out-degrees of a directed graph with loops.

Theorem 11 (Burstein and Rubin [4]). *Consider a degree sequence d with a partition $(a, b) \in \mathsf{BP}(d)$ assuming that a and b have the same length p. Let $\sum a = \sum b = pc$ where c is the average degree. If $a_1 b_1 \leq pc + 1$, then d is realizable by a directed graph with loops.*

In what follows we make use of the following straightforward technical claim which slightly strengthens a similar claim from [2].

Observation 3. *Consider a nonincreasing integer sequence $d = (d_1, \ldots, d_k)$ of total sum $\sum d = D$. Then, $\sum (d[\ell]) \geq \lceil \ell D/k \rceil$, for every $1 \leq \ell \leq k$.*

Proof. Since d is nonincreasing, $\frac{1}{\ell} \sum_{i=1}^{\ell} d_i \geq \frac{1}{k-\ell} \sum_{i=\ell+1}^{k} d_i$. Consequently,

$$D = \sum_{i=1}^{k} d_i = \sum_{i=1}^{\ell} d_i + \sum_{i=\ell+1}^{k} d_i \geq \sum_{i=1}^{\ell} d_i + \left\lceil \frac{k-\ell}{\ell} \sum_{i=1}^{\ell} d_i \right\rceil = \left\lceil \frac{k}{\ell} \sum_{i=1}^{\ell} d_i \right\rceil ,$$

implying the claim.   □

5.1   Bounding the Maximum Multiplicity

Theorem 11 is extended to bipartite multigraphs with bounded maximum multiplicity, i.e., to r-max-bigraphic sequences. The following is a slightly stronger version of Lemma 14 from [2].

Lemma 2. *Let r be a positive integer. Consider a degree sequence d of length n with a partition $(a, b) \in \mathrm{BP}(d)$. If $a_1 \cdot b_1 \leq r \cdot \sum d/2 + r$, then (a, b) is r-max-bigraphic.*

Proof. Let r, d and (a, b) as in the lemma where $a = (a_1, a_2, \ldots, a_p)$ and $b = (b_1, b_2, \ldots, b_q)$. Moreover, let $X = \sum a = \sum b = \sum d/2$. To prove the claim, we assume that $a_1 \cdot b_1 \leq r \cdot X + r$, and show that Equation (7) holds for a fixed index $\ell \in [p]$. The lemma then follows due to Theorem 6.

First, we consider the case where $b_1 \leq \ell r$. Then, $\sum_{i=1}^{q} \min\{\ell r, b_i\} = X \geq \sum_{i=1}^{\ell} a_i$, and Equation (7) holds.

In the following, we assume that $\ell r < b_1$. Note that the conjugate sequence $\tilde{b}$ of b is nonincreasing, and that $\sum_{j=1}^{\ell r} \tilde{b}_j = \sum_{i=1}^{q} \min\{\ell r, b_i\}$. By Observation 3,

$$\sum_{i=1}^{q} \min\{\ell r, b_i\} \geq \lceil \ell r X / b_1 \rceil \geq \lceil \ell(a_1 b_1 - r)/b_1 \rceil = \lceil \ell a_1 - \ell r / b_1 \rceil = \ell a_1 .$$

As a is nonincreasing, we have that $\sum_{i=1}^{\ell} a_i \leq \ell a_1 \leq \sum_{i=1}^{q} \min\{\ell r, b_i\}$. The lemma follows.   □

Lemma 3. *There exists a degree sequence d with a partition $(a, b) \in \mathrm{BP}(d)$, such that $a_1 \cdot b_1 = r \cdot \sum d/2 + r$, which is r-max-bigraphic, but not $(r-1)$-max-bigraphic.*

Proof. Consider the sequence $d = (q^{2k-1}, (q-1)^2)$ for positive integers q, k such that $q = r \cdot k$. This sequence has a unique partition $(a, b) \in \mathrm{BP}(d)$, where $a = b = (q^{k-1}, (q-1))$. One can verify that $a_1 b_1 = q^2$, while

$$r \cdot \sum d/2 + r = r(qk - 1) + r = rqk = q^2 .$$

The partition (a, b) is r-max-bigraphic, but no better.   □

Lemma 2 is stated for a given partition (BDR^P). For BDR, we immediately have the following which is a slight improvement over Corollary 16 form [2].

Corollary 5. *Let r be a positive integer and d be a partitionable degree sequence. If $d_1^2 \leq r \cdot \sum d/2 + r$, then any partition $(a, b) \in \mathrm{BP}(d)$ is r-max-bigraphic.*

5.2   Bounding the Total Multiplicity

In this section, we establish results for total multiplicity analogous to those obtained in the previous section for the maximum multiplicity.

Lemma 4. *Let t be a positive integer. Consider a degree sequence d of length n with a partition $(a,b) \in \mathrm{BP}(d)$. If $a_1 \cdot b_1 \leq \sum d/2 + t + 1$, then (a,b) is t-tot-bigraphic.*

Proof. Let t, d and (a,b) as in the lemma where $a = (a_1, a_2, \ldots, a_p)$ and $b = (b_1, b_2, \ldots, b_q)$, and let $X = \sum a = \sum b = \sum d/2$. To prove the claim, we assume that $a_1 \cdot b_1 \leq X + t + 1$, and show that Equation (8) holds for every index $\ell \in [p]$. The lemma then follows due to Theorem 8.

First, consider the case where $\ell \geq b_1$. In this case,

$$\sum_{i=1}^{q} \min\{\ell, b_i\} = \sum b = X \geq \sum_{i=1}^{\ell} a_i \ ,$$

and Equation (8) holds.

Next, assume that $\ell < b_1$. Note that the conjugate sequence $\tilde{b}$ of b is nonincreasing, and that $\sum_{j=1}^{\ell} \tilde{b}_j = \sum_{i=1}^{q} \min\{\ell, b_i\}$. By Observation 3,

$$\sum_{i=1}^{q} \min\{\ell, b_i\} + t \geq \left\lceil \frac{\ell X}{b_1} \right\rceil + t \geq \left\lceil \frac{\ell(a_1 b_1 - t - 1)}{b_1} \right\rceil + t = \left\lceil \ell a_1 - \frac{\ell(t+1)}{b_1} \right\rceil + t \geq \ell a_1 \ .$$

As a is nonincreasing, we have that $\sum_{i=1}^{\ell} a_i \leq \ell a_1 \leq \sum_{i=1}^{q} \min\{\ell, b_i\} + t$. The lemma follows. $\qquad\square$

The following lemma shows that the above bound it tight.

Lemma 5. *There exists a degree sequence d with a partition $(a,b) \in \mathrm{BP}(d)$, such that $a_1 \cdot b_1 = \sum d/2 + t + 2$, and (a,b) is not t-tot-bigraphic.*

Proof. Consider the sequence $d = (k^{2(k-1)}, 1^2)$, for a positive integer $k > 1$. This sequence has only one partition $(a,b) \in \mathrm{BP}(d)$, where $a = b = (k^{k-1}, 1)$. Observe that $a_1 b_1 = k^2$, while $\sum d/2 = k(k-1) + 1$.

Assume that $t = k - 2$. Hence, $a_1 b_1 = \sum d/2 + t + 1$. For every $\ell < k$, we have that

$$\sum_{i=1}^{k} \min\{\ell, b_i\} + t = k + (\ell - 1)(k - 1) + k - 2 = \ell k - \ell - 1 + k \geq \ell k = \sum_{i=1}^{\ell} a_i \ .$$

For $\ell = k$, we have $\sum_{i=1}^{k} \min\{\ell, b_i\} + t \geq \sum d/2 = \sum_{i=1}^{k} a_i$.

Now assume that $t = k - 3$. Hence, $a_1 b_1 = \sum d/2 + t + 2$. For every $\ell < k$, we have that

$$\sum_{i=1}^{k} \min\{\ell, b_i\} + t = k + (\ell - 1)(k - 1) + k - 3 = \ell k - \ell - 2 + k \ .$$

If $\ell = k - 1$, we get that

$$\sum_{i=1}^{k} \min\{\ell, b_i\} + t = (k - 1)k - (k - 1) - 2 + k = (k - 1)k - 1 < \sum_{i=1}^{\ell} a_i \ ,$$

which means that (a,b) is not t-tot-bigraphic. $\qquad\square$

Similar to above, Lemma 5 (stated for BDR^P) implies the following for BDR.

Corollary 6. *Let t be a positive integer and d be a partitionable degree sequence. If $d_1^2 \leq \sum d/2 + t + 1$, then any partition $(a, b) \in \mathtt{BP}(d)$ is t-tot-bigraphic.*

References

1. Bar-Noy, A., Böhnlein, T., Peleg, D., Perry, M., Rawitz, D.: Relaxed and approximate graph realizations. In: Flocchini, P., Moura, L. (eds.) IWOCA 2021. LNCS, vol. 12757, pp. 3–19. Springer, Cham (2021). https://doi.org/10.1007/978-3-030-79987-8_1
2. Bar-Noy, A., Böhnlein, T., Peleg, D., Rawitz, D.: On realizing a single degree sequence by a bipartite graph. In: 18th SWAT, vol. 227 of LIPIcs, pp. 1:1–1:17 (2022)
3. Bar-Noy, A., Böhnlein, T., Peleg, D., Rawitz, D.: On the role of the high-low partition in realizing a degree sequence by a bipartite graph. In: 47th MFCS, vol. 241 of LIPIcs, pp. 14:1–14:15 (2022)
4. Burstein, D., Rubin, J.: Sufficient conditions for graphicality of bidegree sequences. SIAM J. Discret. Math. **31**(1), 50–62 (2017)
5. Chungphaisan, V.: Conditions for sequences to be r-graphic. Discr. Math. **7**(1–2), 31–39 (1974)
6. Erdös, P., Gallai, T.: Graphs with prescribed degrees of vertices [Hungarian]. Mat. Lapok (N.S.) **11**, 264–274 (1960)
7. Gale, D.: A theorem on flows in networks. Pacific J. Math **7**(2), 1073–1082 (1957)
8. Garey, M.R., Johnson, D.S.: Computers and Intractability: A Guide to the Theory of NP-Completeness. Freeman, San Francisco, CA (1979)
9. Hakimi, S.L.: On realizability of a set of integers as degrees of the vertices of a linear graph -I. SIAM J. Appl. Math. **10**(3), 496–506 (1962)
10. Havel, V.: A remark on the existence of finite graphs [in Czech]. Casopis Pest. Mat. **80**, 477–480 (1955)
11. Hulett, H., Will, T.G., Woeginger, G.J.: Multigraph realizations of degree sequences: maximization is easy, minimization is hard. Oper. Res. Lett. **36**(5), 594–596 (2008)
12. Miller, J.W.: Reduced criteria for degree sequences. Discr. Math. **313**(4), 550–562 (2013)
13. Owens, A.B., Trent, H.M.: On determining minimal singularities for the realizations of an incidence sequence. SIAM J. Appl. Math. **15**(2), 406–418 (1967)
14. Rao, S.B.: A survey of the theory of potentially P-graphic and forcibly P-graphic degree sequences. In: Rao, S.B. (ed.) Combinatorics and Graph Theory. LNM, vol. 885, pp. 417–440. Springer, Heidelberg (1981). https://doi.org/10.1007/BFb0092288
15. Ryser, H.J.: Combinatorial properties of matrices of zeros and ones. Can. J. Math. **9**, 371–377 (1957)
16. Tripathi, A., Vijay, S.: A note on a theorem of Erdös & Gallai. Discr. Math. **265**(1–3), 417–420 (2003)
17. Will, T.G., Hulett, H.: Parsimonious multigraphs. SIAM J. Discr. Math. **18**(2), 241–245 (2004)
18. Zverovich, I.E., Zverovich, V.E.: Contributions to the theory of graphic sequences. Discr. Math. **105**(1–3), 293–303 (1992)

Thirty Years of SIROCCO A Data and Graph Mining Comparative Analysis of Its Temporal Evolution

Pierluigi Crescenzi(✉)🔾

Gran Sasso Science Institute, Viale Francesco Crispi 7, 67100 L'Aquila, Italy
pierluigi.crescenzi@gssi.it
https://pilucrescenzi.it

Abstract. In this paper, we study the temporal evolution of SIROCCO and of other sixteen theoretical computer science conferences. Our goal is to try to understand the evolution of these conferences and to answer several different research questions, related to the number of authors, number of papers, size of collaborations, sex inclusion, research topics, network characteristics, and author centrality. The tentative answer to these questions is given by performing a comparative analysis between the entire set of conferences. Even though the paper focuses on SIROCCO and on a specific set of theoretical computer science conferences, the software used to perform our analysis can be easily used to perform similar analysis in the case of conferences in different computer science research areas.

Keywords: Computer science conference · Data mining · DBLP dataset · Graph mining · Research topic · Temporal network mining

1 Introduction

Computer science is a *young* discipline with a *long history*. According to the corresponding page of Wikipedia [22], the term 'computer science' appears for the first time in 1959 [9] but "algorithms for performing computations have existed since antiquity." Thousands of computer science conferences are now active: indeed, the DBLP computer science bibliography [21] provides information on 6199 computer science conference proceedings (at March 28, 2023). Many of these conferences are celebrating or have celebrated special anniversaries, such as the *IEEE Annual Symposium on Foundations of Computer Science* (60 years in 2019), the *International Colloquium on Automata, Languages and Programming* (50 years in 2022), the *Symposium on the Theory of Computing* (50 years in 2018), and the *Colloquium on Structural Information and Communication Complexity* (in short *SIROCCO*, 30 years in 2023). This implies that sufficient

P. Crescenzi—Part of this work has been done while visiting COATI, INRIA d'Université Côte d'Azur, Sophia Antipolis, France.

data are now available for performing a data and graph mining analysis not only on the entire DBLP dataset (as has often happened in the last two decades), but also on one specific conference or a small set of specific conferences (if we are interested in a comparative analysis).

This paper presents an open-source software resource for doing such an analysis in order to try to provide answers to questions like the following. How do the conferences compare in terms of the evolution of the number of authors, number of papers, and number of collaboration sizes? How much were the conferences more or less open to new entries? Has the percentage of female authors evolved differently over time? Did the conferences present similar densification, diameter shrinking, and small-world phenomenon? What are the most influential authors of the conferences?

As a case study, in this paper we analyse and compare the evolution of SIROCCO and of other sixteen among the most popular *Theoretical Computer Science* (in short, TCS) conferences throughout their temporal evolution. The paper is accompanied by a web page containing all the figures in their full and "interactive" version: we hope that members of the SIROCCO community will enjoy playing with them. It goes without saying that "the data analysis we present has to be taken with a huge pinch of salt and is only meant to provide an overview of" the evolution of some among the most popular TCS conferences and "to be food for thought for" the computer science theory community [1].

Even though in this paper we focus on a relatively small set of TCS conferences, our software allows us to perform similar analysis on any (set of) conferences whose data are included in the DBLP dataset and whose domain might range from artificial intelligence to database systems, from computer networks to bio-informatics, from computer vision to data mining.

1.1   Our Dataset

The DBLP dataset "provides open bibliographic information on major computer science journals and proceedings" [13,21]. The entire dataset is released under the CC0 1.0 Public Domain Dedication license [5]. Each month, a persistent snapshot dump of the DBLP XML dataset is published on the web site: in this paper, we refer to the dump of March 2023. In addition to SIROCCO, of all conferences archived in the DBLP bibliography, in this paper we will use data corresponding to the following sixteen TCS conferences: *International Conference on Computer Aided Verification* (in short, CAV), *Annual International Cryptology Conference* (in short, CRYPTO), *Annual Conference for Computer Science Logic* (in short, CSL), *International Symposium on Distributed Computing* (in short, DISC), *European Symposium on Algorithms* (in short, ESA), *European Symposium on Programming* (in short, ESOP), *International Conference on the Theory and Application of Cryptographic Techniques* (in short, ECRYPT), *IEEE Annual Symposium on Foundations of Computer Science* (in short, FOCS), *International Colloquium on Automata, Languages and Programming* (in short, ICALP), *ACM/IEEE Symposium on Logic in Computer Science* (in short, LICS), *ACM SIGACT-SIGOPS Symposium on Principles of*

Table 1. Basic statistics concerning the editions, the number of authors, and the number of papers of SIROCCO and the other sixteen TCS conferences. Note that the following conferences in some specific years had no edition: CSL in 2019, DISC in 1988, ESOP in 1987, 1989, 1991, 1993, 1995, and 1997, EUROCRYPT in 1983, ICALP in 1973 and 1975. Moreover POPL became a journal issue since 2018.

Conference	Number of authors				Number of papers			
	First	Last	Minimum	Maximum	First	Last	Minimum	Maximum
SIROCCO (1994–2022)	24	48	24 (1994)	92 (2018)	11	16	11 (1994)	32 (2015)
CAV (1990–2022)	77	205	63 (1995)	319 (2021)	38	53	32 (1992)	84 (2021)
CRYPTO (1981–2022)	52	260	46 (1983)	291 (2021)	40	99	29 (1983)	104 (2021)
CSL (1987–2022)	38	85	34 (1990)	165 (2014)	24	37	22 (1993)	80 (2014)
DISC (1987–2022)	58	146	45 (1996)	173 (2021)	30	52	22 (1993)	63 (2021)
ESA (1993–2022)	85	276	73 (1997)	276 (2022)	38	92	38 (1993)	92 (2022)
ESOP (1986–2022)	47	83	37 (1998)	109 (2018)	27	21	18 (1998)	36 (2017)
EUROCRYPT (1982–2022)	34	263	34 (1982)	263 (2022)	26	85	26 (1982)	85 (2022)
FOCS (1960–2022)	10	297	10 (1960)	343 (2020)	9	110	9 (1960)	127 (2020)
ICALP (1972–2022)	65	369	41 (1976)	442 (2018)	49	133	31 (1976)	165 (2018)
LICS (1986–2022)	71	157	64 (1987)	241 (2021)	42	64	37 (1987)	95 (2018)
PODC (1982–2022)	55	163	42 (1989)	229 (2010)	30	64	22 (1988)	92 (2008)
POPL (1973–2017)	36	192	34 (1975)	206 (2016)	22	66	20 (1976)	66 (2017)
SODA (1990–2022)	121	405	98 (1992)	466 (2020)	54	148	53 (1991)	183 (2019)
STACS (1984–2022)	38	178	38 (1984)	178 (2022)	30	60	26 (1990)	71 (1993)
STOC (1969–2022)	40	365	26 (1971)	375 (2021)	31	134	23 (1971)	151 (2021)
TACAS (1995–2022)	37	244	37 (1995)	276 (2019)	13	67	13 (1995)	73 (2019)

Distributed Computing (in short, PODC), *ACM-SIGACT Symposium on Principles of Programming Languages* (in short, POPL), *ACM-SIAM Symposium on Discrete Algorithms* (in short, SODA), *Symposium on Theoretical Aspects of Computer Science* (in short, STACS), *Symposium on the Theory of Computing* (in short, STOC), and *International Conference on Tools and Algorithms for Construction and Analysis of Systems* (in short, TACAS). The year of the first edition of each of the sixteen conferences and the basic statistics concerning the number of authors and the number of papers are summarised in Table 1.

2   A Data Mining Comparative Analysis

In this section we perform some data mining analysis in order to compare the evolution of SIROCCO and of the other sixteen TCS conferences. To provide a framework for the study, we define several research questions.

How do the Conferences Compare in Terms of (The Evolution of) their Author Sets? FOCS, ICALP, STACS, and STOC are general TCS conferences, ESA and SODA are mostly devoted to algorithms, CAV and TACAS to automated verification, CRYPTO and EUROCRYPT to cryptography, DISC and PODC to distributed computation, CSL and LICS to logic, and ESOP and POPL to

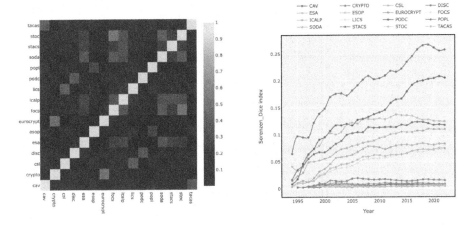

Fig. 1. The similarity between the sixteen TCS conferences, measured by using the Sorensen-Dice index of the corresponding author sets (left), and the evolution of the Sorensen-Dice index between SIROCCO and the other sixteen TCS conferences (right).

programming languages. This rough classification is confirmed by the heat map shown in the left part of Fig. 1, where the similarity between two conferences is computed by using the Sorensen-Dice index [8,20] with respect to the sets of all authors who published at least one paper in a conference (recall that, given two sets X and Y, the Sorensen-Dice index of X and Y is defined as $\mathsf{sd}(X,Y) = \frac{2\mathsf{jac}(X,Y)}{1+\mathsf{jac}(X,Y)}$, where $\mathsf{jac}(X,Y) = \frac{|X \cap Y|}{|X \cup Y|}$ is the Jaccard index of X and Y [10]).[1]

In order to compare the similarity between SIROCCO and the other sixteen TCS conferences, we show in the right part of Fig. 1 the evolution over time of the Sorensen-Dice index between SIROCCO and the other sixteen TCS conferences (always with respect to the sets of all authors who published at least one paper in the corresponding conference). Not surprisingly, DISC and PODC are the conferences more similar to SIROCCO, followed by ESA and STACS. It is worth noting that while DISC and PODC are almost every year more similar to SIROCCO than the previous year, the general TCS and algorithms conferences seem to converge to a stable degree of similarity. Finally, SIROCCO seems to be very different from cryptography, logic, and programming languages conferences. We will further develop this observation while trying to assign to SIROCCO a measure of membership to different ICALP communities.

How do the Conferences Compare in Terms of the Evolution of the Number of Papers? For what concerns the number of papers, three different kinds of evolution can be noted (Fig. 2). There are conferences such as ICALP in which the number of papers increases in a quite regular way, conferences such as STACS

---

[1] Other similarity indices could have been used, such as, for example, the one proposed in [11]. Even if we did not check it, we believe that the results would be very similar.

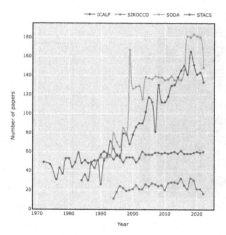

Fig. 2. The evolution of the number of papers for SIROCCO and for a sample of the sixteen TCS conferences.

in which the number of papers remains almost constant, and conferences such as SODA which are a combination of the previous two behaviours (that is, a sudden increase followed by a period of constant value). SIROCCO seems to belong to the second group of conferences: this might be due to the relative small size of SIROCCO but also to a more "rigid" policy concerning the number of accepted papers.

How do the Conferences Compare in Terms of (The Evolution of) the Number Of Collaboration Sizes? Even in the case of the average size of a paper author set, we can note different behaviours. In particular, there are conferences, such as CSL, in which the percentage of single-author papers is significantly higher than the other percentages, while for the majority of the other conferences the percentage curve has a bell shape (see the left part of Fig. 3). In this latter case, there is still a difference for what concerns sizes up to four: indeed, there are conferences, such as ICALP, in which the peak is reached at the value 2 and the percentage of single-author papers is the second one, and there are conferences, such as TACAS, in which the peak is reached at the value 3 and even the percentage of papers with five authors is larger than the percentage of single-author papers (note that the maximum size of an author set for these three conferences is 7, 15, and 24, respectively). SIROCCO seems to be between ICALP and TACAS, since the peak of its curve is at 2, but single-author papers are more frequent than papers with five authors (the maximum size is 13). It is also worth noting that these percentages have evolved over time. For example, in the case of SIROCCO, during the first eight years the number of papers with two authors is significantly higher than the other percentages. While time passing, however, papers with three authors become more and more popular: indeed, during the last seven years these kind of papers are more than one third of all the papers. Moreover, in the period between 2010 and 2017, large coauthor set

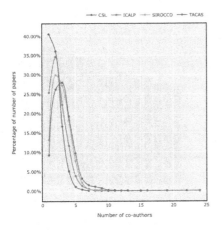
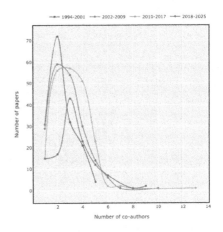

Fig. 3. The average size of a paper author set for SIROCCO and for a sample of the sixteen conferences (left) and its evolution in the case of SIROCCO (right).

have been quite popular: indeed, even the number of papers with five authors is larger than the number of single-author papers (during this period the maximum author set size is also reached). It is worth observing that a similar behaviour can be observed in the case of the other sixteen TCS conferences, thus implying that, in general, theoretical computer scientists tend to collaborate more as time goes by (it would be interesting to compare this observation with other computer science research areas or even with other scientific disciplines).

How much Have the Conferences been more or Less Open to New Entries? All conferences have a similar evolution in terms of the percentage of new authors. Clearly, this percentage starts from 100% and, then, rapidly decreases towards a stable value. This latter value, however, changes from conference to conference (see the left part of Fig. 4), and can range from less than 40% (such as in the case of SODA), between 40% and 50% (such as in the case of ICALP), and above 50% (such as in the case of CAV). In the case of SIROCCO, this percentage significantly oscillates in the last five years, being greater than 55% in 2018, less than 28% in 2019, and around 50% in the last three editions. The difference between the openness to new entries of the analysed conferences can be noted by looking at the average percentage of new authors in the last ten years (see the right part of Fig. 4). The most "conservative" conference is STOC (with an average value approximately equal to 37%), while the most open to new entries is ESOP (with an average value approximately equal to 60%). SIROCCO places itself in the middle with an average percentage approximately equal to 46%. It is worth noting that the situation slightly changes if we consider as new authors only the one whose first paper published in the conference has not been co-authored with an author who already published a paper in the conference: in other words, we can consider as *fully* new authors only the ones which have not been 'introduced' by an author already known in the conference

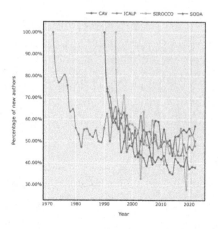
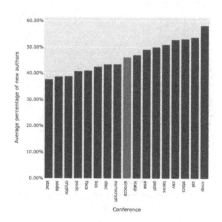

Fig. 4. The evolution of the percentage of new authors for a sample of the sixteen conferences (left) and the average percentage of new authors in the last ten editions of all conferences (right).

community (such as it happens, for instance, in the case of a paper co-authored by a researcher and a student of theirs). By looking at the average percentage of such fully new authors in the last ten years (see Fig. 5), we note that clearly all percentages decrease and that the overall ranking of the conferences changes in several positions. For example, the most "conservative" conference is now CRYPTO (with an average value approximately equal to 5%), while the most open to fully new entries is now CSL (with an average value approximately equal to 22%). SIROCCO is now in fifth position (very close to TACAS) with an average value approximately equal to 15%.

Has the Percentage of Female Authors Evolved Differently Over Time? In order to answer to this question, we first need to "label" each author as male or female. To this aim, we used the web service available at https://genderize.io, which allowed us to assign the labels on the ground of the first name of the authors. Of the 22261 authors who published at least one paper in at least one conference analysed, 1113 could not be assigned a label by using this service (either because the first name was not recognized or because the correctness probability was below 0.5). We then performed a manual search on the web in order to assign the label to these 1113 authors. We were able to do it for 532 authors, which means that 581 authors (approximately, 2.6% of the total) remained unlabeled. Note that almost one third of these authors (precisely, 191) have an *academic age* equal to 1, that is, they published one paper or more in only one year, and then they did not publish any other paper. We expected these authors to be quite difficult to find on the web, and, indeed, they could not be labeled (especially because only 55 of these authors were "active" after 2010). Of the remaining 390 authors, one third published their last paper before 2010 and, hence, they were also difficult to find on the web and could not be labeled. In the left part of Fig. 6

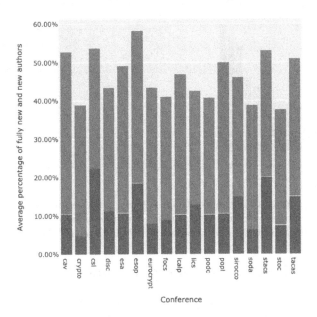

Fig. 5. The average percentage of new and fully new authors in the last ten editions of all conferences.

we show, for each conference, the percentage of unlabeled authors. As we can see, the highest percentages are the ones corresponding to CRYPTO, EURO-CRYPT, and FOCS, which are the only conferences whose percentage is higher than the global percentage. In the right part of Fig. 6 we show the evolution of the percentage of the female authors for the four conferences CRYPTO, DISC, SIROCCO, and SODA. While the first two conferences show a slightly increasing trend (which is more or less shared by the majority of all the other conferences), the second two conferences seems to behave in a different way, showing a quite constant percentage. Moreover, while CRYPTO and DISC manage to repeatedly break the "wall" of 20%, the percentage of female authors for SIROCCO and SODA is still far below this threshold (apart from the 1996 edition of SIROCCO). This latter behavior is, unfortunately, the most frequent one among all the analysed conferences and, indeed, only three other conferences (that is, PODC and TACAS) reached, at least once, a percentage greater than 20%. Finally, it is also worth observing how these percentages are overall "better" than the ones derived in [7] while analyzing the entire computer science research community. In that case, indeed, the percentage of female authors started from 7.3% in the time frame 1991–1995 and ended to 12.9% in the time frame 2016–2020. This seems to suggest that the theoretical computer science community promotes sex inclusion more than the entire computer science community.

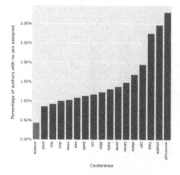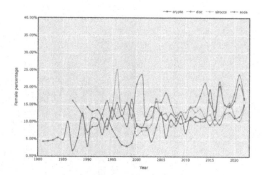

Fig. 6. The percentage of authors without a (sex) label assigned for SIROCCO and for each of the sixteen TCS conferences (left), and the evolution of the percentage of female authors for SIROCCO and for a sample of the sixteen TCS conferences (right).

3   A Graph Mining Comparative Analysis

In this section we perform some graph mining analysis in order to compare the evolution of SIROCCO and of the other sixteen TCS conferences. In particular, for each conference, we consider both the "static" collaboration graph G_s, in which the nodes are the authors who presented at least one paper at the conference and there is an edge (a_1, a_2) if a_1 and a_2 co-authored at least one paper (not necessarily presented at the conference), and the "temporal" collaboration graph G_t, with the same set of nodes and in which there is a "temporal" edge (a_1, a_2, y) if a_1 and a_2 co-authored at least one paper (not necessarily presented at the conference) in year y (for more details about temporal graphs, see [17]). For each conference, the basic statistics of these two collaboration graphs are shown in Table 2, where the density is defined as $\frac{2m_s}{n(n-1)}$, n is the number of nodes, m_s is the number of edges, and the LCC size denotes the fraction of nodes belonging to the largest connected component of the static collaboration graph. As expected, all static graphs contain a huge connected component and are quite sparse.

Did the Conferences Differently Densify Over Time? In [12] the authors observed that many real-world graphs "densify" over time. That is, the number of edges in the graph increases more than linearly with respect to the number of nodes. Formally, suppose that $m_s \propto n^\alpha$ for some constants α with $1 \leq \alpha \leq 2$: the hypothesis is that α is significantly greater than 1. In order to verify this statement, we can execute a linear regression of the log-log relationship between the number of nodes and the number of edges in the static graph (by assuming that when there are no nodes, then there are no edges). The resulting values of α are shown in the last column of Table 2. For these values we can, indeed, conclude that there exists a densification phenomenon, which is more evident in the case of FOCS and STOC (SIROCCO seems to be among the conferences with the lowest level of densification).

Table 2. Basic statistics concerning the static and temporal collaboration graphs of SIROCCO and of the other sixteen TCS conferences.

Conference	# nodes	# edges	# temporal edges	Density	LCC size	α
CAV	2733	17077	48209	0.0046	0.98	1.31
CRYPTO	1988	14323	36460	0.0073	0.95	1.40
CSL	1455	5560	23863	0.0053	0.95	1.27
DISC	1602	8165	28989	0.0064	0.97	1.29
ESA	2868	22761	73502	0.0055	0.99	1.28
ESOP	1367	6368	21689	0.0068	0.97	1.35
EUROCRYPT	1821	12552	31328	0.0076	0.94	1.44
FOCS	3346	27021	82493	0.0048	0.96	1.63
ICALP	4837	37511	123999	0.0032	0.98	1.47
LICS	1953	9886	37650	0.0052	0.98	1.28
PODC	2393	13000	44763	0.0045	0.96	1.30
POPL	1979	9691	32636	0.005	0.96	1.34
SIROCCO	923	4858	19303	0.0114	0.98	1.28
SODA	4173	36097	106273	0.0041	0.99	1.28
STACS	2740	16469	62082	0.0044	0.97	1.32
STOC	3192	27199	81887	0.0053	0.98	1.53
TACAS	2239	13901	41452	0.0055	0.98	1.26

Did the Diameter and the Degrees of Separation Differently Evolve Over Time?
In [12] the authors also observed that, in many real-world graphs, the diameter "shrinks" over time, contrary to the "folklore" hypothesis that it should increase in a logarithmic way with respect to the number of nodes. In the left part of Fig. 7 we show the evolution of the diameter for SIROCCO and for a sample of three others TCS conferences. It can be indeed verified that, after an initial increase, the diameter shrinks, even though, after a while, it tends to converge to a quite small value (between 8 and 11). In the case of SIROCCO, for example, this latter value is among the smallest ones (only ESA converges to a smaller value) and the maximum diameter value is reached in 1994 (that is, the year of its first edition). It is also worth noting that a similar behavior is observed in the case of the so-called "effective" diameter, that is, the value such that, for at least 90% of the pairs of nodes, their distance is at most equal to it. Clearly, in this latter case, the value to which the effective diameter converges is significantly lower than the one corresponding to the diameter (more precisely, either 4 or 5). Another well-studied parameter of large graphs is the average distance between all nodes in the largest connected component, also called "degrees of separation" (inspired by the well-known "small-world" experiment of [18]). As shown in the right part of Fig. 7, all the analysed conference present the small-world phenomenon, that is, their average distance is quite low (between 3.5 and 4.1), which is also

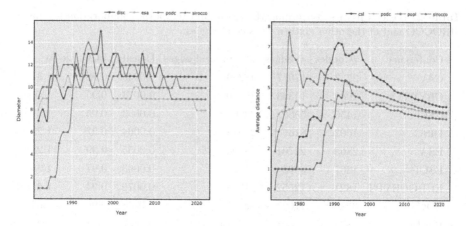

Fig. 7. The evolution of the diameter (left) and of the degrees of separation (right) for SIROCCO and for a sample of the sixteen TCS conferences.

consistent with the more recent experimental results obtained while analysing the degrees of separation of the huge graph of Facebook friendships [2]. It is worth observing that while some conferences became small worlds almost immediately (like PODC) or never reached high degrees of separation (like SIROCCO), there are conferences which had, at the beginning, more than six degrees of separation (actually, more than seven), and then became more and more small worlds.

How the Conferences Compare with Respect to their Most Temporally Central Nodes? Several definitions of temporal centrality in temporal graphs have been proposed in the last few years (see, for example, [4,19]). Here, we are going to use the analogue of the closeness centrality in static graphs, introduced in [6]. According to this definition, the centrality of a node is, intuitively, related to the evolution of its average distance to the other nodes, where in the case of temporal graphs the path has to respect natural temporal constraints (that is, each edge in the path has to appeared after the edge preceding it in the path) and the length of a path is given by the number of its temporal edges. By using this centrality measure, we identified the top-50 authors of SIROCCO and of the other sixteen TCS conferences, and we computed the intersection of these sets of authors (see Table 3) As expected, there is (almost) no intersection between the set of SIROCCO authors and the set of conferences in formal methods and programming languages. Moreover, the biggest intersections are with DISC and PODC (28 and 19 authors, respectively), followed by STACS (13 authors). The biggest intersection in the entire set of conferences is between FOCS and STOC (46 authors). This analysis can be viewed as a refinement of the heat map shown in Fig. 1, since we are now comparing only the most "central" nodes of each conference, and not the entire set of authors who published at least one paper in the conference.

Table 3. The intersection between the sets of the top-50 nodes with respect to the temporal closeness defined in [6]

Conference	C2	C3	C4	C5	C6	C7	C8	C9	C10	C11	C12	C13	C14	C15	C16	C17
C1. CAV	0	12	0	0	6	0	0	1	19	0	16	0	0	0	0	36
C2. CRYPTO		0	3	0	0	43	12	6	0	8	0	1	1	0	12	0
C3. CSL			0	0	5	0	0	1	28	0	8	0	0	0	0	13
C4. DISC				7	0	3	6	14	0	34	2	28	5	10	8	0
C5. ESA					0	0	11	23	0	8	0	6	30	25	10	0
C6. ESOP						0	0	0	9	0	23	0	0	0	0	5
C7. EUROCRYPT							7	4	0	8	0	1	0	0	7	0
C8. FOCS								29	2	17	0	3	22	7	46	0
C9. ICALP									3	21	1	9	29	19	30	1
C10. LICS										1	16	0	1	1	2	18
C11. PODC											2	19	11	10	20	0
C12. POPL												1	0	0	0	12
C13. SIROCCO													4	13	5	0
C14. SODA														19	21	0
C15. STACS															7	0
C16. STOC																0
C17. TACAS																

4   The SIROCCO Membership to ICALP Communities

As a part of the presentation for the fifty years celebration of ICALP, we developed an HTML tool assigning to each ICALP author a "probability" of being a member of four distinct communities, that is, the algorithm, the cryptography, the distributed computing, and the formal methods communities. This tool produces, for every researcher who presented at least one paper at ICALP, the pie chart representing the similarity of the researcher with one of the four communities (see Fig. 8). The tool makes use of ideas similar to the ones described in [16], by associating to each author a document formed by the titles of all the papers published by the author and to each conference a document formed by the titles of all the papers published in the conference. After selecting, for each community, two or more "anchor" conferences (similarly to what has been done in [14] and is suggested in [15]), the tool simply computes the similarity of the document corresponding to an author with the documents corresponding to the anchor conferences of a community. By using this classification of ICALP authors, we can analyse the degree of membership of the entire SIROCCO conference (represented by all the authors who have published a paper both in ICALP and in SIROCCO) to each of the four communities. This analysis is summarised in Fig. 9. As it can be seen, SIROCCO as a conference mostly belongs to the algorithm community, even though its membership to the distributed computing community is also very high. As Andrea Marino, an old friend of ours, said, it seems that SIROCCO is the place where algorithm researchers like to play the role of distributed computing researchers.

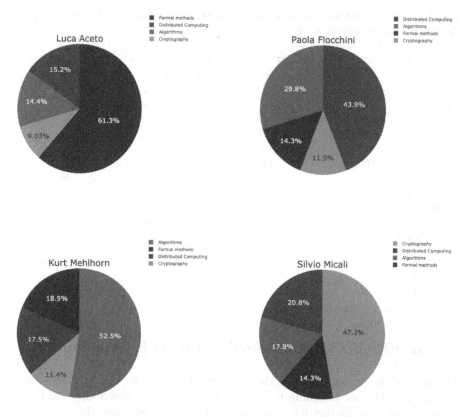

Fig. 8. The similarity of an author with each of the four communities of ICALP, for a sample of four authors.

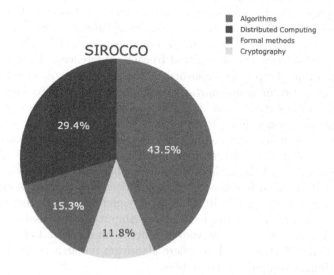

Fig. 9. The global similarity of SIROCCO with each of the four communities of ICALP.

5   Concluding Remarks

In this paper we have analysed the temporal evolution of SIROCCO and of other sixteen theoretical computer science conferences, trying to answer several different research questions, related to the number of authors, number of papers, size of collaborations, sex inclusion, research topics, network characteristics, and author centrality. The tentative answer to these questions has been given by performing a comparative analysis between the entire set of conferences. Apart from performing similar analysis with different sets of computer science conferences, we believe that one promising research line would be to integrate our tool with more sophisticated data and graph mining approaches (even based on machine learning techniques). Just to give an example, we believe it would be interesting to explore the application of time series mining techniques (see, for example, [3]) to the context of computer science conferences, in order to mathematically discover similarities about their quantitative evolution. Finally, even if the tool has been developed in order to analyse conferences included in the DBLP dataset, it should not be difficult to extend its application to journals included in the DBLP dataset or even to different datasets (such as the Scopus dataset). In this latter case, indeed, it should be sufficient to "export" the dataset into the format used by our tool, which could be extended in order to also integrate information concerning the citations between papers and authors.

Code Availability. The code used to perform the comparative analysis is available at

https://github.com/piluc/ConferenceMining

Starting from this website, it is also possible to reach the web page containing the interactive version of all the figures included in this paper, the presentation made for celebrating the fifty years of ICALP in 2022, and an example of a web page that can be produced, by using our tool, in order to analyse one specific conference.

Acknowledgements. This work has been partially supported by the French government through the UCAJEDI. I would like to thank Luca Aceto for proposing me to perform a data and graph mining analysis of ICALP. This was the main motivation for developing the tool which now allows the user to perform a comparative analysis of any set of conferences in the DBLP dataset. I would also like to thanks Daniele Carnevale, with whom I collaborated while computing the community membership algorithm. Finally, I would like to thank Filippos Christodoulou and Oleksandr Skorupskyy, for several discussions at the beginning of this project (during the lectures of the graph mining course at the Gran Sasso Science Institute), and the COATI team for several discussions during my visit at INRIA, Sophia Antipolis.

References

1. Aceto, L., Crescenzi, P.: Concur through time. Bullettin EATCS **138**(October), 1–9 (2022)
2. Backstrom, L., Boldi, P., Rosa, M., Ugander, J., Vigna, S.: Four degrees of separation. In: Proceedings of the 4th ACM Conference on Web Science, pp. 33–42. ACM (2012)
3. Brockwell, P.J., Davis, R.A.: Introduction to Time Series and Forecasting. Springer, New York (2002). https://doi.org/10.1007/0-387-21657-X_8
4. Buß, S., Molter, H., Niedermeier, R., Rymar, M.: Algorithmic aspects of temporal betweenness. In: The 26th ACM SIGKDD Conference on Knowledge Discovery and Data Mining. ACM (2020)
5. Creative Commons: Cc0 1.0 universal (cc0 1.0) public domain dedication (2022). https://creativecommons.org/publicdomain/zero/1.0/
6. Crescenzi, P., Magnien, C., Marino, A.: Finding top-k nodes for temporal closeness in large temporal graphs. Algorithms **13**, 211 (2020)
7. Demetrescu, C., Finocchi, I., Ribichini, A., Schaerf, M.: On computer science research and its temporal evolution. Scientometrics **127**, 4913–4938 (2022)
8. Dice, L.R.: Measures of the amount of ecologic association between species. Ecology **26**(3), 297–302 (1945)
9. Fein, L.: The role of the university in computers, data processing and related fields. Commun. ACM **2**(9), 7–14 (1959)
10. Jaccard, P.: The distribution of the flora in the alpine zone. New Phytol. **11**(2), 37–50 (1912)
11. Kuhn, M., Wattenhofer, R.: The theoretic center of computer science. SIGACT News **38**(4), 54–63 (2007)
12. Leskovec, J., Kleinberg, J., Faloutsos, C.: Graph evolution: densification and shrinking diameters. ACM Trans. Knowl. Discov. Data 1, Article 2 (2007)
13. Ley, M.: DBLP - some lessons learned. Proc. VLDB Endow. **2**(2), 1493–1500 (2009)
14. Lotker, Z.: Voting algorithm in the play Julius Caesar. In: IEEE/ACM International Conference on Advances in Social Networks Analysis and Mining, pp. 848–855 (2015)
15. Lotker, Z.: Analyzing Narratives in Social Networks. Springer, Berlin (2021). https://doi.org/10.1007/978-3-030-68299-6
16. Mathieu, F., Tixeuil, S.: Fun with FUN. In: 11th International Conference on Fun with Algorithms. LIPIcs, vol. 226, pp. 1–13 (2022)
17. Michail, O.: An introduction to temporal graphs: an algorithmic perspective. Internet Math. **12**(4), 239–280 (2016)
18. Milgram, S.: The small world problem. Psychol. Today **1**(1), 61–67 (1967)
19. Oettershagen, L., Mutzel, P., Kriege, N.M.: Temporal walk centrality: ranking nodes in evolving networks. In: The ACM Web Conference. ACM (2022)
20. Sørensen, T.: A method of establishing groups of equal amplitude in plant sociology based on similarity of species and its application to analyses of the vegetation on danish commons. Kongelige Danske Videnskabernes Selskab **5**(4), 1–34 (1948)
21. The DBLP team: DBLP computer science bibliography: monthly snapshot release, March 2023. https://dblp.org/xml/release/dblp-2023-01-01.xml.gz
22. Wikipedia contributors: Computer science – Wikipedia, the free encyclopedia (2022). https://en.wikipedia.org/w/index.php?title=Computer_science&oldid=11 21763527

About Informatics, Distributed Computing, and Our Job: A Personal View

Michel Raynal$^{(\boxtimes)}$

IRISA, Inria, CNRS, Univ Rennes, 35042 Rennes, France
raynal@irisa.fr

Abstract. This article (written for the celebration of the 30th Anniversary of the SIROCCO conference series) is a non-technical article that presents a personal view of what are Informatics, Distributed Computing, and our Job. While it does not pretend to objectivity, its aim is not to launch a controversy on the addressed topics. More modestly it intends to encourage readers to form their own view on these important topics.

Keywords: Algorithms · Informatics · Distributed computing · History of sciences · Personal point of view · Research · Teaching

1 A Short Historical Perspective: What Is Informatics

Preliminary Remark As in a lot of European and Latin countries this article uses the term *informatics* instead of *computer science*. This is due to the fact that, to prevent ambiguity, a scientific domain must be defined by one and only one word as done for physics, biology, chemistry, mathematics, astrophysics, geology (and also in non-scientific domains such as law, history, philosophy, arts, literature, etc.)[1].

Using two words (or more!) is nearly always ambiguous as, in any sentence, words are not equal. As said by Hal Abelson [1] "Computer science is no more about computers, than astronomy is about telescopes or biology is about microscopes..."[2]. *Informatics* is (informally) defined here as the science of computations that can be mechanically done (let us remind that computations are not restricted to numbers).

1.1 A Short History

At the Very Beginning. Let us look at the picture in Fig. 1, known as Plimpton tablet 322 (1800 BC). On this very famous tablet are engraved, in the sexagesimal base, the first fifteen Pythagorean triplets, i.e. the triplets (a, b, c) such that $a^2 + b^2 = c^2$.

So it seems that computing was born before or at the same time as writing. It was mainly used to compute field areas, inheritance transmission, and interest rates. This "proves" that Sumer people were able to run non-trivial computations. So they were able to design "algorithms". Unfortunately, no tablet describing these algorithms has yet been discovered, and we do not know if they were able to prove their "algorithms".

---

[1] See also the first chapter in [27].

[2] This citation is sometimes falsely attributed to Dijkstra.

S. Rajsbaum et al. (Eds.): SIROCCO 2023, LNCS 13892, pp. 33–45, 2023.
https://doi.org/10.1007/978-3-031-32733-9_3

Fig. 1. Plimpton tablet (1800 BC) from Wikipedia

From Recipes to Algorithms. The big step is due to Ancient Greeks, mainly Euclid (300 BC), who proposed axioms, mechanical constructions (algorithms) built from a set of well-defined basic geometric operations, accompanied with their proofs [19,27–29,38].

As a simple example let us consider the bisection of an angle with two basic operations, namely, a ruler to draw lines and a compass to draw arcs and measure the length of segments. The construction (algorithm) is easy, see Fig. 2.

- First, with the compass, draw an arc of a circle centered at the angle vertex denoted A. Observe that the two segments AB and AC are equal.
- From the points B and C draw two circles with the same radius such that these circles intersect. Let D be a point where these circles intersect.
- Draw the line AD.
- Claim: the angles $\angle BAD$ and $\angle CAD$ are equal.

The proof of the claim follows from the following trivial observations. By construction we have $|AB| = |AC| = r_1$, and $|BD| = |CD| = r_2$. Hence, the triangles ABD and ACD are equal, from which we get that the corresponding angles are equal, and we get $\angle BAD = \angle CAD$. QED.

A natural question is then: how to trisect an angle with a ruler and a compass. This problem (also known as "squaring of a circle") remained open until 1837, when it was proved to be impossible by P.L. Wantzell [57]. This shows that, as far as geometric constructions are concerned, the operations provided by a ruler and ta compass are not universal. (This simple example –on what is possible/impossible in a given computation model– can be understood by anyone.)

And Finally. As far as sequential computing is concerned, the currently accepted conjecture, named Church-Turing's thesis, is that anything that can be mechanically computed can be computed with a Turing machine [56]. So, the fundamental results of sequential computing are:

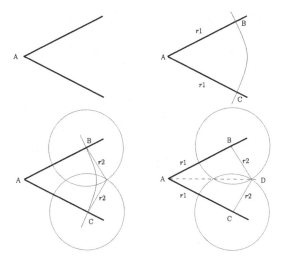

Fig. 2. Bisecting an angle

- Church-Turing's thesis,
- The hierarchy: Finite state automata $\subset$ Pushdown automata $\subset$ Turing machine,
- Computation is symbol manipulation,
- Universality of data representation (at the most basic level, text, video, audio, etc., are sequences of bits).

1.2 Algorithms are (at) the Heart of Informatics

As we have seen, given a computing model (set of primitive operations), a sequential *algorithm* is a sequence of the operations (provided by the model), solving a problem that has been precisely defined (as seen previously with the bisection with a ruler and a compass).

On the Origin of the Word Algorithm. The word *algorithm* comes from the name of the mathematician M. Ibn Musa Al Khwârizmî (700 Khiva – 850 Bagdad), who was working in the *House of Wisdom* created by the Caliph Hârun ar-Rachîd ben Muhammad ben al-Mansûr (Abbasid dynasty 750–1258)[3]. Al Khwârizmî worked on many scientific topics, mainly astronomy and algebra. His today celebrity is related to the computation of the roots of what is today called quadratic equations and written $ax^2 + bx + c$ [23,50,51].

His name started to be known in Europe since the 12th century, and was then disseminated mainly by Leonardo da Pisa (1175–1250, also known as "figlio di Bonacci" shortened to Fibonacci) and Luca Paccioli (1445–1517). Al Khwârizmî was using "algorithms" to describe its methods of calculation and proved their correction using

---

[3] It is worth noticing that the book *The thousand and one nights* was written during this caliphate.

geometry constructions. At the 12th century "computing people" were named *abacist* or *algorithmist* according to the way they were doing their computations; the first ones were using an abacus, the other ones were using Indian numbers (named today Arabic numbers).

On Data and Algorithms. At the center of informatics reside algorithms [24]. This is a direct consequence of the fact that, if we suppress algorithms there is no more computation and so there is no more informatics. A schematic view is presented in Fig. 3. *Informatics* is a science of abstractions (as stated in [21]), And –at its center– *algorithmics* consists then in building higher and higher level computing abstractions (with appropriate data representation).[4]

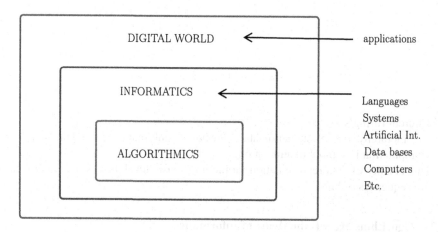

Fig. 3. From algorithms to applications

From Equations to Algorithms. When looking at the past evolution, the main resource until World War II was the pair matter/energy. Now, the main resource is data, which, once processed, provides us with information. As matter/energy, data/information can be collected (extracted), consumed, transformed, stored, carried, etc. But there a fundamental difference between matter/energy and data/information. Differently from matter/energy, data/information is abstract, it does not burn and can be copied at "zero cost".

Moreover, while the aim of science was to "put the world into equations", we know today that everything cannot be captured by a formula. So, the aim of today science is to *put the world into equations and algorithms.* Said in another way, Galileo Galilei said

---

[4] Some people consider algorithmics as a part of mathematics. This is questionable. In the Middle Ages, logic was a part of philosophy, which was part of Rhetoric. Algorithms were born a long time ago (see the abacus-based computations [49,50]), and thanks to the father of modern informatics, A. Turing, Informatics has got its autonomy as a new science. The same appeared for physics with I. Newton, for chemistry with A.-L. Lavoisier., etc.

"the great book of nature is written in mathematical language". This is true for physics, not for other sciences such as life sciences. Today, the pair made up of *mathematics + informatics* seems to be the language of all sciences [8].

A Few Differences Between Informatics and Mathematics. In addition to being the domain of algorithms, informatics and mathematics present an important difference, namely mathematics does not have the touch "run" which transforms a text describing an algorithm (written in some programming language) into an execution! This has a fundamental consequence, namely informatics evolves in the field of "finite". As an example, while mathematics consider the real number π as a number with an infinite number of digits [6], any algorithm that uses π in in is restricted to consider it as defined by a finite number of digits. More generally, any algorithm uses a finite memory and a finite time.

To illustrate the previous claim let us consider the following two objects: the object denoted "1+2" and the object denoted"3" with their usual meaning, and let us ask the question: "are these two objects the same object or are they different objects?". Nearly all mathematicians answer "it is the same object", while nearly all informaticians answer "they are different objects". This is due to the simple observation that informaticians see "1+2" as an algorithm which need to be *executed* to obtain a result, namely the value 3.

2 What is Distributed Computing (or Beyond Turing Machine)

2.1 Parallel Computing Vs Distributed Computing

The Turing world is the world of sequential computing. While parallel computing is an extension of it, distributed computing is not.

Parallel Computing. Parallel computing is a natural extension of sequential computing in the sense that the aim of parallel computing is to detect and exploit *data independence* to obtain efficient programs: once identified, independent sets of data can be processed independently from each other on a multiprocessor. It is nevertheless important to notice that, while independent data can be processed in parallel, any parallel program could be executed on a single processor with an appropriate scheduler (the corresponding sequential execution could be of course highly inefficient!).

Distributed Computing. The nature of distributed computing is totally different from the one of parallel computing. Namely, distributed computing is characterized by the fact that there is a set of predefined (and physically distributed) computing entities (processes) that are *imposed* to the programmers and these entities need to *cooperate to a common goal.* Moreover the behavior of the underlying infrastructure (also called environment) on which the distributed application is executed is not under the control of the programmers who have to consider it as an *hidden input.* Asynchrony and failures are the most frequent phenomenons produced by the environment that create a "context uncertainty" distributed computing has to cope with. In short, distributed computing is characterized by the fact that, in any distributed run, *the run itself is one of its entries* [44,46].

In a Few Words. To summarize, parallel computing is the exploitation of the independence of input data to obtain efficient algorithms (programs), while the aim of distributed computing is to allow predefined computing entities to cooperate to a common goal in a consistent way in the presence of adversaries (mainly, asynchrony, failures). In this sense, *distributed computing is the science of cooperation.* It aims at providing users with high level communication abstractions and cooperation abstractions.

A Famous Quote on the Nature of Distributed Computing.

> *A distributed system is one in which the failure of a computer you didn't even know existed can render your own computer unusable.*   L. Lamport [3]

This quote is nothing else that an humorous statement of the many impossibility results encountered in distributed computing such as the most famous one on consensus in asynchronous crash-prone systems formalized in [20] and known under the acronym FLP. The interested reader will find in [5] a book entirely devoted to unsolvability and lower bound results encountered in distributed computing.

2.2   Sequential Vs Distributed Computing

On the Distributed Computing Problems Side. An important issue in sequential and distributed computing is to understand what is a *computing problem.* Roughly speaking, sequential computing focuses on functions from input strings to output strings [4]. The situation is different in distributed computing: both the input and the output are distributed, and consequently, in the well-studied context of distributed decision problems, the input (resp., output) is a vector with one entry per process. A distributed computing problem is then a mapping from the set of allowed input vectors to the set of allowed output vectors. This gave birth to the notion of *task* [9,36,40]. An in depth study of the solvability of distributed decision problems through the lens of combinatorial topology is presented in [25].

While tasks are appropriate for vector-based input/output specifications (decision problems), the case of distributed objects defined by a non-sequential specification has been investigated in [11, 12].

On the Computing Side. When looking at their impossibility results, the difference between sequential computing and distributed computing is captured by this sentence:

> *In sequential systems, computability is understood through the Church-Turing Thesis: anything that can be computed, can be computed by a Turing Machine. In distributed systems, where computations require coordination among multiple participants, computability questions have a different flavor. Here, too, there are many problems which are not computable, but these limits to computability reflect the difficulty of making decisions in the face of ambiguity, and have little to do with the inherent computational power of individual participants.*   M. Herlihy, S. Rajsbaum, and M. Raynal [26]

This means that, even if the computability power of each computing entity was stronger than the one of the Turing machine, some problems would remain impossible to solve.

2.3   The Two "Most Famous" Distributed Computing Problems

Mutual Exclusion. From a historical point of view, the "very first" distributed computing problem seems to be *mutual exclusion*. This well-known problem was introduced by E.W. Dijkstra in 1965 [17]. Considering a set of asynchronous processes (imposed to the programmer) that share some resource (physical or logical), the mutual exclusion problem consists in ensuring that at most one process at a time accesses the resource. Since Dijkstra's algorithm, plenty of mutual exclusion algorithms have been proposed (see the textbooks and surveys in [41,44,47,54,55]). In the context where the processes communicate through read/write registers, one of the most famous algorithms is Lamport's Bakery algorithm [30]. Differently from nearly all the other mutual exclusion algorithms that consider that the underlying read/write registers are atomic, the Bakery algorithm does not. It is based on safe read/write registers[5].

Consensus. The second "most famous" distributed computing problem is *consensus*, which, despite asynchrony and process crashes and assuming each process propose a value, allows the processes that do not crash to decide on the same output value, that must be one of the input values. As shown in [20] for asynchronous message-passing systems, and in [35] for read/write systems, this problem cannot be solved. So, solving consensus in the presence of asynchrony and even a single process crash, requires to enrich the system with additional computability power (e.g. failure detectors [13], random numbers [7,37], weak synchrony assumptions [10,18]).

Mutual Exclusion vs Consensus. Actually, it seems that these two distributed computing problems are the two sides of the same coin, namely the construction of a total order on the operations applied to an object. Mutual exclusion is for physical objects (e.g. an external device), while consensus is for logical objects (the ones that can be represented as sequences of bits). The table below depicts the main features of this claim, which is developed in [45].[6]

Nature of the object	Poss. underlying replication	Total order obtained from	Underlying Coordination
Physical	No	Mutex	strong
Logical	Yes	Consensus	weak

When looking at the "Underlying coordination" column, it is worth noticing that the weakest information on failures that allows mutex to be solved includes a perpetual

---

[5] A safe register is a register that can be written by a single process and read by any number of processes. A write defines the new value of the register. A read whose execution is not concurrent with a write returns the last value written in the register. A read concurrent with a write returns *any value* that the register can contain (so it can return a value that has never been written in the register!).

[6] Let us also notice that Lamport presented recently in [33] a derivation from his Bakery algorithm described in [30] to his state machine replication algorithm described in [31].

property [15,16], while that the weakest information on failures needed to solve consensus needs to satisfy an eventual property only [13]. This is related to the underlying nature of the object (physical vs logical).

2.4   The Interplay Between Computing Entities and Communication

When considering message-passing communication, an important point of distributed computing lies in the interplay between computing entities and communication (which are key words of the SIROCCO conference series).[7] This interplay and its impact on what can be computed and on efficiency has been mainly considered in the context of synchronous systems in which the processes are connected by a communication graph that is not a clique. The interested reader may consult [22,34,48,53].

3   A Few Personal Points of View

3.1   An Anecdote on the Notion of "First Paper that ..."

As an invited speaker for the SIROCCO Award in 2015 [42], I chose to give a talk on the"patterns" encountered in distributed algorithms. Before moving to distributed computing, I have chosen as an introductory example the well known sequential computing pattern proposed by William George Horner (1786–1837) to efficiently compute the value of a polynomial. Browsing on the web (thanks to Wikipedia) I discovered that this method had been previously proposed by the Italian philosopher and mathematician Paolo Ruffini (1765–1822). A more incisive bibliographic search led me to discover that Horner's method was used by Zhu Shijie (1270–1330), who named it *fan fa* in his book titled *Jade Mirror of the Four Unknowns* (1203), a book in which Pascal's triangle is also described! Books on history of science give many other examples. As a scientist humorously put it: the value of a result could be measured by the number of times it has been rediscovered!

In the same vein, if we ask a researcher which is her most important paper, in a lot of cases, she will answer by presenting the paper she is currently working on.

3.2   On Scientific Competition

There is a big difference between industry competition and scientific competition. Roughly speaking, industry competition means "win or die". In a university context the situation is radically different. Nearly always the competition is a friendly helping. Nothing remains hidden, and (non-virtual) conferences are the playground where we exchange ideas and opinions. Coffee breaks are as important as talks (in some cases even more), and lead to joint work and articles, and, in the long run benefit to the society.

---

[7] A similar interplay was investigated a long time ago in parallel computing, namely the notion of sorting network [2].

3.3   On the Notion of "Best Paper"

As all of you, I have been asked the following question many times (mainly from PhD students): What is a good paper?

At the very beginning (when I was younger, i.e. in the previous millennium!) my answer was mainly based on an objective numerical criterion, namely, a good paper is "a paper with numerous citations". Later, I was saying "a paper that won the best paper award in a top conference". Still later I was saying "a paper that won a prize devoted to more than ten years old papers", etc.

But over time, none of these integer-based definitions fully satisfied me, and I started thinking about the papers that I myself consider as very important papers ... and I discovered that those were papers I was a little bit *kindly jealous* ... not to be a co-author! This was because, those are papers I like to read (and reread) because they are nicely written, their content goes beyond their technical content, they introduce new ideas in a simple and efficient way, and have a very strong impact on the community. This is the effect of good papers: everyone makes them"theirs", assimilating them and passing their essence to students.

3.4   On the Evolution of Informatics

In 1936 A A. Turing laid the foundations of (modern) sequential computing [56], and there is a conjecture on what can be mechanically computed and what cannot not. At this time theory was preceding applications. Today, the schema is reversed. There are more and more applications, but very few fertilize theory and the world of applications seems to become a map-missing megalopolis of applications.

4   By Way of Conclusion: What is Our Job?

Research is THE *raison d'être* of universities, and this is independent of the time and the geographical location[8]. Research is the fuel of university education. It is both a personal and collective adventure of an intellectual nature. By "adventure" I mean that research is not defined by a roadmap. Its goal is to understand the world that surrounds us and the artifacts that we create in it.

Research pretends to a certain universality (let us not forget that the words Universe, Universality, and University derive from the same root). Research relies on curiosity and obstinacy, sagacity and the personal knowledge of its actors, as well as on serendipity (in the sense of an unexpected encounter between personal knowledge, state of mind, context, and... chance... linked to the context in which it is realized).

Tenure positions are crucial regardless of local cultures, social behaviors, and economical structures. This is not accidental, and goes back to the very first Chinese mandarins[9]. The permanence of faculty positions is a necessary condition to establish a long-term approach, without being constrained by hiccups and vagaries of the short

---

[8] The text that follows is a digest of an article that appeared (in French) in [43].

[9] The permanence of their position guaranteed them time for thinking and preventing impulsive and precipitated judgments.

term "time to market" must not be the definitive diktat, nor the criterion for defining research topics, nor the analysis grid presiding over their evaluation.

The sentence "No one enters here if he is not a geometer" engraved on the pediment of the Platonic Academy, summarizes a state of mind of scientific research. While there is a continuum from a part of research to its directly visible impact in society, for an academic researcher applications are more important for the new questions they raise than for their direct impact on the economy. As said by H. Poincaré in [39]:

A science made only with a view to applications is impossible; truths are only fruitful if they are linked to each other. If one attaches oneself only to those from which one expects an immediate result, the intermediate rings will disappear and there will be no chain.

Like yin and yang, at the graduate level, research and teaching are inseparable. On the one hand, research is the domain of uncertainty. One tackles a problem without knowing what will come out of it and, once published, a result is part of the past, piously preserved in DBLP and CVs[10]. When we consider the teaching/research duality, teaching is mainly on the side of certainties and questioning aimed at developing students' critical thinking skills. We teach things, which we partially modify from one year to the next, depending on the topic and what one has assimilated from the research results related to concerned curriculum.

Teaching requires to awake the intellectual curiosity of students and give them the desire to learn (which may seem obvious but is far from easy). The student's desire to learn is far more important than the pedagogical form used to teach. (Too many times we are as a doctor for whom illness is more important than the patient...)

Defining the content of a curriculum requires thinking in terms of several components. The first one is the one that allows students to be operational when they start working. This is important for two reasons: they must have confidence in themselves and their employers must have confidence in their abilities. This requires basic courses both on concepts and knowledge of tools and technology. The second component is to give students the knowledge that will give them the confidence and hindsight that, in thirty years from now, they will still have a job and will not be thrown in the garbage can where is already present the technology they used in their first steps. Finally, the third component consists in transmitting a scientific culture which, without being directly useful, will allow them to better understand current science, innovations and the world around them.

Although immediacy, speed and competition are experienced in today's society as (market) values, it is important to note that we do not think any faster today than twenty-five centuries ago, when emerged, in the Pericles' century, the founding ideas of History, Democracy (in a particular form), Geography, Mathematics, Algorithmics (*with* their proofs), Tragedy, Medicine, etc.

There were (some kind of) universities a thousand years ago, there will be universities a thousand years from now (in one form or another). We are aware that we belong

---

[10] Unfortunately results known as "negative" are too often considered as second-class citizens, whereas they often shed light on an obscure face of some positive results [14].

to a big family. We feel at home in almost any university in the world, and feel solidarity with our present colleagues and those who have gone before us, regardless of their specific beliefs and cultures. We are with them in a logic of exchanges and knowledge and not a logic of economic competitiveness.

To conclude, two teaching-related citations:

- *Teaching is not an accumulation of facts*, from Lamport [32],
- *Teaching is thinking aloud in front of students*, from H. Lebesgue.

Acknowledgments. I want to thank Sergio Rajsbaum (Program Chair of SIROCCO 2023), Alkida Balliu and Dennis Olivetti (Chairs of the 30th Anniversary of SIROCCO ceremony) for their invitation to give a talk at SIROCCO 2023. I also want to thank Gérard Berry for discussions we had and his talks at Collège de France [8], and J. Sifakis for his book [52] and fruitful discussions on the nature of what is Informatics. I also want to thank reviewers for their careful reading that help improve the presentation of this article.

References

1. https://quoteinvestigator.com/2021/04/02/computer-science/
2. https://en.wikipedia.org/wiki/Sorting-network
3. Email message sent to a DEC SRC bulletin board at 12:23:29 PDT on 28 May 1987
4. Aho, A.V.: Computation and computational thinking. Comput. J. **55**(7), 832–835 (2012)
5. Attiya, H., Ellen, F.: Impossibility results for distributed computing. Morgan Claypool Synth. Lect. Distrib. Comput. Theory, 146 (2014)
6. Beckmann, P.: A Hisytory of π, p. 201. St Martin's Press, New York (1971)
7. Ben-Or, M.: Another advantage of free choice: completely asynchronous agreement protocols. In: Proceedings of the 2nd ACM Symposium on Principles of Distributed Computing (PODC 1983), pp. 27–30. ACM Press (1983)
8. Berry, G.: Talks at Collège de France (2012–2019)
9. Biran, O., Moran, S., Zaks, S.: A combinatorial characterization of the distributed 1-solvable tasks. J. Algorithms **11**(3), 420–440 (1990)
10. Bouzid, Z., Mostéfaoui, A., Raynal, M.: Minimal synchrony for Byzantine consensus. In: Proceedings of the 34th ACM Symposium on Principles of Distributed Computing (PODC 2015), pp. 461–470. ACM Press (2015)
11. Castanñeda, A., Rajsbaum, S., Raynal, M.: Unifying concurrent objects and distributed tasks: interval-linearizability. J. ACM **65**(6), 42 (2018). Article 45
12. Castanñeda, A., Rajsbaum, S., Raynal, M.: A linearizability-based hierarchy for concurrent specifications. Commun. ACM **66**(1), 86–97 (2023)
13. Chandra, T.D., Hadzilacos, V., Toueg, S.: The weakest failure detector for solving consensus. J. ACM **43**(4), 685–722 (1996)
14. Chevassus-au-Louis, N.: Malscience: De la fraude dans les labos (in French). Editions du Seuil, p. 216 (2016)
15. Delporte-Gallet, C., Fauconnier, H., Guerraoui, R., Kouznetsov, P.: Mutual exclusion in asynchronous systems with failure detectors. J. Parallel Distrib. Comput. **65**, 492–505 (2005)
16. Delporte-Gallet, C., Fauconnier, H., Raynal, M.: On the weakest information on failures to solve mutual exclusion and consensus in asynchronous crash prone read/write systems. J. Parallel Distrib. Comput. **153**, 110–118 (2021)

17. Dijkstra, E.W.: Solution of a problem in concurrent programming control. Commun. ACM **8**(9), 569 (1965)
18. Dolev, D., Dwork, C., Stockmeyer, L.: On the minimal synchronism needed for distributed consensus. J. ACM **34**(1), 77–97 (1987)
19. Ershov, A.P., Knuth, D.: Algorithms in modern mathematics and computer science. LNCS, vol. 122, p. 487. Springer, Heidelberg (1979). https://doi.org/10.1007/3-540-11157-3
20. Fischer, M.J., Lynch, N.A., Paterson, M.S.: Impossibility of distributed consensus with one faulty process. J. ACM **32**(2), 374–382 (1985)
21. Fischer, M.J., Merritt, M.: Appraising two decades of distributed computing theory research. Distrib. Comput. **16**, 239–247 (2003)
22. Fraigniaud, P., Korman, A., Peleg, D.: Towards a complexity theory for local distributed computing. J. ACM **60**(5), 16 (2013). Article 35
23. Gavin, J., Schärlig, A.: Et l'algèbre fut: de l'*al-jabr* au 9ème siècle au signe *egal* en 1557, 166 (in French). EPFL Press (2020)
24. Harel, D., Feldman, Y.: Algorithmics: The Spirit of Computing, 3rd edn., p. 572. Springer, Cham (2012)
25. Herlihy, M., Kozlov, G., Rajsbaum, S.: Distributed Computing Through Combinatorial Topology, p. 319. Morgan Kaufmann, Burlington (2013)
26. Herlihy, M., Rajsbaum, S., Raynal, M.: Power and limits of distributed computing shared memory models. Theoret. Comput. Sci. **509**, 3–24 (2013)
27. Knuth, D.E.: Selected Papers on Computer Science, p. 274. Cambridge Univertity Press, Cambridge (1996)
28. Knuth, D.E.: Ancient Babylonian algorithms. Commun. ACM **15**(7), 671–677 (1972)
29. Kramer, S.N.: History Begins at Sumer: Thirty-nine Firsts in Man's Recorded History, p. 416. University of Pennsylvania Press, Pennsylvania (1956)
30. Lamport, L.: A new solution of Dijkstra's concurrent programming problem. Commun. ACM **17**(8), 453–455 (1974)
31. Lamport, L.: Time, clocks, and the ordering of events in a distributed system. Commun. ACM **21**(7), 558–565 (1978)
32. Lamport, L.: Teaching concurrency. ACM SIGACT News **40**(1), 58–62 (2009)
33. Lamport, L.: Deconstructing the Bakery to build a distributed state machine. Commun. ACM **65**(9), 58–66 (2022)
34. Linial, N.: Locality in distributed graph algorithms. SIAM J. Comput. **21**(1), 193–201 (1992)
35. Loui, M., Abu-Amara, H.: Memory requirements for agreement among unreliable asynchronous processes. Adv. Comput. Res. 4(163–183), 31. JAI Press Inc. (1987)
36. Moran, S., Wolfsthal, Y.: extended impossibility results for asynchronous complete networks. Inf. Process. Lett. **26**(3), 145–151 (1987)
37. Mostéfaoui, A., Moumen, H., Raynal, M.: Signature-free asynchronous binary Byzantine consensus with $t<n/3$, $O(n^2)$ messages, and $O(1)$ expected time. J. ACM **62**(4), 21 (2015). Article 31
38. Neugebauer, O.E.: The exact sciences in Antiquity, 2nd edition. Princeton University Press (1952), Brown University Press (1957), Reprint: Dover publications (1969)
39. Poincaré, H.: La science et l'hypothèse. Flammarion, p. 322 (1902). English translation: Science and hypothesis, Walter Scott Publishing Company (1905)
40. Rajsbaum, S.: Distributed decision problems: concurrent specifications beyond binary problems. In: Proceedings of the 23rd Int'l Conference on Concurrency Theory (COCUR 2022), LIPIcs, vol. 243, p. 13 (2022). Article 36
41. Raynal, M.: Concurrent Programming: Algorithms, Principles and Foundations, p. 515. Springer, Cham (2013). ISBN: 978-3-642-32026-2

42. Raynal, M.: Communication patterns and input patterns in distributed computing. In: Scheideler, C. (ed.) SIROCCO 2014. LNCS, vol. 9439, pp. 1–15. Springer, Cham (2015). https://doi.org/10.1007/978-3-319-25258-2_1

43. Raynal, M.: Réflexions désordonnées. Bull. 1024 de la Société Informatique de France **9**, 115–122 (2016)

44. Raynal, M.: Fault-Tolerant Message-passing Distributed Systems: An Algorithmic Approach, 550. Springer, Cham (2018)

45. Raynal, M.: Mutual exclusion vs consensus: both Sides of the Same Coin?. Bull. Eur. Assoc. Theoret. Comput. Sci. **140**, 15 (2023)

46. Raynal, M.: A short visit to distributed computing where simplicity is considered a first class property. In: Chapter of the Book The French School of Programming. Springer (2023)

47. Raynal, M., Taubenfeld, G.: A visit to mutual exclusion in seven dates. Theoret. Comput. Sci. **919**, 47–65 (2022)

48. Santoro, N., Widmayer, P.: Time is not a healer. In: Monien, B., Cori, R. (eds.) STACS 1989. LNCS, vol. 349, pp. 304–313. Springer, Heidelberg (1989). https://doi.org/10.1007/BFb0028994

49. Schärlig, A.: Du zéro à la virgule (in French), p. 441. EPFL Press (2012)

50. Schärlig, A.: Un portrait de Gerbert d'Aurillac (in French), p. 133. EPFL Press (2010)

51. Sesiano, J.: Une introduction à l'hstoire de l'algèbre (in French), p. 169. Presses Polytechniques et Universitaires Romandes (2010)

52. Sifakis, J.: Understanding and Changing the World: From Information to Knowledge and Intelligence, p. 158. Springer, Cham (2022)

53. Suomela, J.: Survey of local algorithms. ACM Comput. Surv. **45**(2), 40 (2013). Article 24

54. Taubenfeld, G., Synchronization Algorithms and Concurrent Programming, p. 423. Pearson Education/Prentice Hall (2006). ISBN: 0-131-97259-6

55. Taubenfeld, G.: Concurrent Programming, Mutual Exclusion, pp. 421–425 . Springer Encyclopedia of Algorithms (2016)

56. Turing, A.M.: On computable numbers with an application to the Entscheidungsproblem. Proc. London Math. Soc. **42**, 230–265 (1936)

57. Wantzell, P.L.: Recherches sur les moyens de reconnaître si un problème de géométrie peut se résoudre avec la règle et le compas. J. Math. Pures Appl. **1**(2), 366–372 (1837)

Learning Hierarchically-Structured Concepts II: Overlapping Concepts, and Networks with Feedback

Nancy Lynch[1] and Frederik Mallmann-Trenn[2]([⊠])

[1] MIT, Cambridge, USA
[2] King's College London, London, UK
frederik.mallmann-Trenn@kcl.ac.uk

Abstract. We continue our study from Lynch and Mallmann-Trenn (Neural Networks, 2021), of how concepts that have hierarchical structure might be represented in brain-like neural networks, how these representations might be used to recognize the concepts, and how these representations might be learned. In Lynch and Mallmann-Trenn (Neural Networks, 2021), we considered simple tree-structured concepts and feed-forward layered networks. Here we extend the model in two ways: we allow limited overlap between children of different concepts, and we allow networks to include feedback edges. For these more general cases, we describe and analyze algorithms for recognition and algorithms for learning.

Keywords: Brain · Learning · Hierarchical Concepts

1 Introduction

We continue our study, begun in [4], of how concepts that have hierarchical structure might be represented in brain-like neural networks, how these representations might be used to recognize the concepts, and how these representations might be learned. In [4], we considered only simple tree-structured concepts and simple feed-forward layered networks. Here we consider two important extensions: we allow our data model to include limited *overlap* between the sets of children of different concepts, and we extend the network model to allow some *feedback edges*. We consider these extensions both separately and together. In all cases, we consider both algorithms for recognition and algorithms for learning. Where we can, we quantify the effects of these extensions on the costs of recognition and learning algorithms.

In this paper, as in [4], we consider *robust recognition*, which means that recognition of a concept is guaranteed even in the absence of some of the lowest-level parts of the concept. In [4], we considered both noise-free learning and learning in the presence of random noise. Here we emphasize noise-free learning, but include some ideas for extending the results to the case of noisy learning.

© The Author(s), under exclusive license to Springer Nature Switzerland AG 2023
S. Rajsbaum et al. (Eds.): SIROCCO 2023, LNCS 13892, pp. 46–86, 2023.
https://doi.org/10.1007/978-3-031-32733-9_4

Motivation: This work is inspired by the behavior of the visual cortex, and by algorithms used for computer vision. As described in [4], we are interested in the general problem of how concepts that have structure are represented in the brain. What do these representations look like? How are they learned, and how do the concepts get recognized after they are learned? We draw inspiration from experimental research on computer vision in convolutional neural networks (CNNs) by Zeiler and Fergus [9] and Zhou, et al. [10]. This research shows that CNNs learn to represent structure in visual concepts: lower layers of the network represent basic concepts and higher layers represent successively higher-level concepts. This observation is consistent with neuroscience research, which indicates that visual processing in mammalian brains is performed in a hierarchical way, starting from primitive notions such as position, light level, etc., and building toward complex objects; see, e.g., [1–3].

In [4], we considered only tree-structured concepts and feed-forward layered networks. Here we allow overlap between sets of children of different concepts, and feedback edges in the network. Overlap may be important, for example, in a complicated visual scene in which one object is part of more than one higher-level object, like a corner board being part of two sides of a house. Feedback is critical in visual recognition, since once we recognize a particular higher-level object, we can often fill in lower-level details that were not easily recognized without the help of the context provided by the object. For example, once we recognize that we are looking at a dog, based on seeing some of its parts, we can recognize other parts that are less visible, such as a partially-occluded leg.

Paper Contents: We begin in Sect. 2 by extending our formal concept hierarchy model of [4]. We continue in Sect. 3 with definitions of our networks, both feed-forward and with feedback. Next, in Sect. 4, we define the robust recognition and noise-free learning problems. Section 5 contains algorithms for robust recognition in feed-forward networks, for both tree hierarchies and general hierarchies. Section 6 contains algorithms for robust recognition in networks with feedback, for both tree hierarchies and general hierarchies. In Sect. F, we describe noise-free learning algorithms in feed-forward networks, which produce edge weights for the upward edges that suffice to support robust recognition. In Sect. G, we extend the learning algorithms for feed-forward networks to accommodate feedback.

2   Data Models

We use two types of data models in this paper. One is the same type of tree hierarchy as in [4]. The other allows limited overlap in the sets of children of different concepts. As before, a concept hierarchy is supposed to represent all the concepts that are learned in the "lifetime" of an organism, together with parent/child relationships between them.

We also include two definitions for the notion of "supported", which are used to describe the set of concepts whose recognition should be triggered by a given set of basic concepts. One definition is for the case where information flows only upwards, from children to parents, while the other also allows downward flow, from parents to children. These definitions capture the idea that recognition is robust, in the sense that a certain fraction of neighboring (child and parent) concepts should be enough to support recognition of a given concept.

2.1 Preliminaries

We start by defining some parameters: Let ℓ_{max}, be a positive integer, representing the maximum level number for the concepts we consider. Let n be a positive integer, representing the total number of lowest-level concepts. Let k be a positive integer, representing the number of top-level concepts in a concept hierarchy, and the number of sub-concepts for each concept that is not at the lowest level in the hierarchy. Let r_1, r_2 be reals in $[0, 1]$ with $r_1 \leq r_2$; these represent thresholds for robust recognition. Let o be a real in $[0, 1]$, representing an upper bound on overlap. Let f be a nonnegative real, representing strength of feedback. We assume a universal set D of *concepts*, partitioned into disjoint sets $D_\ell, 0 \leq \ell \leq \ell_{max}$. We refer to any particular concept $c \in D_\ell$ as a *level ℓ concept*, and write $level(c) = \ell$. Here, D_0 represents the most basic concepts and $D_{\ell_{max}}$ the highest-level concepts. We assume that $|D_0| = n$.

2.2 Concept Hierarchies

We define a general notion of a concept hierarchy, which allows overlap. We will refer to our previous notion from [4] as a "tree concept hierarchy"; it can be defined by a simple restriction on the general definition.

A *(general) concept hierarchy* $\mathcal{C}$ consists of a subset $C \subseteq D$, together with a *children* function, satisfying the constraints below. For each ℓ, $0 \leq \ell \leq \ell_{max}$, we define C_ℓ to be $C \cap D_\ell$, that is, the set of level ℓ concepts in $\mathcal{C}$. For each concept $c \in C_0$, we assume that $children(c) = \emptyset$. For each concept $c \in C_\ell$, $1 \leq \ell \leq \ell_{max}$, we assume that $children(c)$ is a nonempty subset of $C_{\ell-1}$. We call each element of $children(c)$ a *child* of c. We extend the *children* notation recursively, namely, we define concept c' to be a *descendant* of a concept c if either $c' = c$, or c' is a child of a descendant of c. We write $descendants(c)$ for the set of descendants of c. Let $leaves(c) = descendants(c) \cap C_0$, that is, the set of all level 0 descendants of c. Also, we call every concept c' for which $c \in children(c')$ a *parent* of c, and write $parents(c)$ for the set of parents of c. Since we allow overlap, the set $parents(c)$ might contain more than one element. If a concept c has only one parent, we write $parent(c)$. We assume the following properties:

1. $|C_{\ell_{max}}| = k$. That is, the number of top-level concepts in the hierarchy is exactly k.[1]
2. For any $c \in C_\ell$, where $1 \leq \ell \leq \ell_{max}$, we have that $|children(c)| = k$. That is, the number of children of any non-leaf concept is exactly k.

---

[1] This assumption and the next are just for uniformity, to reduce the number of parameters and simplify the math.

3. *Limited overlap:* Let $c \in C_\ell$, where $1 \leq \ell \leq \ell_{\max}$. Let $C' = \bigcup_{c' \in C_\ell - \{c\}} children(c')$; that is, C' is the union of the sets of children of all the other concepts in C_ℓ, other than c. Then $|children(c) \cap C'| \leq o \cdot k$.

   To define a *tree hierarchy*, we replace the limited overlap property with the stronger property:

4. *No overlap:* For any two distinct concepts c and c' in C_ℓ, where $1 \leq \ell \leq \ell_{\max}$, we have that $children(c) \cap children(c') = \emptyset$. That is, the sets of children of different concepts at the same level are disjoint. This property is equivalent to Property 3 with $o = 0$. Since children of level ℓ concepts are at level $\ell - 1$, by assumption, there cannot be overlap across different levels.

Properties 1, 2, and 4 are the same as in [4]. Property 3 is new here: we replace the no-overlap condition assumed in [4] with a condition that limits the overlap between the set of children of a concept c and the sets of children of all other concepts at the same level. We require this overlap to be less than a designated fraction $o \cdot k$ of the children of c. See Fig. 1 for an example.

2.3   Support

In this subsection, we fix a particular concept hierarchy $\mathcal{C}$, with its concept set C, partitioned into $C_0, C_1, \ldots, C_{\ell_{\max}}$. We assume that $\mathcal{C}$ satisfies the limited-overlap property. We give two definitions, one that expresses only upward information flow and one that also expresses downward information flow.

Support with only Upward Information Flow. For any given subset B of the general set D_0 of level 0 concepts, and any real number $r \in [0, 1]$, we define a set $supp_r(B)$ of concepts in C. This represents the set of concepts $c \in C$, at any level, such that B contains enough leaves of c to support recognition of c. The notion of "enough" here is defined recursively, in terms of a level parameter ℓ. This definition is equivalent to the corresponding one in [4].

Definition 1 ($supp_r(B)$). *Given $B \subseteq D_0$, define the sets of concepts $S(\ell)$, for $0 \leq \ell \leq \ell_{\max}$: $S(0) = B \cap C_0$ and for $1 \leq \ell \leq \ell_{max}$, $S(\ell)$ is the set of all concepts $c \in C_\ell$ such that $|children(c) \cap S(\ell - 1)| \geq rk$. Define $supp_r(B)$ to be $\bigcup_{0 \leq \ell \leq \ell_{\max}} S(\ell)$. We also write $supp_r(B, \ell)$ for $S(\ell)$, when we want to make the parameters r and B explicit.*

Support with both Upward and Downward Information Flow. Our second definition, which captures information flow both upward and downward in the concept hierarchy, is a bit more complicated. It is expressed in terms of a generic "time parameter" t, in addition to the level parameter ℓ. Here, f is a nonnegative real, as specified at the start of Sect. 2.1.

Definition 2 ($supp_{r,f}(B)$). *Given $B \subseteq D_0$, define the sets of concepts $S(\ell, t)$, for $0 \leq \ell \leq \ell_{\max}$ and $t \geq 0$: 1) $\ell = 0$ and $t \geq 0$: Define $S(0, t) = B$. B is initially*

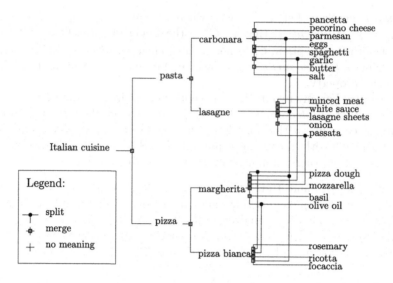

Fig. 1. The figure above illustrates overlapping concepts. In this example, salt is used in all dishes. Moreover, each concept's children intersects with at most five children of other same-level concepts' children taken together. The authors apologize to all Italians for reducing their cuisine to pasta and pizza, and the artistic liberties w.r.t. the recipes.

supported and continues to be supported, and no level 0 concept other than those in B ever gets supported. 2) $1 \leq \ell \leq \ell_{\max}$ and $t = 0$: Define $S(\ell, 0) = \emptyset$. No concepts at levels higher than 0 are initially supported. 3) $1 \leq \ell \leq \ell_{\max}$ and $t \geq 1$: Define $S(\ell, t) = S(\ell, t - 1) \cup \{c \in C_\ell : |children(c) \cap S(\ell - 1, t - 1)| + f |parents(c) \cap S(\ell + 1, t - 1)| \geq rk\}$. Thus, concepts that are supported at time $t - 1$ continue to be supported at time t. In addition, new level ℓ concepts can get supported at time t based on a combination of children and parents being supported at time $t - 1$, with a weighting factor f used for parents.

Define $supp_{r,f}(B)$ to be $\bigcup_{\ell, t} S(\ell, t)$. We sometimes also write $supp_{r,f}(B, \ell, t)$ for $S(\ell, t)$, when we want to make the parameters r, f, and B explicit.

We also use the abbreviations $supp_{r,f}(B, *, t)$ for $\bigcup_\ell S(\ell, t)$, $supp_{r,f}(B, \ell, *)$ for $\bigcup_t S(\ell, t)$, and $supp_{r,f}(B, *, *)$ for $\bigcup_{\ell, t} S(\ell, t)$, Notice that each of these three unions must converge to a finite set since all the sets $S(\ell, t)$ are subsets of the single finite set C_ℓ of concepts. Now we have two monotonicity results, for r and f:

Lemma 1. *Let C be any concept hierarchy satisfying the limited-overlap property, and let $B \subseteq D_0$. Consider r, r', where $0 \leq r \leq r' \leq 1$, and arbitrary f. Then $supp_{r', f}(B) \subseteq supp_{r, f}(B)$.*

Lemma 2. *Let C be any concept hierarchy satisfying the limited-overlap property, and let $B \subseteq D_0$. Consider f, f', where $0 \leq f \leq f'$, and arbitrary $r \in [0, 1]$. Then $supp_{r, f}(B) \subseteq supp_{r, f'}(B)$.*

Also note that the second *supp* definition with $f = 0$ corresponds to the first definition:

Lemma 3. *Let $\mathcal{C}$ be any concept hierarchy satisfying the limited-overlap property, and $B \subseteq D_0$. Then $supp_{r,0}(B) = supp_r(B)$. Moreover, for every ℓ, $0 \leq \ell \leq \ell_{\max}$, $supp_{r,f}(B, \ell, *) = supp_r(B, \ell)$.*

Time Bounds. We would like upper bounds on the time by which the sets $S(\ell, t)$ in the second *supp* definition stabilize to their final values. Specifically, for each value of ℓ, we would like to find a value t^* such that $S(\ell, t^*) = supp_{r,f}(B, \ell, *)$. It follows that, for every $t \geq t^*$, $S(\ell, t) = supp_{r,f}(B, \ell, *)$.

In general, we have only a large (exponential in $\ell_{\max}$) upper bound, based on the fact that $\mathcal{C}$ contains only a bounded number of concepts. However, we have better results in two special cases. The first result is for the case where $f = 0$, that is, where there is no feedback from parents. In this case, for every ℓ, the sets $S(\ell, t)$ stabilize within time ℓ, as the support simply propagates upwards. This can be proven by induction on ℓ.

Proposition 1. *Let $\mathcal{C}$ be any concept hierarchy satisfying the limited-overlap property, and let $B \subseteq D_0$. Then for any ℓ, $0 \leq \ell \leq \ell_{\max}$, we have $supp_{r,0}(B, \ell, *) = supp_{r,0}(B, \ell, \ell)$.*

It follows that $supp_{r,0}(B) = \bigcup_\ell supp_{r,0}(B, \ell, \ell)$. Since $\ell_{\max}$ is the maximum value of ℓ, we get that $supp_{r,0}(B) = \bigcup_\ell supp_{r,0}(B, \ell, \ell_{\max}) = supp_{r,0}(B, *, \ell_{\max})$. This means that all the sets $S(\ell, t)$ stabilize to their final values by time $t = \ell_{\max}$.

The second result is for the special case of a tree hierarchy, with any value of f. In this case, the support may propagate both upwards and downwards. This propagation may follow a complicated schedule, but the total time is bounded by $2\,\ell_{\max}$. To prove this, we use a lemma saying that, if a concept c gets put into an $S(\ell, t)$ set before its parent does, then c is supported by just its descendants. To state the lemma, we here abbreviate $supp_{r,f}(B, *, t)$ by simply $S(t)$. Thus, $S(t)$ is the set of concepts at all levels that are supported by input set B, by step t of the recursive definition of the $S(\ell, t)$ sets. The lemma says that, if a concept c is in $S(t)$ and its parent is not, then it must be that c is supported by just its descendants:

Lemma 4. *Let $\mathcal{C}$ be any tree concept hierarchy, and let $B \subseteq D_0$. Let t be any nonnegative integer. If $c \in S(t)$ and $parent(c) \notin S(t)$ then $c \in supp_{r,0}(B)$.*

Now we can state our time bound result. It says that, for the case of tree hierarchies, the sets $S(\ell, t)$ stabilize within time $2\,\ell_{\max} - \ell$.

Theorem 1. *Let $\mathcal{C}$ be any tree concept hierarchy, and let $B \subseteq D_0$. Then for any ℓ, $0 \leq \ell \leq \ell_{\max}$, we have $supp_{r,f}(B, \ell, *) = supp_{r,f}(B, \ell, 2\,\ell_{\max} - \ell)$.*

3 Network Models

We consider two types of network models, for feed-forward networks and for networks with feedback. The model for feed-forward networks is the same as the network model in [4], with upward edges between consecutive layers. The model for networks with feedback includes edges in both directions between consecutive layers.

We define the following parameters: Let ℓ'_{max}, a positive integer, representing the maximum number of a layer in the network. Let n, a positive integer, representing the number of distinct inputs the network can handle; this is intended to match up with the parameter n in the data model. Let f, a nonnegative real, representing strength of feedback; this is intended to match up with the parameter f in the data model.

Let τ, a positive real number, representing the firing threshold for neurons. Let η, a positive real, representing the learning rate for Oja's rule.

3.1 Network Structure

A network $\mathcal{N}$ consists of a set N of neurons, partitioned into disjoint sets N_ℓ, $0 \le \ell \le \ell'_{max}$, which we call *layers*. We assume that each layer contains exactly n neurons, that is, $|N_\ell| = n$ for every ℓ. We refer to the n neurons in layer 0 as *input neurons*. For feed-forward networks, we assume that each neuron in N_ℓ, $0 \le \ell \le \ell'_{max} - 1$, has an outgoing "upward" edge to each neuron in $N_{\ell+1}$, and that these are the only edges in the network. For networks with feedback, we assume that, in addition to these upward edges, each neuron in $N_\ell, 1 \le \ell \le \ell'_{max}$, has an outgoing "downward" edge to each neuron in $N_{\ell-1}$.

We assume a one-to-one mapping $rep : D_0 \to N_0$, where $rep(c)$ is the input neuron corresponding to level 0 concept c. That is, rep is a mapping from the full set of level 0 concepts to the full set of layer 0 neurons. This allows the network to receive an input corresponding to any level 0 concept, using a simple unary encoding.

3.2 Feed-Forward Networks

Here we describe the contents of neuron states and the rules for network operation, for feed-forward networks.

Neuron States. Each input neuron $u \in N_0$ has just one state component: *firing*, with values in $\{0, 1\}$; $firing = 1$ indicates that the neuron is firing, and $firing = 0$ indicates that it is not firing. We denote the *firing* component of neuron u at time t by $firing^u(t)$. We assume that the value of $firing^u(t)$, for every time t, is set by some external input signal and not by the network.

Each non-input neuron $u \in N_\ell, 1 \le \ell \le \ell'_{max}$, has three state components: *firing*, with values in $\{0, 1\}$, *weight*, a length n vector with entries that are reals in the interval $[0, 1]$, and *engaged*, with values in $\{0, 1\}$.

The *weight* component keeps track of the weights of incoming edges to u from all neurons at the previous layer. The *engaged* component is used to indicate whether neuron u is currently prepared to learn new weights. We denote the three components of non-input neuron u at time t by $firing^u(t)$, $weight^u(t)$, and $engaged^u(t)$, respectively. The initial values of these components are specified as part of an algorithm description. The later values are determined by the operation of the network, as described below.

Potential and Firing. Now we describe how to determine the values of the state components of any non-input neuron u at time $t \geq 1$, based on u's state and on firing patterns for its incoming neighbors at the previous time $t - 1$.

In general, let $x^u(t)$ denote the column vector of *firing* values of u's incoming neighbor neurons at time t. Then the value of $firing^u(t)$ is determined by neuron u's *potential* for time t and its *activation function*. Neuron u's potential for time t, $pot^u(t)$, is given by the dot product of its *weight* vector, $weight^u(t-1)$, and its input vector, $x^u(t-1)$, at time $t-1$: $pot^u(t) = \sum_{j=1}^{n} weight_j^u(t-1) \, x_j^u(t-1)$. The activation function, which determines whether or not neuron u fires at time t, depends on, τ is firing threshold: $firing^u(t) = 1$ if $pot^u(t) \geq \tau$, and 0 otherwise.

Edge Weight Modifications. We assume that the value of the *engaged* flag of u is controlled externally; that is, for every t, the value of $engaged^u(t)$ is set by an external input signal.[2] We assume that each neuron that is engaged at time t determines its weights at time t according to Oja's learning rule [7] with learning rate η. That is, if $engaged^u(t) = 1$, then (using vector notation for $weight^u$ and x^u): *Oja's rule:* $weight^u(t) = weight^u(t-1) + \eta \, pot^u(t) \, (x^u(t-1) - pot^u(t) \, weight^u(t-1))$.

Network Operation. During operation, the network proceeds through a series of *configurations*, each of which specifies a state for every neuron in the network. As described above, the *firing* values for the input neurons and the *engaged* values for the non-input neurons are determined by external sources. The other state components, which are the *firing* and *weight* values for the non-input neurons, are determined by the initial network specification at time $t = 0$, and by the activation and learning functions described above for all times $t > 0$.

3.3   Networks with Feedback

Now we describe the neuron states and rules for network operation for networks with feedback.

---

[2] This is a departure from our usual models [4–6], in which the network determines all the values of the state components for non-input neurons. We expect that the network could be modeled as a composition of sub-networks. One of the sub-networks—a *Winner-Take-All (WTA)* sub-network— would be responsible for setting the *engaged* state components. We will discuss the behavior of the WTA sub-network in Sect. F, but a complete compositional model remains to be developed.

Neuron States. Each neuron in the network has a state component *firing*, with values in $\{0, 1\}$. In addition, each non-input neuron $u \in N_\ell$, $1 \leq \ell < \ell'_{max}$, has state components: *uweight*, a length n vector with entries that are reals in the interval $[0, 1]$; these represent weights on "upward" edges, i.e., those from incoming neighbors at level $\ell - 1$, and *ugaged*, with values in $\{0, 1\}$, representing whether u is engaged for learning of *uweights*.

In addition each neuron $u \in N_\ell$, $0 \leq \ell \leq \ell'_{max} - 1$, has state components: *dweight*, a length n vector with entries that are reals in the interval $[0, f]$; these represent weights on "downward" edges, i.e., those from incoming neighbors at level $\ell + 1$, and *dgaged*, with values in $\{0, 1\}$, representing whether u is engaged for learning of *dweights*. We denote the components of neuron u at time t by $firing^u(t)$, $uweight^u(t)$, $ugaged^u(t)$, $dweight^u(t)$, and $dgaged^u(t)$. As before, the initial values of these components are specified as part of an algorithm description, and the later values are determined by the operation of the network.

Potential and Firing. For a neuron u at level ℓ, $1 \leq \ell \leq \ell'_{max} - 1$, the values of the state components of u at time $t \geq 1$ are determined as follows.

In general, let $ux^u(t)$ denote the vector of *firing* values of u's incoming layer $\ell - 1$ neighbor neurons at time t, and let $dx^u(t)$ denote the vector of *firing* values of u's incoming layer $\ell + 1$ neighbor neurons at time t. Then, as before, the value of $firing^u(t)$ is determined by neuron u's potential for time t and its activation function. The potential at time t is now a sum of two potentials, $upot^u(t)$ coming from layer $\ell - 1$ neurons and $dpot^u(t)$ coming from layer $\ell + 1$ neurons. We define $upot^u(t) = \sum_{j=1}^{n} uweight_j^u(t-1) \, ux_j^u(t-1)$, $dpot^u(t) = \sum_{j=1}^{n} dweight_j^u(t-1) \, dx_j^u(t-1)$ and $pot^u(t) = upot^u(t) + dpot^u(t)$. The activation function is then defined by: $firing^u(t) = 1$ if $pot^u(t) \geq \tau$ and 0 otherwise. For a neuron u at level ℓ'_{max}, the values of the state components of u at time $t \geq 1$ are determined similarly, but using only *uweights* and *ux*.

Edge Weight Modifications. We assume that the values of the *ugaged* and *dgaged* flags of u are controlled externally; that is, for every t, the values of $ugaged^u(t)$ and $dgaged^u(t)$ are set by an external input signal.

For updating the weights, we will use two different rules, one for the *uweights* and one for the *dweights*. The *uweights* are modified as before, using Oja's rule based on the previous *uweights*, the *upot*, and the firing pattern of the layer $\ell - 1$ neurons. Specifically, if $ugaged^u(t) = 1$, then $uweight^u(t) = uweight^u(t-1) + \eta \, (upot^u(t)) \, (ux^u(t-1) - upot^u(t) \, uweight^u(t-1))$, where η is the learning rate. For the *dweights*, we will use a different Hebbian-style learning rule, which we describe in Sect. G.

Network Operation. During operation, the network proceeds through a series of *configurations*, each of which specifies a state for every neuron in the network. As before, the configurations are determined by the initial network specification for time $t = 0$, and the activation and learning functions.

4   Problem Statements

In this section, we define the problems we will consider in the rest of this paper— problems of concept recognition and concept learning. Throughout this section, we use the notation for a concept hierarchy and a network that we defined in Sects. 2 and 3. We assume that the concept hierarchy satisfies the limited-overlap property. We consider both feed-forward networks and networks with feedback, but the notation we specify here is common to both.

Thus, we consider a concept hierarchy $\mathcal{C}$, with concept set C and maximum level $\ell_{\max}$, partitioned into $C_0, C_1, \ldots, C_{\ell_{\max}}$. We use parameters n, k, r_1, r_2, o, and f, according to the definitions for a concept hierarchy in Sect. 2.2. We also consider a network $\mathcal{N}$, with maximum layer ℓ'_{max}, and parameters n, f, τ, and η as in the definitions for a network in Sect. 3. The maximum layer number ℓ'_{max} for $\mathcal{N}$ may be different from the maximum level number $\ell_{\max}$ for $\mathcal{C}$, but the number n of input neurons is the same as the number of level 0 items in $\mathcal{C}$, and the feedback strength f is the same for both $\mathcal{C}$ and $\mathcal{N}$.

We begin with a definition describing how a particular set of level 0 concepts is "presented" to the network. This involves firing exactly the input neurons that represent these level 0 concepts.

Definition 3 (Presented). *If $B \subseteq D_0$ and t is a non-negative integer, then we say that B is* presented *at time t (in some particular network execution) exactly if the following holds. For every layer 0 neuron u: 1) If u is of the form $rep(b)$ for some $b \in B$, then u fires at time t. 2) If u is not of this form, for any b, then u does not fire at time t.*

4.1   Recognition Problems

Here we define what it means for network $\mathcal{N}$ to recognize concept hierarchy $\mathcal{C}$. After learning the concept hierarchy, each concept $c \in C$, at every level, has a unique representing neuron $rep(c)$. We have two definitions, for feed-forward networks and networks with feedback.

Recognition in Feed-forward Networks. The first definition assumes that $\mathcal{N}$ is a feed-forward network. In this definition, we specify not only which neurons fire, but also the precise time when they fire, which is just the time for firing to propagate to the neurons, step by step, through the network layers.

Definition 4 (Recognition problem for feed-forward networks). *Network $\mathcal{N}$ (r_1, r_2)-recognizes $\mathcal{C}$ provided that, for each concept $c \in C$, there is a unique neuron $rep(c)$ such that the following holds. Assume that $B \subseteq C_0$ is presented at time t. Then: 1)When $rep(c)$ must fire: If $c \in supp_{r_2}(B)$, then $rep(c)$ fires at time $t+layer(rep(c))$ and 2) When $rep(c)$ must not fire: If $c \notin supp_{r_1}(B)$, then $rep(c)$ does not fire at time $t + layer(rep(c))$.*

The special case where $r_1 = r_2 = 1$ has a simple characterization:

Lemma 5. *If network $\mathcal{N}$ $(1,1)$-recognizes C, then for each concept $c \in C$, there is a unique neuron $rep(c)$ such that the following holds. If $B \subseteq C_0$ is presented at time t, then $rep(c)$ fires at time $t + layer(rep(c))$ if and only if $leaves(c) \subseteq B$.*

Recognition in Networks with Feedback. The second definition assumes that $\mathcal{N}$ is a network with feedback. For this, the timing is harder to pin down, so we formulate the definition a bit differently. We assume here that the input is presented continually from some time t onward, and we allow flexibility in when the $rep(c)$ neurons are required to fire.

Definition 5 (Recognition problem for networks with feedback). *Network $\mathcal{N}$ (r_1, r_2, f)-recognizes C provided that, for each concept $c \in C$, there is a unique neuron $rep(c)$ such that the following holds. Assume that $B \subseteq C_0$ is presented at all times $\geq t$. Then: 1) When $rep(c)$ must fire: If $c \in supp_{r_2, f}(B)$, then $rep(c)$ fires at some time after t. 2) When $rep(c)$ must not fire: If $c \notin supp_{r_1, f}(B)$, then $rep(c)$ does not fire at any time after t.*

4.2   Learning Problems

In our learning problems, the same network $\mathcal{N}$ must be capable of learning any concept hierarchy C. The definitions are similar to those in [4], but now we extend them to concept hierarchies with limited overlap. As before, we assume that the concepts are shown in a bottom-up manner, though interleaving is allowed for incomparable concepts.[3]

Definition 6. Showing a concept: *Concept c is shown at time t if the set $leaves(c)$ is presented at time t, that is, if for every input neuron u, u fires at time t if and only if $u \in \{rep(c') \mid c' \in leaves(c)\}$.*

Definition 7 (Training schedule). *A training schedule for C is any finite sequence $c_0, c_1, \ldots, c_m$ of concepts in C, possibly with repeats. A training schedule is σ-bottom-up, where σ is a positive integer, provided that the following conditions hold: 1) Each concept in C appears in the list at least σ times. 2) No concept in C appears before each of its children has appeared at least σ times.*

A training schedule $c_0, c_1, \ldots, c_m$ generates a corresponding sequence $B_0, B_1, \ldots, B_m$ of sets of level 0 concepts to be presented to the network in a learning algorithm. Namely, B_i is defined to be $\{rep(c') \mid c' \in leaves(c_i)\}$.

We have two definitions for learning, for networks with and without feedback. The difference is just the type of recognition that is required to be achieved. Each definition makes sense with or without overlap.

---

[3] We might also consider interleaved learning of higher-level concepts and their descendants. The idea is that partial learning of a concept c can be enough to make $rep(c)$ fire, which can help in learning parents of c. We mention this as future work, in Sect. H.

Definition 8 (Learning problem for feed-forward networks). *Network $\mathcal{N}$ (r_1, r_2)-learns concept hierarchy $\mathcal{C}$ with σ repeats provided that the following holds. After a training phase in which all the concepts in $\mathcal{C}$ are shown to the network according to a σ-bottom-up training schedule, $\mathcal{N}$ (r_1, r_2)-recognizes $\mathcal{C}$.*

Definition 9 (Learning problem for networks with feedback). *Network $\mathcal{N}$ (r_1, r_2, f)-learns concept hierarchy $\mathcal{C}$ with σ repeats provided that the following holds. After a training phase in which all the concepts in $\mathcal{C}$ are shown to the network according to a σ-bottom-up training schedule, $\mathcal{N}$ (r_1, r_2, f)-recognizes $\mathcal{C}$.*

5   Recognition Algorithms for Feed-Forward Networks

In this and the following section, we describe and analyze our algorithms for recognition of concept hierarchies (possibly with limited overlap); we consider feed-forward networks in this section and introduce feedback edges in Sect. 6. Throughout both sections, we consider an arbitrary concept hierarchy $\mathcal{C}$ with concept set C, partitioned into $C_0, C_1, \ldots, C_{\ell_{\max}}$. We use the notation n, k, r_1, r_2, o, and f as before. We assume that $r_2 > 0$.

We begin in Sect. 5.1 by defining a basic network, with weights in $\{0, 1\}$, and proving that it (r_1, r_2)-recognizes $\mathcal{C}$. To prove this result, we use a new lemma that relates the firing behavior of the network precisely to the support definition, then obtain the main recognition theorem as a simple corollary. In Sect. 5.3, we extend the main result by allowing weights to be approximate, within an interval of uncertainty. In Sect. 5.4, we extend the results further by allowing scaling of weights and thresholds. In Appendix A, we discuss what happens in a different version of the model, where we replace thresholds by stochastic firing decisions.

5.1   Basic Recognition Results

We define a feed-forward network $\mathcal{N}$ that is specially tailored to recognize concept hierarchy $\mathcal{C}$. We assume that $\mathcal{N}$ has $\ell'_{max} = \ell_{\max}$ layers. Since $\mathcal{N}$ is a feed-forward network, the edges all go upward, from neurons at any layer ℓ to neurons at level $\ell + 1$. We assume the same value of n as in the concept hierarchy $\mathcal{C}$. The edge weights and the threshold τ are defined below. The construction is similar to the corresponding construction in [4]. The earlier paper considered only tree hierarchies; here, we generalize to allow limited overlap. The strategy is simply to embed the digraph induced by $\mathcal{C}$ in the network $\mathcal{N}$. For every level ℓ concept c of $\mathcal{C}$, we assume a unique representative $rep(c)$ in layer ℓ of the network. Let R be the set of all representatives, that is, $R = \{rep(c) \mid c \in C\}$. Let rep^{-1} denote the corresponding inverse function that gives, for every $u \in R$, the corresponding concept $c \in C$ with $rep(c) = u$. Define $\mathcal{E}_{u,v} = u, v \in R$ and $rep^{-1}(u) \in children(rep^{-1}(v))$. We define the weights of the edges as follows. If u is any layer ℓ neuron, $0 \le \ell \le \ell_{\max} - 1$, and v is any layer $\ell + 1$ neuron, then we define the edge weight $weight(u, v)$ to be: $weight(u, v) = 1$ if $\mathcal{E}_{u,v}$ and 0 otherwise.

We would like the threshold τ for every non-input neuron to be a real value in the closed interval $[r_1 k, r_2 k]$; to be specific, we use $\tau = \frac{(r_1+r_2)k}{2}$. Since $r_2 > 0$, we know that $\tau > 0$. Finally, we assume that the initial firing status for all non-input neurons is 0. This completely defines network $\mathcal{N}$, and determines its behavior. The network has been designed in such a way that its behavior directly mirrors the $supp_r$ definition, where $r = \frac{\tau}{k}$. We capture this precisely in the following two lemmas. The first says that, when a subset of C_0 is presented, only *reps* of concepts in C fire at their designated times.

Lemma 6. *Assume C is any concept hierarchy satisfying the limited-overlap property, and $\mathcal{N}$ is the feed-forward network defined above, based on C. Assume that $B \subseteq C_0$ is presented at time t. If u is a neuron that fires at time $t+layer(u)$, then $u \in R$, that is, $u = rep(c)$ for some concept $c \in C$.*

The proof can be found in the appendix. The second lemma says that the *rep* of a concept c fires at time $t + level(c)$ if and only if c is supported by B.

Lemma 7. *Assume C is any concept hierarchy satisfying the limited-overlap property, and $\mathcal{N}$ is the feed-forward network defined above, based on C. Let $r = \frac{\tau}{k}$, where τ is the firing threshold for the non-input neurons of $\mathcal{N}$.*

Assume that $B \subseteq C_0$ is presented at time t. If c is any concept in C, then $rep(c)$ fires at time $t + level(c)(= t + layer(rep(c))$ if and only if $c \in supp_r(B)$.

The proof can be found in the appendix. Using Lemma 7, the basic recognition theorem follows easily:

Theorem 2. *Assume C is any concept hierarchy satisfying the limited-overlap property, and $\mathcal{N}$ is the feed-forward network defined above, based on C. Then $\mathcal{N}$ (r_1, r_2)-recognizes C.*

Recall that the definition of recognition, Definition 4, gives a firing requirement for each individual concept c in the hierarchy. For a concept c, the definition specifies that neuron $rep(c)$ fires at time $t + layer(rep(c)) = t + level(c)$, where t is the time at which the input is presented.

5.2   An Issue Involving Overlap

A new issue arises as a result of allowing overlap: Consider two concepts c and c', with $level(c) = level(c')$. Is it possible that showing concept c' can cause $rep(c)$ to fire? Specifically, suppose that concept c' is shown at some time t, according to Definition 6. That is, the set $leaves(c')$ is presented at time t. Can this cause firing of $rep(c)$ at the designated time $t + level(c)$? We obtain the following negative result whose proof is in the appendix. For this, we assume that the amount of overlap is smaller than the lower bound for recognition.

Theorem 3. *Assume C is any hierarchy satisfying the limited-overlap property, and $\mathcal{N}$ is the feed-forward network defined above, based on C. Suppose that $o < r_1$.*

Let c, c' be two distinct concepts with $level(c') = level(c)$. Suppose that c' is shown at time t. Then $rep(c)$ does not fire at time $t + level(c)$.

5.3   Approximate Weights

So far in this section, we have been considering a simple set of weights, for a network representing a particular concept hierarchy C: $weight(u, v) = 1$ if $\mathcal{E}_{u,v}$ and 0 otherwise. Now we generalize by allowing the weights to be specified only approximately, within some interval. This is useful, for example, when the weights result from a noisy learning process. Here, we assume $0 \leq w_1 \leq w_2$, and allow b to be any positive integer. $weight(u, v) \in [w_1, w_2]$ if $\mathcal{E}_{u,v}$ and $weight(u, v) \in [0, \frac{1}{k^{\ell_{\max}+b}}]$ otherwise.

Again, we set threshold $\tau = \frac{(r_1+r_2)k}{2}$. And we add the (extremely trivial) assumption that $\tau > 1/k^{b-1}$. For this case, we prove the following recognition result. It relies on two inequalities involving the recognition bounds and the weight bounds.

Theorem 4. *Assume C is any concept hierarchy satisfying the limited-overlap property, and $\mathcal{N}$ is the feed-forward network defined above, based on C. Assume that $\frac{(r_1+r_2)k}{2} > \frac{1}{k^{b-1}}$. Suppose that r_1 and r_2 satisfy the following inequalities: $r_2(2w_1 - 1) \geq r_1$ and $r_2 \geq r_1(2w_2 - 1) + \frac{2}{k^b}$. Then, $\mathcal{N}$ (r_1, r_2)-recognizes C.*

To prove Theorem 4, we follow the general pattern of the proof of Theorem 2. We use versions of Lemma 6 and 7: namely Lemma 10 and Lemma 11. The proof can be found in the appendix.

5.4   Scaled Weights and Thresholds

Our recognition results in Sect. 5.3 assume a firing threshold of $\frac{(r_1+r_2)k}{2}$ and bounds w_1 and w_2 on weights on edges from children to parents. The form of the two inequalities in Theorem 4 suggests that w_1 and w_2 should be close to 1, because in that case the two parenthetical expressions are close to 1 and the constraints on the values of r_1 and r_2 are weak. On the other hand, the noise-free learning results in [4] assume a threshold of $\frac{(r_1+r_2)\sqrt{k}}{2}$, that is, our threshold in Sect. 5.3 is multiplied by $\frac{1}{\sqrt{k}}$. Also, in [4], the weights on edges from children to parents approach $\frac{1}{\sqrt{k}}$ in the limit rather than 1, because of the behavior induced by Oja's rule.

We would like to view the results of noise-free learning in terms of achieving a collection of weights that suffice for recognition. That is, we would like a version of Theorem 4 for the case where the threshold is $\frac{(r_1+r_2)\sqrt{k}}{2}$ and the weights are: $weight(u, v) \in [\frac{w_1}{\sqrt{k}}, \frac{w_2}{\sqrt{k}}]$ if $\mathcal{E}_{u,v}$ and $weight(u, v) \in [0, \frac{1}{k^{\ell_{\max}+b}}]$ otherwise. Here, we assume that $w_1 \leq w_2$ and both are close to 1 and recall that $\mathcal{E}_{u,v} = u, v \in R$ and $rep^{-1}(u) \in children(rep^{-1}(v))$. More generally, we can scale by multiplying the threshold and weights by a constant *scaling factor* s, $0 < s < 1$, in place of $\frac{1}{\sqrt{k}}$, giving a threshold of $\frac{(r_1+r_2)ks}{2}$ and weights of: $weight(u, v) \in [w_1 s, w_2 s]$ if $\mathcal{E}_{u,v}$ and $weight(u, v) \in [0, \frac{1}{k^{\ell_{\max}+b}}]$ otherwise. For this general case, we get a new version of Theorem 4:

Theorem 5. *Assume $\mathcal{C}$ is any concept hierarchy satisfying the limited-overlap property, and $\mathcal{N}$ is the feed-forward network defined above (with weights scaled by an arbitrary s). Assume that $\frac{(r_1+r_2)ks}{2} > \frac{1}{k^{b-1}}$. Suppose that r_1 and r_2 satisfy the following inequalities: $r_2(2w_1 - 1) \geq r_1$ and $r_2 \geq r_1(2w_2 - 1) + \frac{2}{k^b s}$. Then, $\mathcal{N}$ (r_1, r_2)-recognizes $\mathcal{C}$.*

Theorem 5 can be proved by slightly adjusting the proofs of Lemma 11 and Theorem 4. Using Theorem 5, we can decompose the proof of correctness for noise-free learning in [4], first showing that the learning algorithm achieves the weight bounds specified above, and then invoking the theorem to show that correct recognition is achieved. For this, we use weights $w_1 = \frac{1}{1+\epsilon}$ and $w_2 = 1$ and scaling factor $s = \frac{1}{\sqrt{k}}$. The value of ϵ is an arbitrary element of $(0, 1]$; the running time of the algorithm depends on ϵ.

6   Recognition Algorithms for Networks with Feedback

In this section, we assume that our network $\mathcal{N}$ includes downward edges, from every neuron in any layer to every neuron in the layer below. We begin in Sect. 6.1 with recognition results for a basic network, with upward weights in $\{0, 1\}$ and downward weights in $\{0, f\}$. Again, we prove these using a new lemma that relates the firing behavior of the network precisely to the support definition.

Recall that for networks with feedback, unlike feed-forward networks, the recognition definition does not specify precise firing times for the *rep* neurons. Therefore, in Sects. E.1 and E.2, we prove time bounds for recognition; these bounds are different for tree hierarchies vs. general hierarchies. Finally, in Sect. 6.2, we extend the main recognition result by allowing weights to be approximate, within an interval of uncertainty. Extension to scaled weights and thresholds should also work in this case.

6.1   Basic Recognition Results

As before, we define a network $\mathcal{N}$ that is specially tailored to recognize concept hierarchy $\mathcal{C}$. We assume that $\mathcal{N}$ has $\ell'_{max} = \ell_{\max}$. We assume the same values of n and f as in $\mathcal{C}$. As before, concept hierarchy $\mathcal{C}$ is embedded, one level per layer, in the network $\mathcal{N}$. Now we define edge weights for both upward and downward edges. Let u be any layer ℓ neuron and v any layer $\ell + 1$ neuron. We define the weight for the upward edge (u, v) as before: $weight(u, v) = 1$ if $\mathcal{E}_{u,v}$ and 0 otherwise. For the downward edge (v, u), we define: $weight(v, u) = f$ if $\mathcal{E}_{u,v}$ and 0 otherwise. Thus, the weight of f on the downward edges corresponds to the weighting factor of f in the $supp_{r,f}$ definition. As before, we set the threshold τ for every non-input neuron to be a real value in the closed interval $[r_1 k, r_2 k]$, specifically, $\tau = \frac{(r_1+r_2)k}{2}$. Again, we assume that the initial firing status of all non-input neurons is 0.

Theorem 6. *Assume C is any concept hierarchy satisfying the limited-overlap property, and N is the network with feedback defined above, based on C. Then N $(r_1, r_2, f)-$ recognizes C.*

In Sect. E, in the appendix, we present time bounds for general and tree hierarchies.

6.2   Approximate Weights

Now, as in Sect. 5.3, we allow the weights to be specified only approximately. We assume that $0 \leq w_1 \leq w_2$, as before. Here we also assume that $b \geq 2$ and $k \geq 2$. Let u be any layer ℓ neuron and v any layer $\ell + 1$ neuron. We define the weight for the upward edge (u, v) by: $weight(u, v) \in [w_1, w_2]$ if $\mathcal{E}_{u,v}$ and otherwise, $weight(u, v) \in [0, k^{\ell_{\max} + b}]$. For the downward edge (v, u), we define: $weight(v, u) \in [fw_1, fw_2]$ if $\mathcal{E}_{u,v}$ and otherwise, $weight(v, u) \in [0, k^{\ell_{\max} + b}]$. As before, we set the threshold $\tau = \frac{(r_1 + r_2)k}{2}$. We also use the trivial assumption that $\tau > \frac{1}{k^{b-2}}$. For this case, we prove:

Theorem 7. *Assume C is any concept hierarchy satisfying the limited-overlap property, and N is the feed-forward network defined above, based on C. Assume that $\frac{(r_1 + r_2)k}{2} \geq \frac{1}{k^{b-2}}$. Suppose that r_1 and r_2 satisfy the following inequalities: $r_2(2w_1 - 1) \geq r_1$ and $r_2 \geq r_1(2w_2 - 1) + \frac{2}{k^{b-1}}$. Then, N (r_1, r_2, f)-recognizes C.*

This theorem follows directly from Lemma 16 and Lemma 17 in the appendix.

Acknowledgement. This work was in part supported by the National Science Foundation awards CCF-1810758 and CCF-2139936, as well as by EPSRC EP/W005573/1.

A   Recognition in Networks with Stochastic Firing

Another type of uncertainty, besides approximate weights, arises when neuron firing is determined stochastically, for example, using a standard sigmoid function. See [5] for an example of a model that uses this strategy. In this case, we cannot make absolute claims about recognition, but we would like to assert that correct recognition occurs with high probability. Here we consider this type of uncertainty in the situation where weights are exactly 1 or 0, as in Sect. 5.1. Extension to allow approximate weights, as well as to networks with feedback, is left for future work.

Following [5], we assume that the potential incoming to a neuron u, pot^u, is adjusted by subtracting a real-valued *bias* value, and the resulting adjusted potential, x, is fed into a standard sigmoid function with temperature parameter λ, in order to determine the firing probability p. Specifically, we have:

$$p(x) = \frac{1}{1 + e^{-x/\lambda}},$$

where $x = pot^u - bias$.

Let δ be a small target failure probability. In terms of our usual parameters n, f, k, and $\ell_{\max}$, and the new parameters λ and δ, our goal is to determine values of r_1 and r_2 so that the following holds: Let $\mathcal{C}$ be any concept hierarchy satisfying the limited-overlap property. Assume that $B \subseteq C_0$ is presented at time t. Then:

1. If $c \in supp_{r_2}(B)$, then with probability at least $1 - \delta$, $rep(c)$ fires at time $t + level(c)$.
2. If $c \notin supp_{r_1}(B)$ then with probability at least $1 - \delta$, $rep(c)$ does not fire at time $t + level(c)$.

In order to determine appropriate values for r_1 and r_2, we start by considering the given sigmoid function. We determine real values b_1 and b_2, $-\infty < b_1 < b_2 < \infty$, that guarantee the following:

1. If the adjusted potential x is $\geq b_2$, then the probability $p(x)$ of firing is $\geq 1 - \delta'$, and
2. If the adjusted potential x is $< b_1$, then the probability $p(x)$ of firing is $\leq \delta'$.

Here, we take δ' to be a small fraction of the target failure probability δ, namely, $\delta' = \frac{\delta}{k^{\ell_{\max}}+1}$. We choose b_1 such that $p(b_1 + bias) = \frac{1}{1+e^{-(b_1+bias)/\lambda}} = \delta'$ and b_2 such that $p(b_2+bias) = \frac{1}{1+e^{-(b_2+bias)/\lambda}} = 1-\delta'$. In other words, $b_1 = \lambda \, log(\frac{1-\delta'}{\delta'}) - bias$, and $b_2 = \lambda \, log(\frac{\delta'}{1-\delta'}) - bias$.

Next, we compute values for r_1 and r_2 based on the values of b_1 and b_2. The values of b_1 and b_2 are adjusted potentials, whereas r_1 and r_2 are fractions of the population of children. To translate, we use $r_1 = \frac{b_1+bias}{k}$ and $r_2 = \frac{b_2+bias}{k}$. This makes sense because having $r_2 k$ children firing yields a potential of $r_2 k$ and an adjusted potential of $r_2 k - bias = b_2$, and not having $r_1 k$ children firing means that the potential is strictly less than $r_1 k$ and the adjusted potential is strictly less than $r_1 k - bias = b_1$.

Note that the requirements on r_1 and r_2 impose some constraints on the value of $bias$. Namely, since we require $r_1 \geq 0$, we must have $b_1 + bias \geq 0$. And since we require $r_2 \leq 1$, we must have $b_2 + bias \leq k$. Within these constraints, different values of $bias$ will yield different values of r_1 and r_2.

With these definitions, we can prove:

Theorem 8. *Let $\mathcal{C}$ be any concept hierarchy satisfying the limited-overlap property. Let δ be a small target failure probability. Let r_1 and r_2 be determined as described above.*

Assume that $B \subseteq C_0$ is presented at time t. Then:

1. *If $c \in supp_{r_2}(B)$, then with probability at least $1 - \delta$, $rep(c)$ fires at time $t + level(c)$.*
2. *If $c \notin supp_{r_1}(B)$ then with probability at least $1 - \delta$, $rep(c)$ does not fire at time $t + level(c)$.*

Proof. (Sketch:) Fix any concept c in C, the set of concepts in $\mathcal{C}$. Note that c has at most $k^{\ell_{\max}+1}$ descendants in C.

For Property 1, suppose that $c \in supp_{r_2}(B)$. Then for every descendant c' of c with $level(c) \geq 1$ and $c' \in supp_{r_2}(B)$, $rep(c')$ fires at time $t + level(c')$ with probability at least $1 - \delta'$, assuming that for each of its children $c'' \in supp_{r_2}(B)$, $rep(c'')$ fires at time $t + level(c'')$. Using a union bound for all such c', we obtain that, with probability at least $1 - k^{\ell_{max}+1} \delta' = 1 - \delta$, $rep(c)$ fires at time $t + level(c)$.

In a bit more detail, for each descendant c' of c with $level(c') \geq 1$ and $c' \in supp_{r_2}(B)$, let $S_{c'}$ denote the set of executions in which $rep(c')$ does not fire at time $t + level(c')$, but for each of its children $c'' \in supp_{r_2}(B)$, $rep(c'')$ fires at time $t + level(c'')$. Then $\bigcup_{c'} S_{c'}$ includes all of the executions in which $rep(c)$ does not fire at time $level(c)$.

Moreover, we claim that the probability of each $S_{c'}$ is at most δ': The fact that $c' \in supp_{r_2}(B)$ imply that at least $r_2 k$ children c'' are in $supp_{r_2}(B)$. Since we assume that all of these $rep(c')$ fire at time $t + level(c'')$, this implies that the potential incoming to c' for time $t + level(c')$ is at least $r_2 k$, and the adjusted potential is at least $r_2 k - bias = b_2$. Then the first property of b_2 yields probability $\leq \delta'$ of c' not firing at time $t + level(c')$.

For Property 2, suppose that $c \notin supp_{r_1}(B)$. Then for every descendant $c' \notin supp_{r_1}(B)$, $rep(c')$ does not fire at time $t + level(c')$, with probability at least $1 - \delta'$, assuming that for each of its children $c'' \notin supp_{r_1}(B)$, $rep(c'')$ does not fire at time $t + level(c'')$. Using a union bound for all $c' \notin supp_{r_1}(B)$, we obtain that, with probability at least $1 - k^{\ell_{max}+1} \delta' = 1 - \delta$, does not fire at time $t + level(c)$.

In a bit more detail, for each descendant c' of c with $level(c') \geq 1$ and $c' \notin supp_{r_1}(B)$, let $S_{c'}$ denote the set of executions in which $rep(c')$ fires at time $t + level(c')$, but for each of its children $c'' \notin supp_{r_1}(B)$, $rep(c'')$ does not fire at time $t + level(c'')$. Then $\bigcup_{c'} S_{c'}$ includes all of the executions in which $rep(c)$ fires at time $level(c)$, and the probability for each $S_{c'}$ is at most δ'; the argument is similar to that for Property 1.

This has been only a sketch of how to analyze stochastic firing, in the simple case of feed-forward networks and exact weights. We leave the complete details of this case, as well as extensions to include approximate weights and networks with feedback, for future work.

B   Missing Statements and Proofs of Section 2

The following monotonicity lemma says that increasing the value of the parameter r can only decrease the supported set.[4]

Lemma 8. *Let C be any concept hierarchy satisfying the limited-overlap property, and let $B \subseteq D_0$. Consider r, r', where $0 \leq r \leq r' \leq 1$. Then $supp_{r'}(B) \subseteq supp_r(B)$.*

---

[4] The mention of the limited-overlap property is just for emphasis, since all of the concept hierarchies of this paper satisty this property.

The following lemma says that any concept c is supported by its entire set of leaves. It can be proven by induction on $level(c)$.

Lemma 9. *Let C be any concept hierarchy satisfying the limited-overlap property. If c is any concept in C, then $c \in supp_1(leaves(c))$.*

Proof (of Lemma 4). We proceed by induction on t. The base case is $t = 0$. In this case $c \in S(0)$, which implies that $c \in supp_{r,0}(B)$, as needed.

For the inductive step, assume that $c \in S(t+1)$ and $parent(c) \notin S(t+1)$. If $c \in S(t)$, then since $parent(c) \notin S(t)$, the inductive hypothesis tells us that $c \in supp_{r,0}(B)$. So assume that $c \notin S(t)$. Then since $parent(c) \notin S(t)$, c is included in $S(t+1)$ based only on its children who are in $S(t)$. Since $c \notin S(t)$, the parent of each of these children is not in $S(t)$. Then by inductive hypothesis, all of c's children that are in $S(t)$ are in $supp_{r,0}(B)$. Since they are enough to support c's inclusion in $supp_{r,0}(B)$, we have that $c \in supp_{r,0}(B)$.

Proof (Proof of Theorem 1). This theorem works because, for each ℓ, the $S(\ell, t)$ sets stabilize after support has had a chance to propagate upwards from level 0 to level $\ell_{\max}$, and then propagate downwards from level $\ell_{\max}$ to level ℓ. Because the concept hierarchy has a simple tree structure, there is no other way for a concept to get added to the $S(\ell, t)$ sets.

Formally, we use a backwards induction on ℓ, from $\ell = \ell_{\max}$ to $\ell = 0$, to prove the claim: If $c \in supp_{r,f}(B)$ and $level(c) = \ell$, then $c \in S(2\,\ell_{\max} - \ell)$. For the base case, consider $\ell = \ell_{\max}$. Since c has no parents, we must have $c \in supp_{r,0}(B)$, so Proposition 1 implies that $c \in S(\ell_{\max}) \subseteq S(2\,\ell_{\max})$, as needed.

For the inductive step, suppose that $c \in supp_{r,f}(B)$ and $level(c) = \ell - 1$. If $c \in supp_{r,0}(B)$ then again Proposition 1 implies that $c \in S(\ell_{\max})$, which suffices. So suppose that $c \in supp_{r,f}(B) - supp_{r,0}(B)$. Then c's inclusion in the set $supp_{r,f}(B)$ relies on its (unique) parent first being included, that is, there is some t for which $c \notin S(t)$ and $parent(c) \in S(t)$. Since $parent(c) \in supp_{r,f}(B)$ and $level(parent(c)) = \ell$, the inductive hypothesis yields that $parent(c) \in S(2\,\ell_{\max} - \ell)$.

Moreover, all the children that c relies on for its inclusion in $supp_{r,f}(B)$ are in $supp_{r,0}(B)$, by Lemma 4. Therefore, by Proposition 1, they are in $S(\ell - 2) \subseteq S(2\,\ell_{\max} - \ell)$. Thus, we have enough support for c in $S(2\,\ell_{\max} - \ell)$ to ensure that $c \in S(2\,\ell_{\max} - \ell + 1)$, as needed.

C   Missing Proofs of Section 5

Proof (Proof of Theorem 2). Let $r = \frac{\tau}{k}$, where τ is the firing threshold for the non-input neurons of $\mathcal{N}$. Assume that $B \subseteq C_0$ is presented at time t. We prove the two parts of Definition 4 separately.

1. If $c \in supp_{r_2}(B)$ then $rep(c)$ fires at time $t + level(c)$.

   Suppose that $c \in supp_{r_2}(B)$. By assumption, $\tau \leq r_2 k$, so that $r = \frac{\tau}{k} \leq r_2$. Then Lemma 8 implies that $c \in supp_r(B)$. Then by Lemma 7, $rep(c)$ fires at time $t + level(c)$.

2. If $c \notin supp_{r_1}(B)$ then $rep(c)$ does not fire at time $t + level(c)$.

   Suppose that $c \notin supp_{r_1}(B)$. By assumption, $\tau \geq r_1 k$, so that $r = \frac{\tau}{k} \geq r_1$. Then Lemma 8 implies that $c \notin supp_r(B)$. Then by Lemma 7, $rep(c)$ does not fire at time $t + level(c)$.

Proof (of Lemma 6). If $layer(u) = 0$, then u fires at time t exactly if $u = rep(c)$ for some $c \in B$, by assumption. So consider u with $layer(u) \geq 1$. We show the contrapositive. Assume that $u \notin R$. Then u has no positive weight incoming edges, by definition of the weights. So u cannot receive enough incoming potential for time $t + layer(u)$ to meet the positive firing threshold τ.

Proof (of Lemma 7). We prove the two directions separately.

1. If $c \in supp_r(B)$ then $rep(c)$ fires at time $t + level(c)$.

   We prove this using induction on $level(c)$. For the base case, $level(c) = 0$, the assumption that $c \in supp_r(B)$ means that $c \in B$, which means that $rep(c)$ fires at time t, by the assumption that B is presented at time t.

   For the inductive step, assume that $level(c) = \ell + 1$. Assume that $c \in supp_r(B)$. Then by definition of $supp_r$, c must have at least rk children that are in $supp_r(B)$. By inductive hypothesis, the *reps* of all of these children fire at time $t + \ell$. That means that the total incoming potential to $rep(c)$ for time $t + \ell + 1$, $pot^{rep(c)}(t + \ell + 1)$, reaches the firing threshold $\tau = rk$, so $rep(c)$ fires at time $t + \ell + 1$.

2. If $c \notin supp_r(B)$ then $rep(c)$ does not fire at time $t + level(c)$.

   Again, we use induction on $level(c)$. For the base case, $level(c) = 0$, the assumption that $c \notin supp_r(B)$ means that $c \notin B$, which means that $rep(c)$ does not fire at time t by the assumption that B is presented at time t.

   For the inductive step, assume that $level(c) = \ell + 1$. Assume that $c \notin supp_r(B)$. Then c has strictly fewer than rk children that are in $supp_r(B)$, and therefore, strictly more than $k - rk$ children that are not in $supp_r(B)$. By inductive hypothesis, none of the *reps* of the children in this latter set fire at time $t + \ell$, which means that the *reps* of strictly fewer than rk children of c fire at time $t + \ell$. So the total incoming potential to $rep(c)$ from *reps* of c's children is strictly less than rk. Since only *reps* of children of c have positive-weight edges to $rep(c)$, that means that the total incoming potential to $rep(c)$ for time $t + \ell + 1$, $pot^{rep(c)}(t + \ell + 1)$, is strictly less than the threshold $\tau = rk$ for $rep(c)$ to fire at time $t + \ell + 1$. So $rep(c)$ does not fire at time $t + \ell + 1$.

Proof (of Theorem 3) Fix c, c' as above, and assume that c' is shown at time t.

Claim: For any descendant d of c that is not also a descendant of c', $rep(d)$ does not fire at time $t + level(d)$.

Proof of Claim: By induction on $level(d)$. For the base case, $level(d) = 0$. We know that $rep(d)$ does not fire at time t because only *reps* of descendants of c' fire at time t.

   For the inductive step, assume that d is a descendant of c that is not also a descendant of c'. By the limited-overlap assumption, d has at most $o \cdot k < r_1 k$

children that are also descendants of c'. By inductive hypothesis, the *reps* of all the other children of d do not fire at time $t + level(d) - 1$. So the number of children of d whose *reps* fire at time $t + level(d) - 1$ is strictly less than $r_1 k$. That is not enough to meet the firing threshold $\tau \geq r_1 k$ for $rep(d)$ to fire at time $t + level(d)$.

End of proof of Claim.

Applying the Claim with $d = c$ yields that $rep(c)$ does not fire at time $t + level(c)$.

Lemma 10. *Assume C is any concept hierarchy satisfying the limited-overlap property, and $\mathcal{N}$ is the feed-forward network defined above, based on C. Assume that $\frac{(r_1 + r_2)k}{2} > \frac{1}{k^{b-1}}$.*

Assume that $B \subseteq C_0$ is presented at time t. If u is a neuron that fires at time $t + layer(u)$, then $u = rep(c)$ for some concept $c \in C$.

Proof. The proof is slightly more involved than the one for Lemma 6. This time we proceed by induction on $layer(u)$. For the base case, If $layer(u) = 0$, then u fires at time t exactly if $u = rep(c)$ for some $c \in B$, by assumption.

For the inductive step, consider u with $layer(u) = \ell + 1$. Assume for contradiction that u is not of the form $rep(c)$ for any $c \in C$. Then the weight of each edge incoming to u is at most $\frac{1}{k^{\ell_{max}+b}}$. By inductive hypothesis, the only layer ℓ incoming neighbors that fire at time $t + \ell$ are *reps* of concepts in C. There are at most $k^{\ell_{max}+1}$ such concepts, hence at most $k^{\ell_{max}+1}$ level ℓ incoming neighbors fire at time $t + \ell$, yielding a total incoming potential for u for time $t + \ell + 1$ of at most $\frac{k^{\ell_{max}+1}}{k^{\ell_{max}+b}} = \frac{1}{k^{b-1}}$. Since the firing threshold $\tau = \frac{(r_1 + r_2)k}{2}$ is strictly greater than $\frac{1}{k^{b-1}}$, u cannot receive enough incoming potential to meet the threshold for time $t + \ell + 1$. $\qquad\blacksquare$

Lemma 11. *Assume C is any concept hierarchy satisfying the limited-overlap property and $\mathcal{N}$ is the feed-forward network defined above, based on C. Assume that $\frac{(r_1 + r_2)k}{2} > \frac{1}{k^{b-1}}$. Suppose that r_1 and r_2 satisfy the following inequalities: $r_2(2w_1 - 1) \geq r_1$ and $r_2 \geq r_1(2w_2 - 1) + \frac{2}{k^b}$. Assume that $B \subseteq C_0$ is presented at time t. If c is any concept in C, then*

1. *If $c \in supp_{r_2}(B)$ then $rep(c)$ fires at time $t + level(c)$.*
2. *If $c \notin supp_{r_1}(B)$ then $rep(c)$ does not fire at time $t + level(c)$.*

Proof. 1. If $c \in supp_{r_2}(B)$ then $rep(c)$ fires at time $t + level(c)$.

We prove this using induction on $level(c)$. For the base case, $level(c) = 0$, the assumption that $c \in supp_{r_2}(B)$ means that $c \in B$, which means that $rep(c)$ fires at time t, by the assumption that B is presented at time t.

For the inductive step, assume that $level(c) = \ell + 1$. Assume that $c \in supp_{r_2}(B)$. Then c must have at least $r_2 k$ children that are in $supp_{r_2}(B)$. By inductive hypothesis, the *reps* of all of these children fire at time $t + \ell$.

We claim that the total incoming potential to $rep(c)$ for time $t + \ell + 1$, $pot^{rep(c)}(t+\ell+1)$, reaches the firing threshold $\tau = \frac{(r_1 + r_2)k}{2}$, so $rep(c)$ fires at time $t + \ell + 1$. To see this, note that the total potential induced by the firing *reps* of children of c is at least $r_2 k w_1$, because the weight of the edge from each firing child to $rep(c)$ is at least w_1. This quantity is $\geq \frac{(r_1+r_2)k}{2}$ because of the assumption that $r_2(2w_1 - 1) \geq r_1$.

2. If $c \notin supp_{r_1}(B)$ then $rep(c)$ does not fire at time $t + level(c)$.

   Again we use induction on $level(c)$. For the base case, $level(c) = 0$, the assumption that $c \notin supp_{r_1}(B)$ means that $c \notin B$, which means that $rep(c)$ does not fire at time t, by the assumption that B is presented at time t.

   For the inductive step, assume that $level(c) = \ell + 1$. Assume that $c \notin supp_{r_1}(B)$. Then c has strictly fewer than $r_1 k$ children that are in $supp_{r_1}(B)$, and therefore, strictly more than $k - r_1 k$ children that are not in $supp_{r_1}(B)$. By inductive hypothesis, none of the *reps* of the children in this latter set fire at time $t + \ell$, which means that the *reps* of strictly fewer than $r_1 k$ children of c fire at time $t + \ell$. Therefore, the total incoming potential to $rep(c)$ from *reps* of c's children is strictly less than $r_1 k w_2$, since the weight of the edge from each firing child to $rep(c)$ is at most w_2.

   In addition, some potential may be contributed by other neurons at level ℓ that are not children of c but fire at time $t + \ell - 1$. By Lemma 10, these must all be *reps* of concepts in C. There are at most $k^{\ell_{max}+1}$ of these, each contributing potential of at most $\frac{1}{k^{\ell_{max}+b}}$, for a total potential of at most $\frac{1}{k^{b-1}}$ from these neurons.

   Therefore, the total incoming potential to $rep(c)$ for time $t+\ell+1$, $pot^{rep(c)}(t+\ell+1)$, is strictly less than $r_1 k w_2 + \frac{1}{k^{b-1}}$. This quantity is $\leq \frac{(r_1+r_2)k}{2}$, because of the assumption that $r_2 \geq r_1(2w_2 - 1) + \frac{2}{k^b}$. This means that the total incoming potential to $rep(c)$ for time $t+\ell+1$ is strictly less than the threshold $\tau = \frac{(r_1+r_2)k}{2}$ for $rep(c)$ to fire at time $t+\ell+1$. So $rep(c)$ does not fire at time $t + \ell + 1$.

Proof (of Theorem 4). (Of Theorem 4:) The proof is similar to that of Theorem 5.3, but now using Lemma 11 in place of Lemma 7. Let $r = \frac{\tau}{k}$, where τ is the firing threshold for the non-input neurons of $\mathcal{N}$. Assume that $B \subseteq C_0$ is presented at time t. We prove the two parts of Definition 4 separately.

1. If $c \in supp_{r_2}(B)$ then $rep(c)$ fires at time $t + level(c)$.

   Suppose that $c \in supp_{r_2}(B)$. By assumption, $\tau \leq r_2 k$, so that $r = \frac{\tau}{k} \leq r_2$. Then Lemma 8 implies that $c \in supp_r(B)$. Then Lemma 11 implies that $rep(c)$ fires at time $t + level(c)$.

2. If $c \notin supp_{r_1}(B)$ then $rep(c)$ does not fire at time $t + level(c)$.

   Suppose that $c \notin supp_{r_1}(B)$. By assumption, $\tau \geq r_1 k$, so that $r = \frac{\tau}{k} \geq r_1$. Then Lemma 8 implies that $c \notin supp_r(B)$. Then Lemma 11 implies that $rep(c)$ does not fire at time $t + level(c)$.

D   Missing Statements and Proofs of Section 6

As before, we have:

Lemma 12. *Assume C is any concept hierarchy satisfying the limited-overlap property, and $\mathcal{N}$ is the network defined above, based on C. Assume that $B \subseteq C_0$ is presented at time t. If u is a neuron that fires at some time after t, then $u \in R$, that is, $u = rep(c)$ for some concept $c \in C$.*

The following preliminary lemma says that the firing of rep neurons is persistent, assuming persistent inputs (as we do in the definition of recognition for networks with feedback).

Lemma 13. *Assume C is any concept hierarchy satisfying the limited-overlap property, and $\mathcal{N}$ is the network with feedback defined above, based on C. Let $r = \frac{\tau}{k}$, where τ is the firing threshold for the non-input neurons of $\mathcal{N}$.*
Assume that $B \subseteq C_0$ is presented at all times $\geq t$. Let c be any concept in C. Then for every $t' \geq t$, if $rep(c)$ fires at time t', then it fires at all times $\geq t'$.

Proof. We prove this by induction on t'. The base case is $t' = t$. The neurons that fire at time t are exactly the input neurons that are *reps* for concepts in B. By assumption, these same inputs continue for all times $\geq t$.

For the inductive step, consider a neuron $rep(c)$ that fires at time t', where $t' \geq t+1$. If $level(c) = 0$ then $c \in B$ and $rep(c)$ continues firing forever. So assume that $level(c) \geq 1$. Then $rep(c)$ fires at time t' because the incoming potential it receives from its children and parents who fire at time $t' - 1$ is sufficient to reach the firing threshold τ. By inductive hypothesis, all of the neighbors of $rep(c)$ that fire at time $t' - 1$ also fire at all times $\geq t' - 1$. So that means that they provide enough incoming potential to $rep(c)$ to make $rep(c)$ fire at all times $\geq t'$. □

Next we have a lemma that is analogous to Lemma 7, but now in terms of eventual firing rather than firing at a specific time. Similarly to before, this works because the network's behavior directly mirrors the $supp_{r,f}$ definition, where $r = \frac{\tau}{k}$.

Lemma 14. *Assume C is any concept hierarchy satisfying the limited-overlap property, and $\mathcal{N}$ is the network with feedback defined above, based on C. Let $r = \frac{\tau}{k}$, where τ is the firing threshold for the non-input neurons of $\mathcal{N}$.*
Assume that $B \subseteq C_0$ is presented at all times $\geq t$. If c is any concept in C, then $rep(c)$ fires at some time $\geq t$ if and only if $c \in supp_{r,f}(B)$.

To prove Lemma 14, it is convenient to prove a more precise version that takes time into account. As before, in Sect. 2.3, we use the abbreviation $S(t) = supp_{r,f}(B, *, t)$. Thus, $S(t)$ is the set of concepts at all levels that are supported by input B by step t of the recursive definition of the $S(\ell, t)$ sets.

Lemma 15. *Assume C is any concept hierarchy satisfying the limited-overlap property, and $\mathcal{N}$ is the network with feedback defined above, based on C. Let $r = \frac{\tau}{k}$, where τ is the firing threshold for the non-input neurons of $\mathcal{N}$.*

Assume that $B \subseteq C_0$ is presented at all times $\geq t$. Let t' be any time $\geq t$. If c is any concept in C, then $rep(c)$ fires at time t' if and only if $c \in S(t' - t)$.

Proof. As for Lemma 7, we prove the two directions separately. But now we use induction on time rather than on $level(c)$.

1. If $c \in S(t' - t)$, then $rep(c)$ fires at time t'.

   We prove this using induction on t', $t' \geq t$. For the base case, $t' = t$, the assumption that $c \in S(0)$ means that c is in the input set B, which means that $rep(c)$ fires at time t.

   For the inductive step, assume that $t' \geq t$ and $c \in S((t'+1)-t)$. If $level(c) = 0$ then again $c \in B$, so c fires at time t, and therefore at time t' by Lemma 13. So assume that $level(c) \geq 1$. If $c \in S(t' - t)$ then $rep(c)$ fires at time t' by the inductive hypothesis, and therefore also at time $t' + 1$ by Lemma 13. Otherwise, enough of c's children and parents must be in $S(t' - t)$ to include c in $S((t' - t) + 1) = S((t' + 1) - t)$; that is, $|children(c) \cap S(t' - t)| + f |parents(c) \cap S(t' - t)| \geq rk$.

   Then by inductive hypothesis, all of the *reps* of the children and parent concepts mentioned in this expression fire at time t'. Therefore, the upward potential incoming to $rep(c)$ for time $t' + 1$, $upot^{rep(c)}(t' + 1)$, is at least $|children(c) \cap S(t' - t)|$, and the downward potential incoming to $rep(c)$ for time $t' + 1$, $dpot^{rep(c)}(t' + 1)$, is at least $f |parents(c) \cap S(t' - t)|$ (since the weight of each downward edge is f). So $pot^{rep(c)}(t' + 1)$, which is equal to $upot^{rep(c)}(t' + 1) + dpot^{rep(c)}(t' + 1)$, is $\geq |children(c) \cap S(t' - t)| + f |parents(c) \cap S(t' - t)| \geq rk$. That reaches the firing threshold $\tau = rk$ for $rep(c)$ to fire at time $t' + 1$.

2. If $rep(c)$ fires at time t', then $c \in S(t' - t)$.

   We again use induction on t', $t' \geq t$. For the base case, $t' = t$, the assumption that $rep(c)$ fires at time t means that c is in the input set B, hence $c \in S(0)$. For the inductive step, suppose that $t' \geq t$ and $rep(c)$ fires at time $t' + 1$. Then it must be that enough of the *reps* of c's children and parents fire at time t' to reach the firing threshold $\tau = rk$ for $rep(c)$ to fire at time $t' + 1$. That is, $upot^{rep(c)}(t'+1)+dpot^{rep(c)}(t'+1) \geq rk$. In other words, the number of *reps* of children of c that fire at time t' plus f times the number of *reps* of parents of c that fire at time t' is $\geq rk$ (since the weight of each downward edge is f).

   By inductive hypothesis, all of these children and parents of c are in $S(t'-t)$. Therefore, $|children(c) \cap S(t' - t)| + f |parents(c) \cap S(t' - t)| \geq rk$. Then by definition of $supp_{r,f}(B)$, we have that $c \in S((t' - t) + 1) = S((t' + 1) - t)$, as needed.

Lemma 14 follows immediately from Lemma 15. Then, as in Sect. 6.1, the main recognition theorem follows easily.

Proof (of Theorem 6). Let $r = \frac{\tau}{k}$, where τ is the firing threshold for the non-input neurons of $\mathcal{N}$. Assume that $B \subseteq C_0$ is presented at all times $\geq t$. We prove the two parts of Definition 5 separately.

1. If $c \in supp_{r_2,f}(B)$ then $rep(c)$ fires at some time $\geq t$.
   Suppose that $c \in supp_{r_2,f}(B)$. By assumption, $\tau \leq r_2 k$, so that $r = \frac{\tau}{k} \leq r_2$.
   Then Lemma 1 implies that $c \in supp_{r,f}(B)$. Then by Lemma 14, $rep(c)$ fires
   at some time $\geq t$.
2. If $c \notin supp_{r_1,f}(B)$ then $rep(c)$ does not fire at any time $\geq t$.
   Suppose that $c \notin supp_{r_1,f}(B)$. By assumption, $\tau \geq r_1 k$, so that $r = \frac{\tau}{k} \geq r_1$.
   Then Lemma 1 implies that $c \notin supp_{r,f}(B)$. Then by Lemma 14, $rep(c)$ does
   not fire at any time $\geq t$.

Lemma 16. *Assume C is any concept hierarchy satisfying the limited-overlap
property, and $\mathcal{N}$ is the network defined above, based on C. Assume that $\frac{(r_1+r_2)k}{2} >
\frac{1}{k^{b-2}}$. Assume that $B \subseteq C_0$ is presented at all times $\geq t$. If u is a neuron that
fires at any time $t' \geq t$, then $u = rep(c)$ for some concept $c \in C$.*

Proof. By induction on the time $t' \geq t$, we show: If u is a neuron that fires at
time t', then $u = rep(c)$ for some concept $c \in C$. For the base case, $t' = t$, if u
fires at time t then $u = rep(c)$ for some $c \in B$, by assumption.

For the inductive step, consider any neuron u that fires at time $t' + 1$, where
$t' \geq t$. Assume for contradiction that u is not of the form $rep(c)$ for any $c \in C$.
Then the weight of each edge incoming to u is at most $k^{\ell_{\max}+b}$. By inductive
hypothesis, the only incoming neighbors that fire at time t' are $reps$ of concepts
in C. There are at most $k^{\ell_{\max}+1} + k^{\ell_{\max}-1}$ concepts at the two layers above
and below $layer(u)$, hence at most $k^{\ell_{\max}+1} + k^{\ell_{\max}-1}$ neighbors of u that fire
at time t', yielding a total incoming potential for u for time $t' + 1$ of at most
$\frac{k^{\ell_{\max}+1}+k^{\ell_{\max}-1}}{k^{\ell_{\max}+b}} = \frac{1}{k^{b-1}} + \frac{1}{k^{b+1}}$. Since $k \geq 2$, this bound on potential is at most
$\frac{1}{k^{b-2}}$. Since the threshold $\tau = \frac{(r_1+r_2)k}{2}$ is assumed to be strictly greater than
$\frac{1}{k^{b-2}}$, u does not receive enough incoming potential to meet the firing threshold
for time $t' + 1$.

Lemma 17. *Assume C is any concept hierarchy satisfying the limited-overlap
property, and $\mathcal{N}$ is the network with feedback as defined above, based on C.
Assume that $\frac{(r_1+r_2)k}{2} > \frac{1}{k^{b-2}}$. Also suppose that r_1 and r_2 satisfy the following inequalities: $r_2(2w_1 - 1) \geq r_1$ and $r_2 \geq r_1(2w_2 - 1) + \frac{2}{k^{b-1}}$. Assume that
$B \subseteq C_0$ is presented at all times $\geq t$. If c is any concept in C, then:*

1. *If $c \in supp_{r_2,f}(B)$ then $rep(c)$ fires at some time $\geq t$.*
2. *If $rep(c)$ fires at some time $\geq t$ then $c \in supp_{r_1,f}(B)$.*

Proof. The proof follows the general outline of the proof of Lemma 14, based on
Lemma 15. As in those results, the proof takes into account both the upward
potential *upot* and the downward potential *dpot*. As before, we split the cases
up and use two inductions based on time. However, now the two inductions
incorporate the treatment of variable weights used in the proof of Lemma 11.

1. If $c \in S(t' - t)$ then $rep(c)$ fires at time t'. Here the set $S(t' - t)$ is defined in
   terms of $supp_{r_2,f}(B)$.

We prove this using induction on t', $t' \geq t$. For the base case, $t' = t$, the assumption that $c \in S(0)$ means that c is in the input set B, which means that $rep(c)$ fires at time t'.

For the inductive step, assume that $t' \geq t$ and $c \in S((t'+1)-t)$. If $level(c) = 0$ then again $c \in B$, so c fires at time t'. So assume that $level(c) \geq 1$. Since $c \in S((t'+1)-t)$, we get that $|children(c) \cap S(t'-t)| + f\,|parents(c) \cap S(t'-t)| \geq r_2 k$.

By the inductive hypothesis, the $reps$ of all of these children and parents fire at time t'. Therefore, the upward potential incoming to $rep(c)$ for time $t'+1$, $upot^{rep(c)}(t'+1)$, is at least $|children(c) \cap S(t'-t)|\, w_1$, and the downward potential incoming to $rep(c)$ for time $t'+1$, $dpot^{rep(c)}(t'+1)$, is at least $f\,|parents(c) \cap S(t'-t)|\, w_1$. Adding these two potentials, we get that the total incoming potential to $rep(c)$ for time $t'+1$, $pot^{rep(c)}(t'+1)$, is at least $(|children(c) \cap S(t'-t)| + f\,|parents(c) \cap S(t'-t)|)\, w_1 \geq r_2 k w_1$. This is at least $\frac{(r_1+r_2)k}{2}$, because of the assumption that $r_2\,(2w_1-1) \geq r_1$. So the incoming potential to $rep(c)$ for time $t'+1$ is enough to reach the firing threshold $\tau = \frac{(r_1+r_2)k}{2}$, so $rep(c)$ fires at time $t'+1$.

2. If $rep(c)$ fires at time t', then $c \in S(t'-t)$. Here the set $S(t'-t)$ is defined in terms of $supp_{r_1,f}(B)$.

We again use induction on t', $t' \geq t$. For the base case, $t' = t$, the assumption that $rep(c)$ fires at time t means that c is in the input set B, hence $c \in S(0)$.

For the inductive step, assume that $rep(c)$ fires at time $t'+1$. Then it must be that $pot^{rep(c)}(t'+1) = upot^{rep(c)}(t'+1) + dpot^{rep(c)}(t'+1)$ reaches the firing threshold $\tau = \frac{(r_1+r_2)k}{2}$ for c to fire at time $t'+1$. Arguing as in the proof of Lemma 16, the total incoming potential to $rep(c)$ from neurons at levels $level(c)-1$ and $level(c)+1$ that are not $reps$ of children or parents of c is at most $\frac{1}{k^{b-2}}$. So the total incoming potential to $rep(c)$ from firing $reps$ of its children and parents must be at least $\frac{(r_1+r_2)k}{2} - \frac{1}{k^{b-2}}$.

By inductive hypothesis, all of these children and parents of c are in $S(t'-t)$. Therefore, $(|children(c) \cap S(t'-t)| + f\,|parents(c) \cap S(t'-t)|)\, w_2 \geq \frac{(r_1+r_2)k}{2} - \frac{1}{k^{b-2}}$. By the assumption that $r_2 \geq r_1(2w_2-1) + \frac{2}{k^{b-1}}$, we get that $|children(c) \cap S(t'-t)| + f\,|parents(c) \cap S(t'-t)| \geq r_1 k$. (In more detail, let $E = |children(c) \cap S(t'-t)| + f\,|parents(c) \cap S(t'-t)|$. So we know that $E w_2 \geq \frac{(r_1+r_2)k}{2} - \frac{1}{k^{b-2}}$. Assume for contradiction that $E < r_1 k$. Then $E w_2 < r_1 k w_2$. But $r_1 k w_2 \leq (r1+r2)k/2 - 1/k^{b-2}$, because of the assumption that $r_2 \geq r_1(2w_2-1) + \frac{1}{k^{b-1}}$. So that means that $E w_2 < (r1+r2)k/2 - 1/k^{b-2}$, which is a contradiction.) Then by definition of $supp_{r_1,f}(B)$, we have that $c \in S((t'-t)+1) = S((t'+1)-t)$, as needed.

The results of this section are also extendable to the case of scaled weights and thresholds, as in Sect. 5.4.

E   Time Bounds for Networks with Feedback

We first give the time bounds for tree hierarchies then for general ones.

E.1   Time Bounds for Tree Hierarchies in Networks with Feedback

It remains to prove time bounds for recognition for hierarchical concepts in networks with feedback. Now the situation turns out to be quite different for tree hierarchies and hierarchies that allow limited overlap. In this section, we consider the simpler case of tree hierarchies.

For a tree network, one pass upward and one pass downward is enough to recognize all concepts, though that is a simplification of what actually happens, since much of the recognition activity is concurrent. Still, for tree hierarchies, we can prove an upper bound of twice the number of levels:

Theorem 9. *Assume C is a tree hierarchy and $\mathcal{N}$ is the network with feedback defined above, based on C. Let $r = \frac{\tau}{k}$, where τ is the firing threshold for the non-input neurons of $\mathcal{N}$.*

Assume that $B \subseteq C_0$ is presented at all times $\geq t$. If $c \in supp_{r,f}(B)$, then $rep(c)$ fires at some time $\leq t + 2\,\ell_{\max}$.

Proof. Assume that $c \in supp_{r,f}(B)$. By Lemma 1, we have that $c \in S(2\,\ell_{\max})$. Then Lemma 15 implies that $rep(c)$ fires at time $t + 2\,\ell_{\max}$.

And this result extends to larger thresholds:

Corollary 1. *Assume C is a tree hierarchy and $\mathcal{N}$ is the network with feedback as defined above, based on C. Assume that $B \subseteq C_0$ is presented at all times $\geq t$. If $c \in supp_{r_2,f}(B)$, then $rep(c)$ fires at some time $\leq t + 2\,\ell_{\max}$.*

Proof. By Theorem 9 and Lemma 1.

E.2   Time Bounds for General Hierarchies in Networks with Feedback

The situation gets more interesting when the hierarchy allows overlap. We use the same network as before. Each neuron gets inputs from its children and parents at each round, and fires whenever its threshold is met. As noted in Lemma 15, this network behavior follows the definition of $supp_{r_2,f}(B)$.

In the case of a tree hierarchy, one pass upward followed by one pass downward suffice to recognize all concepts, though the actual execution involves more concurrency, rather than separate passes. But with overlap, more complicated behavior can occur. For example, an initial pass upward can activate some *rep* neurons, which can then provide feedback on a downward pass to activate some other *rep* neurons that were not activated in the upward pass. So far, this is as for tree hierarchies. But now because of overlap, these newly-recognized concepts can in turn trigger more recognition on another upward pass, then still more on another downward pass, etc. How long does it take before the network is guaranteed to stabilize?

Here we give a simple upper bound and an example that yields a lower bound. Work is needed to pin the bound down more precisely.

Upper Bound. We give a crude upper bound on the time to recognize all the concepts in a hierarchy.

Theorem 10. *Assume $\mathcal{C}$ is any hierarchy satisfying the limited-overlap property, and $\mathcal{N}$ is the network with feedback defined above, based on $\mathcal{C}$. Assume that $B \subseteq C_0$ is presented at all times $\geq t$. If $c \in supp_{r,f}(B)$, then $rep(c)$ fires at some time $\leq t + k^{\ell_{\max}+1}$.*

Proof. All the level 0 concepts in B start firing at time 0. We consider how long it might take, in the worst case, for the *reps* of all the concepts in $supp_{r,f}(B)$ with levels ≥ 1 to start firing.

The total number of concepts in C with levels ≥ 1 is at most $k^{\ell_{\max}+1}$; therefore, the number of concepts in $supp_{r,f}(B)$ with levels ≥ 1 is at most $k^{\ell_{\max}+1}$.

By Lemma 12, the *rep* neurons are the only ones that ever fire. Therefore, the firing set stabilizes at the first time t' such that the sets of *rep* neurons that fire at times t' and time $t'+1$ are the same. Since there are at most $k^{\ell_{\max}+1}$ *rep* neurons with levels ≥ 1, the worst case is if one new *rep* starts firing at each time. But in this case the firing set stabilizes by $t + k^{\ell_{\max}+1}$, as claimed. $\blacksquare$

The bound in Theorem 10 may seem very pessimistic. However, the example in the next subsection shows that it is not too far off, in particular, it shows that the time until all the *reps* fire can be exponential in $\ell_{\max}$.

Lower Bound. Here we present an example of a concept hierarchy $\mathcal{C}$ and an input set B for which the time for the *rep* neurons for all the supported concepts to fire is exponential in $\ell_{\max}$. This yields a lower bound, in Theorem 11.

The concept hierarchy $\mathcal{C}$ has levels $0, \ldots, \ell_{\max}$ as usual. We assume here that $r_1 = r_2 = r$. We assume that the overlap bound o satisfies $o \cdot k \geq 2$, that is, the allowed overlap is at least 2. We take $f = 1$.

The network $\mathcal{N}$ embeds $\mathcal{C}$, as described earlier in this section. As before, we assume that $\ell'_{max} = \ell_{\max}$, and the threshold τ for the non-input nodes in the network is rk. Now we assume that the weights are 1 for both upward and downward edges between *reps* of concepts in C, which is consistent with our choice of $f = 1$ in the concept hierarchy.

We assume that hierarchy $\mathcal{C}$ has overlap only at one level—in the sets of children of level 2 concepts. The upper portion of $\mathcal{C}$, consisting of levels $2, \ldots \ell_{\max}$, is a tree, with no overlap among the sets $children(c)$, $3 \leq level(c) \leq \ell_{\max}$. There is also no overlap among the sets of children of level 1 concepts.

We order the children of each concept with $level \geq 3$ in some arbitrary order, left-to-right. This orients the upper portion of the concept hierarchy, down to the level 2 concepts. Let C' be the set of all the level 2 concepts that are leftmost children of their parents. Since there are $k^{\ell_{\max}-2}$ level 3 concepts, it follows that $|C'| = k^{\ell_{\max}-2}$. Number the elements of C' in order left-to-right as $c_1, \ldots, c_{k_{\ell_{\max}-2}}$. Also, for every concept c_i in C', order its k children in some arbitrary order, left-to-right, and number them 1 through k.

Now we describe the overlap between the sets of children of the level 2 concepts c_i, $1 \leq i \leq k_{\ell_{\max}-2}$. The first $k-1$ children of c_1 are unique to c_1, whereas its kth child is shared with c_2. For $i = k_{\ell_{max}-2}$, the last $k-1$ children of c_i are unique to c_i, whereas its first child is shared with c_{i-1}. For each other index i, the middle $k-2$ children of c_i are unique to c_i, whereas its first child is shared with c_{i-1}, and its kth child is shared with c_{i+1}. There is no other sharing in $\mathcal{C}$.

Next, we define the set B of level 0 concepts to be presented to the network. B consists of the following grandchildren of the level 2 concepts in C':

1. Grandchildren of c_1:
   (a) All the (level 0) children of the children of c_1 numbered $1, \ldots, \lceil rk \rceil$, and
   (b) $\lceil rk \rceil - 1$ of the (level 0) children of the kth child of c_1, which is also the first child of c_2.
2. Grandchildren of each c_i, $2 \le i \le k^{\ell_{max}-2} - 1$:
   (a) $\lceil rk \rceil - 1$ of the (level 0) children of the first child of c_i, which is also the kth child of c_{i-1} (this has already been specified, just above),
   (b) All the (level 0) children of the children of c_i numbered $2, \ldots, \lceil rk \rceil$, and
   (c) $\lceil rk \rceil - 1$ of the (level 0) children of the kth child of c_i, which is also the first child of c_{i+1}.
3. Grandchildren of c_i, $i = k^{\ell_{max}-2}$:
   (a) $\lceil rk \rceil - 1$ of the (level 0) children of the first child of c_i, which is also the kth child of c_{i-1} (this has already been specified, just above), and
   (b) All the (level 0) children of the children of c_i numbered $2, \ldots, \lceil rk \rceil$.

Figure 2 illustrates a sample overlap pattern, for level 2 neurons $c_1, c_2, c_3, \ldots c_m$, where $m = k^{\ell_{max}-2}$. Here we use $k = 4$, $r = 3/4$, and $o = 1/2$.

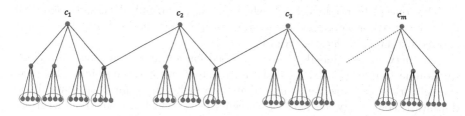

Fig. 2. Concept hierarchy with overlap, and input set.

Theorem 11. *Assume $\mathcal{C}$ is the concept hierarchy defined above, and $\mathcal{N}$ is the network with feedback defined above, based on $\mathcal{C}$. Let B be the input set just defined, and assume that B is presented at all times $\ge t$. Then the time required for the rep neurons for all concepts in $supp_{r,f}(B)$ to fire is at least $2(k^{\ell_{max}} - 2)$.*

Proof. The network behaves as follows:

- Time 0: Exactly the *reps* of concepts in B fire.

- Time 1: The *reps* of the (level 1) children of c_1 numbered $1, \ldots, \lceil rk \rceil$ begin firing. Also, for every c_i, $2 \leq i \leq k^{\ell_{max}-2}$, the *reps* of the (level 1) children numbered $2, \ldots, \lceil rk \rceil$ begin firing. This is because all of these neurons receive enough potential from the *reps* of their (level 0) children that fired at time 0, to trigger firing at time 1. No other neuron receives enough potential to begin firing at time 1.
- Time 2: Neuron $rep(c_1)$ begins firing, since it receives enough potential from the *reps* of its first $\lceil rk \rceil$ children. No other neuron receives enough potential to begin firing at time 2.
- Time 3: Now that $rep(c_1)$ has begun firing, it begins contributing potential to the *reps* of its children, via feedback edges. This potential is enough to trigger firing of the *rep* of the (level 1) k^{th} child of c_1, when it is added to the potential from the *reps* of that child's own level 0 children. So, at time 3, the *rep* of the kth child of c_1 begins firing. No other neuron receives enough potential to begin firing at time 3.
- Time 4: The kth child of c_1 is also the first child of c_2. So its *rep* contributes potential to $rep(c_2)$. This is enough to trigger firing of $rep(c_2)$, when added to the potential from the *reps* of c_2's already-firing children. So, at time 4, $rep(c_2)$ begins firing. No other neuron receives enough potential to begin firing at time 4.
- Time 5: Now that $rep(c_2)$ has begun firing, it contributes potential to the *reps* of its children, via feedback edges. This is enough to trigger firing of the *rep* of the (level 1) k^{th} child of c_2, when added to the potential from the *reps* of that child's own level 0 children. So, at time 5, the *rep* of the kth child of c_2 begins firing. No other neuron begins firing at time 3.
- Time 6: In analogy with that happens at time 4, neuron $rep(c_3)$ begins firing at time 6, and no other neuron begins firing.
- ...
- Time $2(k^{\ell_{max}} - 2)$: Continuing in the same pattern, neuron $rep(c_{k^{\ell_{max}-2}})$ begins firing at time $2(k^{\ell_{max}} - 2)$.

Thus, the time to recognize concept $c_{k^{\ell_{max}-2}}$ is exactly $2(k^{\ell_{max}} - 2)$, as claimed.

F   Learning Algorithms for Feed-Forward Networks

Now we address the question of how concept hierarchies (with and without overlap) can be learned in layered networks. In this section, we consider learning in feed-forward networks, and in Sect. G we consider networks with feedback.

For feed-forward networks, we describe noise-free learning algorithms, which produce edge weights for the upward edges that suffice to support robust recognition. These learning algorithms are adapted from the noise-free learning algorithm in [4], and work for both tree hierarchies and general concept hierarchies. We show that our new learning algorithms can be viewed as producing approximate, scaled weights as described in Sect. 5, which serves to decompose the correctness proof for the learning algorithms. We also discuss extensions to noisy learning.

F.1   Tree Hierarchies

We begin with the case studied in [4], tree hierarchies in feed-forward networks.

Overview of Previous Noise-free Learning Results. In [4], we set the threshold τ for every neuron in layers ≥ 1 to be $\tau = \frac{(r_1+r_2)\sqrt{k}}{2}$. We assumed that the network starts in a state in which no neuron in layer ≥ 1 is firing, and the weights on the incoming edges of all such neurons is $\frac{1}{k^{\ell_{\max}+1}}$. We also assume a Winner-Take-All sub-network satisfying Assumption 13 below, which is responsible for engaging neurons at layers ≥ 1 for learning. These assumptions, together with the general model conventions for activation and learning using Oja's rule, determine how the network behaves when it is presented with a training schedule as in Definition 7. Our main result, for noise-free learning, is (paraphrased slightly)[5]:

Theorem 12 ((r_1, r_2)-Learning Tree concepts). *Let $\mathcal{N}$ be the network described above, with maximum layer ℓ'_{max}, and with learning rate $\eta = \frac{1}{4k}$. Let r_1, r_2 be reals in $[0, 1]$ with $r_1 < r_2$. Let $\epsilon = \frac{r_2-r_1}{r_1+r_2}$. Let $\mathcal{C}$ be any concept hierarchy, with maximum level $\ell_{\max} \leq \ell'_{max}$. Assume that the concepts in $\mathcal{C}$ are presented according to a σ-bottom-up training schedule as defined in Sect. 4.2, where σ is $O\left(\ell_{\max}\log(k) + \frac{1}{\epsilon}\right)$. Then $\mathcal{N}$ (r_1, r_2)-learns $\mathcal{C}$.*

Specifically, we show that the weights for the edges from children to parents approach $\frac{1}{\sqrt{k}}$ in the limit, and the weights for the other edges approach 0.

The Winner-Take-All Assumption. Theorem 12 depends on Assumption 13 below, which hypothesizes a *Winner-Take-All (WTA)* module with certain abstract properties. This module is responsible for selecting a neuron to represent each new concept. It puts the selected neuron in a state that prepares it to learn the concept, by setting the *engaged* flag in that neuron to 1. It is also responsible for engaging the same neuron when the concept is presented in subsequent learning steps. In more detail, while the network is being trained, example concepts are "shown" to the network, according to a σ-bottom-up schedule as described in Sect. 4.2. We assume that, for every example concept c that is shown, exactly one neuron in the appropriate layer gets engaged; this layer is the one with the same number as the level of c in the concept hierarchy. Furthermore, the neuron in that layer that is engaged is one that has the largest incoming potential pot^u:

Assumption 13 (Winner-Take-All Assumption) *If a level ℓ concept c is shown at time t, then at time $t + \ell$, exactly one neuron u in layer ℓ has its engaged state component equal to 1, that is, $engaged^u(t + \ell) = 1$. Moreover, u is chosen so that $pot^u(t + \ell)$ is the highest potential at time $t + \ell$ among all the layer ℓ neurons.*

---

[5] We use O notation here instead of giving actual constants. We omit a technical assumption involving roundoffs.

Thus, the WTA module selects the neuron to "engage" for learning. For a concept c that is being shown for the first time, we showed that a new neuron is selected to represent c—one that has not previously been selected. This is because, if a neuron has never been engaged in learning, its incoming weights are all equal to the initial weight $w = \frac{1}{k^{\ell_{\max}+1}}$, yielding a total incoming potential of kw. On the other hand, those neurons in the same layer that have previously been engaged in learning have incoming weights for all of c's children that are strictly less than the initial weight w, which yields a strictly smaller incoming potential. Also, for a concept c that is being shown for a second or later time, we showed that the already-chosen representing neuron for c is selected again. This is because the total incoming potential for the previously-selected neuron is strictly greater than kw (as a result of previous learning), whereas the potential for other neurons in the same layer is at most kw.

In a complete network for solving the learning problem, the WTA module would be implemented by a sub-network, but we treated it abstractly in [4], and we continue that approach in this paper.

Connections with Our New Results. Here we consider how we might use our scaled result in Sect. 5.4 to decompose the proof of Theorem 12 in [4]. A large part of the proof in [4] consists of proving that the edge weights established as a result of a σ-bottom-up training schedule, for sufficiently large σ, are within certain bounds. If these bounds match up with those in Sect. 5.4, we can use the results of that section to conclude that they are adequate for recognition.

The general definitions in Sect. 5.4 use a threshold of $\frac{(r_1+r_2)ks}{2}$ and weights given by: $weight(u,v) \in [w_1 s, w_2 s]$ if $\mathcal{E}_{u,v}$ and $weight(u,v) \in [0, \frac{1}{k^{\ell_{\max}+b}}]$ otherwise. To make the results of [4] fit the constraints of Sect. 5.4, we can simply take $w_1 = \frac{1}{1+\epsilon}$, $w_2 = 1$, and the scaling factor $s = \frac{1}{\sqrt{k}}$. The two constraints $r_2(2w_1 - 1) \geq r_1$ and $r_2 \geq r_1(2w_2 - 1) + \frac{2}{k^b s}$ now translate into $r_2(\frac{1-\epsilon}{1+\epsilon}) \geq r_1$ and $r_2 \geq r_1 + \frac{2}{k^{b-\frac{1}{2}}}$. The first of these, $r_2(\frac{1-\epsilon}{1+\epsilon}) \geq r_1$, follows from the assumption in [4] that $\epsilon = \frac{r_2-r_1}{r_2+r_1}$. The second inequality is similar to a roundoff assumption in [4] that we have omitted here.[6]

Noisy Learning. In [4], we extended our noise-free learning algorithm to the case of "noisy learning". There, instead of presenting all leaves of a concept c at every learning step, we presented only a subset of the leaves at each step. This subset is defined recursively with respect to the hierarchical concept structure of c and its descendants. The subset varies, and is chosen randomly at each learning step. Similar results hold as for the noise-free case, but with an increase in learning time.[7] The result about noisy learning in [4] assumes a parameter

---

[6] In any case, we can made the decomposition work by adding our new, not-very-severe, inequality as an assumption.

[7] The extension to noisy learning is the main reason that we used the incremental Oja's rule. If the concepts were presented in a noise-free way, we could have allowed learning to occur all-at-once.

p giving the fraction of each set of children that are shown; a larger value of p yields a correspondingly shorter training time. The target weight for learned edges is $\bar{w} = \frac{1}{\sqrt{pk+1-p}}$. The threshold is $r_2 k(\bar{w} - \delta)$, where $\delta = \frac{(r_2 - r_1)\bar{w}}{25}$.

The main result says that, after a certain time σ (larger than the σ used for noise-free learning) spent training for a tree concept hierarchy $\mathcal{C}$, with high probability, the resulting network achieves (r_1, r_2)-recognition for $\mathcal{C}$. Here, a key lemma asserts that, with high probability, after time σ, the weights are as follows: $weight(u, v) \in [\bar{w} - \delta, \bar{w} + \delta]$ if $\mathcal{E}$, and otherwise $weight(u, v) \in [0, \frac{1}{k^2 \ell_{\max}}]$. To make these results fit the constraints of Sect. 5.4, it seems that we should modify the threshold slightly, by using the similar but simpler threshold $(\frac{(r_1 + r_2)k}{2})\bar{w}$ in place of $r_2 k(\bar{w} - \delta)$. The weights can remain the same as above, but in case of $\mathcal{E}_{u,v}$ we have $weight(u, v) \in [(1 - \frac{r_2 - r_1}{25})\bar{w}, (1 + \frac{r_2 - r_1}{25})\bar{w}]$. Thus, we have scaled the basic threshold $\frac{(r_1 + r_2)k}{2}$ by multiplying it by $\bar{w} = \frac{1}{\sqrt{pk+1-p}}$. To make the results fit the constraints of Sect. 5.4, we can take $s = \bar{w}$, $w_1 = 1 - \frac{r_2 - r_1}{25}$, $w_2 = 1 + \frac{r_2 - r_1}{25}$, and $b = \ell_{\max}$. One can easily verify that the new thresholds still fulfill the requirements for recognition. We do this in the full version.

F.2   General Concept Hierarchies

The situation for general hierarchies, with limited overlap, in feed-forward networks is similar to that for tree hierarchies. The same learning algorithm, based on Oja's rule, still works in the presence of overlap, with little modification to the proofs. The only significant new issue to consider is how to choose an acceptable neuron to engage in learning, at each learning step. We continue to encapsulate this choice within a separate WTA service. As before, the WTA should always select an unused neuron (in the right layer) for a concept that is being shown for the first time. And for subsequent times when the same concept is shown, the WTA should choose the same neuron as it did the first time.

An Issue with the Previous Approach. Assumption 13, which we used for tree hierarchies, no longer suffices. For example, consider two concepts c and c' with $level(c') = level(c)$, and suppose that there is exactly one concept d in the intersection $children(c) \cap children(c')$. Suppose that concept c has been fully learned, so a $rep(c)$ neuron has been defined, and then concept c' is shown for the first time. Then when c' is first shown, $rep(c)$ will receive approximately $\frac{1}{\sqrt{k}}$ of total incoming potential, resulting from the firing of $rep(d)$. On the other hand, any neuron that has not previously been engaged in learning will receive potential $\frac{k}{k^{\ell_{\max}+1}} = \frac{1}{k^{\ell_{\max}}}$, based on k neurons each with initial weight $\frac{1}{k^{\ell_{\max}+1}}$, which is smaller than $\frac{1}{\sqrt{k}}$. Thus, Assumption 13 would select $rep(c)$ in preference to any unused neuron. One might consider replacing Oja's learning rule with some other rule, to try to retain Assumption 13, which works based just on comparing potentials. Another approach, which we present here, is to use a "smarter" WTA, that is, to modify Assumption 13 so that it takes more information into account when engaging a neuron.

Approach Using a Modified WTA Assumption. In the assumption below, w denotes the initial weight, $\frac{1}{k^{\ell_{\max}+1}}$. N_ℓ denotes the set of layer ℓ neurons. We make the trivial assumption that $o < 1$ for the noise-free case; for the noisy case, we strengthen that to $o < p$, where p is the parameter indicating how many child concepts are chosen.

Assumption 14 (Revised Winner-Take-All Assumption). *If a level ℓ concept c is shown at time t, then at time $t + \ell$, exactly one neuron $u \in N_\ell$ has its engaged state component equal to 1, that is, $engaged^u(t+\ell) = 1$. Moreover, u is chosen so that $pot^u(t+\ell)$ is the highest potential at time $t+\ell$ among the layer ℓ neurons that have strictly more than $o \cdot k$ incoming neighbors that contribute potential that is $\geq w$.*

Thus, we are assuming that the WTA module is "smart enough" to select the neuron to engaged based on a combination of two criteria: First, it rules out any neuron that has just a few incoming neighbors that contribute potential $\geq w$. This is intended to rule out neurons that have already started learning, but for a different concept. Second, it uses the same criterion as in Assumption 13, choosing the neuron with the highest potential from among the remaining candidate neurons.

We claim that using Assumption 14 in the learning protocol yields appropriate choices for neurons to engage, as expressed by Lemma 19 below. Showing these properties depends on a characterization of the incoming weights for a neuron $u \in N_\ell$ at any point during the learning protocol, as expressed by Lemma 18.

Lemma 18. *During execution of the learning protocol, at a point after any finite number of concept showings, the following properties hold:*

1. *If u has not previously been engaged for learning, then all of u's incoming weights are equal to the initial weight w.*
2. *If u has been engaged for learning a concept c, and has never been engaged for learning any other concept, then all of u's incoming weights for reps of concepts in children(c) are strictly greater than w, and all of its other incoming weights are strictly less than w.*

Proof. Property 1 is obvious—if a neuron is never engaged for learning, its incoming weights don't change. Property 2 follows from Oja's rule.

Lemma 19. *During execution of the learning protocol, the following properties hold for any concept showing:*

1. *If a concept c is being shown for the first time, the neuron that gets engaged for learning c is one that was not previously engaged.*
2. *If a concept c is being shown for the second or later time, the neuron that gets engaged for learning c is the same one that was engaged when c was shown for the first time.*

Proof. We prove Properties 1 and 2 together, by strong induction on the number m of the concept showing. For the inductive step, suppose that concept c is being shown at the mth concept showing. By (strong) induction, we can see that, for each concept that was previously shown, the same neuron was engaged in all of its showings. Therefore, the weights described in Lemma 18, Property 2, hold for all neurons that have been engaged in showings $1, \ldots, m - 1$.

Claim: Consider any neuron u with $layer(u) = level(c)$ that was previously engaged for learning a different concept $c' \neq c$. Then u has at most $o \cdot k$ incoming neighbors that contribute potential to u that is $\geq w$, and so, is not eligible for selection by the WTA.

Proof of Claim: Lemma 18, Property 2, implies that all the incoming weights to neuron u for *reps* of concepts in $children(c')$ are strictly greater than w, and all of its other incoming weights are strictly less than w. Since $|children(c) \cap children(c')| \leq o \cdot k$, u has at most $o \cdot k$ incoming neighbors that contribute potential that is $\geq w$, as claimed.

End of proof of Claim

Now we prove Properties 1 and 2:

1. If concept c is being shown for the first time, the neuron u that gets engaged for learning c is one that was not previously engaged.

   Assume for contradiction that the chosen neuron u was previously engaged. Then it must have been for a different concept $c' \neq c$, since this is the first time c is being shown. Then by the Claim, u is not eligible for selection by the WTA. This is a contradiction.

2. If concept c is being shown for the second or later time, the neuron u that gets engaged for learning c is the same one that was engaged when c was shown for the first time.

   Arguing as for Property 1, again using the Claim, we can see that u cannot have been previously engaged for a concept $c' \neq c$. So the only candidates for u are neurons that were not previously engaged, as well as the (single) neuron that was previously engaged for c. The given WTA rule chooses u from among these candidate based on highest incoming potential.

   For neurons that were not previously engaged, Lemma 18, Property 1, implies that the incoming potential is exactly kw. For the single neuron that was previously engaged for c, Lemma 18, Property 2 implies that the incoming potential is strictly greater than kw. So the WTA rule selects the previously-engaged neuron.

With the new WTA assumption, the learning analysis for general hierarchies follows the same pattern as the analysis for tree hierarchies in [4], and yields the same time bound.

Theorem 15 ((r_1, r_2)-Learning General Hierarchies). *Let $\mathcal{N}$ be the network described above, with maximum layer ℓ'_{max}, and with learning rate $\eta = \frac{1}{4k}$. Let r_1, r_2 be reals in $[0, 1]$ with $r_1 < r_2$. Let $\epsilon = \frac{r_2 - r_1}{r_1 + r_2}$. Let $\mathcal{C}$ be any* **general** *concept hierarchy, with maximum level $\ell_{\max} \leq \ell'_{max}$. Assume the* *revised WTA (Assumption 14) and that the concepts in $\mathcal{C}$ are presented accord-ing to a σ-bottom-up training schedule as defined in Sect. 4.2, where σ is $O\left(\ell_{\max} \log(k) + \frac{1}{\epsilon}\right)$. Then $\mathcal{N}$ (r_1, r_2)-learns $\mathcal{C}$.*

Implementing Assumption 14 will require some additional mechanism, in addition to the mechanisms that are used to implement the basic WTA satisfying Assumption 13. Such a mechanism could serve as a pre-processing step, before the basic WTA. The new mechanism could allow a layer ℓ neuron u to fire (and somehow reflect its incoming potential) exactly if u has strictly more than $o \cdot k$ incoming neighbors that contribute potential $\geq w$ to u.[8]

G   Learning Algorithms for Networks with Feedback

Now we consider how concept hierarchies, with and without overlap, can be learned in layered networks with feedback. The learning algorithms described in Sect. F set the weights on the directed edges from each layer ℓ to the next higher layer $\ell + 1$, that is, the "upward" edges. Now the learning algorithm must also set the weights on the directed edges from each layer ℓ to the next lower layer $\ell - 1$, i.e., the "downward" edges.

One reasonable approach is to separate matters, first learning the weights on the upward edges and then the weights on the downward edges. Fortunately, we can rely on Lemma 9, which says that, if c is any concept in a concept hierarchy $\mathcal{C}$, then $c \in supp_1(leaves(c))$. That is, any concept is supported based only on its descendants, without any help from its parents. This lemma implies that learning of upward edges can proceed bottom-up, as in Sect. F. We give some details below.

G.1   Noise-Free Learning

As in Sect. F, we assume that the threshold is $\frac{(r_1+r_2)\sqrt{k}}{2}$, the initial weight for each upward edge is $w = \frac{1}{k^{\ell_{\max}+1}}$, and $\epsilon = \frac{r_2-r_1}{r_1+r_2}$. Here we also assume that the initial weight for each downward edge is w.[9] We assume that the network starts in a state in which no neuron in layer ≥ 1 is firing.

Our main result, for noise-free learning, is:

---

[8]   For instance, each layer $\ell - 1$ neuron v might have an outgoing edge to a special threshold element that fires exactly if the potential produced by v on the edge (v, u) is at least w, i.e., if v fires and $weight(v, u) \geq w$. Then another neuron associated with u might collect all this firing information from all layer $\ell - 1$ neurons v and see if the number of firing neurons reaches the threshold $\lfloor o \cdot k \rfloor + 1$, which is equivalent to saying that the number of firing neurons is strictly greater than $o \cdot k$. If this special neuron fires, it excites u to act as an input to the basic WTA, but if it does not, u should drop out of contention.

[9]   We are omitting mention here of some trivial technical assumptions, like small lower bounds on τ.

Theorem 16. *Let $\mathcal{N}$ be the network defined in this section, with maximum layer ℓ'_{max}, and with learning rate $\eta = \frac{1}{4k}$. Let r_1, r_2 be reals in $[0, 1]$ with $r_1 \leq r_2$. Let $\epsilon = \frac{r_2 - r_1}{r_1 + r_2}$.*

Let $\mathcal{C}$ be any concept hierarchy, with maximum level $\ell_{max} \leq \ell'_{max}$. Assume that the algorithm described in this section is executed: On the first pass, the concepts in $\mathcal{C}$ are presented according to a σ-bottom-up presentation schedule, where σ is $O\left(\ell_{max} \log(k) + \frac{1}{\epsilon}\right)$. The second pass is as described in Sect. G.1. Then $\mathcal{N}$ (r_1, r_2, f)-learns $\mathcal{C}$.

First Learning Pass. As a first pass, we carry out the learning protocol from Sect. F for all the concepts in the concept hierarchy $\mathcal{C}$, working bottom-up. Learning each concept involves applying Oja's rule for that concept, for enough steps to ensure that the weights of the upward edges end up within the bounds described in Sect. 5.4.

Consider the network after the first pass, when the weights of all the upward edges have reached their final values. At that point, we have that the network (r_1, r_2)-recognizes the given concept hierarchy $\mathcal{C}$, as described in Sect. F. Moreover, we obtain:

Lemma 20. *The weights of the edges after the completion of the first learning pass are as follows:*

1. *The weights of the upward edges from reps of children to reps of their parents are in the range $[\frac{1}{(1+\epsilon)\sqrt{k}}, \frac{1}{\sqrt{k}})$, and the weights of the other upward edges are in the range $[0, \frac{1}{2^{l_{max}+b}}]$.*
2. *The weights of all downward edges are $w = \frac{1}{k^{\ell_{max}+1}}$.*

As a consequence of these weight settings, we can prove the following about the network resulting from the first pass:

Lemma 21. *The following properties hold of the network that results from the completion of the first learning pass:*

1. *Suppose c is any concept in C. Suppose that c is shown (that is, the set $leaves(c)$ is presented) at time t, and no inputs fire at any other times. Then $rep(c)$ fires at time $t + level(c) = t + layer(rep(c))$, and does not fire at any earlier time.*
2. *Suppose c is any concept in C. Suppose that c is shown at time t, and no inputs fire at any other times. Suppose c' is any other concept in C with $level(c') = level(c)$. Then $rep(c')$ does not fire at time $t + level(c)$.*
3. *Suppose that u is a neuron in the network that is not a rep of any concept in C. Suppose that precisely the set of level 0 concepts in C is presented at time t, and no inputs fire at any other times. Then neuron u never fires.*

Proof. Property 1 follows from the analysis in [4], plus the fact that $level(c)$ is the time it takes to propagate a wave of firing from the inputs to $layer(rep(c)$. Property 2 can be proved by induction on $level(c)$, using the limited-overlap property. Property 3 can be proved by induction on the time $t' \geq t$, analogously to Lemma 16.

Second Learning Pass. The second pass sets all the weights for the downward edges. Here, to be simple, we set the weight of each edge to its final value in one learning step, rather than proceeding incrementally.[10] We aim to set the weights of all the "important" downward edges, that is, those that connect the *rep* of a concept to the *rep* of any of its children, to $\frac{f}{\sqrt{k}}$, and the weights of all other downward edges to 0.

We first set the weights on the "important" downward edges. For this, we proceed level by level, from 1 to $\ell_{\max}$. The purpose of the processing for level ℓ is to set the weights on all the "important" downward edges from layer ℓ to layer $\ell - 1$ to $\frac{f}{\sqrt{k}}$, while leaving the weights of the other downward edges equal to the initial weight w.

For each particular level ℓ, we proceed sequentially through the level ℓ concepts in C, one at a time, in any order. For each such level ℓ concept c, we carry out the following three steps:

1. Show concept c, that is, present the set $leaves(c)$, at some time t. By Theorem 20, the *reps* of all children of c fire at time $t + \ell - 1$, and $rep(c)$ fires at time $t + \ell$.
2. Engage all the layer $\ell - 1$ neurons to learn their incoming *dweights* at time $t + \ell + 1$, by setting their *dgaged* flags.
3. *Learning rule:* At time $t + \ell + 1$, each *dgaged* layer $\ell - 1$ neuron u that fired at time $t + \ell - 1$ sets the weights of any incoming edges from layer ℓ neurons that fired at time $t + \ell$ (and hence contributed potential to u) to $\frac{f}{\sqrt{k}}$. Neuron u does not modify the weights of other incoming edges.

Note that, in Step 3, each neuron u that fired at time $t + \ell - 1$ will set the weight of at most one incoming downward edge to $\frac{f}{\sqrt{k}}$; this is the edge from $rep(c)$, in case u is the *rep* of a child of c.

Also note that u does not reduce the weights of other incoming downward edges during this learning step. This is to allow u to receive potential from other layer ℓ neurons when those concepts are processed. This is important because u may represent a concept with multiple parents, and must be able to receive potential from all parents when they are processed.

Finally, note that, to implement this learning rule, we need some mechanism to engage the right neurons at the right times. For now, we just treat this abstractly, as we did for learning in feed-forward networks in Sect. F.

At the end of the second pass, each neuron u resets the weights of all of its incoming downward edges that still have the initial weight w, to 0. The neurons can all do this in parallel.

Lemma 22. *The weights of the edges after the completion of the second learning pass are as follows:*

---

[10] This seems reasonable since we are not considering noise during this second pass. Of course, that might be interesting to consider at some point.

1. *The weights of the upward edges from reps of children to reps of their parents are in the range $[\frac{1}{(1+\epsilon)\sqrt{k}}, \frac{1}{\sqrt{k}})]$, and the weights of the other upward edges are in the range $[0, \frac{1}{2^{l_{max}+b}}]$.*
2. *The weights of the downward edges from reps of parents to reps of their children are $\frac{f}{\sqrt{k}}$, and the weights of the other downward edges are 0.*

Proof. Property 1 follows from Lemma 20 and the fact that the weights of the upward edges are unchanged during the second pass.

For Property 2, the second pass is designed to set precisely the claimed weights. This depends on the neurons firing at the expected times. This follows from Lemma 21, once we note that the three claims in that lemma remain true throughout the second pass. (The first two properties depend on upward weights only, which do not change during the second pass. Property 3 follows because only *rep* neurons have their incoming weights changing during the second pass.)

With these weights, we can now prove the main theorem:

Proof. (Of Theorem 16:) We use a scaled version of Theorem 7. Here we use $w_1 = \frac{1}{1+\epsilon}$, $w_2 = 1$, and a scaling factor $s = \frac{1}{\sqrt{k}}$.[11]

G.2   Noisy Learning

We can extend the results of the previous section to allow noisy learning in the first pass. For this, we use a threshold of $(\frac{(r_1+r_2)k}{2})\bar{w}$ and retain the initial edge weights of $w = \frac{1}{k^{\ell_{max}}+1}$. We define $w_1 = 1 - \frac{r_2-r_1}{25}$, $w_2 = 1 - \frac{r_2-r_1}{25}$, and the scaling factor s to be $\bar{w} = \frac{1}{\sqrt{pk+1-p}}$.

The ideas are analogous to the noise-free case. The differences are:

1. The first phase continues long enough to complete training for the weights of the upward edges using Oja's rule.
2. The weights of the upward edges from *reps* of children to *reps* of their parents are in the range $[(1 - \frac{r_2-r_1}{25})\bar{w}, (1 + \frac{r_2-r_1}{25})\bar{w}]$, and the weights of the other upward edges are in the range $[0, \frac{1}{k^2\,\ell_{max}}]$.
3. The weights of the downward edges are set to $f\bar{w}$.

With these changes, we can obtain a theorem similar to Theorem 16 but with a larger training time, yielding (r_1, r_2, f)-learning of $\mathcal{C}$.

H   Future Work

There are many possible directions for extending the work. For example:

---

[11] The scaled case isn't actually worked out in Sect. 6 but should follow as a natural extension of the un-scaled results.

Concept Recognition: It would be interesting to study *recognition behavior after partial learning.* The aim of the learning process is to increase weights sufficiently to guarantee recognition of a concept when partial information about its leaves is presented. Initially, even showing all the leaves of a concept c should not be enough to induce $rep(c)$ to fire, since the initial weights are very low. At some point during the learning process, after the weights increase sufficiently, presenting all the leaves of c will guarantee that $rep(c)$ fires. As learning continues, fewer and fewer of the leaves will be needed to guarantee firing. It would be interesting to quantify the relationship between amount of learning and the number of leaves needed for recognition.

Also, the definition of robustness that we have used in this paper involves just omitting some inputs. In would be interesting to also consider other types of noise, such as *adding extraneous inputs.* How well do our recognition algorithms handle this type of noise?

Another type of noise arises if we replace the deterministic threshold elements with neurons that fire stochastically, based on some type of sigmoid function of the incoming potential. How well do our recognition algorithms handle this type of noise? Some initial ideas in this direction appear in Appendix A, but more work is needed.

Learning of Concept Hierarchies: Our learning algorithms depend heavily on Winner-Take-All subnetworks. We have treated these abstractly in this paper, by giving formal assumptions about their required behavior. It remains to *develop and analyze networks implementing the Winner-Take-All assumptions.*

Another interesting issue involves possible *flexibility in the order of learning concepts.* In our algorithms, incomparable concepts can be learned in any order, but children must be completely learned before we start to learn their parents. We might also consider some interleaving in learning children and parents. Specifically, in order to determine $rep(c)$ for a concept c, according to our learning algorithms, we would need for the *reps* of all of c's children to be already determined, and for the children to be learned sufficiently so that their *reps* can be made to fire by presenting "enough" of their leaves.

But this does not mean that the child concepts must be completely learned, just that they have been learned sufficiently that it is possible to make them fire (say, when all, or almost all, of their leaves are presented). This suggests that it is possible to allow some interleaving of the learning steps for children and parents. This remains to be worked out.

Another issue involves *noise in the learning process.* In Sects. F and G, we have outlined results for noisy learning of weights of upward edges, in the various cases studied in this paper, but full details remain to be worked out. The approach should be analogous to that in [4], based on presenting randomly-chosen subsets of the leaves of a concept being learned. The key here should be to articulate simple lemmas about achieving approximate weights with high probability. It also remains to consider noise in the learning process for weights of downward edges.

Finally, our work on learning of weights of upward edges has so far relied on Oja's learning rule. It would be interesting to consider *different learning rules* as well, comparing the guarantees that are provided by different rules, with respect to speed of learning and robustness to noise during the learning process.

Different Data Models, Different Network Assumptions, Different Representations: One can consider many variations on our assumptions. For example, what is the impact of loosening the very rigid assumptions about the shape of concept hierarchies? What happens to the results if we have limited connections between layers, rather than all-to-all connections? Such connections might be randomly chosen, as in [8]. Also, we have been considering a simplified representation model, in which each concept is represented by precisely one neuron; can the results be extended the to accommodate more elaborate representations?

Experimental Work in Computer Vision: Finally, it would be interesting to try to devise experiments in computer vision that would reflect some of our theoretical results. For example, can the high-latency recognition behavior that we identified in Sect. E.2, involving extensive information flow up and down the hierarchy, be exhibited during recognition of visual scenes?

References

1. Felleman, D.J., Van Essen, D.C.: Distributed hierarchical processing in the primate cerebral cortex. Cereb. cortex (New York, NY: 1991) **1**(1), 1–47 (1991)
2. Hubel, D., Wiesel, T.: Receptive fields, binocular interaction, and functional architecture in the cat's visual cortex. J. Physiol. **160**, 106–154 (1962)
3. Hubel, D.H., Wiesel, T.N.: Receptive fields of single neurones in the cat's striate cortex. J. Physiol. **148**(3), 574–591 (1959). https://doi.org/10.1113/jphysiol.1959.sp006308
4. Lynch, N., Mallmann-Trenn, F.: Learning hierarchically structured concepts. Neural Netw. **143**, 798–817 (2021)
5. Lynch, N., Musco, C.: A basic compositional model for spiking neural networks, April 2021. arXiv:1808.03884v2. Also, submitted for publication
6. Lynch, N., Musco, C., Parter, M.: Winner-take-all computation in spiking neural networks, April 2019. arXiv:1904.12591
7. Oja, E.: Simplified neuron model as a principal component analyzer. J. Math. Biol. **15**(3), 267–273 (1982)
8. Leslie, G.: Valiant. Circuits of the Mind. Oxford University Press, Oxford (2000)
9. Zeiler, M.D., Fergus, R.: Visualizing and understanding convolutional networks. In: Fleet, D., Pajdla, T., Schiele, B., Tuytelaars, T. (eds.) ECCV 2014. LNCS, vol. 8689, pp. 818–833. Springer, Cham (2014). https://doi.org/10.1007/978-3-319-10590-1_53
10. Zhou, B., Bau, D., Oliva, A., Torralba, A.: Interpreting deep visual representations via network dissection. IEEE Trans. Pattern Anal. Mach. Intell. **41**(9), 2131–2145 (2019). https://doi.org/10.1109/TPAMI.2018.2858759

SIROCCO Main Track

Distributed Coloring of Hypergraphs

Duncan Adamson[1]📧, Magnús M. Halldórsson[2]📧, and Alexandre Nolin[3](✉)📧

[1] University of Göttingen, Göttingen, Germany
`duncan.adamson@cs.uni-goettingen.de`
[2] Reykjavik University, Reykjavík, Iceland
`mmh@ru.is`
[3] CISPA Helmholtz Center for Information Security, Saarbrücken, Germany
`alexandre.nolin@cispa.de`

Abstract. For any integer $r \geq 2$, a linear r-uniform hypergraph is a generalization of ordinary graphs, where edges contain r vertices and two edges intersect in at most one node. We consider the problem of coloring such hypergraphs in several constrained models of computing, i.e., computing a partition such that no edge is fully contained in the same class. In particular, we give a $\mathrm{poly}(\log \log n)$-round randomized LOCAL algorithm that computes a $O(\Delta^{1/(r-1)})$-coloring w.h.p. This is tight up to polynomial factors of the time complexity as $\Omega(\log_\Delta \log n)$ distributed rounds are necessary for even obtaining a Δ-coloring, where Δ is the maximum degree, and tight up to logarithmic factors of the number of colors, as $\Theta((\Delta/\log \Delta)^{1/(r-1)})$ colors are necessary for existence. We also give simple algorithms that run in $O(1)$-rounds of the CONGESTED CLIQUE model and in a single-pass of the semi-streaming model.

Keywords: Hypergraph coloring · Distributed computing · LOCAL model · Congested Clique

1 Introduction

In a seminal work [33], Linial opened the field of distributed computing with upper and lower bounds on the problem of finding a coloring of a distributed network. Since then, a large body of work has been committed to finding colorings faster and with fewer colors [5,12,25,26]. The goal is usually to find a $\Delta + 1$-coloring where Δ is the maximum degree of the network, a number which ensures that the graph can be colored without any monochromatic edge.

One direction that has been left largely unexplored is that of finding colorings of *hypergraphs*. Hypergraphs generalize traditional graphs, replacing edges with *hyperedges*. Each hyperedge corresponds to a set of nodes without a fixed upper bound on the size of the set. A hypergraph is said to be r-*uniform* or of (uniform) *rank* r if its hyperedges all have cardinality r. A hypergraph is called *linear* (or *simple*) if no pair of hyperedges shares more than a single node[1]. As in graphs,

---

[1] Note that linear hypergraphs generalize (ordinary) graphs: a graph is a 2-uniform linear hypergraph.

S. Rajsbaum et al. (Eds.): SIROCCO 2023, LNCS 13892, pp. 89–111, 2023.
https://doi.org/10.1007/978-3-031-32733-9_5

a valid coloring is one without monochromatic edges. Erdős and Lovász showed that every r-uniform hypergraph of maximum degree Δ has a chromatic number of no more than $O(\Delta^{1/(r-1)})$, and further that this bound is tight [17]. Frieze and Mubayi [21] obtained an improved (and also tight) bound of $O((\Delta/\log \Delta)^{1/(r-1)})$ colors for linear hypergraphs of rank $r \geq 3$. This research leaves a large gap between the $\Delta+1$-coloring that can be computed in the traditional graph model, and what is theoretically possible.

In this paper, we focus on the following question; given an r-uniform linear hypergraph in the LOCAL model, what is the fewest number of rounds needed to compute a $\tilde{\Theta}(\Delta^{1/(r-1)})$-coloring[2]? More generally, what is the round complexity of k-coloring, for each value of k at least $\Theta(\Delta^{1/(r-1)})$ and at most Δ?

> **Main result (informal):**
> *The randomized round complexity of k-coloring r-uniform linear hypergraphs is $\log^{\Theta(1)} \log n$, for every $k \in \{\Theta(\Delta^{1/(r-1)}), \ldots, \Delta\}$.*

One well-known approach to solving this problem would be an application of the Lovász Local Lemma (LLL). Indeed, LLL was first introduced in a paper of Erdős and Lovász as a tool for coloring hypergraphs [17]. Starting with the work of Moser and Tardos [35], a series of successive works provide constructive, distributed techniques giving an $\log^{O(1)} n$-round algorithm in both the randomized [14,35] and deterministic [11,14,38] setting. Our result provides a rare case where a poly$(\log \log n)$-round algorithm is known for a problem with a strict LLL formulation, even on high-degree graphs.

1.1   Our Results

Our main result is a randomized $\log^{O(1)} \log n$-round distributed algorithm in the LOCAL model for finding a $O(\Delta^{1/(r-1)})$-coloring, w.h.p. This provides a significant improvement over the $O(\log n)$ round algorithm given using the LLL based approach. It comes close to the recent lower bound of $\Omega(\log_{r\Delta} \log n)$ for coloring with Δ colors, due to Balliu et al. [4].

It is worth noting the large gap between the $\tilde{\Theta}(\Delta^{1/(r-1)})$-coloring that is achieved by our algorithm, and the lower bound of $\Omega(\log_{\Delta} \log n)$ on finding a Δ-coloring. This suggests that the complexity of hypergraph coloring "plateaus" between Δ and $\tilde{\Theta}(\Delta^{1/(r-1)})$, which contrasts with the significant gap in complexity between Δ-coloring and Δ^2-coloring in traditional graphs. The lower bound for Δ-coloring is particularly surprising as, unlike in the graph case, r-uniform hypergraphs are $O(\sqrt{\Delta})$-colorable when $r \geq 3$, leading to a significant gap.

We also give simple algorithms for $o(\Delta^{1/(r-1)})$-coloring in two models: an $O(1)$-round algorithm in the CONGESTED CLIQUE model, and a single-pass streaming algorithm using $\tilde{O}(n)$ space. For completeness, we supplement the distributed results with two additional results: An $O(\log^* n)$ round deterministic

---

[2] Here and throughout the paper, $\tilde{O}(x) = x \log^{O(1)}(x)$, $\tilde{\Omega}(x) = x/\log^{O(1)}(x)$, and $\tilde{\Theta}(x) = \tilde{O}(x) \cap \tilde{\Omega}(x)$.

algorithm for computing an $\tilde{O}(r \cdot \Delta^{r/(r-1)})$-coloring, and the tightness of that time bound. These results build on the $O(\log^* n)$-round algorithm for $O(\Delta^2)$-coloring graphs and matching lower bound due to Linial [33].

1.2 Related Work

The related problem of $\Delta + 1$-coloring graphs has been studied extensively. The current best randomized LOCAL algorithm is due to Chang et al. [12], and uses $O(\log^3 \log n)$ rounds when using the recent poly$(\log n)$-round deterministic algorithm of [25] as a subroutine. The best known complexity in terms of Δ is $O(\sqrt{\Delta \log \Delta} + \log^* n)$ [5,34].

As for distributed symmetry-breaking in hypergraphs, poly$(\log n)$ algorithms are known for Maximal Independent Set (MIS) [28,30,31]. Hypergraph maximal matching (HMM) has received more attention, in part due to existing reductions from some graph problems to HMM, notably edge-coloring [20]. $\tilde{O}(\text{poly}(\log n, r))$-round deterministic and $\tilde{O}(\text{poly}(\log \log n, \log \Delta, r))$-round randomized algorithms are known for HMM [20,24,29].

Brandt et al. recently showed that finding w.h.p. a Δ-coloring in graphs requires $\Omega(\log_{\Delta} \log n)$ rounds [8]. This uses the round elimination framework, which was automatized by Brandt [7] and applied in numerous works. Much of the work has been on graph problems, but the basic formulation also applies to hypergraph problems. In fact, Balliu et al. recently proved lower bounds for hypergraph coloring, strong coloring, MIS and maximal matching [4]. Most relevant to this paper, they showed that randomized hypergraph Δ-coloring requires $\Omega(\log_{r\Delta} \log n)$ rounds in LOCAL. Their proof involves finding a fixed point for the round elimination technique, a method previously applied by the same authors to the graph Δ-coloring problem [3].

In the streaming setting, there are recent single-pass randomized algorithms for $\Delta + 1$-coloring [1] and Δ-coloring graphs [2] using $O(n \text{ poly}(\log n))$ bits of memory. The only related work on hypergraphs is on the 2-coloring problem [37] (a.k.a., Property B), where the randomized algorithm matches the sequential results but deterministic algorithms are shown to be too weak. In the CONGESTED CLIQUE model, $O(1)$-round algorithms are known for $\Delta + 1$-coloring graphs, both randomized [9] and deterministic [15], but we are not aware of any similar hypergraph results.

The Lovász local lemma technique was first introduced in [17] and applied there specifically to the hypergraph coloring problem studied here. It has had an outsize importance to numerous combinatorial problems. A general and efficient algorithm was given by Moser and Tardos [35]. It is highly parallel in nature, which allows for efficient implementations in distributed computing.

The LLL is particularly important in distributed computing due to the discovery of a time hierarchy for a large class of problems known as LCLs (Locally Checkable Labeling problems). An LCL task consists of assigning labels to node and edges satisfying some locally checkable property (e.g., in coloring, adjacent nodes must receive distinct labels). On bounded degree graphs, every LCL that runs in $o(\log n)$ distributed (randomized) rounds can be sped up to $O(T_{\text{LLL}})$

rounds, where T_{LLL} is the round complexity of LLL [13]. The Moser-Tardos algorithm runs in $O(\log^2 n)$-rounds of LOCAL, later improved to $O(\log n \cdot \log \Delta)$ [14,22] in general and to $O(\log n)$ [14] for weaker forms of LLL. In particular, it gives an $O(\log^2 n)$-round algorithm for $O(\Delta^{1/(r-1)})$-coloring (general) hypergraphs. It is known that $T_{\mathsf{LLL}} = \Omega(\log_\Delta \log n)$ [8], but there is currently a huge gap for large Δ. There have been many attempts at obtaining improved algorithms. Fischer and Ghaffari [19] gave an $O(\Delta^2 + \mathrm{poly}(\log \log n))$-round algorithm, which largely answers the question for low-degree graphs. Their algorithm (and the ones that follow) only work for LLLs with polynomially-weakened criterion, a weaker form insufficient for hypergraph coloring with an optimal number of colors. There are recent results on still weaker forms of LLL [16], certain splitting problems [27], or restricted classes of graphs [10], but we are not aware of other distributed results that yield $\mathrm{poly}(\log \log n)$-round solutions to strict forms of LLL.

Outline. The next section provides definitions and some key results from the literature. Section 3 provides a simple partitioning approach, enabling algorithms for streaming, CONGESTED CLIQUE, and LOCAL. An improved LOCAL algorithm is given in Sect. 4. Additional results follow in Sect. 5.

2   Preliminaries

Let $G = (V, E)$ be a hypergraph. For any node $v \in V$, let d_v be the degree of v, defined as the number of edges in E containing v. Let Δ be the maximum degree of G and let n be the number of nodes in G. We assume that every node in G knows the values of both Δ and n.

Definition 1 (Underlying Graph). *The* underlying graph *of a hypergraph* $G = (V, E)$ *is the graph* $G' = (V, E')$ *formed by replacing each hyperedge of rank* r *with the* r-clique, *i.e.,* $E' = \{(u, v) \in \binom{V}{2} \mid \exists e \in E, \{u, v\} \subseteq e\}$ *(Fig. 1, left).*

Note that the underlying graph has maximum degree $r\Delta$.

Definition 2 (Induced Subhypergraph). *Let* $V' \subseteq V$, *the* subhypergraph *$G[V']$ induced by V' is the hypergraph* $G' = (V', E')$ *formed by only keeping edges whose endpoints are all in* V', *i.e.,* $E' = \{e \in E \mid e \subseteq V'\}$ *(Fig. 1, right).*

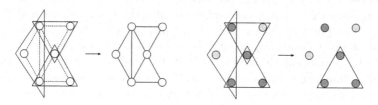

Fig. 1. (Left) a hypergraph and its underlying graph. (Right) a colored hypergraph and the subhypergraph induced by the blue vertices. (Color figure online)

Importantly, the induced subhypergraph is defined s.t. an edge disappears if at least one of its vertices is not part of V', rather than becoming an edge of rank $r' < r$. Thus, an induced subhypergraph of a rank r hypergraph is also rank r.

Definition 3 (Hypergraph coloring). *A coloring of a rank r hypergraph $G = (V, E)$ is a assignment ψ of colors to the nodes such that every edge $e \in E$, contains a pair of vertices $v_i, v_j \in e$, $\psi(v_i) \neq \psi(v_j)$ (Fig. 2).*

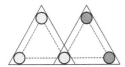 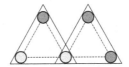 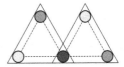

Fig. 2. (Left) an invalid coloring, containing a monochromatic hyperedge. (Center) a valid coloring which is not a strong coloring. (Right) a strong coloring.

We say an edge is *monochromatic* if $\psi(v_i) = \psi(v_j)$ for all $v_i, v_j \in e$. A valid coloring is a coloring without any monochromatic edge. It may be equivalently defined as a coloring such that the subhypergraph induced by each color class is *empty*, i.e., containing no hyperedges. Note that coloring a hypergraph with edges of minimum cardinality r is no harder than coloring a r-uniform hypergraph – it trivially reduces to it. A *strong*[3] coloring of a hypergraph $G = (V, E)$ is a coloring of the vertices such that no two adjacent vertices share a color. A strong coloring may be defined as a proper coloring of the underlying graph.

Problem 1. The distributed c-coloring problem for rank r hypergraphs

Input. A hypergraph $G = \{V, E\}$ with minimum rank r and an integer c.
Output. A coloring of G using at most c colors.

2.1   Communication Model

In this paper we consider two models of communication, LOCAL and CONGESTED CLIQUE. In both, communication is done over synchronous rounds, and each node has some globally unique $\Theta(\log n)$-bit ID. In LOCAL, in each round, each node can only send messages to its neighbors in the graph, but the messages can be arbitrarily large. CONGESTED CLIQUE removes the locality constraint and adds congestion. In this model, nodes may only send messages of size $O(\log n)$ bits, but their recipients can be all other nodes in the graph. I.e., while in LOCAL the graph of communication is also the input graph, in CONGESTED CLIQUE nodes are restricted in bandwidth but not who they can talk to.

---

[3] Prior work sometimes refer to hypergraph coloring as hypergraph *weak* coloring, by opposition to strong coloring. We do not use this terminology here, to avoid confusion with the graph weak coloring problem, which only asks that each node has at least one non-monochromatic edge.

2.2   An LLL for Hypergraph Coloring

In this section we provide an overview of the key results underpinning the application of the Lovász Local Lemma to the problem of hypergraph coloring.

Lemma 1 (Lovász Local Lemma, [17]). *Consider a set $\mathcal{E}$ of events such that for each $A \in \mathcal{E}$:*

1. $\Pr[A] \leq p < 1$, and
2. $A \in \mathcal{E}$ is mutually independent of a set of all but at most d of the other events.

If $4pd \leq 1$ then with positive probability, none of the events in $\mathcal{E}$ occur.

Informally, Lemma 1 states that given a set of events that are sufficiently independent, with a low enough probability of failure, then there is a positive probability of global success. As LLL instances are locally checkable, they are a natural tool for use in distributed algorithms. Hypergraph coloring was the first problem to be solved using Lemma 1 [17].

Lemma 2 ([17]). *Any hypergraph G with maximum degree Δ and rank r can be colored with $(4r \cdot \Delta)^{1/(r-1)}$ colors.*

Proof. Let $k = \lceil (4r \cdot \Delta)^{1/(r-1)} \rceil$. Consider the uniform probability distribution over the set of k colors. Further, as an edge has at least r vertices, the probability p of an edge being monochromatic is at most $p \leq \frac{k}{k^r} \leq \frac{1}{4r \cdot \Delta}$. As each hyperedge is adjacent to at most $d \leq r \cdot \Delta$ other edges, the event of an edge being monochromatic is dependent on at most $r \cdot \Delta$ other events. Hence, by Lemma 1 as $4pd \leq 4\frac{1}{4r\Delta}r\Delta \leq 1$, there exists a k-coloring of G.

Lemma 2 provides an immediate method of computing a $(4r \cdot \Delta)^{1/(r-1)}$-coloring by brute force. The breakthrough work by Moser and Tardos [35] provided a $O(\log^2 n)$ randomized LOCAL algorithm for LLL, later improved to $O(\log n \cdot \log d)$ rounds [14,22].

Theorem 1. *There is a poly$(\log n)$-round deterministic LOCAL algorithm that computes a $(4r \cdot \Delta)^{1/(r-1)}$-coloring of a hypergraph with n vertices, rank r and maximum degree Δ. This holds even if the nodes' IDs are of order $\exp(\text{poly}(n))$.*

Proof. There is an LLL formulation of hypergraph k-coloring, for $k = (4r \cdot \Delta)^{1/(r-1)}$, by Lemma 2. Thus the distributed LLL algorithm of [35] (and [14]) gives a randomized poly$(\log n)$-round algorithm to find such a coloring. This is a locally checkable labeling problem, which implies by the network decomposition result of [38] that there exists a poly$(\log n)$-round *deterministic* algorithm for the problem. To handle large IDs, one can either use the improved network decomposition algorithm of [23] or run Linial's algorithm to reduce IDs to poly(n) [33].

2.3   Shattering and Concentration Bounds

In the shattering technique, a randomized algorithm is first used to solve a large subset of the graph so that the unsolved parts of the graph induce small connected components.

Lemma 3 (Lemma 4.1 of [12]). *Consider a randomized procedure that generates a subset* Bad $\subseteq V$ *of vertices. Suppose that for each* $v \in V$, *we have* $\Pr[v \in \text{Bad}] \leq \Delta^{-3c}$, *and each event* $v \in$ Bad *is determined by the random choices within distance* c *of* v. *W.p.* $1 - n^{-\Omega(c')}$, *each connected component in* $G[\text{Bad}]$ *has size at most* $(c'/c)\Delta^{2c} \log_\Delta n$.

Lemma 4 (Chernoff bounds). *Let* $\{X_i\}_i$ *be a family of independent random variables taking values in* $[0,1]$, *and let* $X = \sum_i X_i$.

$$\Pr[X \geq (1+\delta)\,\mathbb{E}[X]] \leq \exp(-\min(\delta, \delta^2)\,\mathbb{E}[X]/3), \qquad \forall \delta > 0, \quad (1)$$

$$\Pr[X \leq (1-\delta)\,\mathbb{E}[X]] \leq \exp(-\delta^2\,\mathbb{E}[X]/2), \qquad \forall \delta \in (0,1). \quad (2)$$

As corollary of Eq. (1) *when* $\mathbb{E}[X] > 0$, $\forall t \geq 2\,\mathbb{E}[X]$, $\Pr[X \geq t] \leq \exp(-t/6)$.

3   Simple Splitting Primitive and Its Applications

We first consider a simple zero-round randomized primitive (see Algorithm 1) for splitting the vertex set and apply it in three different models. The splitting forms a defective coloring, defined as follows.

Given a coloring of the vertices, the *defect* $\text{def}(v)$ of a node $v \in V$ is the number of monochromatic edges incident to v, i.e., the number of incident edges whose nodes all have the same color as v. A coloring is d-defective if $\text{def}(v) \leq d$ for all nodes v. A 0-defective coloring is a normal valid coloring.

Algorithm 1. SPLIT(Vertex set V, maximum degree Δ, integer x)

Input: A hypergraph on V of degree Δ, and a parameter $x \geq 1$.
Output: Partition of V into $V_1, V_2, \ldots, V_x$ and V_{Bad}.
Assign each $v \in V$ a value r_v in $[x]$, u.a.r., partitioning V into $V_1, V_2, \ldots, V_x$.
Move into V_{Bad} the nodes with defect at least $\eta = 2\Delta/x^{r-1}$.

Lemma 5. *Consider the partition computed by* SPLIT(V, Δ, x). *The probability that a given node is in* V_{Bad} *is at most* $\exp(-\eta/6) = \exp(-\Delta/(3x^{r-1}))$.

Proof. Observe first that for any edge e incident to v to be monochromatic, every vertex other than v in e must pick the same color as v. As there are x colors, the probability of any set of $r - 1$ nodes choosing the same specified color is $1/x^{r-1}$. The expected defect of v is therefore $\mathbb{E}[\text{def}(v)] = d_v/x^{r-1}$. Since the hypergraph is linear, edges incident on v share no other vertex. Therefore, once conditioned on v's choice of random color, whether each edge incident on v is monochromatic is independent from whether other edges incident on v are monochromatic. Using that $\eta \geq 2d_v/x^{r-1} = 2\,\mathbb{E}[\mathrm{def}(v)]$, and applying Lemma 4 (Chernoff), we get that $\Pr[\mathrm{def}(v) \geq \eta] \leq \exp(-\eta/6)$.

Lemma 6. *Suppose we run* SPLIT(V, Δ, x) *with* $x \leq (\Delta/(24\log(r\Delta)))^{1/(r-1)}$. *Then,* $G[V_{\mathsf{Bad}}]$ *consists of connected components of size* $O(\Delta^2 \log n)$ *(i.e., the graph is shattered), w.h.p. If* $x \leq (\Delta/(6\log n))^{1/(r-1)}$, *then* V_{Bad} *is empty, w.h.p.*

Proof. The claim that $V_{\mathsf{Bad}} = \emptyset$ when $x \leq (\Delta/\log n)^{1/(r-1)}$ follows from Lemma 5. Otherwise, the set V_{Bad} consists of the nodes of defect at least $\eta = 2\Delta/x^{r-1} \geq 48\log(r\Delta)$. By Lemma 5, the probability that $\mathrm{def}(v) \geq 48\log(r\Delta)$ is less than $(r\Delta)^{-8}$. Each event $v \in V_{\mathsf{Bad}}$ is fully determined by the random choices of v and its at most $r\Delta$ neighbors. The lemma now follows from Lemma 3.

Lemma 5 immediately implies a single-round randomized algorithm to produce an $O((\Delta/\log n)^{1/(r-1)})$ coloring with a defect of $O(\log n)$, w.h.p., when $\Delta = \Omega(\log n)$. It suffices then to solve the coloring problem on hypergraphs of degree $\Delta = O(\log n)$. Further, we observe the following.

Lemma 7. *Suppose we run* SPLIT(V, Δ, x) *and that we further color each subgraph* $H = G[X]$ *where* $X \in \{V_i\}_i \cup V_{\mathsf{Bad}}$ *with* $O(\Delta(H)^{1/(r-1)})$ *colors. The coloring of* G *obtained by concatenating these colorings uses* $O(\Delta^{1/(r-1)})$ *colors.*

Proof. SPLIT produces x subgraphs of degree $\Delta(H) \leq \eta = 2\Delta/x^{r-1}$ and one subgraph of small size. By assumption, the total number of colors is on the order of

$$\sum_{i=1}^{x} \Delta(H)^{1/(r-1)} + \Delta^{1/(r-1)} \leq x \cdot \eta^{1/(r-1)} + \Delta^{1/(r-1)} \leq 3\Delta^{1/(r-1)}.$$

3.1   Streaming Algorithm

We first give a simple application of the SPLIT algorithm to the *semi-streaming* model. Introduced in [18,36], the *semi-streaming* model is a model of computation for solving problems on massive graphs with an $O(n\,\mathrm{poly}\log n)$ amount of storage space. The goal of the semi-streaming model is to provide an algorithm that can compute a solution to graph problems without needing to store the complete graph explicitly.

In the semi-streaming model, the input graph $G = (V, E)$ is given as a stream of edge changes (insertions and deletions). For hypergraphs, the stream is instead an ordered list of hyperedge changes. After each edge change, a semi-streaming algorithm is given some amount of time to process the change, with the restriction that no more than $O(n\,\mathrm{poly}\log n)$ space is used at any given time. The order that the edge changes are assigned is assumed to be determined by an *oblivious adversary*. Such an adversary has access to the algorithm being used and is capable of simulating it, but does not have access to any source of randomness being used.

Theorem 2. *There exists a single-pass semi-streaming randomized algorithm for $O((\Delta/\sigma)^{1/(r-1)})$-coloring linear hypergraphs against an oblivious adversary, where $\sigma = \min\{\log \Delta, \log \log n\}$.*

Proof. When $\Delta = O(\log n)$, the full graph is represented using $O(rn \log^2 n)$ bits: we store it fully and color it with $O((\Delta/\log \Delta)^{1/(r-1)})$ colors using the method of [21]. Otherwise, we apply $\text{SPLIT}(V, \Delta, x)$ with $x = O((\Delta/\log n)^{1/(r-1)})$. By Lemma 5, $V_{\text{Bad}} = \emptyset$ and each V_i is $O(\log n)$-defective. Thus, $O(rn \log^2 n)$ bits suffice to represent all the subgraphs in the partition. Applying the algorithm of [21] on each of them, we use a total of $x \cdot O((\log n/\log \log n)^{1/(r-1)}) = O(\Delta^{1/(r-1)}/\log \log n)$ colors.

3.2   Congested Clique Algorithm

In the CONGESTED CLIQUE, $O(\log n)$-bit messages can be sent between any pair of vertices, not just the adjacent ones. It does not matter for our argument whether the hyperedges are represented by a separate node (as in the client-server model), or if we are given the underlying graph representation. We propose an algorithm that partitions the hypergraph into a collection of hypergraphs that can be represented in small space. Each of these can then be gathered at a single node and colored separately. Recall that the bound obtained is best possible for linear hypergraphs [21].

Theorem 3. *There is a $O(1)$-round randomized algorithm in the CONGESTED CLIQUE model for $O((\Delta/\log \Delta)^{1/(r-1)})$-coloring r-uniform linear hypergraphs.*

Proof. Apply $\text{SPLIT}(V, \Delta, x)$ with $x = \Delta^{1/r}$ to obtain a partition of V into $V_1, V_2, \ldots, V_x$ and V_{Bad}, where each V_i is at most η-defective, $\eta = 2\Delta/x^{r-1} = 2\Delta^{1/r}$. For each $i \in [x]$, send all monochromatic edges within V_i to node i, which then computes an $O((\eta/\log \eta)^{1/(r-1)})$-coloring of V_i. Similarly, send the edges within V_{Bad} to a single node and let it $O((\Delta/\log \Delta)^{1/(r-1)})$-color V_{Bad} [21]. Concatenate these colorings to obtain a coloring of G using

$$x \cdot O((\eta/\log \eta)^{1/(r-1)}) + O((\Delta/\log \Delta)^{1/(r-1)}) = O((\Delta/\log \Delta)^{1/(r-1)})$$

colors. It remains to explain how to achieve this communication in $O(1)$ rounds.

By Lemma 4 (Chernoff), each V_i contains at most $2n/x$ vertices, w.h.p., and by definition each node has degree at most η (within V_i). Thus, the number of monochromatic edges in each part is at most $2n/x \cdot \eta = 2\Delta n/x^r = 2n/r$, w.h.p. They are represented in $r \cdot 2n/r = 2n$ space (of $O(\log n)$-bit words), and can be forwarded to a single node using Lenzen routing in $O(1)$ rounds [32].

3.3 Simple LOCAL Algorithm

Suppose $\Delta = O(\log n)$ for now, leaving the $\Delta \in \Omega(\log n)$ case for later. We apply $\text{SPLIT}(V, \Delta, x)$, where $x = (\Delta/(24 \log(r\Delta)))^{1/(r-1)}$ to partition V into $V_1, V_2, \ldots, V_x$ and V_{Bad}. We color V_{Bad} with $(4r\Delta)^{1/(r-1)}$ colors using the deterministic LLL algorithm of Theorem 1. We then color each V_i by replacing each r-edge e of $G[V_i]$ by an arbitrary 2-edge $e' \subseteq e$ and applying the $\Delta + 1$-coloring algorithm of [12] on each obtained (standard) graph, in parallel. Each of these steps runs in $\text{poly}(\log \log n)$ rounds. We use $\eta + 1 = O(\log(r\Delta))$ colors on each V_i, for a total of $x(\eta + 1) = O((\Delta \log^{r-2}(r\Delta))^{1/(r-1)})$.

When $\Delta \in \Omega(\log n)$, we reduce the problem to coloring $O(\log n)$-degree instance by applying $\text{SPLIT}(V, \Delta, x)$ with $x = (\Delta/(6 \log n))^{1/(r-1)}$.

4 Improved LOCAL Algorithm

We give an improved algorithm that uses $O(r^2 \Delta^{1/(r-1)})$ colors, using different techniques. The main component of our method is an algorithm for triangle-free (girth 4) hypergraphs, i.e., when there are no vertices x, y, z where each pair belongs to a distinct edge.

Theorem 4. *There is an $O(\text{poly}(\log \log n))$-round randomized LOCAL algorithm to color a triangle-free hypergraph of rank r with $O(\Delta^{1/(r-1)})$ colors, w.h.p. When $\Delta \geq 4^{r-1}(18 \log n)^{(r-1)^2}$, the algorithm takes $O(\log \log \Delta + \log^* n)$ rounds.*

The algorithm and the proof of the theorem are given in the next subsection. We then give in Sect. 4.2 a reduction of the general coloring problem (of linear hypergraphs) to the triangle-free case.

4.1 Triangle-Free Hypergraphs

We consider the following simple method GEOMETRICTRIALS (Algorithm 2), in which the nodes try random colors from geometrically decreasing palettes. In this algorithm, all nodes initially participate in trying colors, and across $(i_{\text{last}} + 1) \in O(\log \log \Delta)$ successive iterations, they progressively either get colored or quit the process (joining a shattered subinstance), thereby reducing competition for other nodes. The nodes still active in iteration i induce an hypergraph of maximum degree Δ_i, where the sequence $\Delta_0 \ldots \Delta_{i_{\text{last}}}$ decreases doubly exponentially in i and $\Delta_{i_{\text{last}}+1} < \Delta_{i_{\text{last}}}$ is set to $\Delta_{\text{goal}} = \Delta^{1/(r-1)}$. The quitters in iteration i are those that both fail to color themselves and whose degree remains above Δ_{i+1}.

By using geometrically shrinking palettes of initial size $K = 4\Delta^{1/(r-1)}$ and shrinking factor $\alpha = 1/2$, clearly, at most $O(K/(1 - \alpha)) = O(\Delta^{1/(r-1)})$ distinct colors are used by GEOMETRICTRIALS. We show that nodes left uncolored by GEOMETRICTRIALS can also be colored using $O(\Delta^{1/(r-1)})$ colors.

By definition, the nodes still active after the last iteration of the algorithm induce a graph of maximum degree $\Delta_{\text{goal}} = \Delta^{1/(r-1)}$, which can be efficiently

$O(\Delta_{\mathsf{goal}})$-colored by an algorithm from the distributed graph coloring literature. What remains to be proved is that quitters can also be efficiently colored with $O(\Delta^{1/(r-1)})$ colors, even though they might induce a hypergraph of degree $\omega(\Delta^{1/(r-1)})$. We resolve this by a shattering argument: quitting the process early occurs with sufficiently low probability, and with sufficient independence between the nodes, that early quitters form connected components of size $\mathrm{poly}(\log n)$ which can be handled by the deterministic LLL algorithm of Theorem 1. The triangle-free property is crucial to the analysis of this probability: it allows us to argue that what happens in each edge incident on a node is somewhat independent from what happens in other incident edges, and so the degree decreases as needed with a high (enough) probability.

Algorithm 2. GEOMETRICTRIALS(Integer C, $\alpha \in [0,1)$) (on hypergraph G of maximum degree Δ)

For all $i \geq 0$, let $C_i = C^{2^i}$, $K_i = \alpha^i \cdot C \cdot \Delta^{1/(r-1)}$, and $a_i = \sum_{j<i} K_j$.
Let $\Delta_{\mathsf{goal}} = \Delta^{1/(r-1)}$. For all $i \geq 0$, let $\Delta_i = \max\{\Delta_{\mathsf{goal}}, (K_i/C_i)^{r-1}\}$.
for $i \leftarrow 0$ to $i_{\mathsf{last}} = \max\{i \mid \Delta_i > \Delta_{\mathsf{goal}}\}$ **do**
    Each live uncolored node ($v \in V^{(i)}$) picks a color u.a.r. in $[a_i, a_{i+1})$.
    Each node part of a monochromatic edge drops its temporary color.
    Nodes who kept their temporary color make it permanent (join $V_{\mathsf{Good}}^{(i)}$).
    Remove every partially colored edge from the graph, update degrees.
    Every $v \in V^{(i)} \setminus V_{\mathsf{Good}}^{(i)}$ of current degree $d_v > \Delta_{i+1}$ quits the process (joins $V_{\mathsf{Quit}}^{(i)}$).
    Remove all edges containing a node in $V_{\mathsf{Quit}}^{(i)}$.
end for

Notation. The algorithm executes a loop for $i_{\mathsf{last}} + 1$ iterations, where $i_{\mathsf{last}} \in O(\log \log \Delta)$. For any $i \in [0, i_{\mathsf{last}}]$, we denote by $V^{(i)}$ the set of uncolored nodes trying a color in iteration i in Algorithm 2. We denote by $V_{\mathsf{Good}}^{(i)}$ the nodes of $V^{(i)}$ that successfully color themselves in iteration i, and $V_{\mathsf{Quit}}^{(i)}$ the nodes that abandon the process in iteration i. The remaining nodes form $V^{(i+1)}$, i.e., $V^{(i+1)} = V^{(i)} \setminus (V_{\mathsf{Good}}^{(i)} \sqcup V_{\mathsf{Quit}}^{(i)})$.

We also consider the sets $V_{\mathsf{Good}} = \bigsqcup_{i=0}^{i_{\mathsf{last}}} V_{\mathsf{Good}}^{(i)}$, $V_{\mathsf{Quit}} = \bigsqcup_{i=0}^{i_{\mathsf{last}}} V_{\mathsf{Quit}}^{(i)}$, and $V_{\mathsf{Low}} = V \setminus (V_{\mathsf{Good}} \cup V_{\mathsf{Quit}})$. V_{Good} are all the nodes that got colored by the process, V_{Quit} are all the quitters, and V_{Low} are the nodes that remained active through the whole process and thus have had their degree reduced. We have as initial condition $V^{(0)} = V$.

Degree Reduction. First, we analyze how the degree of a node behaves when all nodes in a hypergraph of maximum degree Δ try a color u.a.r. from a set of size $K \gg \Delta^{1/(r-1)}$, as in GEOMETRICTRIALS. There are two ways for an edge e incident on v to survive: either e itself was monochromatic, or each node of e was part of a monochromatic edge other than e. Note that these are not mutually exclusive. The first type of event is analyzed when considering splitting (Sect. 3). The following lemma analyzes the second type of event.

Lemma 8. *Let $C \geq 2^{1/(r-2)}$ be a constant, and $G = (V, E)$ be a triangle-free graph of rank r and maximum degree Δ. Let each node v try a random color in $[K] = [C \cdot \Delta^{1/(r-1)}]$ and uncolor itself if it is part of a monochromatic edge. Let $s \in [2\Delta^{1/(r-1)}/C^{r-2}, K]$ and $t \geq 2(s/C)^{r-1}$. Then w.p. at least $1 - 2\exp(-t/6) - (r-1)\Delta \exp(-s/6)$, v's degree (number of fully uncolored incident edges) becomes at most $2t$.*

Proof (Proof sketch). We consider an edge e incident on v and bound the probability that each node in e other than v is part of a monochromatic edge. The survivals of edges incident on v are not necessarily independent due to 4-cycles: for two edges e, e' incident on v, there might be a vertex u at distance 2 from v that is connected to both a vertex in $w \in e$ and a vertex $w' \in e'$. We handle this non-independence by arguing independence once the colors at distance 2 from v have been fixed. When fixing those colors, some edges incident on a neighbor of v might be monochromatic on the already selected $(r-1)$ colors. We bound the probability that this occurs on too many edges, so as to argue that nodes in an edge incident on v are unlikely to all pick a color that makes an edge they're part of monochromatic. We defer the full proof to Appendix A.1

Lemma 9. *Let Δ_i be the maximum degree of active nodes in the i-th round of* GEOMETRICTRIALS, *K_i the number of colors to choose from in that round, and $C_i = \Delta_i^{1/(r-1)}/K_i$. Then, for each live node v in the i-th round of* GEOMETRICTRIALS:

- *v gets colored w.p. at least $1 - \Delta_i/K_i^{r-1}$.*
- *Let d'_v be the degree of v after this round. For any $t \geq 4\Delta_i/C_i^{(r-1)^2}$,*

$$\Pr[d'_v \leq t] \geq 1 - 3(r-1)\Delta_i \exp(-(t/4)^{1/(r-1)}C_i/6).$$

Proof. For the first item, we use that each edge incident on v is monochromatic w.p. $1/K_i^{r-1}$. By union bound, v is part of no monochromatic edge w.p. at least $1 - \Delta_i/K_i^{r-1}$, in which case it gets colored. For the second item, we apply Lemma 8 with $K = K_i$, $C = C_i$, $\Delta = \Delta_i$, and $s = C \cdot (t/4)^{1/(r-1)}$. Note that though the lemma may not be applied with this s when $t \geq 4\Delta$, the result still holds since v's degree is always less than Δ.

We now analyze the probability that a node quits during GEOMETRICTRIALS.

Lemma 10. *For each node v, the probability that v is in V_{Quit} is at most*

$$3(r-1)\Delta(\log\log\Delta) \cdot \exp(-(\Delta_{\mathsf{goal}}/4)^{1/(r-1)}/6).$$

Proof (Proof sketch). Armed with previous technical lemmas, this proof only consists of a union bound over all iterations, summing the probabilities that v's degree does not decrease sufficiently when it fails to get colored. The full proof is deferred to Appendix A.2.

The Full Algorithm. Our algorithm and its analysis follow naturally once GEOMETRICTRIALS has been introduced and analyzed. We first run GEOMETRICTRIALS. Uncolored nodes now fall into two categories: quitters and non-quitters. The non-quitters induce a graph of maximum degree $\Delta_{\mathsf{goal}} = \Delta^{1/(r-1)}$. When Δ is a large enough poly($\log n$), w.h.p., there are no quitters. For smaller Δ, w.h.p., the graph induced by quitters is shattered; more precisely, it consists of connected components of size $\log^{O(\log \log \log n)} n$. All that remains is to apply results from the literature to color those connected components, and to color the remaining other uncolored nodes of degree $O(\Delta^{1/(r-1)})$.

Algorithm 3. COLOR(Triangle-free hypergraph G, maximum degree Δ, rank r)

GEOMETRICTRIALS$(4, 1/2)$.
Color V_{Quit} by the deterministic LLL algorithm of Theorem 1 (if $V_{\mathsf{Quit}} \neq \emptyset$).
Color V_{Low} by an algorithm for graph coloring using $O(\Delta_{\mathsf{goal}})$ colors.

Theorem 4. *There is an $O(\mathrm{poly}(\log \log n))$-round randomized LOCAL algorithm to color a triangle-free hypergraph of rank r with $O(\Delta^{1/(r-1)})$ colors, w.h.p. When $\Delta \geq 4^{r-1}(18 \log n)^{(r-1)^2}$, the algorithm takes $O(\log \log \Delta + \log^* n)$ rounds.*

Proof. Recall that $\Delta_{\mathsf{goal}} = \Delta^{1/(r-1)}$. After GEOMETRICTRIALS, each uncolored node is either in V_{Quit} or V_{Low}, and only $O(\Delta^{1/(r-1)})$ colors were used. We color V_{Quit} and V_{Low} each with their own set of $O(\Delta^{1/(r-1)})$ colors, for a total number of colors of the same order of magnitude. Note that we can color them in parallel since they use distinct colors. We split the analysis depending on the value of Δ, starting with Δ large (at least some poly($\log n$)).

When $\Delta \geq 4^{r-1}(18 \log n)^{(r-1)^2}$, by Lemma 10, w.h.p., there are no nodes in V_{Quit}. This means that the only nodes that remain to be colored are in V_{Low}, so we can skip the costly application of Theorem 1 that we otherwise use to color V_{Quit}. The hypergraph induced by V_{Low} has maximum degree $O(\Delta^{1/(r-1)})$. We project each hyperedge e of $G[V_{\mathsf{Low}}]$ to an arbitrary 2-edge uv, $\{u, v\} \subseteq e$. The resulting graph also has maximum degree $O(\Delta^{1/(r-1)})$, and we color it in $O(\log^* n)$ rounds with $\Theta(\Delta^{1/(r-1)}) = \Omega(\log^{r-1} n)$ colors by the algorithm for $O(\Delta + \log^{1+1/\log^* n} n)$-coloring from [39]. This gives the complexity of $O(\log \log \Delta + \log^* n)$ rounds for large Δ.

For smaller Δ, we color V_{Low} with the poly($\log \log n$) algorithm of [6] for $(\Delta + 1)$-coloring. To color the nodes of V_{Quit} efficiently, we argue that the hypergraph they induce is shattered, w.h.p. More precisely, we show that $G[V_{\mathsf{Quit}}]$ consists of connected components of size $O(\log^{1+2\log \log \Delta} n) = \log^{O(\log \log \log n)} n$, w.h.p.

Let $c = \log \log \Delta > i_{\mathsf{last}}$. Whether a node quits GEOMETRICTRIALS is determined by the random choices within distance c from v during the process, and it occurs with probability at most $3(r-1)\Delta(\log \log \Delta) \exp(-(\Delta_{\mathsf{goal}}/4)^{1/(r-1)}/6) = \exp(-\Theta(\Delta^{1/(r-1)^2})) \leq (r\Delta)^{-3c}$ by Lemma 10, for Δ larger than some sufficiently big constant (constant degree hypergraphs can be colored with $O(\Delta^2) = O(1) = $

$O(\Delta^{1/(r-1)})$ colors in $O(\log^* n)$ by [33]). By Lemma 3, the graph induced by V_{Quit} has connected components of size $O((r\Delta)^{2c} \log n) = \log^{(r^{O(1)} \log\log\log n)} n$. Theorem 1 therefore colors V_{Quit} in poly($\log\log n$) rounds, for a total complexity of poly($\log\log n$) LOCAL rounds, and using at most $O(\Delta^{1/(r-1)})$ colors in total.

4.2   Reduction to the Triangle-Free Case

We now reduce the problem of coloring general linear hypergraphs to that of coloring triangle-free ones. We partition the hypergraph into hypergraphs that are either triangle-free or of polylogarithmic size. The former are solved by the algorithm of the preceding section, while the latter are solved by Theorem 1. This reduction is adapted from the work of Frieze and Mubayi [21], and modified only slightly for a distributed context, in particular avoiding a degeneracy-based coloring (that is known to require $\Omega(\sqrt{\log n})$-rounds [33]).

The following two lemmas are stated existentially in [21], but the statements below follow immediately from the proofs of their lemmas in [21]. When splitting nodes into subsets $V_1, \ldots, V_m$, inducing subhypergraphs $H_1, \ldots, H_m$, where v ends in H_{i_v}, a pair of nodes $x, y \in N_H(v)$ is said to be *covered* if

- the edges $S, S' \in E$ s.t. $\{v, x\} \subseteq S$ and $\{v, y\} \subseteq S'$ are both in H_{i_v};
- there exists an edge $S'' \in E$ that contains both x and y but not v ($\{x, y\} \subseteq S$ and $v \notin S''$), and $S'' \in H_{i_v}$.

Intuitively, x, y is a covered pair of v if v, x, and y form a triangle that survived splitting.

Lemma 11 (Lemma 5 [21]). *Let H be a linear rank r hypergraph of maximum degree Δ. Let $m = \lceil \Delta^{2/(3r-4)} - \varepsilon \rceil$. Suppose we partition the nodes u.a.r. into subsets $V_1, V_2, \ldots, V_m$, inducing subhypergraphs $H_1, H_2, \ldots, H_m$. Then, for each $i = 1, \ldots, m$ and each $v \in V_i$,*

1. *v has degree more than $2\Delta/m^{r-1}$ in V_i w.p. at most Δ^{-5}.*
2. *The H_i neighborhood of v $N_{H_i}(v)$ contains more than $r^2\Delta^2/m^{3r-4}$ covered pairs w.p. at most Δ^{-5}.*

Lemma 12 (Lemma 6 [21]). *Let δ be a sufficiently small positive constant depending on r. Let L be a linear rank r hypergraph of maximum degree at most d. Suppose that each vertex neighborhood $N_L(v)$ contains at most d^δ covered pairs. Let $\ell = d^{1/(r-1)-\delta}$. Suppose we partition the nodes u.a.r. into subsets $W_1, W_2, \ldots, W_\ell$, inducing subhypergraphs $L_1, L_2, \ldots, L_\ell$. Then, for each $j = 1, \ldots, \ell$ and each $v \in W_j$,*

1. *v has more than $2d/\ell^{r-1}$ neighbors in W_j is w.p. at most d^{-5}.*
2. *v belongs to more than $400r^2$ triangles within L_j is w.p. at most d^{-5}.*

Theorem 5. *There is a randomized algorithm for $O(r^2\Delta^{1/(r-1)})$-coloring r-uniform linear hypergraphs in poly($\log\log n$) LOCAL rounds, w.h.p.*

Proof. We reduce the problem to the case where the maximum degree is $O(\log n)$. When $\Delta > \log n$, we apply SPLIT(V, Δ, x) with $x = (\Delta/(6 \log n))^{1/(r-1)}$. By Lemma 6, each of the obtained vertex sets induces a subhypergraph of maximum degree $\Delta' \in O(\log n)$, w.h.p. Applying a coloring algorithm that uses only $O(r^2(\Delta')^{1/(r-1)})$ on each of them results in an overall coloring that uses $O(r^2 \Delta^{1/(r-1)})$ colors in total, by Lemma 7. The $\Delta \in \Omega(\log n)$ degree case thus reduces to the $O(\log n)$ degree case.

By combining Lemma 11 and Lemma 12, we reduce the problem to the coloring of a collection of triangle-free hypergraphs. Nodes that fail the first (second) condition of Lemma 11 are moved to the set V_{Bad}^1 (V_{Bad}^2), and those that fail the first (second) condition of Lemma 12 are moved to V_{Bad}^3 (V_{Bad}^4), respectively. By Lemma 3, each V_{Bad}^i is shattered, inducing components of size at most $N = \text{poly}(\log n)$. Thus, we can color each V_{Bad}^i with $(4r\Delta)^{1/(r-1)}$ colors in $\text{poly}(\log N) = \text{poly}(\log \log n)$ rounds, by the LLL algorithm of [14], for a total of $O(\Delta^{1/(r-1)})$ colors.

The rest of the nodes are partitioned into sets $V_1, \ldots, V_m$, each of which is $d = 2\Delta'/m^{r-1}$-defective. Each such V_i is partitioned into $W_1^i, \ldots, W_\ell^i$, which are $q = 2d/\ell^{r-1}$-defective, where $\ell = d^{1/(r-1)-\delta}$. Crucially, each node in the subhypergraph L_j^i induced by each W_j^i is part of at most $400r^2$ triangles. Consider the underlying graph of L_j^i, and focus on the subgraph M_j^i consisting of the edges (v, x), (v, y) involved in a covered pair x, y (with some node v). This M_j^i has maximum degree $O(r^2)$, by the bound on triangle participation in L_j^i. We color this graph with $O(r^2)$ colors using a $\text{poly}(\log \log n)$-round randomized algorithm for $\Delta + 1$-coloring graphs [6,38]. Let $W_j^{i,1}, \ldots, W_j^{i,c}$ be the vertices of each of the c color classes, where $c \in O(r^2)$. Each node that is not in any class joins one at random. For each $k \in [c]$ let $L_j^{i,k}$ be subhypergraph induced by $W_j^{i,k}$.

Note that each $L_j^{i,k}$ is triangle-free (girth 4), and like L_j^i has maximum degree at most $q = 2d/\ell^{r-1}$. We apply the algorithm of Theorem 4 to each $L_j^{i,k}$ in parallel, which uses $O(q^{1/(r-1)}) = O(d^{1/(r-1)}/\ell)$ colors. So, the total number of colors used on each V_i is $\ell \cdot c \cdot O(d^{1/(r-1)}/\ell) = O(r^2 d^{1/(r-1)})$. By the same token, the total number of colors used on all the classes $V_1, \ldots, V_m$ is $m \cdot O(r^2 d^{1/(r-1)}) = m \cdot O(r^2 \Delta^{1/(r-1)}/m) = O(r^2 \Delta^{1/(r-1)})$, as desired.

5   Additional Results

We also provide two results lifting the classic $O(\Delta^2)$-coloring algorithm and $\Omega(\log^* n)$ lower bound due to Linial to the hypergraph setting. Contrary to our main results, these apply to general hypergraphs. We defer their proofs to Appendix B.

Theorem 6. *There is a deterministic $O(\log^* n)$-round CONGEST algorithm to $O(r \cdot \Delta^{r/(r-1)} \log(r\Delta))$-color hypergraphs of maximum degree Δ and rank r.*

Theorem 7. *For any pair of constants r and c, no LOCAL algorithm can find a $O(\Delta^c/r)$ coloring of a rank r hypergraph in fewer than $\Omega((\log^* n)/r)$ rounds.*

Acknowledgements. This project was supported by Icelandic Research Fund grant no. 217965. Part of the work was done while D. Adamson and A. Nolin were with the CS Department of Reykjavik University.

A   Missing Proofs of Main Algorithm

A.1   Proof of Lemma 8 (Degrees Decrease in GeometricTrials)

Lemma 8. *Let $C \geq 2^{1/(r-2)}$ be a constant, and $G = (V, E)$ be a triangle-free graph of rank r and maximum degree Δ. Let each node v try a random color in $[K] = [C \cdot \Delta^{1/(r-1)}]$ and uncolor itself if it is part of a monochromatic edge. Let $s \in [2\Delta^{1/(r-1)}/C^{r-2}, K]$ and $t \geq 2(s/C)^{r-1}$. Then w.p. at least $1-2\exp(-t/6)-(r-1)\Delta\exp(-s/6)$, v's degree (number of fully uncolored incident edges) becomes at most $2t$.*

Proof. As explained in the main text, an edge e incident on a node v has two ways of surviving this process: by being monochromatic itself, or by having each of its nodes be part of a monochromatic edge distinct from e.

The number of monochromatic edges incident on v corresponds to its defect in the tentative coloring, which we previously analyzed. As in the proof of Lemma 5, at most t edges survive that way, w.p. $1 - \exp(-t/6)$.

We turn to the second type of surviving edges. Let us say an edge e incident on v is *forbidding* to v if the nodes in e other than v all have the same color. We say that a color c is *forbidden* to a node v if v has an incident forbidding edge whose nodes other than v are all colored c. Let $F(v)$ be the set of edges forbidding to v. The second type of surviving edge occurs when each of its nodes selects a forbidden colors. We show that $|F(v)|$ is concentrated.

Claim. For a node v and integers $x > 0$, $t \geq 2d_v/x^{r-2}$, $\Pr[|F(v)| \geq t] \leq \exp(-t/6)$.

Proof (Proof of Appendix A.1). An edge e is forbidding to $v \in e$ with probability $1/x^{r-2}$. Therefore, $\mathbb{E}[|F(v)|] = d_v/x^{r-2}$, and $t \geq 2\mathbb{E}[|F(v)|]$. Because v's neighborhood is triangle-free, edges incident on v share no other vertex. Therefore, whether each edge is forbidding to v is independent of whether other edges are, and by Lemma 4 (Chernoff bound), the probability ensues.

We now bound the probability that many edges survive due to all of its nodes picking a forbidden color.

Recall $s \geq 2\Delta^{1/(r-1)}/C^{r-2}$. By Appendix A.1, a node u has less than s forbidden colors w.p. $1 - \exp(-s/6)$. Therefore, all the neighbors of v have less than s forbidden colors w.p. $1-(r-1)\Delta\exp(-s/6)$. In the rest of the argument, we condition on the event that nodes at distance 2 from v forbade at most s colors to each direct neighbor of v, and fix the random choices of nodes at distance 2 from v to a specific assignment satisfying this conditioning.

Consider an edge e incident on v with vertices $v, u_1, \ldots, u_{r-1}$. The probability that u_i picks a color forbidden by the distance 2 neighbors of v it is adjacent to is at most s/K. The probability that the $r-1$ u_i's do so is at most $(s/K)^{r-1}$ by the independence of their choices. Therefore, the expected number of edges that remain uncolored due to this second argument is at most $\Delta(s/K)^{r-1}$.

Finally, for each e incident on v let X_e be the indicator random variable of the event that all its nodes other than v picked a color forbidden by the nodes at distance 2 from v. Let X be the sum $\sum_{e \ni v} X_e$. The X_e's are all independent once the random choices of nodes at distance 2 are fixed. Therefore, by Lemma 4 (Chernoff bound), for $t \geq 2\Delta(s/K)^{r-1} = 2(s/C)^{r-1}$, $\Pr[X \geq t] \leq 1 - \exp(-t/6)$.

Putting everything together, w.p. at least $1 - 2\exp(-t/6) - (r - 1)\Delta \exp(-s/6)$, each of the two sources of surviving edges contributes at most t edges, for a total of at most $2t$.

A.2   Proof of Lemma 10

Lemma 10. *For each node v, the probability that v is in V_{Quit} is at most*

$$3(r-1)\Delta(\log\log\Delta) \cdot \exp(-(\Delta_{\text{goal}}/4)^{1/(r-1)}/6).$$

Proof. Recall the values of variables which dictate how nodes behave in GEO-METRICTRIALS,

- $C = 4, \alpha = 1/2, \Delta_{\text{goal}} = \Delta^{1/(r-1)}$,
- $C_i = C^{2^i} = 4^{2^i}, K_i = \alpha^i \Delta^{1/(r-1)}$,
- $\Delta_i = \max\{(K_i/C_i)^{r-1}, \Delta_{\text{goal}}\}, i_{\text{last}} = \max\{i \mid \Delta_i > \Delta_{\text{goal}}\}$.

Let us analyze the probability that a live node v in the i-th iteration of GEOMETRICTRIALS decreases its degree to less than Δ_{i+1} (or gets colored). By Lemma 9, if $\Delta_{i+1} \geq 4\Delta_i/C_i^{(r-1)^2}$, then v's degree decreases to less than Δ_{i+1} with probability

$$1 - 3(r-1)\Delta_i \exp(-(\Delta_{i+1}/4)^{1/(r-1)}C_i/6).$$

We verify that $\Delta_{i+1} \geq 4\Delta_i/C_i^{(r-1)^2}$ indeed holds. Note that, by definition,

$$\Delta_{i+1} \geq (K_{i+1}/C_{i+1})^{r-1} = \alpha^{(i+1)(r-1)}KC^{-(r-1)2^{i+1}} = \Delta_i \cdot (\alpha^{r-1}C^{-(r-1)2^i})$$
$$= \Delta_i \cdot 4^{-(r-1)/2-(r-1)2^i} \geq 4\Delta_i/C_i^{(r-1)^2}.$$

For each node active in iteration $i \leq i_{\text{last}}$ (which has therefore degree at most Δ_i), by Lemma 9, the probability that its degree fails to decrease to Δ_{i+1} or less after each live node tries a color is at most

$$3(r-1)\Delta_i e^{-(\Delta_{i+1}/4)^{1/(r-1)}/6}.$$

Summing over all the rounds, the probability that a node joins V_{Quit} during the $i_{\text{last}} + 1 \leq \log\log\Delta$ loop iterations of GEOMETRICTRIALS is at most

$$3(r-1)\sum_{i=0}^{i_{\text{last}}}\Delta_i e^{-(\Delta_{i+1}/4)^{1/(r-1)}/6} \leq 3(r-1)\Delta(\log\log\Delta)e^{-(\Delta_{\text{goal}}/4)^{1/(r-1)}/6}.$$

B   Missing Proofs for the $\Theta(\log^* n)$ Algorithm and Lower Bound

We give two results on deterministic algorithms. Firstly, we give an $O(\log^* n)$ rounds algorithm for $\tilde{O}(\Delta^{r/(r-1)})$-coloring any r-uniform hypergraph. Secondly, we complement this algorithm with a lower bound of $\Omega\left(\frac{\log^* n}{r}\right)$ on finding such a coloring.

B.1   Finding a $\tilde{O}(\Delta^{r/(r-1)})$-Coloring

In this section, we give a one round algorithm for transforming a strong $\tilde{O}(r^2\Delta^2)$-coloring into a weak $\tilde{O}(r \cdot \Delta^{r/(r-1)})$-coloring for an r-regular hypergraph. This result can be viewed as an addendum to Linial's algorithm for finding a $O(\Delta^2)$-coloring in $O(\log^* n)$ rounds. This reduction is performed via a combinatorial argument extending the notion of a Δ-cover free family to an r-weak Δ-cover free family. Note that a $\tilde{O}(r^2\Delta^2)$ strong coloring can be found in $O(\log^* n)$ rounds by using Linial's algorithm on the underlying graph. In order to obtain such a coloring, it is useful to introduce r-weak Δ-cover free families of sets. This generalization of Δ-cover free families serves to relax coloring constraint to the problem of finding a weak coloring.

Definition 4 (r-weak Δ-cover free families). *Let $\mathcal{F}$ be a family of sets. The family $\mathcal{F}$ is a r-weak Δ-cover free family if, for every set $S_0 \in \mathcal{F}$, and Δ subfamilies $\mathcal{S}_j = \{S_{j,1}, \ldots, S_{j,r-1}\} \subseteq \mathcal{F} \setminus \{S_0\}$, each of size $r - 1$, the following holds:*

$$S_0 \not\subseteq \bigcup_{j=1}^{\Delta} \bigcap_{k=1}^{r-1} S_{j,k}$$

Note that a 2-weak Δ-cover free family is equivalent to the classical definition of an Δ-cover free family.

Lemma 13 (Lower bound on the size of r-weak Δ-cover free families). *For three integers $n, \Delta, r \in \mathbb{N}$ such that $n \geq r \geq 2$ and $\Delta \geq 1$, there exists an r-weak Δ-cover free family $\mathcal{F}$ of size n, where each $S \in \mathcal{F}$ is a subset of a ground set $[m]$, $m = 5\lceil r\Delta^{r/(r-1)} \ln(n) \rceil$.*

Proof. In the proof that follows, we use that $e^{-s} \geq \left(1 - \frac{s}{r}\right)^r$ for all $1 \leq s \leq r$. For some m, consider a random collection $\mathcal{F} = \{S_1, \ldots, S_n\}$ of subsets of $[m]$ constructed the following way: for every element $x \in [m]$ and index $i \in [n]$, x belongs to S_i with some fixed probability p, independently of every other pair $(x', i') \neq (x, i)$. For any given element x, index $i_0 \in [n]$, and Δ sets of $r - 1$ indices $\{i_{j,1}, \ldots, i_{j,r-1}\} \subseteq [n] \setminus \{i_0\}$, the probability of x being in the set S_{i_0} but out of $\bigcup_{j=1}^{\Delta}\left(\bigcap_{k=1}^{r-1} S_{i_{j,k}}\right)$ is:

$$\Pr\left[x \in S_{i_0} \setminus \bigcup_{j=1}^{\Delta}\left(\bigcap_{k=1}^{r-1} S_{i_{j,k}}\right)\right] \geq \Pr[x \in S_{i_0}]\left(1 - \sum_{j=1}^{\Delta}\Pr\left[x \in \bigcap_{k=1}^{r-1} S_{i_{j,k}}\right]\right)$$

$$= p(1 - \Delta p^{r-1})$$

Setting $p = (2\Delta)^{-1/(r-1)}$, this probability is at least $\frac{1}{4\Delta^{1/(r-1)}}$. Therefore the probability that for every $x \in [m]$, $x \notin S_{i_0} \setminus \bigcup_{j=1}^{\Delta}\left(\bigcap_{k=1}^{r-1} S_{i_{j,k}}\right)$ is no more than $\left(1 - \frac{1}{4\Delta^{1/(r-1)}}\right)^m \leq e^{-m/(4\Delta^{1/(r-1)})} \leq n^{-5r\Delta/4}$. The probability that valid multiset of indices $i_0, i_{1,1}, \ldots, i_{\Delta,(r-1)}$ exists such that $S_{i_0} \subseteq \bigcup_{j=1}^{\Delta}\left(\bigcap_{k=1}^{r-1} S_{i_{j,k}}\right)$ is no more than $n^{(r-1)\Delta+1}n^{-5r\Delta/4} < 1$. Therefore, an r-weak Δ-cover free family of n sets with no such indices exists.

Theorem 6. *There is a deterministic $O(\log^* n)$-round* CONGEST *algorithm to $O(r \cdot \Delta^{r/(r-1)} \log(r\Delta))$-color hypergraphs of maximum degree Δ and rank r.*

Proof. Let ϕ be a strong $O(r^2\Delta^2)$-coloring of the graph, computed using Linial's algorithm on the underlying graph. We show that ϕ can be converted in to a weak $\tilde{O}(r \cdot \Delta^{r/(r-1)})$-coloring in a single round. From Lemma 13, there must be an r-weak Δ-cover free family $\mathcal{F}$ of $O(r^2\Delta^2)$ sets from a universe of $O(r \cdot \Delta^{r/(r-1)} \log(r\Delta))$ elements. By indexing these sets in some universal order, each vertex v can choose the set S_v at index $\phi(v)$. As the set is a member of $\mathcal{F}$, following Definition 4 there must exist at least one element $c_v \in S_v$, such that for every edge e incident to v, $c_v \notin \bigcap_{u \in e \setminus \{v\}} S_u$. Therefore, coloring v with c_v, v can not be incident to any monochromatic edge. As computing the value of c_v only requires the color of each neighbor in the $O(r^2\Delta^2)$-coloring, this can be done in a single round from the $O(r^2\Delta^2)$-strong coloring. Hence the total round complexity of this process is $O(\log^* n)$, dominated by the process of finding the initial strong coloring. Further, as c_v is selected from a universe of size $O(r \cdot \Delta^{r/(r-1)} \log(r\Delta))$, the coloring of G from this process corresponds to a weak $O(r \cdot \Delta^{r/(r-1)} \log(r\Delta)) = \tilde{O}(r \cdot \Delta^{r/(r-1)})$-coloring of G.

B.2 Lower Bounds on Finding Polynomial Colorings

We show that there exists a lower bound of $\Omega\left(\frac{\log^* n}{r}\right)$ for finding on finding a poly(Δ)-coloring, by generalizing the classic lower bound due to Linial [33]. Rather than using a simple n-cycle, we construct a *strongly connected n-hyper-cycle*. A strongly connected n-hyper-cycle with minimum rank r can be derived from an n-cycle C by constructing an edge for each connected component of size r in C. Note that the degree of a vertex in such a graph is $2(r-1)$. We provide a lower bound using this construction be reduction from the problem of $O(\Delta^c)$-coloring an n-cycle.

Theorem 7. *For any pair of constants r and c, no* LOCAL *algorithm can find a $O(\Delta^c/r)$ coloring of a rank r hypergraph in fewer than $\Omega((\log^* n)/r)$ rounds.*

Proof. For the sake of contradiction, let $\mathcal{A}$ be an algorithm that can weakly color a strongly connected n-hyper-cycle with $O(\Delta^c)$ colors for some pair of constants r and c. Let $T_{\mathcal{A}}$ be its complexity. Let $G = (V, E)$ be a cycle graph with n vertices. It is known that no algorithm can find a $O(\Delta^2)$-coloring on G in fewer than $\Omega(\log^* n)$ rounds. We show that an algorithm for coloring G in $O(T_{\mathcal{A}})$ rounds exists, implying the lower bound on $T_{\mathcal{A}}$.

Let $G' = (V, H)$ be an r-uniform hypergraph constructed from G, with edge set $H = \{(v_1, \ldots, v_r) \in V^r \mid (v_i, v_{i+1}) \in E, \forall i \in 1, 2, \ldots, r-1\}$. Note that G' corresponds to a strongly connected n-cycle. Observe that any algorithm on G' can be simulated on G in at most a factor of r additional rounds. Let ϕ be the coloring on V after running $\mathcal{A}$. Given any vertex $v \in V$, let $h_1, h_2 \in H$ be the pair of hyperedges incident to v such that $N(v) = h_1 \cup h_2$. In other words, h_1 and h_2 are the hyperedges that include v and the two vertices at a distance of $r-1$ from v in G. As ϕ is a weak coloring of G', there must exists two vertices $u_1, u_2 \in h_1 \times h_2$ such that $\phi(v) \notin \{\phi(u_1), \phi(u_2)\}$.

Let $C = (V', E')$ be a connected component in G such that $\forall v, u \in V', \phi(v) = \phi(u)$. Following the above observation, the maximum length of such a component is $2r - 3$. As the number of colors assigned by ϕ is at most $O(r^c)$ for some pair of constants r and c, ϕ can be turned into a proper coloring of G by going through each color class and coloring the nodes in each component in order of decreasing ID. Therefore ϕ can be transformed into a proper coloring of G in at most $O(r^{c+1})$ rounds, and hence $r \cdot (T_{\mathcal{A}} + O(r^{c+1})) \in \Omega(\log^* n)$, i.e. $\mathcal{A}$ must take at least $\Omega(\log^* n)$ rounds, since r and c are constants. $\square$

Note that the hypergraph used for the lower bound is not linear, i.e., the theorem does not rule out the existence of an $o(\log^* n)$ round algorithm whose scope is limited to linear hypergraphs.

References

1. Assadi, S., Chen, Y., Khanna, S.: Sublinear algorithms for $(\Delta+1)$ vertex coloring. In: Proceedings of the ACM-SIAM Symposium on Discrete Algorithms (SODA), pp. 767–786 (2019). https://doi.org/10.1137/1.9781611975482.48. Full version at arXiv:1807.08886
2. Assadi, S., Kumar, P., Mittal, P.: Brooks' theorem in graph streams: a single-pass semi-streaming algorithm for Δ-coloring. In: Proceedings of the ACM Symposium on Theory of Computing (STOC), pp. 234–247. ACM (2022). https://doi.org/10.1145/3519935.3520005
3. Balliu, A., Brandt, S., Kuhn, F., Olivetti, D.: Distributed Δ-coloring plays hide-and-seek. In: Proceedings of the ACM Symposium on Theory of Computing (STOC), pp. 464–477. ACM (2022). https://doi.org/10.1145/3519935.3520027
4. Balliu, A., Brandt, S., Kuhn, F., Olivetti, D.: Distributed maximal matching and maximal independent set on hypergraphs. In: Proceedings of the ACM-SIAM Symposium on Discrete Algorithms (SODA), pp. 2632–2676 (2023). https://doi.org/10.1137/1.9781611977554.ch100

5. Barenboim, L.: Deterministic ($\Delta + 1$)-coloring in sublinear (in Δ) time in static, dynamic, and faulty networks. J. ACM **63**(5), 47:1–47:22 (2016). https://doi.org/10.1145/2979675

6. Barenboim, L., Elkin, M., Pettie, S., Schneider, J.: The locality of distributed symmetry breaking. J. ACM **63**(3), 20:1–20:45 (2016). https://doi.org/10.1145/2903137

7. Brandt, S.: An automatic speedup theorem for distributed problems. In: Proceedings of the ACM Symposium on Principles of Distributed Computing (PODC), pp. 379–388. ACM (2019). https://doi.org/10.1145/3293611.3331611

8. Brandt, S., et al.: A lower bound for the distributed Lovász local lemma. In: Proceedings of the ACM Symposium on Theory of Computing (STOC), pp. 479–488. ACM (2016). https://doi.org/10.1145/2897518.2897570

9. Chang, Y., Fischer, M., Ghaffari, M., Uitto, J., Zheng, Y.: The complexity of ($\Delta+1$) coloring in congested clique, massively parallel computation, and centralized local computation. In: Proceedings of the ACM Symposium on Principles of Distributed Computing (PODC), pp. 471–480. ACM (2019). https://doi.org/10.1145/3293611.3331607

10. Chang, Y.J., He, Q., Li, W., Pettie, S., Uitto, J.: Distributed edge coloring and a special case of the constructive Lovász local lemma. ACM Trans. Algorithms (TALG) **16**(1), 1–51 (2019). https://doi.org/10.1145/3365004

11. Chang, Y.J., Kopelowitz, T., Pettie, S.: An exponential separation between randomized and deterministic complexity in the LOCAL model. SIAM J. Comput. **48**(1), 122–143 (2019). https://doi.org/10.1137/17M1117537

12. Chang, Y.J., Li, W., Pettie, S.: Distributed $\Delta + 1$-coloring via ultrafast graph shattering. SIAM J. Comput. **49**(3), 497–539 (2020). https://doi.org/10.1137/19M1249527

13. Chang, Y.J., Pettie, S.: A time hierarchy theorem for the LOCAL model. SIAM J. Comput. **48**(1), 33–69 (2019). https://doi.org/10.1137/17M1157957

14. Chung, K.-M., Pettie, S., Su, H.-H.: Distributed algorithms for the Lovász local lemma and graph coloring. Distrib. Comput. **30**(4), 261–280 (2016). https://doi.org/10.1007/s00446-016-0287-6

15. Czumaj, A., Davies, P., Parter, M.: Simple, deterministic, constant-round coloring in congested clique and MPC. SIAM J. Comput. **50**(5), 1603–1626 (2021). https://doi.org/10.1137/20M1366502

16. Davies, P.: Improved distributed algorithms for the Lovász local lemma and edge coloring. In: Proceedings of the ACM-SIAM Symposium on Discrete Algorithms (SODA), pp. 4273–4295 (2023). https://doi.org/10.1137/1.9781611977554.ch163

17. Erdős, P., Lovász, L.: Problems and results on 3-chromatic hypergraphs and some related questions. In: Colloquia Mathematica Societatis Janos Bolyai 10. Infinite and Finite Sets, Keszthely, Hungary (1973)

18. Feigenbaum, J., Kannan, S., McGregor, A., Suri, S., Zhang, J.: On graph problems in a semi-streaming model. Theor. Comput. Sci. **348**(2–3), 207–216 (2005). https://doi.org/10.1016/j.tcs.2005.09.013

19. Fischer, M., Ghaffari, M.: Sublogarithmic distributed algorithms for Lovász local lemma, and the complexity hierarchy. In: Proceedings of the International Symposium on Distributed Computing (DISC) (2017). https://doi.org/10.4230/LIPIcs.DISC.2017.18

20. Fischer, M., Ghaffari, M., Kuhn, F.: Deterministic distributed edge-coloring via hypergraph maximal matching. In: Proceedings of the Symposium on Foundations of Computer Science (FOCS), pp. 180–191 (2017). https://doi.org/10.1109/FOCS.2017.25

21. Frieze, A., Mubayi, D.: Coloring simple hypergraphs. J. Comb. Theory Ser. B **103**(6), 767–794 (2013)
22. Ghaffari, M.: An improved distributed algorithm for maximal independent set. In: Proceedings of the ACM-SIAM Symposium on Discrete Algorithms (SODA), pp. 270–277. SIAM (2016). https://doi.org/10.1137/1.9781611974331.ch20
23. Ghaffari, M., Grunau, C., Rozhoň, V.: Improved deterministic network decomposition. In: Proceedings of the ACM-SIAM Symposium on Discrete Algorithms (SODA), pp. 2904–2923. SIAM (2021). https://doi.org/10.1137/1.9781611976465.173
24. Ghaffari, M., Harris, D.G., Kuhn, F.: On derandomizing local distributed algorithms. In: Proceedings of the Symposium on Foundations of Computer Science (FOCS), pp. 662–673. IEEE Computer Society (2018). https://doi.org/10.1109/FOCS.2018.00069
25. Ghaffari, M., Kuhn, F.: Deterministic distributed vertex coloring: simpler, faster, and without network decomposition. In: Proceedings of the Symposium on Foundations of Computer Science (FOCS), pp. 1009–1020. IEEE (2021). https://doi.org/10.1109/FOCS52979.2021.00101
26. Halldórsson, M.M., Kuhn, F., Nolin, A., Tonoyan, T.: Near-optimal distributed degree+1 coloring. In: Proceedings of the ACM Symposium on Theory of Computing (STOC), pp. 450–463. ACM (2022). https://doi.org/10.1145/3519935.3520023
27. Halldórsson, M.M., Maus, Y., Nolin, A.: Fast distributed vertex splitting with applications. In: Proceedings of the International Symposium on Distributed Computing (DISC). LIPIcs, vol. 246, pp. 26:1–26:24 (2022). https://doi.org/10.4230/LIPIcs.DISC.2022.26
28. Harris, D.G.: Derandomized concentration bounds for polynomials, and hypergraph maximal independent set. ACM Trans. Algorithms (TALG) **15**(3), 1–29 (2019). https://doi.org/10.1145/3326171
29. Harris, D.G.: Distributed local approximation algorithms for maximum matching in graphs and hypergraphs. SIAM J. Comput. **49**(4), 711–746 (2020). https://doi.org/10.1137/19M1279241
30. Kuhn, F., Zheng, C.: Efficient distributed computation of MIS and generalized MIS in linear hypergraphs. arXiv preprint arXiv:1805.03357 (2018)
31. Kutten, S., Nanongkai, D., Pandurangan, G., Robinson, P.: Distributed symmetry breaking in hypergraphs. In: Kuhn, F. (ed.) DISC 2014. LNCS, vol. 8784, pp. 469–483. Springer, Heidelberg (2014). https://doi.org/10.1007/978-3-662-45174-8_32
32. Lenzen, C.: Optimal deterministic routing and sorting on the congested clique. In: Proceedings of the ACM Symposium on Principles of Distributed Computing (PODC), pp. 42–50. ACM (2013). https://doi.org/10.1145/2484239.2501983
33. Linial, N.: Locality in distributed graph algorithms. SIAM J. Comput. **21**(1), 193–201 (1992). https://doi.org/10.1137/0221015
34. Maus, Y., Tonoyan, T.: Local conflict coloring revisited: linial for lists. In: Proceedings of the International Symposium on Distributed Computing (DISC). LIPIcs, vol. 179, pp. 16:1–16:18. LZI (2020). https://doi.org/10.4230/LIPIcs.DISC.2020.16
35. Moser, R.A., Tardos, G.: A constructive proof of the general Lovász local lemma. J. ACM **57**(2), 11:1–11:15 (2010). https://doi.org/10.1145/1667053.1667060
36. Muthukrishnan, S., et al.: Data streams: algorithms and applications. Found. Trends® Theor. Comput. Sci. **1**(2), 117–236 (2005). https://doi.org/10.1561/0400000002
37. Radhakrishnan, J., Shannigrahi, S., Venkat, R.: Hypergraph two-coloring in the streaming model. arXiv preprint arXiv:1512.04188 (2015)

38. Rozhoň, V., Ghaffari, M.: Polylogarithmic-time deterministic network decomposition and distributed derandomization. In: Proceedings of the ACM Symposium on Theory of Computing (STOC), pp. 350–363. ACM (2020). https://doi.org/10.1145/3357713.3384298

39. Schneider, J., Wattenhofer, R.: A new technique for distributed symmetry breaking. In: Proceedings of the ACM Symposium on Principles of Distributed Computing (PODC), pp. 257–266. ACM (2010). https://doi.org/10.1145/1835698.1835760

Lockless Blockchain Sharding
with Multiversion Control

Ramesh Adhikari$^{(\boxtimes)}$ and Costas Busch

Augusta University, Augusta, GA 30912, USA
{radhikari,kbusch}@augusta.edu

Abstract. Sharding is used to address the performance and scalability issues of the blockchain protocols, which divides the overall transaction processing costs among multiple clusters of nodes. Shards require less storage capacity and communication and computation cost per node than the existing whole blockchain networks, and they operate in parallel to maximize performance. However, existing sharding solutions use locks for transaction isolation which lowers the system throughput and may introduce deadlocks. In this paper, we propose a lockless transaction method for ensuring transaction isolation without using locks, which improves the concurrency and throughput of the transactions. In our method, transactions are split into subtransactions to enable parallel processing in multiple shards. We use versions for the transaction accounts to implement consistency among the shards. We provide formal proof for liveness and correctness. We also evaluate experimentally our proposed protocol and compare the execution time and throughput with lock-based approaches. The experiments show that the transaction execution time is considerably shorter than the lock-based time and near to the ideal (no-lock) execution time.

Keywords: Blockchains · Blockchain Sharding · Lockless Transactions · Transaction Conflicts · Parallel Commits

1 Introduction

The popularity of blockchains has grown due to their numerous benefits in decentralized applications. They have several special features such as fault tolerance, transparency, non-repudiation, and immutability [25]. To maximize bandwidth usage, every transaction is hashed with a cryptographic function and multiple transactions are divided into blocks [9]. After that, a ledger is created by chaining all the blocks together using a consensus mechanism to append blocks. Assuming that nobody else can be trusted, every node is in charge of keeping its own copy of the distributed ledger. As a result, if someone or some system attempts to alter or restore a portion of these transactions it will be detected, which provides assurances of data integrity and finality.

The distributed cryptocurrency blockchain system known as Bitcoin [22] is one of the first and most well-known instances of how blockchain was originally

© The Author(s), under exclusive license to Springer Nature Switzerland AG 2023
S. Rajsbaum et al. (Eds.): SIROCCO 2023, LNCS 13892, pp. 112–131, 2023.
https://doi.org/10.1007/978-3-031-32733-9_6

designed for the reliable exchange of digital goods. A permissionless blockchain allows anyone to join or leave the network without having to reveal their true identity. No participant can be truly trusted in such situations. Due to the lack of identity, a computationally intensive consensus process called proof-of-work that is based on cryptography is required. On the other hand, in permissioned systems the environment is more controlled and allows for more power-efficient consensus protocols based on Byzantine agreement [1]; nevertheless, even these blockchain protocols do not scale well due to large communication overhead.

Unfortunately, conventional blockchain applications have a fully replicated architecture where each node stores a copy of the whole blockchain and processes every transaction which causes scalability issues in contemporary very big data-based applications [23]. When the number of transactions and storage nodes increases, the blockchain system not only takes a longer time to achieve a consensus among nodes but also takes more time to process the transaction; therefore, it reduces the overall performance of the system. To mitigate the scalability issue of the blockchain, several blockchain protocols like Elastico [20], OmniLedger [18], RapidChain [27], and ByShard [14] has proposed to introduce sharding to provide scalability which divides the whole replicated single blockchain system to multiple shards and each shard processes its own transactions independently.

The blockchain nodes are divided into clusters of nodes called shards. Subsets in each shard may contain Byzantine nodes. We presume that each shard employs a BFT (Byzantine Fault Tolerant) consensus algorithm with authentication, such as PBFT [7]. Existing sharding solutions such as, Elastico [20], OmniLedger [18], and RapidChain [27] are tailored for supporting open (or permissionless) cryptocurrency applications and are not easily generalizable to other systems. To address system-specific specialized approaches towards sharding, Hellings *et al.* [14], introduced ByShard, and combine two conventional sharded database concepts, two-phase commit and two-phase locking for atomicity and isolation of transactions in a Byzantine environment. However, their sharding solutions are not optimal as the locks are expensive, and when a process locks a data set for reading or writing, all other processes attempting to access the same data set are blocked until the lock is released, which lowers system throughput.

In this paper, we propose a different method for ensuring transaction isolation without using locks, which improves the transaction processing time. We propose a novel algorithm for ensuring atomicity and isolation of the transactions in the distributed environment.

1.1   Contributions

To our knowledge, we give the first lockless approach to blockchain sharding. We provide the following contributions:

- We provide a lockless protocol for sharded blockchains. Our protocol is based on multi-version concurrency control of the various shared objects (accounts) that the transactions access. A transaction is first split into subtransactions

that execute in parallel in multiple shards. Using object versioning, the subtransactions can detect whether there is a conflict with other subtransactions that attempt to access the same shared objects concurrently. In case of a conflict, a transaction may need to restart and attempt to commit again.
– We provide correctness proofs for the safety and liveness of our proposed protocol. We also evaluate our protocol experimentally through simulations and we observe that the transaction execution time is considerably faster than the lock-based approaches and also the throughput of the transactions is improved with an increasing number of shards.

Paper Organization: The rest of this paper is structured as follows: Sect. 2 provides previous related works. Section 3 describes the preliminaries for this study and the sharding model. Section 4 discusses our proposed lockless sharding protocol. In Sect. 5 we provide the correctness analysis. Section 6 discusses the performance evaluation of our work, experimental setup, and experimental results. Finally, we give our conclusions in Sect. 7.

2   Related Work

Several proposals have come forward to address the blockchain scalability issue in the consensus layer [10,12,15–17]. Although these protocols have addressed the scalability issues to some extent, the system still cannot maintain good performance as the network size grows too large (thousand or more node participants). The sharding technique has been used to further improve the scalability of a blockchain network. Sharding is a fundamental concept in databases which has been recently used to improve the efficiency of blockchains [14,27].

The way that conventional big database systems achieve scalability is by separating the whole database into shards (or partitions) [2], which increases the efficiency of the system by dividing the workload among the shards. To ensure ACID characteristics [11] of the database transactions, coordination is needed among the multiple shards if the transaction access multiple shards objects. In the distributed database system two-phase commit (2PC), and two-phase locking (2PL) [8], are used for atomicity, concurrency control, and isolation for the transactions. And we achieved these characteristics in our model in a different way without using locks.

Similarly, the blockchain network can be split up into smaller committees using the well-researched and tested technique of sharding, which also serves to scale up databases and lower the overhead of consensus algorithms. Elastico [20], OmniLedger [18], and RapidChain [27] are a few examples of sharded blockchains. These methods are not generalizable to other applications since they concentrate on a simple data model, that is the unspent transaction output (UTXO) model [13]. In addition, these methods use locks for the isolation of transactions. As in databases, a blockchain transaction must be isolated since it interacts with the global state. In reality, it is necessary to avoid dirty, phantom, or unrepeatable reads [3]. Additionally, transactions must comply with all of the ACID properties [11]. Typically, two-phase locking [8] is used to accomplish optimistic concurrency control [19], serializable snapshot isolation [6,24].

To mitigate the system-specific specialized approaches towards sharding, Hellings *et al.* [14], propose ByShard. It uses a two-phase commit to ensure the atomicity of the transaction and two-phase locking for isolation of the transaction in a Byzantine environment of the blockchain system. However, locking is expensive because when a process locks a data set for reading or writing, all other processes attempting to access the same data-set are blocked until the lock is released, which lowers system throughput. An innovative lock-free method for ensuring transaction isolation is presented by Hagar Meir *et al.* [21], In order to construct version-based snapshot isolation, it takes advantage of the key-value pair versioning that already exists in the database and is mostly utilized at the validation phase of the transaction to detect the read-write [21]. However, they are not addressing their solution in a sharding-based blockchain model.

In our solution, we are using a lockless approach to achieve transaction isolation with sharding. We use multiversion concurrency control [5], as we describe in our proposed model in Sect. 4.

3   Preliminaries and Sharding Model

Shards: We assume that the system consists of a set of N (replica) nodes, where $n = |N|$. We design a sharded system as a partitioning of the N nodes into w shards $S_1, S_2, \ldots, S_w$, where $N = \cup_i S_i$ and each $S_i \subseteq N$ is a subset of the nodes such that $S_i \cap S_j = \emptyset$, for $i \neq j$. Let $n_i = |S_i|$ represent the number of replica nodes in shard S_i, and f_i represent the number of Byzantine nodes in shard S_i. Similar to related work [27], to achieve Byzantine fault tolerance in each shard we assume that $n_i > 3f_i$ within each shard. Hence, we focus on the consistency aspects, assuming there is an underlying consensus protocol in each shard.

Let $\mathcal{O}$ be a set of shared objects that are accessed by the transactions. Similar to related work [14] we assume that every shard is responsible for a subset of the shared objects (accounts) that are accessed by the transactions. Namely, $\mathcal{O}$ is partitioned into subsets $\mathcal{O}_1, \ldots, \mathcal{O}_w$, where $\mathcal{O}_i$ is the set of objects handled by shard S_z. Every shard S_i maintains its own local ledger (local chain) L_i and runs a local consensus algorithm to achieve this (e.g. PBFT [7]). The shard S_i processes subtransactions related to the object set $\mathcal{O}_i$ (see below for the description of subtransactions). The local chains define implicitly the global blockchain, that is, the global order of all transactions is implied by the order of their respective subtransactions in the local chains.

Timing Assumptions: We consider a semi-synchronous setting where communication delay is upper bounded by some time Δ_1, which means that every message is guaranteed to be delivered within Δ_1 time. Our sharding protocol does not require knowledge of Δ_1. We assume that every transaction has a unique ID based on its generation timestamp, hence IDs grow over time. Due to the semi-synchronous model, since local clocks are not perfectly synchronized, we assume a new timestamp (generated at any node in N) will be strictly larger than any timestamp generated $c \cdot \Delta_1$ time earlier (where the constant c depends on the system). Hence, we assume that for a transaction T_i that arrives at time t, any other transaction T_j that arrives after $t + c \cdot \Delta_1$ will have always a higher ID than T_i.

For guaranteeing liveness in our protocol, we assume that each Δ_2 time each shard sends the lowest transaction ID from its transaction pool to other shards. Here Δ_2 is known to each shard (in order to periodically perform the lowest ID transmission) but is not related to Δ_1. In this way, each shard maintains the set of lowest transaction IDs which are periodically updated with new lowest ID information from each shard. The transaction which has the global lowest ID gets within a bounded time high priority and is eventually added to the blockchain. The process of propagating the lowest IDs is running in the background while the normal execution phases take place.

Similar to previous works [14], we assume that each shard runs locally a PBFT-style [7] consensus algorithm in every phase of our algorithm which takes bounded time Δ_3 for decisions (e.g. in [14] it is assumed that $\Delta_3 = 30\,\text{ms}$). Our protocol does not need to know Δ_3.

Subtransactions: We model each transaction T_i which consist of subtransactions $T_{i,k_1}, \ldots, T_{i,k_j}$, such that:

- Subtransaction T_{i,k_l} uses only objects in $\mathcal{O}_{k_l}$ in shard S_{k_l}. We also say that the subtransaction T_{i,k_l} belongs to shard S_{k_l}.
- The subtransactions of a transaction T_i do not depend on each other and can be executed in parallel in any relative order.
- A subtransaction consists of two parts: (i) condition checking, where various explicit conditions on the objects are checked, and (ii) updates on the objects.

Example 1. Consider a transaction (T_1) consisting of read-write operations on the accounts with several conditions.

$T_1 = $ "Transfer 2000 from Rock account to Asma account, if Rock has 3000 and Asma has 500 and Mark has 200". We split this transaction into three subtransactions, where shards r, a, and m are responsible for the respective accounts of Rock, Asma, and Mark:

$T_{1,r}$: "Check Rock has 3000" $T_{1,a}$: "Check Asma has 500"
    : "Remove 2000 from Rock   : "Add 2000 to Asma account"
    account"

$T_{1,m}$: "Check Mark has 200"

After splitting the transaction T_1 into its subtransactions we send each subtransaction to its respective shard associated with that account. If the conditions are satisfied (for example in $T_{1,r}$ check if Rock has 3000) and the transaction is valid (for example in $T_{1,r}$ Rock has indeed 2000 in the account to be removed) then the destination shards are ready to commit the subtransactions which imply that T_1 will commit as well. Otherwise, if any of the conditions in the subtransactions is not satisfied or the subtransactions are invalid, then the corresponding subtransactions abort, which results in T_1 aborting as well. In this case, all subtransactions of T_1 will also be forced to abort.

4 Sharding Algorithm

Our sharding protocol consists of two parts, the leader shard algorithm (Algorithm 1), and the destination shard algorithm (Algorithm 2).

Every transaction has a designated *leader shard*, which will handle its processing. Each leader shard has a transaction pool for all the transactions that have it as their leader. The job of the leader shard is to pick a transaction from the transaction pool and split it into subtransactions and send them to *destination shards*. The leader shard interacts with the destination shards through a protocol with seven phases which decide whether the subtransactions they receive are able to commit locally or not. The leader shard picks the transaction from its transaction pool on the basis of the priority of the transactions so that the earliest transaction (i.e. with lower ID) proceeds first. Whereas the destination shard checks each received subtransaction and if it is valid then it commits it and appends it to its local ledger, otherwise, it aborts the subtransaction and sends the corresponding message to its leader shard.

To achieve transaction isolation, we use multi-version concurrency control [5] in each destination shard, which saves multiple versions of each object (account) so that data can be safely read and modified simultaneously. When a destination shard processes a subtransaction, it takes a snapshot of the current version of each object that the subtransaction will access. When the subtransaction is about to commit, it compares the latest version with the recorded snapshot version. If these are the same then the subtransaction is eligible to commit; otherwise, the subtransaction cannot commit. The leader shard is informed accordingly from the destination shards. If all subtransactions are eligible to commit then the whole transaction will commit and is removed from the leader shard pool. If however, a subtransaction is not eligible to commit, the whole transaction will restart and is reinserted back into its pool.

In our algorithm, each transaction whose conditional statements are satisfied will eventually commit (as we show in the correctness proofs). Our algorithm may attempt to commit the transaction multiple times by restarting it in case of conflicts with other transactions. However, if the condition of a transaction is not satisfied then the transaction is aborted and will not restart (is removed completely from the pool). Using the object versions the algorithm guarantees safety, as it does not allow conflicting transactions to commit concurrently. To ensure liveness, the algorithm prioritizes earlier generated transactions.

4.1 Detailed Algorithm

We now describe the details of our protocol in Algorithms 1 and 2. Our combined protocol consists of seven phases. As mentioned in Sect. 3, to ensure liveness, periodically every Δ_2 time, each leader shard sends the lowest ID from its transaction pool to every other shard. So that in case of conflict, priority is given to the transaction with the smallest known ID. In this way, each destination shard maintains in T_l'' (Algorithm 2) the lowest known ID that it received from all leader shards. If a subtransaction realizes that it belongs to a transaction with the smallest ID in the system then it gets the highest priority and enforces itself to commit. This is further achieved with the help of a rollback mechanism that we discuss below.

Now we describe each phase of our algorithm. For simplicity of presentation, we assume that each subtransaction accesses a single object in each destination shard. However, our algorithms can be generalized for the case where each subtransaction accesses multiple objects.

Phase 1: (Algorithm 1) the leader shard (S_k) picks a transaction with the lowest transaction ID from its transaction pool (P_k) and splits that transaction T_i into its subtransaction $T_{i,j}$ and sends each $T_{i,j}$ to corresponding destination shards (S_j) in parallel.

Phase 2: (Algorithm 2) after receiving the subtransaction $T_{i,j}$ in destination shard (S_j) accessing an object, say O_d, it takes the latest version (say V_d) of the object O_d. After that, it checks the conditions (constraints) of the subtransaction $T_{i,j}$. If the constraints match (means subtransaction is eligible to commit) then, it adds the $T_{i,j}$ to the read set $R(O_d)$ and if $T_{i,j}$ will also write to O_d then it also adds $T_{i,j}$ to write set $W(O_d)$ and sends a "commit vote" to the leader shard S_k. Otherwise, it sends a "abort vote" to the leader shard.

Phase 3: (Algorithm 1) the leader shard S_k collects the votes from all the destination shards, and if it gets all "commit vote", (that means constraints are matched in all respective destination shards) then it sends the "commit" message to the corresponding destination shards. Similarly, if it gets any "abort vote" then it sends an "abort" message to all respective destination shards.

Phase 4: (Algorithm 2) if the destination shard receives a "commit" message from a leader shard then, it checks the read set $(R(O_d))$ and write set $(W(O_d))$ of the accessing object and also checks the version of the object. If the subtransaction $T_{i,j}$ is only reading the object O_d and the latest version of the object O_d is still the same (i.e. V_d) then the shard appends this subtransaction to its local ledger L_j and sends "committed" message to the leader. Similarly, if $T_{i,j}$ is trying to update object O_d and the read set only contains $T_{i,j}$ (i.e. ($T_{i,j} \in W(O_d)$ and $R(O_d)\backslash\{T_{i,j}\} = \emptyset$)) and the latest version of the object O_d is still same as the previously taken version (i.e. V_d) (that means the object is not modified by other transactions) then it does the necessary update operation and adds the subtransaction $T_{i,j}$ to its local chain and sends the "committed" message to the leader shard. Moreover, if the transaction ID of subtransaction $T_{i,j}$ is equal to the lowest known transaction ID (T_l''), that means the current subtransaction $T_{i,j}$ is the earliest subtransaction among all and it has a higher priority to execute. So it appends the subtransaction $T_{i,j}$ to its local chain and sends a "committed" message to its leader shard, and also sends "force rollback" to the

---

Algorithm 1: Leader Shard S_k

---

1 Let P_k be the pool of pending transactions in shard S_k;
   // Periodically, every Δ_2 time the transaction with lowest ID in P_k is sent to every other shard
2 Let C_k be the committed transaction pool;

   // Phase 1
3 Pick transaction T_i with lowest ID from P_k and remove it from P_k;
4 Split T_i into subtransactions;
5 Let $S(T_i)$ be the set of destination shards for the subtransactions of T_i;
6 Send each subtransaction $T_{i,j}$ to the corresponding destination shard S_j (in parallel for all subtransactions of T_i);

   // Phase 3
7 **if** "commit vote" *message is received from all shards in* $S(T_i)$ **then**
8 | Send "commit" message to all shards in $S(T_i)$;
9 **else if** "abort vote" *message is received from any shard in* $S(T_i)$ **then**
10 | Send "abort" message to all shards in $S(T_i)$;

   // Phase 5
11 **if** "committed" *message is received from all shards in* $S(T_i)$ **then**
12 | Send "release" to all shards in $S(T_i)$;
13 **else if** "restart vote" *message is received from any shard in* $S(T_i)$ **then**
14 | Send "restart" message to all shards in $S(T_i)$;
15 **else if** "aborted" *message is received from all shards in* $S(T_i)$ **then**
16 | Transaction T_i is discarded;

   // Phase 7
17 **if** "released" *message is received from all shards in* $S(T_i)$ **then**
18 | Transaction T_i has completed;
19 | Add T_i to C_k;
20 **else if** "restarted" *message is received from all shards in* $S(T_i)$ **then**
21 | Transaction T_i is reinserted into the pool P_k to be processed again;

   // Handling Force Rollback Messages
22 **if** "force rollback $T'_{x,j}$" *message is received and* S_k *is the leader shard of the respective transaction* T' **then**
23 | **if** $T' \in C_k$ **then**
24 | Get subtransaction information from C_k;
25 | Send respective "force rollback $T'_{x,z}$" to all destination shards of T';
26 **if** "rollbacked $T'_{x,z}$" *message is received from all destination shards of* T' *and* S_k *is the leader shard of the transaction* T' **then**
27 | Insert T' back in the pool P_k;
28 | **if** $T' \in C_k$ **then**
29 | Remove T' from C_k;

---

leader of the subtransaction which is in the write set $(W(O_d))$. Otherwise, it sends a "restart vote", which means there is another higher-priority transaction accessing the object O_d and not released yet. Similarly, if it receives the "abort" message then it sends an "aborted" message to its leader shard.

Algorithm 2: Destination Shard S_j

---

1 T_l'' ← the lowest transaction ID from the IDs propagation process;

   // Phase 2

2 Subtransaction $T_{i,j}$ from leader shard S_k is received;

3 Suppose $T_{i,j}$ accesses object O_d;

4 Let $R(O_d)$ and $W(O_d)$ be a set of transactions that will respectively read or write O_d;

5 Let V_d be the latest version of object O_d;

6 **if** *constraint match* **then**

7 | Add $T_{i,j}$ to $R(O_d)$;

8 | **if** $T_{i,j}$ *will write to* O_d **then**

9 | Add $T_{i,j}$ to $W(O_d)$;

10 | Send "commit vote" message to S_k;

11 **else**

12 | Send "abort vote" message to S_k;

   // Phase 4

13 **if** "commit" *message is received from* S_k **then**

14 | **if** $((((W(O_d)\backslash\{T_{i,j}\} = \emptyset)$ *or* $(T_{i,j} \in W(O_d)$ *and* $R(O_d)\backslash\{T_{i,j}\} = \emptyset))$ *and* *(the latest version of object O_d is still V_d))* *or* $(T_{i,j} = T_l'')$ **then**

15 | Append transaction $T_{i,j}$ to local chain L_j;

16 | Send "committed" message to S_k;

17 | **if** $T_{i,j} = T_l''$ // $T_{i,j}$ **has the lowest ID in the system**

18 | **then**

19 | For each $T_{x,j}' \in W(O_d)$ send "force rollback" message to its respective leader shard;

20 | **else**

21 | send "restart vote" message to S_k;

22 **else if** "abort" *message is received from* S_k **then**

23 | Send "aborted" message to S_k;

   // Phase 6

24 **if** "restart" *message is received from* S_k **then**

25 | Remove $T_{i,j}$ from $R(O_d)$ and $W(O_d)$ and from local chain L_j;

26 | Send "restarted" message to the leader S_k;

27 **else if** "release" *message is received from* S_k **then**

28 | **if** $T_{i,j}$ in $W(O_d)$ **then**

29 | Create new version $V_d + 1$ for the object O_d;

30 | Remove $T_{i,j}$ from $R(O_d)$ and $W(O_d)$;

31 | Send *"released"* message to S_k;

   // Handling Force Rollback Messages

32 **if** "force rollback $T_{x,j}'$" *message is received* **then**

33 | Remove $T_{x,j}'$ from $R(O_d)$ and $W(O_d)$;

34 | Let Z be the suffix in local chain L_j starting from $T_{x,j}'$;

35 | Remove from L_j the suffix Z and send "rollbacked $T_{x,j}'$" message to its leader shard;

36 | For each subtransaction $T_{x,j}'$ in Z, send "force rollback $T_{x,j}'$" message to the leader shard of $T_{x,j}'$;

Phase 5: (Algorithm 1) if it receives a "committed" message from all destination shards (means that eligible to commit subtransactions are added to their local chain) then it sends a "release" message to the respective destination shards to release the subtransactions from their read set, write set and also to update the version of the object if required. Similarly, if the leader receives any "restart vote" message from any shards then it sends the "restart" message to the respective destination shards because some of the shards may have appended the subtransaction to their local chain so that should be removed, and restart should be consistent in all shards. Moreover, if it receives an "aborted" message from all destination shards, then transaction T_i is discarded.

Phase 6: (Algorithm 2) if the destination shard receives a "restart" message from the leader shard, then it removes the transaction $T_{i,j}$ from $R(O_d)$ and $W(O_d)$ and also removes $T_{i,j}$ from its local chain L_k if it already added and sends "restarted" message to the leader. Similarly, if it receives a "release" message and $T_{i,j}$ is already in $W(O_d)$ that means it updated the object O_d so it creates the new version of the object as $V_d + 1$. After that, it removes $T_{i,j}$ from $R(O_d)$ and $W(O_d)$ and sends a "released" message to its leader shard.

Phase 7: (Algorithm 1) if the leader shard receives a "released" message from all destination shards that means the transaction T_i is completed, so it adds T_i to the pool of committed transactions (C_k) so that it can get all the subtransaction information of T_i in case of rollback. Similarly, if it receives a "restarted" message from all destination shards, then this transaction needs to be processed again, and is reinserted into the transaction pool P_k.

Handling Rollbacks: This part of our protocol executes only in the special case (i.e. when the current transaction has the highest priority to execute than the already running transaction accessing the same object). After receiving the "force rollback" message from destination shards, the leader shard checks whether that subtransaction belongs to the committed transaction pool (C_k) or not. If the transaction of that subtransaction is in C_k then it gets the other subtransaction information from C_k otherwise it has the information about the currently running transaction, then it sends a "force rollback" message to all respective destination shards because it should be rollbacked in all the shards to be consistent. So if the destination shard receives the "force rollback $T'_{x,j}$" message from the leader then it rollbacks $T'_{x,j}$ from its shards and sends "rollbacked $T'_{x,j}$" message to its leader. Furthermore, if there exists some depending subtransaction T' on $T'_{x,j}$ accessed the version of the object added by $T'_{x,j}$ then all depending transactions should be rollbacked and it sends "rollback T'" to its leader shard so this function executes recursively to rollback all the transactions which read the version of object added by T'. The leader shard collects the "rollbacked" messages from all the destination shards, and after receiving all the "rollbacked" messages from all the respective shards for the transaction T', it adds T' to its transaction pool to be processed again and removes T' from the committed pool (C_k) if T' is already in C_k.

Example 2. Consider two conflicting transactions T_1 and T_2 consisting of read-write operations on the accounts. We explain how our protocol handles these transactions.

$T_1 =$ "Transfer 2000 from Rock account to Asma account, if Rock has 3000 and Asma has 500 and Mark has 200".

$T_2 =$ "Transfer 500 from Asma account to Bob account, if Asma has 5000".

Suppose leader shard (S_{k_1}) handles transaction T_1 and splits it into three sub-transactions, where shards S_r, S_a, and S_m are responsible for the respective accounts of Rock, Asma, and Mark. Similarly, the leader shard (S_{k_2}) handles transaction T_2 and splits it into two subtransactions where shard S_a is responsible for the Asma account and shard S_b is responsible for the Bob account.

$T_{1,r}$: "Check Rock has 3000"
      : "Remove 2000 from Rock account"
	$T_{2,a}$: "Check Asma has 5000"
	      : "Remove 500 from Asma account"

$T_{1,a}$: "Check Asma has 500" : "Add 2000 to Asma account"	$T_{2,b}$: "Add 500 to Bob account"

$T_{1,m}$: "Check Mark has 200"

Let us consider both leader shards $(S_{k_1}$ and $S_{k_2})$ are trying to execute their transaction in parallel. At this condition, there are two subtransactions accessing the same account of Asma (i.e. $T_{1,a}$, $T_{2,a}$) so there will be a conflict on the respective subtransactions. So, in our algorithm, each destination shard takes the version of every account and checks that version at the time of commit. In case of conflicts, the transaction that updated earlier the read and write sets (R and W) at the destination shard will have a chance to commit.

5   Correctness Analysis

Consider a set of transactions $\mathcal{T} = \{T_1, T_2, \ldots, T_\zeta\}$. The objective is to arrange all transactions in $\mathcal{T}$ in a sequence $B = T_{i_1}, T_{i_2}, \ldots, T_{i_\zeta}$, which is agreed upon by all non-faulty nodes in N. We also write $T_{i_l} \prec_B T_{i_{l'}}$ for $l < l'$ to denote the relative order between two transactions in the sequence B. The sharding system does not maintain the actual B as a single blockchain (or ledger) explicitly, but rather, the blockchain consists of a collection of local chains which if combined they jointly give the whole blockchain B.

Each shard S_α maintains a local chain L_α of the sub-transactions $T_{i,\alpha}$ that it receives. The subtransactions are appended in L_α according to the order that they commit in S_α. If $T_{i,\alpha} \prec_{L_\alpha} T_{j,\alpha}$, and $T_{i,\alpha}$ conflicts with $T_{j,\alpha}$ (the two subtransactions conflict if they access the same object in S_α and one of the two is updating the object), then we say that $T_{i,\alpha}$ *causes* $T_{j,\alpha}$ and we write $T_{i,\alpha} \to_{L_\alpha} T_{j,\alpha}$.

We define the *local chain system* as the tuple $L = (L_1, \ldots, L_w)$ consisting of local chains in shards. If $T_{i,\alpha} \to_{L_\alpha} T_{j,\alpha}$, we can also simply write $T_{i,\alpha} \to_L T_{j,\alpha}$. The casual relation $\to$ can be extended across two local chains L_α and L_β, $\alpha \neq \beta$, in the following way.

- If $T_{i,\alpha} \to_{L_\alpha} T_{j,\alpha}$ and T_j has a subtransaction $T_{j,\beta}$.
- If T_i has subtransactions $T_{i,\alpha}$ and $T_{i,\beta}$ such that $T_{i,\beta} \to_{L_\beta} T_{j,\beta}$.

In both cases, we say that $T_{i,\alpha}$ causes $T_{j,\beta}$, and we write $T_{i,\alpha} \to_L T_{j,\beta}$. Consider from now on the transitive closure of the causal relation $\to_L$.

We say that the local chain system L is *valid* if there is no subtransaction $T_{i,\alpha}$ such that $T_{i,\alpha} \to_L T_{i,\alpha}$. That is, L is a valid local chain system if there is no cyclic (transitive) causal relationship of a subtransaction to itself.

We say that a sequence B is a *valid serialization* of the local chain system L if B is a sequence of all the subtransactions which preserves the causal relationship of L. Namely, if $T_{i,\alpha} \to_L T_{j,\beta}$ then $T_{i,\alpha} \prec_B T_{j,\beta}$. We say that a sequence B is a *blockchain serialization* of L if B is a valid serialization of L, and for each transaction T_i its subtransactions $T_{i,j_1}, \ldots, T_{i,j_k}$ appear consecutively in B (without being interleaved by subtransactions of other transactions).

The goal is to show that our sharding protocol generates a local chain system L that has a blockchain serialization B. We introduce the *shard-coherence property* which we will use to prove the existence of B.

Definition 1 (Shard-coherence). *We say that transactions T_i and T_j are shard-coherent with respect to local chain system L if whenever two of their subtransactions are casually related as $T_{i,\alpha} \to_L T_{j,\beta}$, then for any two of their conflicting subtransactions $T_{i,\gamma}$ and $T_{j,\gamma}$ it holds that $T_{i,\gamma} \prec_{L_\gamma} T_{j,\gamma}$. The local chain system L is shard-coherent if every pair of transactions are shard-coherent.*

The following result shows that in order to build a blockchain serialization B from a chain system L, it suffices to prove that L is shard-coherent. (The proof of Proposition 1 is in Appendix A.1.)

Proposition 1. *If a local chain system L is shard-coherent, then L has a blockchain serialization B.*

Next, we continue to show that in our sharding protocol two transactions that conflict in the same shard, they cannot have some of their phases interleave. (The proof of Lemma 1 is in Appendix A.2.)

Lemma 1. *If two transactions T_i and T_j conflict in a destination shard S_γ, and their respective subtransactions are processed concurrently by S_γ so that they both go past phase 2 in S_γ concurrently, then at least one of the two transactions will restart or rollback.*

Theorem 1 (Safety). *The local chain system L produced by our protocol has a blockchain serialization B.*

Proof. From Proposition 1, we only need to prove that L is shard-coherent.

Consider any two transactions T_i and T_j such that $T_{i,\alpha} \to_L T_{j,\beta}$. Suppose that $T_{i,\gamma}$ and $T_{j,\gamma}$ conflict in shard S_γ because they access at least one common object O_d and one of the two subtransactions updates O_d. It suffices to show that $T_{i,\gamma} \prec_{L_\gamma} T_{j,\gamma}$.

Since $T_{i,\alpha} \to_L T_{j,\beta}$, from the definition of the $\to_L$ relation, we have that there is a sequence of transactions $T_{k_1}, T_{k_2}, \ldots, T_{k_z}$ with $T_{k_1} = T_i$, $T_{k_z} = T_j$ and $T_{k_i} \to T_{k_{i+1}}$, for $1 \le i < z$, such that any pair of consecutive transactions T_{k_l} and $T_{k_{l+1}}$ have respective conflicting subtransactions $T_{k_l,\delta}$ and $T_{k_{l+1},\delta}$ on some common shard S_δ such that $T_{k_l,\delta} \prec_{L_\delta} T_{k_{l+1},\delta}$.

Since $T_{k_l,\delta}$ and $T_{k_{l+1},\delta}$ are appended in the local chain L_δ, while they both conflict, we have from Lemma 1 that they cannot go past phase 2 concurrently without one of them restarting or rolling back. Therefore, $T_{k_l,\delta}$ finishes phase 6, before $T_{k_{l+1},\delta}$ enters phase 4.

This implies that phase 5 of T_{k_l} (at its leader shard) finishes before phase 5 of $T_{k_{l+1}}$ starts (at its leader shard). Therefore, by induction, we can easily show that the end of phase 5 of T_i (at its leader shard) occurs earlier than the beginning of phase 5 of T_j (at its leader shard).

Suppose now that $T_{j,\gamma} \prec_{L_\gamma} T_{i,\gamma}$. Since T_i and T_j commit in S_γ and also conflict in S_γ by sharing the same object, then from Lemma 1, phase 6 of T_j ends before phase 4 of T_i starts in S_γ. Therefore, phase 5 of T_j ends before phase 5 of T_i starts (at their respective home shards). This is a contradiction. Therefore, $T_{i,\gamma} \prec_{L_\gamma} T_{j,\gamma}$, as needed.

Theorem 2 (Liveness). *Our protocol guarantees that every issued transaction will eventually commit.*

Proof. Consider the timing assumptions for Δ_1, Δ_2, and Δ_3 as described in Sect. 3. Consider a transaction T_i with ID $ID(T_i)$ generated at time t. In the worst case, T_i will execute when its ID is the lowest in the system, through force rollback messages.

After $c \cdot \Delta_1$ time steps, every new transaction generated will have a larger ID than $ID(T_i)$, and hence lower precedence than T_i. It takes additional time Δ_2 to propagate $ID(T_i)$.

Let $ID'_{\min}$ be the smallest ID of all transactions considering all the pools of all shards at time t. Let q be the number of transactions which at time $t + \Delta_2 + c \cdot \Delta_1$ have ID at least $ID'_{\min}$ and less than $ID(T_i)$. In the worst case, all of these q transactions may commit before T_i. As we have 7 phases in our protocol, for each committed transaction, the combined upper bound for communication and consensus delay time is $7(\Delta_1 + \Delta_3)$. Hence, it takes at most $q \cdot 7(\Delta_1 + \Delta_3)$ time to commit the q transactions. Therefore, by time $t + q \cdot 7(\Delta_1 + \Delta_3) + \Delta_2 + c \cdot \Delta_1$ transaction T_i will be committed in the blockchain.

6   Performance Evaluation

We set up our experiments in a virtual machine in M1 MAC PC with a 10-core CPU and 32-core GPU, including 32 GB RAM. We used Python programming language for the experiments which supports multiprocessing and multithreading. We virtually created multiple shards within a machine and conducted the experiment with different numbers of shards. For the communication between the shards, we use socket programming in Python, which enables the communication between shards by message passing. Same as previous work [14], we also assume that each shard runs the consensus algorithm and takes 30 ms (say Δ_3) for decisions.

We generate 1000 accounts randomly by using the combination of the English alphabet letters and assigned an initial balance of 3000 to each account. Moreover, we generate 1500 transactions by randomly selecting the account from 1000 accounts. Each transaction includes the read and writes operations with some constraints. The generated 1500 transactions are divided with respect to the number of shards and randomly assigned to the transaction pool of each leader shard.

We show the experimental results in three categories. Firstly, optimal (no lock), means there is no transaction isolation; concurrent transactions can access the accounts and update those accounts without any consideration of the data consistency. Secondly, We used the concept of exclusive lock protocol to ensure transaction isolation and concurrency control. This approach acquires a lock on an object (account), at the time of accessing it and releases the lock after the transaction completes [4]. This prevents other transactions from accessing the same object until the lock is released, ensuring that transactions do not interfere with each other. When a transaction acquires an exclusive lock on a data item, no other transaction can read or modify that data item until the lock is released, providing exclusive access to the data. In our implementation, when an object is locked, other transactions attempting to access the same object wait until the lock is released. This guarantees that the transaction holding the lock has exclusive access to the data and can modify it without interference from other transactions. Finally, we used our protocol to achieve transaction isolation and concurrency control without using a lock, which takes a snapshot of each object (account) and if there is conflict occurs then priority to access the object is given to the earliest transaction and other transaction are restart and rollback to re-execute again.

Experimental Results: In the first experiment, we evaluate the average throughput of the transactions using 1500 generated transactions, in which each transaction checks whether the account has sufficient balance or not before transferring from its own account to another account, and the other three constraints. If the transaction is valid and satisfies all the conditions then the transaction is executed by removing the balance from one account and adding that balance to another account (i.e. two write operations). To measure the average throughput of transactions, we initialize the start time at the beginning of the transaction processing and capture the final time after processing all the 1500 transactions. The average throughput of the transaction with respect to the number of shards is shown in Fig. 1, where we measure the average throughput of the transactions by varying the number of shards. From the experiment, we observe that the throughput increases with the number of shards. From Fig. 1 we can see that the transaction throughput of our protocol is better than the lock-based protocol and quite close to the no-lock protocol.

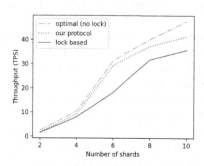

Fig. 1. Average transaction throughput with the number of shards

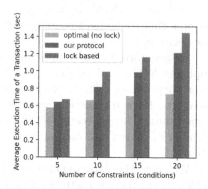

Fig. 2. Average execution time of a transaction with numbers of constraints

In the second experiment, we set up the environment with four shards and calculate the average execution time of a transaction with respect to the number of conditions in each transaction. We increases the constraints of the transactions and recorded the execution of the transactions. In each experiment, we found that the average execution time of our protocol is less than the lock-based protocol also shown in Fig. 2. From the experimental result, we see that as the number of conditions to execute the transaction increased, the commit process takes a long time. As a result, in the lock-based protocol, the lock is kept for a long period, which adds a lot of overhead and takes more time for the execution of transactions than in our proposed protocol.

7   Conclusion

In this research work, we presented a lockless transaction scheduling protocol for blockchain sharding. Our protocol is based on taking a snapshot version of the various shared objects (accounts) that the transactions access in each shard. We provide a correctness proof with the safety and liveness properties of our protocol. We also evaluate our protocol experimentally through simulations and we observe that the transaction execution time is considerably faster than the lock-based approaches and also the throughput of the transactions is improved with an increasing number of shards.

This study still has some room for improvement. One possible extension could be a study on efficient communication between leader shards and destination shards. Introducing a formal performance analysis for blockchain sharding is another interesting topic for future work, which will quantify the performance based on parameters of the blockchain, such as the number of shards and the sizes of the shards.

In recent literature, Schwarzmann [26] reviewed several requirements that need to be satisfied by electronic poll book systems, such as ensuring correctness, security, integrity, fault-tolerance, consistent distributed storage, etc. Our proposed protocol can be used to address some of these issues because it not only provides blockchain features but also offers scalability and better performance for recording transactions. Overall, our protocol may offer many unique features for electronic check-in poll book systems, including decentralization, immutability, and consensus.

Acknowledgements. This paper is supported by NSF grant CNS-2131538.

A   Appendix

A.1   Proof of Proposition 1

The proof of Proposition 1 follows directly from Corollary 1 and Lemma 4 given below. In the results below consider a local chain system $L = (L_1, \ldots, L_w)$ for transactions $T = \{T_1, T_2, \ldots, T_\zeta\}$.

Lemma 2. *If L is a valid local chain system, then L has a valid serialization.*

Proof. Consider the sequence A of subtransactions which is the concatenation of sequences $L_1, L_2, \ldots, L_w$. Suppose that $A = a_1, a_2, \ldots, a_\delta$, where $a_\sigma = T_{i_\sigma, j_\sigma}$ where T_{i_σ, j_σ} is a subtransaction of transaction $T_{i_\sigma} \in T$.

From A we incrementally build a sequence A' which is a valid serialization of L. Let A'_σ denote the sequence that we obtain after we appropriately insert (as explained below) the σth element of A into A'. We prove by induction that A'_σ is a valid serialization of the involved subtransactions of the respective induced subsystem L^σ of L that consists of the σ subtransactions of L under consideration (the subsystem L^σ keeps from each L_i the involved subtransactions; note that the subsystem is valid). The main claim follows when we consider $\sigma = \zeta$ which gives $A' = A'_\sigma$.

For the basis case $\sigma = 1$, and $A'_1 = a_1$ which is trivially a valid serialization of the single subtransaction. Suppose that we built A'_σ which is a valid serialization of the first σ subtransactions in A, where $\sigma < \zeta$.

In order to build $A'_{\sigma+1}$ we take $a_{\sigma+1}$ and insert it into A'_σ, as follows. Suppose that $A'_\sigma = a'_1, \ldots, a'_\sigma$. If there is no $a'_i \in A'_\sigma$ such that $a_{\sigma+1} \to_L a'_i$, then append $a_{\sigma+1}$ at the end of A'_σ, to obtain $A'_{\sigma+1}$, which is clearly a valid serialization.

Otherwise, let a'_i be the earliest subtransaction in A'_σ (i is the smallest index within A'_σ) such that $a_{\sigma+1} \to_L a'_i$, and let a'_j be the latest subtransaction in A'_σ (j is the largest index within A'_σ) such that $a'_j \to_L a_{\sigma+1}$. We examine two cases:

- $j < i$: in this case we append $a_{\sigma+1}$ just before a'_i (and clearly after a'_j) in A'_σ to obtain $A'_{\sigma+1}$, which gives a valid serialization.
- $i < j$: we examine three sub-cases as follows.
  - $a'_i \to_L a'_j$: this case is impossible since this would create a cycle $a_{\sigma+1} \to_L a_{\sigma+1}$ in the causal relation $\to_L$, and hence, L would not be valid, contradicting the assumptions.
  - $a'_j \to_L a'_i$: since $a'_i \prec_{A'_\sigma} a'_j$ this would imply that in A'_σ is not a valid serialization of the involved subtransactions of A, which contradicts the induction hypothesis.
  - a'_j and a'_i are not related by $\to_L$ to one another: consider the subsequence s of A'_σ from a'_i to a'_j (including a'_i and a'_j). Let s_1 be the subsequence of s that includes all a'_q such that $a'_i \to_L a'_q$; let s_2 be the subsequence of s that includes all a'_q such that $a'_q \to_L a'_j$; let s_3 be the remaining elements of s. Note that s_1 and s_2 are disjoint, since otherwise $a'_i \to_L a'_j$. Next, we move all the elements in the sequence s_2 (keeping their relative order) to be before the first element in s_1. Moreover, add $a_{\sigma+1}$ between the last element of s_2 and the first element of s_1. The resulting sequence $A'_{\sigma+1}$ is clearly a valid serialization of the involved subtransactions.

Lemma 3. *If the local chain system L has a valid serialization, then L is blockchain serializable.*

Proof. Let A be a valid serialization of L. Suppose that $A = a_1, a_2, \ldots, a_\delta$, where $a_\sigma = T_{i_\sigma, j_\sigma}$ and T_{i_σ, j_σ} is a subtransaction of $T_{i_\sigma} \in \mathcal{T}$.

We will rearrange the subtransactions in A to a new sequence A' such that each transaction T_i has its subtransactions consecutively in A'. We will show how to do the transformation for a single transaction, and this can repeat for the remaining transactions.

For a transaction T_i let $T_{i,j_1}, \ldots, T_{i,j_q}$ denote its subtransactions, with respective positions $a_{s_1}, \ldots, a_{s_q}$ in A.

From the validity of A and transitivity of $\to_L$, we have that if for some $l \in [q]$, $a_j \to_L a_{s_l}$, then $a_j \prec_A a_{s_{l'}}$, for every $l' \in [q]$. Hence, if a_{s_l} is the earliest subtransaction of T_i in A (i.e. s_l has the smallest index among those with $l \in [q]$), any a_j that causes (through $\to_L$) any of the subtransactions of T_i must appear in A before a_{s_l}. Therefore, we can move the subtransactions of T_i and arrange them to appear consecutively starting at the position of a_{s_l}, so that a_{s_1} will take the place of a_{s_l}, a_{s_2} will appear immediately after a_{s_1}, and so on, until a_{s_q}.

Let A' be the resulting sequence after we rearrange the subtransactions of T_i. Clearly, this transformation of A has preserved its validity and also the subtransactions of T_i appear consecutively in A'. By repeating this process for each remaining transaction we obtain the final A'. By induction (on the number of transactions), it is clear that the final A' is a blockchain serialization of L.

From Lemmas 2 and 3 we obtain the following corollary.

Corollary 1. *A valid local chain system L is blockchain serializable.*

Lemma 4. *If a local chain system L is shard-coherent, then L is valid.*

Proof. Suppose that L is shard-coherent. Suppose for the sake of contradiction that there is subtransaction $T_{i,j}$ such that $T_{i,j_k} \to_L T_{i,j_k}$ (that is, there is a cycle in L with respect to causal relation $\to_L$).

Let $p = a_1, a_2, \ldots, a_\ell$ be a transitive "relation path", where each node in a_i is a subtransaction of some transaction in T and $a_1 = a_\ell = T_{i,j_k}$, and $a_i \to_L a_{i+1}$, for each $1 \leq i < \ell$. Among all possible relation paths starting and ending to T_{i,j_k}, let p be the longest (and if there are multiple paths of the same longest length then pick one of them arbitrarily). Note that it has to be $\ell > 2$ since a subtransaction alone by itself cannot create cyclic dependencies.

First, consider the case where each subtransaction in p is in the same shard $S_\alpha = S_{j_k}$ as that of T_{i,j_k}. We consider two sub-cases:

- $a_1 \prec_{L_\alpha} a_2$: let a_r, $1 < r < \ell$, have the largest index r such that $a_1 \prec_{L_\alpha} a_r$. Then clearly, $a_{r+1} \prec_{L_\alpha} a_r$ (note that a_{r+1} exists since we took $r < \ell$ and also it holds $\ell > 2$). However, since $a_r \to_L a_{r+1}$, the shard-coherence property of L is violated between a_r and a_{r+1}, a contradiction.
- $a_2 \prec_{L_\alpha} a_1$: since $a_1 \to_L a_2$, the shard-coherence of L is violated between a_1 and a_2, a contradiction.

Next, consider the case where some subtransaction in p is in a different shard than S_α. Let a_r, where $1 \leq r < \ell$ be the first subtransaction (with the smallest index r) in p which is in a different shard, say S_β, where $\alpha \neq \beta$.

We now show that a_{r+1} must also be in S_β conflicting with a_r. Suppose to the contrary that a_{r+1} is not in S_β. Since $a_r \to_L a_{r+1}$, there must be a subtransaction T' in S_β which conflicts with a_r, such that $a_r \to_{L_\beta} T'$ and $T' \to_L a_{r+1}$. However, this implies that p can be augmented with T', which is a contradiction since p is the longest relation path. Thus, a_{r+1} is in S_β. Moreover, a_{r+1} must be conflicting with a_r, since otherwise we would find as above some other transaction T' that conflicts with a_r which could be inserted into p to increase its length. We examine two cases:

- $a_r \prec_{L_\beta} a_{r+1}$: from the cyclicity of path p we have that $a_{r+1} \to_L a_r$ (going through a_1). Hence, the shard-coherency is violated between a_r and a_{r+1}, a contradiction.
- $a_{r+1} \prec_{L_\beta} a_r$: from p we have that $a_r \to_L a_{r+1}$. Hence, the shard-coherency is violated between a_r and a_{r+1}, a contradiction.

A.2   Proof of Lemma 1

Proof. Suppose transactions T_i and T_j have respective subtransactions $T_{i,\gamma}$ and $T_{j,\gamma}$ in S_γ. Moreover, suppose that these subtransactions conflict in S_γ by accessing the same object O_d and at least one of two of them is updating O_d.

Without loss of generality assume that $T_{i,\gamma}$ is updating O_d. Suppose that $T_{i,\gamma}$ has finished executing phase 2. Hence, $T_{i,\gamma}$ has been added to write set $W(O_d)$. Then when $T_{j,\gamma}$ reaches phase 4, it will observe that $T_{i,\gamma}$ is already in $W(O_d)$ which will force $T_{j,\gamma}$ to restart.

On the other hand, if $T_{j,\gamma}$ has the lowest ID then when $T_{j,\gamma}$ will reach phase 4 it will force $T_{i,j}$ to rollback. In either case, one of the two transactions will either restart or rollback.

References

1. Amiri, M.J., Agrawal, D., El Abbadi, A.: SharPer: sharding permissioned blockchains over network clusters. In: Proceedings of the 2021 International Conference on Management of Data, pp. 76–88 (2021)
2. Bagui, S., Nguyen, L.T.: Database sharding: to provide fault tolerance and scalability of big data on the cloud. Int. J. Cloud Appl. Comput. (IJCAC) 5(2), 36–52 (2015)
3. Berenson, H., Bernstein, P., Gray, J., Melton, J., O'Neil, E., O'Neil, P.: A critique of ANSI SQL isolation levels. ACM SIGMOD Rec. 24(2), 1–10 (1995)
4. Bernstein, P.A., Goodman, N.: Concurrency control in distributed database systems. ACM Comput. Surv. (CSUR) 13(2), 185–221 (1981)
5. Bernstein, P.A., Goodman, N.: Multiversion concurrency control-theory and algorithms. ACM Trans. Database Syst. (TODS) 8(4), 465–483 (1983)
6. Cahill, M.J., Röhm, U., Fekete, A.D.: Serializable isolation for snapshot databases. ACM Trans. Database Syst. (TODS) 34(4), 1–42 (2009)
7. Castro, M., Liskov, B., et al.: Practical Byzantine fault tolerance. In: OsDI, vol. 99, pp. 173–186 (1999)
8. Eswaran, K.P., Gray, J.N., Lorie, R.A., Traiger, I.L.: The notions of consistency and predicate locks in a database system. Commun. ACM 19(11), 624–633 (1976)
9. Friedman, R., Van Renesse, R.: Packing messages as a tool for boosting the performance of total ordering protocols. In: Proceedings of the Sixth IEEE International Symposium on High Performance Distributed Computing (Cat. No. 97TB100183), pp. 233–242. IEEE (1997)
10. Giridharan, N., Howard, H., Abraham, I., Crooks, N., Tomescu, A.: No-commit proofs: defeating livelock in BFT. Cryptology ePrint Archive (2021)
11. Gray, J., Reuter, A.: Transaction Processing: Concepts and Techniques, 1st edn. Morgan Kaufmann Publishers Inc., San Francisco (1992)
12. Gueta, G.G., et al.: SBFT: a scalable and decentralized trust infrastructure. In: 2019 49th Annual IEEE/IFIP International Conference on Dependable Systems and Networks (DSN), pp. 568–580. IEEE (2019)
13. Hellings, J., Hughes, D.P., Primero, J., Sadoghi, M.: Cerberus: minimalistic multi-shard byzantine-resilient transaction processing. arXiv preprint arXiv:2008.04450 (2020)
14. Hellings, J., Sadoghi, M.: ByShard: sharding in a Byzantine environment. Proc. VLDB Endow. 14(11), 2230–2243 (2023). https://doi.org/10.1007/s00778-023-00794-0
15. Jalalzai, M.M., Busch, C.: Window based BFT blockchain consensus. In: iThings, IEEE GreenCom, IEEE (CPSCom) and IEEE SSmartData 2018, pp. 971–979, July 2018

16. Jalalzai, M.M., Busch, C., Richard, G.G.: Proteus: a scalable BFT consensus protocol for blockchains. In: 2019 IEEE International Conference on Blockchain (Blockchain), pp. 308–313. IEEE (2019)
17. Jalalzai, M.M., Feng, C., Busch, C., Richard, G.G., Niu, J.: The Hermes BFT for blockchains. IEEE Trans. Dependable Secure Comput. $19(6)$, 3971–3986 (2021)
18. Kokoris-Kogias, E., Jovanovic, P., Gasser, L., Gailly, N., Syta, E., Ford, B.: OmniLedger: a secure, scale-out, decentralized ledger via sharding. In: 2018 IEEE Symposium on Security and Privacy (SP), pp. 583–598. IEEE (2018)
19. Kung, H.T., Robinson, J.T.: On optimistic methods for concurrency control. ACM Trans. Database Syst. (TODS) $6(2)$, 213–226 (1981)
20. Luu, L., Narayanan, V., Zheng, C., Baweja, K., Gilbert, S., Saxena, P.: A secure sharding protocol for open blockchains. In: Proceedings of the 2016 ACM SIGSAC Conference on Computer and Communications Security, pp. 17–30 (2016)
21. Meir, H., Barger, A., Manevich, Y., Tock, Y.: Lockless transaction isolation in hyperledger fabric. In: 2019 IEEE International Conference on Blockchain (Blockchain), pp. 59–66 (2019). https://doi.org/10.1109/Blockchain.2019.00017
22. Nakamoto, S.: Bitcoin: a peer-to-peer electronic cash system (2009)
23. Pisa, M., Juden, M.: Blockchain and economic development: hype vs. reality. Center for Global Development Policy Paper 107, p. 150 (2017)
24. Ports, D.R., Grittner, K.: Serializable snapshot isolation in PostgreSQL. arXiv preprint arXiv:1208.4179 (2012)
25. Sankar, L.S., Sindhu, M., Sethumadhavan, M.: Survey of consensus protocols on blockchain applications. In: 2017 4th International Conference on Advanced Computing and Communication Systems (ICACCS), pp. 1–5. IEEE (2017)
26. Schwarzmann, A.A.: Towards a robust distributed framework for election-day voter check-in. In: Johnen, C., Schiller, E.M., Schmid, S. (eds.) SSS 2021. LNCS, vol. 13046, pp. 173–193. Springer, Cham (2021). https://doi.org/10.1007/978-3-030-91081-5_12
27. Zamani, M., Movahedi, M., Raykova, M.: RapidChain: scaling blockchain via full sharding. In: Proceedings of the 2018 ACM SIGSAC Conference on Computer and Communications Security, pp. 931–948 (2018)

Self-adjusting Linear Networks
with Ladder Demand Graph

Vitaly Aksenov[1]([envelope]) [ID], Anton Paramonov[2], Iosif Salem[3] [ID],
and Stefan Schmid[3] [ID]

[1] ITMO University, St. Petersburg, Russia
aksenov.vitaly@gmail.ru
[2] EPFL, Lausanne, Switzerland
[3] TU Berlin, Berlin, Germany
iosif.salem@inet.tu-berlin.de, stefan.schmid@tu-berlin.de

Abstract. Self-adjusting networks (SANs) have the ability to adapt
to communication demand by dynamically adjusting the workload (or
demand) embedding, i.e., the mapping of communication requests into
the network topology. SANs can reduce routing costs for frequently communicating node pairs by paying a cost for adjusting the embedding. This
is particularly beneficial when the demand has structure, which the network can adapt to. Demand can be represented in the form of a demand
graph, which is defined by the set of network nodes (vertices) and the
set of pairwise communication requests (edges). Thus, adapting to the
demand can be interpreted by embedding the demand graph to the network topology. This can be challenging both when the demand graph is
known in advance (offline) and when it revealed edge-by-edge (online).
The difficulty also depends on whether we aim at constructing a static
topology or a dynamic (self-adjusting) one that improves the embedding
as more parts of the demand graph are revealed. Yet very little is known
about these self-adjusting embeddings.

In this paper, the network topology is restricted to a line and the
demand graph to a ladder graph, i.e., a $2 \times n$ grid, including all possible
subgraphs of the ladder. We present an online self-adjusting network that
matches the known lower bound asymptotically and is 12-competitive in
terms of request cost. As a warm up result, we present an asymptotically optimal algorithm for the cycle demand graph. We also present an
oracle-based algorithm for an arbitrary demand graph that has a constant overhead.

Keywords: Ladder graph · Self-adjusting networks · Traffic patterns ·
Online algorithms

Supported partially by the Austrian Science Fund (FWF) project I 4800-N (ADVISE)
and the European Research Council (ERC) under the European Union's Horizon 2020
research and innovation programme, grant agreement No. 864228 "Self-Adjusting Networks (Adjust-Net)".

S. Rajsbaum et al. (Eds.): SIROCCO 2023, LNCS 13892, pp. 132–148, 2023.
https://doi.org/10.1007/978-3-031-32733-9_7

1   Introduction

Traditional networks are static and demand-oblivious, i.e., designed without considering the communication demand. While this might be beneficial for all-to-all traffic, it doesn't take into account temporal or spatial locality features in demand. That is, sets of nodes that temporarily cover the majority of communication requests may be placed diameter-away from each other in the network topology. This is a relevant concern as studies on datacenter network traces have shown that communication demand is indeed bursty and skewed [3].

Self-adjusting networks (SANs) are optimized towards the traffic they serve. SANs can be static or dynamic, depending on whether it is possible to reconfigure the embedding (mapping of communication requests to the network topology) in between requests, and offline or online, depending on whether the sequence of communication requests is known in advance or revealed piece-wise. In the online case, we assume that the embedding can be adjusted in between requests at a cost linear to the added and deleted edges, thus, bringing closer frequently communicating nodes. Online algorithms for SANs aim to reduce the sum of routing and reconfiguration (re-embedding) costs for any communication sequence.

We can express traffic in the form of a demand graph that is defined by the set of nodes in the network and the set of pairwise communication requests (edge set) among them. Knowing the structure of the demand graph could allow us to further optimize online SANs, even though the demand is still revealed online. That is, by re-embedding the demand graph to the network we optimize the use of network resources according to recent patterns in demand.

To the best of our knowledge, the only work on demand graph re-embeddings to date is [2], where the network topology is a line and the demand graph is also a line. The authors presented an algorithm that serves $m = \Omega(n^2)$ requests at cost $O(n^2 \log n + m)$ and showed that this complexity is the lower bound. The problem is inspired by the Itinerant List Update Problem [12] (ILU). To be more precise, the problem in [2] appears to be the restricted version of the online Dynamic Minimum Linear Arrangement problem, which is another reformulation of ILU.

Contributions. In this work, we take the next step towards optimizing online SANs for more general demand graphs. We restrict the network topology to a line, but assume that the demand graph is a ladder, i.e., a $2 \times n$ grid. We assume that before performing a request, we can re-adjust the line topology by performing several swaps of two neighbouring nodes, paying one for each swap. We present a 12-competitive online algorithm that embeds a ladder demand graph to the line topology, thus, asymptotically matching the lower bound in [2]. This algorithm can be applied to any demand graph that is a subgraph of the ladder graph and that when all edges of the demand graph are revealed the topology is optimal and no more adjustments are needed. We also optimally solve the case of cycle demand graphs, which is a simple generalization of the line demand graph, but is not a subcase of the ladder due to odd cycles. Finally, we provide a generic algorithm for arbitrary demand graphs, given an oracle that computes an embedding with the cost of requests bounded by the bandwidth.

A solution for the ladder is the first step towards the $k \times n$ grid demand graph where k is an arbitrary constant. Moreover, a ladder (and a cycle) has a constant *bandwidth*, i.e., a minimum value over all embeddings in a target line topology of a maximal path between the ends of an edge (request). It can be shown that given a demand graph G the best possible complexity per request is the bandwidth.

Related Work. Avin et al. [2], consider a fixed line (host) network and a line demand graph. Their online algorithm re-embeds the demand graph to the host line topology with minimum number of swaps on the embedding. Both [1,6] present constant-competitive online algorithms for a fixed and complete binary tree, where nodes can swap and the demand is originating only from the source. However, these two works do not consider a specific demand graph. Moreover, [5] studied optimal but static and bounded-degree network topologies, when the demand is known. Self-adjusting networks have been formally organized and surveyed in [7]. Other existing online SAN algorithms consider different models. The most distinct difference is our focus on online re-embedding while keeping a fixed host graph (i.e., a line) compared to other works that focus on changing the network topology. The latter is what, for example, SplayNet [14] is proposing, where tree rotations change the form of the binary search tree network, without optimizing for a specific family of demand patterns.

Online demand graph re-embedding also relates to dynamically re-allocating network resources to follow traffic patterns. In [4], the authors consider a fixed set of clusters of bounded size, which contain all nodes and migrate nodes online according to the communication demand. But more broadly, [8] assumes a fixed grid network and migrates tasks according to their communication patterns.

Online embedding of metric spaces is studied in [11]. Authors consider the problem in which elements of some metric space are exposed one after another and the goal is to map them into another metric space while preserving the smallest expansion possible. There are several differences with our problem: 1) they care about all pairs of elements, while we consider a special demand graph; 2) nodes can not be re-embedded after being placed.

Also, relevant problems, from a migration point of view, are the classic list update problem (LU) [15], the related Itinerant List Update (ILU) problem [12], and the Minimum Linear Arrangement (MLA) problem [10]. In contrast to those problems, we study an online problem where requests occur between nodes.

Roadmap. Section 2 describes the model and background. Section 3 contains the summary of our three contributions (ladder, cycle, general demand graph) and their high-level proofs. Section 4 presents the algorithm and the analysis for ladder demand graphs. Some technical details can be found in the technical report [13].

2   Model and Background

Let us introduce the notation that we are going to use throughout the paper. Let $V(H)$ and $E(H)$ be the sets of vertices and edges in graph H, respectively.

Sometimes, we just use V and E if the graph H is obvious from the context. Let $d_H(u, v)$ be the distance between u and v in graph H.

Let N be the network topology and σ be a sequence of pairwise communication requests between nodes in N. Let the demand graph G be the graph built over the nodes in N and the pairs of nodes that appear in σ, i.e. $G = (V(N), \{\sigma_i = (s_i, d_i) \mid \sigma_i \in \sigma\})$. We assume that the demand graph is of a certain type and our overall goal will be to embed the demand graph G in the actual network topology N at a minimum cost. This is non-trivial as requests are selected from G by an online adversary and G is not known in advance. In the following, we formalize demand graph embedding and topology reconfiguration.

A configuration (or an embedding) of G (the demand graph) in a graph N (the host network) is a bijection of $V(G)$ to $V(N)$; $C_{G \to N}$ denotes the set of all such configurations. A configuration $c \in C_{G \to N}$ is said to serve a communication request $(u, v) \in E(G)$ at the cost $d_N(c(u), c(v))$. A finite communication sequence $\sigma = (\sigma_1, \ldots, \sigma_m)$ is served by a sequence of configurations $c_0, c_1, \ldots, c_m \in C_{G \to N}$. The cost of serving σ is the sum of serving each σ_i in c_i plus the reconfiguration cost between subsequent configurations c_i and c_{i+1}. The reconfiguration cost between c_i and c_{i+1} is the number of *migrations* necessary to change from c_i to c_{i+1}; a *migration* swaps the images of two neighbouring nodes u and v under c in N. Moreover, $E_i = \{\sigma_1, \ldots, \sigma_i\}$ denotes the first i requests of σ interpreted as a set of edges on V. We present algorithms for an online self-adjusting linear network: a network whose topology forms a 1-dimensional grid, i.e., a line.

Definition 1 (Working Model). *Let G be the demand graph, n be the number of vertices in G, $N = (\{1, \ldots, n\}, \{(1, 2), (2, 3), \ldots, (n-1, n)\})$ be a line (or list) graph L_n (host network), c be a configuration from $C_{G \to N}$, and σ be a sequence of communication requests. The cost of serving $\sigma_i = (u, v) \in \sigma$ is given by $|c(u) - c(v)|$, i.e., the distance between u and v in N. Migrations can occur before serving a request and can only occur between nodes configured on adjacent vertices in N.*

In the following we introduce notions relevant to our new results.

Definition 2 (Bandwidth). *Given a graph G, the Bandwidth of an embedding $c \in C_{G \to L_n}$ is equal to the maximum over all edges $(u, v) \in E$ of $|c(u) - c(v)|$, i.e., the distance between u and v on L_n. Bandwidth(G) is the minimum bandwidth over all embeddings from $C_{G \to L_n}$.*

Remark 1. *The Bandwidth computation of an arbitrary graph is an NP-hard problem [9].*

To save the space, we typically omit the proofs of lemmas and theorems in this paper and put them in [13, Appendix C]. Here we define the $2 \times n$ grid or ladder graph for which we get the main results of our paper.

Definition 3. *A graph $Ladder_n = (V, E)$ is represented as follows. The vertices V are the nodes of the grid $2 \times n$—$\{(1, 1), (1, 2), \ldots, (1, n), (2, 1), (2, 2), \ldots, (2, n)\}$. There is an edge between vertices (x_1, y_1) and (x_2, y_2) iff $|x_1 - x_2| + |y_1 - y_2| = 1$.*

Lemma 1. Bandwidth($Ladder_n$) = 2.

Proof. The bandwidth is greater than 1, because there are nodes of degree three. The bandwidth of 2 can be achieved via the "level-by-level" enumeration as shown on the figure.

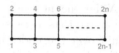

Fig. 1. Optimal ladder numeration.

Lemma 2. *For each subgraph S of a graph G,* Bandwidth(S) $\leqslant$ Bandwidth(G).

2.1 Background

Let us overview the previous results from [2]. In that work, both the demand and the host graph (network topology) were the line L_n on n vertices. It was shown that there exists an algorithm that performs $O(n^2 \log n)$ migrations in total, while serving the requests themselves in $O(1)$. By that, if the number of requests is $\Omega(n^2 \log n)$ then each request has $O(1)$ amortized cost.

Theorem 1 (Avin et al. [2]). *Consider a linear network L_n and a linear demand graph. There is an algorithm such that the total time spent on migrations is $O(n^2 \log n)$, while each request is performed in $O(1)$ omitting the migrations.*

We give an overview of this algorithm. At each moment in time, we know some subgraph of the line demand graph. For each new communication request, there are two cases: 1) the edge from the demand graph is already known— then, we do nothing; 2) the new edge is revealed. In the second case, this edge connects two connected components. We just move the smaller component on the line network closer to the larger component. The move of each node in one reconfiguration does not exceed n. Since, the total number of reconfigurations in which the node participates does not exceed $\log n$, we have $O(n^2 \log n)$ upper bound on the algorithm. From [2], $\Omega(n^2 \log n)$ is also the lower bound on the total cost. Thus, the algorithm is asymptotically optimal in the terms of complexity.

Corollary 1. *If $|\sigma| = \Omega(n^2 \log n)$ the amortized service cost per request is $O(1)$.*

The algorithms are not obliged to perform migrations at all, but the sum of costs for $\Theta(n^2)$ requests can be lower-bounded with $\Omega(n^2 \log n)$.

Theorem 2 (Lower bound, Avin et al. [2]). *For every online algorithm ON there is a sequence of requests σ_{ON} of length $\Theta(n^2)$ with the demand graph being a line, such that $cost(ON(\sigma_{ON})) = \Omega(n^2 \log n)$.*

That implies $\Omega(\log n)$ optimality (or competitive) factor since any offline algorithm knowing the whole request sequence σ in advance can simply reconfigure the network to match the (line) demand graph by paying $\Theta(n^2)$ in the worst case.

3 Summary of Contributions

In this work we present self-adjusting networks with a line topology for a demand graph that is either a cycle, or a $2 \times n$ grid (ladder), or an arbitrary graph. We study offline and online algorithms on how to best embed the demand graph on the line, such that the total cost is minimized. The online case is more challenging, as the demand graph is revealed edge-by-edge and the embedding changes, with a cost. The result for the cycle follows from [2] almost directly. However, the result for the ladder is non-trivial and requires new techniques; it is not simple to reconfigure a subgraph on a $2 \times n$ grid after revealing a new edge in order to get $O(n^2 \log n)$ cost of modifications in total. We give an overview of each case below.

3.1 Cycle Demand Graph

We start with the following observation. Let C_n be a cycle graph on n vertices, i.e., $E(C_n) = \{(1, 2), \ldots, (n-1, n), (n, 1)\}$. Then, Bandwidth$(C_n) = 2$. We give a brief description of how the algorithm works. We start with the same algorithm as for the line (Sect. 2.1): while the number of revealed edges is not more than $n - 1$, we can emulate the algorithm for the line. When the last edge appears we restructure the whole embedding in order to get bandwidth 2, which is the cycle bandwidth. For the last-step restructuring using swaps, we pay no more than $O(n^2)$. This cost is less than the total time spent on the reconstruction $\Omega(n^2 \log n)$.

Theorem 3. *Suppose the demand graph is C_n. There is an algorithm such that the total cost spent on the migrations is $O(n^2 \log n)$ and each request is performed in $O(1)$. In particular, if the number of requests is $\Omega(n^2 \log n)$ each request has $O(1)$ amortized cost.*

The full proof appears in [13, Appendix A]

Remark 2. *The lower bound with $\Omega(n^2 \log n)$ that was presented for a line demand graph still holds in the case of a cycle, since the cycle contains the line as the subgraph. Thus, our algorithm is optimal.*

3.2 Ladder Demand Graph

Now, we state the main result of the paper—the algorithm for the case when the demand graph is a ladder.

Theorem 4. *Suppose a demand graph is a ladder. There is an algorithm such that the total cost spent on the migrations is $O(n^2 \log n)$ and each request is performed in $O(1)$. In particular, if the number of requests is $\Omega(n^2 \log n)$ each request has $O(1)$ amortized cost.*

We provide a brief description of the algorithm. We say that a ladder has n levels from left to right: i.e., the nodes $(1, y)$ and $(2, y)$ are on the same level y (see Fig. 1). On a high-level, we use the same algorithmic approach as in Theorem 1 for the line demand graph. The main difference is that instead of embedding the demand graph right away in the line network, at first, we "quasi-embed" the graph in the $2n$-ladder graph, which then we embed in the line. By "quasi-embedding" we mean a relaxation of the embedding defined earlier: at most **three** vertices of the demand graph are mapped to each level of the ladder.

Suppose for a moment that we have a dynamic algorithm that quasi-embeds the graph in the $2n$-ladder. Given this quasi-embedding we can then embed the $2n$-ladder in the line L_n. We sequentially go through from level 1 to level $2n$ of our ladder and map (at most three) vertices from the level to the line in some order (see Theorem 1). Such a transformation from the ladder to the line costs only a constant factor in bandwidth.

We explain briefly how to design a dynamic quasi-embedding algorithm with the desired complexity. At first, we present a static quasi-embedding algorithm, i.e., we are given a subgraph of the ladder and we need to quasi-embed it. This algorithm consists of three parts: embed a tree, embed a cycle, embed everything together. To embed a tree we find a special path in it, named trunk. We embed this trunk from left to right: one vertex per level. All the subgraphs connected to trunk are pretty simple and can be easily quasi-embedded in parallel to the trunk (see Fig. 2). To embed a cycle we just have to decide which orientation it should have. To simplify the algorithm we embed only the cycles of length at least 6, omitting the cycles of length 4. This decision introduces just the multiplicative constant of the cost. Finally, we embed the whole graph: we construct its cycle-tree decomposition and embed cycles and trees one by one from left to right.

Now, we give a high-level description of our dynamic algorithm. We maintain the invariant that all the components are quasi-embedded. When an already served request (edge) appears, we do nothing. The complication comes from a newly revealed edge-request. There are two cases. The first one is when the edge connects nodes in the same component—thus, there is a cycle. We redo only the part of the quasi-embedding of the component around the new cycle; the rest of the component remains. In the second case, the edge connects two components. We move the smaller component to the bigger one as in Theorem 1. The bigger component does not move and we redo the quasi-embedding of the smaller one.

Now, we briefly calculate the complexity of the dynamic algorithm. For the requests of the first case, if the nodes are on the cycle for the first time (this event happens

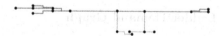

Fig. 2. Quasi-correct embedding of a tree

only once for each node), we pay $O(n)$ for it. Otherwise, there are already nodes in the cycle. In this case we make sure to re-embed the existing cycle in a way that all the nodes are moved for a $O(1)$ distance. As for the neighboring nodes, it can be shown that each node is moved only once as a part of the cycle neighborhood, so we also bound this movement with $O(n)$ cost. This gives us $O(n^2)$ complexity in total—each node is moved by at most $O(n)$. For the requests of the second case, we always move the smaller component and, thus, we pay $O(n^2 \log n)$ in total: each node can be moved by $O(n)$ at most $O(\log n)$ times, i.e., any node can be at most $\log n$ times in the "smaller" component. Our algorithm matches the lower bound, since the ladder contains L_n as a subgraph.

3.3 General Graph

We finish the list of contributions with a general result; the case where the demand graph is an arbitrary graph G. The full proofs are available in [13, Appendix D].

Theorem 5. *Suppose we are given a (demand) graph G and an algorithm B, that for any subgraph S of G outputs an embedding $c \in C_{S \to L_{|V(G)|}}$ with bandwidth less than or equal to $\lambda \cdot \mathrm{Bandwidth}(G)$ for some λ. Then, for any sequence of requests σ with a demand graph G there is an algorithm that serves σ with a total cost of $O(|E(G)| \cdot |V(G)|^2 + \lambda \cdot \mathrm{Bandwidth}(G) \cdot |\sigma|)$. In particular, if the number of requests is $\Omega(|E(G)| \cdot |V(G)|^2)$ each request has $O(\lambda \cdot \mathrm{Bandwidth}(G))$ amortized cost.*

Here we give a brief description of the algorithm. Suppose that the current configuration c_i is the embedding of the current demand graph G_i in $L_{|V(G)|}$ after i requests. Now, we need to serve a new request in $\lambda \cdot \mathrm{Bandwidth}(G_i) \leqslant \lambda \cdot \mathrm{Bandwidth}(G)$. If the corresponding edge already exists in the demand graph, we simply serve the request without the reconfiguration. Now, suppose the request reveals a new edge and we get the demand graph G_{i+1}. Using the algorithm B we get the configuration (embedding) c_{i+1} that has $\lambda \cdot \mathrm{Bandwidth}(G_{i+1}) \leqslant \lambda \cdot \mathrm{Bandwidth}(G)$. Please note that we do not put any constraints on the algorithm B: typically this problem is NP-complete. To serve the request fast, we should rebuild the configuration c_i into the configuration c_{i+1}. By using the swap operations on the line we can get from c_i to c_{i+1} in $O(|V(G)|^2)$ operations: each vertex moves by at most $V(G)$. After the reconfiguration we can serve the request with the desired cost.

A new edge appears at most $|E(G)|$ times while the reconfiguration costs $|V(G)|^2$. Each request is served in $\lambda \cdot \mathrm{Bandwidth}(G)$. Thus, the total cost of requests σ is $O(|E(G)| \cdot |V(G)|^2 + \lambda \cdot \mathrm{Bandwidth}(G) \cdot |\sigma|)$.

Lemma 3. *Given a demand graph G. For each online algorithm ON there is a request sequence σ_{ON} such that ON serves each request from σ_{ON} for a cost of at least $\mathrm{Bandwidth}(G)$.*

4 Embedding a Ladder Demand Graph

We present our algorithms for embedding a demand graph that is a subgraph of the ladder graph ($2 \times n$-grid) on the line. We first present the offline case, where the demand graph is known in advance (Sect. 4.1). Then we present the dynamic case, where requests are revealed online, revealing also the demand graph and thus possibly changing the current embedding (Sect. 4.2). Finally, we discuss the cost of the dynamic case in Sect. 4.3.

Though our final goal is to embed a demand graph into the line, we will first focus on how to embed a partially-known demand graph into $Ladder_N$, where N is large enough to make the embedding possible, i.e., no more than $2n$. When we have such an embedding one might construct an embedding from $Ladder_N$ into $Line_n$, simply composing it with a level by level (see the proof of Lemma 1) embedding of $Ladder_N$ to $Line_{2N}$ and then by omitting empty images we get $Line_n$. Such a mapping of $Ladder_N$ to $Line_{2N}$ enlarges the bandwidth for at most a factor of 2, but significantly simplifies the construction of our embedding.

Definition 4. *A ladder graph l consists of two line-graphs on n vertices l_1 and l_2 with additional edges between the lines: $\{(l_1[i], l_2[i]) \mid i \in [n]\}$, where $l_j[i]$ is the i-th node of the line-graph l_j. We call the set of two vertices, $\{l_1[i], l_2[i]\}$, the i-th level of the ladder and denote it as $level_{Ladder_n}(i)$ or just $level(i)$ if it is clear from the context. We refer to $l_1[i]$ and $l_2[i]$ as $level(i)[1]$ and $level(i)[2]$, respectively. We say that $level\langle v \rangle = i$ for $v \in V(Ladder_n)$ if $v \in level_{Ladder_n}(i)$. We refer to l_1 and l_2 as the sides of the ladder.*

Definition 5. *A correct embedding of a graph A into a graph B is an injective mapping $\varphi : V(A) \to V(B)$ that preserves edges, i.e.*

$$\begin{cases} \forall u, v \in V(A) \text{ with } u \neq v \Rightarrow \varphi(u) \neq \varphi(v) \\ (u, v) \in E(A) \Rightarrow (\varphi(u), \varphi(v)) \in E(B) \end{cases}$$

4.1 Static Quasi-embedding

We start with one of the basic algorithms—how to quasi-embed any graph that can be embedded in $Ladder_n$ onto $Ladder_N$ with large N. We present a tree and cycle embedding and then we show how to combine them in an embedding of a general component (by first doing a cycle-tree decomposition). The whole algorithm is presented in [13, Appendix B.1].

Tree Embedding. In this case, our task is to embed a tree on a ladder graph. We start with some definitions and basic lemmas.

Definition 6. *Consider some correct embedding φ of a tree T into $Ladder_n$. Let $r = \arg\max_{v \in V(T)} level\langle \varphi(v) \rangle$ and $l = \arg\min_{v \in V(T)} level\langle \varphi(v) \rangle$ be the "rightmost" and "leftmost" nodes of the embedding, respectively. The trunk of T is a path in T connecting l and r. The trunk of a tree T for the embedding φ is denoted with $trunk_\varphi(T)$.*

Definition 7. *Let T be a tree and φ be its correct embedding into $Ladder_n$. The level i of $Ladder_n$ is called occupied if there is a vertex $v \in V(T)$ on that level, i.e., $\varphi(v) \in level_{Ladder_n}(i)$.*

Statement 1. *For every occupied level i there is $v \in trunk_\varphi(T)$ such that $v \in level(i)$.*

Proof. By the definition of the trunk, an image goes from the minimal occupied level to the maximal. It cannot skip a level since the trunk is connected and the correct embedding preserves connectivity.

The trunk of a tree in an embedding is a useful concept to define since the following holds for it. The proofs for the lemmas in this section appear in [13, Appendix C].

Lemma 4. *Let T be a tree correctly embedded into $Ladder_n$ by some embedding φ. Then, all the connected components in $T \setminus trunk_\varphi(T)$ are line-graphs.*

Lemma 5. *For the tree T and for each node v of degree three (except for maximum two of them) we can verify in polynomial time if for any correct embedding φ, $trunk_\varphi(T)$ passes through v or not.*

Support nodes are the nodes of two types: either a node of degree three without neighbours of degree three or a node that is located on some path between two nodes with degree three. The path through passing through all support nodes is called *trunk core*. We denote this path for a tree T as $trunkCore(T)$. Intuitively, the trunk core consists of vertices that lie on a trunk of any embedding. It can be proven that the support nodes appear in the trunk of every correct embedding (proof appears in [13]).

Definition 8. *Let T be a tree. All the connected components in $T \setminus trunkCore(T)$ are called* simple-graphs *of tree T.*

Lemma 6. *The simple-graphs of a tree T are line-graphs.*

Definition 9. *The edge between a simple-graph and the trunk core is called a* leg. *The end of a leg in the simple-graph is called a* head *of the simple-graph. The end of a leg in the trunk core is called a* foot *of the simple-graph. If you remove the head of a simple-graph and it falls apart into two connected components, such simple-graph is called* two-handed *and those parts are called its* hands. *Otherwise, the graph is called* one-handed, *and the sole remaining component is called a* hand. *If there are no nodes in the simple-graph but just a head it is called zero-handed. We refer to Fig. 3.*

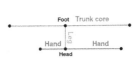

Fig. 3. Hands, Legs, and Trunk core.

Definition 10. *A simple-graph connected to some end node of the trunk core is called* exit-graph. *A simple-graph connected to an inner node of the trunk core is called* inner-graph.

Please note that the next definition is about a much larger ladder graph, $Ladder_N$, rather than $Ladder_n$. Here, N is equal to $2n$ to make sure that we have enough space to embed.

Definition 11. *An embedding* $\varphi : V(G) \rightarrow V(Ladder_N)$ *of a graph* G *into* $Ladder_N$ *is called* quasi-correct *if:*

- $(u,v) \in E(G) \Rightarrow (\varphi(u), \varphi(v)) \in E(Ladder_N)$, *i.e., images of adjacent vertices in G are adjacent in the ladder.*
- *There are no more than* **three** *nodes mapped into each level of $Ladder_N$, i.e., the two ladder nodes on each level are the images of no more than three nodes.*

We can think of a quasi-correct embedding as an embedding into levels of the ladder with no more than three nodes embedded to the same level. Then, we can compose this embedding with an embedding of a ladder into the line which is the enumeration level by level. More formally if a node u is embedded to level i and a node v is embedded to level j and $i < j$ then the resulting number of u on the line is smaller than the number of v, but if two nodes are embedded to the same level, we give no guarantee.

Lemma 7. *Any graph mapped into the ladder graph by the quasi-correct embedding described above can be mapped to the line level by level with the property that any pair of adjacent nodes are embedded at the distance of at most five.*

Assume, we are given a tree T that can be embedded into $Ladder_n$. Furthermore, there are two special nodes in the tree: one is marked as R (right) and another one is marked as L (left). It is known that there exists a correct embedding of T into $Ladder_n$ with R being the rightmost node, meaning no node is embedded more to the right or to the same level, and L being the leftmost node.

We now describe how to obtain a quasi-correct embedding of T in $Ladder_N$ with R being the rightmost node and L being the leftmost one while L is mapped to $ImageL$—some node of the $Ladder_N$. Moreover, our embedding obeys the following invariant.

Invariant 1 (Septum invariant). *For each inner simple-graph, its foot and its head are embedded to the same level and no other node is embedded to that level.*

We embed a path between L and R simply horizontally and then we orient line-graphs connected to it in a way that they do not violate our desired invariant. It can be shown that it is always possible if T can be embedded in $Ladder_n$. The pseudocode is in [13, Appendix, Algorithm 1].

Suppose now that not all information, such as R, L, and $ImageL$, is provided. We explain how we can embed a tree T. We first get the *trunk core* of the given tree. This can be

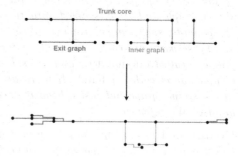

Fig. 4. Example of a quasi-correct embedding

done by following the definition. Now the idea would be to first embed the trunk core and its inner line-graphs using a tree embedding presented earlier with R and L to be the ends of the trunk core. Then, we embed exit-graphs strictly horizontally "away" from the trunk core. That means, that the hands of exit-graphs that are connected to the right of the trunk core are embedded to the right, and the hands of those exit-graphs that are connected to the left of the trunk core are embedded to the left. An example of the quasi-correct embedding is shown in Fig. 4.

If a tree does not have a trunk core, then its structure is quite simple (in particular it has no more than two nodes of degree three). Such a tree can be embedded without conflicts. The pseudocode appears in [13, Appendix, Algorithm 2].

Cycle Embedding. Now, we show how to embed a cycle into $Ladder_N$. First, we give some important definitions and lemmas.

Definition 12. *A maximal cycle C of a graph G is a cycle in G that cannot be enlarged, i.e., there is no other cycle C' in G such that $V(C) \subsetneq V(C')$.*

Definition 13. *Consider a graph G and a maximal cycle C of G. A whisker W of C is a line inside G such that: 1) $V(W) \neq \varnothing$ and $V(W) \cap V(C) = \varnothing$. 2) There exists only one edge between the cycle and the whisker (w, c) for $w \in V(W)$ and $c \in V(C)$. Such c is called a* foot *of W. The nodes of W are enumerated starting from w. 3) W is maximal, i.e., there is no W' in G such that W' satisfies previous properties and $V(W) \subsetneq V(W')$. We refer to Fig. 5.*

Fig. 5. Cycle and its Whiskers.

Definition 14. *Suppose we have a graph G that can be correctly embedded into $Ladder_n$ by φ and a cycle C in G. Whiskers W_1 and W_2 of C are called adjacent (or neighboring) for the embedding φ if $\forall i \leqslant \min(|V(W_1)|, |V(W_2)|)$ $(\varphi(W_1[i]), \varphi(W_2[i])) \in E(Ladder_n)$.*

Lemma 8. *Suppose we have a graph G that can be correctly embedded into $Ladder_n$ and there exists a maximal cycle C in G with at least 6 vertices with two neighbouring whiskers W_1 and W_2 of C, i.e., $(foot(W_1), foot(W_2)) \in E(G)$. Then, W_1 and W_2 are adjacent in any correct embedding of G into $Ladder_N$.*

Definition 15. *Assume, we have a graph G and a maximal cycle C of length at least 6. The frame for C is a subgraph of G induced by vertices of C and $\{W_1[i], W_2[i] \mid i \leqslant \min(|V(W_1)|, |V(W_2)|)\}$ for each pair of adjacent whiskers W_1 and W_2. Adding all the edges $\{(W_1[i], W_2[i]) \mid i \leqslant \min(|V(W_1)|, |V(W_2)|)\}$ for each pair of adjacent whiskers W_1 and W_2 makes a frame completed. We refer to Fig. 6.*

Fig. 6. Cycle, its frame, and edges (dashed) to make the frame completed

Given a cycle C of length at least six and its special nodes $L, R \in V(C)$, we construct a correct embedding of C into $Ladder_N$ with $level\langle L \rangle \leqslant level\langle u \rangle \leqslant level\langle R \rangle$ for all u in $V(C)$, while L is mapped into the node $ImageL$.

We first check if it is possible to satisfy the given constraints of placing the L node to the left and a R node to the right. If it is indeed possible, we place L to the desired place $ImageL$ and then we choose an orientation (clockwise or counterclockwise) following which we could embed the rest of the nodes, keeping in mind that R must stay on the rightmost level. The pseudocode appears in [13, Appendix, Algorithm 3].

Now, suppose that not all the information, such as R, L, and $ImageL$, is provided. We reduce this problem to the case when the missing variables are known. This subtlety might occur since there are inner edges in the cycle. In this case, we choose missing L/R more precisely in order to embed an inner edge vertically. For more intuition, please see Figs. 7a and 7b. A dashed line denotes an inner edge. The pseudocode appears in [13, Appendix, Algorithm 4].

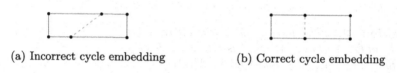

(a) Incorrect cycle embedding   (b) Correct cycle embedding

Fig. 7. Cycle embeddings.

Embedding a Connected Component of the Demand Graph. Combining the previous results, we can now explain how to embed in $Ladder_N$ a connected component S that can be embedded in $Ladder_n$.

Definition 16. *By the cycle-tree decomposition of a graph G we mean a set of maximal cycles $\{C_1, \dots C_n\}$ of G and a set of trees $\{T_1, \dots, T_m\}$ of G such that*

- $\bigcup\limits_{i \in [n]} V(C_i) \cup \bigcup\limits_{i \in [m]} V(T_i) = V(G)$
- $V(C_i) \cap V(C_j) = \varnothing \ \forall i \neq j$
- $V(T_i) \cap V(T_j) = \varnothing \ \forall i \neq j$
- $V(T_i) \cap V(C_j) = \varnothing \ \forall i \in [m], j \in [n]$
- $\forall i \neq j \ \forall u \in V(T_i) \ \forall v \in V(T_j) \ (u, v) \notin E(G)$

We start with an algorithm on how to make a cycle-tree decomposition of S assuming no incomplete frames. To obtain a cycle-tree decomposition of a graph: 1) we find a maximal cycle; 2) we split the graph into two parts by logically removing the cycle; 3) we proceed recursively on those parts, and, finally, 4) we combine the results together maintaining the correct order between cycle and two parts (first, the result for one part, then the cycle, and then the result for the second part). Since we care about the order of the parts, we say that it is a *cycle-tree decomposition chain*. The decomposition pseudocode appears in [13, Appendix, Algorithm 5].

We describe how to obtain a quasi-correct embedding of S. We preprocess S: 1) we remove one edge from cycles of size four; 2) we complete incomplete frames with vertical edges. Then, we embed parts of S from the cycle-tree decomposition chain one by one in the relevant order using the corresponding algorithm (either for a cycle or for a tree embedding) making sure parts are glued together correctly. The pseudocode appears in [13, Appendix, Algorithm 6].

4.2   Online Quasi-embedding

In the previous subsection, we presented an algorithm on how to quasi-embed a static graph. Now, we will explain how to operate when the requests are revealed in an online manner. The full version of the algorithm is presented in [13, Appendix B.2].

There are two cases: a known edge is requested or a new edge is revealed. In the first case the algorithm does nothing since we already know how to quasi-correctly embed the current graph and, thus, we already can embed into the line network with constant bandwidth. Thus, further, we consider only the second case.

We describe how one should change the embedding of the graph after the processing of a request in an online scenario. At each moment some edges of the demand graph $Ladder_n$ are already revealed, forming connected components. After an edge reveal we should reconfigure the target line topology. For that, instead of line reconfiguration we reconfigure our embedding onto $Ladder_N$ that is then embedded to the line level by level and introduces a constant factor. So, we can consider the reconfiguration only of $Ladder_N$ and forget about the target line topology at all. When doing the reconfiguration of an embedding we want to maintain the following invariants:

1. The embedding of any connected component is quasi-correct.
2. For each tree in the cycle-tree decomposition its embedding respects Septum Invariant 1.
3. There are no maximal cycles of length 4.
4. Each cycle frame is completed with all "vertical" edges even if they are not yet revealed.
5. There are no conflicts with cycle nodes, i.e., each cycle node is the only node mapped to its image in the embedding to $Ladder_N$.

For each newly revealed edge there are two cases: either it connects two nodes from one connected component or not. We are going to discuss both of them.

Edge in One Component. The pseudocode appears in [13, Appendix, Algorithm 8]. If the new edge is already known or it forms a maximal cycle of length four, we simply ignore it. Otherwise, it forms a cycle of length at least six, since two connected nodes are already in one component. We then perform the following steps:

1. Get the completed frame of a (possibly) new cycle.
2. Logically "extract" it from the component and embed maintaining the orientation (not twisting the core that was already embedded in some way).
3. Attach two components appeared after an extraction back into the graph, maintaining their relative order.

Edge Between Two Components. The pseudocode appears in [13, Appendix, Algorithm 9]. In order to obtain an amortization in the cost, we always "move" the smaller component to the bigger one. Thus, the main question here is how to glue a component to the existing embedding of another component. The idea is to consider several cases of where the smaller component will be connected to the bigger one. There are three possibilities:

1. *It connects to a cycle node.* In this case, there are again two possibilities. Either it "points away" from the bigger component meaning that the cycle to which we connect is the one of the ends in the cycle-tree decomposition of the bigger component. Here, we just simply embed it to the end of the cycle-tree decomposition while possibly rotating a cycle at the end.

Or, the smaller component should be placed somewhere between two cycles in the cycle-tree decomposition. Here, it can be shown that this small graph should be a line-graph, and we can simply add it as a whisker, forming a larger frame.

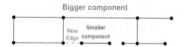

2. *It connects to a trunk core node of a tree in the cycle-tree decomposition.* It can be shown that in this case the smaller component again must be a line-graph. Thus, our only goal is to orient it and possibly two of its inner simple-graphs neighbours to maintain the Septum Invariant 1 for the corresponding tree from the decomposition.

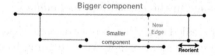

3. *It connects to an exit graph node of an end tree of the cycle-tree decomposition.* In this case, we straightforwardly apply a static embedding algorithm of this tree and the smaller component from scratch. Please, note that only the exit graphs of the end tree will be moved since the trunk core and its inner graphs will remain.

4.3   Complexity of the Online Embedding

Now, we calculate the cost of our online algorithm (a more detailed discussion on the cost of the algorithm appears in [13, Appendix C.5]): how many swaps we should do and how much we should pay for the routing requests. Recall that we first apply the reconfiguration and, then, the routing request.

We start with considering the routing requests. Their cost is $O(1)$ since they lie pretty close on the target line network, i.e., by no more than 12 nodes apart. This bound holds because the nodes are quasi-correctly embedded on $Ladder_N$, two adjacent nodes at G are located not more than four levels apart (in the worst case, when we remove an edge of a cycle with length four) where each level of the quasi-correct embedding has at most three images of nodes of G. Thus, on the target line, if we enumerate level by level, the difference between any two adjacent nodes of G is at most 12.

Then, we consider the reconfiguration. We count the total cost of each case of the online algorithm before all the edges are revealed.

In the first case, we add an edge in one component. By that, either a new frame is created or some frame was enlarged. In both cases, only the nodes, that appear on some frame for the first time, are moved. Since, a node can be moved only once to be mapped to a frame and it is swapped at most $N = O(n)$ times to move to any position, the total cost of this type of reconfiguration is at most $O(n^2)$. Also, there are several adjustments that could be done: 1) the "old" frame can rotate by one node, and 2) possibly, we should flip the first inner-graphs of two components connected to the frame. In the first modification, each node at the frame can only be "rotated" once, thus, paying $O(n)$ cost in total. In the second modification, inner-graph can change orientation at most once in order to satisfy the Septum invariant (Invariant 1), thus, paying $O(n^2)$ cost in total—each node can move by at most $N = O(n)$.

In the second case, we add an edge in between two components. At first, we calculate the time spent on the move of the small component to the bigger one: each node is moved at most $O(\log n)$ times since the size of the component always grows at least two times, the number of swaps of a vertex is at most $N = O(n)$ to move to any place, thus, the total cost is $O(n^2 \log n)$. Secondly, there are two more modification types: 1) a rotation of a cycle, and 2) some simple-graphs can be reoriented. The cycle can be rotated only once, thus, we should pay at most $O(n)$ there. At the same time, each simple-graph can be reoriented at most once to satisfy the Septum invariant (Invariant 1), thus, the total cost is $O(n^2)$ for that type of a reconfiguration.

To summarize, the total cost of requests σ is $O(n^2 \log n)$ for the whole reconfiguration plus $O(|\sigma|)$ per requests. This matches the lower bound that was obtained for the line demand graph. The same result holds for any demand graph that is the subgraph of the ladder of size n.

Theorem 6. *The online algorithm for embedding the ladder demand graph of size n on the line has total cost $O(n^2 \log n + |\sigma|)$ for a sequence of communication requests σ.*

5   Conclusion

We presented methods for statically or dynamically re-embedding a ladder demand graph (or a subgraph of it) on a line, both in the offline and online case. As side results, we also presented how to embed a cycle demand graph and a meta-algorithm for a general demand graph. Our algorithms for the cycle and the ladder cases match the lower bounds. Our work is a first step towards a tight bound on dynamically re-embedding more generic demand graphs, such as arbitrary $k \times n$ grids.

References

1. Avin, C., Bienkowski, M., Salem, I., Sama, R., Schmid, S., Schmidt, P.: Deterministic self-adjusting tree networks using rotor walks. In: 2022 IEEE 42nd International Conference on Distributed Computing Systems (ICDCS), pp. 67–77. IEEE (2022)
2. Avin, C., van Duijn, I., Schmid, S.: Self-adjusting linear networks. In: Ghaffari, M., Nesterenko, M., Tixeuil, S., Tucci, S., Yamauchi, Y. (eds.) SSS 2019. LNCS, vol. 11914, pp. 368–382. Springer, Cham (2019). https://doi.org/10.1007/978-3-030-34992-9_29
3. Avin, C., Ghobadi, M., Griner, C., Schmid, S.: On the complexity of traffic traces and implications. Proc. ACM Measur. Anal. Comput. Syst. 4(1), 1–29 (2020)
4. Avin, C., Loukas, A., Pacut, M., Schmid, S.: Online balanced repartitioning. In: Gavoille, C., Ilcinkas, D. (eds.) DISC 2016. LNCS, vol. 9888, pp. 243–256. Springer, Heidelberg (2016). https://doi.org/10.1007/978-3-662-53426-7_18
5. Avin, C., Mondal, K., Schmid, S.: Demand-aware network design with minimal congestion and route lengths. IEEE/ACM Trans. Netw. 30(4), 1838–1848 (2022)
6. Avin, C., Mondal, K., Schmid, S.: Push-down trees: optimal self-adjusting complete trees. IEEE/ACM Trans. Netw. 30(6), 2419–2432 (2022)
7. Avin, C., Schmid, S.: Toward demand-aware networking: a theory for self-adjusting networks. ACM SIGCOMM Comput. Commun. Rev. 48(5), 31–40 (2019)
8. Batista, D.M., da Fonseca, N.L.S., Granelli, F., Kliazovich, D.: Self-adjusting grid networks. In: 2007 IEEE International Conference on Communications, pp. 344–349. IEEE (2007)
9. Díaz, J., Petit, J., Serna, M.: A survey of graph layout problems. ACM Comput. Surv. (CSUR) 34(3), 313–356 (2002)
10. Hansen, M.D.: Approximation algorithms for geometric embeddings in the plane with applications to parallel processing problems. In: 30th Annual Symposium on Foundations of Computer Science, pp. 604–609. IEEE Computer Society (1989)
11. Newman, I., Rabinovich, Y.: Online embedding of metrics. arXiv preprint arXiv:2303.15945 (2023)
12. Olver, N., Pruhs, K., Schewior, K., Sitters, R., Stougie, L.: The itinerant list update problem. In: Epstein, L., Erlebach, T. (eds.) WAOA 2018. LNCS, vol. 11312, pp. 310–326. Springer, Cham (2018). https://doi.org/10.1007/978-3-030-04693-4_19
13. Paramonov, A., Salem, I., Schmid, S., Aksenov, V.: Self-adjusting linear networks with ladder demand graph. arXiv preprint arXiv:2207.03948 (2022)
14. Schmid, S., Avin, C., Scheideler, C., Borokhovich, M., Haeupler, B., Lotker, Z.: Splaynet: towards locally self-adjusting networks. IEEE/ACM Trans. Netw. 24(3), 1421–1433 (2016)
15. Sleator, D.D., Tarjan, R.E.: Amortized efficiency of list update and paging rules. Commun. ACM 28(2), 202–208 (1985)

Compatibility of Convergence Algorithms for Autonomous Mobile Robots (Extended Abstract)

Yuichi Asahiro[1(✉)] and Masafumi Yamashita[2]

[1] Kyushu Sangyo University, Fukuoka, Japan
asahiro@is.kyusan-u.ac.jp
[2] Kyushu University, Fukuoka, Japan
masafumi.yamashita@gmail.com

Abstract. We investigate autonomous mobile robots in the Euclidean plane. A robot has a function called *target function* to decide the destination from the robots' positions, and operates in Look-Compute-Move cycles, i.e., identifies the robots' positions, computes the destination by the target function, and then moves there. Robots can have different target functions. Let Φ and Π be a set of target functions and a problem, respectively. If the robots whose target functions are chosen from Φ always solve Π, we say that Φ is *compatible* with respect to Π. Suppose that Φ is compatible with respect to Π. Then two swarms controlled by (possibly different) target functions in Φ can merge to form a larger swarm, and a broken robot can be replaced with another robot with any target function in Φ, keeping the correctness of solving Π. We investigate the convergence, the gathering, and some fault tolerant convergence problems, assuming crash failures, from the view point of compatibility.

Keywords: Autonomous mobile robot · Compatibility · Convergence · Crash fault · Gathering

1 Introduction

Over the last three decades, swarms of autonomous mobile robots have obtained much attention in a variety of contexts. Among them is understanding solvable problems by a swarm consisting of many simple and identical robots in a distributed manner, which has been constantly attracting researchers in distributed computing society [1–3,5,6,8–20].

Many of the works mentioned above adopt the following robot model. The robots look identical and indistinguishable. Each robot is represented by a point that moves in the Euclidean plane. It lacks identifier and communication devices,

---

Due to the space limitation, we omit most of the proofs and some contributions. The full version of the paper [4] contains them.

© The Author(s), under exclusive license to Springer Nature Switzerland AG 2023
S. Rajsbaum et al. (Eds.): SIROCCO 2023, LNCS 13892, pp. 149–164, 2023.
https://doi.org/10.1007/978-3-031-32733-9_8

and operates in Look-Compute-Move cycles. When a robot starts a cycle, it identifies the multiset of the robots' positions in its local x-y coordinate system such that it is right-handed, and its origin is always the position of the robot, computes the destination point using a target function[1] based only on the multiset identified, and then moves towards the target position. If each cycle starts at a time t and finishes, reaching the target position, before (not including) $t + 1$, for some integer t, the scheduler is said to be *semi-synchronous* ($\mathcal{SSYNC}$). If cycles can start and end any time (even on the way to the target point), it is *asynchronous* ($\mathcal{ASYNC}$).

This paper investigates several *convergence problems*, e.g., [2, 8–10, 13, 14, 16, 18]. The simplest convergence problem requires the robots to converge to a single point. For the $\mathcal{SSYNC}$ model, the problem is solvable for robots with unlimited visibility [18], and is also solvable for robots with limited visibility [2].

Under the $\mathcal{ASYNC}$ model, it is solvable by a target function called CoG, which always outputs the center of gravity of the robots' positions [8]. In [8], the authors also showed that CoG correctly works under the sudden-stop model, under which the movement of a robot towards the center of gravity might stop on the way after traversing at least some fixed distance. This implies that the robots can correctly converge to a point, even when they are controlled by different target functions as long as they always move robots towards the current center of gravity over distance at least some fixed constant. This idea is extended in [10]: The authors proposed the δ-*inner* property[2] of target functions, and showed that the robot system converges to a point if all robots take δ-inner target functions, provided $\delta \in (0, 1/2]$. Finally [16] gives a convergence algorithm for robots with limited visibility under the $\mathcal{ASYNC}$ model.

Consider a problem Π and a set of target functions Φ. If the robots whose target functions are chosen from Φ always solve Π, we say that Φ is *compatible with respect to* Π. For example, every (non-empty) set of $(1/3)$-inner functions is compatible with respect to the convergence problem [10].

If a singleton $\{\phi\}$ is compatible with respect to Π, we abuse to say that target function ϕ is an *algorithm*[3] for Π. If a set Φ of target functions is compatible with respect to Π, every target function $\phi \in \Phi$ is an algorithm for Π. (The converse is not always true.) Thus there is an algorithm for Π, if and only if there is a compatible set Φ with respect to Π. We say that a problem Π is *solvable*, if there is a compatible set Φ with respect to Π, meaning that there is an algorithm for Π.

---

[1] Roughly, a target function is a function from $(R^2)^n$ to R^2, where R is the set of real numbers and n is the number of robots, i.e., given a snapshot in $(R^2)^n$, it returns a destination point in R^2. Later, we define a target function a bit more carefully.

[2] Let P, D, and o be the multiset of robots' positions, the axes aligned minimum box containing P, and its center, respectively. Define $\delta * D = \{(1 - 2\delta)x + 2\delta o : x \in D\}$. A function ϕ is δ-inner, if $\phi(P)$ is included in $\delta * D$ for any P.

[3] Here, we abuse term "algorithm," since an algorithm must have a finite description. A target function may not. To compensate the abuse, we insist on giving a finite procedure when we show the existence of a target function.

We would like to find a large compatible set Φ with respect to Π. That Π has a large compatible set with respect to Π implies that Π has many algorithms. The difficulty of problems might be compared by the sizes of their compatible sets. A problem Π with a large compatible set Φ seems to have some practical merits. Two swarms both of which are controlled by target functions in Φ (which may be produced by different makers) can merge to form a larger swarm, keeping the correctness of solving Π. When a robot breaks down, we can safely replace it with another robot, as long as it is controlled by a target function in Φ.

Fault Tolerant Convergence Problems. This paper investigates three fault-tolerant convergence problems, besides the convergence and the gathering problems. We consider only crash faults: A faulty robot can stop functioning at any time, becoming permanently inactive. A faulty robot may not cause a malfunction, forever. We cannot distinguish such a robot from non-faulty ones. Let n and $f(\leq n-1)$ be the number of robots and the number of faulty robots.

The *fault-tolerant (n, f)-convergence problem* (FC(f)) is the problem to find an algorithm which ensures that, as long as at most f robots are faulty, *all non-faulty robots* converge to a single point. The *fault-tolerant (n, f)-convergence problem to f points* (FC(f)-PO) is the problem to find an algorithm which ensures that, as long as at most f robots are faulty, *all robots* (including faulty ones) converge to at most f points. All non-faulty robots need not converge to the same point. If f faulty robots have crashed at different positions, each non-faulty robot must converge to one of the faulty robots. The *fault-tolerant (n, f)-convergence problem to a convex f-gon* (FC(f)-CP) is the problem to find an algorithm which ensures that, as long as at most f robots are faulty, the convex hull of the positions of *all robots* (including faulty ones) converges to a convex h-gon CH for some $h \leq f$, in such a way that, for each vertex of CH, there is a robot that converges to the vertex.

Since an algorithm for FC(1)-PO solves FC(1), the former is not easier than the latter. (Note that for $f \geq 2$, an algorithm for FC(f)-PO may not solve FC(f).) Since an algorithm for FC(f)-PO solves FC(f)-CP, again the former is not easier than the latter. In [8], the authors showed that, for all $f \leq n-2$, CoG is an algorithm for FC(f) under the $\mathcal{ASYNC}$ model. As far as we know, FC(f)-PO and FC(f)-CP have not been investigated so far.

Gathering Problem. The *gathering problem* requires the robots to gather in the exactly the same location. For $\mathcal{SSYNC}$, the gathering problem is not solvable if $n = 2$. If $n > 2$, it is solvable, provided that all robots initially occupy distinct positions [18]. For $\mathcal{ASYNC}$, the same results hold [7]. The gathering problem has been investigated under a variety of assumptions [1,5,7,11–14].

Contributions. Let R be the set of real numbers. Formally, a *target function* ϕ is a function from $(R^2)^n$ to $R^2 \cup \{\perp\}$ for all $n \geq 1$ such that $\phi(P) = \perp$ if and only if $(0,0) \notin P$. Here, $\perp$ is a special symbol to denote that $(0,0) \notin P$. Suppose that a robot r identifies a multiset P of n points, which are the positions of the

robots in its local x-y coordinate system Z, in Look phase. Then $(0,0) \in P$.[4]
Using its target function ϕ, r computes the target point $\boldsymbol{x} = \phi(P)$ in Compute
phase. Then it moves to $\boldsymbol{x}$ $(\neq \perp)$ in Z in Move phase.

Let the convex hull and the center of gravity of P be $CH(P)$ and $g(P)$,
respectively. For any $0 \leq d$, let $d * CH(P) = \{d\boldsymbol{x} + (1 - d)g(P) : \boldsymbol{x} \in CH(P)\}$.
The *scale* $\alpha(\phi)$ of a target function ϕ is defined by

$$\alpha(\phi) = \sup_{P \in (R^2)^n} \alpha(\phi, P),$$

where $\alpha(\phi, P)$ is the smallest d satisfying $\phi(P) \in d * (CH(P))$.[5] Then the scale
of a set Φ of target functions is defined by

$$\alpha(\Phi) = \sup_{\phi \in \Phi} \alpha(\phi).$$

The only target function ϕ satisfying $\alpha(\phi) = 0$ is CoG. Thus the set Φ of
target functions satisfying $\alpha(\Phi) = 0$ is a singleton $\{\text{CoG}\}$. The idea of scale is
similar to that of the δ-inner property in [10], and more directly embodies the
idea behind the δ-inner target function.

Our contributions are summarized in Table 1. For example, the entry of Convergence and $\alpha(\phi) = 0$ is A. Thus $\{\text{CoG}\}$ is compatible with respect to the
convergence problem, or CoG is an algorithm for the convergence problem, as
[8] shows. Not only the case $\alpha(\phi) = 0$, but also the case $0 < \alpha(\Phi) < 1$, every Φ
is compatible with respect to the convergence problem.

Organization. After introducing the robot model in Sect. 2, we investigate the
convergence problem in Sect. 3. In Sect. 4, we discuss the compatibilities of two
convergence problems FC(1) and FC(1)-PO. Sections 5 and 6 respectively investigate the compatibilities of FC(f) and FC(f)-CP for $f \geq 2$. Section 7 first
shows that a target function ϕ is an algorithm for FC(f)-PO for $f \geq 2$, only if
$\alpha(\phi) \geq 1$. We then present an algorithm $\psi_{(n,2)}$ for FC(2)-PO. Section 8 investigates the gathering problem to show the difference between this and the convergence problems. We conclude the paper in Sect. 9.

2   The Model

Consider a robot system $\mathcal{R}$ consisting of n robots $r_1, r_2, \ldots, r_n$. Each robot r_i
has its own unit of length, and a local compass defining an local x-y coordinate
system Z_i, which is assumed to be right-handed and self-centric, i.e., its origin
$(0,0)$ is always the position of r_i. We also assume that r_i has the strong multiplicity detection capability, i.e., it can count the number of robots resides at
a point. Given a target function ϕ_i, each robot $r_i \in \mathcal{R}$ repeatedly executes a
Look-Compute-Move cycle:

---

[4] That $(0,0) \notin P$ means an error of eye sensor, which we assume will not occur, in
this paper.
[5] For the sake of completeness, we assume that $\alpha(\phi, P) = 0$ when $\phi(P) = \perp$.

Table 1. The compatibility of a set Φ of target functions with respect to a problem Π, taking its scale $\alpha(\Phi)$ as a parameter. Each entry contains the status A, E, N, or ? of the compatibility of Φ with respect to Π (and the theorem/corollary/observation/citation number establishing the result in parentheses). Letter 'A' means that every Φ such that $\alpha(\Phi)$ is in the range is compatible with respect to Π. Letter 'N' means that any Φ such that $\alpha(\Phi)$ is in the range is not compatible with respect to Π, which indicates the absence of an algorithm. Letter 'E' means that some Φ is compatible, while some other is not, which indicates the existence of an algorithm. Letter '?' means that the answer is unknown.

problem Π	scale $\alpha(\Phi)$		
	$\alpha(\Phi) = 0$	$0 < \alpha(\Phi) < 1$	$\alpha(\Phi) = 1$
Convergence	A (Theorem 1 [8])	A (Theorem 2)	E (Theorem 3)
FC(1)	A ([8])	A (Cor. 1)	E (Theorem 5)
FC(1)-PO	A (Theorem 4)	A (Theorem 4)	E (Theorem 5)
FC(f) ($f \geq 2$)	A (Theorem 6 [8])	E (Theorem 7)	E (Corollary 2)
FC(f)-CP ($f \geq 2$)	A (Theorem 8)	A (Theorem 8)	E (Trivial)
FC(2)-PO	N (Theorem 9)	N (Theorem 9)	E (Theorem 10)
FC(f)-PO ($f \geq 3$)	N (Theorem 9)	N (Theorem 9)	?
Gathering	N (Theorem 12)	N (Theorem 12)	E (Theorem 11 [18])

Look: Robot r_i identifies the multiset P of the robots' positions (including the one of r_i) in Z_i. Since r_i has the strong multiplicity detection capability, it can identify P not only distinct positions of P.

Compute: Robot r_i computes $\boldsymbol{x}_i = \phi_i(P)$. (We do not mind even if ϕ_i is not computable. We simply assume that $\phi_i(P)$ is given by an oracle.)

Move: Robot r_i moves to $\boldsymbol{x}_i$. We assume that r_i always reaches $\boldsymbol{x}_i$ before this Move phase ends.

We assume a discrete time $0, 1, \ldots$. At each time $t \geq 0$, the scheduler activates some unpredictable subset (that may be none or all) of robots. Then activated robots execute a cycle which starts at t and ends before (not including) $t + 1$ (unless it has crashed), i.e., the scheduler is semi-synchronous ($\mathcal{SSYNC}$). Let Z_0 be the global x-y coordinate system, which is right-handed and is not accessible by any robot r_i. The coordinate transformation from Z_i to Z_0 is denoted by γ_i. We use Z_0 and γ_i just for the purpose of explanation.

The position of robot r_i at time t in Z_0 is denoted by $\boldsymbol{x}_t(r_i)$. Then $P_t = \{\boldsymbol{x}_t(r_i) : 1 \leq i \leq n\}$ is a multiset representing the positions of all robots at time t, and is called the *configuration* at t.

Given an initial configuration P_0, an assignment $\mathcal{A}$ of a target function ϕ_i to each robot r_i, and an $\mathcal{SSYNC}$ activation schedule, the execution of $\mathcal{R}$ is a sequence $\mathcal{E} : P_0, P_1, \ldots, P_t, \ldots$ of configurations starting from P_0. Here, for all r_i and $t \geq 0$, if r_i is not activated at t, $\boldsymbol{x}_{t+1}(r_i) = \boldsymbol{x}_t(r_i)$. Otherwise, if it is activated, r_i identifies $Q_t^{(i)} = \gamma_i^{-1}(P_t)$ in Z_i, computes $\boldsymbol{y} = \phi_i(Q_t^{(i)})$, and moves to y in Z_i.[6] Then $x_{t+1}(r_i) = \gamma_i(y)$. We assume that the scheduler is fair: It activates every robot infinitely many times. Throughout the paper, we regard the scheduler as an adversary.

We introduce several notations. Let $P \in (R^2)^n$. The distinct points of P is denoted by $\overline{P}$. Then $|P|$ (resp. $|\overline{P}|$) denotes the number of points (resp. the number of distinct points) in P. Let $CH(P)$ be the convex hull of P. We sometimes denote $CH(P)$ by a sequence of vertices of $CH(P)$ appearing on the boundary counter-clockwise. The center of gravity $g(P)$ of P is defined by $g(P) = \sum_{x \in P} x/n$. For two points x and y in R^2, $dist(x, y)$ denotes the Euclidean distance between x and y. For a set $B(\subseteq R^2)$ of points and a point $a \in R^2$, $dist(a, B) = \min_{x \in B} dist(a, x)$. Finally, let $\mathcal{P} = \{P \in (R^2)^n : (0,0) \in P, n \geq 1\}$. We regard $\mathcal{P}$ as the domain of target functions.

3   Convergence Problem

We investigate the convergence problem, provided that all robots are non-faulty. For any $0 \leq \alpha \leq 1$, consider a target function CoG_α defined by $\text{CoG}_\alpha(P) = (1 - \alpha)g(P)$, for any $P \in \mathcal{P}$. The scale of CoG_α is α, and $\text{CoG}_0 = \text{CoG}$. The following theorem holds, since CoG works correctly under the sudden-stop model.

Theorem 1 ([8]). *For any $0 \leq \alpha < 1$, let $\Phi_\alpha = \{\text{CoG}_\alpha\}$. Then Φ_α is compatible with respect to the convergence problem, or equivalently, CoG_α is an algorithm for the convergence problem.*

We extend Theorem 1 to have the following theorem.

Theorem 2. *Let Φ be a set of target functions such that $0 \leq \alpha(\Phi) < 1$. Then Φ is compatible with respect to the convergence problem.*

Proof (Sketch). Let $\phi_i \in \Phi$ be the target function taken by robot r_i for $i = 1, 2, \ldots, n$. Let $\alpha(\phi_i) = \alpha_i$ and $\alpha = \max_{1 \leq i \leq n} \alpha_i$. Then $\alpha \leq \alpha(\Phi) < 1$. Consider any execution $\mathcal{E} : P_0, P_1, \ldots$ starting from any initial configuration P_0. We show that P_t converges to a point.

Suppose that $P_t = \{x, x, \ldots, x\}$ at some time t, i.e., $|\overline{P_t}| = 1$. Since $g_t = g(P_t) = x$, $P_{t+1} = P_t$. Thus convergence has already been achieved. We assume without loss of generality that $|\overline{P_t}| \geq 2$ for all $t \geq 0$.

Let $A_t \subseteq \mathcal{R}$ be the set of robots activated at time t. If $x_t(r) = g_t$ for all $r \in A_t$, $P_{t+1} = P_t$ holds. However, there is a robot r such that $x_t(r) \neq g_t$ since $|\overline{P_t}| \geq 2$, and r is eventually activated by the fairness of scheduler. Thus, without loss of generality, we assume that there is a robot $r \in A_t$ such that $x_t(r) \neq g_t$, and that $P_{t+1} \neq P_t$ holds for all $t \geq 0$.

We denote $CH(P_t)$ by CH_t. Since $\alpha < 1$, $CH_{t+1} \subseteq CH_t$, which implies that CH_t converges to a convex k-gon CH (including a point and a line segment) for some positive integer k. We show that CH is a point, i.e., $k = 1$.

---

[6] Since $(0,0) \in Q_t^{(i)}$ by definition, $y \neq \perp$.

Let $p_0, p_1, \ldots, p_{k-1}$ be the vertices of CH aligned counter-clockwise on the boundary. To derive a contradiction, we assume that $k \geq 2$. For any pair (i, j) $(0 \leq i < j \leq k-1)$, let $L_{(i,j)} = dist(p_i, p_j)$, and $L = \min_{0 \leq i < j \leq k-1} L_{(i,j)}$. Since CH_t converges to CH, for any $0 < \epsilon \ll (1-\alpha)L/n$, there is a time instant t_0 such that, for all $t \geq t_0$, $CH \subseteq CH_t \subseteq N_\epsilon(CH)$. For any vertex p of CH, $dist(p, \alpha * CH_t) > (1-\alpha)(L/n - \epsilon) - \epsilon \gg \epsilon$, since $dist(p, g_t) > L/n - \epsilon$.

Suppose that a robot r is activated at some time $t \geq t_0$. Then $x_{t+1}(r) \in \alpha * CH_t$, which implies that $x_{t+1}(r) \notin N_\epsilon(p)$, for any vertex p of CH. If r is reactivated at some time $t' > t$ for the first time after t, since $CH_{t'} \subseteq CH_t$ and $x_{t'+1}(r) \in \alpha * CH_{t'}$, $x_{t'+1}(r) \notin N_\epsilon(p)$, for any vertex p of CH. Therefore, for any $t' > t$ and any vertex p of CH, $x_{t'}(r) \notin N_\epsilon(p)$.

On the other hand, all robots will be activated infinitely many times after time t, by the fairness of scheduler. It is a contradiction to the assumption that CH_t converges to CH, since there is a time instant $t' > t$ such that for any robot r and any vertex p of CH, $x_{t'}(r) \notin N_\epsilon(p)$ holds. □

Let Φ and Φ' be two sets of target functions. If $\alpha(\Phi) < 1$ and $\alpha(\Phi') < 1$, Φ, Φ', and $\Phi \cup \Phi'$ are all compatible with respect to the convergence problem by Theorem 2. However, the following claim does not hold:

If both of Φ and Φ' are compatible with respect to the convergence problem, so is $\Phi \cup \Phi'$.

To observe this fact, examine two target functions ϕ_T and ϕ_S. For a configuration P, define a condition Ψ as follows:

Ψ: $|P| = 7$, $(0,0) \in P$, $P = T \cup S$, T is an equilateral triangle, S is a square, T and S have the same side length, and T and S do not overlap.

[**Target function ϕ_T**]

1. If P satisfies Ψ:
   (a) If $(0,0) \in T$, $\phi_T(P)$ is the middle point on the line segment connecting $(0,0)$ and $g(T)$.
   (b) If $(0,0) \in S$, $\phi_T(P) = g(P)$.
2. If P does not satisfy Ψ: $\phi_T(P) = g(P)$.

[**Target function ϕ_S**]

1. If P satisfies Ψ:
   (a) If $(0,0) \in S$, $\phi_S(P)$ is the middle point on the line segment connecting $(0,0)$ and $g(S)$.
   (b) If $(0,0) \in T$, $\phi_S(P) = g(P)$.
2. If P does not satisfy Ψ: $\phi_S(P) = g(P)$.

Recall that $g(P)$, $g(T)$, and $g(S)$ are the centers of gravity of P, T, and S, respectively, and that when a robot identifies P in Look phase, $(0,0)$ always in P, which corresponds to its current position.

Let us observe that $\alpha(\phi_T) = 1$. Since $\phi_T(P) \in CH(P)$ for all P, $\alpha(\phi_T) \leq 1$. To see that $\alpha(\phi_T) \geq 1$, consider any number $0 < a < 1$. It is easy to construct a P satisfying Ψ such that $\frac{dist((0,0),g(T))}{dist((0,0)),g(P))} < a$, which implies that $\alpha(\phi_T) > 1 - a$. Thus $\alpha(\phi_T) = 1$ by the definition of α. By the same argument, $\alpha(\phi_S) = 1$.

Theorem 3. *Both $\Phi_T = \{\phi_T\}$ and $\Phi_S = \{\phi_S\}$ are compatible with respect to the convergence problem, but $\Phi = \Phi_T \cup \Phi_S$ is not.*

4   Convergence When at Most One Robot Crashes

We investigate the fault-tolerant $(n, 1)$-convergence problem (FC(1)) and the fault-tolerant $(n, 1)$-convergence problem to a point (FC(1)-PO). There is an algorithm for FC(1) [8], but FC(1)-PO is not easier than FC(1). We have the following theorem, which implies the existence of an algorithm for FC(1)-PO.

Theorem 4. *Let Φ be a set of target functions such that $0 \leq \alpha(\Phi) < 1$. Then Φ is compatible with respect to FC(1)-PO.*

Corollary 1. *Let Φ be a set of target functions such that $0 \leq \alpha(\Phi) < 1$. Then Φ is compatible with respect to FC(1).*

Next we reconsider the target functions ϕ_T and ϕ_S.

Theorem 5. *Both $\Phi_T = \{\phi_T\}$ and $\Phi_S = \{\phi_S\}$ are compatible with respect to FC(1)-PO. However, $\Phi = \Phi_T \cup \Phi_S$ is not. Recall that $\alpha(\Phi_T) = \alpha(\Phi_S) = \alpha(\Phi) = 1$.*

5   FC(f) for $f \geq 2$

We go on the fault tolerant (n, f)-convergence problem (FC(f)) for $f \geq 2$. Since CoG is an algorithm for FC(f) [8], the next theorem holds.

Theorem 6 ([8]). *Suppose that $f \leq n - 1$. The set $\Phi_0 = \{\text{CoG}\}$ is compatible with respect to FC(f), or equivalently, a set Φ of target functions is compatible with respect to FC(f), if $\alpha(\Phi) = 0$.*

Corollary 1 states that every set Φ of target functions such that $0 \leq \alpha(\Phi) < 1$ is compatible with respect to FC(1). In contrast, for any $2 \leq f \leq n - 1$ and $0 < \alpha < 1$, there is a set Φ of two target functions such that (1) $\alpha(\Phi) = \alpha$, (2) each target function in Φ is compatible with respect to FC(f), but (3) Φ is not compatible with respect to FC(f). We use target functions $\xi_{(\alpha,n)}$ and $\xi'_{(\alpha,n)}$. Let $\ell = \lfloor \frac{n-2}{2} \rfloor$ and $\ell' = \lceil \frac{n-2}{2} \rceil$. Thus $\ell + \ell' = n - 2$. For a configuration P, define a condition Ψ^+ by a conjunction of two conditions (i) and (ii).

Ψ^+:

   (i) $P = \{p_1, p_2, \ldots, p_n\} \subseteq \overline{p_1 p_n}$, where $p_1, p_{\ell+1}, p_{\ell+2}, p_{\ell+3}$ are distinct and aligned on $\overline{p_1 p_n}$ in this order, $p_1 = p_2 = \cdots = p_\ell$, i.e., the multiplicity of p_1 is ℓ, $p_{\ell+3} = p_{\ell+4} = \cdots = p_n$, i.e., the multiplicity of $p_{\ell+3}$ is ℓ'.
   (ii) Let $L = dist(p_1, p_n)$. Then $dist(p_1, p_{\ell+1}) = \frac{1}{2}L$. If n is even, $dist(p_{\ell+1}, p_{\ell+2}) = \frac{\alpha n}{2(\alpha+n-1)}L$; otherwise, if it is odd, $dist(p_{\ell+1}, p_{\ell+2}) = \frac{(2\alpha-1)n+(1-\alpha)}{2(\alpha(n+1)-1)}L$.

[**Target function** $\xi_{(\alpha,n)}$]

1. If P satisfies Ψ^+, $\xi_{(\alpha,n)}(P)$ is
    (a) $\boldsymbol{p}_{\ell+2}$, if $\boldsymbol{p}_{\ell+1} = (0,0)$,
    (b) $(0,0)$, if $\boldsymbol{p}_{\ell+2} = (0,0)$, and
    (c) $g(P)$, otherwise.
2. If P does not satisfy Ψ^+, $\xi_{(\alpha,n)}(P) = g(P)$.

[**Target function** $\xi'_{(\alpha,n)}$]

1. If P satisfies Ψ^+, $\xi'_{(\alpha,n)}(P)$ is
    (a) $(0,0)$, if $\boldsymbol{p}_{\ell+1} = (0,0)$,
    (b) $\alpha\boldsymbol{p}_1 + (1-\alpha)g(P)$, if $\boldsymbol{p}_{\ell+2} = (0,0)$, and
    (c) $g(P)$, otherwise.
2. If P does not satisfy Ψ^+, $\xi'_{(\alpha,n)}(P) = g(P)$.

Theorem 7. *For any $2 \leq f \leq n-1$ and $0 < \alpha < 1$, (1) $\alpha(\xi_{(\alpha,n)}) = \alpha(\xi'_{(\alpha,n)}) = \alpha$, (2) both of $\Phi = \{\xi_{(\alpha,n)}\}$ and $\Phi' = \{\xi'_{(\alpha,n)}\}$ are compatible with respect to $FC(f)$, but (3) $\Phi \cup \Phi'$ is not.*

Before closing this section, we examine the case $\alpha = 1$.

Corollary 2. *For any $2 \leq f \leq n-1$, there are two target functions $\xi_{(1,n)}$ and $\xi'_{(1,n)}$ such that (1) $\alpha(\xi_{(1,n)}) = \alpha(\xi'_{(1,n)}) = 1$, (2) both of $\Phi = \{\xi_{(1,n)}\}$ and $\Phi' = \{\xi'_{(1,n)}\}$ are compatible with respect to $FC(f)$, but (3) $\Phi \cup \Phi'$ is not.*

6   FC(f)-CP for $f \geq 2$

We next investigate the fault tolerant (n,f)-convergence problem to a convex f-gon (FC(f)-CP). FC(1)-CP is FC(1)-PO. FC(f)-CP seems to be substantially easier than FC(f) (and FC(f)-PO), since the convergence of CH_t to a convex f-gon does not always mean the convergence of P_t. We have the following theorem.

Theorem 8. *Let Φ be any set of target functions such that $0 \leq \alpha(\Phi) < 1$. Then Φ is compatible with respect to $FC(f)$-CP for any $2 \leq f \leq n-1$.*

Let Φ and Φ' be any sets of target functions such that $\alpha(\Phi) < 1$ and $\alpha(\Phi') < 1$ hold. Then all of Φ, Φ', and $\Phi \cup \Phi'$ are compatible with respect to FC(f)-CP for all $2 \leq f \leq n-1$, since $\alpha(\Phi \cup \Phi') < 1$, by Theorem 8. However, we cannot extend this observation to include the case $\alpha = 1$. Consider the following two target functions τ and τ' for three robots.

[**Target function** τ]

1. If $P = \{\boldsymbol{p}_1, \boldsymbol{p}_2, \boldsymbol{p}_3\}$ is a triangle such that $\angle\boldsymbol{p}_1 < \angle\boldsymbol{p}_2 < \angle\boldsymbol{p}_3$, where $\angle\boldsymbol{p}_i$ is the angle of vertex $\boldsymbol{p}_i$ of the triangle, $\tau(P)$ is
    (a) $g(P)$ if $\boldsymbol{p}_1 = (0,0)$, and
    (b) $\boldsymbol{p}_1$, otherwise.

2. Otherwise, $\tau(P) = g(P)$.

[**Target function** τ']

1. If $P = \{p_1, p_2, p_3\}$ is a triangle such that $\angle p_1 < \angle p_2 < \angle p_3$, where $\angle p_i$ is the angle of vertex p_i of the triangle, then $\tau'(P) = p_1$.
2. Otherwise, $\tau'(P) = g(P)$.

Let $\Phi = \{\tau\}$ and $\Phi' = \{\tau'\}$. Then $\alpha(\Phi) = \alpha(\Phi') = 1$. Sets Φ is compatible with respect to the fault tolerant $(3, 2)$-convergence problem to a line segment, but Φ is not.

7   FC(f)-PO for $f \geq 2$

This section investigates the fault tolerant (n, f)-convergence problem to f points (FC(f)-PO) for $f \geq 2$. At a glance, FC(f)-PO looks to have properties similar to FC(f), and readers might consider that the former would be easier than the latter, since in the former, the non-faulty robots are not requested to converge to a point. On the contrary, we shall see that FC(f)-PO is a formidable problem even if $f = 2$.

7.1   Compatibility

We show a difference between FC(f) and FC(f)-PO for $f \geq 2$.

Theorem 9. *Let $f \geq 2$. Any target function ϕ is not an algorithm for FC(f)-PO, if $0 \leq \alpha(\phi) < 1$, or equivalently, Φ is not compatible with respect to FC(f)-PO, if $0 \leq \alpha(\Phi) < 1$.*

Recall that $\Phi = \{\xi_{(\alpha,n)}\}$ (or $\Phi' = \{\xi'_{(\alpha,n)}\}$) is compatible with respect to FC(f) for all $2 \leq f \leq n - 1$ and $0 \leq \alpha < 1$ by Theorem 7. Since $\alpha(\Phi) = \alpha$, by Theorem 9, we have:

Corollary 3. *Neither Φ nor Φ' is compatible with respect to FC(f)-PO, for all $f \geq 2$ and $0 \leq \alpha < 1$.*

7.2   Algorithm for FC(2)-PO

In Sect. 7.1, we showed that, for any $f \geq 2$, there is no FC(f)-PO algorithm whose scale is less than 1. It is a clear difference between FC(f)-PO and FC(f), which is solved, e.g., by CoG$_\alpha$ for any $0 \leq \alpha < 1$. This section proposes an algorithm $\psi_{(n,2)}$ with $\alpha(\psi_{(n,2)}) = 1$ for FC(2)-PO. Unfortunately, proposing an algorithm for FC(f)-PO for $f \geq 3$ is left as a future work.

Algorithm $\psi_{(n,2)}$. We propose an algorithm $\psi_{(n,2)}$ to solve FC(2)-PO. Since the case $n = 3$ is easy, we assume $n \geq 4$. Algorithm $\psi_{(n,2)}$ calls another algorithm LN$_{(n,2)}$, which solves FC(2)-PO when an initial configuration P_0 is linear, i.e., when $CH(P_0)$ is a line segment.

To compare positions p and q, we frequently use a lexicographic order $>$. It has however a drawback for our purpose: Suppose that p (resp. q) in Z_0 is $p^{(i)}$ (resp. $q^{(i)}$) in Z_i. Then $p^{(i)} < q^{(i)}$ and $p^{(j)} > q^{(j)}$ can happen for some $i \neq j$. In $\psi_{(n,2)}$, we introduce an order $\succ$ that all robots can consistently compute.

Let $P = \{p_1, p_2, \ldots, p_n\}$ be a multiset of n points, and $\overline{P} = \{q_1, q_2, \ldots, q_m\}$ be the set of distinct points in P. The multiplicity of a point $q \in \overline{P}$ is denoted by $\mu_P(q)$. In the definition of $\succ_P$, it is convenient to treat $\mu_P(q)$ points q in P as a point q with a label $\mu_P(q)$. We thus identify P with a pair $(\overline{P}, \mu_P)$, where μ_P is a labeling function to associate label $\mu_P(q)$ with each point $q \in \overline{P}$. Let o_P be the center of the smallest enclosing circle C_P of P.

Let G_P be the rotation group $G_{\overline{P}}$ of $\overline{P}$ about o_P preserving μ_P. The order $|G_P|$ of G_P is denoted by k_P. Note that k_P does not depend on $\mu_P(o_P)$. It is similar to the symmetricity $\sigma(P)$ of P defined in [18], but $k_P \neq \sigma(P)$ in general. Let $\Gamma_P(q) \subseteq \overline{P}$ be the orbit of G_P through $q \in \overline{P}$. Then $|\Gamma_P(q)| = k_P$ for all $q \in \overline{P} \backslash \{o_P\}$, and $\mu_P(q') = \mu_P(q)$ for all $q' \in \Gamma_P(q)$. If $o_P \in \overline{P}$, $\Gamma_P(o_P) = \{o_P\}$. Let $\Gamma_P = \{\Gamma_P(q) : q \in \overline{P}\}$ be the set of all orbits. Then Γ_P is a partition of $\overline{P}$.

To define $\succ_P$, we need the concept of *view*. Define an x-y coordinate system Ξ_q for any point $q \in \overline{P} \backslash \{o_P\}$. The origin of Ξ_q is q, the unit distance is the radius of C_P, and the x-axis is taken so that it goes through o_P in its positive side. Finally, it is right-handed. Let γ_q be the coordinate transformation from Ξ_q to Z_0. Then the view $V_P(q)$ of q is defined to be $\gamma_q^{-1}(P)$. That is, $\gamma_q(V_P(q)) = P$, i.e., P in Z_0 is $V_P(q)$ in Ξ_q. By definition, $V_P(q) = V_P(q')$ if and only if $q' \in \Gamma_P(q)$. Let $View_P = \{V_P(q) : q \in \overline{P} \backslash \{o_P\}\}$.

To compare two views in $View_P$, we arbitrarily choose and fix a total order $\sqsupset$ on the set of multisets of n points. We define a total order $\succ_P$ on Γ_P as follows. For any two distinct orbits $\Gamma_P(q)$ and $\Gamma_P(q')$ in Γ_P, $\Gamma_P(q) \succ_P \Gamma_P(q')$, if one of the following conditions hold: (1) $\mu_P(q) > \mu_P(q')$, (2) $\mu_P(q) = \mu_P(q')$ and $dist(q, o_P) < dist(q', o_P)$, or (3) $\mu_P(q) = \mu_P(q'), dist(q, o_P) = dist(q', o_P)$, and $V_P(q) \sqsupset V_P(q')$. When $o_P \in \overline{P}$, we assume that $\Gamma_P(q) \succ_P \Gamma_P(o_P)$ for all $q \neq o_P$. Now $\succ_P$ is a total order on Γ_P. If $k_P = 1$, since $\Gamma_P(q) = \{q\}$ for all $q \in \overline{P}$, we regard $\succ_P$ as a total order on $\overline{P}$ (by identifying $\Gamma_P(q)$ with q).

We partition the set of all multisets $P = \{p_1, p_2, \ldots, p_n\}$ for all $n \geq 4$ into six types G, L, T, I, S, and Z. Let $m_P = |\overline{P}|$.

G(oal): $m_P \leq 2$.

L(ine): $CH(P)$ is a line segment.

T(riangle): $m_P = 3$ and $CH(P)$ is a triangle.

I(nside): $m_P = 4$, $CH(P)$ is a triangle, and $o_P \in P$.

S(ide): $m_P = 4$, $CH(P)$ is a triangle, and $M_P \in P$, where M_P is the middle point of a longest side of $CH(P)$.

Z: P does not belong to the above five types.

We define a target function $\psi_{(n,2)}$.

[Target function $\psi_{(n,2)}$]

1. When P is type Z:

(a) If $k_P \geq 2$, $\psi_{(n,2)}(P) = o_P$.

(b) If $k_P = 1$, $\psi_{(n,2)}(P) = a_P$, where $a_P \in \overline{P}$ is the largest point with respect to $\succ_P$, which is well-defined since $k_P = 1$.

2. If P is type L, invoke $LN(n,2)$.

3. When P is type T, let $\overline{P} = \{a, b, c\}$.

(a) If triangle abc is equilateral, $\psi_{(n,2)}(P) = o_P$.

(b) If triangle abc is not equilateral, $\psi_{(n,2)}(P) = M_P$, where M_P is the middle point of the longest side. If there are two longest sides, M_P is the middle point of the side next to the shortest side counter-clockwise.

4. If P is type I, $\psi_{(n,2)}(P) = o_P$.

5. If P is type S, $\psi_{(n,2)}(P) = M_P$ (which is defined in the definition of type S).

Algorithm $LN_{(n,2)}$. We present target function $LN_{(n,2)}$. Let $P = \{p_1, p_2, \ldots, p_n\} \in \mathcal{P}$ be a configuration of type L, which may be a configuration that a robot identifies in Look phase. We identify a point p_i in R^2 with a point in R: Since $(0,0) \in P$, we rotate P about $(0,0)$ counter-clockwise so that the resultant P becomes the multiset of points in the x-axis. Then we denote $(p, 0)$ by p. In what follows in this section, a configuration P is thus regarded as a multiset of n real numbers, including at least one 0. We assume $p_1 \leq p_2 \leq \cdots \leq p_n$. By $\overline{P} = \{b_1, b_2, \ldots, b_{m_P}\}$, we denote the set of distinct real numbers in P, where m_P is the size $|\overline{P}|$ of $\overline{P}$, and $b_1 < b_2 < \cdots < b_{m_P}$. The length of $CH(P)$ is denoted by $L_P = b_{m_P} - b_1 = p_n - p_1$. Let $\lambda_P = \max_{p \in P} \min\{p - p_1, p_{m_P} - p\} \leq L_P/2$. Define j^* by $b_{j^*} = 0$. (Thus the current position of a robot r_i who identifies P in Look phase is b_{j^*} in Z_i.) Since P is type L, $k_P \leq 2$. We denote the middle point of x and y by M_{xy}, i.e., $M_{xy} = (x + y)/2$.

Like $\psi_{(n,2)}$, we consider 10 types, which we define as follows:

G: $m_P \leq 2$.

B$_3$: $m_P = 3$ and $k_P = 2$.

B$_4$: $m_P = 4$ and $k_P = 2$.

B$_5$: $m_P = 5$ and $k_P = 2$.

B$_6$: $m_P = 6$ and $k_P = 2$.

B: $m_P \geq 7$ and $k_P = 2$.

U$_3$: $m_P = 3$ and $k_P = 1$.

W: $m_P = 4$, $k_P = 1$, and $\overline{P} = \{b_1, b_2, b_3, b_4\}(b_1 < b_2 < b_3 < b_4)$ satisfies either (a) $2(b_2 - b_1) = b_3 - b_2$ and $b_3 \leq M_{b_1 b_4}$, or (b) $2(b_4 - b_3) = b_3 - b_2$ and $b_2 \geq M_{b_1 b_4}$.

U$_4$: $m_P = 4$, $k_P = 1$, and P is not type W.

U: $m_P \geq 5$ and $k_P = 1$.

We now give the target function $LN_{(n,2)}$.

[Target function $LN_{(n,2)}$]

1. If P is type G, $LN_{(n,2)}(P) = 0$.

2. When P is type B: If $j^* \leq \lceil m_P/2 \rceil$, $LN_{(n,2)}(P) = b_1$. Otherwise if $j^* > \lceil m_P/2 \rceil$, $LN_{(n,2)}(P) = b_{m_P}$.

3. When P is type B_3: $m_P = 3$. If $j^* \leq 2$, $\mathrm{LN}_{(n,2)}(P) = M_{b_1 b_2}$. Otherwise if $j^* = 3$, $\mathrm{LN}_{(n,2)}(P) = M_{b_2 b_3}$.

4. When P is type B_4: $m_P = 4$. If $j^* \leq 2$, $\mathrm{LN}_{(n,2)}(P) = M_{b_1 b_2}$. Otherwise if $j^* \geq 3$, $\mathrm{LN}_{(n,2)}(P) = M_{b_3 b_4}$.

5. When P is type B_5: $m_P = 5$. If $j^* \leq 3$, $\mathrm{LN}_{(n,2)}(P) = b_2$. Otherwise if $j^* \geq 4$, $\mathrm{LN}_{(n,2)}(P) = b_4$.

6. When P is type B_6: $m_P = 6$. If $j^* \leq 3$, $\mathrm{LN}_{(n,2)}(P) = b_2$. Otherwise if $j^* \geq 4$, $\mathrm{LN}_{(n,2)}(P) = b_5$.

7. When P is type U: Since $k_P = 1$, either $b_1 \succ_P b_{m_P}$ or $b_{m_P} \succ_P b_1$ holds. If $b_1 \succ_P b_{m_P}$, then $\mathrm{LN}_{(n,2)}(P) = b_1$. Otherwise if $b_{m_P} \succ_P b_1$, $\mathrm{LN}_{(n,2)}(P) = b_{m_P}$.

8. When P is type U_3: Since $k_P = 1$ and $m_P = 3$, if $b_2 = M_{b_1 b_3}$, then $\mu_P(b_1) \neq \mu_P(b_3)$. If $b_2 < M_{b_1 b_3}$ or $(b_2 = M_{b_1 b_3}) \wedge (\mu_P(b_1) > \mu_P(b_3))$, then $\mathrm{LN}_{(n,2)}(P) = (2b_1 + b_2)/3$. Otherwise, if $b_2 > M_{b_1 b_3}$ or $(b_2 = M_{b_1 b_3}) \wedge (\mu_P(b_1) < \mu_P(b_3))$, then $\mathrm{LN}_{(n,2)}(P) = (b_2 + 2b_3)/3$.

9. When P is type W: $k_P = 1$, $m_P = 4$, and P satisfies either condition (a) or (b) (of the definition of type W).
   (a) If $2(b_2 - b_1) = b_3 - b_2$ and $b_3 \leq M_{b_1 b_4}$, then $\mathrm{LN}_{(n,2)}(P) = b_2$.
   (b) If $2(b_4 - b_3) = b_3 - b_2$ and $b_2 \geq M_{b_1 b_4}$, then $\mathrm{LN}_{(n,2)}(P) = b_3$.

10. When P is type U_4: $k_P = 1$, $m_P = 4$, and P is not type W. Suppose that $\mu_P(b_1) \geq \mu_P(b_4)$ holds. (The case P satisfies $\mu_P(b_1) < \mu_P(b_4)$ is symmetric, and we omit it.)
   (a) If $\mu_P(b_1) \geq \mu_P(b_3)$, then $\mathrm{LN}_{(n,2)}(P) = b_1$.
   (b) If $(\mu_P(b_1) < \mu_P(b_3)) \wedge (\mu_P(b_3) \geq 3)$, $\mathrm{LN}_{(n,2)}(P) = b_1$, if $b_3 = 0$, and $\mathrm{LN}_{(n,2)}(P) = 0$, otherwise if $b_3 \neq 0$.
   (c) Otherwise if $(\mu_P(b_1) < \mu_P(b_3)) \wedge (\mu_P(b_3) < 3)$, $\mu_P(b_1) = \mu_P(b_4) = 1$ and $\mu_P(b_3) = 2$. $\mathrm{LN}_{(n,2)}(P) = b_1$, if $(b_2 = 0) \vee (b_3 = 0)$, and $\mathrm{LN}_{(n,2)}(P) = 0$, otherwise if $(b_1 = 0) \vee (b_4 = 0)$.

We have the following theorem:

Theorem 10. *Target function $\psi_{(n,2)}$, which satisfies $\alpha(\psi_{(n,2)}) = 1$, is an algorithm for FC(2)-PO.*

8 Gathering Problem

We finally investigate the gathering problem, provided that there are no faulty robots, to emphasize that the gathering and the convergence problems have completely different properties from the viewpoint of compatibility. Since the gathering problem is not solvable if $n = 2$ [18], we assume $n \geq 3$ in this section. Moreover, we assume that the robots initially occupy distinct points. There are many gathering algorithms. The following algorithm GAT [18] is one of them.

[Target function GAT]

1. If there is a unique $p \in P$ such that $\mu_P(p) > 1$, $\mathrm{GAT}(P) = p$.
2. Otherwise, if $\mu_P(p) = 1$ for all $p \in P$:

(a) If $k_P = 1$, $\mathrm{GAT}(P) = p$, where p is the largest point in P with respect to $\succ_P$.

(b) If $k_P > 1$, $\mathrm{GAT}(P) = o_P$.

Observe that $\alpha(\mathrm{GAT}) = 1$. We can modify Step 2(a) of GAT to obtain another algorithm GAT'. For example, $\mathrm{GAT}'(P)$ can be the smallest point p' in P with respect to $\succ_P$, instead of p. Then indeed GAT' is also a gathering algorithm with $\alpha(\mathrm{GAT}') = 1$, but obviously $\varPhi = \{\mathrm{GAT}, \mathrm{GAT}'\}$ is not compatible with respect to the gathering problem. Let us summarize.

Theorem 11 [18]. *Let $\varPhi = \{\mathrm{GAT}\}$ and $\varPhi' = \{\mathrm{GAT}'\}$. Then $\varPhi$ and $\varPhi'$ are compatible with respect to the gathering problem, but $\varPhi \cup \varPhi'$ is not. Here $\alpha(\varPhi) = \alpha(\varPhi') = \alpha(\varPhi \cup \varPhi') = 1$.*

Theorem 12. *Any target function ϕ is not a gathering algorithm if $\alpha(\phi) < 1$, or equivalently, any set $\varPhi$ of target functions such that $\alpha(\varPhi) < 1$ is not compatible with respect to the gathering problem.*

9   Conclusions

We introduced the concept of compatibility and investigated the compatibilities of several convergence problems. A compatible set $\varPhi$ of target functions with respect to a problem $\varPi$ is an extension of an algorithm ϕ for $\varPi$, in the sense that every target function $\phi \in \varPhi$ is an algorithm for $\varPhi$.

The problems we investigated are the convergence problem, the fault tolerant (n, f)-convergence problem $(\mathrm{FC}(f))$, the fault tolerant (n, f)-convergence problem to a convex f-gon $(\mathrm{FC}(f)\text{-}\mathrm{CP})$, and the fault tolerant (n, f)-convergence problem to f points $(\mathrm{FC}(f)\text{-}\mathrm{PO})$, for crash faults. The gathering problem was also investigated. The results are summarized in Table 1. Main observations we would like to emphasize are:

1. The convergence, $\mathrm{FC}(1)$, $\mathrm{FC}(1)\text{-}\mathrm{PO}$, and $\mathrm{FC}(f)\text{-}\mathrm{CP}$ share the same property: Every set $\varPhi$ of target functions is compatible, if $0 \le \alpha(\varPhi) < 1$.
2. The gathering problem and $\mathrm{FC}(f)\text{-}\mathrm{PO}$ for $f \ge 2$ share the same property: Any set $\varPhi$ of target functions is **not** compatible, if $0 \le \alpha(\varPhi) < 1$.
3. $\mathrm{FC}(f)$ $(f \ge 2)$ is in between $\mathrm{FC}(f)\text{-}\mathrm{CP}$ and $\mathrm{FC}(f)\text{-}\mathrm{PO}$.

Before closing the paper, we list some open problems:

1. Extend Table 1 to contain the results for $\alpha(\varPhi) > 1$.
2. Suppose that ϕ and ϕ' are algorithms for the convergence problem. Find a necessary and/or a sufficient condition for $\varPhi = \{\phi, \phi'\}$ to be compatible with respect to the convergence problem.
3. Investigate the compatibility of $\mathrm{FC}(f)\text{-}\mathrm{PO}$ for $f \ge 2$ under the $\mathcal{FSYNC}$ model.
4. Investigate the compatibility of convergence problems under the $\mathcal{ASYNC}$ model.

5. Investigate the compatibility of convergence problems in the presence of Byzantine failures.
6. Investigate the compatibility of fault tolerant gathering problems.
7. Find interesting problems with a large compatible set.

Acknowledgments. This work is supported in part by JSPS KAKENHI Grant Numbers JP17K00024 and JP22K11915.

References

1. Agmon, N., Peleg, D.: Fault-tolerant gathering algorithms for autonomous mobile robots. In: Proceedings of the 15th Annual ACM-SIAM Symposium on Discrete Algorithms, pp. 1063–1071 (2004)
2. Ando, H., Oasa, Y., Suzuki, I., Yamashita, M.: A distributed memoryless point convergence algorithm for mobile robots with limited visibility. IEEE Trans. Robot. Autom. **15**, 818–828 (1999)
3. Asahiro, Y., Suzuki, I., Yamashita, M.: Monotonic self-stabilization and its application to robust and adaptive pattern formation. Theor. Comput. Sci. **934**, 21–46 (2022)
4. Asahiro, Y., Yamashita, M.: Compatibility of convergence algorithms for autonomous mobile robots. arXiv:2301.10949 (2023)
5. Bouzid, Z., Das, S., Tixeuil, S.: Gathering of mobile robots tolerating multiple crash faults. In: Proceedings of the IEEE 33rd International Conference on Distributed Computing Systems, pp. 337–346 (2013)
6. Buchin, K., Flocchini, P., Kostitsyana, I., Peters, T., Santoro, N., Wada, K.: On the computational power of energy-constrained mobile robots: algorithms and cross-model analysis. In: Parter, M. (ed.) SIROCCO 2022. LNCS, vol. 13298, pp. 42–61. Springer, Cham (2022). https://doi.org/10.1007/978-3-031-09993-9_3
7. Cieliebak, M., Flocchini, P., Prencipe, G., Santoro, N.: Distributed computing by mobile robots: gathering. SIAM J. Comput. **41**, 829–879 (2012)
8. Cohen, R., Peleg, D.: Convergence properties of the gravitational algorithm in asynchronous robot systems. SIAM J. Comput. **34**, 1516–1528 (2005)
9. Cohen, R., Peleg, D.: Convergence of autonomous mobile robots with inaccurate sensors and movements. SIAM J. Comput. **38**, 276–302 (2008)
10. Cord-Landwehr, A., et al.: A new approach for analyzing convergence algorithms for mobile robots. In: Aceto, L., Henzinger, M., Sgall, J. (eds.) ICALP 2011. LNCS, vol. 6756, pp. 650–661. Springer, Heidelberg (2011). https://doi.org/10.1007/978-3-642-22012-8_52
11. Das, S., Flocchini, P., Santoro, N., Yamashita, M.: Forming sequences of geometric patterns with oblivious mobile robots. Distrib. Comput. **28**, 131–145 (2015). https://doi.org/10.1007/s00446-014-0220-9
12. Défago, X., Potop-Butucaru, M., Tixeuil, S.: Fault-tolerant mobile robots. In: Flocchini, P., Prencipe, G., Santoro, N. (eds.) Distributed Computing by Mobile Entities. LNCS, vol. 11340, pp. 234–251. Springer, Cham (2019). https://doi.org/10.1007/978-3-030-11072-7_10
13. Flocchini, P.: Gathering. In: Flocchini, P., Prencipe, G., Santoro, N. (eds.) Distributed Computing by Mobile Entities. LNCS, vol. 11340, pp. 63–82. Springer, Cham (2019). https://doi.org/10.1007/978-3-030-11072-7_4

14. Flocchini, P., Prencipe, G., Santoro, N.: Distributed computing by oblivious mobile robots. In: Synthesis Lectures on Distributed Computing Theory 10. Morgan & Claypool Publishers (2012)
15. Izumi, T., et al.: The gathering problem for two oblivious robots with unreliable compasses. SIAM J. Comput. **41**, 26–46 (2012)
16. Katreniak, B.: Convergence with limited visibility by asynchronous mobile robots. In: Kosowski, A., Yamashita, M. (eds.) SIROCCO 2011. LNCS, vol. 6796, pp. 125–137. Springer, Heidelberg (2011). https://doi.org/10.1007/978-3-642-22212-2_12
17. Prencipe, G.: Pattern formation. In: Flocchini, P., Prencipe, G., Santoro, N. (eds.) Distributed Computing by Mobile Entities. LNTCS, vol. 11340, pp. 37–62. Springer, Cham (2019). https://doi.org/10.1007/978-3-030-11072-7_3
18. Suzuki, I., Yamashita, M.: Distributed anonymous mobile robots - formation and agreement problems. SIAM J. Comput. **28**, 1347–1363 (1999)
19. Yamashita, M., Suzuki, I.: Characterizing geometric patterns formable by oblivious anonymous mobile robots. Theor. Comput. Sci. **411**, 2433–2453 (2010)
20. Yamauchi, Y., Uehara, T., Kijima, S., Yamashita, M.: Plane formation by synchronous mobile robots in the three-dimensional Euclidean space. J. ACM **64**, 1–43 (2017)

FNF-BFT: A BFT Protocol with Provable Performance Under Attack

Zeta Avarikioti[1]([✉]), Lioba Heimbach[2], Roland Schmid[2], Laurent Vanbever[2], Roger Wattenhofer[2], and Patrick Wintermeyer[2]

[1] TU Wien, Vienna, Austria
georgia.avarikioti@tuwien.ac.at
[2] ETH Zürich, Zürich, Switzerland
roschmi@ethz.ch

Abstract. We introduce FNF-BFT, the first partially synchronous BFT protocol with performance guarantees under truly byzantine attacks during stable networking conditions. At its core, FNF-BFT parallelizes the execution of requests by allowing all replicas to act as leaders independently. Leader parallelization distributes the load over all replicas. Consequently, FNF-BFT fully utilizes all correct replicas' processing power and increases throughput by overcoming the single-leader bottleneck.

We prove lower bounds on FNF-BFT's efficiency and performance in synchrony: the amortized communication complexity is linear in the number of replicas and thus competitive with state-of-the-art protocols; FNF-BFT's amortized throughput with less than $\frac{1}{3}$ byzantine replicas is at least $\frac{16}{27}$th of its best-case throughput. We also provide a proof-of-concept implementation and preliminary evaluation of FNF-BFT.

Keywords: BFT · SMR · parallel leaders · byzantine-resilient performance

1 Introduction

Byzantine fault tolerance has been the gold standard for making distributed systems more robust. Instead of modeling every single failure scenario, the byzantine failure model considers *arbitrarily malicious* actors that may infiltrate the system, thus covering many unforeseeable failure scenarios. This failure model has been broadly applied to *state machine replication (SMR)*. In SMR, a set of distributed replicas aims to agree on a unique ordering of client requests, even though a subset of the replicas, the byzantine failures, tries to disrupt the protocol. Therefore, the primary objectives of a protocol are the system's correctness (*safety*) and continuous progress (*liveness*). SMR protocols that offer these guarantees, i.e., are resilient against byzantine failures while continuing system operation, are known as *byzantine fault-tolerant (BFT)* protocols.

The first practical BFT system, PBFT [9], was introduced more than two decades ago and has inspired numerous other BFT systems, e.g., [18,22,38]. However, even today, BFT protocols do not scale well with an increasing number of replicas, making large-scale deployment of BFT systems a challenge.

© The Author(s), under exclusive license to Springer Nature Switzerland AG 2023
S. Rajsbaum et al. (Eds.): SIROCCO 2023, LNCS 13892, pp. 165–198, 2023.
https://doi.org/10.1007/978-3-031-32733-9_9

Often, the origin of this issue stems from the *single-leader bottleneck*: most BFT protocols rest the responsibility of uniquely ordering client requests on a single leader instead of distributing this task amongst the replicas [35]. More recently, protocols tackling the single-leader bottleneck through *parallelization* emerged, demonstrating a staggering performance increase over state-of-the-art sequential-leader protocols [11,19,20,26,34–36]. Similar to most of their single-leader counterparts, these works only consider non-malicious faults for their performance analysis. However, malicious attacks may lead to significant performance losses that are not evaluated. While these systems exhibit promising performance with simple faults, they *fail to provide lower bounds on their performance under attack.*

Our Contribution. In this work, we introduce the first parallel-leader BFT protocol *with a provable performance guarantee under truly byzantine attack in stable network conditions*, which we term FAST'N'FAIR-BFT (FNF-BFT). To formally capture this performance guarantee, we define the *byzantine-resilient performance* of a BFT protocol as the ratio between its worst-case and its best-case throughput, i.e., the effective utilization. For FNF-BFT, we bound this ratio to be constant, meaning that the throughput of our protocol under byzantine faults is lower-bounded by a constant fraction of its best-case throughput where no faults are present. Concretely, we show that FNF-BFT achieves byzantine-resilient performance with a ratio of 16/27 while maintaining safety and liveness.

In short, FNF-BFT is the first BFT protocol that *provably* achieves all the following properties in the partially synchronous communication model, i.e., where after some unknown *global stabilization time (GST)*, messages are delivered within a known bound Δ.

- **Optimistic Performance:** After GST, the best-case throughput is $\Omega(n)$ times higher than the throughput of sequential-leader protocols.
- **Byzantine-Resilient Performance:** After GST, the worst-case throughput of the system is at least a constant fraction of its best-case throughput.
- **Efficiency:** After GST, the amortized authenticator complexity of reaching consensus is $\Theta(n)$.

FNF-BFT achieves these properties by combining two key insights: First, by enabling all replicas to continuously act as leaders in parallel – sharing the load of clients' requests. Second, FNF-BFT does not replace leaders upon failure but configures each leader's load based on the leader's past performance. With this combination, we guarantee a *fair* distribution of requests according to each replica's capacity, which in turn results in *fast* processing of requests.

The rest of this paper is structured as follows: First, we present our formal model, an overview of the protocol, and define the protocol goals (Sect. 2). We then explain the design of FNF-BFT (Sect. 3), and analyze its security and performance formally (Sect. 4). We conclude with a related work section (Sect. 5). Proofs omitted in Sect. 4 can be found in Appendix A, where we present the complete analysis of FNF-BFT. A description and preliminary evaluation of our proof-of-concept implementation of FNF-BFT based on HotStuff [38] can be found in Appendix B.

2   FnF-BFT Overview

Model. The system consists of $n = 3f + 1$ authenticated replicas and a set of clients. We index replicas by $i \in [n] = \{1, 2, \ldots, n\}$. Throughout an execution, at most f unique replicas in the system are *byzantine*, i.e., they are controlled by an adversary with full information on their internal state. All other replicas are *correct*, i.e., following the protocol. The adversary cannot intercept the communication between two correct replicas. Any number of clients may be byzantine.

Communication Model: We assume a *partially synchronous communication model*, i.e., a known bound Δ on message transmission will hold between any two correct replicas after some unknown global stabilization time (GST). We show that FnF-BFT is safe in asynchrony, that is, when messages between correct replicas may arrive in arbitrary order after any finite delay. We evaluate all other properties of the system after GST, thus assuming a synchronous network.

Cryptographic Primitives: We make the usual cryptographic assumptions: the adversary is computationally bounded, and cryptographically-secure communication channels, computationally secure hash functions, (threshold) signatures, and encryption schemes exist. Similar to other BFT algorithms [5,18,38], FnF-BFT makes use of threshold signatures. In an (l, n) threshold signature scheme, there is a single public key held by all replicas and clients. Additionally, each replica u holds a distinct private key allowing the generation of a partial signature $\sigma_u(m)$ of any message m. Any set of l distinct partial signatures for the same message, $\{\sigma_u(m) \mid u \in U, |U| = k\}$ can be combined (by any replica) into a unique signature $\sigma(m)$. The combined signature can be verified using the public key. We assume that the scheme is *robust*, i.e., any verifier can easily filter out invalid signatures from malicious replicas. In this work, we set $l = 2f + 1$.

Authenticator Complexity: Message complexity has long been considered the main throughput-limiting factor in BFT protocols [18,38]. In practice, however, the throughput of a BFT protocol is limited by both its computational footprint (mainly caused by cryptographic operations), as well as its message complexity. Hence, to assess the performance and efficiency of FnF-BFT, we adopt a complexity measure called authenticator complexity [38]. An *authenticator* is any (partial) signature. We define the *authenticator complexity* of a protocol as the average number of all computations or verifications of any authenticator by replicas during the protocol execution per request. Note that the authenticator complexity also captures the message complexity of a protocol if, like in FnF-BFT, each message can be assumed to contain at least one signature. Unlike [38], where only the number of received signatures is considered, our definition allows to capture the load handled by each individual replica more accurately. Note that authenticator complexities according to the two definitions only differ by a constant factor. We only analyze the authenticator complexity after GST, as it is impossible for a BFT protocol to ensure deterministic progress and safety at the same time in an asynchronous network [15].

Protocol Overview. The FNF-BFT protocol implements a state machine (cf. Sect. 2) that is replicated across all replicas in the system. Clients broadcast requests to the system. Given client requests, replicas decide on the order of request executions and deliver commit-certificates to the clients.

Our protocol moves forward in *epochs*. In an epoch, each replica u is responsible for ordering a set of up to C_u client requests that are independent of all requests ordered by other replicas in the epoch. Every replica in the system simultaneously acts as both a leader and a backup to the other leaders. The number of assigned client requests C_u is based on u's past performance as a leader. The client space is rotated between replicas between epochs to guarantee liveness. More precisely, during the epoch-change, a designated replica acting as primary: (a) ensures that all replicas have a consistent view of the past leader and primary performance, (b) deduces non-overlapping sequence numbers for each leader, and (c) assigns parts of the client space to leaders.

An epoch-change occurs when requested by more than two-thirds of replicas. Replicas requesting an epoch-change immediately stop participating in the previous epoch. The primary in charge of the epoch-change is selected through periodic rotation based on performance history. Replicas request an epoch-change if: (a) all replicas have exhausted their requests, (b) their local epoch timeout is exceeded, (c) not enough progress by other leaders is observed, or (d) enough other replicas request an epoch-change. Hence, epochs have bounded-length.

Protocol Goals. FNF-BFT achieves scalable and byzantine fault-tolerant *state machine replication (SMR)*. In SMR, a group of replicas decide on a growing log of client requests. Clients are provided with cryptographically secure certificates which prove the commitment of their request. The protocol ensures:

1. **Safety:** If any two correct replicas commit a request with the same sequence number, they both commit the same request.
2. **Liveness:** If a correct client broadcasts a request, then every correct replica eventually commits the request.

Thus, FNF-BFT will eventually make progress, and valid client requests cannot be censored. Additionally, FNF-BFT guarantees low overhead in reaching consensus. Unlike other protocols limiting the worst-case efficiency for a single request, we analyze the amortized authenticator complexity per request after GST. We find this to be the relevant throughput-limiting factor:

3. **Efficiency:** After GST, the amortized authenticator complexity of reaching consensus is $\Theta(n)$.

Furthermore, FNF-BFT achieves competitive performance under both optimistic and pessimistic adversarial scenarios:

4. **Optimistic Performance:** After GST, the best-case throughput is $\Omega(n)$ times higher than the throughput of sequential-leader protocols.
5. **Byzantine-Resilient Performance:** After GST, the worst-case throughput of the system is at least a constant fraction of its best-case throughput.

Hence, unlike many other BFT systems, FNF-BFT *guarantees that byzantine replicas cannot arbitrarily slow down the system when the network is stable.*

3   FNF-BFT Architecture

FNF-BFT executes client requests on a state machine replicated across n replicas. We advance FNF-BFT in epochs – identified by monotonically increasing *epoch numbers*. Replicas in the system act as leaders and backups concurrently. As a leader, a replica is responsible for ordering client requests within its jurisdiction. Each leader v is assigned a predetermined number of requests C_v to execute during an epoch. To deliver a client request, v starts by picking the next available sequence number and shares the request with the backups. Leader v must collect $2f + 1$ signatures from replicas in the leader prepare and commit phase (Algorithm 1) to commit the request. We employ threshold signatures for the signature collection – allowing us to achieve linear authenticator complexity for reaching consensus on a request. Additionally, we use low and high watermarks for each leader to represent a range of request sequence numbers that each leader can propose concurrently to boost individual leaders' throughput.

Each epoch has a unique primary responsible for the preceding epoch-change, i.e., moving the system into the epoch. The primary changes every epoch and its selection is based on the system's history. A replica calls for an epoch-change in any of the following cases: (a) the replica has locally committed requests for all sequence numbers available in the epoch, (b) the maximum epoch time expired, (c) the replica has not seen sufficient progress, or (d) the replica has observed at least $f + 1$ epoch-change messages from other replicas.

FNF-BFT generalizes PBFT [9] and Mir-BFT [35] to the n leader setting. Additionally, we avoid PBFT's expensive all-to-all communication during epoch operation similar to Linear-PBFT [18]. Throughout this section, we discuss the various components of the protocol in further detail.

3.1   Client

Each client has a unique identifier. A client c requests the execution of an operation r by sending a $\langle request, r, t, c \rangle$ to all leaders. Here, timestamp t is a monotonically increasing sequence number used to order the requests from one client. By using watermarks, we allow clients to have more than one request in flight. Client watermarks, low and high, represent the range of timestamp sequence numbers which the client can propose concurrently. Thus, we require t to be within the low and high watermarks of client c. The client watermarks are advanced similarly to the leader watermarks (cf. Sect. 3.6). Upon executing operation r, replica u responds to the client with $\langle reply, e, d, u \rangle$, where e is the epoch number and d is the request digest (cf. Sect. 3.5)[1]. The client waits for $f + 1$ such responses from the replicas.

---

[1] Instead of committing client request independently, the protocol could be adapted to process client requests in batches – a standard BFT protocol improvement [22,35,38].

Algorithm 1. Committing a request proposed by leader v

1: *Leader prepare phase*
2:    **as** replica u:
3:       **upon** receiving a valid $\langle request, r, t, c \rangle$ from client c:
4:          map client request to hash bucket
5:    **as** leader v:
6:       accept $\langle request, r, t, c \rangle$ assigned to one of v's buckets
7:       pick next assigned sequence number sn
8:       broadcast $\langle pre\text{-}prepare, sn, e, h(r), v \rangle$
9: *Backup prepare phase*
10:    **as** backup w:
11:       accept $\langle pre\text{-}prepare, sn, e, h(r), v \rangle$
12:       **if** the pre-prepare message is valid:
13:          compute partial signature $\sigma_w(d)$, where $d = h(sn\|e\|r)$
14:          send $\langle prepare, sn, e, \sigma_w(d) \rangle$ to leader v
15:    **as** leader v:
16:       compute partial signature $\sigma_v(d)$
17:       **upon** receiving $2f$ prepare messages:
18:          compute $(2f + 1, n)$ threshold signature $\sigma(d)$
19:          broadcast $\langle prepared\text{-}certificate, sn, e, \sigma(d) \rangle$
20: *Commit phase*
21:    **as** backup w:
22:       accept $\langle prepared\text{-}certificate, sn, e, \sigma(d) \rangle$
23:       compute partial signature $\sigma(\sigma_w(d))$
24:       $\langle commit, sn, e, \sigma_w(\sigma(d)) \rangle$ to leader v
25:    **as** leader v:
26:       compute partial signature $\sigma(\sigma_v(d))$
27:       **upon** receiving $2f$ commit messages:
28:          compute $(2f + 1, n)$ threshold signature $\sigma(\sigma(d))$
29:          broadcast $\langle commit\text{-}certificate, sn, e, \sigma(\sigma(d)) \rangle$

3.2 Sequence Number Distribution

We distribute sequence numbers to leaders for the succeeding epoch during the epoch-change. While we commit requests from each leader in order, the requests from different leaders are committed independently of each other in our protocol. Doing so allows leaders to continue making progress in an epoch, even though other leaders might have stopped working. Otherwise, a natural attack for byzantine leaders is to stop working and force the system to an epoch-change. Such attacks are possible in other parallel-leader protocols such as Mir-BFT [35].

To allow leaders to commit requests independently of each other, we need to allocate sequence numbers to all leaders during the epoch-change. Thus, we must also determine the number of requests each leader is responsible for before the epoch. The number of requests for leader v in epoch e is denoted by $C_v(e)$. It can be computed deterministically by all replicas in the network, based on the known history of the system (cf. Sect. 3.7).

When assigning sequence numbers, we first automatically yield to each leader $v \in [n]$ the sequence numbers of the $O_v(e)$ existing hanging requests from previous epochs in the assigned bucket(s). The remaining $C_v(e) - O_v(e)$ sequence numbers for each leader are distributed to them one after the other according to their ordering from the set of available sequence numbers. Note that $O_v(e)$ cannot exceed $C_v(e)$. For each leader v the assigned sequence numbers are mapped to local sequence numbers $1_{v,e}, 2_{v,e}, \ldots, C_v(e)_{v,e}$ in epoch e. These sequence numbers are later used to simplify checkpoint creation (cf. Sect. 3.6).

3.3 Hash Space Division

The client hash space is partitioned into buckets to avoid duplication. Each bucket is assigned to a single leader every epoch. We consider the client identifier to be the request input and hash the client identifier ($h_c = h(c)$) to map requests into buckets. The hash space partition ensures that no two conflicting requests will be assigned to different leaders[2].

Thus, the requests served by different leaders are independent of each other. Additionally, the bucket assignment is rotated round-robin across epochs, preventing request censoring. The hash space is portioned into $m \cdot n$ non-intersecting buckets of equal size, where $m \in \mathbb{Z}^+$ is a configuration parameter. Each leader v is then assigned $m_v(e)$ buckets in epoch e according to their load $C_v(e)$ (cf. Sect. 3.7). Leaders can only include requests from their active buckets.

When assigning buckets to leaders, the protocol ensures that every leader is assigned at least one bucket, as well as distributing the buckets according to the load handled by the leaders. Precisely, the number of buckets leader v is assigned in epoch e is given by $m_v(e) = \left\lfloor \frac{C_v(e)}{\sum_{u \in [n]} C_u(e)} (m - 1) \cdot n \right\rfloor + 1 + \tilde{m}_v(e)$, where $\tilde{m}_v(e) \in \{0, 1\}$ distributes the remaining buckets to the leaders – ensuring $\sum_{u \in [n]} m_u(e) = m \cdot n$. The remaining buckets are allocated to leaders v with the biggest value: $\left\lfloor \frac{C_v(e)}{\sum_{u \in [n]} C_u(e)} (m - 1) \cdot n \right\rfloor + 1 - \frac{C_v(e)}{\sum_{u \in [n]} C_u(e)} \cdot m \cdot n$.

Note that the system will require a sufficiently long stability period for all correct leaders to be working at their capacity limit, i.e., $C_v(e)$ matching the performance of leader v in epoch e. Once correct leaders operate at capacity, the number of buckets they serve matches that. The hash buckets are distributed to leaders through a deterministic rotation such that each leader repeatedly serves each bucket under $f + 1$ unique primaries, i.e., preventing byzantine replicas from censoring specific hash buckets. For the remaining paper, we assume that there are always client requests pending in each bucket. Since we aim to optimize throughput, this assumption is in-sync with our protocol goals.

---

[2] Note that in case the requests are transactions with multiple inputs, the hash space division is more challenging to circumvent double-spending attacks. In such cases, we can employ well-known techniques [21,39] with no performance overhead as long as the average number of transactions' inputs remains constant [7].

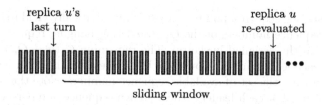

Fig. 1. FNF-BFT primary rotation in a system with $n = 10$ replicas. In blue, we show epochs led by primaries elected based on their performance. Epochs shown in yellow are led by replicas re-evaluated once their last turn as primary falls out of the sliding window. (Color figure online)

3.4 Primary Rotation

While all replicas are tasked with being a leader at all times, only a single replica, the primary, initiates an epoch. FNF-BFT assigns primaries periodically, exploiting the performance of good primaries and being reactive to network changes. The primary rotation consists of two core building blocks. First, FNF-BFT repeatedly rotates through the $2f + 1$ best primaries – exploiting their performance. Second, FNF-BFT explores every primary at least once within a sliding window. The sliding window consists of $g \in \mathbb{Z}$ epochs, and we set $g \geq 3f + 1$ to allow the exploration of all primaries throughout a sliding window. We depict a sample rotation in Fig. 1.

Throughout the protocol, all replicas record the performance of each primary. We measure performance as the number of requests successfully committed under a primary in an epoch. Performance can thus be determined during the succeeding epoch-change by each replica (cf. Sect. 3.7). To deliver a reactive system, we update a replica's primary performance after each turn.

We rotate through the best $2f + 1$ primaries repeatedly. After every $2f + 1$ primaries, the best $2f + 1$ primaries are redetermined and subsequently elected as primary in order of the time passed since their last turn as primary. The primary that has not been seen for the longest time is elected first. Cycling through the best primaries maximizes system performance. Simultaneously, basing performance solely on a replica's preceding primary performance strips byzantine primaries from the ability to misuse a good reputation. Every so often, we interrupt the continuous exploitation of the best $2f + 1$ primaries to revisit replicas that fall out of the sliding window. If replica u's last turn as primary occurred in epoch $e - g$ by the time epoch e rolls around, replica u would be re-explored as primary in epoch e. The exploration allows us to re-evaluate all replicas as primaries periodically and ensures that FNF-BFT is reactive to network changes.

The protocol starts by exploring all primaries ordered by their identifiers. Note that only one primary can fall out of the sliding window at any time after the first exploration. Thus, we always know which primary will be re-evaluated.

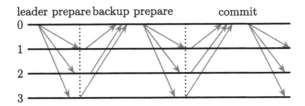

leader prepare backup prepare   commit

Fig. 2. Schematic message flow for one request.

3.5   Epoch Operation

To execute requests, we use a leader-based adaption of PBFT, similar to Linear-PBFT [18]. Threshold signatures are commonly used to reduce the complexity of the backup prepare and commit phases of PBFT. The leader of a request is used as a collector of partial signatures to create a $(2f + 1, n)$ threshold signature in the intermediate stages of the backup prepare and commit phases. We visualize the schematic of the message flow for one request led by replica 0 in Fig. 2 and summarize the protocol executed locally by replicas in Algorithm 1.

Leader Prepare Phase. Upon receiving $\langle request, r, t, c \rangle$ from a client, each replica computes the hash of the client identifier c. If the request falls into one of leader v's active buckets, v verifies $\langle request, r, t, c \rangle$. The request is discarded if either it has already been prepared or it is already pending. Once verified, leader v broadcasts $\langle pre\text{-}prepare, sn, e, h(r), v \rangle$, where sn is the sequence number, e the current epoch, $h(r)$ is the hash digest of request r and v represents the leader's signature. The cryptographic hash function h maps an arbitrary-length input to a fixed-length output. We can use the digest $h(r)$ as a unique identifier for a request r, as we assume the hash function to be collision-resistant.

Backup Prepare Phase. Backup w accepts $\langle pre\text{-}prepare, sn, e, h(r), v \rangle$ from leader v, if (a) the epoch number matches its local epoch number, (b) w has not prepared another request with the same sequence number sn in epoch e, (c) leader v leads sequence number sn, (d) sn lies between the low and high watermarks of leader v, (e) r is in the active bucket of v, and (f) r was submitted by an authorized client. Upon accepting $\langle pre\text{-}prepare, sn, e, h(r), v \rangle$, w computes $d = h(sn\|e\|r)$ where h is a hash function. Additionally, w signs d by computing a verifiable partial signature $\sigma_w(d)$. Then w sends $\langle prepare, sn, e, \sigma_w(d) \rangle$ to leader v. Upon receiving $2f$ prepare messages for sn in epoch e, leader v forms a combined signature $\sigma(d)$ from the $2f$ prepare messages and its own signature. Leader v then broadcasts $\langle prepared\text{-}certificate, sn, e, \sigma(d) \rangle$ to all backups.

Commit Phase. Backup w accepts the *prepared-certificate* and replies with $\langle commit, sn, e, \sigma_w(\sigma(d)) \rangle$ to leader v. After collecting $2f$ commit messages, v creates a combined signature $\sigma(\sigma(d))$ using the signatures from the collected commit messages and its own signature. Once the combined signature is prepared,

v continues by broadcasting $\langle$commit-certificate, $sn, e, \sigma(\sigma(d))\rangle$. Upon receiving the *commit-certificate*, replicas execute r after delivering all preceding requests led by v, and send replies to the client.

3.6   Checkpointing

Similar to PBFT [9], we periodically create checkpoints to prove the correctness of the current state. Instead of requiring a costly round of all-to-all communication to create a checkpoint, we add an intermediate phase and let the respective leader collect partial signatures to generate a certificate optimistically. Additionally, we expand the PBFT checkpoint protocol to run for n parallel leaders.

For each leader v, we repeatedly create checkpoints to clear the logs and advance the watermarks of leader v whenever the local sequence number $sn_{v,e,k}$ is divisible by a constant $k \in \mathbb{Z}^+$. Recall that when a replica u delivers a request for leader v with local sequence number $sn_{v,e,k}$, this implies that all requests led by v with local sequence number lower than $sn_{v,e,k}$ have been locally committed at replica u. Hence, after delivering the request with local sequence number $sn_{v,e,k}$, replica u sends $\langle$checkpoint, $sn_{v,e,k}, h(sn'_{v,e,k}), u\rangle$ to leader v. Here, $sn'_{v,e,k}$ is the last checkpoint and $h(sn'_{v,e,k})$ is the hash digest of the requests with sequence number sn_v in the range $sn'_{v,e,k} \leq sn_v \leq sn_{v,e,k}$. Leader v proceeds by collecting $2f + 1$ checkpoint messages (including its own) and generates a *checkpoint-certificate* by creating a combined threshold signature. Then, leader v sends the checkpoint-certificate to all other replicas. If a replica sees the checkpoint-certificate, the checkpoint is *stable* and the replica can discard the corresponding messages from its logs, i.e., for sequence numbers belonging to leader v lower than $sn_{v,e,k}$.

We use checkpointing to advance low and high watermarks. In doing so, we allow several requests from a leader to be in flight. The low watermark L_v for leader v is equal to the sequence number of the last stable checkpoint, and the high watermark is $H_v = L_v + 2k$. We set k to be large enough such that replicas do not stall. Given its watermarks, leader v can only propose requests with a local sequence number between low and high watermarks.

Calling Epoch-Change. Replicas call an epoch-change by broadcasting an epoch-change message in four cases:

1. Replica u triggers an epoch-change in epoch e, once it has committed everyone's assigned requests locally.
2. Replica u calls for an epoch-change when its *epoch timer* expires. The value of the epoch timer T is set to ensure that after GST, correct replicas can finish at least $C_{\min}$ requests during an epoch. $C_{\min} \in \Omega(n^2)$ is the minimum number of requests assigned to leaders.
3. Replicas call epoch-changes upon observing inadequate progress. Each replica u has individual *no-progress timers* for all leaders. The no-progress timer is initialized with the same value T_p for all leaders. Initially, replicas set all no-progress timers for the first time after 5Δ in the epoch – accounting for the

message transmission time of the initial requests. A replica resets the timer for leader v every time it receives a commit-certificate from v. In case the replica has already committed C_v requests for leader v, the timer is no longer reset. Upon observing no progress timeouts for $b \in [f + 1, 2f + 1]$ different leaders, a replica calls an epoch-change. Requiring at least $f + 1$ leaders to make progress ensures that a constant fraction of leaders makes progress, and at least one correct leader is involved. On the other hand, we demand no more than $2f + 1$ leaders to make progress such that byzantine leaders failing to execute requests cannot stop the epoch early. We let $b = 2f + 1$ and set the no-progress timer such that it does not expire for correct leaders and simultaneously ensures sufficient progress, i.e., $T_p \in \Theta(T/C_{\min})$.

4. Finally, replica u calls an epoch-change if it sees that $f + 1$ other replicas have called an epoch-change for an epoch higher than e. Replica u picks the smallest epoch in the set such that byzantine replicas cannot advance the protocol an arbitrary number of epochs.

After sending an epoch-change message, the replica will only start its epoch-change timer, upon seeing at least $2f + 1$ epoch-change messages. We will discuss the epoch-change timer in more detail later.

3.7   Epoch-Change

At high level, in FnF-BFT's epoch-change protocol, we modify the PBFT view-change protocol as follows: we use threshold signatures to reduce the message complexity and extend the view-change message to include information about all leaders. Similar to Mir-BFT [35], we introduce a round of reliable broadcast to share information needed to determine the configuration of the next epoch(s). We determine the load assigned to each leader in the next epoch, based on their past performance, and also record the performance of the preceding primary.

Starting Epoch-Change (Algorithm 2, Steps 1–5). To move the system to epoch $e + 1$, replica u sends $\langle$epoch-change, $e + 1, \mathcal{S}, \mathcal{C}, \mathcal{P}, \mathcal{Q}, u\rangle$ to all replicas in the system. Here, $\mathcal{S}$ is a vector of sequence numbers sn_v of the last stable checkpoints S_v $\forall v \in [n]$ known to u for each leader v. $\mathcal{C}$ is a set of checkpoint-certificates proving the correctness of S_v $\forall v \in [n]$, while $\mathcal{P}$ contains sets $\mathcal{P}_v$ $\forall v \in [n]$. For each leader v, $\mathcal{P}_v$ contains a prepared-certificate for each request r that was prepared at u with sequence number higher than sn_v, if replica v does not possess a commit-certificate for r. Similarly, $\mathcal{Q}$ contains sets $\mathcal{Q}_v$ $\forall v \in [n]$. $\mathcal{Q}_v$ consists of a commit-certificate for each request r that was prepared at u with sequence number higher than sn_v.

Reliable Broadcast (Algorithm 2, Steps 6–11). The primary of epoch $e + 1$ (p_{e+1}) waits for $2f$ epoch-change messages for epoch e. Upon receiving a sufficient number of messages, the primary performs a classical 3-phase reliable broadcast. During the broadcast, the primary informs leaders on the number of requests assigned to each leader in the next epoch and the identifiers of the replicas which send epoch-change messages. The number of requests assigned to a

Algorithm 2. Epoch-change protocol for epoch $e + 1$

1: *Starting epoch-change*
2:    **as** replica u:
3:       broadcast $\langle$epoch-change, $e + 1, \mathcal{S}, \mathcal{C}, \mathcal{P}, \mathcal{Q}, u\rangle$
4:       **upon** receiving $2f$ epoch-change messages for $e + 1$:
5:          start epoch-change timer T_e
6: *Reliable broadcast*
7:    **as** primary p_{e+1}:
8:       compute $C_v(e + 1)$ (Algorithm 3) for all leaders $v \in [n]$
9:       perform 3-phase reliable broadcast sharing configuration details of epoch $e + 1$ and the performance of primary p_e
10:   **as** replica u:
11:      participate in reliable broadcast initiates by p_{e+1}
12: *Starting epoch*
13:   **as** primary p_{e+1}:
14:      broadcast $\langle$new-epoch, $e + 1, \mathcal{V}, \mathcal{O}, p_{e+1}\rangle$
15:      enter epoch $e + 1$
16:   **as** replica u:
17:      accept $\langle$new-epoch, $e + 1, \mathcal{V}, \mathcal{O}, p_{e+1}\rangle$
18:      enter epoch $e + 1$

---

leader is computed deterministically (Algorithm 3). Through the reliable broadcast, we ensure that the primary cannot share conflicting information regarding the sequence number assignment and, in turn, the next epoch's sequence number distribution. In addition to sharing information about the epoch configuration, the primary also broadcasts the total number of requests committed during the previous epoch. This information is used by the network to evaluate primary performance and determine epoch primaries.

Starting Epoch (Algorithm 2, Steps 12–18). The primary p_{e+1} multicasts $\langle$new-epoch, $e + 1, \mathcal{V}, \mathcal{O}, p_{e+1}\rangle$. Here, the set $\mathcal{V}$ contains sets $\mathcal{V}_u$, which carry the valid epoch-change messages of each replica u of epoch e received by the primary of epoch $e + 1$, plus the epoch-change message the primary of epoch $e + 1$ would have sent. $\mathcal{O}$ consists of sets $\mathcal{O}_v \; \forall v \in [n]$ containing pre-prepare messages and commit-certificates.

$\mathcal{O}_v$ is computed as follows. First, the primary determines the sequence number $S_{\min}(v)$ of the latest stable checkpoint in $\mathcal{V}$ and the highest sequence number $S_{\max}(v)$ in a prepare message in $\mathcal{V}$. For each sequence number sn_v between $S_{\min}(v)$ and $S_{\max}(v)$ of all leaders $v \in [n]$ there are three cases: (a) there is at least one set in $\mathcal{Q}_v$ of some epoch-change message in $\mathcal{V}$ with sequence number sn_v, (b) there is at least one set in $\mathcal{P}_v$ of some epoch-change message in $\mathcal{V}$ with sequence number sn_v and none in $\mathcal{Q}_v$, or (c) there is no such set. In the first case, the primary simply prepares a commit-certificate it received for sn_v. In the second case, the primary creates a new message $\langle$pre-prepare, $sn_v, e + 1, d, p_{e+1}\rangle$, where d is the request digest in the pre-prepare message for sequence number sn_v with the highest epoch number in $\mathcal{V}$. In the third case, the primary creates a

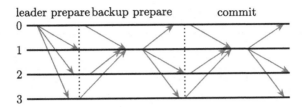

Fig. 3. Schematic of message flow for hanging requests. In this example, the primary is replica 0, and the request falls into the bucket of replica 1.

new pre-prepare message $\langle \text{pre-prepare}, sn_v, e + 1, d^{null}, p_{e+1} \rangle$, where d^{null} is the digest of a special null request; a null request goes through the protocol like other requests, but its execution is a no-op. If there is a gap between $S_{\max}(v)$ and the last sequence number assigned to leader v in epoch e, these sequence numbers will be newly assigned in the next epoch.

Next, the primary appends the messages in $\mathcal{O}$ to its log. If $S_{\min}(v)$ is greater than the sequence number of its latest stable checkpoint, the primary also inserts the proof of stability (the checkpoint with sequence number $S_{\min}(v)$) in its log. Then it enters epoch $e + 1$; at this point, it can accept messages for epoch $e + 1$.

A replica accepts a new-epoch message for epoch $e + 1$ if: (a) it is signed properly, (b) the epoch-change messages it contains are valid for epoch $e + 1$, (c) the information in $\mathcal{V}$ matches the new request assignment, and (d) the set $\mathcal{O}$ is correct. The replica verifies the correctness of $\mathcal{O}$ by performing a computation similar to the one previously used by the primary. Then, the replica adds the new information contained in $\mathcal{O}$ to its log and decides all requests for which a commit-certificate was sent. Replicas rerun the protocol for messages with sequence numbers between $S_{\min}(v)$ and $S_{\max}(v)$ without a commit-certificate. They do not execute client requests again (they use their stored information about the last reply sent to each client instead). As request messages and stable checkpoints are not included in new-epoch messages, a replica might not have some of them available. In this case, the replica can easily obtain the missing information from other replicas in the system.

Hanging Requests. While the primary sends out the pre-prepare message for all hanging requests, replicas in whose buckets the requests fall, are responsible for computing prepared- and commit-certificates of the individual requests. In the example shown in Fig. 3, the primary of epoch $e + 1$, replica 0, sends a pre-prepare message for a request in a bucket of replica 1, contained in the new-epoch message, to everyone. Replica 1 is then responsible for prepared- and commit-certificates, as well as collecting the corresponding partial signatures.

The number of request $C_v(e + 1)$ assigned to leader v in epoch $e + 1$ is determined deterministically based on its past performance (Algorithm 3). By $c_v(e)$ we denote the number of requests committed under leader v in epoch e. Each leader is re-evaluated during the epoch-change. If a leader successfully committed all assigned requests in the preceding epoch, we double the number of requests this leader is given in the following epoch. Else, it is assigned the maximum number of requests it committed within the last $f + 1$ epochs.

Algorithm 3. Configuration adjustment

1: initially $C_v(1) = C_{\min}$ for all replicas v
2: **if** $c_v(e) < C_v(e)$:
3:     $C_v(e + 1) = \max \left(C_{\min}, \max_{i \in \{0,\dots,f\}} c_v(e - i) \right)$
4: **else:**
5:     $C_v(e + 1) = 2 \cdot c_v(e)$

Epoch-Change Timer. A replica sets an epoch-change timer T_e upon entering the epoch-change for epoch $e + 1$. By default, we configure the timer T_e such that a correct primary can successfully finish the epoch-change after GST. If the timer expires without seeing a valid new-epoch message, the replica requests an epoch-change for epoch $e + 2$. If a replica has experienced at least f unsuccessful consecutive epoch-changes previously, the replica doubles the timer's value. It continues to do so until it sees a valid new-epoch message. We only start doubling the timer after f unsuccessful consecutive epoch-changes to avoid having f byzantine primaries in a row, i.e., the maximum number of subsequent byzantine primaries possible, purposely increasing the timer value exponentially and, in turn, decreasing the system throughput significantly. As soon as replicas witness a successful epoch-change, they reduce T_e to its default again.

Assignment of Requests. Finally, the number of requests assigned to each leader is updated during the epoch-change. We limit the number of requests that can be processed by each leader per epoch to assign the sequence numbers ahead of time and allow leaders to work independently of each other.

We assign sequence numbers to leaders according to their abilities. As soon as we see a leader outperforming their workload, we double the number of requests they are assigned in the following epoch. Additionally, leaders operating below their expected capabilities are allocated requests according to the highest potential demonstrated in the past $f + 1$ rounds. By looking at the previous $f + 1$ epochs, we ensure that there is at least one epoch with a correct primary in the leader set. In this epoch, the leader had the chance to display its capabilities. Thus, basing a leader's performance on the last $f + 1$ rounds allows us to see its ability independent of the possible influence of byzantine primaries.

4   Analysis

We show that FNF-BFT satisfies the properties specified in Sect. 2. A detailed analysis can be found in Appendix A.

Safety. We prove FNF-BFT is safe under asynchrony. FNF-BFT generalizes Linear-PBFT [18], an adaptation of PBFT [9] that reduces its authenticator complexity during epoch operation. We thus rely on similar arguments to prove FNF-BFT's safety in Theorem 1.

Liveness. We show FNF-BFT makes progress after GST (Theorem 2). FNF-BFT's epoch-change uses the following techniques to ensure that correct replicas become synchronized (Definition 1) after GST: (1) A replica in epoch e observing epoch-change messages from $f + 1$ other replicas calling for any epoch(s) greater than e issues an epoch-change message for the smallest such epoch e'. (2) A replica only starts its epoch-change timer for epoch e' after receiving $2f$ other epoch-change messages for epoch e', thus ensuring that at least $f + 1$ correct replicas have broadcasted an epoch-change message for epoch e'. Hence, all correct replicas start their epoch-change timer for an epoch e' within at most 2 message delay. After GST, this amounts to at most 2Δ. (3) Byzantine replicas are unable to impede progress by calling frequent epoch-changes, as an epoch-change will only happen if at least $f + 1$ replicas call it. A byzantine primary can hinder the epoch-change from being successful. However, there can only be f byzantine primaries in a row.

Efficiency. To demonstrate FNF-BFT's efficiency, we analyze the authenticator complexity for reaching consensus during an epoch. Like Linear-PBFT [18], using each leader as a collector for partial signatures in the backup prepare and commit phase, allows FNF-BFT to achieve linear complexity during epoch operation. We continue by calculating the authenticator complexity of an epoch-change. Intuitively speaking, we reduce PBFT's view-change complexity from $\Theta(n^3)$ to $\Theta(n^2)$ by employing threshold signatures. However, as FNF-BFT allows for n simultaneous leaders, we obtain an authenticator complexity of $\Theta(n^3)$ as a consequence of sharing the same information for n leaders during the epoch-change. Finally, we argue that after GST, there is sufficient progress by correct replicas to compensate for the high epoch-change cost (Theorem 3).

Optimistic Performance. We assess FNF-BFT's optimistic performance, i.e., we theoretically evaluate its best-case throughput, assuming all replicas are correct and the network is synchronous. We further assume that the best-case throughput is limited by the available computing power of each replica – mainly required for the computation and verification of cryptographic signatures – and that the available computing power of each correct replica is the same. In this model, we demonstrate that FNF-BFT achieves higher throughput than sequential-leader protocols by the means of leader parallelization. To show the speed-up gained through parallelization, we first analyze the *optimistic epoch throughput* of FNF-BFT, i.e., the throughput of the system during stable networking conditions in the best-case scenario with $3f + 1$ correct replicas (Lemma 6). Later, we consider the repeated epoch changes and show that

FnF-BFT's throughput is dominated by its authenticator complexity during the epochs. To that end, observe that for $C_{min} \in \Omega(n^2)$, every epoch will incur an authenticator complexity of $\Omega(n^3)$ per replica and thus require $\Omega(n^3)$ time units. We show that after GST, an epoch-change under a correct primary requires $\Theta(n^2)$ time units (Lemma 7). We conclude our analysis by quantifying FnF-BFT's overall best-case throughput. Specifically, we prove that the speed-up gained by moving from a sequential-leader protocol to a parallel-leader protocol is proportional to the number of leaders (Theorem 4).

Byzantine-Resilient Performance. While many BFT protocols present practical evaluations of their performance that ignore byzantine adversarial behavior [9,18,35,38], we provide a novel, theoretical byzantine-resilience guarantee. We first analyze the impact of byzantine replicas in an epoch under a correct primary. We consider the replicas' roles as backups and leaders separately. On the one hand, for a byzantine leader, the optimal strategy is to leave as many requests hanging, while not making any progress (Lemma 9). On the other hand, as a backup, the optimal byzantine strategy is not helping other leaders to make progress (Lemma 10). In conclusion, we observe that byzantine replicas have little opportunity to reduce the throughput in epochs under a correct primary. Specifically, we show that after GST, the effective utilization under a correct primary is at least $\frac{8}{9}$ for $n \to \infty$ (Theorem 5).

Next, we discuss the potential strategies of a byzantine primary trying to stall the system. We first show that under a byzantine primary, an epoch is either aborted quickly or $\Omega(n^3)$ new requests become committed (Lemma 11). Then, we prove that rotating primaries across epochs based on primary performance history reduces the control of the byzantine adversary on the system. In particular, byzantine primaries only have one turn as primary throughout any sliding window in a stable network. Combining all the above, we conclude that FnF-BFT's byzantine-resilient utilization is asymptotically $\frac{8}{9} \cdot \frac{g-f}{g} > \frac{16}{27}$ for $n \to \infty$ (Theorem 7), where g is the fraction of byzantine primaries in the system's stable state, while simultaneously dictates how long it takes to get there after GST.

5   Related Work

Lamport et al. [23] first discussed the problem of reaching consensus in the presence of byzantine failures. Following its introduction, byzantine fault tolerance was initially studied in the synchronous network setting [12,13,30]. Dwork et al. [14] proposed the concept of partial synchrony and demonstrated the feasibility of reaching consensus in partially synchronous networks. Subsequently, Reiter [32,33] introduced Rampart, an early protocol tackling byzantine fault tolerance for state machine replication in asynchrony. Then, with PBFT, Castro and Liskov [9] devised the first efficient protocol for state machine replication that tolerates byzantine failures. The leader-based protocol requires $\mathcal{O}(n^2)$ communication to reach consensus, as well as $\mathcal{O}(n^3)$ for leader replacement. While widely deployed, PBFT does not scale well when the number of replicas increases.

Kotla et al. [22] were the first to achieve $\mathcal{O}(n)$ complexity with Zyzzyva. The complexity of leader replacement in Zyzzyva remains $\mathcal{O}(n^3)$, and safety violations were later exposed [2]. Later, SBFT [18], improved the complexity of exchanging leaders to $\mathcal{O}(n^2)$. While reducing the overall complexity, both Zyzzyva and SBFT suffer from the single-leader bottleneck.

Developed by Yin et al. [38], leader-based HotStuff matches the $\mathcal{O}(n)$ complexity of Zyzzyva and SBFT. HotStuff rotates the leader with every request and is the first to achieve $\mathcal{O}(n)$ for leader replacement. However, HotStuff offers little parallelization due to its sequential proposal of requests, and experiments have revealed high complexity in practice [35]. Recently, Gelashvili et al. [16] improved on HotStuff's latency while adding an asynchronous fallback to enhance its performance during epoch synchronization. Although this work improves the overall performance of HotStuff, requests are still processed sequentially. In contrast, FNF-BFT enables n parallel leaders to propose requests simultaneously.

Parallel Leaders. Leveraging parallel leaders to overcome the single-leader bottleneck was initially introduced by Mao et al. [26, 28] with Mencius and BFT-Mencius. Mencius maps client requests to the closest leader, and in turn, requests can become censored. However, no de-duplication measures are in place to handle the re-submission of censored client requests. FNF-BFT addresses this problem by periodically rotating leaders over the client space.

Later, Stathakopoulou et al. [35] proposed Mir-BFT that significantly improved throughput compared to sequential-leader approaches. Mir runs instances of PBFT on a set of leaders, updating the leader set as soon as a single leader in the set stops making progress. Hence, we expect Mir's performance to drop significantly in the presence of byzantine replicas, as it allows byzantine leaders to repeatedly end epochs early. This is despite its high throughput demonstrated in the presence of faults. In a follow-up work, Stathakopoulou et al. [36] addressed Mir's temporary loss of throughput during epoch changes, but their protocol still offers no guarantees under attack, unlike FNF-BFT that maintains a constant fraction of its best-case throughput under byzantine attacks.

In parallel, Gupta et al. [19] proposed RCC protocol-agnostic approach to parallelize existing BFT protocols. While allowing multiple instances to each run an individual request, the protocol requires instances to unify after each request, creating a significant overhead. Further, RCC relies on failure detection, which is only possible in synchronous networks [24]. With FNF-BFT, we allow leaders to make progress independently of each other without relying on failure detection.

Another paradigm that has recently gained traction and enables replicas to operate in parallel to increase throughput is DAG-based consensus. The core idea is that the client requests are spread reliably as fast as the network permits and replicas accumulate them in a DAG. Subsequently, the replicas extract the total ordering of the accumulated requests from their local DAG without exchanging additional messages. DAG-based protocols, initially introduced as consensus systems for data-center replication, precede blockchains [3,8,27,29,31]. These initial DAG protocols are, however, very complex and have high latency. Lately, several DAG-based consensus protocols have been proposed this time in the context of blockchains. Some are purely DAG-based [1] while others employ the DAG structure as transportation means for unconfirmed requests, e.g., [11], and on top execute a randomized BFT protocol [11,17,20,25,34]. The state-of-the-art Bullshark [34] achieves (minimum) constant latency with linear communication complexity in the partially synchronous model, similarly to FNF-BFT. While all these works achieve staggering throughput, none of them provide any provable guarantees on their throughput under Byzantine attacks.

Byzantine Resilience. Byzantine resilience was initially explored by Clement et al. [10] with Aardvark. Aardvark is an adaptation of PBFT with frequent view-changes: a leader only stays in its position when displaying an increasing throughput level. This approach comes with significant performance cuts in networks without failures. Parallel leaders allow FNF-BFT to be byzantine-resilient without accepting significant performance losses in an ideal setting.

Prime, proposed by Amir et al. [4], aims to maximize performance in malicious environments. Besides adding delay constraints that further confine the partially synchronous network model, Prime restricts its evaluation to delay attacks, i.e., a byzantine leader adds as much delay to the protocol as possible. Similarly, Veronese et al. [37] only evaluated their proposed protocol, Spinning, in the presence of delay attacks – not fully capturing possible byzantine attacks. Consequently, the maximum performance degradation Spinning and Prime can incur under byzantine faults is at least 78% [6]. We analyze FNF-BFT theoretically to capture the entire spectrum of possible byzantine attacks.

Aublin et al. [6] further explored the performance of BFT protocols under byzantine attacks with RBFT. RBFT runs f backup instances on the same set of client requests as the master instance to discover whether the master instance is byzantine. Thus, RBFT incurs quadratic communication complexity for every request. To the contrary, FNF-BFT achieves a communication complexity of $\mathcal{O}(n)$ and further increases performance through parallelization – allowing byzantine-resilience without the added burden of detecting byzantine leaders.

Acknowledgments. The work was partially supported by the Austrian Science Fund (FWF) through the project CoRaF (grant agreement 2020388) and by the European Research Council (ERC) under the ERC Starting Grant (SyNET) 851809.

A   Analysis

We show that FNF-BFT satisfies the properties specified in Sect. 2. In particular, we prove the safety and liveness of FNF-BFT, argue that it is efficient, and evaluate its resilience to byzantine attacks in stable network conditions.

A.1 Safety

FNF-BFT generalizes Linear-PBFT [18], which is an adaptation of PBFT [9] that reduces its authenticator complexity during epoch operation. We thus rely on similar arguments to prove FNF-BFT's safety in Theorem 1.

Theorem 1. *If any two correct replicas commit a request with the same sequence number, they both commit the same request.*

Proof. We start by showing that if $\langle$prepared-certificate, $sn, e, \sigma(d)\rangle$ exists, then $\langle$prepared-certificate, $sn, e, \sigma(d')\rangle$ cannot exist for $d' \neq d$. Here, $d = h(sn\|e\|r)$ and $d' = h(sn\|e\|r')$. Further, we assume the probability of $r \neq r'$ and $d = d'$ to be negligible. The existence of $\langle$prepared-certificate, $sn, e, \sigma(d)\rangle$ implies that at least $f + 1$ correct replicas sent a pre-prepare message or a prepare message for the request r with digest d in epoch e with sequence number sn. For $\langle$prepared-certificate, $sn, e, \sigma(d')\rangle$ to exist, at least one of these correct replicas needs to have sent two conflicting prepare messages (pre-prepare messages in case it leads sn). This is a contradiction.

Through the epoch-change protocol we further ensure that correct replicas agree on the sequence of requests that are committed locally in different epochs. The existence of $\langle$prepared-certificate, $sn, e, \sigma(d)\rangle$ implies that there cannot exist $\langle$prepared-certificate, $sn, e', \sigma(d')\rangle$ for $d' \neq d$ and $e' > e$. Any correct replica only commits a request with sequence number sn in epoch e if it saw the corresponding commit-certificate. For a commit-certificate for request r with digest d and sequence number sn to exist a set R_1 of at least $f + 1$ correct replicas needs to have seen $\langle$prepared-certificate, $sn, e, \sigma(d)\rangle$. A correct replica will only accept a pre-prepare message for epoch $e' > e$ after having received a new-epoch message for epoch e'. Any correct new-epoch message for epoch $e' > e$ must contain epoch-change messages from a set R_2 of at least $f + 1$ correct replicas. As there are $2f + 1$ correct replicas, R_1 and R_2 intersect in at least one correct replica u. Replica u's epoch-change message ensures that information about request r being prepared in epoch e is propagated to subsequent epochs, unless sn is already included in the stable checkpoint of its leader. In case the prepared-certificate is propagated to the subsequent epoch, a commit-certificate will potentially be propagated as well. If the new-epoch message only includes the prepared-certificate for sn, the protocol is redone for request r with the same sequence number sn. In the two other cases, the replicas commit sn locally upon seeing the new-epoch message and a correct replica will never accept a request with sequence number sn again.   □

A.2 Liveness

One cannot guarantee safety and liveness for deterministic BFT protocols in asynchrony [15]. We will, therefore, show that FNF-BFT eventually makes progress after GST. In other words, we consider a stable network when discussing liveness. Furthermore, we assume that after an extended period without progress, the time required for local computation in an epoch-change is negligible. Thus, we focus on analyzing the network delays for liveness.

Definition 1. *Two replicas are called synchronized, if they start their epoch-change timer for an epoch e within at most 2Δ.*

Similar to PBFT [9], FNF-BFT's epoch-change employs three techniques to ensure that correct replicas become synchronized (Definition 1) after GST:

1. A replica in epoch e observing epoch-change messages from $f+1$ other replicas calling for any epoch(s) greater than e issues an epoch-change message for the smallest such epoch e'.
2. A replica only starts its epoch-change timer after receiving $2f$ other epoch-change messages, thus ensuring that at least $f+1$ correct replicas have broadcasted an epoch-change message for the epoch (or higher). Hence, all correct replicas start their epoch-change timer for an epoch e' within at most 2 message delay. After GST, this amounts to at most 2Δ.
3. Byzantine replicas are unable to impede progress by calling frequent epoch-changes, as an epoch-change will only happen if at least $f+1$ replicas call it. A byzantine primary can hinder the epoch-change from being successful. However, there can only be f byzantine primaries in a row.

Lemma 1. *After GST, correct replicas eventually become synchronized.*

Proof. Let u be the first correct replica to start its epoch-change timer for epoch e at time t_0. Following (2), this implies that u received at least $2f$ other epoch-change messages for epoch e (or higher). Of these $2f$ messages, at least f originate from other correct replicas. Thus, together with its own epoch-change message, at least $f+1$ correct replicas broadcasted epoch-change messages by time t_0. These $f+1$ epoch-change messages are seen by all correct replicas at the latest by time $t_0 + \Delta$. Thus, according to (1), at time $t_0 + \Delta$ all correct replicas broadcast an epoch-change message for epoch e. Consequently, at time $t_0 + 2\Delta$ all correct replicas have received at least $2f$ other epoch-change messages and will start the timer for epoch e.   □

Lemma 2. *After GST, all correct replicas will be in the same epoch long enough for a correct leader to make progress.*

Proof. From Lemma 1, we conclude that after GST, all correct replicas will eventually enter the same epoch if the epoch-change timer is sufficiently large. Once the correct replicas are synchronized in their epoch, the duration needed for a correct leader to commit a request is bounded. Note that all correct replicas will be in the same epoch for a sufficiently long time as the timers are configured accordingly. Additionally, byzantine replicas are unable to impede progress by calling frequent epoch-changes, according to (3).   □

Theorem 2. *If a correct client c broadcasts request r, then every correct replica eventually commits r.*

Proof. Following Lemmas 1 and 2, we know that all correct replicas will eventually be in the same epoch after GST. Hence, in any epoch with a correct primary, the system will make progress. Note that a correct client will not issue invalid requests. It remains to show that an epoch with a correct primary and a correct leader assigned to hash bucket $h(c)$ will occur. We note that this is given by the bucket rotation, which ensures that a correct leader repeatedly serves each bucket in a correct primary epoch.   □

A.3   Efficiency

To demonstrate that FnF-BFT is efficient, we first analyze the authenticator complexity for reaching consensus during an epoch. Like Linear-PBFT [18], using each leader as a collector for partial signatures in the backup prepare and commit phase allows FnF-BFT to achieve linear complexity during epoch operation.

Lemma 3. *The authenticator complexity for committing a request during an epoch is $\Theta(n)$.*

Proof. During the leader prepare phase, the authenticator complexity is at most n. The primary computes its signature to attach it to the pre-prepare message. This signature is verified by no more than $n - 1$ replicas.

Furthermore, the backup prepare and commit phase's authenticator complexity is less than $3n$ each. Initially, at most $n - 1$ backups, compute their partial signature and send it to the leader, who, in turn, verifies $2f$ of these signatures. The leader then computes its partial signature, as well as computing the combined signature. Upon receiving the combined signature, the $n - 1$ backups need to verify the signature.

Overall, the authenticator complexity committing a request during an epoch is thus at most $7n + o(n) \in \Theta(n)$.   □

Next, we analyze the authenticator complexity of an epoch-change. Intuitively speaking, we reduce PBFT's view-change complexity from $\Theta(n^3)$ to $\Theta(n^2)$ by employing threshold signatures. However, as FnF-BFT allows for n simultaneous leaders, we obtain an authenticator complexity of $\Theta(n^3)$ as a consequence of sharing the same information for n leaders during the epoch-change.

Lemma 4. *The authenticator complexity of an epoch-change is $\Theta(n^3)$.*

Proof. The epoch-change for epoch $e + 1$ is initiated by replicas sending epoch-change messages to the primary of epoch $e + 1$. Each epoch-change message holds n authenticators for each leader's last checkpoint-certificates. As there are at most $2k$ hanging requests per leader, further $\mathcal{O}(n)$ authenticators for prepared- and commit-certificates of the open requests per leader are included in the message. Additionally, the sending replica also includes its signature. Each replica newly computes its signature to sign the epoch-change message, the remaining authenticators are already available and do not need to be created by the replicas. Thus, a total of no more than n authenticators are computed for the epoch-change messages. Note that epoch-change messages contain $\Theta(n)$ authenticators. Thus, the number of authenticators received by each replica is $\Theta(n^2)$.

After the collection of $2f + 1$ epoch-change messages, the primary performs a classical 3-phase reliable broadcast. The primary broadcasts the same signed message to start the classical 3-phase reliable broadcast. While the primary computes 1 signature, at most $n - 1$ replicas verify this signature. In the two subsequent rounds of all-to-all communication, each participating replica computes 1 and verifies $2f$ signatures. Thereby, the authenticator complexity of each round of all-to-all communication is at most $(2f + 1) \cdot n$. Thus, the authenticator complexity of the 3-phase reliable broadcast is bounded by $(4f + 3) \cdot n \in \Theta(n^2)$.

After successfully performing the reliable broadcast, the primary sends out a new-epoch message to every replica in the network. The new-epoch message contains the epoch-change messages held by the primary and the required pre-prepare messages for open requests. There are $\mathcal{O}(n)$ such pre-prepare messages, all signed by the primary. Finally, each new-epoch message is signed by the primary. Thus, the authenticator complexity of the new-epoch message is $\Theta(n^2)$. However, suppose a replica has previously received and verified an epoch-change from replica u whose epoch-change message is included in the new-epoch message. In that case, the replica no longer has to check the authenticators in u's epoch-change message again. For the complexity analysis, it does not matter when the replicas verify the signature. We assume that all replicas verify the signatures contained in the epoch-change messages before receiving the new-epoch messages. Thus, the replicas only need to verify the $\mathcal{O}(n)$ new authenticators contained in the new-epoch message.

Overall, the authenticator complexity of an epoch-change is $\Theta(n^3)$.   □

Finally, we argue that after GST, there is sufficient progress by correct replicas to compensate for the high epoch-change cost.

Theorem 3. *After GST, the amortized authenticator complexity of committing a request is $\Theta(n)$.*

Proof. To find the amortized authenticator complexity of committing a request, we consider an epoch and the following epoch-change. After GST, the authenticator complexity of committing a request for a correct leader is $\Theta(n)$. The timeout value is set such that a correct worst-case leader creates at least $C_{\min}$ requests in each epoch initiated by a correct primary. Thus, there are $\Theta(n)$ correct replicas, each committing $C_{\min}$ requests. By setting $C_{\min} \in \Omega(n^2)$, we guarantee that at least $\Omega(n^3)$ requests are created during an epoch given a correct primary.

Byzantine primaries can ensure that no progress is made in epochs they initiate, simply by withholding the new-epoch messages. However, at most a constant fraction of epochs lies in the responsibility of byzantine primaries. We conclude that, on average, $\Omega(n^3)$ requests are created during an epoch.

Following Lemma 4, the authenticator complexity of an epoch-change is $\Theta(n^3)$. Note that the epoch-change timeout T_e is set so that correct primaries can successfully finish the epoch-change after GST. Not every epoch-change will be successful immediately, as byzantine primaries might cause unsuccessful epoch-changes. Specifically, byzantine primaries can purposefully summon an unsuccessful epoch-change to decrease efficiency.

In case of an unsuccessful epoch-change, replicas initiate another epoch-change – and continue doing so – until a successful epoch-change occurs. However, we only need to start $\mathcal{O}(1)$ epoch-changes on average to be successful after GST, as the primary rotation ensures that at least a constant fraction of primaries is correct. Hence, the average cost required to reach a successful epoch-change is $\Theta(n^3)$.

We find the amortized request creation cost by adding the request creation cost to the ratio between the cost of a successful epoch-change and the number of requests created in an epoch, that is, $\Theta(n) + \frac{\Theta(n^3)}{\Omega(n^3)} = \Theta(n)$.   □

A.4   Optimistic Performance

Throughout this section, we make the following optimistic assumptions: all replicas are considered correct, and the network is stable and synchronous. We employ this model to assess the optimistic performance of FNF-BFT, i.e., theoretically evaluating its best-case throughput. Note that this scenario is motivated by practical applications, as one would hope to have functioning hardware at hand, at least initially. Additionally, we assume that the best-case throughput is limited by the available computing power of each replica – predominantly required for the computation and verification of cryptographic signatures. We further assume that the available computing power of each correct replica is the same, which we believe is realistic as the same hardware will often be employed for each replica in practice. Without loss of generality, each leader can compute/verify one authenticator per time unit. As *throughput*, we define the number of requests committed by the system per time unit. Finally, we assume that replicas only verify the authenticators of relevant messages. For example, a leader receiving $3f$ prepare messages for a request will only verify $2f$ authenticators. Similarly, pre-prepare messages outside the leaders' watermarks will not be processed by backups. Note that we will carry all assumptions into the next section. There they will, however, only apply to correct replicas.

Sequential-Leader Protocols. We claim that FNF-BFT achieves higher throughput than sequential-leader protocols due to its leader parallelization. To support this claim, we compare FNF-BFT's throughput to that of a generic sequential-leader protocol. The generic sequential-leader protocol serves as an asymptotic characterization of several sequential-leader protocols, e.g., [9,18,38].

A *sequential-leader protocol* characteristically relies on a unique leader at any point in time (Fig. 4). During its reign, the leader is responsible for serving all client requests. The leader can be rotated repeatedly or only upon failure.

Lemma 5. *A sequential-leader protocol requires at least $\Omega(n)$ time units to process a client request.*

Proof. In sequential-leader protocols, a unique replica is responsible for serving all client requests at any point in time. This replica must verify $\Omega(n)$ signatures to commit a request while no other replica leads requests simultaneously. Thus, a sequential-leader protocol requires $\Omega(n)$ time units to process a request.   □

Fig. 4. Sequential leader example with four leaders taking turns in serving client requests. Leader changes are indicated by vertical lines.

FNF-BFT Epoch. With FNF-BFT, we propose a *parallel-leader protocol* that divides client requests into $m \cdot n$ independent hash buckets. Each hash bucket is assigned to a unique leader at any time (Fig. 5). The hash buckets are rotated between leaders across epochs to ensure liveness (cf. Sect. 3.3). Within an epoch, a leader is only responsible for committing client requests from its assigned hash bucket(s). Overall, this parallelization leads to a significant speed-up.

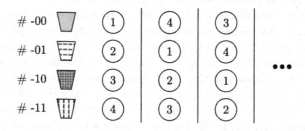

Fig. 5. Parallel leader example with four leaders and four hash buckets. In each epoch, leaders are only responsible for serving client requests in their hash bucket. Epoch-changes are indicated by vertical lines.

To show the speed-up gained through parallelization, we first analyze the *optimistic epoch throughput* of FNF-BFT, i.e., the throughput of the system during stable networking conditions in a best-case scenario with $3f + 1$ correct replicas. Furthermore, we assume the number of requests included in a checkpoint to be sufficiently large, such that no leader must ever stall when waiting for a checkpoint to be created. Finally, we analyze the effects of epoch-changes and compute the overall best-case throughput of FNF-BFT in the aforementioned optimistic setting.

Lemma 6. *After GST, the best-case epoch throughput with $3f + 1$ correct replicas is* $\dfrac{k \cdot (3f + 1)}{k \cdot (19f + 3) + (8f + 2)}$.

Proof. In the optimistic setting, all epochs are initiated by correct primaries, and thus all replicas will be synchronized after GST.

In FNF-BFT, n leaders work on client requests simultaneously. Similar to sequential-leader protocols, each leader needs to verify at least $\mathcal{O}(n)$ signatures to commit a request. A leader needs to compute 3 and verify $4f$ authenticators precisely to commit a request it proposes during epoch operation. Thus, leaders need to process a total of $4f+3 \in \Theta(n)$ signatures to commit a request. With the help of threshold signatures, backups involved in committing a request only need to compute 2 and verify 3 authenticators. We follow that a total of $4f+3+5\cdot3f = 19f + 3$ authenticators are computed/verified by a replica for one of its own requests and $3f$ requests of other leaders.

After GST, each correct leader v will quickly converge to a C_v such that it will make progress for the entire epoch-time, hence, working at its full potential. We achieve this by rapidly increasing the number of requests assigned to each leader outperforming its assignment and never decreasing the assignment below what the replica recently managed.

Checkpoints are created every k requests and add to the computational load. A leader verifies and computes a total of $2f + 2$ messages to create a checkpoint, and the backups are required to compute 1 partial signature and verify 1 threshold signature. The authenticator cost of creating $3f + 1$ checkpoints, one for each leader, is, therefore, $8f + 2$ per replica.

Thus, the best-case throughput of the system is

$$\frac{k \cdot (3f + 1)}{k \cdot (19f + 3) + (8f + 2)}.$$

$\square$

Note that it would have been sufficient to show that the epoch throughput is $\Omega(1)$ per time unit, but this more precise formula will be required in the next section. Additionally, we would like to point out that the choice of k does not influence the best-case throughput asymptotically.

FNF-BFT Epoch-Change. As FNF-BFT employs bounded-length epochs, repeated epoch-changes have to be considered. In the following, we will show that FNF-BFT's throughput is dominated by its authenticator complexity during the epochs. To that end, observe that for $C_{\min} \in \Omega(n^2)$, every epoch will incur an authenticator complexity of $\Omega(n^3)$ per replica and thus require $\Omega(n^3)$ time units.

Lemma 7. *After GST, an epoch-change under a correct primary requires $\Theta(n^2)$ time units.*

Proof. Following Lemma 4, the number of authenticators computed and verified by each replica for all epoch-change messages is $\Theta(n^2)$. Each replica also processes $\Theta(n)$ signatures during the reliable broadcast, and $\mathcal{O}(n)$ signatures for the new-epoch messages. Overall, each replica thus processes $\Theta(n^2)$ authenticators during the epoch-change. Subsequently, this implies that the epoch-change requires $\Theta(n^2)$ time units, as we require only a constant number of message delays to initiate and complete the epoch-change protocol. Recall that we assume the throughput to be limited by the available computing power of each replica.   □

Theoretically, one could set C_{min} even higher such that the time the system spends with epoch-changes becomes negligible. However, there is a trade-off for practical reasons: increasing C_{min} increases the minimal epoch-length, allowing a byzantine primary to slow down the system for a longer time. Note that the guarantee for byzantine-resilient performance (cf. Sect. A.5) would still hold.

FnF-BFT Optimistic Performance. Ultimately, it remains to quantify FnF-BFT's overall best-case throughput.

Lemma 8. *After GST, and assuming all replicas are correct, FnF-BFT requires $\mathcal{O}(n)$ time units to process n client requests on average.*

Proof. Under a correct primary, each correct leader will commit at least $C_{min} \in \Omega(n^2)$ requests after GST. Hence, FnF-BFT will spend at least $\Omega(n^3)$ time units in an epoch, while only requiring $\Theta(n^2)$ time units for an epoch-change (Lemma 7). Thus, following Lemma 6, FnF-BFT requires an average of $\mathcal{O}(n)$ time units to process n client requests.   □

Following Lemmas 5 and 8, the speed-up of a parallel-leader protocol over a sequential-leader protocol is proportional to the number of leaders.

Theorem 4. *If the throughput is limited by the (equally) available computing power at each replica, the speed-up for equally splitting requests between n parallel leaders over a sequential-leader protocol is at least $\Omega(n)$.*

A.5   Byzantine-Resilient Performance

While many BFT protocols present practical evaluations of their performance that neglect truly byzantine adversarial behavior [9,18,35,38], we provide a novel, theory-based byzantine-resilience guarantee. We first analyze the impact of byzantine replicas in an epoch under a correct primary. Next, we discuss the potential strategies of a byzantine primary trying to stall the system. And finally, we conflate our observations into a concise statement.

Correct Primary Throughput. To gain insight into the byzantine-resilient performance, we analyze the optimal byzantine strategy. In epochs led by correct primaries, we will consider their roles as backups and leaders separately. On the one hand, for a byzantine leader, the optimal strategy is to leave as many requests hanging, while not making any progress (Lemma 9).

Lemma 9. *After GST and under a correct primary, the optimal strategy for a byzantine leader is to leave 2k client requests hanging and commit no request.*

Proof. Correct replicas will be synchronized as a correct primary initiates the epoch. Thus, byzantine replicas' participation is not required to make progress.

Byzantine leaders can follow the protocol accurately (at any chosen speed), send messages that do not comply with the protocol, or remain unresponsive.

Hanging requests reduce the throughput as they increase the number of authenticators shared during the epoch and the epoch-change. Hence, byzantine leaders leave the maximum number of requests hanging, i.e., $2k$ requests as all further prepare messages would be discarded by correct replicas.

While byzantine replicas cannot hinder correct leaders from committing requests, committing any request can only benefit the throughput of FnF-BFT. To that end, note that after GST, each correct leader v will converge to a C_v such that it will make progress during the entire epoch-time; hence, prolonging the epoch-time is impossible. The optimal strategy for byzantine leaders is thus to stall progress on their assigned hash buckets.

Finally, note that we assume the threshold signature scheme to be robust and can, therefore, discard any irrelevant message efficiently. □

On the other hand, as a backup, the optimal byzantine strategy is not helping other leaders to make progress (Lemma 10).

Lemma 10. *Under a correct primary, the optimal strategy for a byzantine backup is to remain unresponsive.*

Proof. Byzantine participation in the protocol can only benefit the correct leaders' throughput. Invalid messages can simply be ignored, while additional authenticators are not verified and thus do not reduce the system throughput. □

In conclusion, we observe that byzantine replicas have little opportunity to reduce the throughput in epochs under a correct primary.

Theorem 5. *After GST, the effective utilization under a correct primary is at least $\frac{8}{9}$ for $n \to \infty$.*

Proof. Moving from the best-case scenario with $3f + 1$ correct leaders to only $2f + 1$ correct leaders, each correct leader still processes $4f + 3$ authenticators per request, and 5 authenticators for each request of other leaders. We know from Lemma 9 that only the $2f + 1$ correct replicas are committing requests and creating checkpoints throughout the epoch. The authenticator cost of creating $2f + 1$ checkpoints, one for each correct leader, becomes $6f + 2$ per replica.

Byzantine leaders can open at most $2k$ new requests in an epoch. Each hanging request is seen at most twice by correct replicas without becoming committed. Thus, each correct replica processes no more than $8k$ authenticators for requests purposefully left hanging by a byzantine replica in an epoch. Thus, the utilization is reduced at most by a factor $\left(1 - \frac{8kf}{T}\right)$, where T is the maximal epoch length. While epochs can finish earlier, this will not happen after GST as soon as each correct leader v works at its capacity C_v.

Hence, the byzantine-resilient epoch throughput becomes

$$\frac{k \cdot (2f+1)}{k \cdot (14f+3) + (6f+2)} \cdot \left(1 - \frac{8kf}{T}\right).$$

By comparing this to the best-case epoch throughput from Lemma 6, we obtain a maximal throughput reduction of

$$\frac{(2f+1)(k \cdot (19f+3) + (8f+2))}{(3f+1)(k \cdot (14f+3) + (6f+2))} \cdot \left(1 - \frac{8kf}{T}\right).$$

Observe that the first term decreases and approaches $\frac{8}{9}$ for $n \to \infty$:

$$\frac{(2f+1)(k \cdot (19f+3) + (8f+2))}{(3f+1)(k \cdot (14f+3) + (6f+2))} \stackrel{n \to \infty}{=} \frac{16 + 38k}{18 + 42k} \geq \frac{8}{9}.$$

We follow that the epoch time is $T \in \Omega(n^3)$, as we set $C_{\min} \in \Omega(n^2)$ and each leader requires $\Omega(n)$ time units to commit one of its requests. Additionally, we know that $8kf \in \mathcal{O}(n)$, and thus: $\left(1 - \dfrac{8kf}{T}\right) \stackrel{n \to \infty}{=} 1$.

For $n \to \infty$, the throughput reduction byzantine replicas can impose on the system during a synchronized epoch is therefore bounded by a factor $\dfrac{8}{9}$.   □

Byzantine Primary Throughput. A byzantine primary, evidently, aims to perform the epoch-change as slow as possible. Furthermore, a byzantine primary can impede progress in its assigned epoch entirely, e.g., by remaining unresponsive. We observe that there are two main byzantine strategies to be considered.

Lemma 11. *Under a byzantine primary, an epoch is either aborted quickly or $\Omega(n^3)$ new requests become committed.*

Proof. A byzantine adversary controlling the primary of an epoch has three options. Following the protocol and initiating the epoch for all $2f+1$ correct replicas will ensure high throughput and is thus not optimal. Alternatively, initiating the epoch for $s \in [f+1, 2f]$ correct replicas will allow the byzantine adversary to control the progress made in the epoch, as no correct leader can make progress without a response from at least one byzantine replica. However, slow progress can only be maintained as long as at least $2f+1$ leaders continuously make progress. By setting the no-progress timeout $T_p \in \Theta(T/C_{\min})$, $\Omega(n^3)$ new requests per epoch can be guaranteed. In all other scenarios, the epoch will be aborted after at most one epoch-change timeout T_e, the initial message transmission time 5Δ, and one no-progress timeout T_p.

Note that we do not increase the epoch-change timer T_e for f unsuccessful epoch-changes in a row. In doing so, we prevent f consecutive byzantine primaries from increasing the epoch-change timer exponentially; thus potentially reducing the system throughput significantly.   □

FnF-BFT Primaries. Primaries rotate across epochs based on their performance history to reduce the control of the byzantine adversary on the system.

Lemma 12. *After a sufficiently long stable time period, the performance of a byzantine primary can only drop below the performance of the worst correct primary once throughout the sliding window.*

Proof. The network is considered stable for a sufficiently long time when all leaders work at their capacity limit, i.e., the number of requests they are assigned in an epoch matches their capacity, and primaries have subsequently been explored once. As soon as all leaders are working at their capacity limit, we observe the representative performance of all correct primaries, at least.

FnF-BFT repeatedly cycles through the $2f+1$ best primaries. A primary's performance is based on its last turn as primary. Consequently, a primary is removed from the rotation as soon as its performance drops below one of the f remaining primaries. We conclude that a byzantine primary will only be nominated beyond its single exploration throughout the sliding window if its performance matches at least the performance of the worst correct primary. □

As its successor determines a primary's performance, the successor can influence the performance slightly. However, this is bounded by the number of open requests – $\mathcal{O}(n)$ many – which we consider being well within natural performance variations, as $\Omega(n^3)$ requests are created in an epoch under a correct primary. Thus, we will disregard possible performance degradation originating at the succeeding primary.

From Lemma 12, we easily see that the optimal strategy for a byzantine primary is to act according to Lemma 11 – performing better would only help the system. In a stable network, byzantine primaries will thus only have one turn as primary throughout any sliding window. In the following, we consider a primary to be behaving byzantine if it performs worse than all correct primaries.

Theorem 6. *After the system has been in stability for a sufficiently long time period, the fraction of byzantine behaving primaries is $\frac{f}{g}$.*

Proof. Following from Lemma 12, we know that a primary can only behave byzantine once in the sliding window. There are a total of g epochs in a sliding window, and the f byzantine replicas in the network can only act byzantine in one epoch included in the sliding window. We see that the fraction of byzantine behaving primaries is $\frac{f}{g}$. □

The configuration parameter g determines the fraction of byzantine primaries in the system's stable state, while simultaneously dictating how long it takes to get there after GST. Setting g to a small value ensures that the system quickly recovers from asynchrony. On the other hand, setting g to larger values provides near-optimal behavior once the system is operating at its optimum.

FnF-BFT Byzantine-Resilient Performance. Combining the byzantine strategies from Theorem 5, Lemma 11 and Theorem 6, we obtain the following.

Theorem 7. *After GST, the effective utilization is asymptotically $\frac{8}{9} \cdot \frac{g-f}{g}$ for $n \to \infty$.*

Proof. To estimate the effective utilization, we only consider the throughput within epochs. That is because the time spent in correct epochs dominates the time for epoch-changes, as well as the time for failed epoch-changes under byzantine primaries, as the number of replicas increases (Lemma 7). Without loss of generality, we consider no progress to be made in byzantine primary epochs. We make this assumption, as we cannot guarantee asymptotically significant throughput. From Theorem 5, we know that in an epoch initiated by a correct primary, the byzantine-resilient effective utilization is at least $\frac{8}{9}$ for $n \to \infty$. Further, at least $\frac{g-f}{g}$ of the epochs are led by correct primaries after a sufficiently long time period in stability and thus obey this bound (Theorem 6). In the limit for $n \to \infty$ the effective utilization is $\frac{8}{9} \cdot \frac{g-f}{g}$.   □

B   Implementation and Preliminary Evaluation

Features. FnF-BFT's proof-of-concept implementation is directly based on the code of HotStuff's open-source prototype implementation `libhotstuff`.[3] We implement the basic functionality of FnF-BFT including the epoch-change and watermarks, while only changing ≈ 2000 lines of code and maintaining the general framework and experiment setup. In addition, we extend both implementations to support BLS threshold signatures.[4]

Threshold Signatures. Note that while HotStuff is designed with threshold signatures in mind and relies on them for its theoretic performance analysis [38], `libhotstuff` uses sets of $2f + 1$ signatures instead of real threshold signatures. While this workaround maintains a complexity of $\mathcal{O}(n)$ for creation of such a "threshold signature", it comes at the expense of a verification complexity of $\mathcal{O}(n)$ as well. In HotStuff, this additional overhead affects mainly the non-primary replicas, which would otherwise be idle in the HotStuff protocol. However, the design of FnF-BFT ensures that all replicas' computational resources are utilized at all times. Since the originally used `secp256k1` cryptographic signatures appear to be more optimized than the BLS threshold signatures, and to ensure a fair comparison, we thus compare FnF-BFT's throughput and latency to HotStuff using the identical BLS threshold signature implementation.

Limitations. As in the theoretical analysis (see Sect. 3.1), we did not implement a batching process for client requests. Hence, each block contains only a single

---

[3] https://github.com/hot-stuff/libhotstuff.
[4] https://github.com/herumi/bls.

client request. For this reason, the expected throughput (and typically reported number for BFT protocols) in practical deployments is much higher.

While FNF-BFT's design inherently allows to utilize concurrent threads during epoch operation, e.g., to interact with all parallel leaders, our proof-of-concept implementation currently only supports single-threaded operation.

Setup and Methodology. We compare FNF-BFT's single-threaded implementation to both single- and multi-threaded HotStuff (with 12 threads per replica) with respect to best-case performance, i.e., throughput and latency when all n replicas operate correctly. Experiments are repeated for $n \in \{4, 7, 10, 16, 31, 61\}$ replicas. We deploy both protocols on Amazon EC2 using c5.4xlarge AWS cloud instances. Each replica is assigned to a dedicated VM instance with 16 CPU cores powered by Intel Xeon Platinum 8000 processors clocked at up to 3.6 GHz, 32 GB of RAM, and a maximum network bandwidth of 10 Gigabits per second.

We measure the average throughput and latency over multiple epochs to include the expected drop in performance for FNF-BFT during the epoch-change. For each experiment, we run both protocols for at least three minutes and measure their average performance accordingly. We divide the hash space into n buckets, resulting in one bucket per replica. For generating requests, we run the `libhotstuff` client with its default settings, meaning that the payload of each request is empty. Clients generate and broadcast four requests in parallel, and issue a new request whenever one of their requests is answered. For the throughput measurements, we launch sufficiently many clients until we observe that no buckets are idle. For the latency measurement, we run a single client instance such that the system does not operate at its throughput limit. In general, we use the same settings for both protocols wherever applicable (Table 1).

Table 1. Experiment parameters used for FNF-BFT and HotStuff, if applicable.

Parameter	Value	Parameter	Value
Requests per block	1	No progress timeout	2s
Threads per replica	1	Blocks per checkpoint (K)	50
Threads per client	4	Watermark window size $(2 * K)$	100
Epoch timeout	30s	Initial epoch watermark bounds	10000

Performance. Figure 6 depicts a best-case operation of FNF-BFT over five epochs and demonstrates its consistently high throughput.[5] As expected, the throughput of our protocol stalls during an epoch-change. However, in comparison to HotStuff (Fig. 7), FNF-BFT's average throughput remains on top over

---

[5] Note that a rate of 200 batches per second with a typical batch size of 500 commands per batch translates to a throughput of 100,000 requests per second.

multiple epochs (i.e., including the epoch-change gap). As HotStuff's throughput decreases for increasing number of replicas, FNF-BFT showcases its superior scalability. Specifically, FNF-BFT handles large amounts of requests up to 4.7× faster than multi-threaded and 16× faster than single-threaded HotStuff.

Figure 8 depicts the average latency of both FNF-BFT and HotStuff, showing that they scale similarly with the number of replicas. As latency expresses the time between a request being issued and committed, both protocols exhibit very fast finality for requests on average, even with many replicas. In combination, Fig. 7 and Fig. 8 demonstrate the high performance and competitiveness of FNF-BFT with HotStuff, especially when scaling to many replicas.

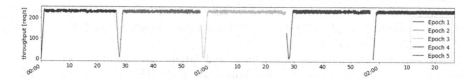

Fig. 6. Throughput of FNF-BFT with $n = 4$ replicas over 5 epochs.

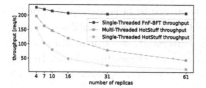

Fig. 7. Average Throughput Comparison

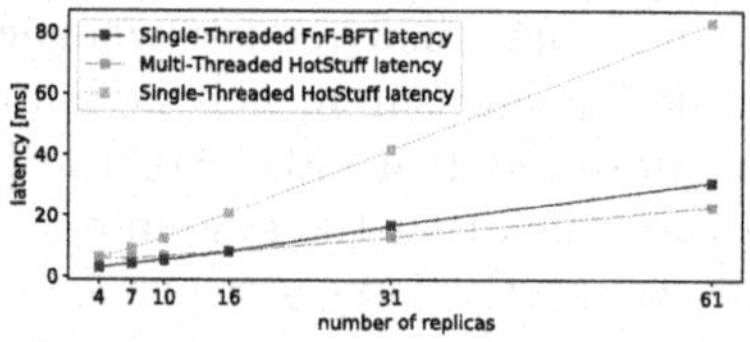

Fig. 8. Average Latency Comparison

References

1. The swirdls hashgraph consensus algorithm: fair, fast, byzantine fault tolerance. https://www.swirlds.com/downloads/SWIRLDS-TR-2016-01.pdf. Accessed 30 Jan 2023
2. Abraham, I., Gueta, G., Malkhi, D., Alvisi, L., Kotla, R., Martin, J.P.: Revisiting fast practical byzantine fault tolerance (2017)
3. Amir, Y., Dolev, D., Kramer, S., Malki, D.: Transis: a communication subsystem for high availability. In: 1992 Digest of Papers. FTCS-22: The Twenty-Second International Symposium on Fault-Tolerant Computing, pp. 76–84 (1992). https://doi.org/10.1109/FTCS.1992.243613
4. Amir, Y., Coan, B., Kirsch, J., Lane, J.: Prime: byzantine replication under attack. IEEE Trans. Dependable Secure Comput. **8**(4), 564–577 (2010)

5. Amir, Y., et al.: Steward: scaling byzantine fault-tolerant replication to wide area networks. IEEE Trans. Dependable Secure Comput. **7**(1), 80–93 (2008)

6. Aublin, P., Mokhtar, S.B., Quéma, V.: RBFT: redundant byzantine fault tolerance. In: ICDCS, pp. 297–306 (2013)

7. Avarikioti, G., Kokoris-Kogias, E., Wattenhofer, R.: Divide and scale: formalization of distributed ledger sharding protocols (2019)

8. Birman, K., Joseph, T.: Exploiting virtual synchrony in distributed systems. In: Proceedings of the Eleventh ACM Symposium on Operating Systems Principles, SOSP 1987, pp. 123–138. Association for Computing Machinery, New York (1987). https://doi.org/10.1145/41457.37515

9. Castro, M., Liskov, B.: Practical byzantine fault tolerance and proactive recovery. ACM Trans. Comput. Syst. (TOCS) **20**(4), 398–461 (2002)

10. Clement, A., Wong, E.L., Alvisi, L., Dahlin, M., Marchetti, M.: Making byzantine fault tolerant systems tolerate byzantine faults. In: NSDI, pp. 153–168 (2009)

11. Danezis, G., Kokoris-Kogias, L., Sonnino, A., Spiegelman, A.: Narwhal and tusk: a DAG-based mempool and efficient BFT consensus. In: Proceedings of the Seventeenth European Conference on Computer Systems, EuroSys 2022, pp. 34–50. Association for Computing Machinery, New York (2022). https://doi.org/10.1145/3492321.3519594

12. Dolev, D., Reischuk, R.: Bounds on information exchange for byzantine agreement. J. ACM (JACM) **32**(1), 191–204 (1985)

13. Dolev, D., Strong, H.R.: Polynomial algorithms for multiple processor agreement. In: ACM STOC, pp. 401–407 (1982)

14. Dwork, C., Lynch, N., Stockmeyer, L.: Consensus in the presence of partial synchrony. J. ACM (JACM) **35**(2), 288–323 (1988)

15. Fischer, M.J., Lynch, N.A., Paterson, M.S.: Impossibility of distributed consensus with one faulty process. J. ACM (JACM) **32**(2), 374–382 (1985)

16. Gelashvili, R., Kokoris-Kogias, L., Sonnino, A., Spiegelman, A., Xiang, Z.: Jolteon and Ditto: network-adaptive efficient consensus with asynchronous fallback. In: Eyal, I., Garay, J. (eds.) FC 2022. LNCS, vol. 13411, pp. 296–315. Springer, Cham (2022). https://doi.org/10.1007/978-3-031-18283-9_14

17. Gągol, A., Leundefinedniak, D., Straszak, D., undefinedwiundefinedtek, M.: Aleph: efficient atomic broadcast in asynchronous networks with byzantine nodes. In: Proceedings of the 1st ACM Conference on Advances in Financial Technologies, AFT 2019, pp. 214–228. Association for Computing Machinery, New York (2019). https://doi.org/10.1145/3318041.3355467

18. Gueta, G.G., et al.: SBFT: a scalable decentralized trust infrastructure for blockchains (2018)

19. Gupta, S., Hellings, J., Sadoghi, M.: RCC: resilient concurrent consensus for high-throughput secure transaction processing. In: 2021 IEEE 37th International Conference on Data Engineering (ICDE), pp. 1392–1403. IEEE (2021)

20. Keidar, I., Kokoris-Kogias, E., Naor, O., Spiegelman, A.: All you need is DAG. In: Proceedings of the 2021 ACM Symposium on Principles of Distributed Computing, PODC 2021, pp. 165–175. Association for Computing Machinery, New York (2021). https://doi.org/10.1145/3465084.3467905

21. Kokoris-Kogias, E., Jovanovic, P., Gasser, L., Gailly, N., Syta, E., Ford, B.: Omniledger: a secure, scale-out, decentralized ledger via sharding. In: IEEE SP, pp. 19–34 (2018)

22. Kotla, R., Alvisi, L., Dahlin, M., Clement, A., Wong, E.: Zyzzyva: speculative byzantine fault tolerance. SIGOPS Oper. Syst. Rev. **41**(6), 45–58 (2007)

23. Lamport, L., Shostak, R., Pease, M.: The byzantine generals problem. ACM Trans. Program. Lang. Syst. **4**(3), 382–401 (1982)
24. Lynch, N.A.: Distributed Algorithms. Elsevier (1996)
25. Malkhi, D., Szalachowski, P.: Maximal extractable value (MEV) protection on a DAG (2022). https://doi.org/10.48550/ARXIV.2208.00940
26. Mao, Y., Junqueira, F.P., Marzullo, K.: Mencius: building efficient replicated state machines for wans. In: USENIX OSDI, pp. 369–384 (2008)
27. Melliar-Smith, P., Moser, L., Agrawala, V.: Broadcast protocols for distributed systems. IEEE Trans. Parallel Distrib. Syst. **1**(1), 17–25 (1990). https://doi.org/10.1109/71.80121
28. Milosevic, Z., Biely, M., Schiper, A.: Bounded delay in byzantine-tolerant state machine replication. In: IEEE SRDS, pp. 61–70 (2013)
29. Moser, L.E., Melliar-Smith, P.M.: Byzantine-resistant total ordering algorithms. Inf. Comput. **150**(1), 75–111 (1999). https://doi.org/10.1006/inco.1998.2770
30. Pease, M., Shostak, R., Lamport, L.: Reaching agreement in the presence of faults. J. ACM (JACM) **27**(2), 228–234 (1980)
31. Peterson, L.L., Buchholz, N.C., Schlichting, R.D.: Preserving and using context information in interprocess communication. ACM Trans. Comput. Syst. **7**(3), 217–246 (1989). https://doi.org/10.1145/65000.65001
32. Reiter, M.K.: Secure agreement protocols: reliable and atomic group multicast in rampart. In: ACM CCS, pp. 68–80 (1994)
33. Reiter, M.K.: The Rampart toolkit for building high-integrity services. In: Birman, K.P., Mattern, F., Schiper, A. (eds.) Theory and Practice in Distributed Systems. LNCS, vol. 938, pp. 99–110. Springer, Heidelberg (1995). https://doi.org/10.1007/3-540-60042-6_7
34. Spiegelman, A., Giridharan, N., Sonnino, A., Kokoris-Kogias, L.: Bullshark: DAG BFT protocols made practical. In: Proceedings of the 2022 ACM SIGSAC Conference on Computer and Communications Security, CCS 2022, pp. 2705–2718. Association for Computing Machinery, New York (2022). https://doi.org/10.1145/3548606.3559361
35. Stathakopoulou, C., David, T., Vukolić, M.: Mir-BFT: high-throughput BFT for blockchains (2019)
36. Stathakopoulou, C., Pavlovic, M., Vukolić, M.: State machine replication scalability made simple. In: Proceedings of the Seventeenth European Conference on Computer Systems, pp. 17–33 (2022)
37. Veronese, G.S., Correia, M., Bessani, A.N., Lung, L.C.: Spin one's wheels? Byzantine fault tolerance with a spinning primary. In: IEEE SRDS, pp. 135–144 (2009)
38. Yin, M., Malkhi, D., Reiter, M.K., Gueta, G.G., Abraham, I.: HotStuff: BFT consensus with linearity and responsiveness. In: ACM PODC, pp. 347–356 (2019)
39. Zamani, M., Movahedi, M., Raykova, M.: Rapidchain: scaling blockchain via full sharding. In: ACM CCS, pp. 931–948 (2018)

Divide & Scale: Formalization and Roadmap to Robust Sharding

Zeta Avarikioti[1(✉)], Antoine Desjardins[2], Lefteris Kokoris-Kogias[2,4], and Roger Wattenhofer[3]

[1] TU Wien, Vienna, Austria
georgia.avarikioti@tuwien.ac.at
[2] ISTA, Klosterneuburg, Austria
[3] ETH Zürich, Zürich, Switzerland
[4] Mysten Labs, San Francisco, USA

Abstract. Sharding distributed ledgers is a promising on-chain solution for scaling blockchains but lacks formal grounds, nurturing skepticism on whether such complex systems can scale blockchains securely. We fill this gap by introducing the first formal framework as well as a roadmap to robust sharding. In particular, we first define the properties sharded distributed ledgers should fulfill. We build upon and extend the Bitcoin backbone protocol by defining *consistency* and *scalability*. Consistency encompasses the need for atomic execution of cross-shard transactions to preserve safety, whereas scalability encapsulates the speedup a sharded system can gain in comparison to a non-sharded system.

Using our model, we explore the limitations of sharding. We show that a sharded ledger with n participants cannot scale under a fully adaptive adversary, but it can scale up to m shards where $n = c'm \log m$, under an epoch-adaptive adversary; the constant c' encompasses the trade-off between security and scalability. This is possible only if the sharded ledgers create succinct proofs of the valid state updates at every epoch. We leverage our results to identify the sufficient components for robust sharding, which we incorporate in a protocol abstraction termed Divide & Scale. To demonstrate the power of our framework, we analyze the most prominent sharded blockchains (Elastico, Monoxide, OmniLedger, RapidChain) and pinpoint where they fail to meet the desired properties.

Keywords: Blockchains · Sharding · Scalability · Formalization

1 Introduction

A promising solution to scaling blockchain protocols is sharding, e.g. [33,36,53,55]. Its high-level idea is to employ multiple blockchains in parallel, the *shards*, that operate using the same consensus protocol. Different sets of participants run consensus and validate transactions, so that the system "scales out".

However, there is no formal definition of a robust sharded ledger (similar to the definition of what a robust transaction ledger is [24]), which leads to multiple problems. First, each protocol defines its own set of goals, which tend to

S. Rajsbaum et al. (Eds.): SIROCCO 2023, LNCS 13892, pp. 199–245, 2023.
https://doi.org/10.1007/978-3-031-32733-9_10

favor the protocol design presented. These goals are then confirmed achievable by experimental evaluations that demonstrate their improvements. Additionally, due to the lack of robust comparisons (which cannot cover all possible Byzantine behaviors), sharding is often criticized as some believe that the overhead of transactions between shards cancels out the potential benefits. In order to fundamentally understand sharding, one must formally define what sharding really is, and then see whether different sharding techniques live up to their promise.

Related Work. Recently, a few systemizations of knowledge on sharding [52], consensus [8], and cross-shard communication [56] which have also discussed part of sharding, have emerged. These works, however, do not define sharding in a formal fashion to enable an "apples-to-apples" comparison of existing works nor do they explore its limitations.

There are very few works that lay formal foundations for blockchain protocols. In particular, the Bitcoin backbone protocol [24] was the first to formally define and prove a blockchain protocol, specifically Bitcoin, in a PoW setting. Later, Pass et al. [41] showed that there is no PoW protocol that can be robust under asynchrony. With Ouroboros [30] Kiayias et al. extended the ideas of backbone to the Proof-of-Stake (PoS) setting, where they showed that it is possible to have a robust transaction ledger in a semi-synchronous environment as well [21]. However, all of these works consider only non-sharded ledgers but can be used as stepping stones to the formalization of sharded ledgers.

Our Contribution. In this work, we take up the challenge of providing formal "common grounds" under which we can capture the sharding limitations, determine the necessary components of a sharding system, and fairly compare different sharding solutions. We achieve this by defining a formal sharding framework as well as formal bounds of what a sharded transaction ledger can achieve.

To maintain compatibility with the existing models of a robust transaction ledger, we build upon the work of Garay et al. [24]. We generalize the transaction ledger properties, originally introduced in [24], namely *Persistence* and *Liveness*, to also apply to sharded ledgers. Persistence expresses the agreement between honest parties on the transaction order, while liveness encompasses that a transaction will eventually be processed and included in the transaction ledger. Further, we extend the model to capture what sharding offers to blockchain systems by defining *Consistency* and *Scalability*. Consistency is a security property that conveys the atomic property of *cross-shard transactions* (transactions that span multiple shards and should either abort or commit in all shards). Scalability, on the other hand, is a performance property that encapsulates the resource gains per party (in bandwidth, storage, and computation) in a sharded system compared to a non-sharded system.

Once we define the properties, we explore the limitations of sharding protocols that satisfy them. We identify a *trade-off between the bandwidth requirements and how adaptive the adversary is*, i.e., how "quickly" the adversary can change the corrupted parties. Specifically, with a fully adaptive adversary, scalable and secure sharding is impossible in our model. With a slowly-adaptive adversary, however, sharding can scale securely with up to m shards, where $n = c'm \log m$. The constant c' encompasses the trade-off between scalability and security: if the overall and per-shard adversarial thresholds are close to each other, then c' must be large to ensure security within each hard. Furthermore, scaling against a somewhat adaptive adversary is only possible under two conditions: first, the parties of a shard *cannot be light clients* to other shards to scale storage. Second, shards must periodically *compact the state* updates in a verifiable and succinct manner (e.g., via checkpoints [33], cryptographic accumulators [10], zero-knowledge proofs [9,39] or other techniques [15,26–28]); else eventually the bandwidth resources per party will exceed those of a non-sharded blockchain.

Once we provide solid bounds on the design of sharding protocols, we identify seven components that are critical to designing a robust permissionless sharded ledger: (a) a core consensus protocol for each shard, (b) a protocol to partition transactions in shards, (c) an atomic cross-shard communication protocol that enables transferring of value across shards, (d) a Sybil-resistance mechanism that forces the adversary to commit resources in order to participate, (e) a process that guarantees honest and adversarial nodes are appropriately dispersed to the shards to defend security against adversarial adaptivity, (f) a distributed randomness generation protocol, (g) a process to occasionally compact the state in a verifiable manner. We then employ these components to introduce a protocol abstraction, termed *Divide & Scale*, that achieves robust sharding in our model. We explain the design rationale, provide security proofs, and identify which components affect the scalability and throughput of our protocol abstraction.

To demonstrate the power of our framework, we further describe, abstract, and analyze the most well-established permisionless sharding protocols: Elastico [36] (inspiration of Zilliqa), OmniLedger [33] (inspiration of Harmony), Monoxide [53], and RapidChain [55]. We demonstrate that all sharding systems fail to meet the desired properties in our model. Elastico and Monoxide do not actually (asymptotically) improve on storage over non-sharded blockchains according to our model. OmniLedger is susceptible to a liveness attack where the adaptive adversary can simply delete a shard's state effectively preventing the system's progress. Albeit, with a simple fix, OmniLedger satisfies all the desired properties in our model. Last, we prove RapidChain meets the desired properties but only in a weaker adversarial model. For all protocols, we provide elaborate proofs while for OmniLedger and RapidChain we further estimate how much they improve over their blockchain substrate. To that end, we define and use a theoretical performance metric, termed *throughput factor*, which expresses the average number of transactions that can be processed per round under the worst possible Byzantine behavior. We show that both OmniLedger and RapidChain scale optimally with m shards where $n = O(m \log m)$.

In summary, the contribution of this work is the following:

– We introduce a framework where sharded transaction ledgers are formalized and the necessary properties of sharding protocols are defined. Further, we define a throughput factor to estimate the transaction throughput improvement of sharding blockchains over non-sharded blockchains (Sect. 2).

- We explore the limitations of secure and efficient sharding protocols under our model (Sect. 3, Appendix A).
- We identify the critical and sufficient ingredients for designing a robust sharded ledger, which we incorporate into a protocol abstraction for robust sharding, termed Divide & Scale (Sect. 4, Appendix B).
- We evaluate Elastico, Monoxide, OmniLedger, and RapidChain. We pinpoint where the former three fail to satisfy our properties, whereas the latter satisfies them all only under a weaker adversarial model (Sect. 5, Appendix C).

2   The Sharding Framework

In this section, we define the desired security and performance properties of a secure and efficient distributed sharded ledger, extending the work of Garay et al. [24]. We further define a theoretical performance metric, the transaction throughput. To assist the reader, we provide a glossary of the most frequently used parameters in Table 2 (Section Figures).

2.1   The Model

Network Model. We analyze blockchain protocols assuming a synchronous communication network. In particular, a protocol proceeds in *rounds*, and at the end of each round the participants of the protocol are able to synchronize, and all messages are delivered. A set of R consecutive rounds $E = \{r_1, r_2, \ldots, r_R\}$ defines an *epoch*. We consider a fixed number of participants in the system denoted by n. However, this number might not be known to the parties.

Threat Model. The adversary is slowly-adaptive, meaning that the adversary can corrupt parties on the fly at the beginning of each epoch but cannot change the corrupted set during the epoch, i.e., the adversary is static during each epoch. In addition, in any round, the adversary decides its strategy after receiving all honest parties' messages. The adversary can change the order of the honest parties' messages but cannot modify or drop them. Furthermore, the adversary is computationally bounded and can corrupt at most f parties during each epoch. This bound f holds strictly at every round of the protocol execution. Note that depending on the specifications of each protocol, i.e., which Sybil-attack-resistant mechanism is employed, the value f represents a different manifestation of the adversary's power (e.g., computational power, stake in the system).

Transaction Model. We assume transactions consist of inputs and outputs that can only be spent as a whole. Each transaction input is an unspent transaction output (UTXO). Thus, a transaction takes UTXOs as inputs, destroys them and creates new UTXOs, the outputs. A transaction ledger that handles such transactions is UTXO-based, similarly to Bitcoin [40]. Most protocols considered in this work are UTXO-based. Transactions can have multiple inputs and outputs. We define the *average size of a transaction*, i.e., the average number of inputs and outputs of a transaction in a transaction set, as a parameter v. This way v correlates to the number of shards a transaction is expected to affect; the actual size in bytes is proportional to v but unimportant for measuring scalability. Further, we assume a transaction set T follows a distribution D_T (e.g. D_T is the uniform distribution if the sender(s) and receiver(s) of each transaction are chosen uniformly at random from all possible users).

2.2 Sharded Transaction Ledgers

In this section, we introduce the necessary properties a sharding blockchain protocol must satisfy in order to maintain a robust sharded transaction ledger. We build upon the definition of a robust transaction ledger introduced in [24].

A sharded transaction ledger is defined with respect to a set of valid[1] transactions T and a collection of transaction ledgers for each shard $S = \{S_1, S_2, \ldots, S_m\}$. In each shard $i \in [m] = \{1, 2, \ldots, m\}$, a transaction ledger is defined with respect to a set of valid ledgers[2] S_i and a set of valid transactions. Each set possesses an efficient membership test. A ledger $L \in S_i$ is a vector of sequences of transactions $L = \langle x_1, x_2, \ldots, x_l \rangle$, where $tx \in x_j \Rightarrow tx \in T, \forall j \in [l]$.

In a sharding blockchain protocol, a sequence of transactions $x_i = tx_1 \ldots tx_e$ is inserted in a block which is appended to a party's local chain C in a shard. A chain C of length l contains the ledger $L_C = \langle x_1, x_2, \ldots, x_l \rangle$ if the input of the j-th block in C is x_j. The *position* of transaction tx_j in the ledger of a shard L_C is the pair (i, j) where $x_i = tx_1 \ldots tx_j \ldots tx_e$ (i.e., the block that contains the transaction). Essentially, a party *reports* a transaction tx_j in position i only if one of their shards' local ledger includes transaction tx_j in the i-th block. We assume that a block has constant size, i.e., there is a maximum constant number of transactions included in each block[3].

Furthermore, we define a symmetric relation on T, denoted by $M(\cdot, \cdot)$, that indicates if two transactions are conflicting, i.e., $M(tx, tx') = 1 \Leftrightarrow tx, tx'$ are conflicting. Note that valid ledgers can never contain conflicting transactions. Similarly, a valid sharded ledger cannot contain two conflicting transactions even across shards. In our model, we assume there exists a verification oracle denoted by $V(T, S)$, which instantly verifies the validity of a transaction with respect to a ledger. In essence, the oracle V takes as input a transaction $tx \in T$ and a valid ledger $L = \langle x_1, x_2, \ldots, x_l \rangle \in S$ and checks whether the transaction is valid and not conflicting in this ledger; formally, $V(tx, L) = 1 \Leftrightarrow \exists tx' \in L$ s.t. $M(tx, tx') = 1$ or $L' = \langle x_1, x_2, \ldots, x_l, tx \rangle$ is an invalid ledger.

Next, we introduce the security and performance properties a blockchain protocol must uphold to maintain a robust and efficient sharded transaction ledger: persistence, consistency, liveness, and scalability. Intuitively, *persistence* expresses the agreement between honest parties on the transaction order,

---

[1] Validity depends on the application using the ledger.

[2] Only one of the ledgers is actually committed as part of the shard's ledger, but before commitment there are multiple potential ledgers.

[3] To scale in bandwidth, the block size cannot depend on the parties or transactions.

whereas *consistency* conveys that cross-shard transactions are either committed or aborted atomically (in all shards). *Liveness* indicates that transactions will eventually be included in a shard, i.e., the system makes progress. Last, *scalability* encapsulates the speedup of a sharded system in comparison to a non-sharded system: The blockchain's throughput limitation stems from the need for data propagation, maintenance, and verification by every party. Thus, to scale via sharding, each party must broadcast, maintain and verify mainly local information.

Definition 1 (Persistence). *Parameterized by $k \in \mathbb{N}$ ("depth" parameter), if in a certain round an honest party reports a shard that contains a transaction tx in a block at least k blocks away from the end of the shard's ledger (such transaction will be called "stable"), then whenever tx is reported by any honest party it will be in the same position in the shard's ledger.*

Definition 2 (Consistency). *Parametrized by $k \in \mathbb{N}$ ("depth" parameter), there is no round r in which there are two honest parties P_1, P_2 reporting transactions tx_1, tx_2 respectively as stable (at least in depth k in the respective shards), such that $M(tx_1, tx_2) = 1$.*

Both persistence and consistency are necessary properties because one may fail while the other holds. For instance, if a party double-spends across two shards without reverting a stable transaction (e.g., due to a badly designed mechanism to process cross-shard transactions), consistency fails while persistence holds.

We further note consistency depends on the average size of transactions $v \in \mathbb{N}$ as well as the distribution of the input set of transactions D_T. For example, if all transactions are intra-shard, consistency is trivially satisfied due to persistence.

To evaluate the system's progress, we assume that the block size is sufficiently large, thus a transaction will never be excluded due to space limitations.

Definition 3 (Liveness). *Parameterized by u ("wait time") and k ("depth" parameter), provided that a valid transaction is given as input to all honest parties of a shard continuously for the creation of u consecutive rounds, then all honest parties will report this transaction at least k blocks from the end of the shard, i.e., all report it as stable.*

Scaling distributed ledgers depends on three vectors: *communication, space,* and *computation*. In particular, to allow high transaction throughput, the bandwidth and computation required per party should ideally be constant and independent of the number of parties while the storage requirements per party should decrease with the number of parties. Such a system can scale optimally because an increased transaction load, e.g. double, can be processed with the same storage resources if the parties increase proportionally, e.g. double, as well as the same communication and computation resources per node. To measure scalability, i.e., the resource requirements per node, we define three scaling factors, namely the communication, space, and computation factor.

We define the **communication factor** ω_m as the communication complexity of the system (per transaction) scaled over the number of participants. In essence, ω_m represents the average amount of sent or received data (bandwidth) required per party to include a transaction in the ledger. ω_m expresses the worst communication complexity of all the subroutines of the system, incorporating the bandwidth requirements of the protocols both within an epoch (i. e., within and across shards communication), as well as during epoch transitions (amortized over the epoch's length). The latter becomes the bottleneck for scalability in the long run as rotating parties must bootstrap to new shards and download the ever-growing shard ledgers.

We next introduce the **space factor** ω_s that estimates how much data each party stores in the system. To do so, we count the amount of data stored in total by all the parties scaled over the number of parties and the transaction load. When ω_s is constant, $\Theta(1)$, each node stores all transactions equivalently to a central database, e.g., Bitcoin. On the contrary, a perfectly scalable system allows parties to share the transaction load equally, $\omega_s = c/n$, c constant; as a result, if parties increase proportionally to the transaction load the space resources per party remain the same.

To define the space factor we introduce the notion of average-case analysis. Typically, sharding protocols scale well when the analysis is optimistic, that is, for transaction inputs that contain neither cross-shard nor multi-input (multi-output) transactions. However, in practice transactions are both cross-shard and multi-input/output. For this reason, we define the space factor as a random variable dependent on an input set of transactions T drawn uniformly at random from a distribution D_T.

We assume T is given well in advance as input to all parties. To be specific, we assume every transaction $tx \in T$ is given at least for u consecutive rounds to all parties of the system. Hence, from the liveness property, all transaction ledgers held by honest parties will report all transactions in T as stable. Further, we denote by $L^{\lceil k}$ the vector L where the last k positions are "pruned", while $|L^{\lceil k}|$ denotes the number of transactions contained in this "pruned" ledger. We note that a similar notation holds for a chain C where the last k positions map to the last k blocks. Each party P_j maintains a collection of ledgers $SL_j = \{L_1, L_2, \ldots, L_s\}, 1 \le s \le m$. We may now define the *space factor* for a sharding protocol with input T as the number of stable transactions included in every party's collection of transaction ledgers over the number of parties n and the number of input transactions T[4], $\omega_s(T) = \sum_{\forall j \in [n]} \sum_{\forall L \in SL_j} |L^{\lceil k}|/(n|T|)$.

Lastly, we consider the verification process which can be computationally expensive. In our model, we focus on the average verification cost per transaction. We assume a constant computational cost per verification, i. e., a party's running time of verifying if a transaction is invalid or conflicting with a ledger is considered constant because this process can always speed up using efficient data structures (e.g. trees allow for logarithmic lookup time). Thus, the computational cost of a party is defined by the number of times the party executes the verification process. For this purpose, we employ a verification oracle V. Each

---

[4] Without loss of generality, we assume all transactions are valid and thus are eventually included in all honest parties' ledgers.

party calls the oracle to verify transactions, pending or included in a block. We denote by q_i the number of times party P_i calls oracle V in a protocol execution. The **computational factor** ω_c reflects the total number of times all parties call the verification oracle in a protocol execution scaled over the number of transactions T, $\omega_c(T) = \sum_{\forall i \in [n]} q_i / |T|$.

An ideal sharding system only involves a constant number of parties to verify each transaction, $\omega_c = \Theta(1)$, while both a typical BFT-based protocol and Bitcoin demand all nodes to verify all transactions, $\omega_c = \Theta(n)$. Furthermore, the computational factor is a random variable, hence the objective is to calculate the expected value of ω_c, i.e., the probability-weighted average of all possible values, where the probability is taken over the input transactions T.

Intuitively, scaling means processing more transactions with similar (i.e., not proportionally increasing) resources per party. If parties share the transaction load, e.g., space scales $\omega_s = c/n$, increased transactions can be processed by increasing the number of parties. Subsequently, the communication and computational costs must not increase proportionally to the number of parties, i.e., $\omega_c = o(n)$ and $\omega_m = o(n)$, else the system cannot truly scale the transaction load. We observe, however, that in practice protocols may scale well in one dimension but fail in another. A notable example is the Bitcoin protocol which has minimal communication overhead but does not scale in space and computation. To ensure overall scaling capabilities, we define the scalability property of sharded ledgers below; we say that a sharded ledger satisfies scalability if and only if the system scales in all the aforementioned dimensions.

Definition 4 (Scalability). *Parameterized by n (number of participants), $v \in \mathbb{N}$ (average size of transactions), D_T (distribution of the input set of transactions), the communication, space and computational factors of a sharding blockchain protocol are $\omega_m = o(n)$, $\omega_s = o(1)$, and $\omega_c = o(n)$, respectively.*

In order to adhere to standard security proofs from now on we say that the protocol Π satisfies property Q in our model if Q holds with overwhelming probability (in a security parameter). Note that a probability p is overwhelming if $1 - p$ is negligible. A function $negl(k)$ is negligible if for every $c > 0$, there exists an $N > 0$ such that $negl(k) < 1/k^c$ for all $k >> N$. Furthermore, we denote by $\mathbb{E}(\cdot)$ the expected value of a random variable.

Definition 5 (Robust Sharded Transaction Ledger). *A protocol that satisfies the properties of persistence, consistency, liveness, and scalability maintains a robust sharded transaction ledger.*

2.3   (Sharding) Blockchain Protocols

In this section, we adopt the definitions and properties of [24] for blockchain protocols, while we slightly change the notation to fit our model. In particular, we assume the parties of a shard of any sharding protocol maintain a chain (ledger) to achieve consensus. This means that every shard internally executes a blockchain (consensus) protocol that has three properties as defined by [24]: chain growth, chain quality, and common prefix. Each consensus protocol satisfies these properties with different parameters.

In this work, we will use the properties of the shards' consensus protocol to prove that a sharding protocol maintains a robust sharded transaction ledger. In addition, we will specifically use the shard growth and shard quality parameters to estimate the transaction throughput of a sharding protocol. The following definitions follow closely Definitions 3, 4 and 5 of [24].

Definition 6 (Shard Growth Property). *Parametrized by $\tau \in \mathbb{R}$ and $s \in \mathbb{N}$, for any honest party P with chain C, it holds that for any s rounds there are at least $\tau \cdot s$ blocks added to chain C of P.*

Definition 7 (Shard Quality Property). *Parametrized by $\mu \in \mathbb{R}$ and $l \in \mathbb{N}$, for any honest party P with chain C, it holds that for any l consecutive blocks of C the ratio of honest blocks in C is at least μ.*

Definition 8 (Common Prefix Property). *Parametrized by $k \in \mathbb{N}$, for any pair of honest parties P_1, P_2 adopting chains C_1, C_2 (in the same shard) at rounds $r_1 \leq r_2$ respectively, it holds that $C_1^{\lceil k} \preceq C_2$, where $\preceq$ denotes the prefix relation.*

Next, we define the **degree of parallelism** (DoP) of a sharding protocol, denoted m'. To evaluate the DoP of a protocol with input T, we need to determine how many shards are affected by each transaction on average; essentially, estimate how many times we run consensus for each valid transaction until it is stable. This is determined by the mechanism that handles the cross-shard transactions. To that end, we define $m_{i,j} = 1$ if the j-th transaction of set T has either an input or an output that is assigned to the i-th shard; otherwise $m_{i,j} = 0$. Then, the DoP of a protocol execution over a set of transactions T is defined as follows: $m' = \frac{T \cdot m}{\sum_{j=1}^{T} \sum_{i=1}^{m} m_{i,j}}$. The DoP of a protocol execution depends on the distribution of transactions D_T, the average size of transactions v, and the number of shards m. For instance, assuming a uniform distribution D_T, the expected DoP is $\mathbb{E}(m') = m/v$.

We can now define an efficiency metric, the *transaction throughput* of a sharding protocol. Considering constant block size, we have:

Definition 9 (Throughput). *The expected transaction throughput in s rounds of a sharding protocol with m shards is $\mu \cdot \tau \cdot s \cdot m'$. We define the throughput factor of a sharding protocol $\sigma = \mu \cdot \tau \cdot m'$.*

Intuitively, the throughput factor expresses the average number of blocks that can be processed per round by a sharding protocol. Thus, the transaction throughput (per round) can be determined by the block size multiplied by the throughput factor. The block size is considered constant; however, it cannot be arbitrarily large. The limit on the block size is determined by the bandwidth of the "slowest" party within each shard. At the same time, the constant block size guarantees low latency. If the block size is very large or depends on the number of shards or the number of participants, *bandwidth or latency* becomes the performance bottleneck. As our goal is to estimate the efficiency of the transactions' parallelism in a protocol, other factors like cross-shard communication latency are omitted.

3   Limitations of Sharding Protocols

In this section, we present a summary of our analysis on the limitations of sharding protocols in our framework (cf. Appendix A).

First, we focus on the limitations that stem from the nature of the transaction workload. In particular, sharding protocols are affected by two characteristics of the input transaction set: the transaction size v (number of inputs and outputs of each transaction), and more importantly the number of cross-shard transactions.

The average size of transactions is fairly small in practice, e.g., an average Bitcoin transaction has 2 inputs and 3 outputs with a small deviation [1]. We thus assume a fixed number of UTXOs participating in each transaction, meaning the transaction size v is a small constant. Furthermore, as v increases, more shards are affected by each transaction on expectation, hence the number of cross-shard transactions increases. To meaningfully lower bound the ratio of cross-shard transactions, we thus consider the minimum transaction size $v = 2$. If a transaction has more UTXOs, its chance of being cross-shard only increases.

The number of cross-shard transactions depends on the distribution of the input transactions D_T, as well as the process that partitions transactions into shards. First, we assume each ledger interacts (i.e., shares a cross-shard transaction) with γ other ledgers on average, γ being a function dependent on the number of shards m. We examine protocols where parties maintain information on shards other than their own and derive an upper bound for the expected value of γ such that scalability holds. Leveraging that, we prove the following:

Theorem 10. *There is no protocol maintaining a robust sharded transaction ledger against an adaptive adversary in our model controlling $f \geq n/m$, where m is the number of shards, and n is the number of parties.*

Next, we extend our results assuming, similarly to most sharding systems, that the UTXO space is partitioned uniformly at random into shards. In particular, we first show that a constant fraction of transactions is expected to be cross-shard. Using that we demonstrate there is no sharded ledger that satisfies scalability if parties store any information on ledgers (other than their own) involved in cross-shard transactions, i.e., are light clients on other shards [18]. We stress that our results hold for *any distribution* where the expected number of cross-shard transactions is *proportional* to the number of shards.

Theorem 11. *There is no protocol that maintains a robust sharded transaction ledger in our model under uniform space partition when parties are light nodes on the shards involved in cross-shard transactions.*

We further identify a concrete trade-off between security and scalability, that stems from the way parties are partitioned into shards. In particular, when parties are randomly permuted among shards, which is a common practice in sharding, e.g., [33,36], sharding scales almost linearly. The trade-off is now captured by the constant c': if the overall and per-shard adversarial thresholds are close to each other, then c' must be large to ensure security within each shard.

Theorem 12. *Any protocol that maintains a robust sharded transaction ledger in our model under uniformly random partition of the state and parties, can scale at most by a factor of m, where $n = c'm \log m$ and the constant c' encompasses the trade-off between security and scalability.*

Finally, we demonstrate the importance of periodical compaction of the valid state-updates in sharding protocols: we prove that any sharding protocol that satisfies scalability in our model, when the state is uniformly partitioned and the parties are periodically shuffled among shards, requires a state-compaction process such as checkpoints [33], cryptographic accumulators [10], zero-knowledge proofs [9], non-interactive proofs of proofs-of-work [15,28], proof of necessary work [27], erasure codes [26], etc. Intuitively, parties must be periodically shuffled among shards to maintain security against adaptivity. Subsequently, the parties must occasionally bootstrap to the new ever-increasing blockchains, leading to bandwidth or storage overheads that exceed those of a non-sharded blockchain in the long run. We stress that this result holds even if the parties are not randomly shuffled among the shards, as long as a significant fraction of parties changes shards from epoch to epoch.

Theorem 13. *Any protocol that maintains a robust sharded transaction ledger in our model, under uniformly random partition of the state and parties, employs verifiable compaction of the state.*

4   Divide & Scale

In this section, we discuss our design rationale for robust sharding; using the bounds of Sect. 3, we deduce some sufficient components for robust sharding in our model. We leverage these components to introduce a *protocol abstraction* for robust sharding, termed *Divide & Scale*, in Algorithm 1. We prove Divide & Scale is secure in our model (assuming the components are secure) and evaluate its efficiency depending on the choices of the individual components in Appendix B.

Sharding Components. We explain our design rationale and introduce the ingredients of a protocol that maintains a robust sharded ledger.

(a) **Consensus protocol of shards or Consensus:** A sharding protocol either runs consensus in every shard separately (multi-consensus) or provides a single total ordering for all the blocks generated in each shard (uni-consensus [2,43]). Since uni-consensus takes polynomial cost per block, such a protocol can only scale if the block size is also polynomial (e.g., includes $\Omega(n)$ transactions [43]). However, in such a case, the resources of each node generating an $\Omega(n)$-sized block must also grow with n, and therefore scalability cannot be satisfied[5]. For this reason, in our protocol abstraction, we chose the multi-consensus approach.

The consensus protocol run per shard must satisfy the properties of Garay et al. [24]: *common prefix, chain quality, and chain growth*. These properties are necessary (but not sufficient) to ensure persistence, liveness, and consistency.

(b) **Cross-shard mechanism or `CrossShard`:** The cross-shard mechanism is the protocol that handles the transactions that span across multiple shards. It is critical for the security of the sharding system, as it guarantees consistency, as well as scalability; a naively designed cross-shard mechanism may induce high storage or communication overhead on the nodes when handling several cross-shard transactions. To that end, the limitations of Sect. 3 apply. The cross-shard mechanism should provide the *ACID properties* (as in database transactions). Durability and Isolation are provided directly by the blockchains of the shards, hence, the cross-shard mechanism should provide Consistency, i.e., every transaction that commits produces a semantically valid state, and Atomicity, i.e., transactions are committed and aborted atomically (all or nothing). Typically the cross-shard mechanism runs hand in hand with the consensus protocol to guarantee consistency across shards.

(c) **Sybil-resistance mechanism or `Sybil`:** The Sybil-resistance mechanism enables the participants of a permissionless setting to reach a global consensus on a set of fairly-selected valid identities. Its fair selection, i.e., assigning valid identities to each party proportionally to its spent resources, guarantees the security bounds of the consensus protocol (e.g., $f < 1/3$ for BFT). To ensure fairness against slowly-adaptive adversaries, the Sybil-resistance mechanism must have access to unknown unbiasable randomness (see below DRG). The exact protocol (e.g. PoW, PoS) is irrelevant to our analysis as long as it guarantees (i) *correctness*: all parties can verify a valid identity, (ii) *fairness*: each party is selected with probability proportional to its resources, and (iii) *unpredictability*: no party can predict beforehand the valid set of identities (for the new epoch).

(d) **`StatePartition`:** This protocol determines how the state (e.g. transactions) is partitioned into shards. A naive design may violate consistency but there are several secure solutions to employ, e.g. [33,55]. We perform our analysis assuming all transactions are cross-shard, because any secure protocol that performs well in the pessimistic case, also performs well when transactions are intra-shard. Moreover, in the latter case, scaling is not challenging as the transaction throughput can be processed securely in blockchains that work in parallel.

---

[5] Due to their inherent inability to asymptotically scale, we believe uni-consensus systems are categorized as performance optimizations of consensus, e.g., [5,7,19,49, 50].

(e) **Division of nodes to shards or** `Divide2Shards`**:** This is the protocol that determines how parties are assigned to shards. It is crucial for security against slowly adaptive adversaries as a fully corrupted shard may result in the loss of all three security properties. It is also the reason that sharding cannot tolerate fully-adaptive adversaries in our model (Theorem 10). Note that static adversaries are an easier subcase of the slowly adaptive one.

In particular, to ensure transaction finality (i.e., liveness and persistence), either the consensus security bounds must hold for each shard, or the protocol must guarantee that if the adversary compromises a shard then the security violation will be restored within a specific (small) number of rounds. Specifically, if an adversary completely or partially compromises a shard, effectively violating the consensus bounds, then the adversary can double spend within the shard (violates persistence), as well as across shards (because nodes cannot verify cross-shard transactions from Lemma 15). Therefore, the transactions included in these blocks can only be executed when honest parties have verified them. Partial solutions towards this direction have been proposed such as proofs of fraud that allow an honest party to later prove misbehavior. Another challenge of this approach is to guarantee data availability.

Due to the complexity of such solutions and their implications on the transactions' finality, we design Divide & Scale assuming *the security bounds of consensus are maintained* when parties are divided into shards. Specifically, the parties are shuffled at the beginning of each epoch so that the threat model holds. A secure shuffling process requires an ubiasable source of randomness (see below DRG). When assigned to a shard, the nodes update their local state with the state of the new shard they are asked to secure, which in turn affects scalability. *The frequency of shuffling is thereby incorporating the trade-off between scalability and adaptive security.*

(f) **Randomness generation protocol or** `DRG`**:** The DRG protocol *provides unpredictable unbiasable randomness* [11, 14, 16, 20, 22, 34, 45, 51] such that both `Sybil` and `Divide2Shards` result in shards that maintain the security bounds for the consensus protocol. Given a slowly-adaptive adversary, the DRG protocol must be executed (at least) once per epoch; its high communication complexity can be amortized over the rounds of an epoch such that the system scales.

(g) **Verifiable compaction of state or** `CompactState`**:** `CompactState` guarantees that periodically state updates can be verifiably compacted. This protocol is necessary for scaling sharding systems in the long run, as it ensures that new parties can bootstrap with minimal effort (Theorem 13). The compacted state must be broadcasted to all parties, e.g. via reliable broadcast [13], to ensure data integrity and data availability; else a slowly-adaptive adversary can corrupt an entire shard after an epoch transition, violating liveness. Any protocol that ensures data binding and data availability can be used. In summary, this protocol must guarantee (i) *verifiable asymptotic compression* (more than constant), and (ii) *data integrity and* *availability*, i.e., the ledgers' history is available and can be retrieved. To satisfy scalability, the protocol must also ensure (iii) *efficient communication complexity* with respect to the epoch size (in rounds).

5   Evaluation of Sharding Protocols

To showcase the wide applicability and value of our framework, we evaluate in our model the well-established sharding protocols Elastico, Monoxide, OmniLedger, and RapidChain, and discuss Chainspace. We refer the reader to Appendix C for the complete analysis where we identify each protocol's sharding components as defined in Sect. 4 which we use to prove or disprove the desired properties of Sect. 2, often leveraging the bounds of Sect. 3. Due to space limitations, we only discuss here the final results of our analysis, also illustrated in Table 1, with key insights on how each protocol fails to meet some of the properties. We include in the evaluation the "permissionless" and "slowly-adaptive" properties to fairly compare the protocols. In our analysis, we evaluate the cross-shard communication protocols considering the fixes of [48] against replay attacks.

We first show that **Elastico** does not satisfy *consistency* in our model because the adversary may double-spend across shards when multi-input transactions are allowed (Theorem 25). Additionally, Elastico does not satisfy *scalability* by design regardless of the transaction distribution – even with a few cross-shard transactions (Theorem 27). Specifically, all epoch-transition protocols are executed for every block while parties maintain a global hash chain. Thus, transactions are only compressed by a constant factor, the block size, resulting in space and communication growing proportionally to the number of parties.

We then show that **Monoxide** does not satisfy *scalability* because miners must mine in parallel in all shards, verifying and storing all transactions to ensure security (Theorem 30). Due to its design rationale, Monoxide cannot scale even with optimistic transaction distributions with no cross-shard transactions.

Third, we prove that **OmniLedger** satisfies all properties but *liveness* (Theorems 35, 37, 36, 41). Specifically, OmniLedger checkpoints the UTXO pool at each epoch transition, but the state is not broadcasted to the network. Hence, a slowly adaptive adversary can corrupt a shard from the previous epoch before the new nodes of the shard bootstrap to the state in epoch transition. This attack violates liveness but simply adding a reliable broadcast step after checkpointing restores the liveness since all other components satisfy it already. The overhead of reliable broadcast can be amortized over the rounds of the epoch hence the overall scalability is not affected.

Fourth, we prove **RapidChain** maintains a robust sharded ledger but only under a *weaker model* than the one defined in Sect. 2 (Theorems 47, 49, 48, 53). Specifically, the protocol only allows a constant number of parties to join or leave and the adversary can at most corrupt a constant number of additional parties with each epoch transition. Another shortcoming of RapidChain is the synchronous consensus mechanism it employs. In case of temporary loss of synchrony in the network, the consensus of cross-shard transactions is vulnerable, hence consistency might break [55]. However, most of these drawbacks can be

---

Protocol Abstraction 1: Divide & Scale

Data: N_0 nodes are participating in the system at round 0 (genesis block). $m(N_E)$ denotes the function that determines the number of shards in epoch E. The transactions of epoch E are T_E. i denotes the block round (its relation to the communication rounds depends on the employed components).

Result: Shard state $T = \{T_0, T_1, \dots\}$.

```
   /* Initialization                                                    */
1  i ← 1
2  E ← 0
   /* Beginning of epoch: retrieve identities from Sybil resistant
      protocol, execute the DRG protocol to create the new epoch
      randomness, and assign nodes to shards                            */
3  if i mod R = 1 :
4      E ← E + 1
5      if i ≠ 1 :
6          N_E ← Sybil(r_{E-1})
7      r_E ← DRG(N_E)
8      Call Divide2Shards(N_E, m(N_E), r_E)
   /* End of epoch: compact the state of the shard                      */
9  elif i mod R = 0 :
10     Call CompactState(i)
   /* During epoch: run the consensus protocol for intra-shard and
      cross-shard transactions                                          */
11 else:
12     if If transaction t ∈ T_E is cross-shard :
13         Call CrossShard(t) ;      // Invokes Consensus in multiple shards
14     else:
15         Call Consensus(t)
16 i ← i + 1
17 Go to step 3
```

---

addressed with simple solutions, such as changing the consensus protocol (tradeoff performance with security), replacing the epoch transition process with one similar to (fixed) OmniLedger, etc. Although OmniLedger (with the proposed fix) maintains a robust sharded ledger in a stronger model (as defined in Sect. 2), RapidChain introduces practical speedups on specific components of the system. These improvements are not asymptotically important – and thus not captured by our framework – but might be significant for the performance of deployed sharding protocols.

Finally, we include in the comparison **Chainspace**, which maintains a robust sharded transaction ledger but only in the permissioned setting against a static adversary. Chainspace could be secure in our model in the permissioned setting if it adopts OmniLedger's epoch transition protocols and the proposed fix for data availability in the verifiable compaction of state. We omit the security proofs for Chainspace since they are either included in [3] or are similar to OmniLedger.

Table 1. Summarizing sharding protocol properties under our model

Protocol	Persistence	Consistency	Liveness	Scalability	Permissionless	S.-adaptive
Elastico	✓	✗	✓	✗	✓	✓
Monoxide	✓	✓	✓	✗	✓	✓
OmniLedger	✓	✓	✗	✓	✓	✓
RapidChain	✓	✓	✓	✓	✓	~
Chainspace	✓	✓	✓	✓	✗	✗

Discussion. Although we restrict our evaluation to the most impactful (so far) sharding proposals, we stress that the power of our framework and the bounds we provide are not limited to these works. For instance, we observe that Chainweb [37], a recently deployed sharding proposal, does not scale because it violates Theorem 11. We believe our framework is general enough to cover most sharding approaches, and we aspire it will be established as a tool for proving the security of future sharding protocols.

Figures

Table 2. (Glossary) The parameters in our analysis.

n	number of parties
f	number of Byzantine parties
m	number of shards
v	average transaction size (number of inputs and outputs)
E	epoch, i.e., a set of consecutive rounds
T	set of transactions (input)
k	"depth" security parameter (persistence)
u	"wait" time (liveness)
ω_m	communication factor
ω_s	space factor
ω_c	computational factor
σ	throughput factor
μ	chain quality parameter
τ	chain growth parameter
v	average transaction size
m'	degree of parallelism
γ	average number of a shard's interacting shards (cross-shard)

Acknowledgments. The work was partially supported by the Austrian Science Fund (FWF) through the project CoRaF (grant agreement 2020388).

A   Limitations of Sharding Protocols

A.1   General Bounds

First, we prove there is no robust sharded transaction ledger that has a constant number of shards. Then, we show that there is no protocol that maintains a robust sharded transaction ledger against an adaptive adversary.

Lemma 14. *In any robust sharded transaction ledger the number of shards (parametrized by n) is $m = \omega(1)$.*

Proof. Suppose there is a protocol that maintains a constant number m of sharded ledgers, denoted by $x_1, x_2, \ldots, x_m$. Let n denote the number of parties and T the number of transactions to be processed (wlog assumed to be valid). A transaction is processed only if it is stable, i.e. is included deep enough in a ledger (k blocks from the end of the ledger where k a security parameter). Each ledger will include T/m transactions on expectation. Now suppose each party participates in only one ledger (best case), thus broadcasts, verifies, and stores the transactions of that ledger only. Hence, every party stores T/m transactions on expectation. The expected space factor is $\omega_s = \sum_{\forall i \in [n]} \sum_{\forall x \in L_i} |x^{\lceil k \rceil}|/(n|T|) = \sum_{\forall x \in L_i} \frac{T}{nmT} = \frac{n}{nm} = \Theta(\frac{1}{m}) = \Theta(1)$, when m in constant. Thus, scalability is not satisfied.

Suppose a party is participating in shard x_i. If the party maintains information (e.g. the headers of the chain for verification purposes) on the chain of shard x_j, we say that the party is a *light node* for shard x_j. In particular, *a light node for shard x_j maintains information at least proportional to the length of the shard's chain x_j.* This holds because blocks must be of constant size to be able to scale in bandwidth (aka communication), and thus storing all the headers of a shard is asymptotically similar in overhead to storing the entire shard with the block content. Sublinear light clients [15, 28] verifiably compact the shard's state, thus are not considered light nodes but are discussed later. We next prove that if parties act as light clients to all shards involved in cross-shard transactions, then the sharded ledger can scale only if each shard does not interact with all the other shards (or a constant fraction thereof).

Lemma 15. *For any robust sharded transaction ledger that requires every participant to be a light node for all the shards affected by cross-shard transactions, it holds $\mathbb{E}(\gamma) = o(m)$.*

Proof. We assumed that every ledger interacts on average with γ different ledgers, i. e., the cross-shard transactions involve γ many different shards on expectation. The block size is considered constant, meaning each block includes at most e transactions where e is constant. Thus, each party maintaining a ledger and being a light node to γ other ledgers must store on expectation $(1 + \frac{\gamma}{e})\frac{T}{m}$ information. Hence, the expected space factor is

$$\mathbb{E}(\omega_s) = \sum_{\forall i \in [n]} \sum_{\forall x \in L_i} |x^{\lceil k}|/(n|T|) = n\frac{(1 + \frac{\gamma}{e})\frac{T}{m}}{nT} = \Theta\left(\frac{\gamma}{m}\right)$$

where the second equation holds due to linearity of expectation. To satisfy scalability, we demand $\mathbb{E}(\omega_s) = o(1)$, thus $\gamma = o(m)$.

Next, we show that there is no protocol that maintains a robust transaction ledger against an adaptive adversary in our model. We highlight that our result holds because we assume *any node is corruptible* by the adversary. If we assume more restrictive corruption sets, e.g. each shard has at least one honest well-connected node, sharding against an adaptive adversary may be possible if we employ other tools, such as fraud and data availability proofs [4].

Theorem 10. *There is no protocol maintaining a robust sharded transaction ledger against an adaptive adversary in our model controlling $f \geq n/m$, where m is the number of shards, and n is the number of parties.*

Proof (Towards contradiction). Suppose there exists a protocol Π that maintains a robust sharded ledger against an adaptive adversary that corrupts $f = n/m$ parties. From the pigeonhole principle, there exists at least one shard x_i with at most n/m parties (independent of how shards are created). The adversary is adaptive, hence at any round can corrupt all parties of shard x_i. In a malicious shard, the adversary can perform arbitrary operations, thus can spend the same UTXO in multiple cross-shard transactions. However, for a cross-shard transaction to be executed it needs to be accepted by the output shard, which is honest. Now, suppose Π allows the parties of each shard to verify the ledger of another shard. For Lemma 15 to hold, the verification process can affect at most $o(m)$ shards. Note that even a probabilistic verification, i.e., randomly select some transactions to verify, can fail due to storage requirements and the fact that the adversary can perform arbitrarily many attacks. Therefore, for each shard, there are at least 2 different shards that do not verify the cross-shard transactions (since Lemma 15 essentially states they cannot all be verified). Thus, the adversary can simply attempt to double-spend the same UTXO across every shard and will succeed in the shards that do not verify the validity of the cross-shard transaction. Hence, consistency is not satisfied.

A.2   Bounds Under Uniform Shard Creation

In this section, we assume that the creation of shards is UTXO-dependent; transactions are assigned to shards independently and uniformly at random. This assumption is in sync with the proposed protocols in the literature. In a non-randomized process of creating shards, the adversary can precompute and thus bias the process in a permissionless system. Hence, all sharding proposals employ a random process for shard creation. Furthermore, all shards validate approximately the same amount of transactions; otherwise the efficiency of the protocol would depend on the shard that validates most transactions. For this reason, we assume the UTXO space is partitioned to shards uniformly at random. Note that we consider UTXOs to be random strings.

Under this assumption, we prove a constant fraction of transactions are cross-shard on expectation. As a result, we prove no sharding protocol can maintain a robust sharded ledger when participants act as light clients on all shards involved in cross-shard transactions. Our observations hold for any transaction distribution D_T that results in a constant fraction of cross-shard transactions.

Lemma 16. *The expected number of cross-shard transactions is $\Theta(|T|)$.*

Proof. Let Y_i be the random variable that shows if a transaction is cross-shard; $Y_i = 1$ if $tx_i \in T$ is cross-shard, and 0 otherwise. Since UTXOs are assigned to shards uniformly at random, $Pr[i \in x_k] = \frac{1}{m}$, for all $i \in v$ and $k \in [m] = \{1, 2, \ldots, m\}$. The probability that all UTXOs in a transaction $tx \in T$ belong to the same shard is $\frac{1}{m^{v-1}}$ (where v is the cardinality of UTXOs in tx). Hence, $Pr[Y_i = 1] = 1 - \frac{1}{m^{v-1}}$. Thus, the expected number of cross-shard transactions is $\mathbb{E}(\sum_{\forall tx_i \in T} Y_i) = |T|(1 - \frac{1}{m^{v-1}})$. Since, $m(n) = \omega(1)$ (Lemma 14) and v constant, the expected cross-shard transactions converges to T for n sufficiently large.

Lemma 17. *For any protocol that maintains a robust sharded transaction ledger, it holds $\gamma = \Theta(m)$.*

Proof. We assume each transaction has a single input and output, hence $v = 2$. This is the worst-case input for evaluating how many shards interact per transaction; if $v \gg 2$ then each transaction would most probably involve more than two shards and thus each shard would interact with more different shards for the same set of transactions.

For $v = 2$, we can reformulate the problem as a graph problem. Suppose we have a random graph G with m nodes, each representing a shard. Now let an edge between nodes u and w represent a transaction between shards u and w. Note that in this setting we allow self-loops, which represent the intra-shard transactions. We create the graph G with the following random process: We choose an edge independently and uniformly at random from the set of all possible edges including self-loops, denoted by E'. We repeat the process independently $|T|$ times, i.e., as many times as the cardinality of the transaction set. We note that each trial is independent and the edges chosen uniformly at random due to the corresponding assumptions concerning the transaction set and the shard creation. We will now show that the average degree of the graph is $\Theta(m)$, which immediately implies the statement of the lemma.

Let the random variable Y_i represent the existence of edge i in the graph, i.e., $Y_i = 1$ if edge i was created at any of the T trials, 0 otherwise. The set of all possible edges in the graph is E, $|E| = \binom{m}{2} = \frac{m(m-1)}{2}$. Note that this is not

the same as set E' which includes self-loops and thus $|E'| = \binom{m}{2} + m = \frac{m(m+1)}{2}$.
For any vertex u of G, it holds

$$\mathbb{E}[deg(u)] = \frac{2\mathbb{E}[\sum_{\forall i \in E} Y_i]}{m}$$

where $deg(u)$ denotes the degree of node u. We have,

$$Pr[Y_i = 1] = 1 - Pr[Y_i = 0]$$

$$= 1 - Pr[Y_i = 0 \text{ at trial } 1] Pr[Y_i = 0 \text{ at trial } 2] \dots$$

$$Pr[Y_i = 0 \text{ at trial } T] = 1 - \left(1 - \frac{2}{m(m+1)}\right)^{|T|}$$

Thus,

$$\mathbb{E}[deg(u)] = \frac{2m(m-1)}{2}\left[1 - \left(1 - \frac{2}{m(m+1)}\right)^{|T|}\right]$$

$$= (m-1)\left[1 - \left(1 - \frac{2}{m(m+1)}\right)^{|T|}\right]$$

Therefore, for many transactions we have $|T| = \omega(m^2)$ and consequently
$\mathbb{E}[deg(u)] = \Theta(m)$.

Theorem 11. *There is no protocol that maintains a robust sharded transaction ledger in our model under uniform space partition when parties are light nodes on the shards involved in cross-shard transactions.*

Proof. Immediately follows from Lemmas 15 and 17.

A.3   Bounds Under Random Permutation of Parties to Shards

In this section, we assume parties are periodically randomly shuffled among shards, using a random permutation of their IDs. Any other shard assignment strategy yields equivalent or worse guarantees since we have no knowledge of which parties are Byzantine. Our goal is to upper bound the number of shards for a protocol that maintains a robust sharded transaction ledger in our security model. To satisfy the security properties, we demand each shard to contain at least a constant fraction of honest parties $1 - a$ ($< 1 - \frac{f}{n}$), where a is the tolerance of the shards. This is due to classic lower bounds of consensus protocols [35].

The *size* of a shard is the number of the parties assigned to the shard. We say shards are *balanced* if all shards have approximately the same size. In what follows, we assume shards to be balanced (this can be done by drawing uniformly at random a balanced partition of parties). We denote by $p = f/n$ the (constant) fraction of the Byzantine parties. A shard is *a-honest* if at least a fraction of $1 - a$ parties in the shard are honest.

The following lemma, proven by Raab and Steger [42] will be useful later:

Lemma 18. *Let M be the random variable that counts the number of balls in any bin if we throw pn balls independently and uniformly at random into m bins. Then $Pr[M > k_\alpha] = o(1)$ if $\alpha > 1$ and $Pr[M > k_\alpha] = 1 - o(1)$ if $0 < \alpha < 1$, where*

$$
k_\alpha = \begin{cases}
\frac{\log m}{\log \frac{m \log m}{pn}} * (1 + \alpha \frac{\log^{(2)} \frac{m \log m}{pn}}{\log \frac{m \log m}{pn}}) & \text{if } \frac{m}{polylog(m)} \leq pn \ll m \log m, \\
(d_c - 1 + \alpha) \log m & \text{if } pn = cm \log m \text{ for some constant } c, \\
\frac{pn}{m} + \alpha \sqrt{2 \frac{pn}{m} \log m} & \text{if } m \log m \ll pn \leq mpolylog(m), \\
\frac{pn}{m} + \sqrt{\frac{2pn \log m}{m}} (1 - \frac{\log^{(2)} m}{2\alpha \log m}) & \text{if } pn \gg m(\log m)^3
\end{cases}
$$

$$(1)$$

Lemma 19. *Given n parties are assigned uniformly at random to m shards of constant size $s = \frac{n}{m}$ and the adversary corrupts at most $f = pn$ parties, all shards are a-honest (p, a are constants with p the proportion of corrupted parties and a the tolerance of the model) with probability $1 - o(1)$ if and only if the number of shards is at most $n = cmlog(m)/p$, where c is a constant and p/a is small enough depending only on the value of c.*

Proof. We start by reformulating the problem in order to show it is equivalent to the well-know Generalized Birthday Paradox.

Assuming we build m shards of equivalent size $s = \frac{n}{m}$ using a random permutation with uniform probability. Then this is equivalent to distributing the Byzantine processes to shards at random following a uniform law, but with the shards being of maximum size s. In other words, we throw $f = pn$ balls in m bins of limited capacity s. We would like to know the probability that the maximum load of the bins be greater or equal to a.

Reformulated as the Birthday paradox, what is the probability that, in a room of n people whose birthdays are spread uniformly at random over m days, a people share the same birthday? We denote that probability by $f(pn, m, a)$.

Notice that our reformulation as the Birthday Paradox does not take into account the limited size of the possible birthdays (no more than s people can have the same birthday). Both problems are however equivalent, as we can reconstruct that probability easily using Bayes' formula:

$$
P(A|B) = \frac{P(B|A) * P(A)}{P(B)}
$$

Where $A =$ "the maximum load is $\leq as$", $B =$ "the maximum load is $\leq s$" and $A|B = C =$"all shards are $a - honest$. $P(B|A) = 1$ since $a < 1$ so

$$
P(C) = \frac{P(A)}{P(B)}
$$

hence solving the Birthday Paradox solves our problem with very little additional calculation. Our calculation will actually be conducted using $A' =$ "the maximum load is $\geq as$" and $B' =$ "the maximum load is $\geq s$"

$$P(C) = \frac{1 - P(A')}{1 - P(B')}$$

Since $\frac{1 - o(1)}{1 - o(1)} \geq 1 - o(1)$, it is sufficient for $P(C) = 1 - o(1)$ that $P(A') = o(1)$ and $P(B') = o(1)$. The problem is sometimes denoted as the Cell Occupancy Problem [23].

We then use Lemma 18 (beware, in the original paper [42] n and m are reverse when compared with our notation). We want $\alpha > 1$, $k_\alpha = \frac{an}{m}$.

When applying this, we immediately get impossible equations for the third and fourth values of k_α, hence it is not possible to have m in that range of values compared to n ($m \gg n\log(n)$):

$$\frac{an}{m} = \frac{pn}{m} + \alpha\sqrt{2\frac{pn}{m}\log m}$$

$$\frac{(a - p)n}{m} = \alpha\sqrt{2\frac{pn}{m}\log m}$$

$$\frac{n}{m} = \frac{\alpha\sqrt{2p}}{(a - p)}\sqrt{\frac{n}{m}\log m}$$

$$\sqrt{n} = \frac{\alpha\sqrt{2p}}{(a - p)}\sqrt{m\log m}$$

$$n = \frac{\alpha^2 2p}{(a - p)^2}m\log m$$

As we can see, we also violate the hypothesis that $pn \gg m\log m$, which is absurd. For the fourth equation, we can simply notice that since $\alpha > 1$, $(1 - \frac{\log^{(2)} m}{2\alpha\log m}) \leq 1$ hence reusing the calculation made for the third case n will be even smaller when compared with $m\log m$, thus the hypothesis $pn \gg m(\log m)^3$ is broken.

The equations however is correct under the hypothesis that $pn = cm\log m$ (see calculation below). This indicates that this is as high a value of m we can use while keeping the shards safe with overwhelming probability.

$$\frac{an}{m} = (d_c - 1 + \alpha)\log m$$

$$n = \frac{1}{a}(d_c - 1 + \alpha)m\log m$$

We can see already that we are indeed verifying the hypothesis $pn = cm\log m$ for some constant c (the constant d_c is a scalar not dependant on either n or m). If $k_\alpha = \frac{n}{m}$, then $n = (d_c - 1 + \alpha)m\log m$ and the hypothesis is also verified.

We now need to make sure that $\alpha > 1$ for both cases.

Since, by hypothesis, $pn = cm\log m$, we identify that $c = \frac{p}{a}(d_c - 1 + \alpha)$, where $d_c \geq c$. In order to obtain $\alpha > 1$, it is necessary that $c > \frac{p}{a}d_c$ where $p < a$. d_c is a function of c with $d_c > c$, hence for a given c it is always possible to enforce $\alpha > 1$ if p/a is small enough.

for the case $k_\alpha = \frac{n}{m}$, the previous result holds trivially with $a = 1$.

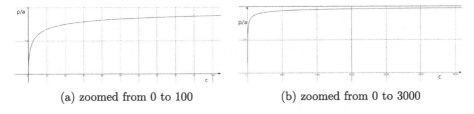

(a) zoomed from 0 to 100   (b) zoomed from 0 to 3000

Fig. 1. $p/a = g(c)$ as described in Corollary 20. p is the proportion of corrupted parties in the system, while $1 - a$ is the maximum proportion of corrupted parties allowed per shard.

Using the previous calculations, we can exhibit the trade-off between security and scalability in a mathematical formulation in Corollary 20. A systems designer may choose to adjust either parameter p/a or c, one being computed thanks to the chosen value of the other. Since the expression is not mathematically intuitive, we provide a plotting of the increasing function $p/a = g(c)$ in Fig. 1.

Corollary 20. *In a sharding protocol maintaining a robust sharded transaction ledger against an adversary, the trade-off between scalability (low value of c) and security (high value of p/a) is described by $\frac{c}{d_c} > \frac{p}{a}$. c is the multiplicative constant in the relation $pn = cm\log(m)$, d_c is a function of c, while p and $1 - a$ are the proportion of corrupted parties in the system and per shard, respectively.*

Proof. According to Lemma 19, the constant d_c is a real number dependant only on c and

$$\frac{c}{d_c} > \frac{p}{a}$$

which means the value of p/a is ceiled by the value of c/d_c.

As explained in [42], d_c is the solution to the equation $1 + x(log(c) - log(x) + 1) - c = 0$ that is greater than c. Thus we have the exact mathematical expression of the well-known security/scalability trade-off.

Corollary 21. *In a sharding protocol maintaining a robust sharded transaction ledger against an adversary, m is upper-bounded by $f(n) = \frac{n}{c'\log(\frac{n}{c'\log(n)})}$ with $c' = \frac{c}{p}$ and c a constant as described in Corollary 20.*

Proof. Because of Lemma 19, $cm\log(m) = pn$. using $m = \frac{n}{c'\log(m)}$ (a), we obtain $m = \frac{n}{c'\log(\frac{n}{c'\log(m)})}$ and since $n \geq m$, an upper-bound is $f(n) = \frac{n}{c'\log(\frac{n}{c'\log(n)})}$. Note we could build a tighter but more complex upper bound by replacing m by its expression (a) instead of n as many times as desired.

Next, we prove that any sharding protocol may scale at most by an $n/\log n$ factor. This bound refers to independent nodes. If, for instance, we "shard" per authority, but all authorities represented in each shard, the bound of the theorem does not hold and the actual system should be considered sharded since every authority holds all the data.

Theorem 12. *Any protocol that maintains a robust sharded transaction ledger in our model under uniformly random partition of the state and parties, can scale at most by a factor of m, where $n = c'm \log m$ and the constant c' encompasses the trade-off between security and scalability.*

Proof. In our security model, the adversary can corrupt $f = pn$ parties, p constant. Hence, from Corollary 20, $m = O(\frac{n}{\log m})$. Each party stores at least T/m transactions on average and thus the expected space factor is $\omega_s \geq n\frac{T/m}{T} = \frac{n}{m}$. Therefore, any sharding protocol can scale at most $O(\frac{n}{\log m})$.

Next, we show that any sharding protocol that satisfies scalability requires some process of *verifiable compaction of state* such as checkpoints [33], cryptographic accumulators [10], zero-knowledge proofs [9], non-interactive proofs of proofs-of-work [15,28], proof of necessary work [27] or erasure codes [26]. Such a process allows the state of the distributed ledger (e.g., stable transactions) to be compressed significantly while users can verify the correctness of the state. Intuitively, in any sharding protocol secure against a slowly adaptive adversary parties must periodically shuffle in shards. To verify new transactions the parties must receive a verifiably correct UTXO pool for the new shard without downloading the full shard history; otherwise the communication overhead of the bootstrapping process eventually exceeds that of a non-sharded blockchain. Although existing evaluations typically ignore this aspect with respect to bandwidth, we stress its importance in the long-term operation: *the bootstrap cost will eventually become the bottleneck due to the need for nodes to regularly shuffle.*

Theorem 13. *Any protocol that maintains a robust sharded transaction ledger in our model, under uniformly random partition of the state and parties, employs verifiable compaction of the state.*

Proof (Towards contradiction). Suppose there is a protocol that maintains a robust sharded ledger without employing any process that verifiably compacts the blockchain. To guarantee security against a slowly-adaptive adversary, the parties change shards at the end of each epoch. At the beginning of each epoch, the parties must process a new set of transactions. To check the validity of this new set of transactions, each (honest) shard member downloads and maintains the corresponding ledger. Note that even if the party only maintains the hash-chain of a ledger, the cost is equivalent to maintaining the list of transactions given that the block size is constant. We will show that the communication factor increases with time, eventually exceeding that of a non-sharded blockchain; thus scalability is not satisfied from that point on.

In each epoch transition, a party changes shards with probability $1 - 1/m$, where m is the number of shards. As a result, a party changing a shard in epoch k must download the shard's ledger of size $\dfrac{k \cdot T}{m}$. Therefore, the expected communication factor of bootstrapping during the k-th epoch transition is $\dfrac{k \cdot T}{m} \cdot (1 - \dfrac{1}{m})$. We observe the communication overhead grows with the number of epochs k, hence it will eventually become the scaling bottleneck. For instance, for $k > m \cdot n$, the communication factor is greater than linear to the number of parties in the system n, thus the protocol does not satisfy scalability.

Theorem 13 holds even if parties are not assigned to shards uniformly at random but follow some other shuffling strategy like in [43]. *As long as a significant fraction of honest parties change shards from epoch to epoch, verifiable compaction of state is necessary* to restrict the bandwidth requirements during bootstrapping in order to satisfy scalability.

B   Analysis

We show that Divide & Scale is secure in our model (i.e., satisfies persistence, consistency, and liveness), while its efficiency (i.e., scalability and throughput factor) depends on the chosen subprotocols. For the purpose of our analysis, we assume all employed subprotocols satisfy liveness.

Theorem 22. *Divide & Scale satisfies persistence in our system model assuming at most f Byzantine nodes.*

Proof. Assuming `Sybil` guarantees the fair distribution of identities (Sybil, property iv), and `Divide2Shard` maintains the distribution within the desired limits to guarantee the securities bounds of `Consensus` (Divide2Shard, property iii), the common prefix property is satisfied in each shard, so persistence is satisfied.

Theorem 23. *Divide & Scale satisfies consistency in our system model assuming at most f Byzantine nodes.*

Proof. Transactions can either be intra-shard (all UTXOs within a single shard) or cross-shard. Consistency is satisfied for intra-shard transactions as long as `Sybil` and `Divide2Shard` result in a distribution that respects the security bounds of `Consensus`, hence the common prefix property is satisfied. Furthermore, consistency is satisfied for cross-shard transactions from the `CrossShard` protocol as long as it correctly provides atomicity.

Theorem 24. *Divide & Scale satisfies liveness in our system model assuming at most f Byzantine nodes.*

Proof. Follows from the assumption that all subprotocols satisfy liveness, as well as the `CompactState` protocol that ensures data availability between epochs.

Scalability. The scalability of Divide & Scale depends on the worse scaling factor, i.e., communication, space, computation, of all the components it employs. The maximum scaling factor for `DRG`, `Divide2Shards`, `Sybil`, and `CompactState` can be amortized over the rounds of an epoch because these protocols are executed once per epoch. Thus, the size of an epoch is critical for scalability. Intuitively, this implies that *if the size of the epoch is small, hence the adversary highly-adaptive, sharding is not that beneficial as the protocols that are executed on the epoch transaction are as resource demanding as the consensus in a non-sharded system.*

Throughput Factor. Similarly to scalability, the throughput factor also depends on the chosen subroutines, and in particular, Consensus and CrossShards. To be specific, the throughput factor depends on the shard growth and shard quality parameters which are determined by Consensus. In addition, given a transaction input, the degree of parallelism, which is the last component of the throughput factor, is determined by the maximum number of shards possible and the way cross-shard transactions are handled. The maximum number of shards depends on Consensus and Divide2Shards, while CrossShard determines how many shards are affected by a single transaction. For instance, if the transactions are divided in shards uniformly at random, Divide & Scale can scale at most by $n/\log n$ as stated in Corollary 20. We further note that the minimum number of affected shards for a specific transaction is the number of UTXOs that map to different shards; otherwise security cannot be guaranteed.

We demonstrate in Appendix C how to calculate the scaling factors and the throughput factor for OmniLedger and RapidChain.

C   Evaluation of Existing Protocols

In this section, we evaluate existing sharding protocols in our model with respect to the desired properties defined in Sect. 2.2. A summary of our evaluation can be found in Table 1 in Sect. 5.

The analysis is conducted in the synchronous model and thus any details regarding performance on periods of asynchrony are discarded. The same holds for other practical refinements that do not asymptotically improve performance.

C.1   Elastico

Overview. Elastico is the first distributed blockchain sharding protocol introduced by Luu et al. [36]. The protocol lies in the intersection of traditional BFT protocols and the Nakamoto consensus. The protocol is synchronous and proceeds in epochs. The setting is permissionless, and during each epoch, the participants create valid identities for the next epoch by producing proof-of-work (PoW) solutions. The adversary is slowly-adaptive (see Sect. 2) and controls at most 25% of the computational power of the system or equivalently $f < \frac{n}{4}$ out of n valid identities in total.

At the beginning of each epoch, parties are partitioned into small shards (committees) of constant size c. The number of shards is $m = 2^s$, where s is a small constant such that $n = c \cdot 2^s$. A shard member contacts its directory committee to identify the other members of the same shard. For each party, the directory committee consists of the first c identities created in the epoch in the party's local view. Transactions are randomly partitioned in disjoint sets based on the hash of the transaction input (in the UTXO model); hence, each shard only processes a fraction of the total transactions in the system. The shard members execute a BFT protocol to validate the shard's transactions and then send the validated transactions to the final committee. The final committee consists of all members with a fixed s-bit shards identity, and is in charge of two operations: (i) computing and broadcasting the final block, which is a digital signature on the union of all valid received transactions[6] (via executing a BFT protocol), and (ii) generating and broadcasting a bounded exponential biased random string to be used as a public source of randomness in the next epoch (e.g. for the PoW).

Consensus: Elastico does not specify the consensus protocol but instead can employ any standard BFT protocol, like PBFT [17].

CrossShard & StatePartition: Each transaction is assigned to a shard according to the hash of the transaction's inputs. Every party maintains the entire blockchain, thus each shard can validate the assigned transaction independently, i. e., there are no cross-shard transactions. Note that Elastico assumes that transactions have a single input and output, which is not the case in cryptocurrencies as discussed in Sect. 3. To generalize Elastico's transaction assignment method to multiple inputs, we assume each transaction is assigned to the shard corresponding to the hash of all its inputs. Otherwise, if each input is assigned to a different shard according to its hash value, an additional protocol is required to guarantee the atomicity of transactions and hence the security (consistency) of Elastico.

Sybil: Participants create valid identities by producing PoW solutions using the randomness of the previous epoch.

Divide2Shards & CompactState: The protocol assigns each identity to a random shard in 2^s, identified by an s-bit shard identity. At the end of each epoch, the final committee broadcasts the final block that contains the Merkle hash root of every block of all shards' block. The final block is stored by all parties in the system. Hence, when the parties are re-assigned to new shards they already have the hash-chain to confirm the shard ledger and future transactions. Essentially, an epoch in Elastico is equivalent to a block generation round.

DRG: In each epoch, the final committee (of size c) generates a set of random strings R via a commit-and-XOR protocol. First, all committee members generate an r-bit random string r_i and send the hash $h(r_i)$ to all other committee members. Then, the committee runs an interactive consistency protocol to agree on a single set of hash values S, which they include on the final block. Later, each (honest) committee member broadcasts its random string r_i to all parties in the network. Each party chooses and XORs $c/2 + 1$ random strings for which the corresponding hash exists in S. The output string is the party's randomness for the epoch. Note that $r > 2\lambda + c - \log(c)/2$, where λ is a security parameter.

---

[6] The final committee in Elastico broadcasts only the Merkle root for each block. However, this is asymptotically equivalent to including all transactions since the block size is constant. Furthermore, the final committee does not check if the received transactions are conflicting but merely verifies the presence of signatures.

Analysis. Elastico's threat model allows for adversaries that can drop or modify messages, and send different messages to honest parties, which is not allowed in our model. However, we show that even under a more restrictive adversarial model, Elastico fails to meet the desired sharding properties. Specifically, we prove Elastico does not satisfy *scalability* and *consistency*. From the security analysis of [36], it follows that Elastico satisfies persistence and liveness in our system model.

Theorem 25. *Elastico does not satisfy consistency in our system model.*

Proof. Suppose a party submits two valid transactions, one spending input x and another spending input x and input y. Note that the second is a single transaction with two inputs. In this case, the probability that both hashes (transactions), $H(x, y)$ and $H(x)$, land in the same shard is $1/m$. Hence, the probability of a successful double-spending in a set of T transactions is almost $1 - (1/m)^T$, which converges to 0 as T grows, for any value $m > 1$. However, $m > 1$ is necessary to satisfy scalability (Lemma 14). Therefore, there will be almost surely a round in which two parties report two conflicting transactions. Since the final committee does not verify the validity of transactions but only checks the appropriate signatures are present, consistency is not satisfied.

Lemma 26. *The communication and space factors of Elastico are $\omega_m = \Theta(n)$ and $\omega_s = \Theta(1)$.*

Proof. At the end of each epoch, which corresponds to the generation of one block per shard, the final committee broadcasts the final block to the entire network. All parties download and store the final block. hence all parties maintain the entire input set of transactions. Since the block size is considered constant, downloading and storing the final block which consists of the hash-chains of all shards is equivalent to downloading and storing all the shards' ledgers. It follows that the space factor is $\omega_s = \Theta(1)$ as all parties store a constantly-compressed version of the input T, regardless of the nature of the input set T. Similarly, it follows that the communication factor is $\omega_m = \Theta(n)$ as the broadcast of the final block takes place regularly at the generation of one block per shard, i. e., Elastico's epoch.   □

Theorem 27. *Elastico does not satisfy scalability in our system model.*

Proof. Immediately follows from Definition 4 and Lemma 26.   □

C.2   Monoxide

Overview. Monoxide [53] is an asynchronous proof-of-work protocol, where the adversary controls at most 50% of the computational power of the system. The protocol uniformly partitions the space of user addresses into shards (zones) according to the first k bits. Every party is permanently assigned to a shard uniformly at random. Each shard employs the GHOST [47] consensus protocol.

Participants are either full-nodes that verify and maintain the transaction ledgers, or miners investing computational power to solve PoW puzzles for profit in addition to being full-nodes. Monoxide introduces a new mining algorithm, called Chu-ko-nu, that enables miners to mine in parallel for all shards. The Chu-ko-nu algorithm aims to distribute the hashing power to protect individual shards from an adversarial takeover. Successful miners include transactions in blocks. A block in Monoxide is divided into two parts: the chaining block that includes all metadata (Merkle root, nonce for PoW, etc.) creating the hash-chain, and the transaction-block that includes the list of transactions. All parties maintain the hash-chain of every shard in the system.

Furthermore, all parties maintain a distributed hash table for peer discovery and identifying parties in a specific shard. This way the parties of the same shard can identify each other and cross-shard transactions are sent directly to the destination shard. Cross-shard transactions are validated in the shard of the payer and verified from the shard of the payee via a relay transaction and the hash-chain of the payer's shard.

Consensus: The consensus protocol of each shard is GHOST [47]. GHOST is a DAG-based consensus protocol similar to Nakamoto consensus [40], but the consensus selection rule is the heaviest subtree instead of the longest chain.

StatePartition: Monoxide is account-based hence all transactions are single input and single output.

CrossShard: An input shard is a shard that corresponds to the address of a sender of a transaction (payer) while an output shard one that corresponds to the address of a receiver of a transaction (payee). Each cross-shard transaction is processed in the input shard, where an additional relay transaction is created and included in a block. The relay transaction consists of all metadata needed to verify the validity of the original transaction by only maintaining the hash-chain of a shard (i. e. for light nodes). The miner of the output shard verifies that the relay transaction is stable and then includes it in a block in the output shard. Note that in case of forks in the input shard, Monoxide invalidates the relay transactions and rewrites the affected transaction ledger to maintain consistency.

Sybil: In a typical PoW election scheme, the adversary can create many identities and target its computational power to specific shards to gain control over more than half of the shard's participants. In such a case, the security of the protocol fails (both persistence and consistency properties do not hold). To address this issue, Monoxide introduces a new mining algorithm, Chu-ko-nu, that allows parallel mining on all shards. Specifically, a miner can batch valid transactions from all shards and use the root of the Merkle tree of the list of chaining headers in the batch as input to the hash, alongside with the nonce (and some configuration data). Thus, when a miner successfully computes a hash lower than the target, the miner adds a block to every shard.

Divide2Shards: Parties are permanently assigned to shards uniformly at random according the first k bits of their address.

DRG: The protocol uses deterministic randomness (e.g. hash function) and does not require any random source.

CompactState: No compaction of state is used in Monoxide.

Analysis. We prove that Monoxide satisfies persistence, liveness, and consistency, but *does not satisfy scalability*. The same result is also immediately derived from our impossibility result stated in Theorem 11 as Monoxide demands each party to verify cross-shard transactions by acting as a light node to all shards; effectively demonstrating the effectiveness of our framework and the usability of our results.

Theorem 28. *Monoxide satisfies persistence and liveness in our system model for $f < n/2$.*

Proof. From the analysis of Monoxide, it holds that if all honest miners follow the Chu-ko-nu mining algorithm, then honest majority within each shard holds with high probability for any adversary with $f < n/2$ (Sect. 5.3 [53]).

Assuming honest majority within shards, persistence depends on two factors: the probability a stable transaction becomes invalid in a shard's ledger, and the probability a cross-shard transaction is reverted after being confirmed. Both these factors solely depend on the common prefix property of the shards' consensus mechanism. Monoxide employs GHOST as the consensus mechanism of each shard, hence the common prefix property is satisfied if we assume that invalidating the relay transaction does not affect other shards [29]. Suppose common prefix is satisfied with probability $1 - p$ (which is overwhelming on the "depth" security parameter k). Then, the probability none of the outputs of a transaction are invalidated is $(1-p)^{(v-1)}$ (worst case where $v - 1$ outputs – relay transactions – link to one input). Thus, a transaction is valid in a shard's ledger after k blocks with probability $(1 - p)^v$, which is overwhelming in k since v is considered constant. Therefore, persistence is satisfied.

Similarly, liveness is satisfied within each shard. Furthermore, this implies liveness is satisfied for cross-shard transactions. In particular, both the initiative and relay transactions will be eventually included in the shards' transaction ledgers, as long as chain quality and chain growth are guaranteed within each shard [29]. □

Theorem 29. *Monoxide satisfies consistency in our system model for $f < n/2$.*

Proof. The common prefix property is satisfied in GHOST [30] with high probability. Thus, intra-shard transactions satisfy consistency with high probability (on the "depth" security parameter). Furthermore, if a cross-shard transaction output is invalidated after its confirmation, Monoxide allows rewriting the affected transaction ledgers. Hence, consistency is restored in case of cross-transaction failure. Thus, overall, consistency is satisfied in Monoxide. □

Note that allowing to rewrite the transaction ledgers in case a relay transaction is invalidated strengthens the consistency property but weakens the persistence and liveness properties.

Intuitively, to satisfy persistence in a sharded PoW system, the adversarial power needs to be distributed across shards. To that end, Monoxide employs a new mining algorithm, Chu-ko-nu, that incentivizes honest parties to mine in parallel on all shards. However, this implies that a miner needs to verify transactions on all shards and maintain a transaction ledger for all shards. Hence, the computation and space factors are proportional to the number of (honest) participants and the protocol does not satisfy scalability.

Theorem 30. *Monoxide does not satisfy scalability in our system model for* $f < n/2$.

Proof. Let m denote the number of shards (zones), m_p the fraction of mining power running the Chu-ko-nu mining algorithm and m_d the rest of the mining power ($m_p + m_d = 1$). Additionally, suppose m_s denotes the mining power of one shard. The Chu-ko-nu algorithm enforces the parties to verify transactions that belong to all shards, hence the parties store all sharded ledgers. To satisfy scalability, the space factor of Monoxide can be at most $o(1)$. Similarly, it follows that the verification overhead expressed through the computational factor must be bounded by $o(n)$. Thus, at most $o(n)$ parties can run the Chu-ko-nu mining algorithm, hence $nm_p = o(n)$. We note that the adversary will not participate in the Chu-ko-nu mining algorithm as distributing the hashing power is to the adversary's disadvantage.

To satisfy persistence, every shard running the GHOST protocol [47] must satisfy the common prefix property. Thus, the adversary cannot control more than $m_a < m_s/2$ hash power, where $m_s = \frac{m_d}{m} + m_p$. Consequently, we have $m_a < \frac{m_s}{2(m_d+m_p)} = \frac{1}{2} - \frac{m_d(m-1)}{2m(m_d+m_p)}$. For n sufficiently large, m_p converges to 0; hence $m_a < \frac{1}{2} - \frac{(m-1)}{2m} = \frac{1}{2m}$. From Lemma 14, $m = \omega(1)$, thus the adversarial power $m_a < 0$ for sufficiently large n. We conclude that Monoxide does not satisfy scalability in our model. Moreover, we identify in Monoxide a clear trade-off between security and scaling storage and verification.   □

C.3   OmniLedger

Overview. OmniLedger [33] proceeds in epochs, assumes a partially synchronous model within each epoch (to be responsive), synchronous communication channels between honest parties (with a large maximum delay), and a slowly-adaptive computationally-bounded adversary that can corrupt up to $f < n/4$ parties.

The protocol bootstraps using techniques from ByzCoin [31]. The core idea is that there is a global identity blockchain that is extended once per epoch with Sybil resistant proofs (proof-of-work, proof-of-stake, or proof-of-personhood [12]) coupled with public keys. At the beginning of each epoch a sliding window mechanism is employed to define the eligible validators as the ones with identities in the last W blocks, where W depends on the adaptivity of the adversary. For our definition of slowly adaptive, we set $W = 1$. The UTXO space is partitioned uniformly at random into m shards, each shard maintaining its own ledger.

At the beginning of each epoch, a new common random value is created via a distributed randomness generation (DRG) protocol. The DRG protocol employs verifiable random functions (VRF) to elect a leader who runs RandHound [51] to create the random value. The random value is used as a challenge for the next epoch's identity registration and as a seed to assigning identities of the current epoch into shards.

Once the participants for this epoch are assigned to shards and bootstrap their internal states, they start validating transactions and updating the shards' transaction ledgers by operating ByzCoinX, a modification of ByzCoin [31]. When a transaction is cross-shard, a protocol that ensures the atomic operation of transactions across shards called *Atomix* is employed. Atomix is a client-driven atomic commit protocol secure against Byzantine adversaries.

Consensus: OmniLedger suggests the use of a strongly consistent consensus in order to support Atomix. This modular approach means that any consensus protocol [17,25,31,32,44] works with OmniLedger as long as the deployment setting of OmniLedger respects the limitations of the consensus protocol. In its experimental deployment, OmniLedger uses a variant of ByzCoin [31] called ByzCoinX [32] in order to maintain the scalability of ByzCoin and be robust as well. We omit the details of ByzCoinX as it is not relevant to our analysis.

StatePartition: The UTXO space is partitioned uniformly at random into m shards.

CrossShard (Atomix): Atomix is a client-based adaptation of two-phase atomic commit protocol running with the assumption that the underlying shards are correct and never crash. This assumption is satisfied because of the random assignment of parties to shards, as well as the Byzantine fault-tolerant consensus of each shard.

In particular, Atomix works in two steps: First, the client that wants the transaction to go through requests a proof-of-acceptance or proof-of-rejection from the shards managing the inputs, who log the transactions in their internal blockchain. Afterwards, the client either collects proof-of-acceptance from all the shards or at least one proof-of-rejection. In the first case, the client communicates the proofs to the output shards, who verify the proofs and finish the transaction by generating the necessary UTXOs. In the second case, the client communicates the proofs to the input shards who revert their state and abort the transaction. Atomix, has a subtle replay attack, hence we analyze OmniLedger with the proposed fix [48].

Sybil: A global identity blockchain with Sybil resistant proofs coupled with public keys is extended once per epoch.

Divide2Shards: Once the parties generate the epoch randomness, the parties can independently compute the shard they are assigned to for this epoch by permuting (*mod* n) the list of validators (available in the identity chain).

`DRG`: The DRG protocol consists of two steps to produce unbiasable randomness. On the first step, all parties evaluate a VRF using their private key and the randomness of the previous round to generate a "lottery ticket". Then the parties broadcast their ticket and wait for Δ to be sure that they receive the ticket with the lowest value whose generator is elected as the leader of RandHound.

This second step is a partially-synchronous randomness generation protocol, meaning that even in the presence of asynchrony safety is not violated. If the leader is honest, then eventually the parties will output an unbiasable random value, whereas if the leader is dishonest there are no liveness guarantees. To recover from this type of fault the parties can view-change the leader and go back to the first step in order to elect a new leader.

This composition of randomness generation protocols (leader election and multiparty generation) guarantees that all parties agree on the final randomness (due to the view-change) and the protocol remains safe in asynchrony. Furthermore, if the assumed synchrony bound (which can be increasing like PBFT [17]) is correct, an honest leader will be elected in a constant number of rounds.

Note, however, that the DRG protocol is modular, thus any other scalable distributed randomness generation protocol with similar guarantees, such as Hydrand [45] or Scrape [16], can be used.

`CompactState`: A key component that enables OmniLedger to scale is the epoch transition. At the end of every epoch, the parties run consensus on the state changes and append the new state (e.g. UTXO pool) in a state-block that points directly to the previous epoch's state-block. This is a classic technique [17] during reconfiguration events of state machine replication algorithms called checkpointing. New validators do not replay the actual shard's ledger but instead, look only at the checkpoints which help them bootstrap faster.

In order to guarantee the continuous operation of the system, after the parties finish the state commitment process, the shards are reconfigured in small batches (at most 1/3 of the parties in each shard at a time). If there are any blocks committed after the state-block, the validators replay the state-transitions directly.

Analysis. In this section, we prove OmniLedger satisfies persistence, consistency, and scalability (on expectation) but *fails to satisfy liveness*. Nevertheless, we estimate the efficiency of OmniLedger by providing an upper bound on its throughput factor.

Lemma 31. *At the beginning of each epoch, OmniLedger provides an unbiased, unpredictable, common to all parties random value (with overwhelming probability in t within t rounds).*

Proof. If the elected leader that orchestrates the distributed randomness generation protocol (RandHound or equivalent) is honest the statement holds. On the other hand, if the leader is Byzantine, the leader cannot affect the security of the protocol, meaning the leader cannot bias the random value. However, a

Byzantine leader can delay the process by being unresponsive. We show that there will be an honest leader, hence the protocol will output a random value, with overwhelming probability in the number of rounds t.

The adversary cannot pre-mine PoW puzzles, because the randomness of each epoch is used in the PoW calculation of the next epoch. Hence, the expected number of identities the adversary will control (number of Byzantine parties) in the next epoch is $f < n/4$. Hence, the adversary will have the smallest ticket – output of the VRF – and thus will be the leader that orchestrates the distributed randomness generation protocol (RandHound) with probability $1/2$. Then, the probability there will be an honest leader in t rounds is $1 - \frac{1}{2^t}$, which is overwhelming in t.

The unpredictability is inherited by the properties of the employed distributed randomness generation protocol. □

Lemma 32. *The distributed randomness generation protocol has $O(\frac{n \log^2 n}{R})$ amortized communication complexity, where R is the number of rounds in an epoch.*

Proof. The DRG protocol inherits the communication complexity of RandHound, which is $O(c^2 n)$ [45]. In [51], the authors claim that c is constant. However, the protocol requires a constant fraction of honest parties (e.g. $n/3$) in each of the n/c partitions of size c against an adversary that can corrupt a constant fraction of the total number of parties (e.g. $n/4$). Hence, from Lemma 19, we have $c = \Omega(\log n)$, which leads to communication complexity $O(n \log^2 n)$ for each epoch. Assuming each epoch consist of R rounds, the amortized per round communication complexity is $O(\frac{n \log^2 n}{R})$. □

Corollary 33. *In each epoch, the expected size of each shard is n/m.*

Proof. Due to Lemma 31, the n parties are assigned independently and uniformly at random to m shards. Hence, the expected number of parties in a shard is n/m. □

Lemma 34. *In each epoch, all shards are $\frac{1}{3}$-honest for $m \leq f(n)$ with $f(n)$ as described in Corollary 21.*

Proof. Due to Lemma 31, the n parties are assigned independently and uniformly at random to m shards. Since $a = 1/3 > p = 1/4$, both a, p constant, the statement holds from Lemma 19 and Corollary 21. □

Note that the bound is theoretical and holds for a large number of parties since the probability tends to 1 as the number of parties grows. For practical bounds, we refer to OmniLedger's analysis [33].

Theorem 35. *OmniLedger satisfies persistence in our system model for $f < n/4$.*

Proof. From Lemma 34, each shard has an honest supermajority $\frac{2}{3}\frac{n}{m}$ of participants. Hence, persistence holds by the common prefix property of the consensus protocol of each shard. Specifically, for ByzCoinX, persistence holds for depth parameter $k = 1$ because ByzCoinX guarantees finality. □

Theorem 36. *OmniLedger does not satisfy liveness in our system model for* $f < n/4$.

Proof. To estimate the liveness of the protocol, we need to examine all the subprotocols: (i) `Consensus`, (ii) `CrossShard` or Atomix, (iii) `DRG`, (iv) `CompactState`, and (v) `Divide2Shards`.

`Consensus`: From Lemma 34, each shard has an honest supermajority $\frac{2}{3}\frac{n}{m}$ of participants. Hence, in this stage liveness holds by chain growth and chain quality properties of the underlying blockchain protocol (an elaborate proof can be found in [24]). The same holds for `CompactState` as it is executed similarly to `Consensus`.

`CrossShard`: Atomix guarantees liveness since the protocol's efficiency depends on the consensus of each shard involved in the cross-shard transaction. Note that liveness does not depend on the client's behavior; if the appropriate information or some part of the transaction is not provided in multiple rounds to the parties of the protocol then the liveness property does not guarantee the inclusion of the transaction in the ledger. Furthermore, if some other party wants to continue the process it can collect all necessary information from the ledgers of the shards.

`DRG`: During the epoch transition, the DRG protocol provides a common random value with overwhelming probability within t rounds (Lemma 31). Hence, liveness is satisfied in this subprotocol as well.

`Divide2Shrds`: Liveness is not satisfied in this protocol. The reason is that a slowly-adaptive adversary can select who to corrupt during epoch transition, and thus can corrupt a shard from the previous epoch. Since the compact state has not been disseminated in the network, the adversary can simply delete the shard's state. Thereafter, the data unavailability prevents the progress of the system. □

Theorem 37. *OmniLedger satisfies consistency in our system model for* $f < n/4$.

Proof. Each shard is $\frac{1}{3}$-honest (Lemma 34). Hence, consistency holds within each shard, and the adversary cannot successfully double-spend. Nevertheless, we need to guarantee consistency even when transactions are cross-shard. OmniLedger employs Atomix, a protocol that guarantees cross-shard transactions are atomic. Thus, the adversary cannot validate two conflicting transactions across different shards.

Moreover, the adversary cannot revert the chain of a shard and double-spend an input of a cross-shard transaction after the transaction is accepted in all relevant shards because persistence holds (Theorem 35). Suppose persistence holds with probability p. Then, the probability the adversary breaks consistency in a cross-shard transaction is the probability of successfully double-spending in one of the relevant to the transaction shards, $1 - p^v$, where v is the average size of transactions. Since v is constant, consistency holds with high probability, given that persistence holds with high probability. $\qquad\square$

To prove OmniLedger satisfies scalability (on expectation) we need to evaluate the scaling factors in the following subprotocols of the system: (i) Consensus, (ii) CrossShard, (iii) DRG, and (iv) Divide2Shards. Note that CompactState is merely an execution of Consensus.

Lemma 38. *The scaling factors of Consensus are $\omega_m = O(n/m)$, $\omega_s = O(1/m)$, and $\omega_c = O(n/m)$.*

Proof. From Corollary 33, the expected number of parties in a shard is n/m. ByzCoin has quadratic to the number of parties' worst-case communication complexity, hence the communication factor of the protocol is $O(n/m)$. The verification complexity collapses to the communication complexity. The space factor is $O(1/m)$, as each party maintains the ledger of the assigned shard for the epoch. $\qquad\square$

Lemma 39. *The communication factor of Atomix (CrossShard) is $\omega_m = O(v\frac{n}{m})$, where v is the average size of transactions.*

Proof. In a cross-shard transaction, Atomix allows the participants of the output shards to verify the validity of the transaction's inputs without maintaining any information on the input shards' ledgers. This holds due to persistence (see Theorem 35).

Furthermore, the verification process requires each input shard to verify the validity of the transaction's inputs and produce a proof-of-acceptance or proof-of-rejection. This corresponds to one query to the verification oracle for each input. In addition, each party of an output shard must verify that all proofs-of-acceptance are present and no shard rejected an input of the cross-shard transaction. The proof-of-acceptance (or rejection) consists of the signature of the shard which is linear to the number of parties in the shard. The relevant parties have to receive all the information related to the transaction from the client (or leader), hence the communication factor is $O(v\frac{n}{m})$.

So far, we considered the communication complexity of Atomix. However, each input must be verified within the corresponding input shard. From Lemma 38, we get that the communication factor at this step is $O(v\frac{n}{m})$.

Lemma 40. *The communication factor of Divide2Shards is $\omega_m = O(\frac{n}{mR})$, while the space factor is $\omega_s = O(1/R)$, where R is size of an epoch.*

Proof. During the epoch transition each party is assigned to a shard uniformly at random and thus most probably needs to bootstrap to a new shard, meaning the party must store the new shard's ledger. At this point, within each shard OmniLedger introduces checkpoints, the state blocks that summarize the state of the ledger (CompactState). Therefore, when a party syncs with a shard's ledger, it does not download and store the entire ledger but only the active UTXO pool corresponding to the previous epoch's state block.

For security reasons, each party that is reassigned to a new shard must receive the state block of the new shard by $O(n/m)$ parties. Thus, the communication complexity of the protocol is $O(\frac{n}{mR})$ amortized per round, where R is the number of rounds in an epoch.

The space complexity is constant but amortized over the epoch length since the state block has a constant size and is broadcast once per epoch, $\omega_s = O(1/R)$. There is no verification process at this stage. □

Theorem 41. *OmniLedger satisfies scalability in our system model for $f < n/4$ with communication and computational factor $O(n/m)$ and space factor $O(1/m)$, where $n = O(m \log m)$.*

Proof. To evaluate the scalability of OmniLedger, we need to estimate the dominating scaling factors of all the subprotocols of the system: (i) Consensus, (ii) CrossShard, (iii) DRG, and (iv) Divide2Shards.

The scaling factors of Consensus are $\omega_m = O(n/m)$, $\omega_s = O(1/m)$, and $\omega_c = O(n/m)$ (Lemma 38), while Atomix (CrossShard) has expected communication factor $O(v\frac{n}{m})$ (Lemma 39) where the average size of transaction v is constant (see Sect. 3).

The epoch transition consists of the DRG, CompactState, and Divide2Shards protocols. We assume a large enough epoch in rounds, $R = \Omega(n \log n)$, in order to amortize the communication-heavy protocols that are executed only once per epoch. CompactState has the same overhead as Consensus hence it is not critical. For $R = \Omega(n \log n)$, DRG has an expected amortized communication factor $O(\log n)$ (Lemma 32), while Divide2Shards has an expected amortized communication factor of $\omega_m = O(\frac{1}{m \log n})$ and an amortized space factor of $\omega_s = O(1/R) = O(\frac{1}{n \log n})$ (Lemma 40).

Overall, considering the worst of the aforementioned scaling factors for OmniLedger, we have expected communication and computational factors $O(n/m)$ and space factor $O(1/m)$, where $n = O(m \log m)$ (see Lemma 14 and Lemma 34). □

Theorem 42. *In OmniLedger, the throughput factor is $\sigma = \mu \cdot \tau \cdot \dfrac{m}{v} < \dfrac{\mu \cdot \tau \cdot f(n)}{v}$ where $f(n) = \dfrac{n}{c' \log(\frac{n}{c' \log(n)})}$ with $c' = \frac{c}{p}$ and c a constant as described in Corollary 20.*

Proof. In Atomix, at most v shards are affected per transaction, thus $m' < m/v^7$. From Lemma 19 and Corollary 21, $n \leq f(n)$. Therefore, $\sigma < \frac{\mu \cdot \tau \cdot f(n)}{v}$ □

---

[7] Note that if v is constant, a more elaborate analysis could yield a lower upper bound on m' better than m/v (depending on D_T). However, if v is not constant but approximates the number of shards m, then m' is also bounded by the scalability of the Atomix protocol (Lemma 39), and thus the throughput factor can be much lower.

The parameter v depends on the input transaction set. The parameters μ, τ, a, p depend on the choice of the consensus protocol. Specifically, μ represents the ratio of honest blocks in the chain of a shard. On the other hand, τ depends on the latency of the consensus protocol, i. e., what is the ratio between the propagation time and the block generation time. Last, a expresses the resilience of the consensus protocol (e.g., $1/3$ for PBFT), while p the fraction of corrupted parties in the system ($f = pn$).

In OmniLedger, the consensus protocol is modular, so we chose to maintain the parameters for a fairer comparison to other protocols.

C.4   RapidChain

Overview. RapidChain [55] is a synchronous protocol and proceeds in epochs. The adversary is slowly-adaptive, computationally-bounded and corrupts less than $1/3$ of the participants ($f < n/3$).

The protocol bootstraps via a committee election protocol that selects $O(\sqrt{n})$ parties – the root group. The root group generates and distributes a sequence of random bits used to establish the reference committee. The reference committee consists of $O(\log n)$ parties, is re-elected at the end of each epoch, and is responsible for: (i) generating the randomness of the next epoch, (ii) validating the identities of participants for the next epoch from the PoW puzzle, and (iii) reconfiguring the shards from one epoch to the next (to protect against single shard takeover attacks).

The parties are divided into shards of size $O(\log n)$ (committees). Each shard handles a fraction of the transactions, assigned based on the prefix of the transaction ID. Transactions are sent by external users to an arbitrary number of active (for this epoch) parties. The parties then use an inter-shard routing scheme (based on Kademlia [38]) to send the transactions to the input and output shards, i. e., the shards handling the inputs and outputs of a transaction, resp.

To process cross-shard transactions, the leader of the output shard creates an additional transaction for every different input shard. Then the leader sends (via the inter-shard routing scheme) these transactions to the corresponding input shards for validation. To validate transactions (i. e., a block), each shard runs a variant of the synchronous consensus of Ren et al. [44] and thus tolerates $1/2$ Byzantine parties.

At the end of each epoch, the shards are reconfigured according to the participants registered in the new reference block. Specifically, RapidChain uses a bounded version of Cuckoo 'rule [46]; the reconfiguration protocol adds a new party to a shard uniformly at random, and also moves a constant number of parties from each shard and assigns them to other shards uniformly at random.

Consensus: In each round, each shard randomly picks a leader. The leader creates a block, gossips the block header H (containing the round and the Merkle root) to the members of the shard, and initiates the consensus protocol on H. The consensus protocol consists of four rounds: (1) The leader gossips ($H, propose$), (2) All parties gossip the received header ($H, echo$), (3) The honest parties that received at least two echoes containing a different header gossip $(H', pending)$, where H' contains the null Merkle root and the round, (4) Upon receiving $\frac{nf}{m} + 1$ echos of the same and only header, an honest party gossips $(H, accept)$ along with the received echoes. To increase the transaction throughput, RapidChain allows new leaders to propose new blocks even if the previous block is not yet accepted by all honest parties.

StatePartition: Each shard handles a fraction of the transactions, assigned based on the prefix of the transaction ID.

CrossShard: For each cross-shard transaction, the leader of the output shard creates one "dummy" transaction for each input UTXO in order to move the transactions' inputs to the output shard, and execute the transaction within the shard. To be specific, assume we have a transaction with two inputs I_1, I_2 and one output O. The leader of the output shard creates three new transactions: tx_1 with input I_1 and output I'_1, where I'_1 holds the same amount of money with I_1 and belongs to the output shard. tx_2 is created similarly. tx_3 with inputs I'_1 and I'_2 and output O. Then the leader sends tx_1, tx_2 to the input shards respectively. In principle, the output shard is claiming to be a trusted channel [6] (which is guaranteed from the assignment), hence the input shards should transfer their assets there and then execute the transaction atomically inside the output shard (or abort by returning their assets back to the input shards).

Sybil: A party can only participate in an epoch if it solves a PoW puzzle with the previous epoch's randomness, submit the solution to the reference committee, and consequently be included in the next reference block. The reference block contains the active parties' identities for the next epoch, their shard assignment, and the next epoch's randomness, and is broadcast by the reference committee at the end of each epoch.

Divide2Shards: During bootstrapping, the parties are partitioned independently and uniformly at random in groups of size $O(\sqrt{n})$ with a deterministic random process. Then, each group runs the DRG protocol and creates a (local) random seed. Every node in the group computes the hash of the random seed and its public key. The e (small constant) smallest tickets are elected from each group and gossiped to the other groups, along with at least half the signatures of the group. These elected parties are the root group. The root group then selects the reference committee of size $O(\log n)$, which in turn partitions the parties randomly into shards as follows: each party is mapped to a random position in $[0, 1)$ using a hash function. Then, the range $[0, 1)$ is partitioned into k regions, where k is constant. A shard is the group of parties assigned to $O(\log n)$ regions.

During epoch transition, a constant number of parties can join (or leave) the system. This process is handled by the reference committee which determines the next epoch's shard assignment, given the set of active parties for the epoch. The reference committee divides the shards into two groups based on each shard's number of active parties in the previous epoch: group A contains the $m/2$ larger in size shards, while the rest comprise group I. Every new node is assigned uniformly at random to a shard in A. Then, a constant number of parties is evicted from each shard and assigned uniformly at random in a shard in I.

DRG: RapidChain uses Feldman's verifiable secret sharing [22] to distributively generate unbiased randomness. At the end of each epoch, the reference committee executes a distributed randomness generation (DRG) protocol to provide the random seed of the next epoch. The same DRG protocol is also executed during bootstrapping to create the root group.

CompactState: No protocol for compaction of the state is used.

Analysis. RapidChain does not maintain a robust sharded transaction ledger under our security model since it assumes a weaker adversary. To fairly evaluate the protocol, we weaken our security model. First, assume the adversary cannot change more than a constant number of Byzantine parties during an epoch transition, which we term *constant-adaptive adversary*. In general, we assume *bounded epoch transitions*, i. e., at most a constant number of leave/join requests during each transition. Furthermore, the number of epochs is asymptotically less than polynomial to the number of parties. In this weaker security model, we prove RapidChain maintains a robust sharded transaction ledger, and provide an upper bound on the throughput factor of the protocol.

Note that in cross-shard transactions, the "dummy" transactions that are committed in the shards' ledgers as valid, spend UTXOs that are not signed by the corresponding users. Instead, the original transaction, signed by the users, is provided to the shards to verify the validity of the "dummy" transactions. Hence, the transaction validation rules change. Furthermore, the protocol that handles cross-shard transactions has no proof of security against Byzantine leaders. For analysis purposes, we assume the following holds:

Assumption 43. *CrossShard satisfies safety even under a Byzantine leader (of the output shard).*

Lemma 44. *The communication factor of DRG is $O(n/m)$.*

Proof. The DRG protocol is executed by the final committee once each epoch. The size of the final committee is $O(n/m) = O(\log n)$. The communication complexity of the DRG protocol is quadratic to the number of parties [22]. Thus, the communication factor is $O(n/m)$. □

Lemma 45. *In each epoch, all shards are $\frac{1}{2}$-honest for $m \leq f(n)$ with $f(n)$ from Corollary 21.*

Proof. During the bootstrapping process of RapidChain (first epoch), the n parties are partitioned independently and uniformly at random into m shards [22]. For $p = 1/3$, the shards are $\frac{1}{2}$-honest only if $m \leq f(n)$ with $f(n)$ from corollary 21. At any time during the protocol, all shards remain $\frac{1}{2}$-honest ([55], Theorem 5). Hence, the statement holds after each epoch transition, as long as the number of epochs is $o(n)$. □

Lemma 46. *In each epoch, the expected size of each shard is $O(n/m)$.*

Proof. During the bootstrapping process of RapidChain (first epoch), the n parties are partitioned independently and uniformly at random into m shards [22]. The expected shard size in the first epoch is n/m. Furthermore, during epoch transition the shards remain "balanced" (Theorem 5 [55]), i.e., the size of each shard is $O(n/m)$. □

Theorem 47. *RapidChain satisfies persistence in our system model for constant-adaptive adversaries with $f < n/3$ and bounded epoch transitions.*

Proof. The consensus protocol in RapidChain achieves safety if the shard has no more than $t < 1/2$ fraction of Byzantine parties ([55], Theorem 2). Hence, the statement follows from Lemma 45. □

Theorem 48. *RapidChain satisfies liveness in our system model for constant-adaptive adversaries with $f < n/3$ and bounded epoch transitions.*

Proof. To estimate the liveness of RapidChain, we need to examine the following subprotocols: (i) `Consensus`, (ii) `CrossShard`, (iii) `DRG`, and (iv) `Divide2Shards`.

The consensus protocol in RapidChain achieves liveness if the shard has less than $\frac{n}{2m}$ Byzantine parties (Theorem 3 [55]). Thus, liveness is guaranteed during `Consensus` (Lemma 45).

Furthermore, the final committee is $\frac{1}{2}$-honest with high probability. Hence, the final committee will route each transaction to the corresponding output shard. We assume transactions will reach all relevant honest parties via a gossip protocol. RapidChain employs IDA-gossip protocol, which guarantees message delivery to all honest parties (Lemma 1 and Lemma 2 [55]). From Assumption 43, the protocol that handles cross-shard transactions satisfies safety even under a Byzantine leader. Hence, all "dummy" transactions will be created and eventually delivered. Since the consensus protocol within each shard satisfies liveness, the "dummy" transactions of the input shards will become stable. Consequently, the "dummy" transaction of the output shard will become valid and eventually stable (consensus liveness). Thus, `CrossShard` satisfies liveness.

During epoch transition, `DRG` satisfies liveness [22]. Moreover, `Divide2Shards` allows only for a constant number of leave/join/move operations and thus terminates in a constant number of rounds. □

Theorem 49. *RapidChain satisfies consistency in our system model for constant-adaptive adversaries with $f < n/3$ and bounded epoch transitions.*

Proof. In every epoch, each shard is $\frac{1}{2}$-honest; hence, the adversary cannot double-spend and consistency is satisfied.

Nevertheless, to prove consistency is satisfied across shards, we need to prove that cross-shard transactions are atomic. `CrossShard` in RapidChain ensures that the "dummy" transaction of the output shard becomes valid only if all "dummy" transactions are stable in the input shards. If a "dummy" transaction of an input shard is rejected, the "dummy" transaction of the output shard will not be executed, and all the accepted "dummy" transactions will just transfer the value of the input UTXOs to other UTXOs that belong to the output shard. This holds because the protocol satisfies safety even under a Byzantine leader (Assumption 43).

Lastly, the adversary cannot revert the chain of a shard and double-spend an input of the cross-shard transaction after the transaction is accepted in all relevant shards because consistency with each shard and persistence (Theorem 35) hold. Suppose persistence holds with probability p. Then, the probability the adversary breaks consistency in a cross-shard transaction is the probability of successfully double-spending in one of the relevant to the transaction shards, hence $1 - p^v$ where v is the average size of transactions. Since v is constant, consistency holds with high probability, given persistence holds with high probability. $\square$

Similarly to OmniLedger, to calculate the scaling factor of RapidChain, we need to evaluate the following protocols of the system: (i) Consensus, (ii) CrossShard, (iii) DRG, and (iv) Divide2Shards.

Lemma 50. *The scaling factors of Consensus are* $\omega_m = O(\frac{n}{m})$, $\omega_s = O(\frac{1}{m})$, *and* $\omega_c = O(\frac{n}{m})$.

Proof. From Lemma 46, the expected number of parties in a shard is $O(n/m)$. The consensus protocol of RapidChain has quadratic to the number of parties' communication complexity. Hence, the communication factor Consensus is $O(\frac{n}{m})$. The verification complexity (computational factor) collapses to the communication complexity. The space factor is $O(\frac{1}{m})$, as each party maintains the ledger of the assigned shard for the epoch.

Lemma 51. *The communication and computational factors of CrossShard are both* $\omega_m = \omega_c = O(v\frac{n}{m})$, *where* v *is the average size of transactions.*

Proof. During the execution of the protocol, the interaction between the input and output shards is limited to the leader, who creates and routes the "dummy" transactions. Hence, the communication complexity of the protocol is dominated by the consensus within the shards. For an average size of transactions v, the communication factor is $O(vn/m + v) = O(vn/m)$ (Lemma 46). Note that this bound holds for the worst case, where transactions have $v - 1$ inputs and a single output while all UTXOs belong to different shards.

For each cross-shard transaction, each party of the input and output shards queries the verification oracle once. Hence, the computational factor is $O(vn/m)$. The protocol does not require any verification across shards, thus the only storage requirement per party is to maintain the ledger of its own shard. $\square$

Lemma 52. *The communication factor of Divide2Shards is* $O(\frac{R \cdot n}{m^2})$.

Proof. The number of join/leave and move operations is constant per epoch, denoted by k. Further, each shard is $\frac{1}{2}$-honest (Lemma 45) and has size $O(\frac{n}{m})$ (Lemma 46); these guarantees hold as long as the number of epochs is $o(n)$.

Each party changing shards receives the new shard's ledger of size T/m by $O(n/m)$ parties in the new shard. Thus the total communication complexity at this stage is $O(\frac{T}{m} \cdot \frac{n}{m})$, hence the communication factor is $O(\frac{T}{m^2}) = O(\frac{R \cdot e}{m^2})$, where R is the number of rounds in each epoch and e the number of epochs since genesis. Since $e = o(n)$, the communication factor is $O(\frac{R \cdot n}{m^2})$.

□

Theorem 53. *RapidChain satisfies scalability in our system model for constant-adaptive adversaries with $f < n/3$ and bounded epoch transitions, with communication and computational factor $O(n/m)$ and space factor $O(1/m)$, where $n = O(m \log m)$, assuming epoch size $R = O(m)$.*

Proof. Consensus has on expectation communication and computational factors bounded by $O(n/m)$ and space factor $O(1/m)$ (Lemma 50). These bounds are similar in CrossShard where the communication and computational factors are bounded by $O(vn/m)$ (Lemma 51), where v is constant (see Sect. 3).

During epoch transitions, the communication factor dominates: In DRG $\omega_m = O(\frac{n}{m})$ (Lemma 44) while in Divide2Shards $\omega_m = O(\frac{n \cdot R}{m^2})$ (Lemma 52). Thus for $R = O(m)$, the communication factor during epoch transitions is $O(n/m)$.

Overall, RapidChain's expected scaling factors are as follows: $\omega_m = \omega_c = O(n/m) = O(\log m)$ and $\omega_s = O(1/m)$, where the equation holds for $n = c'm \log m$ (Lemma 45).

Theorem 54. *In RapidChain, the throughput factor is $\sigma = \mu \cdot \tau \cdot \dfrac{m}{v} < \dfrac{\mu \cdot \tau \cdot f(n)}{v}$ with $f(n) = \dfrac{n}{c' \log(\frac{n}{c' \log(n)})}$ with $c' = \frac{c}{p}$ and constant c from Corollary 20.*

Proof. At most v shards are affected per transaction – when each transaction has $v - 1$ inputs and one output, and all belong to different shards. Therefore, $m' < m/v$. From Lemma 19 and Corollary 21, $m < f(n)$. Therefore, $\sigma < \frac{\mu \cdot \tau \cdot f(n)}{v}$.

□

In RapidChain, the consensus protocol is synchronous and thus not practical. We estimate the throughput factor irrespective of the chosen consensus, to provide a fair comparison to other protocols. We notice that both RapidChain and OmniLedger have the same throughout factor when v is constant.

We provide an example of the throughput factor in case the employed consensus is the one suggested in RapidChain. In this case, we have $a = 1/2$, $p = 1/4$ (hence $p/a = 2/3$), $\mu < 1/2$ (Theorem 1 [55]), and $\tau = 1/8$ (4 rounds are needed to reach consensus for an honest leader, and the leader will be honest every two rounds on expectation [54].). Note that τ can be improved by allowing the next leader to propose a block even if the previous block is not yet accepted by all honest parties; however, we do not consider this improvement. Because of the values of p and a we can compute $c \simeq 2.6$, thus $c' \simeq 10.4$. Hence, for $v = 5$, we have throughput factor:

$$\sigma < \frac{1}{2} \cdot \frac{1}{8} \cdot \frac{1}{5} \cdot \frac{1}{10.4} \frac{n}{\log(\frac{n}{10.4 \log n})} = \frac{n}{832 \log(\frac{n}{10.4 \log n})}$$

C.5   Chainspace

Chainspace is a sharding protocol introduced by Al-Bassam et al. [3] that operates in the permissioned setting. The main innovation of Chainspace is on the application layer. Specifically, Chainspace presents a sharded, UTXO-based distributed ledger that supports smart contracts. Furthermore, limited privacy is enabled by offloading computation to the clients, who need to only publicly provide zero-knowledge proofs that their computation is correct. Chainspace focuses on specific aspects of sharding; epoch transition or reconfiguration of the protocol is not addressed. Nevertheless, the cross-shard communication protocol, namely S-BAC, is of interest as a building block to secure sharding.

S-BAC Protocol. S-BAC is a shard-led cross-shard atomic commit protocol used in Chainspace. In S-BAC, the client submits a transaction to the input shards. Each shard internally runs a BFT protocol to tentatively decide whether to accept or abort the transaction locally and broadcasts its local decision to other shards that take part in the transaction. If the transaction fails locally (e.g., is a double-spend), then the shard generates pre-abort(T), whereas if the transaction succeeds locally the shard generates pre-accept(T) and changes the state of the input to 'locked'. After a shard decides to pre-commit(T), it waits to collect responses from other participating shards, and commits the transaction if all shards respond with pre-accept(T), or aborts the transaction if at least one shard announces pre-abort(T). Once the shards decide, they send their decision (accept(T) or abort(T)) to the client and the output shards. If the decision is accept(T), the output shards generate new 'active' objects and the input shards change the input objects to 'inactive'. If an input shard's decision is abort(T), all input shards unlock the input objects by changing their state to 'active'.

S-BAC, just like Atomix, is susceptible to replay attacks [48]. To address this problem, sequence numbers are added to the transactions, and output shards generate dummy objects during the first phase (pre-commit, pre-abort). More details and security proofs can be found on [48], as well as a hybrid of Atomix and S-BAC called Byzcuit.

References

1. Bitcoin statistics on transaction utxos. https://bitcoinvisuals.com/. Accessed 20 Nov 2020
2. Al-Bassam, M.: LazyLedger: a distributed data availability ledger with client-side smart contracts. arXiv preprint arXiv:1905.09274 (2019)
3. Al-Bassam, M., Sonnino, A., Bano, S., Hrycyszyn, D., Danezis, G.: Chainspace: a sharded smart contracts platform. In: 25th Annual Network and Distributed System Security Symposium (2018)
4. Al-Bassam, M., Sonnino, A., Buterin, V.: Fraud and data availability proofs: maximising light client security and scaling blockchains with dishonest majorities. arXiv preprint arXiv:1809.09044 (2018)

5. Androulaki, E., et al.: Hyperledger fabric: a distributed operating system for permissioned blockchains. In: Proceedings of the 13th EuroSys Conference, pp. 30:1–30:15 (2018)
6. Androulaki, E., Cachin, C., De Caro, A., Kokoris-Kogias, E.: Channels: horizontal scaling and confidentiality on permissioned blockchains. In: Lopez, J., Zhou, J., Soriano, M. (eds.) ESORICS 2018. LNCS, vol. 11098, pp. 111–131. Springer, Cham (2018). https://doi.org/10.1007/978-3-319-99073-6_6
7. Avarikioti, Z., Heimbach, L., Schmid, R., Wattenhofer, R.: FnF-BFT: exploring performance limits of BFT protocols. arXiv preprint arXiv:2009.02235 (2020)
8. Bano, S., et al.: SoK: consensus in the age of blockchains. In: Proceedings of the 1st ACM Conference on Advances in Financial Technologies, pp. 183–198. ACM (2019)
9. Ben-Sasson, E.: A cambrian explosion of crypto proofs (2020). https://nakamoto.com/cambrian-explosion-of-crypto-proofs/
10. Boneh, D., Bünz, B., Fisch, B.: Batching techniques for accumulators with applications to IOPs and stateless blockchains. In: Boldyreva, A., Micciancio, D. (eds.) CRYPTO 2019. LNCS, vol. 11692, pp. 561–586. Springer, Cham (2019). https://doi.org/10.1007/978-3-030-26948-7_20
11. Bonneau, J., Clark, J., Goldfeder, S.: On bitcoin as a public randomness source. IACR Cryptology ePrint Archive, Report 2015/1015 (2015)
12. Borge, M., Kokoris-Kogias, E., Jovanovic, P., Gasser, L., Gailly, N., Ford, B.: Proof-of-personhood: redemocratizing permissionless cryptocurrencies. In: IEEE European Symposium on Security and Privacy Workshops, pp. 23–26 (2017)
13. Bracha, G., Toueg, S.: Asynchronous consensus and broadcast protocols. J. ACM 32(4), 824–840 (1985)
14. Bünz, B., Goldfeder, S., Bonneau, J.: Proofs-of-delay and randomness beacons in ethereum. In: IEEE Security and Privacy on the Blockchain (2017)
15. Bünz, B., Kiffer, L., Luu, L., Zamani, M.: FlyClient: super-light clients for cryptocurrencies. In: IEEE Symposium on Security and Privacy, pp. 928–946 (2020)
16. Cascudo, I., David, B.: SCRAPE: scalable randomness attested by public entities. In: Gollmann, D., Miyaji, A., Kikuchi, H. (eds.) ACNS 2017. LNCS, vol. 10355, pp. 537–556. Springer, Cham (2017). https://doi.org/10.1007/978-3-319-61204-1_27
17. Castro, M., Liskov, B.: Practical byzantine fault tolerance. In: Proceedings of the 3rd USENIX Symposium on Operating Systems Design and Implementation, pp. 173–186 (1999)
18. Chatzigiannis, P., Baldimtsi, F., Chalkias, K.: SoK: blockchain light clients. In: Eyal, I., Garay, J. (eds.) FC 2022. LNCS, vol. 13411, pp. 615–641. Springer, Cham (2022). https://doi.org/10.1007/978-3-031-18283-9_31
19. Danezis, G., Kokoris-Kogias, L., Sonnino, A., Spiegelman, A.: Narwhal and tusk: a DAG-based mempool and efficient BFT consensus. In: Proceedings of the Seventeenth European Conference on Computer Systems, pp. 34–50 (2022)
20. Das, S., Yurek, T., Xiang, Z., Miller, A., Kokoris-Kogias, L., Ren, L.: Practical asynchronous distributed key generation. In: 2022 IEEE Symposium on Security and Privacy (SP), pp. 2518–2534. IEEE (2022)
21. David, B., Gaži, P., Kiayias, A., Russell, A.: Ouroboros praos: an adaptively-secure, semi-synchronous proof-of-stake blockchain. In: Nielsen, J.B., Rijmen, V. (eds.) EUROCRYPT 2018. LNCS, vol. 10821, pp. 66–98. Springer, Cham (2018). https://doi.org/10.1007/978-3-319-78375-8_3
22. Feldman, P.: A practical scheme for non-interactive verifiable secret sharing. In: 28th Annual IEEE Symposium on Foundations of Computer Science, pp. 427–438. IEEE (1987)

23. Fisher, T., Funk, D., Sams, R.: The birthday problem and generalizations. Carlton College, Mathematics Comps Gala (2013). https://d31kydh6n6r5j5.cloudfront.net/uploads/sites/66/2019/04/birthday_comps.pdf

24. Garay, J., Kiayias, A., Leonardos, N.: The bitcoin backbone protocol: analysis and applications. In: Oswald, E., Fischlin, M. (eds.) EUROCRYPT 2015. LNCS, vol. 9057, pp. 281–310. Springer, Heidelberg (2015). https://doi.org/10.1007/978-3-662-46803-6_10

25. Gilad, Y., Hemo, R., Micali, S., Vlachos, G., Zeldovich, N.: Algorand: scaling byzantine agreements for cryptocurrencies. In: Proceedings of the 26th Symposium on Operating Systems Principles, pp. 51–68. ACM (2017)

26. Kadhe, S., Chung, J., Ramchandran, K.: SeF: a secure fountain architecture for slashing storage costs in blockchains. arXiv preprint arXiv:1906.12140 (2019)

27. Kattis, A., Bonneau, J.: Proof of necessary work: succinct state verification with fairness guarantees. IACR Cryptology ePrint Archive, Report 2020/190 (2020)

28. Kiayias, A., Miller, A., Zindros, D.: Non-interactive proofs of proof-of-work. In: Bonneau, J., Heninger, N. (eds.) FC 2020. LNCS, vol. 12059, pp. 505–522. Springer, Cham (2020). https://doi.org/10.1007/978-3-030-51280-4_27

29. Kiayias, A., Panagiotakos, G.: On trees, chains and fast transactions in the blockchain. In: Lange, T., Dunkelman, O. (eds.) LATINCRYPT 2017. LNCS, vol. 11368, pp. 327–351. Springer, Cham (2019). https://doi.org/10.1007/978-3-030-25283-0_18

30. Kiayias, A., Russell, A., David, B., Oliynykov, R.: Ouroboros: a provably secure proof-of-stake blockchain protocol. In: Katz, J., Shacham, H. (eds.) CRYPTO 2017. LNCS, vol. 10401, pp. 357–388. Springer, Cham (2017). https://doi.org/10.1007/978-3-319-63688-7_12

31. Kogias, E.K., Jovanovic, P., Gailly, N., Khoffi, I., Gasser, L., Ford, B.: Enhancing bitcoin security and performance with strong consistency via collective signing. In: 25th USENIX Security Symposium, pp. 279–296 (2016)

32. Kokoris-Kogias, E.: Robust and scalable consensus for sharded distributed ledgers. IACR Cryptology ePrint Archive, Report 2019/676 (2019)

33. Kokoris-Kogias, E., Jovanovic, P., Gasser, L., Gailly, N., Syta, E., Ford, B.: OmniLedger: a secure, scale-out, decentralized ledger via sharding. In: 39th IEEE Symposium on Security and Privacy, pp. 583–598. IEEE (2018)

34. Kokoris-Kogias, E., Malkhi, D., Spiegelman, A.: Asynchronous distributed key generation for computationally-secure randomness, consensus, and threshold signatures. In: 27th ACM SIGSAC Conference on Computer and Communications Security, pp. 1751–1767. ACM (2020)

35. Lamport, L., Shostak, R., Pease, M.: The byzantine generals problem. In: Concurrency: The Works of Leslie Lamport, pp. 203–226 (2019)

36. Luu, L., Narayanan, V., Zheng, C., Baweja, K., Gilbert, S., Saxena, P.: A secure sharding protocol for open blockchains. In: Proceedings of the 25th ACM SIGSAC Conference on Computer and Communications Security, pp. 17–30. ACM (2016)

37. Martino, W., Quaintance, M., Popejoy, S.: Chainweb: a proof-of-work parallel-chain architecture for massive throughput (2018). https://www.kadena.io/whitepapers

38. Maymounkov, P., Mazières, D.: Kademlia: a peer-to-peer information system based on the XOR metric. In: Druschel, P., Kaashoek, F., Rowstron, A. (eds.) IPTPS 2002. LNCS, vol. 2429, pp. 53–65. Springer, Heidelberg (2002). https://doi.org/10.1007/3-540-45748-8_5

39. Meckler, I., Shapiro, E.: Coda: decentralized cryptocurrency at scale (2018). https://cdn.codaprotocol.com/static/coda-whitepaper-05-10-2018-0.pdf

40. Nakamoto, S.: Bitcoin: a peer-to-peer electronic cash system (2008)
41. Pass, R., Seeman, L., Shelat, A.: Analysis of the blockchain protocol in asynchronous networks. In: Coron, J.-S., Nielsen, J.B. (eds.) EUROCRYPT 2017. LNCS, vol. 10211, pp. 643–673. Springer, Cham (2017). https://doi.org/10.1007/978-3-319-56614-6_22
42. Raab, M., Steger, A.: "Balls into bins"—a simple and tight analysis. In: Luby, M., Rolim, J.D.P., Serna, M. (eds.) RANDOM 1998. LNCS, vol. 1518, pp. 159–170. Springer, Heidelberg (1998). https://doi.org/10.1007/3-540-49543-6_13
43. Rana, R., Kannan, S., Tse, D., Viswanath, P.: Free2shard: adaptive-adversary-resistant sharding via dynamic self allocation. arXiv preprint arXiv:2005.09610 (2020)
44. Ren, L., Nayak, K., Abraham, I., Devadas, S.: Practical synchronous byzantine consensus. arXiv preprint arXiv:1704.02397 (2017)
45. Schindler, P., Judmayer, A., Stifter, N., Weippl, E.: HydRand: practical continuous distributed randomness. IACR Cryptology ePrint Archive, Report 2018/319 (2018)
46. Sen, S., Freedman, M.J.: Commensal cuckoo: secure group partitioning for large-scale services. ACM SIGOPS Oper. Syst. Rev. **46**(1), 33–39 (2012)
47. Sompolinsky, Y., Zohar, A.: Secure high-rate transaction processing in bitcoin. In: Böhme, R., Okamoto, T. (eds.) FC 2015. LNCS, vol. 8975, pp. 507–527. Springer, Heidelberg (2015). https://doi.org/10.1007/978-3-662-47854-7_32
48. Sonnino, A., Bano, S., Al-Bassam, M., Danezis, G.: Replay attacks and defenses against cross-shard consensus in sharded distributed ledgers. arXiv preprint arXiv:1901.11218 (2019)
49. Spiegelman, A., Giridharan, N., Sonnino, A., Kokoris-Kogias, L.: Bullshark: DAG BFT protocols made practical. In: Proceedings of the 2022 ACM SIGSAC Conference on Computer and Communications Security, pp. 2705–2718 (2022)
50. Stathakopoulou, C., David, T., Vukolić, M.: Mir-BFT: High-throughput BFT for blockchains. arXiv preprint arXiv:1906.05552 (2019)
51. Syta, E., et al.: Scalable bias-resistant distributed randomness. In: IEEE Symposium on Security and Privacy, pp. 444–460 (2017)
52. Wang, G., Shi, Z.J., Nixon, M., Han, S.: SoK: Sharding on blockchain. In: Proceedings of the 1st ACM Conference on Advances in Financial Technologies, pp. 41–61. ACM (2019)
53. Wang, J., Wang, H.: Monoxide: scale out blockchains with asynchronous consensus zones. In: 16th USENIX Symposium on Networked Systems Design and Implementation, pp. 95–112 (2019)
54. Yin, M., Malkhi, D., Reiter, M.K., Gueta, G.G., Abraham, I.: HotStuff: BFT consensus with linearity and responsiveness. In: Proceedings of the 38th ACM Symposium on Principles of Distributed Computing, pp. 347–356. ACM (2019)
55. Zamani, M., Movahedi, M., Raykova, M.: RapidChain: scaling blockchain via full sharding. In: Proceedings of the 2018 ACM SIGSAC Conference on Computer and Communications Security, pp. 931–948. ACM (2018)
56. Zamyatin, A., et al.: SoK: communication across distributed ledgers. IACR Cryptology ePrint Archive, Report 2019/1128 (2019)

Zero-Memory Graph Exploration with Unknown Inports

Hans-Joachim Böckenhauer[1], Fabian Frei[1]([✉]), Walter Unger[2], and David Wehner[1]

[1] ETH Zürich, Zürich, Switzerland
{hjb,fabian.frei,david.wehner}@inf.ethz.ch
[2] RWTH Aachen University, Aachen, Germany
quax@algo.rwth-aachen.de

Abstract. We study a very restrictive graph exploration problem. In our model, an agent without persistent memory is placed on a vertex of a graph and only sees the adjacent vertices. The goal is to visit every vertex of the graph, return to the start vertex, and terminate. The agent does not know through which edge it entered a vertex. The agent may color the current vertex and can see the colors of the neighboring vertices in an arbitrary order. The agent may not recolor a vertex. We investigate the number of colors necessary and sufficient to explore all graphs. We prove that $n-1$ colors are necessary and sufficient for exploration in general, 3 colors are necessary and sufficient if only trees are to be explored, and $\min(2k-3, n-1)$ colors are necessary and $\min(2k-1, n-1)$ colors are sufficient on graphs of size n and circumference k, where the circumference is the length of a longest cycle. Moreover, we prove that recoloring vertices is very powerful by designing an algorithm with recoloring that uses only 7 colors and explores all graphs.

Keywords: Graph exploration · Mobile agents · Zero memory

1 Introduction

Say you wake up one morning in an unknown hotel with the desire to stroll around and visit every place in the city. Considering your terrible headache, you don't bother to remember anything about which places you have visited, but still, at the end of the day, you want to return to your hotel. You know this is not possible without further aid so you decide to take some crayons with you and color every place you visit. You are endowed with keen eyes and you're able to see the colors of the places around you. All you now need to know is how many colors you have to take along and how you color the places. This paper deals exactly with that situation.

Part of the work by Fabian Frei was done during a visit at Hosei University and supported by grant GR20109 by the Swiss National Science Foundation (SNSF) and the Japan Society for the Promotion of Science (JSPS).

© The Author(s), under exclusive license to Springer Nature Switzerland AG 2023
S. Rajsbaum et al. (Eds.): SIROCCO 2023, LNCS 13892, pp. 246–261, 2023.
https://doi.org/10.1007/978-3-031-32733-9_11

The exploration of an unknown environment by a mobile entity is one of the basic tasks in many areas. Its applications range from robot navigation over image recognition to sending messages over a network. Due to the manifold purposes, there is a great deal of different settings in which exploration has been analyzed. In this paper, we consider the fundamental problem of a single agent, e.g., a robot or a software agent, that has to first explore all vertices of an initially unknown undirected graph G, then return to the start vertex and terminate. By exploring we mean that the agent is located at a vertex and can, in each step, either go to an adjacent vertex or terminate.

If the vertices of G have unique labels and without further restrictions to the agent, this becomes a trivial task. However, in many applications, the environment is unknown and the agent is a simple and inexpensive device. Hence, we consider *anonymous* graphs, that is, there are no unique labels on the vertices or edges. Moreover, the edges have no port labels; the labeling is given implicitly by the order in which the agent sees the edges and can be different at each visit of a vertex. The agent itself is *oblivious*, that is, it has no persistent memory. Such agents are sometimes also called *zero-memory algorithms* or *1-state robots* [1,2].

Clearly, with these restrictions, there does not exist a feasible exploration algorithm. In fact, not even a graph consisting of one single edge could be explored since the algorithm would not know when to terminate. Therefore, in most models with anonymous graphs and oblivious agents, the agent remembers through which port it entered a vertex. We, in contrast, assume that the agent does not know through which edge it entered a vertex. This is sometimes called *unknown inports* or *no inports* [6]. Instead, we allow the agent to color the current vertex, a feature which is also referred to as placing distinguishable pebbles [4] or labeling vertices [1]. In this paper, we prefer the notion of coloring[1]. This notion emphasizes that a colored vertex may never be recolored unless we explicitly allow it and then use the term "recoloring." However, having the ability to color vertices alone is still utterly useless for an oblivious agent that has to return to the start vertex. Therefore, we relax our restrictions by allowing the agent to see the labels (i.e., colors) of the neighboring vertices.

We consider storage efficiency and analyze the minimum amount of colors necessary and sufficient to explore any graph with n vertices. We prove that 3 colors are both necessary and sufficient to explore trees, whereas $n - 1$ colors are necessary and sufficient to explore every graph with n vertices. This striking difference is not limited to planarity, graphs of large treewidth or graphs with a large feedback vertex set; in fact, even planar graphs with treewidth 2 and feedback vertex set number 1 need $\Omega(n)$ colors. We discover that the driving parameter of a graph is its circumference, the length of a longest cycle. We show that $2k - 1$ colors are sufficient and $\min\{2k - 3, n - 1\}$ colors are necessary to explore all graphs of circumference at most k. Finally, we make an ostensibly

---

[1] Note that this coloring is just a normal labeling and has nothing to do with graph coloring such as in 3-COLORING; it is perfectly fine to color adjacent vertices with the same color.

inconspicuous change to our model and analyze the case where we allow recoloring vertices. We show that, in this model, 7 colors are enough to explore all graphs.

This paper is organized as follows. In the rest of this introduction, we consider related work and lay out the basic definitions. In Sect. 2, we present the analysis when the graphs to be explored are trees. In Sect. 3, we analyze the general case before we then turn to graphs of a certain circumference in Sect. 4. Afterwards, we consider recoloring in Sect. 5 before we conclude in Sect. 6. Due to space constraints, some proofs are omitted.

1.1   Related Work

There is a vast body of literature on exploration and navigation problems. A great deal of aspects have been analyzed; in general, they can be categorized along four dimensions: environment, agent, goal, complexity measure. We briefly discuss these dimensions and highlight the setting considered in this paper.

The environment dimension is concerned with whether there is a specific geometric setting or a more *abstract setting* such as a graph or a *graph class*. Sometimes there are special environmental features such as faulty links or, as in our case, *local memory*, sometimes also called storage. In the agent dimension, we find the aspects such as whether there is a *single agent* or whether there are multiple agents; whether the agents are *deterministic* or probabilistic; whether the agents have some restrictions such as *limited memory*, range or *view*; and whether the agents possess special abilities such as *marking of vertices* or teleportation. Some goals include mapping the graph, finding a treasure, meeting, exploring all edges, and *exploring all vertices*. For the latter, most researchers consider one of the following three modes of termination: perpetual exploration, where the agent has to visit every vertex infinitely often; exploration with stop, where the agent has to stop at some point after it has explored everything; and *exploration with return*, where the agent has to return to the starting point after the exploration and then terminate. The main complexity measures are time complexity, space/memory complexity, *storage complexity* and competitive ratio[2].

As highlighted above, in this paper, we focus on vertex exploration by a single agent with local memory or storage. For an overview on this segment of exploration problems, we refer to the excellent overview of Das [2] and the first two chapters of [5] by Gąsieniec and Radzik. We are not aware of any research on the exploration of anonymous graphs by oblivious agents where the labels of the neighboring vertices are visible. However, the models of Cohen et al. [1] and Disser et al. [4] are similar to ours; they analyze graph exploration with anonymous graphs, local port labels, a single oblivious agent, and the ability to label the current vertex.

---

[2] Here, we use "time complexity" as the complexity of the algorithm that calculates the decisions of the agent and "competitive ratio" as the number of time steps of the agent compared to an optimal number of time steps.

In the model of Cohen et al. [1], there is a robot $\mathcal{R}$ with a finite number of states that has to explore all vertices of an unknown undirected graph and then terminate. The robot sees at each vertex the incident edges as *port numbers*. The order of these edges is fixed per vertex, but unknown to $\mathcal{R}$. The robot knows through which port it entered a vertex. In a preprocessing state, the vertices are labeled with pairwise different labels. Cohen et al. analyzed how many labels are necessary to explore all graphs. They proved that a robot with constant memory can explore all graphs with just three labels. Moreover, they showed that for any $d > 4$, an oblivious robot that uses at most $\lfloor \log d \rfloor - 2$ pairwise different labels cannot explore all graphs of maximum degree d.

Their model is different from ours in several ways: In our model, the incoming port number is not known, the order of the port numbering is not fixed, the labels may not be changed, and the algorithm has to return to the start vertex, but, most importantly, the algorithm sees the labels of the adjacent vertices. Even though the models are quite different, we can see that known inports is a much stronger feature than seeing the labels of neighboring vertices: Our general lower bound does not depend on the maximum degree, but on the number of vertices in the graph. We provide a lower bound that is linear in the number of vertices even when the maximum degree is restricted to 3.

The model of Disser et al. [4] is even closer to our model. Here, the labels— they call them distinguishable pebbles—are assigned during the exploration by the agent/algorithm as well. Moreover, the goal is exploration with return. They showed that, for any agent with sub-logarithmic memory, $\Theta(\log \log n)$ labels are necessary and sufficient to explore any graph with n vertices. Moreover, they characterized the trade-off between memory and the number of labels: When the agent has $\Omega(\log(n))$ bits of memory, all graphs on n vertices can be explored without any labels. As soon as the agent only has $\mathcal{O}(\log(n)^{1-\varepsilon})$ bits of memory, $\Omega(\log \log(n))$ labels are needed to explore all graphs on n vertices. However, with that many labels, even a constant amount of memory suffices for the exploration.

As before, this model does not directly compare to ours since neither is contained in the other. Their results seem to support the idea that knowing inports is stronger than seeing the labels of neighboring vertices; however, the focus of their work was on constant or sub-logarithmic memory.

1.2   Basic Definitions

We use the usual notions from graph theory as found for example in the textbook by Diestel [3]. The graph exploration setting considered in this paper is defined as follows. We first describe the setting informally. An agent is placed on a vertex, called the start vertex, of an undirected connected graph and moves along edges, one edge per step. In a step, the agent may use an arbitrary natural number to color the vertex on which it is currently located, if this vertex was uncolored up to now, and then move to a neighbor. As basis for its decisions, the agent may only use the color of the current vertex and the colors of the neighboring vertices. The agent can neither use the identity of the vertices nor any numbering of the edges. The agent has no persistent memory; in particular, the agent does not know through which edge it entered the current vertex (if any; the start vertex is not entered from anywhere). On the basis of the coloring of the current vertex and the neighbors, the agent chooses a neighbor it wants to move to, or it decides to stop. For no decision or choice can the agent distinguish between neighbors that have the same color.

The task is to provide a strategy for the agent—which is called an algorithm ALG—that determines for each situation the action of the agent in such a way that, no matter on which graph and on which start vertex the agent is placed, and no matter which neighbors are used to go to if there is a choice, the agent will visit all vertices, return to the start vertex, and stop there.

As with classical online problems, the concept of an *adversary* is useful to formulate bounds on the number of colors necessary and sufficient to explore all graphs. This adversary makes the decisions that are left open in the informal description above, namely choosing the start vertex and choosing the vertex the agent visits next if there is a choice among neighbors of the same color to which the agent wants to go. For all graphs and for all possible choices of the adversary, the agent must visit all vertices and then stop on the start vertex.

An algorithm in this model is a function that determines what the agent should do when located on a vertex with a certain color structure in the neighborhood. We denote the color structure by a pair (c_0, E), where $c_0 \in \mathbb{N}$ stands for the color of the current vertex and $E: \mathbb{N} \to \mathbb{N}$, where E is 0 almost everywhere, stands for the colors of the neighbors. For a number $c \in \mathbb{N}$, $E(c)$ stands for the number of neighbors of color c. In cases where we consider only a restricted amount of colors, we consider functions c_0 and E of smaller domain, for example colors $c_0 \in \{1, \ldots, n_c\}$, and $E: \{1, \ldots, n_c\} \to \mathbb{N}$. We denote the set of all pairs (c_0, E) by $\mathcal{E}$ and call it the *environment*.

An *algorithm* in this model is a function move: $\mathcal{E} \to \mathbb{N} \times \mathbb{N} \cup \{\text{STOP}\}$, with the restriction that whenever move$(c_0, E) = (c_1, d)$, we must have $E(d) > 0$, that is, the algorithm is not allowed to send the agent to a neighbor that does not exist. When the number of colors allowed is at most n_c, the range of move is $\{1, \ldots, n_c\} \times \{1, \ldots, n_c\} \cup \{\text{STOP}\}$. Moreover, we must have $c_1 = c_0$ unless $c_0 = 0$, that is, the agent may only color the current vertex if it was not colored before. We analyze the model where this last restriction is canceled in Sect. 5.

To carry out an algorithm on a given graph G, we use an adversary, and carry out steps. Initially, all vertices are uncolored, that is, have color 0. The adversary selects a start vertex v_0 and places the agent there. Then, each step works as follows: The agent is located in a vertex v. The coloring in the neighborhood is translated into an element (c_0, f) of $\mathcal{E}$ in the obvious way, by counting the number of neighbors for each occurring color. The value move(c_0, f) determined by the algorithm is then used as follows. If it is STOP, the run of the algorithm stops. If it is a pair (c_1, d), vertex v is colored with color c_1, and the adversary chooses a neighbor of v of color d, to which the agent moves for the next step.

An algorithm *successfully explores all graphs* if for all graphs G and for all adversaries, after finitely many steps, all vertices of G have been visited (they do not need to have a color $c > 0$, but they must have been the "current vertex" in some step), the agent is located on the start vertex v_0, and the decision of the algorithm is STOP. An algorithm is correct on a graph class $\mathcal{C}$ if it successfully explores all graphs $G \in \mathcal{C}$.

Throughout the paper, we denote by n the number of vertices of G and use $[n]$ to denote $\{1, \ldots, n\}$. For a vertex $v \in V$, we write $c(v) \in \mathbb{N}$ for its color; $N(v)$ for the open neighborhood of v, that is, the set of vertices adjacent to v; and $N[v]$ for the closed neighborhood of v, where $N[v] := N(v) \cup \{v\}$. We use mod_1 to denote a modulo operator shifted by 1, i.e., $n \bmod_1 m := ((n-1) \bmod m) + 1$. Instead of having the numbers $0, \ldots, m-1$ as the outcome of the modulo operation, one thus obtains the numbers $1, \ldots, m$. For convenience, we sometimes speak of an algorithm behaving in a particular way and mean by this formulation that the agent of the algorithm behaves in a particular way.

2   Exploration of Trees

We begin by analyzing the problem on trees. We show that only three colors are enough to explore all trees. In line with the research that analyzes graph exploration with pebbles, we do not count 0 as a color. The idea of the algorithm is to alternatingly label the vertices with colors 1, 2, and 3 and then follow a simple depth-first search (DFS) strategy. Since there are no cycles, there is always a unique path back to the start vertex and backtracking is possible. Moreover, the start vertex is recognized by the fact that it is the only vertex with color 1 where all neighbors have color 2. The formalization of this strategy, which we call TREEEXPLORATION, is omitted here; it enables us to use at most 3 colors and successfully explore all graphs, which we note as a first theorem:

Theorem 1. TREEEXPLORATION *is correct on trees and uses at most 3 colors.*

We prove that TREEEXPLORATION is optimal in the sense that it is impossible to use fewer than 3 colors to solve graph exploration in our model on trees. Before doing so, we make the following crucial observation.

Observation 1 (Functional Nature). *Due to its functional nature, once an algorithm is in a vertex v where all vertices in $N[v]$ have been colored and takes a decision upon which it goes to a neighbor w or upon which the adversary chooses neighbor w as next vertex, this choice can be made by the adversary each time the algorithm returns to v.*

Theorem 2. *There is no algorithm that solves graph exploration as in our model on every tree and that uses less than 3 colors.*

Proof. Assume by contradiction that there exists such an algorithm. Consider a path with seven vertices. Denote them by v_1 to v_7 in sequential order. Let v_1 be the start vertex.

For $i \in [5]$, if an agent does not color v_i, it cannot visit v_{i+2}: The agent could continue to v_{i+1}, but then it cannot distinguish between v_i and v_{i+2}. Therefore, the adversary can make the agent go back to v_i. If the agent now does not color v_i,[3] it is caught in an endless loop by Observation 1, which contradicts the assumption that the algorithm works correctly on all trees.

When the agent has reached v_7, it has to return to v_1. It must make a step from v_4 to v_3 at some point. If $c(v_3) = c(v_5)$, the adversary can make the agent go back to v_5 instead, which again opens the way to an endless loop. Hence, $c(v_3) \neq c(v_5)$. The same is true at v_3 and at v_2. Hence, $c(v_4) \neq c(v_2)$ and $c(v_1) \neq c(v_3)$. Without loss of generality, assume $c(v_1) = 1$. Then this implies that either $(c(v_1), \ldots, c(v_5)) = (1, 1, 2, 2, 1)$ or $(c(v_1), \ldots, c(v_5)) = (1, 2, 2, 1, 1)$.

If $(c(v_1), \ldots, c(v_5)) = (1, 1, 2, 2, 1)$, the agent has to go from v_4 to v_3; however, then the agent goes from v_3 back to v_4 because of its functional nature. Similarly, if $(c(v_1), \ldots, c(v_5)) = (1, 2, 2, 1, 1)$, the agent has to go from v_3 to v_2, but then it goes from v_2 back to v_3. □

3   Exploration of General Graphs

How can an agent proceed on graphs that are not necessarily trees? Again, a depth-first search strategy suffices; however, this time, the agent uses almost as many colors as vertices.

Let us first describe the idea. The agent starts at the start vertex and colors it with color 1. This vertex is the root. All other vertices will receive larger colors; hence, the root is easy to recognize. We wish to carry out DFS. It is easy to go whenever possible to an unexplored—or, equivalently, uncolored—neighbor of the current vertex. But how shall the agent find the way back? The idea is to color a new vertex with a color by 1 larger than the largest color in the neighborhood. This ensures that colors increase along paths taken forward in the tree. A notable exception is when the agent arrives at a leaf in the DFS tree, that is, a vertex where all neighbors are colored. Then, the agent can save a color by assigning the largest color in the neighborhood—which is the color of the vertex the agent came from–instead of the largest color in the neighborhood plus one. For backtracking from a vertex v, the agent goes to a vertex whose color is one less than $c(v)$. Such a vertex always exists, except if v is the root. The only thing one has to prove is that this neighbor is unique. The formal description of DEPTHFIRSTSEARCH is omitted here.

Clearly, DEPTHFIRSTSEARCH is well defined and uses at most $n - 1$ colors. To show that DEPTHFIRSTSEARCH explores all vertices, we first prove that whenever DEPTHFIRSTSEARCH goes to a vertex that has been colored before, this vertex is the predecessor[4] of the current vertex.

Theorem 3. DEPTHFIRSTSEARCH *is correct and uses at most* $n - 1$ *colors.*

---

[3] Note that the agent may have colored v_{i+1} and thus the environment of the second visit of v_i may be different from the environment of the first visit, leading to a potentially different decision.

[4] By *predecessor* of a vertex v that is not the root, we mean the neighbor w from which the agent moved to v when v was first visited.

We are going to show in the following that DEPTHFIRSTSEARCH is optimal in our model. Hence, our perspective changes. We now deal with an unknown algorithm of which we only know that it successfully explores all graphs. In order to show a lower bound, we have to define a graph and the decisions of an adversary such that every algorithm with a limited amount of colors fails to explore our graph given the decisions of our adversary.

We start with an easy observation and two technical lemmas, whose proofs are omitted in this extended abstract. These lemmas will be crucial in creating lower bounds for our model. They will allow us to define an instance and then argue that the agent has to take a certain path in this instance and cannot choose a different route.

Observation 2. *If the agent of an algorithm* ALG *that successfully explores all graphs arrives at a colored vertex v and there is an uncolored neighbor, then the agent has to visit an uncolored neighbor next.*

Lemma 1. *Let* ALG *be an algorithm that successfully explores all graphs. If the agent of* ALG *is on a vertex v with only one colored neighbor p, which is its predecessor, and at least one uncolored neighbor u, and if p has an uncolored neighbor $v' \neq v$, then the algorithm has to go to an uncolored neighbor of v.*

Lemma 2. *Let* ALG *be an algorithm that successfully explores all graphs. Assume the agent has just moved from a (now) colored vertex p to an uncolored vertex v, and that p has at least one other uncolored neighbor v'. Then the agent must color v now.*

We now prove that DEPTHFIRSTSEARCH uses the minimal number of colors necessary.

Theorem 4. *For every algorithm* ALG *that successfully explores all graphs and for every natural number n there is a graph G of size n such that* ALG *uses at least $n - 1$ colors to explore G.*

Proof. For n at most 4, we verify the statement by checking all cases. For $n \leq 5$, assume towards contradiction that there is an algorithm ALG that, for some $n_0 \geq 5$, uses at most $n_0 - 2$ colors on every graph of size n_0. We are going to define a family of graphs of size $2(n_0 - 1) - 1$ and prove that there is a graph G_0 in this family such that ALG uses at least $n_0 - 1$ colors on its first $n_0 - 1$ steps in order to successfully explore G_0. Afterwards, we define a graph G_1 of size n_0 which locally looks, for the first $n_0 - 1$ steps, exactly as G_0. Therefore, ALG uses at least $n_0 - 1$ colors on G_0 as well.

We construct our family of graphs as follows. Consider a graph G that consists of a path $v_1, \ldots, v_{n_0-1}$ of length $n_0 - 1$ where, for each $r \in [n_0 - 1]$, a leaf l_r is connected to v_r. For fixed $i, j \in [n_0 - 1]$, we merge the leaves l_i and l_j and call the new vertex $l_{i/j}$ and the graph $G_{n_0,i,j}$. Such a graph is depicted in Fig. 1.

Consider any general algorithm ALG with at most $n_0 - 2$ colors that explores a graph $G_{n_0,r,s}$ for yet to be determined r and s. The adversary makes such decisions that ALG walks from v_1 to v_{n_0-1} without exploring any leaves or

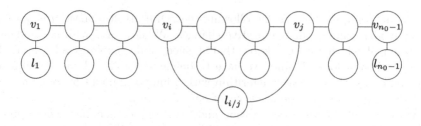

Fig. 1. The graph $G_{n_0,i,j}$

leaving any visited vertex uncolored. This is possible by Lemma 1 and Lemma 2.

By the pigeonhole principle, there are $i < j$ such that ALG assigns the same color to v_i and v_j. Let $r = i$ and $s = j$; in other words, consider the exploration of ALG on $G_{n_0,i,j}$. Without loss of generality, we assume that v_i and v_j are not adjacent since, if that were the case, clearly, all v_k with $k > j$ would be assigned the color of v_j and ALG cannot explore all graphs this way.

On its way back to v_1, ALG reaches v_j at some point. From there, ALG has to visit $l_{i/j}$ by Observation 2. Since ALG cannot distinguish v_i and v_j when located in $l_{i/j}$, the adversary can send ALG to either vertex if ALG wants to visit a neighbor. If ALG decides not to color $l_{i/j}$, then the adversary sends ALG to v_i and ALG will have to go to $l_{i/j}$ by Observation 2 and then either color $l_{i/j}$ or be caught in an endless loop. Hence, ALG colors $l_{i/j}$ and then goes to one of its neighbors. The adversary is going to send ALG to v_i.

When ALG arrives in v_i, the closed neighborhood $N[v_i]$ is colored. If ALG decides to go to $l_{i/j}$, it will then again be sent to v_i and will never terminate. If ALG decides to go to v_{i-1}, it will never visit any leaf between v_i and v_j. If ALG decides to go to v_{i+1}, then ALG can never terminate since all paths from v_{i+1} to v_1 lead through v_i and from there, ALG always chooses v_{i+1}. Therefore, no matter how ALG behaves, it cannot successfully explore $G_{n_0,i,j}$, unless it uses at least $n_0 - 1$ colors on the first $n_0 - 1$ steps.

Consider now the graph G_1 depicted in Fig. 2. G_1 consists of a path of $n_0 - 1$ vertices that are connected to a universal vertex u. Locally, G_1 looks exactly the same as $G_{n_0,i,j}$. An algorithm can only successfully explore $G_{n_0,i,j}$ if it assigns $n_0 - 1$ colors on the first $n_0 - 1$ steps during the exploration of $G_{n_0,i,j}$. Then, however, such an algorithms assigns $n_0 - 1$ colors on the first $n_0 - 1$ steps during

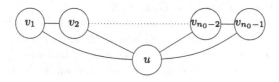

Fig. 2. The graph G_1 for the general lower bound

Algorithm 1. SMALLDFS$_k$

1: **if** $c(v) = 0$ **then**

2:    $c(v) := \left(\left(\max^k_{v' \in N(v)} c(v') \right) + 1 \right) \mod_1 (2k-1)$

3: **if** there is an uncolored neighbor w **then**

4:    go to w

5: **else if** there is a neighbor w with $c(w) = (c(v) - 1) \mod_1 (2k-1)$ **then**

6:    go to w

7: **else**

8:    **terminate**

the exploration of G_1 as well. G_1 has n_0 vertices; hence, we have discovered a graph on n_0 vertices where ALG uses at least $n_0 - 1$ colors. This contradicts our assumption and thus finishes the proof. □

4   Between Trees and General Graphs

We want to discover the parameter that determines the difficulty of graph exploration of a given graph. Since the exploration of trees is quite easy, it is tempting to think that a good parameter to measure the difficulty for graph exploration could be treewidth, feedback vertex set, number of cycles, etc. However, considering the difficult instance in Fig. 1, we see that treewidth and feedback vertex set are no suitable parameters. In fact, even restricting the number of cycles to 1 and the maximum degree to 3 does not remove the linear amount of colors. Bipartite graphs as generalizations of trees might still serve as plausible candidates. After all, the proof of Lemma 2 does not work for bipartite graphs and the graphs $G_{n_0,i,j}$ used in the construction for Theorem 4 are not bipartite if $j - i$ is odd. However, these defects can be remedied and only exploring bipartite graphs is almost as difficult as exploring all graphs.

We now argue why the *circumference* of a graph, i.e., the size of a longest simple cycle, is the right parameter. First, we observe that the construction from the proof of Theorem 4 immediately results in the following corollary:

Corollary 1. *Any algorithm needs at least $k - 2$ colors to color all graphs of circumference at most k.*

Second, we present with SMALLDFS$_k$ an algorithm that uses at most $2k - 1$ colors and successfully explores all graphs of circumference at most $k \geq 3$. This algorithm sorts numbers a, b with respect to whether the distance between $a, \ldots, b$ or the distance between $b, b+1, \ldots, 2k-1, 1, 2, \ldots, a$ is larger. We use the following maximum function for three natural numbers a, b, k with $a \leq b \leq 2k-1$:

$$\max^k\{a, b\} := \begin{cases} a, & \text{if } |b - a + 1| < |2k - b + a| \\ b, & \text{if } |b - a + 1| > |2k - b + a|. \end{cases}$$

Note that, if the two values $(b - a + 1)$ and $(2k - b + a)$ were equal, this would imply $2(b - a) = 2k - 1$, which is impossible for three natural numbers a, b and k. In other words, $\max^k$ is well-defined. With this maximum function, that is, calculating modulo $2k - 1$ shifted by 1, a number is smaller than the next $k - 1$ numbers and larger than the preceding $k - 1$ numbers.

The strategy of SMALLDFS$_k$ is a depth-first search strategy that assigns color $c \bmod_1 (2k - 1)$ to vertices that are visited in distance $c - 1$ (in the depth-first search tree) from the start vertex. This is done by looking at all neighbor colors and then assigning the largest color modulo $(2k - 1)$ to the current vertex. However, in order to be able to assign the color 1 after the color $2k - 1$, the algorithm takes the maximum function defined above to determine the largest color number. Since there are no cycles of length $k + 1$, the maximum function is well-defined. Thus, it is possible, as we are going to prove in a moment, to determine the direct predecessor and backtrack accurately.

We can prove that SMALLDFS$_k$ terminates on the start vertex and only on the start vertex; the proofs are omitted here. We have yet to show that all vertices are indeed explored.

Lemma 3. *On any vertex v, either* SMALLDFS$_k$ *goes to an unvisited vertex or it goes back to the direct predecessor of v.*

Proof. We prove something slightly different, namely that (1) SMALLDFS$_k$ always assigns color $c(v') + 1 \bmod_1 (2k - 1)$ to a vertex v with predecessor v' and that (2) there is never a vertex v with two neighbors w_1, w_2 with $c(w_1) = c(w_2) = c(v) - 1 \bmod_1 (2k - 1)$. This implies that SMALLDFS$_k$ always goes back to the direct predecessor.

Assume towards contradiction that one of these two claims is wrong. Consider the first step in which one of these claims is wrong. We have three cases:

Case 1 In this step, SMALLDFS$_k$ does not assign color $c(v') + 1 \bmod_1 (2k - 1)$ to a vertex v with predecessor v'. This can only happen if there is a neighbor x of v with greater color than v', that is, $\max^k\{c(x), c(v')\} = c(x)$.

Both x and v' are colored before v. Assume as subcase 1.1 that x is colored before v'. Consider the search sequence $(x, v_1, \ldots, v_n, v')$ of SMALLDFS$_k$ between x and v'. The first step when SMALLDFS$_k$ "skips" a color is between v' and v; hence, the colors between vertices next to each other in this search sequence differ by exactly 1. Since $c(x)$ is smaller than the next $k - 1$ numbers and since we have $\max^k\{c(x), c(v')\} = c(x)$, there have to be at least $k - 1$ vertices between x and v'. Thus, we have together with v a cycle of length at least $k + 2$, which is not allowed. Hence, x is not colored first.

Consider subcase 1.2, namely that v' is colored before x. Then SMALLDFS$_k$ went from v' to x, then back to v' and then to v. However, this is not possible since backtracking (applying line 6) is only allowed when there are no uncolored neighbors and thus, SMALLDFS$_k$ cannot backtrack from x, which has v as an uncolored neighbor.

Case 2 In this step, SMALLDFS$_k$ assigns color $c(v)$ to a vertex v with two neighbors w_1, w_2 with $c(w_1) = c(w_2) = c(v) - 1 \bmod_1 (2k - 1)$. Assume without loss of generality that first w_1 is visited (and colored), then w_2, then v.

We can argue exactly as before. The search sequence of SMALLDFS$_k$ between w_1 and w_2 is $(w_1, v_1, \ldots, v_r, w_2)$. SMALLDFS$_k$ could not backtrack from w_1 and thus went to some uncolored vertex v_1. Again, backtracking would only be possible up to w_1; therefore, we can assume without loss of generality that all vertices $v_1, \ldots, v_r$ in the search sequence are visited for the first time. In order for w_2 to obtain the same color as w_1, there have to be at least $2k - 2$ vertices between w_1 and w_2. Since w_1 and w_2 are both adjacent to v, this results in a cycle of size $2k + 1$.

Case 3 In this step, SMALLDFS$_k$ assigns color $c(w_1)$ to a vertex w_1 adjacent to a vertex v with $c(w_1) = c(v) - 1 \bmod_1 (2k - 1)$ and v is adjacent to a vertex $w_2 \neq w_1$ with $c(w_2) = c(w_1)$. Consider the search sequence $(v, \ldots, w_1)$ between v and w_1. Since w_1 receives a greater number than v, there have to be at least $k - 1$ vertices between v and w_1. Together with v and w_1, they form a cycle of size $k + 1$, which is impossible.

We see that the assumption that one of these two claims is wrong leads to a contradiction; therefore, both claims are true, which proves the lemma.   □

It is easy to prove that SMALLDFS$_k$ explores all vertices and we thus obtain the following theorem.

Theorem 5. SMALLDFS$_k$ *uses at most* $2k - 1$ *colors and is correct on graphs with circumference at most* k.

Contrasting Corollary 1 and Theorem 5, we see that SMALLDFS$_k$ uses at most $k + 1$ too many colors. We narrow this gap with the following theorem.

Theorem 6. *An algorithm needs at least* $2k - 3$ *colors to successfully color all graphs of circumference at most* k.

Proof. Let ALG be an algorithm that uses at most $2k - 4$ colors and colors every graph of circumference k. Without loss of generality, assume that ALG uses exactly $2k-4$ colors. The colors ALG assigns have to repeat at some point, that is, if we let ALG explore a path $v_1, \ldots, v_r$ of length $r > 2k-4$, after some initial color assignments $c(v_1), c(v_2), \ldots, c(v_s)$, the colors ALG assigns always follow the same pattern $c(v_s), c(v_{s+1}), \ldots, c(v_t), c(v_s), c(v_{s+1})$ etc. Without loss of generality, we assume this pattern starts with v_1. This behavior can be enforced even if we add some leaves or connect some vertices of the path: Due to Lemma 1, a general algorithm may not go back to a visited vertex if there is an unvisited neighbor. Therefore, we may even add some leaves to the path or connect some vertices of the path and there is still an adversary that ensures that ALG goes from v_1 directly to v_r without visiting any leaf (apart from v_1 and v_r, of course) and without taking any shortcuts.

Consider now the two graphs G_1 and G_2 from Fig. 3, both consisting of a path $v_1, \ldots, v_{2k}$ of length $2k - 1$ and one shortcut edge and some leaves attached to it. Note that both G_1 and G_2 have circumference exactly k.

Consider the exploration of ALG on G_1. As explained, there is an adversary that lets ALG walk from v_1 to v_{2k-2} without taking the shortcut $\{v_1, v_k\}$ and

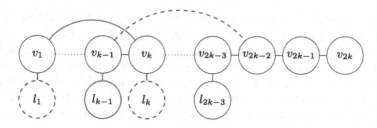

Fig. 3. The graphs G_1 (black and blue) and G_2 (black and red). (Color figure online)

without exploring the leaves. Moreover, by assuming that the repeating pattern of ALG starts with v_1, we obtain in particular that $c(v_1) = c(v_{2k-3})$.

Now, on vertex v_k, ALG receives as input $c(v_k) = 0$ and the colors $c(v_1)$, $c(v_{k-1})$, and 0 and assigns color $c(v_k) > 0$. Then, ALG goes on to explore the rest of the graph. At some point, ALG returns to v_k and receives as input $c(v_k)$ and the colors $c(v_1)$, $c(v_{k-1})$, and $c(v_{k+1})$. Since l_{k-1} has not yet been explored and since upon going to v_1, ALG has to terminate, it is crucial that ALG chooses v_{k-1} as its next vertex.

However, consider now the exploration of ALG on G_2. Since G_2 looks locally exactly like G_1, there is an adversary that lets ALG behave on G_2 exactly as on G_2 for at least the first $2k - 3$ steps. On vertex v_{2k-2}, ALG receives as input $c(v_{2k-2}) = 0$ and the colors $c(v_{2k-3}) = c(v_1)$, $c(v_{k-1})$, and 0. This is the same input as before and because of its functional nature, ALG has to assign color $c(v_{2k-2}) = c(v_k) > 0$ and then has to go to the unvisited neighbor v_{2k-1}. On v_{2k-1}, ALG is in the same situation as during the exploration of G_1 on v_{k+1} and assigns color $c(v_{2k-1}) = c(v_{k+1})$. After then visiting v_{2k}, ALG returns to v_{2k-2}. Now, its input is again $c(v_k) = c(v_{2k-2})$ and the colors $c(v_1) = c(v_{2k-3})$, $c(v_{k-1})$, and $c(v_{2k-1}) = c(v_{k+1})$.

However, this time, since l_{2k-3} has not yet been explored, it is crucial for ALG to choose v_{2k-3} and not v_{k-1}. Since ALG is a function, it cannot output different values on the same input; hence, ALG fails to explore either G_1 or G_2.   □

5   Exploration with Recoloring

In this section, we allow recoloring of already colored vertices.

We prove that, in this case, seven colors are enough to explore any graph. This demonstrates the superior strength of strategies with recoloring. We omit the formal description of our algorithm RECOLORER. Let us describe the basic strategy. We abuse the term "color" and say a vertex either receives label x or a label 1, 2 or 3 together with one of the two colors green and red. The special label x is assigned to *delete* vertices, that is, to mark vertices to which the algorithm must never return. The algorithm performs breadth-first search and labels every vertex in *depth* k, i.e., at distance k from the start vertex with label $k+1 \bmod_1 3$. Unless a vertex is deleted, the labels are never changed, only the colors.

We imagine the start vertex as top vertex and go from top to bottom when moving further away from the start vertex. Since vertices in depth k have only neighbors in depth $k-1$, k, and $k+1$, this labeling provides a sense of direction.

When RECOLORER is on a vertex v that is in depth k, we call neighbors in depth $k-1$ *parents* (of v), neighbors in depth $k+1$ *children* (of v), and neighbors in the same depth as v *siblings*. Note that the agent is going to be able to determine whether labeled neighbors are parents, siblings, or children. To ensure that all vertices in a certain depth are colored, the algorithm colors the vertices from green to red and vice versa, depending on the current phase.

There are green phases and red phases in alternating order.

Initially, the start vertex receives label 1 and color red. Then, a green phase starts. In a green phase, the algorithm follows the red labels from 1 to 2 to 3 to 1 etc. until an unvisited vertex is reached. This vertex is then marked with the next label and colored green. The algorithm then proceeds to a parent p. If p has unvisited neighbors—these are all children, as we will see—, then one of these neighbors is visited. Otherwise, if all children of p are colored, the algorithm goes down to a child of the same color as p until it eventually finds an unvisited vertex, which is then colored green before going up to a parent. As soon as all children of a parent are of another color, the algorithm recolors the parent and then goes up to a grandparent. This process takes place until the start vertex is reached and all children of the start vertex are colored green. Now, the start vertex is marked green as well and a red phase starts, which works analogously.

This process would suffice to achieve perpetual exploration. To ensure that the algorithm terminates, the algorithm deletes a vertex whenever all its children are deleted or if there are no children at all. An example is depicted in Fig. 4.

We have the following theorem, whose proof we omit here.

Theorem 7. *There is an algorithm that never uses more than 7 colors and explores every graph with recoloring.*

We mention that allowing the recoloring of vertices might render it feasible to adapt an algorithm by Cohen et al. [1, Theorem 2.1], which uses 25 colors, to obtain a weaker version of Theorem 7. It is possible to simulate the knowledge of the inport via color-encoded temporary flags in the vertices (each flag-type doubling the number of colors), and it seems that the other extra feature of their model, namely the fixed port numbers, can be compensated in our model by the agent directly seeing the labels of all neighboring vertices. Another small difference is that our agent is required to terminate at the start vertex, whereas in the model by Cohen at al., the agent may terminate as soon as the entire graph is explored. This issue is easy to address, however. We can simulate an algorithm not guaranteeing a termination at the start vertex twice: first once after marking the start vertex with a unique flag, and then a second time, with a different set of colors (effectively squaring the number of required colors), starting from the vertex at which the previous simulation terminated, until the marked start vertex is encountered, prompting the algorithm to terminate. However, even if the simulation as described above is possible, the required number of colors will be in the hundreds or thousands instead of seven.

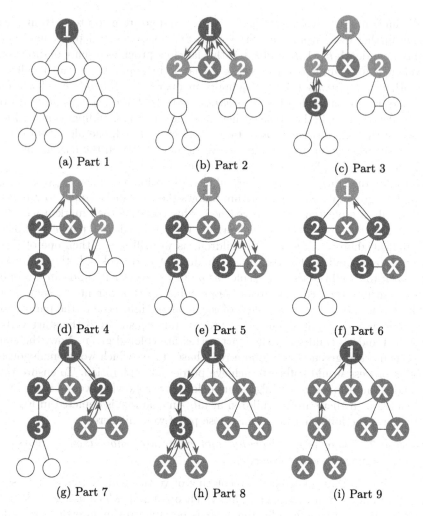

Fig. 4. An example of an exploration by RECOLORER.

6 Conclusion

We investigated graph exploration by a very limited agent and showed tight bounds for the exploration of trees and of general graphs. Essential for our upper bounds was the idea that the algorithm needs a way to uniquely determine the direct predecessor of a vertex. On a high level, our algorithms try to enumerate the number of situations that can occur locally. Surprisingly, we discovered that this number is not determined by the degree of a vertex, but by the circumference of the graph. We showed almost tight lower bounds for graphs of a certain circumference as well.

For our lower bounds, Lemma 1 and Lemma 2 form the basis of our argumentation. They make it possible to construct specific graphs and know exactly—up to renaming of the colors—how a general algorithm behaves on this graph.

Finally, we studied the case where recoloring vertices is allowed. We showed that in this case, seven colors are already sufficient to explore all graphs, which stands in stark contrast to the amount of colors needed in general.

There are at least three ways to continue this research. First, it would be interesting to generalize Lemma 1 and Lemma 2 to certain subgraphs or minors. This would facilitate the analysis of graph exploration in our model for large graph classes. Second, a natural extension of our model would be to allow a constant number of memory bits. This would in particular enable a direct comparison to the existing research on oblivious graph exploration with known inports. Third, it might be interesting to analyze how additional information can help the agent. On the one hand, one could study the classical advice complexity model for similar graph exploration models; on the other hand, one could restrict the advice to vertices. This would allow comparing global advice and local advice, which is much more restricted, but can be accessed exactly when needed.

Acknowledgments. We thank the anonymous reviewers for their useful suggestions for improvement.

References

1. Cohen, R., Fraigniaud, P., Ilcinkas, D., Korman, A., Peleg, D.: Label-guided graph exploration by a finite automaton. ACM Trans. Algorithms **4**(4), 42:1–42:18 (2008). https://doi.org/10.1145/1383369.1383373
2. Das, S.: Graph explorations with mobile agents. In: Flocchini, P., Prencipe, G., Santoro, N. (eds.) Distributed Computing by Mobile Entities, Current Research in Moving and Computing. LNCS, vol. 11340, pp. 403–422. Springer, Cham (2019). https://doi.org/10.1007/978-3-030-11072-7_16
3. Diestel, R.: Graph Theory. Graduate Texts in Mathematics, 5th edn. Springer, Heidelberg (2017). https://doi.org/10.1007/978-3-662-53622-3
4. Disser, Y., Hackfeld, J., Klimm, M.: Tight bounds for undirected graph exploration with pebbles and multiple agents. J. ACM **66**(6), 40:1–40:41 (2019). https://doi.org/10.1145/3356883
5. Gąsieniec, L., Radzik, T.: Memory efficient anonymous graph exploration. In: Broersma, H., Erlebach, T., Friedetzky, T., Paulusma, D. (eds.) WG 2008. LNCS, vol. 5344, pp. 14–29. Springer, Heidelberg (2008). https://doi.org/10.1007/978-3-540-92248-3_2
6. Menc, A., Pajak, D., Uznanski, P.: Time and space optimality of rotor-router graph exploration. Inf. Process. Lett. **127**, 17–20 (2017). https://doi.org/10.1016/j.ipl.2017.06.010

The Energy Complexity of Diameter and Minimum Cut Computation in Bounded-Genus Networks

Yi-Jun Chang[✉][ID]

National University of Singapore, Singapore, Singapore
cyijun@nus.edu.sg

Abstract. This paper investigates the energy complexity of distributed graph problems in multi-hop radio networks, where the energy cost of an algorithm is measured by the maximum number of awake rounds of a vertex. Recent works revealed that some problems, such as broadcast, breadth-first search, and maximal matching, can be solved with energy-efficient algorithms that consume only $\operatorname{poly} \log n$ energy. However, there exist some problems, such as computing the diameter of the graph, that require $\Omega(n)$ energy to solve. To improve energy efficiency for these problems, we focus on a special graph class: bounded-genus graphs. We present algorithms for computing the exact diameter, the exact global minimum cut size, and a $(1 \pm \epsilon)$-approximate s-t minimum cut size with $\tilde{O}(\sqrt{n})$ energy for bounded-genus graphs. Our approach is based on a generic framework that divides the vertex set into high-degree and low-degree parts and leverages the structural properties of bounded-genus graphs to control the number of certain connected components in the subgraph induced by the low-degree part.

Keywords: Energy-aware computation · Radio networks · Diameter

1 Introduction

We consider the multi-hop *radio network* model [16] of distributed computing, where a communication network is modeled as a graph $G = (V, E)$: Each vertex $v \in V$ is a computing device and each edge $\{u, v\} \in E$ indicates that u and v are within the transmission range of each other. The graph topology of the underlying network G is initially unknown to all devices, except that two parameters $n = |V|$ and $\Delta = \max_{v \in V} \deg(v)$ are global knowledge.

Communication proceeds in synchronized rounds. All devices agree on the same start time. In each round, each device can choose to do one of the following three operations: (i) listen to the channel, (ii) transmit a message, or (iii) stay idle. We do not allow a device to simultaneously transmit and listen, and we assume that there is no message size constraint.

Each transmitting or idle device does not receive any feedback from the communication channel, so a transmitting device u does not know whether its

S. Rajsbaum et al. (Eds.): SIROCCO 2023, LNCS 13892, pp. 262–296, 2023.
https://doi.org/10.1007/978-3-031-32733-9_12

message is successfully received by any of its neighbors $N(u)$. A listening device v successfully receives a message from a transmitting device $u \in N(v)$ if u is the only transmitting device in $N(v)$. If the number of transmitting devices in $N(v)$ is zero, then a listening device v hears silence. For the case the number of transmitting devices in $N(v)$ is greater than one, then the feedback that a listening device v receives depends on the underlying model. In the No-CD model (without collision detection), v still hears silence. In the CD model (with collision detection), v hears collision. All our algorithms presented in this paper work in the No-CD model.

We assume that each device has access to an unlimited local random source. We say that an event occurs *with high probability* (w.h.p.) if the event occurs with probability $1 - 1/\text{poly}(n)$. In this paper, we only consider *Monte Carlo* algorithms that succeed w.h.p. If we let each vertex $v \in V$ locally assign themselves $O(\log n)$-bit identifiers $\text{ID}(v)$, then they are distinct with high probability, so we assume that each device has a distinct identifier of length $O(\log n)$.

Complexity Measures. *Time* and *energy* are the two main complexity measures of a distributed algorithm in radio networks. The time complexity of an algorithm is the number of rounds of the algorithm. The energy complexity of an algorithm is the maximum energy cost of a device, where the energy cost of a device v is the number of rounds that v is non-idle. We only consider *worst-case* time and energy complexities. The motivation for studying energy complexity is that energy is a scarce resource in small battery-powered wireless devices, and such devices can save energy by entering a low-power sleep mode.

1.1   Prior Work

Most of the early work on the energy complexity focused on *single-hop* radio networks, which is the special case that $G = (V, E)$ is a complete graph. Over the last two decades, there is a long line of research to optimize the energy complexity of *leader election* and its related problems in single-hop radio networks [6,8,9, 11,13,14,29–32,34,35,39].

This line of research was recently extended to multi-hop radio networks [10, 12,19,20]. Chang et al. [10] considers the problem of *broadcasting* a message from one device to all other devices in a multi-hop radio network. They showed that broadcasting can be done in poly $\log n$ energy. Specifically, they presented randomized broadcasting algorithms for CD and No-CD using energy $O\left(\frac{\log n \log \log \Delta}{\log \log \log \Delta}\right)$ and $O(\log \Delta \log^2 n)$ w.h.p., respectively. They also proved that any algorithm transmitting a message from one endpoint to the other endpoint of an n-vertex path costs $\Omega(\log n)$ energy *in expectation*. The lower bound applies even to the LOCAL model of distributed computing.

Chang et al. [12] showed that *breadth-first search* can be done w.h.p. using $2^{O(\sqrt{\log n \log \log n})}$ energy in No-CD. Their algorithm is based on a hierarchical clustering using the low-diameter decomposition algorithm of Miller, Peng, and Xu [38]. The energy complexity of breadth-first search was recently improved to poly $\log n$ by Dani and Thomas [20]. Combining the polylogarithmic-energy breadth-first search algorithm with the diameter approximation algorithm of Roditty and Williams [41], an approximation $\tilde{D}$ of the diameter D such that $\tilde{D} \in [\lfloor 2D/3 \rfloor, D]$ can be computed with $\tilde{O}(\sqrt{n})$ energy w.h.p. [12]. The notation $\tilde{O}(\cdot)$ suppresses any poly $\log n$ factor.

Dani et al. [19] showed that a *maximal matching* can be computed in $O(\Delta \log n)$ time and $O(\log \Delta \log n)$ energy w.h.p. in No-CD. There exists a family of graphs such that these time and energy bounds are simultaneously optimal up to polylogarithmic factors.

1.2   Our Contribution

Not all problems admit energy-efficient algorithms in multi-hop radio networks. It was shown in [12] that any algorithm that computes a $(1.5 - \epsilon)$-approximation of the diameter requires $\tilde{\Omega}(n)$ energy w.h.p. The lower bound holds even on graphs with arboricity $O(\log n)$ and treewidth $O(\log n)$.

To improve energy efficiency for diameter computation, we focus on the class of bounded-genus graphs. We show that the diameter of the graph can be computed using $\tilde{O}(\sqrt{n})$ energy w.h.p. in bounded-genus graphs.

Theorem 1. *There is an algorithm that computes the diameter in $\tilde{O}(n^{1.5})$ time and $\tilde{O}(\sqrt{n})$ energy w.h.p. for bounded-genus graphs in No-CD.*

Our approach is based on a generic framework that divides the vertex set into high-degree and low-degree parts. We then classify the connected components of the subgraph induced by the low-degree part into several types. We will leverage the structural properties of bounded-genus graphs to upper-bound the number of connected components of one type. For the remaining connected components, we will design energy-efficient algorithms that extract all the necessary information from these connected components for the purpose of diameter computation.

Our approach is sufficiently general so that it is applicable to other problems as well. Using the same approach, we show that the exact global minimum cut size and a $(1 \pm \epsilon)$-approximate s-t minimum cut size can also be computed using $\tilde{O}(\sqrt{n})$ energy w.h.p. in bounded-genus graphs.

Theorem 2. *There is an algorithm that computes the minimum cut size in $\tilde{O}(n^{1.5})$ time and $\tilde{O}(\sqrt{n})$ energy w.h.p. for bounded-genus graphs in No-CD.*

Theorem 3. *There is an algorithm that computes an $(1 \pm \epsilon)$-approximate s-t minimum cut size in $\tilde{O}(n^{1.5}) \cdot \epsilon^{-O(1)}$ time and $\tilde{O}(\sqrt{n}) \cdot \epsilon^{-O(1)}$ energy w.h.p. for bounded-genus graphs in No-CD.*

To complement these algorithmic results, we show that any algorithm that computes the exact size of an s-t minimum cut or a global minimum cut requires $\Omega(n)$ energy. The lower bound for the s-t minimum cut holds even for planar bipartite graphs, so it is necessary that we consider approximation algorithms for this problem. These lower bounds apply to both No-CD and CD.

Theorem 4. *For any randomized algorithm that computes the s–t minimum cut size of a planar bipartite graph w.h.p. in* CD, *the energy complexity of the algorithm is* $\Omega(n)$.

Theorem 5. *For any randomized algorithm that computes the minimum cut size of a unit-disc graph w.h.p. in* CD, *the energy complexity of the algorithm is* $\Omega(n)$.

1.3   Additional Related Work

There are numerous works studying energy-aware distributed computing in multi-hop networks from different perspectives. In radio networks, the power of a signal received is proportional to $O(1/d^\alpha)$, where d is the distance to the sender, and α is a constant related to environmental factors. Kirousis et al. [33] studied the optimization problem of assigning transmission ranges of devices subject to some connectivity and diameter constraints so as to minimize the total power consumption. See [2,17,43] for related work.

There are several works [7,22,42] on the subject of reducing the number of rounds or transmissions required to complete a specific communication task. In the setting of known network topology, Gasieniec et al. [23] designed a randomized protocol for broadcasting in $O(D + kn^{1/(k-2)} \log^2 n)$ rounds such that each device transmits at most k times.

The energy complexity has recently been studied in the well-known LOCAL and CONGEST models of distributed computing [3,5,15,21,27].

There is a large body of research on distributed graph algorithms in planar networks, bounded-genus networks, or more broadly H-minor-free networks: distributed approximation [1,18,36,44], low-congestion shortcuts and its applications [24–26,28], and other planar graph algorithms [37,40].

1.4   Organization

In Sect. 2, we present the basic tools. In Sect. 3, we present our lower bounds. In Sect. 4, we present our decomposition of bounded-genus networks. In Sect. 5 and Appendix A, we present our algorithm for diameter computation. In Appendix B, we present our algorithm for minimum cut computation. In Appendix C, we present the proof details for our basic tools.

2   Tools

2.1   Communication Between Two Sets of Vertices

Let $\mathcal{S}$ and $\mathcal{R}$ be two vertex sets that are not necessarily disjoint. The task SR-comm [10] is defined as follows. Each vertex $u \in \mathcal{S}$ holds a message m_u that it wishes to send, and each vertex $v \in \mathcal{R}$ wants to receive a message from vertices in $N^+(v) \cap \mathcal{S}$, where $N^+(v) = N(v) \cup \{v\}$ is the inclusive neighborhood of v.

Table 1. The time and energy complexities of SR-comm and its variants.

Task	Time	Energy
SR-comm	$O(\log \Delta \log n)$	$O(\log \Delta \log n)$
SR-comm^{all}	$O(\Delta' \log n)$	$O(\Delta' \log n)$
SR-comm^{min}	$O(K \log \Delta \log n)$	$O(\log K \log \Delta \log n)$
SR-comm^{max}		
SR-comm^{apx}	$O(\epsilon^{-6} \log W \log \Delta \log n)$	$O(\epsilon^{-6} \log W \log \Delta \log n)$
SR-comm^{multi}	$O(M \log \Delta \log^2 n)$	$O(M \log \Delta \log^2 n)$

Specifically, the task SR-comm requires that for each vertex $v \in \mathcal{R}$ with $N^+(v) \cap \mathcal{S} \neq \emptyset$, vertex v receives a message m_u from *at least one* vertex $u \in N^+(v) \cap \mathcal{S}$ w.h.p. Several variants of SR-comm are defined as follows.

All messages: SR-comm^{all}. The task SR-comm^{all} requires that each vertex $v \in \mathcal{R}$ receives the message m_u for each $u \in N^+(v) \cap \mathcal{S}$ w.h.p.

Approximate sum: SR-comm^{apx}. Suppose the message m_u sent from each vertex $u \in \mathcal{S}$ is an integer within the range $[W]$. The task SR-comm^{apx} requires each vertex $v \in \mathcal{R}$ computes an $(1 \pm \epsilon)$-factor approximation of the summation $\sum_{u \in N^+(v) \cap \mathcal{S}} m_u$ w.h.p.

Minimum and maximum: SR-comm^{min} and SR-comm^{max}. The message m_u sent from each vertex $u \in \mathcal{S}$ contains a key k_u from the key space $[K] = \{1, 2, \ldots, K\}$. For SR-comm^{min}, it is required that w.h.p., each vertex $v \in \mathcal{R}$ with $N^+(v) \cap \mathcal{S} \neq \emptyset$ receives a message m_u from a vertex $u \in N^+(v) \cap \mathcal{S}$ such that $k_u = \min_{u' \in N^+(v) \cap \mathcal{S}} k_{u'}$. The task SR-comm^{max} is defined analogously by replacing minimum with maximum.

Multiple messages: SR-comm^{multi}. Consider the setting where each vertex $u \in \mathcal{S}$ holds a set of messages $\mathcal{M}_u$. For each message m, all vertices holding the same message m have access to some shared random bits associated with m. We assume that for each $v \in \mathcal{R}$, the number of distinct messages in $\bigcup_{u \in N^+(v) \cap \mathcal{S}} \mathcal{M}_u$ is upper bounded by a number M that is known to all vertices. The task SR-comm^{multi} requires that each vertex $v \in \mathcal{R}$ receives all distinct messages in $\bigcup_{u \in N^+(v) \cap \mathcal{S}} \mathcal{M}_u$ w.h.p.

Table 1 summarizes the time and energy complexities of our algorithms for these tasks. For SR-comm^{all}, the parameter Δ' can be any known upper bound on $|\mathcal{S} \cap N(v)|$, for each $v \in \mathcal{R}$. For example, we may set $\Delta' = \Delta$ if no better upper bound is known. The proofs for these results are left to Appendix C.

2.2   Communication via a Good Labeling

A *good labeling* is a vertex labeling $\mathcal{L} : V(G) \mapsto \{0, \ldots, n-1\}$ such that each vertex v with $\mathcal{L}(v) > 0$ has a neighbor u with $\mathcal{L}(u) = \mathcal{L}(v) - 1$ [10]. A vertex v is called a *layer-i* vertex if $\mathcal{L}(v) = i$. Observe that if there is a unique layer-0 vertex r, then $\mathcal{L}$ represents a tree T rooted at r, so we call r the *root* of the good labeling $\mathcal{L}$. Since a vertex might have multiple choices of the parent, the tree T is not unique in general. The following lemma was proved in [10].

Lemma 1 ([10]). *A good labeling $\mathcal{L}$ with a unique layer-0 vertex can be constructed in $O(n \log \Delta \log^2 n)$ time and $O(\log \Delta \log^2 n)$ energy w.h.p.*

The following lemma shows that a good labeling allows the vertices in the graph to broadcast messages in an energy-efficient manner.

Lemma 2. *Suppose that we are given a good labeling $\mathcal{L}$ with a unique layer-0 vertex. Then we can achieve the following.*

1. *It takes $O(n\Delta \log n)$ time and $O(\Delta \log n)$ energy for each vertex to broadcast a message to the entire network w.h.p.*
2. *It takes $O(nx \log \Delta \log^2 n)$ time and $O(x \log \Delta \log^2 n)$ energy for x vertices to broadcast messages to the entire network w.h.p.*

Proof. Let r be the root of $\mathcal{L}$. For the first task, consider the following algorithm. We relay the message of each vertex to the root r using the following *convergecast* algorithm. For $i = n - 1$ down to 1, do SR-comm^{all} with $\mathcal{S}$ being the set of all layer-i vertices and $\mathcal{R}$ being the set of all layer-$(i-1)$ vertices. For each execution of SR-comm^{all}, each vertex in $\mathcal{S}$ transmits not only its message but also all other messages that it has received so far. Although we perform SR-comm^{all} $n - 1$ times, each vertex only participates at most twice. By Lemma 22, the cost of the convergecast algorithm is $O(n\Delta \log n)$ time and $O(\Delta \log n)$ energy.

At the end of the convergecast algorithm, the root r has gathered all messages sent during the algorithm. After that, the root r then broadcasts this information to all vertices via the following *divergecast* algorithm. For $i = 0$ to $n - 2$, do SR-comm with $\mathcal{S}$ being the set of all layer-i vertices and $\mathcal{R}$ being the set of all layer-$(i + 1)$ vertices. Similarly, although we perform SR-comm for $n - 1$ times, each vertex only participates at most twice. By Lemma 21, the cost of the divergecast algorithm is $O(n \log \Delta \log n)$ time and $O(\log \Delta \log n)$ energy. At the end of the divergecast algorithm, all vertices have received all messages.

For the rest of the proof, we consider the second task. Let X be the set of x vertices that attempt to broadcast a message. We solve this task similarly in two steps:

- We first do a convergecast, using SR-comm^{multi} with $M = x$, to gather all x messages to the root. By Lemma 23, SR-comm^{multi} costs $O(x \log \Delta \log^2 n)$ time, so the convergecast costs $O(nx \log \Delta \log^2 n)$ time and $O(x \log \Delta \log^2 n)$ energy.
- After that, we do a divergecast based on SR-comm to broadcast these messages from root to everyone. The divergecast costs $O(n \log \Delta \log n)$ time and $O(\log \Delta \log n)$ energy.

In order to use SR-comm^{multi}, the initial holder of each message m needs to first generate a sufficient number of random bits and attach them to the message. These random bits serve as the shared randomness associated with the message m, which is needed in the definition of SR-comm^{multi}.   □

Lemma 3. *There is an algorithm that lets all vertices learn the entire graph topology in $O(n\Delta \log n)$ time and $O(\Delta \log n)$ energy w.h.p.*

Proof. We first let each vertex v learn the list of identifiers in $N(v)$ by doing SR-comm$_{\mathrm{all}}$ with $\mathcal{S} = \mathcal{R} = V$, where the message of each vertex v is ID(v). By Lemma 22, this step takes $O(\Delta \log n)$ time and energy. After that, we apply Lemma 1 to construct a good labeling with a unique layer-0 vertex, and then we apply Lemma 2(1) to let all vertices learn the entire network topology by having each v broadcasting ID(v) and the list of identifiers in $N(v)$. This step takes $O(n\Delta \log n)$ time and $O(\Delta \log n)$ energy.   □

3 Lower Bounds

In this section, we prove the two lower bounds: Theorems 4 and 5.

Theorem 4. *For any randomized algorithm that computes the s–t minimum cut size of a planar bipartite graph w.h.p. in CD, the energy complexity of the algorithm is $\Omega(n)$.*

Proof. Suppose that there is a randomized algorithm $\mathcal{A}$ that computes the exact s–t minimum cut size of any planar bipartite graph with high probability and using $o(n)$ energy. Let G be a complete bipartite graph $K_{2,\Delta}$ with the bipartition $\{s, t\}$ and $\{v_1, \ldots, v_\Delta\}$. Set $X = \Delta/5$. We select Δ to be sufficiently large so that it is guaranteed that both s and t use at most X unit of energy in an execution of $\mathcal{A}$ on G.

Let G' be the result of removing v_Δ from G. The size of a s–t minimum cut of G is Δ, and the size of a s–t minimum cut of G' is $\Delta - 1$. Therefore, $\mathcal{A}$ lets s correctly distinguish between G and G' with high probability.

Consider an execution of $\mathcal{A}$ on G. Let S be the subset of $\{v_1, \ldots, v_\Delta\}$ such that $v_i \in S$ if there is a time slot τ where (i) v_i transmits, (ii) the number of vertices in $\{v_1, \ldots, v_\Delta\}$ that transmit is at most 2, and (iii) at least one of s and t listens.

We claim that $|S| \leq 4X = 4\Delta/5$. Let T be the set of all time slots τ such that the above conditions (i), (ii), and (iii) hold for at least one $v_i \in \{v_1, \ldots, v_\Delta\}$. In view of condition (ii), we must have $|T| \geq |S|/2$. In view of condition (iii), if $\tau \in T$, then at least one of s and t must listen at time τ, so the energy cost of one of s and t must be at least $|T|/2 \geq |S|/4$, which implies $X \geq |S|/4$.

Let $\mathcal{E}$ be the event that $v_\Delta \notin S$ in an execution of $\mathcal{A}$ on G. Whether or not $\mathcal{E}$ occurs depends only on the local randomness stored in the vertices $\{s, t\}$ and $\{v_1, \ldots, v_\Delta\}$. Since $|S| \leq 4\Delta/5$, at least $1/5$ fraction of the vertices in $\{v_1, \ldots, v_\Delta\}$ are not in S. Since the probability that $v_i \notin S$ is identical for all $v_i \in \{v_1, \ldots, v_\Delta\}$, we have $\Pr[\mathcal{E}] \geq 1/5$.

Consider the following scenario. All vertices in $\{s, t\}$ and $\{v_1, \ldots, v_\Delta\}$ have decided their random bits in advance. With probability $1/2$, we run $\mathcal{A}$ on G. With probability $1/2$, we run $\mathcal{A}$ on G'. If $\mathcal{E}$ occurs, then the execution of $\mathcal{A}$ on both G and G' is completely identical from the point of view of each vertex, except for v_Δ. Therefore, conditioning on event $\mathcal{E}$, the probability that vertex s correctly decides whether the underlying graph is G or G' is at most $1/2$, as s can only guess randomly.

Since $\Pr[\mathcal{E}] \geq 1/5$, the probability that vertex s fails to correctly decide whether the underlying graph is G or G' is at least $(1/2) \cdot (1/5) = 1/10$, so s fails to correctly calculate the s–t minimum cut with probability at least $1/10$ in the above scenario. This contradicts the assumption that $\mathcal{A}$ is able to compute the s–t minimum cut with high probability. □

Theorem 5. *For any randomized algorithm that computes the minimum cut size of a unit-disc graph w.h.p. in* CD, *the energy complexity of the algorithm is* $\Omega(n)$.

Proof. Consider the case where the underlying graph is K_n with probability $1/2$, and is $K_n - e$ with probability $1/2$, where the edge e is chosen uniformly at random from the set of all edges in K_n. Let $\mathcal{A}$ be any randomized algorithm that computes the size of a minimum cut exactly with high probability. Observe that the size of a minimum cut of K_n is $n - 1$ and the size of a minimum cut of $K_n - e$ is $n - 2$, so $\mathcal{A}$ is able to distinguish between K_n and $K_n - e$ with high probability. It was shown in [12] that any algorithm that distinguishes between K_n and $K_n - e$ with success probability at least $3/4$ necessarily has energy cost $\Omega(n)$ in both CD and No-CD, so the energy complexity of $\mathcal{A}$ is $\Omega(n)$. □

4   Graph Partitioning

The *genus* of a graph G is the minimum number g such that G can be drawn on an oriented surface of g handles without crossing. For example, planar graphs are the graphs with genus zero, and the graphs that can be drawn on a torus without crossing are the graphs with genus at most one. A class of graphs is called *bounded-genus* if the genus of all graphs in the class can be upper bounded by some constant $g = O(1)$. In this section, we consider a classification of the connected components of the subgraph induced by the low-degree vertices in a bounded-genus graph. Our algorithms, which will be presented in subsequent sections, make use of the classification.

Let $G = (V, E)$ be any bounded-genus graph. Let V_H be the set of vertices that have degree at least $\sqrt{n}$. Let $V_L = V \setminus V_H$. Since bounded-genus graphs have arboricity $O(1)$, we have $|E| = O(n)$, which implies $|V_H| = O(\sqrt{n})$.

From now on, we assume $|V_H| \geq 1$, since otherwise G has maximum degree $\Delta \leq \sqrt{n}$, so we can already solve *all* problems using $O(n\Delta \log \Delta \log n)) = \tilde{O}(n^{1.5})$ time and $O(\Delta \log \Delta \log n)) = \tilde{O}(\sqrt{n})$ energy by learning the entire graph topology using the algorithm of Lemma 3.

Given a set of vertices S, we write $G[S]$ to denote the subgraph of G induced by S and write $G^+[S]$ to denote the subgraph of G induced by all edges that have *at least one* endpoint in S. We divide the connected components of $G[V_L]$ into three types.

Type 1. A connected component S of $G[V_L]$ is of type-1 if $|S| < \sqrt{n}$ and $|\bigcup_{w \in S} N(w) \cap V_H| = 1$. For each vertex $u \in V_H$, we write $C(u)$ to denote the set of type-1 components S such that $\bigcup_{w \in S} N(w) \cap V_H = \{u\}$.

Type 2. A connected component S of $G[V_L]$ is of type-2 if $|S| < \sqrt{n}$ and $|\bigcup_{w \in S} N(w) \cap V_H| = 2$. For each pair of two distinct vertices $\{u, v\} \subseteq V_H$, we write $C(u,v)$ to denote the set of type-2 components S such that $\bigcup_{w \in S} N(w) \cap V_H = \{u, v\}$.

Type 3. A connected component S of $G[V_L]$ is of type-3 if it is neither of type-1 nor of type-2.

A connected component S of $G[V_L]$ is of type-3 if $|S| \geq \sqrt{n}$ or $|\bigcup_{w \in S} N(w) \cap V_H| \geq 3$. The number of type-3 components S with $|S| \geq \sqrt{n}$ is clearly at most $|V|/\sqrt{n} = \sqrt{n}$. We will show that the number of type-3 components with $|\bigcup_{w \in S} N(w) \cap V_H| \geq 3$ is also $O(\sqrt{n})$.

Lemma 4. Let $G = (V, E)$ be a bipartite graph of genus at most g. Let $V = X \cup Y$ be the bipartition of G. If $\deg(v) \geq 3$ for each $v \in X$, then $|X| \leq 2|Y| + 4(g-1)$.

Proof. Consider any embedding of G into a surface of genus g, and let F be set of faces. In a bipartite graph, each face has at least four edges, and each edge appears in at most two faces, so $|E| \geq 2|F|$. Combining this inequality with Euler's polyhedral formula $|V| - |E| + |F| \geq 2 - 2g$, we obtain that

$$2V| - |E| \geq 4(1 - g).$$

Since $\deg(v) \geq 3$ for each $v \in X$, we have $|E| \geq 3|X|$, so

$$2V| - |E| = 2(|X| + |Y|) - |E| \leq 2(|X| + |Y|) - 3|X| = 2|Y| - |X|.$$

Combining these upper and lower bounds of $2|V| - |E|$, we obtain that $2|Y| - |X| \geq 4(1 - g)$, so $|X| \leq 2|Y| + 4(g - 1)$, as required. $\qquad\square$

Note that Lemma 4 is precisely the reason that our algorithms only apply to bounded-genus graphs and do not work on an arbitrary H-minor-free graph. Consider a complete bipartite graph with the bipartition X and Y such that $|Y| = 3$. Such a graph does not contain K_5 as a minor, regardless of the size of X. Therefore, K_5-minor-freeness does not allow us to upper bound $|X|$ by any function of $|Y|$. Therefore, the bounded-genus requirement in Lemma 4 cannot be relaxed to H-minor-freeness for an arbitrary H.

Lemma 5. If G is a bounded-genus graph, then the number of type-3 components is $O(\sqrt{n})$.

Proof. A connected component S of $G[V_L]$ is of type-3 if $|S| \geq \sqrt{n}$ or $|\bigcup_{w \in S} N(w) \cap V_H| \geq 3$. As discussed earlier, the number of type-3 components S with $|S| \geq \sqrt{n}$ is at most $\sqrt{n}$, so we just need to prove that the number of type-3 components S with $|\bigcup_{w \in S} N(w) \cap V_H| \geq 3$ is also $O(\sqrt{n})$. Consider a bipartite graph $G^* = (V^*, E^*)$ with the bipartition $V^* = X^* \cup Y^*$ defined as follows.

- X^* is the set of all type-3 components S such that $|\bigcup_{w \in S} N(w) \cap V_H| \geq 3$.
- $Y^* = V_H$.
- For each component $S \in X^*$ and each vertex $v \in Y^*$, $\{S, v\} \in E^*$ if $v \in \bigcup_{w \in S} N(w)$.

Alternatively, G^* can be constructed from G by the following steps.

- Remove all type-1, type-2, and type-3 components S with $|\bigcup_{w \in S} N(w) \cap V_H| \leq 2$.
- For each type-3 component S with $|\bigcup_{w \in S} N(w) \cap V_H| \geq 3$, contract S into a vertex.

As G^* can be obtained from G via a sequence of edge contractions and vertex removals, G^* is a bounded-genus graph. Observe that $\deg(S) \geq 3$ for each $S \in X^*$ in G^*, so we may apply Lemma 4, which shows that the number $|X^*|$ of type-3 components S such that $|\bigcup_{w \in S} N(w) \cap V_H| \geq 3$ satisfies $|X^*| \leq 2|Y^*| + O(1) = 2|V_H| + O(1) = O(\sqrt{n})$.   □

We write G_H to denote the graph defined by the vertex set V_H and the edge set $\{\{u, v\} : |C(u, v)| > 0\}$. The following observation is useful.

Lemma 6. *If G is a bounded-genus graph, then G_H is also a bounded-genus graph, so the number of edges in G_H is $O(\sqrt{n})$ and there exists an edge orientation of G_H such that each vertex has outdegree $O(1)$.*

Proof. The graph G_H can be obtained from G via a sequence of edge contractions and vertex removals, so G_H is a bounded-genus graph. As bounded-genus graphs have arboricity $O(1)$, and so the number of edges in G_H is at most linear in the number of vertices in G_H, which is $O(\sqrt{n})$, and we can orient the edges of G_H in such a way that each vertex has outdegree $O(1)$.   □

5   Diameter

In this section, we show that for bounded-genus graphs, the diameter can be computed using $\tilde{O}(\sqrt{n})$ energy. We begin with discussing the high-level proof idea. First of all, using Lemma 2, learning the entire graph topology of the subgraph induced by V_H and all type-3 components is doable using $\tilde{O}(\sqrt{n})$ energy. Intuitively, this is due to the following facts: (i) $|V_H| = O(\sqrt{n})$, (ii) $\deg(v) = O(\sqrt{n})$ for each $v \in V_L$, and (iii) the number of type-3 components is $O(\sqrt{n})$.

The main difficulty in the diameter computation is dealing with type-1 and type-2 components. For example, a vertex $u \in V_H$ can be connected to $\Theta(n)$ type-1 components in that $|C(u)| = \Theta(n)$. Since we aim for an algorithm with energy complexity $\tilde{O}(\sqrt{n})$, throughout the entire algorithm, u can only receive messages from at most $\tilde{O}(\sqrt{n})$ components in $C(u)$. The challenge here is to show that the diameter can still be calculated with a limited amount of information about type-1 and type-2 components and show that such information can be extracted in an energy-efficient manner in the radio network model.

We will define a set of parameters of type-1 and type-2 components and show that with these parameters, the exact value of the diameter can be calculated. Based on this result, we will define a subgraph G^* of G such that the diameter of G equals the diameter of G^*, and then in Appendix A we will design an energy-efficient algorithm to learn the graph topology of G^*.

In the subsequent discussion, we write eccentricity(u, S) to denote $\max_{v \in S} \text{dist}(u, v)$. By default, all distances are measured in the underlying network G. We use subscripts to describe distances that are measured in a vertex set, an edge set, or a subgraph.

Parameters for Type-1 Components. We first consider the type-1 components in $C(u)$, for any vertex $u \in V_H$.

$(A_i[u], a_i[u])$. Let $A_1[u]$ be a component $S \in C(u)$ that maximizes eccentricity(u, S), and let $A_2[u]$ be a component $S \in C(u) \setminus \{A_1[u]\}$ that maximizes eccentricity(u, S). For $i \in \{1, 2\}$, we write $a_i[u] = $ eccentricity$(u, A_i[u])$.

$(B[u], b[u])$. Let $B[u]$ be a component $S \in C(u)$ that maximizes $\max_{s,t \in S \cup \{u\}} \text{dist}(s, t)$, and we write $b[u] = \max_{s,t \in B[u] \cup \{u\}} \text{dist}(s, t)$.

In the above definitions, ties can be broken arbitrarily if there are multiple choices. Some of the above definitions become undefined when $|C(u)|$ is too small. For example, if $|C(u)| = 1$, then $A_2[u]$ and $a_2[u]$ are undefined. In such a case, we set these parameters to their default values: zero or an empty set. For example, if $|C(u)| = 1$, then we set $A_2[u] = \emptyset$ and $a_2[u] = 0$.

Any path connecting a vertex in $\bigcup_{S \in C(u)} S$ to the rest of the graph must pass the vertex $u \in V_H$, so the amount of information we can afford to extract from $\bigcup_{S \in C(u)} S$ is limited. Intuitively, for the purpose of calculating the diameter, we only need the following information from $\bigcup_{S \in C(u)} S$:

- The longest distance between two vertices in $\bigcup_{S \in C(u)} S \cup \{u\}$, which is $\max\{b[u], a_1[u] + a_2[u]\}$.
- The longest distance between u and a vertex in $\bigcup_{S \in C(u)} S$, which is $a_1[u]$.

Regardless of the size of $C(u)$, we only need to learn $a_1[u]$, $a_2[u]$, and $b[u]$ from the components of $C(u)$. Later we will show that these parameters can be learned efficiently via SR-comm$^{\max}$.

Parameters for Type-2 Components. Next, we consider the type-2 components in $C(u, v)$, for any two distinct vertices $u, v \in V_H$.

$(R[u, v], r[u, v])$. Let $R[u, v]$ be a component $S \in C(u, v)$ that minimizes $\text{dist}_{G+[S]}(u, v)$, and we write $r[u, v] = \text{dist}_{G+[R[u,v]]}(u, v)$. In other words, $R[u, v]$ is a component that contains a shortest path between u and v, among all u–v paths via the vertices in $\bigcup_{S \in C(u,v)} S$.

$(A_i^k[u, v], a_i^k[u, v])$. For each component $S \in C(u, v)$, we write $S^{u,k}$ to denote the set of vertices $\{w \in S : \text{dist}_{G+[S]}(w, v) - \text{dist}_{G+[S]}(w, u) \geq k\}$. In other words, $S^{u,k}$ is the set of all vertices in S whose distance to u in $G^+[S]$ is shorter than that to v by *at least* k.

Let $A_1^k[u,v]$ be a component $S \in C(u,v)$ that maximizes eccentricity$_{G^+[S]}$ $(u, S^{u,k})$, and let $A_2^k[u,v]$ be a component $S \in C(u,v) \backslash \{A_1^k[u,v]\}$ that maximizes eccentricity$_{G^+[S]}(u, S^{u,k})$. We write $a_i^k[u,v] = $ eccentricity$_{G^+[A_i^k[u,v]]}$ $(u, A_i^k[u,v])$. We only consider $k \in \{-\sqrt{n}, \ldots, \sqrt{n}\}$.

$(B^l[u,v], b^l[u,v])$. For a component $S \in C(u,v)$, we write $G^l[S]$ to denote the graph resulting from adding to $G^+[S]$ a path of length l connecting u and v, and we write $\phi^l(S)$ to denote the maximum value of dist$_{G^l[S]}(s,t)$ among all pairs of vertices $s,t \in S \cup \{u,v\}$. A useful observation here is that if dist$_{V \backslash S}(u,v) = l$, then $\phi^l(S)$ equals the maximum value of dist$_G(s,t)$ among all pairs of vertices $s,t \in S \cup \{u,v\}$.

Let $B^l[u,v]$ be a component $S \in C(u,v) \setminus \{R[u,v]\}$ that maximizes $\phi^l(S)$, and we write $b^l(u,v) = \phi^l(B^l[u,v])$. We only consider $l \in \{1, \ldots, \sqrt{n}\}$.

Similar to the parameters of type-1 components, all the above parameters are set to their default values if they are undefined. Note that the definitions of $a_i^k[u,v]$ and $A_i^k[u,v]$ are *asymmetric* in the sense that we might have $a_i^k[u,v] \neq a_i^k[v,u]$ and $A_i^k[u,v] \neq A_i^k[v,u]$. All remaining parameters for type-2 components are symmetric.

We briefly explain how the above parameters can be used in the diameter calculation. Let $P = (s, \ldots, t)$ be an s–t shortest path in G whose length equals the diameter. There are three possible ways that P intersects the vertex set $\bigcup_{S \in C(u,v)} S$.

- Suppose s and t are within $G^+[S]$, for a component $S \in C(u,v)$. In this case, if dist$_{V \backslash S}(u,v) = l$, then the length of P equals $\phi^l(S) = b^l[u,v]$.
- Suppose there is a subpath $P' = (u, \ldots, v)$ of P whose intermediate vertices are all in $\bigcup_{S \in C(u,v)} S$. In this case, the length of P' equals $r[u,v]$.
- Suppose there is a component $S \in C(u,v)$ such that $s \in S$ and $t \notin S \cup \{u,v\}$. Suppose u is the first vertex of P that is not in S. Consider the subpath $P' = (s, \ldots, u)$ of P. If dist$(t,v) - $dist$(t,u) = k$, then we must have $s \in S^{u,k}$, since otherwise dist$(s,v) + $dist$(v,t)$ is smaller than the length of P, violating the assumption that P is an s–t shortest path. If $t \notin A_1^k[u,v]$, then the length of P' equals $a_1^k[u,v]$. If $t \in A_1^k[u,v]$, then the length of P' equals $a_2^k[u,v]$.

Intuitively, the above discussion shows that the parameters described above capture all the necessary information needed to be extracted from the type-2 components for the purpose of diameter computation. We have $O(\sqrt{n})$ parameters for each $C(u,v)$. We will later show that all these parameters can be learned using $O(\sqrt{n})$ energy.

The graph $G^\star$. We define $G^\star$ as the subgraph induced by the union of (i) V_H, (ii) all type-3 components, (iii) $A_1[u]$, $A_2[u]$, and $B[u]$, for each $u \in V_H$, and (iv) $A_i^k[u,v]$, $A_i^k[v,u]$, $B^l[u,v]$, and $R[u,v]$, for each pair of distinct vertices $\{u,v\} \subseteq V_H$, $i \in \{1,2\}$, $k \in \{-\sqrt{n}, \ldots, \sqrt{n}\}$, and $l \in \{1, \ldots, \sqrt{n}\}$. In the subsequent discussion, we prove that the diameter of G equals the diameter of $G^\star$, so the task of computing the diameter of G is reduced to learning the topology of $G^\star$. We will show that the following two statements are correct.

(S1) For each pair of vertices $\{s, t\}$ in graph $G^\star$, we have $\mathrm{dist}_G(s, t) = \mathrm{dist}_{G^\star}(s, t)$.

(S2) For each pair of vertices $\{s, t\}$ in graph G, there exists a pair of vertices $\{s', t'\}$ in graph $G^\star$ satisfying $\mathrm{dist}_G(s, t) \leq \mathrm{dist}_G(s', t')$.

These statements imply that G and $G^\star$ have the same diameter. We first prove that (S1) is true.

Lemma 7. *For any two vertices s and t in $G^\star$, we have $\mathrm{dist}_G(s, t) = \mathrm{dist}_{G^\star}(s, t)$.*

Proof. We choose P to be an s–t path in G whose length is $\mathrm{dist}_G(s, t)$ that uses the minimum number of vertices not in $G^\star$. If P is entirely in $G^\star$, then we are done. For the rest of the proof, we assume that P is not entirely in $G^\star$. Then P contains a subpath $P' = (u, \ldots, v)$ whose intermediate vertices are all within a type-2 component $S \in C(u, v)$ that is not included to $G^\star$. By the definition of $R[u, v]$, the length of P' is at least $r[u, v]$, which is the shortest path length between u and v via $R[u, v]$. Therefore, replacing P' with a shortest u–v path in $R_j[u, v]$, which is entirely in $G^\star$, does not increase the path length. This contradicts our choice of P. Hence P is entirely in $G^\star$. $\square$

Lemma 8. *Let S be a type-1 or type-2 component that is not included in $G^\star$. Let $s \in S$. Let t be any vertex in G that does not belong to $G^+[S]$. Then there exists a vertex s' in $G^\star$ such that $\mathrm{dist}_G(s', t) \geq \mathrm{dist}_G(s, t)$.*

Proof. Let P be an s–t shortest path in G. Suppose that $S \in C(u)$ is of type-1. Because S is not included in $G^\star$, we must have $|C(u)| \geq 3$, so both $A_1[u] \neq S$ and $A_2[u] \neq S$ are not $\emptyset$. Let $i \in \{1, 2\}$ be an index such that t is not in $A_i[u]$. Consider the subpath $\tilde{P} = (s, \ldots, u)$ of P. By the definition of $a_i[u]$ and $A_i[u]$, the length of $\tilde{P}$ is at most $a_i[u]$, and there exists a vertex $s' \in A_i[u]$ such that the length of the shortest path between s' and u equals $a_i[u]$. Thus, we have

$$\mathrm{dist}_G(s', t) = \mathrm{dist}_G(s', u) + \mathrm{dist}_G(u, t) \geq \mathrm{dist}_G(s, u) + \mathrm{dist}_G(u, t) = \mathrm{dist}_G(s, t).$$

Next, consider the case that $S \in C(u, v)$ is of type-2. The path P must contain at least one of u and v. Without loss of generality, assume that u is the first vertex of P that is not in S, so there is a subpath $\tilde{P} = (s, \ldots, u)$ of P such that all vertices in $\tilde{P}$ other than u are in S. The length of P equals $\mathrm{dist}_{G^+[S]}(s, u) + \mathrm{dist}_G(u, t)$.

Let $k = \mathrm{dist}_{G^+[S]}(s, v) - \mathrm{dist}_{G^+[S]}(s, u)$, so $S^{u,k} \supseteq \{s\} \neq \emptyset$. Since S is not of type-3, $|S| < \sqrt{n}$, so $k \in \{-\sqrt{n}, \ldots, \sqrt{n}\}$. Because S is not included in $G^\star$, both $A_1^k[u, v] \neq S$ and $A_2^k[u, v] \neq S$ are not $\emptyset$. At least one of $A_1^k[u, v]$ and $A_2^k[u, v]$ does not contain t. We choose $S' = A_i^k[u, v]$ as any one of them that does not contain t. We choose $s' \in S'$ as a vertex such that $\mathrm{dist}_{G^+[S']}(s', u) = a_i^k[u, v]$

and $\text{dist}_{G+[S']}(s', v) - \text{dist}_{G+[S']}(s', u) \geq k$. The existence of such a vertex s' is guaranteed by the definition of $A_i^k[u, v]$.

Our plan is to show that (i) $a_i^k[u, v] + \text{dist}_G(u, t) \geq \text{dist}_G(s, t)$ and (ii) $\text{dist}_G(s', t) = a_i^k[u, v] + \text{dist}_G(u, t)$. Combining these two inequalities give us the desired result: $\text{dist}_G(s', t) \geq \text{dist}_G(s, t)$.

Proof of (i). By the definition of $A_i^k[u, v]$, we must have

$$\text{dist}_{G+[S']}(s', u) = a_i^k[u, v] \geq \text{dist}_{G+[S]}(s, u),$$

so we have

$$a_i^k[u, v] + \text{dist}_G(u, t) \geq \text{dist}_{G+[S]}(s, u) + \text{dist}_G(u, t) = \text{dist}_G(s, t).$$

Proof of (ii). Suppose that (ii) is not true. Then any shortest path between s' and t must contain a subpath $P' = (s', \ldots, v)$ such that u is not in P', and so we have:

$$\text{dist}_G(s', t) = \text{dist}_{G+[S']}(s', v) + \text{dist}_G(v, t) < \text{dist}_{G+[S']}(s', u) + \text{dist}_G(u, t).$$

Combining this inequality with the known fact $\text{dist}_{G+[S']}(s', v) - \text{dist}_{G+[S']}(s', u) \geq k$, we have:

$$\text{dist}_G(u, t) - \text{dist}_G(v, t) > \text{dist}_{G+[S']}(s', v) - \text{dist}_{G+[S']}(s', u) \geq k,$$

which implies that $\text{dist}_G(v, t) < \text{dist}_G(u, t) - k$ $(\star)$. We calculate an upper bound of $\text{dist}_G(s, t)$:

$$
\begin{aligned}
\text{dist}_G(s, t) &\leq \text{dist}_{G+[S]}(s, v) + \text{dist}_G(v, t) \\
&= (k + \text{dist}_{G+[S]}(s, u)) + \text{dist}_G(v, t) && \text{by definition of } k. \\
&< (k + \text{dist}_{G+[S]}(s, u)) + (\text{dist}_G(u, t) - k) && \text{by } (\star). \\
&= \text{dist}_{G+[S]}(s, u) + \text{dist}_G(u, t).
\end{aligned}
$$

This contradicts the assumption that P is a shortest path between s and t in G, as the length of P equals $\text{dist}_{G+[S]}(s, u) + \text{dist}_G(u, t)$. □

The following lemma shows that (S2) is true.

Lemma 9. *For any two vertices s and t in graph G, there exist two vertices s' and t' in graph $G^\star$ such that $\text{dist}_G(s, t) \leq \text{dist}_G(s', t')$.*

Proof. If both s and t are already in $G^\star$, then we are done by setting $s' = s$ and $t' = t$. In the subsequent discussion, we focus on the case that at least one of s and t is not in $G^\star$. By symmetry, we will assume that s is not in $G^\star$, so there is a type-1 or a type-2 component S that is not included in $G^\star$ such that $s \in S$.

Case 1: t belongs to $G^+[S]$. If $S \in C(u)$ for some $u \in V_H$, then there exist two vertices s' and t' in the component $B[u] \in C(u)$ such that $\text{dist}_G(s', t') = b[u] \geq \text{dist}_G(s, t)$ by the definition of $B[u]$.

The remaining case is that $S \in C(u, v)$ for some $u, v \in V_H$. Let $l = \text{dist}_{V \setminus S}(u, v)$. We observe that $l \leq r[u, v]$. The reason is that the existence of a component $S \neq R[u, v]$ guarantees that $R[u, v] \neq \emptyset$, which implies that $l = \text{dist}_{V \setminus S}(u, v) \leq \text{dist}_{G^+[R[u,v]]}(u, v) = r[u, v]$, as $G^+[R[u, v]]$ is a subgraph of $G[V \setminus S]$.

Since $R[u, v]$ is of type-2, we have $r[u, v] \leq |R[u, v]| + 1 \leq \sqrt{n}$, so $l \in \{1, \ldots, \sqrt{n}\}$ Consider the component $B^l[u, v] \in C(u, v)$. We observe that $l = \text{dist}_{V \setminus B^l[u,v]}(u, v)$, since the shortest u–v path length via $R[u, v]$ is at most the length of any u–v path via S or $B^l[u, v]$, by our choice of $R[u, v]$. More precisely, we have $l = \text{dist}_{V \setminus S}(u, v) = \text{dist}_V(u, v) = \text{dist}_{V \setminus B^l[u,v]}(u, v)$, as the above discussion implies that including S and excluding $B^l[u, v]$ in the subscript does not change the shortest u–v path length. Here we use the fact that $B^l[u, v] \neq R[u, v]$, which is due to the definition of $B^l[u, v]$.

Since $l = \text{dist}_{V \setminus B^l[u,v]}(u, v)$, by the definition of $B^l[u, v]$, there exist two vertices s' and t' in $G^+[B^l[u, v]]$ such that $\text{dist}_G(s', t') \geq \text{dist}_G(s, t)$, since otherwise we would have selected $B^l[u, v] = S$.

Case 2: t does not belong to $G^+[S]$. We apply Lemma 8 to find a vertex s' in $G^\star$ such that $\text{dist}_G(s, t) \leq \text{dist}_G(s', t)$. If t is already in $G^\star$, then we are done. Otherwise, there is a type-1 or a type-2 component S' that is not included in $G^\star$ such that $t \in S'$. There are two sub-cases.

- Suppose s' belongs to $G^+[S']$. Then we may apply the same argument for Case 1 above to find two vertices s'' and t'' in $G^\star$ such that $\text{dist}_G(s, t) \leq \text{dist}_G(s', t) \leq \text{dist}_G(s'', t'')$.
- Suppose s' does not belong to $G^+[S']$. Then we may apply Lemma 8 again to find a vertex t' in $G^\star$ such that $\text{dist}_G(s, t) \leq \text{dist}_G(s', t) \leq \text{dist}_G(s', t')$.

In both sub-cases, we can find two vertices in $G^\star$ whose distance in G is at least $\text{dist}_G(s, t)$.   □

Lemma 10. *The diameter of G equals the diameter of $G^\star$.*

Proof. Lemma 7 shows that (S1) is true. Lemma 9 shows that (S2) is true. These two results together imply that G and $G^\star$ have the same diameter. Statement (S1) implies that the diameter of $G^\star$ is at most the diameter of G. For the other direction, let s and t be two vertices in G such that $\text{dist}(s, t)$ equals the diameter of G. By (S2), there exist two vertices s' and t' in $G^\star$ such that $\text{dist}_G(s, t) \leq \text{dist}_G(s', t')$. By (S1), $\text{dist}_G(s', t') = \text{dist}_{G^\star}(s', t')$, so the diameter of $G^\star$ is at least the diameter of G.   □

In Appendix A, we will design an energy-efficient algorithm to learn the graph topology of $G^\star$. This algorithm, combined with Lemma 10, allows us to prove Theorem 1.

A   Learning the Topology of $G^\star$

By Lemma 10, the task of computing the diameter of a bounded-genus graph G is reduced to computing the diameter of $G^\star$. In this section, we show that all vertices can learn the graph topology of $G^\star$ using $\tilde{O}(\sqrt{n})$ energy.

Recall that G_H is the graph defined by the vertex set V_H and the edge set $\{\{u,v\} : |C(u,v)| > 0\}$. By Lemma 6, we know that $E(G_H) = O(\sqrt{n})$ and there exists an assignment $F : E(G_H) \mapsto V_H$ mapping each pair $\{u,v\} \in E(G_H)$ to one vertex in $\{u,v\}$ such that each $w \in V_H$ is mapped to at most $O(1)$ times. Let $\mathcal{A}'$ be any deterministic centralized algorithm that finds such an assignment F, and we fix $F^\star$ to be the outcome of $\mathcal{A}'$ on the input G_H. If each vertex $v \in V$ already knows the graph G_H, then v can locally calculate $F^\star$ by simulating $\mathcal{A}'$.

To learn $G^\star$, we will let each vertex $u \in V$ learn the following information:

Basic information $\mathcal{I}_0(u)$. For each vertex $u \in V$, $\mathcal{I}_0(u)$ contains the following information: (i) whether $u \in V_H$ or $u \in V_L$, (ii) the list of vertices in $N(u) \cap V_H$, and (iii) the set of all pairs $\{u', v'\} \in E(G_H)$.
   If u is in a connected component S of $G[V_L]$, then $\mathcal{I}_0(u)$ contains the following additional information: (i) the list of vertices in S, and (ii) the topology of the subgraph $G^+[S]$.

Information about type-1 components $\mathcal{I}_1(u)$. For each $u \in V_H$, $\mathcal{I}_1(u)$ contains the graph topology of $G^+[S']$, for each $S' = A_1[u]$, $A_2[u]$, and $B[u]$.

Information about type-2 components $\mathcal{I}_2(u)$. For each $u \in V_H$, $\mathcal{I}_2(u)$ contains the following information. For each pair $\{u,v\} \in E(G_H)$ such that $F^\star(\{u,v\}) = u$, $\mathcal{I}_2(u)$ includes the graph topology of $G^+[S']$, for each $S' = A_i^k[u,v]$, $A_i^k[v,u]$, $B^l[u,v]$, and $R[u,v]$, for each $i \in \{1,2\}$, $k \in \{-\sqrt{n}, \ldots, \sqrt{n}\}$, and $l \in \{1, \ldots, \sqrt{n}\}$.

The information $\mathcal{I}_0(u)$ contains the graph topology of G_H, allowing each vertex u to calculate $F^\star$ locally. Note that $\mathcal{I}_1(u)$ and $\mathcal{I}_2(u)$ contain nothing if $u \in V_L$. The following lemma shows that the graph topology of $G^\star$ can be learned efficiently given that each vertex $u \in V$ already knows $\mathcal{I}_0(u)$, $\mathcal{I}_1(u)$, and $\mathcal{I}_2(u)$.

Lemma 11. *Given that each $u \in V$ already knows $\mathcal{I}_0(u)$, $\mathcal{I}_1(u)$, and $\mathcal{I}_2(u)$, using $\tilde{O}(n^{1.5})$ time and $\tilde{O}(\sqrt{n})$ energy, we can let all vertices in G learn the graph topology of $G^\star$ w.h.p.*

Proof. To learn $G^\star$, it suffices to know the following information: (i) $\mathcal{I}_1(u)$ and $\mathcal{I}_2(u)$ for each $u \in V_H$, (ii) the graph topology of $G^+[S]$ for each type-3 component S, and (iii) the graph topology of the subgraph induced by V_H. For each type-3 component S, let r_S be the smallest ID vertex in S. In view of the above, to let each vertex learn the topology of $G^\star$, it suffices to let the following $O(\sqrt{n})$ vertices broadcast the following information:

- For each $u \in V_H$, u broadcasts $\mathcal{I}_1(u)$, $\mathcal{I}_2(u)$, and the list of vertices $N(u) \cap V_H$, which is contained in $\mathcal{I}_0(u)$.
- For each $u \in V_L$ such that $u = r_S$ for a type-3 component S, u broadcasts the graph topology of $G^+[S]$. Note that each vertex $u \in V_L$ can decide locally using the information in $\mathcal{I}_0(u)$ whether or not u itself is r_S for a type-3 component S.

Since $|V_H| = O(\sqrt{n})$ and the number of type-3 components is also $O(\sqrt{n})$ by Lemma 5, the number of vertices that need to broadcast is $O(\sqrt{n})$. We run the algorithm of Lemma 1 to find a good labeling $\mathcal{L}$ of G, and then we use Lemma 2(2) with $x = O(\sqrt{n})$ to let the above $O(\sqrt{n})$ vertices broadcast their information. This can be done in time $\tilde{O}(n^{1.5})$, and energy $\tilde{O}(\sqrt{n})$. After that, all vertices know the graph topology of $G^\star$. $\qquad\square$

Next, we consider the task of learning the basic information $\mathcal{I}_0(u)$.

Lemma 12. *Using $\tilde{O}(\sqrt{n})$ time and energy, we can let all vertices $v \in V$ learn the following information w.h.p.*

- *Each $v \in V$ learns whether $v \in V_H$ or $v \in V_L$.*
- *If $v \in V_H$, then v also learns the list of vertices in $N(v) \cap V_H$.*
- *If $v \in V_L$, then v also learns the two lists of vertices $N(v) \cap V_L$ and $N(v) \cap V_H$.*

Proof. First, we run SR-comm^{apx} with $W = 1$, $\epsilon = 1$, $\mathcal{S} = \mathcal{R} = V$, and $m_u = 1$, for each $u \in \mathcal{S}$. This step lets each $v \in V$ estimate $\deg(v)$ up to a factor of 2. This step costs poly $\log n$ time, by Lemma 27.

After that, we run SR-comm^{all} with $\mathcal{S} = V$ and $\mathcal{R}$ being the set of all vertices v whose estimate of $\deg(v)$ is at most $2\sqrt{n}$. The message m_v for each vertex v is ID(v), and we use the bound $\Delta' = 4\sqrt{n}$ for SR-comm^{all}. Recall that V_L is the set of vertices of degree at most $\sqrt{n}$, so we must have $V_L \subseteq \mathcal{R}$. The algorithm of SR-comm^{all} allows each vertex $v \in \mathcal{R}$ to calculate $\deg(v)$ precisely. Therefore, after this step, each vertex $v \in V$ has enough information to decide whether $v \in V_H$ or $v \in V_L$. Furthermore, if $v \in V_L$, then v knows the list of all vertices $N(v)$. This step takes $\tilde{O}(\sqrt{n})$ time, by Lemma 22.

In order for each vertex to learn all the required vertex lists, we run SR-comm^{all} again with the following parameters: $\mathcal{S} = V_H$, $\mathcal{R} = V$, and the message m_v for each vertex $v \in \mathcal{S}$ is its ID(v). This time we may use the bound $\Delta' = \sqrt{n} \geq |V_H|$. After the algorithm of SR-comm^{all}, each vertex $v \in V$ knows the list of vertices in $N(v) \cap V_H$. For each $v \in V_L$, since v already knows the list of all vertices $N(v)$, it can locally calculate the list $N(v) \cap V_L$. This step also takes $\tilde{O}(\sqrt{n})$ time. $\qquad\square$

Lemma 13. *Using $\tilde{O}(n^{1.5})$ time and $\tilde{O}(\sqrt{n})$ energy, we can let all vertices v in all connected components S of $G[V_L]$ learn (i) the vertex set S and (ii) the graph topology of $G^+[S]$ w.h.p.*

Proof. First, we apply Lemma 12 to let all vertices $v \in V_L$ learn the two lists $N(v) \cap V_L$ and $N(v) \cap V_H$. To let all vertices learn the required information in the lemma statement, it suffices to let each vertex $v \in S$ broadcast the two lists $N(v) \cap V_L$ and $N(v) \cap V_H$ to all other vertices in S, for all connected components S of $G[V_L]$.

We do the above broadcasting task in parallel, for all connected components S of $G[V_L]$. We use Lemma 1 to let each component S compute a good labeling, and then we use Lemma 2(1) to let each vertex $v \in S$ broadcast the two lists

$N(v) \cap V_L$ and $N(v) \cap V_H$ to all other vertices in S. Recall that the degree of any vertex in V_L is less than $\sqrt{n}$, so the algorithm of Lemma 2(1) costs $\tilde{O}(n^{1.5})$ time and $\tilde{O}(\sqrt{n})$ energy. □

For each connected component S of $G[V_L]$, at the end of the algorithm of Lemma 13, each vertex $w \in S$ is able to determine the type of S. If S is of type-1, w knows the vertex $u \in V_H$ such that $S \in C(u)$. If S is of type-2, w knows the two vertices $u, v \in V_H$ such that $S \in C(u,v)$. Given such information, in the following lemma, we design an algorithm for learning the topology of G_H.

Lemma 14. *Suppose that each vertex in each type-2 component S already knows (i) the vertex set S and (ii) the graph topology of $G^+[S]$. Using $\tilde{O}(n^{1.5})$ time and $\tilde{O}(\sqrt{n})$ energy, all vertices in the graph can learn the set of all pairs $\{u,v\} \in E(G_H)$ w.h.p.*

Proof. First of all, we let all vertices in V_H agree on a fixed ordering $V_H = \{v_1, \ldots, v_{|H|}\}$ as follows. We use Lemma 1 to compute a good labeling of G, and then we use Lemma 2(2) with $x = \sqrt{n}$ to let each vertex $v \in V_H$ broadcast $ID(v)$. After that, we may order $V_H = \{v_1, \ldots, v_{|H|}\}$ by increasing ordering of ID. This step takes $\tilde{O}(n^{1.5})$ time and $\tilde{O}(\sqrt{n})$ energy.

Next, we consider the task of letting each $u \in V_H$ learn the list of all $v \in V_H$ such that $C(u,v) \neq \emptyset$. We solve this task by $|V_H|$ invocations of SR-comm. Given a type-2 component $S \in C(u,v)$, we define $z_{u,S}$ as the smallest-ID vertex in $N(v) \cap S$. The vertex $z_{u,S}$ will be responsible for letting v know that $C(u,v) \neq \emptyset$. For $i = 1$ to $|V_H|$, we do an SR-comm with $\mathcal{R} = V_H$ and $\mathcal{S}$ being the set of all vertices $z_{v_i,S}$ such that S is a type-2 component S with $v_i \in G^+[S]$. Observe that a vertex $u \in V_H$ receives a message during the ith iteration if and only if $C(u, v_i) \neq \emptyset$, i.e., $\{u, v_i\} \in E(G_H)$. By Lemma 21, this step takes $|V_H| \cdot \text{poly} \log n = \tilde{O}(\sqrt{n})$ time.

At the end of the above algorithm, each $u \in V_H$ knows the list of all $v \in V_H$ such that $C(u,v) \neq \emptyset$. In order to let all vertices in G learn the topology of G_H, it suffices to let all $u \in V_H$ broadcast this information. This can be done using Lemma 2(2) with $x = \sqrt{n}$, which costs $\tilde{O}(n^{1.5})$ time and $\tilde{O}(\sqrt{n})$ energy. □

Lemma 15. *In $\tilde{O}(n^{1.5})$ time and $\tilde{O}(\sqrt{n})$ energy, we can let all $u \in V$ learn $\mathcal{I}_0(u)$ w.h.p.*

Proof. This follows from Lemma 13 and Lemma 14. □

Next, we consider the task of learning $\mathcal{I}_1(u)$ and $\mathcal{I}_2(u)$.

Lemma 16. *Suppose that each $v \in V$ knows $\mathcal{I}_0(v)$. Using $\tilde{O}(n^{1.5})$ time and $\tilde{O}(\sqrt{n})$ energy, we can let all vertices $u \in V_H$ learn $\mathcal{I}_1(u)$ and $\mathcal{I}_2(u)$ w.h.p.*

Proof. Consider any vertex $u \in V_H$. For each component $S \in C(u)$, we let $r_{S,u}$ be the smallest-ID vertex in the set $S \cap N(u)$. For each $v \in V_H$ such that $F^\star(\{u,v\}) = u$, and for each component $S \in C(u,v)$, we similarly let $r_{S,u}$ be the smallest-ID vertex in the set $S \cap N(u)$. As we will later see, $r_{S,u}$ will be the vertex in S responsible for sending the graph topology $G^+[S]$ to u in case $G^+[S]$ belongs to $\mathcal{I}_1(u)$ or $\mathcal{I}_2(u)$.

Recall that $\mathcal{I}_1(u)$ and $\mathcal{I}_2(u)$ consist of the graph topology $G^+[S']$ of some selected type-1 and type-2 component S' such that u belongs to $G^+[S']$. We will present a generic approach that lets $u \in V_H$ learn one graph topology in $\mathcal{I}_1(u)$ and $\mathcal{I}_2(u)$. As we will later see, the cost of learning one graph topology is poly $\log n$ time and energy. If the graph topology to be learned is in $C(u)$, then only u and the vertices $r_{S,u}$ for all $S \in C(u)$ need to participate in the algorithm for learning the graph topology. If the graph topology to be learned is in $C(u, v)$, then only u and the vertices $r_{S,u}$ for all $S \in C(u, v)$ need to participate in the algorithm for learning the graph topology. We only describe the algorithms that let $u \in V_H$ learn $A_1[u]$ and $A_2[u]$. The algorithms for learning the remaining graph topologies are analogous.

Learning $A_1[u]$. Recall that $A_1[u]$ is a component $S' \in C(u)$ that maximizes eccentricity(u, S'). To learn $A_1[u]$, we use SR-comm$^{\max}$ with $\mathcal{S} = \{r_{S,u} : S \in C(u)\}$ and $\mathcal{R} = \{u\}$. The message m_v of $v = r_{S,u}$ is the graph topology of $G^+[S]$, and the key of $v = r_{S,u}$ is $k_v =$ eccentricity(u, S). Since each type-1 and type-2 component satisfies $|S| \le \sqrt{n}$, the maximum possible value of eccentricity(u, S) is $\sqrt{n}$, so the size of the key space for SR-comm$^{\max}$ is $K = \sqrt{n}$.

If $|C(u)| > 0$, then the message that u receives from SR-comm$^{\max}$ is the topology of $G^+[S']$, for a component $S' \in C(u)$ that attains the maximum value of eccentricity(u, S') among all components in $C(u)$, so u may set $A_1[u] = S'$. If $|C(u)| = 0$, the vertex u receives nothing from SR-comm$^{\max}$, so u may set $A_1[u] = \emptyset$. By Lemma 24, the cost of SR-comm$^{\max}$ is $O(\log K \log \Delta \log n) =$ poly $\log n$.

Learning $A_2[u]$. The procedure for learning $A_2[u]$ is almost exactly the same as that for $A_1[u]$, with only one difference. Recall that $A_2[u]$ is a component $S' \in C(u) \setminus \{A_1[u]\}$ that maximizes eccentricity(u, S'), so we need to exclude the component $A_1[u]$ from participating. To do so, before we apply SR-comm$^{\max}$, we use one round to let u send $\text{ID}(r_{A_1[u],u})$ to all vertices $\{r_{S,u} : S \in C(u)\}$. This allows each $r_{S,u}$ to learn whether or not $S = A_1[u]$.

For each $u \in V_H$, the number of pairs $\{u, v\}$ such that $F^*(\{u, v\}) = u$ is $O(1)$, so the number of graph topologies needed to be learned in $\mathcal{I}_1(u)$ and $\mathcal{I}_2(u)$ by u is $O(\sqrt{n})$. The total number of graph topologies needed to be learned, for all $u \in V_H$, is at most $|V_H| \cdot O(\sqrt{n}) = O(n)$. We fix any ordering of these learning tasks and solve them sequentially. For each of these tasks, we use the above generic approach to solve the task, so the time and energy cost for learning one graph topology is poly $\log n$. Since there are $O(n)$ tasks, the overall time complexity is $O(n) \cdot$ poly $\log n = \tilde{O}(n)$. Each vertex participates in $O(\sqrt{n})$ tasks, so the overall energy complexity is $O(\sqrt{n}) \cdot$ poly $\log n = \tilde{O}(\sqrt{n})$.   □

Lemma 17. *Using $\tilde{O}(n^{1.5})$ time and $\tilde{O}(\sqrt{n})$ energy, we can let all vertices in G learn the graph topology of $G^\star$ w.h.p.*

Proof. The lemma follows from combining Lemmas 11, 15 and 16.   □

Now we are ready to prove Theorem 1.

Theorem 1. *There is an algorithm that computes the diameter in $\tilde{O}(n^{1.5})$ time and $\tilde{O}(\sqrt{n})$ energy w.h.p. for bounded-genus graphs in* No-CD.

Proof. The theorem follows from combining Lemmas 10 and 17.   □

B   Minimum Cut

In this section, we apply the approach introduced in Sect. 5 to show that (i) the exact global minimum cut size and (ii) an approximate s–t minimum cut size of any bounded-genus graph can be computed in $\tilde{O}(\sqrt{n})$ energy. We follow the same approach introduced in Sect. 5. That is, we still decompose the vertex set into V_H and V_L, and we categorize the connected components of $G[V_L]$ into three types. The only difference here is the information that we extract from type-1 and type-2 components.

Given a cut $\mathcal{C} = (X, V \setminus X)$ of $G = (V, E)$, the two vertex sets $X \neq \emptyset$ and $V \setminus X \neq \emptyset$ are called the two parts of $\mathcal{C}$, and the *cut edges* of $\mathcal{C}$ are defined as $\{\{u, v\} : u \in X, v \in V \setminus X\}$. The *size* of a cut $\mathcal{C}$, which we denote as $|\mathcal{C}|$, is defined as the number of cut edges of $\mathcal{C}$. A *minimum cut* of a graph is a cut $\mathcal{C}$ that minimizes $|\mathcal{C}|$ among all possible cuts. An s–t *minimum cut* of a graph is a cut $\mathcal{C}$ the minimizes $|\mathcal{C}|$ among all possible cuts subject to the constraint that s and t belong to different parts. We consider the following definitions:

$c(S)$. For any type-1 component S, let $c(S)$ be the minimum cut size of $G^+[S]$.

$c'(S)$. For any type-2 component $S \in C(u, v)$, let $c'(S)$ be the u–v minimum cut size of $G^+[S]$.

$c''(S)$. For any type-2 component $S \in C(u, v)$, let $c''(S)$ be the minimum cut size of $G^+[S]$ among all cuts such that both u and v are within the same part of the cut.

We make the following observations.

Lemma 18. *Let $\mathcal{C} = (X, V \setminus X)$ be any minimum cut of G. For any vertex $u \in V_H$, one of the following statements is true:*

- *One part of the cut contains all vertices in $\bigcup_{S \in C(u)} S \cup \{u\}$.*
- *the size of the cut is $\min_{S \in C(u)} c(S)$.*

Proof. Suppose that the first statement is false. Then there exists a component $S' \in C(u)$ such that $S' \cup \{u\}$ intersects both parts of the cut, so $\mathcal{C}' = (X \cap (S' \cup \{u\}), (V \setminus X) \cap (S' \cup \{u\}))$ is a cut of $G^+[S']$. Therefore, $\min_{S \in C(u)} c(S) \leq c(S') \leq |\mathcal{C}'| \leq |\mathcal{C}|$. To prove that the second statement is true, we just need to show that $|\mathcal{C}| \leq \min_{S \in C(u)} c(S)$. This inequality follows from the observation that for any component $S \in C(u)$, any cut of $G^+[S]$ can be extended to a cut of G of the same size by adding all vertices in $V \setminus (S \cup \{u\})$ to the part of the cut that contains u.   □

Lemma 19. *Let $C = (X, V \setminus X)$ be any minimum cut of G. For two distinct vertices $u, v \in V_H$, one of the following statements is true:*

- *One part of the cut contains all vertices in $\bigcup_{S \in C(u,v)} S \cup \{u, v\}$.*
- *The size of the cut is $\min_{S \in C(u,v)} c''(S)$.*
- *u and v belong to different parts of the cut, and the number of cut edges that have at least one endpoint in $\bigcup_{S \in C(u,v)} S'$ is $\sum_{S \in C(u,v)} c'(S)$.*

Proof. Suppose that the first statement is false. We first focus on the case where u and v belong to the same part of the cut C. In this case, there exists a component $S' \in C(u, v)$ such that $S' \cup \{u, v\}$ intersects both parts of the cut, so $C' = (X \cap (S' \cup \{u, v\}), (V \setminus X) \cap (S' \cup \{u, v\}))$ is a cut of $G^+[S]$ such that u and v belong to the same part of the cut. Therefore, $\min_{S \in C(u,v)} c''(S) \leq c''(S') \leq |C'| \leq |C|$. Similar to the proof of Lemma 18, we also have $|C| \leq \min_{S \in C(u,v)} c''(S)$, as any cut of $G^+[S]$ such that u and v belong to the same part of the cut can be extended to a cut of G of the same size. Therefore, we must have $|C| = \min_{S \in C(u,v)} c''(S)$, that is, the second statement is true.

For the rest of the proof, we consider the case where u and v belong to different parts of the cut C. For each component $S \in C(u, v)$, we write Z_S to denote the number of cut edges of C that have at least one endpoint in S. Then we must have $Z_S = c'(S)$, since otherwise C is not a minimum cut. Therefore, the number of cut edges that have at least one endpoint in $\bigcup_{S \in C(u,v)} S'$ is $\sum_{S \in C(u,v)} c'(S)$, that is, the third statement is true. □

Using the above two observations, we prove Theorem 2.

Theorem 2. *There is an algorithm that computes the minimum cut size in $\tilde{O}(n^{1.5})$ time and $\tilde{O}(\sqrt{n})$ energy w.h.p. for bounded-genus graphs in* No-CD.

Proof. Bounded-genus graphs have bounded arboricity. The minimum degree of any graph of arboricity α is at most $2\alpha - 1$. The minimum cut size of any graph is at most the minimum degree of the graph. Therefore, there is a constant λ_0 such that the minimum cut size of G is at most λ_0.

Define the graph $G^\diamond$ as the result of applying the following operations to G:

- Remove all type-1 components.
- For each pair $\{u, v\}$ of distinct vertices in V_H with $|C(u, v)| > 0$, replace $C(u, v)$ with $\min\{\lambda_0, \sum_{S \in C(u,v)} c'(S)\}$ multi-edges between u and v.

By Lemmas 18 and 19, the minimum cut size of G is the minimum of the following numbers:

- The minimum value of $\min_{S \in C(u)} c(S)$ among all $u \in V_H$ such that $|C(u)| > 0$.
- The minimum value of $\min_{S \in C(u,v)} c''(S)$ among all $u, v \in V_H$ such that $|C(u, v)| > 0$.
- The minimum cut size of $G^\diamond$.

For each vertex $u \in V$, we define $\mathcal{I}_0^\diamond(u)$, $\mathcal{I}_1^\diamond(u)$, and $\mathcal{I}_2^\diamond(u)$ as follows.

- $\mathcal{I}_0^\diamond(u)$ is the same as the basic information $\mathcal{I}_0(u)$ defined in Sect. 5.
- $\mathcal{I}_1^\diamond(u)$ contains the number $\min_{S \in C(u)} c(S)$.
- $\mathcal{I}_2^\diamond(u)$ contains $\min_{S \in C(u,v)} c''(S)$ and $\min\{\lambda_0, \sum_{S \in C(u,v)} c'(S)\}$, for all pairs $\{u,v\} \in E(G_H)$ such that $F^\star(\{u,v\}) = u$.

Similar to the proof of Theorem 1, $\mathcal{I}_1(u)$ and $\mathcal{I}_2(u)$ contain nothing if $u \in V_L$.

As $\mathcal{I}_0^\diamond(u) = \mathcal{I}_0(u)$, we may use the algorithm of Lemma 15 to let all vertices $u \in V$ learn the information $\mathcal{I}_0^\diamond(u)$ using $\tilde{O}(n^{1.5})$ time and $\tilde{O}(\sqrt{n})$ energy.

The algorithm of Lemma 16 can be modified to allow all vertices $u \in V_H$ learn the information $\mathcal{I}_1^\diamond(u)$ and $\mathcal{I}_2^\diamond(u)$. Specifically, the number $\min_{S \in C(u)} c(S)$ can be learned by the same algorithm for learning $A_1[u]$ described in the proof of Lemma 15 by replacing SR-comm^{max} with SR-comm^{min} letting $v = r_{S_u}$ use the key $k_v = c(S)$. The algorithm for learning $\min_{S \in C(u,v)} c''(S)$ is similar.

For each pair $\{u,v\} \in E(G_H)$ such that $F^\star(\{u,v\}) = u$, to let u learn $\min\{\lambda_0, \sum_{S \in C(u,v)} c'(S)\}$, we use SR-comm^{apx} with the following parameters:

- $\mathcal{S} = \{r_{S,u} : S \in C(u,v)\}$.
- $\mathcal{R} = \{u\}$.
- $\epsilon = 1/(2\lambda_0 + 1)$.
- $W = \lambda_0$.
- For each $S \in C(u,v)$, the message m_v of the representative $v = r_{S,u}$ of S is $\min\{\lambda_0, c'(S)\}$.

After the algorithm of SR-comm^{apx}, u learns an $(1 \pm \epsilon)$-approximation of

$$\sum_{v \in N^+(u) \cap \mathcal{S}} m_v = \sum_{S \in C(u,v)} \min\{\lambda_0, c'(S)\}.$$

We claim that this allows u to calculate $\min\{\lambda_0, \sum_{S \in C(u,v)} c'(S)\}$ precisely. To prove this claim, we break the analysis into two cases. Let x be the approximation of $\sum_{S \in C(u,v)} \min\{\lambda_0, c'(S)\}$ computed by SR-comm^{apx}.

If $\min\{\lambda_0, \sum_{S \in C(u,v)} c'(S)\} = \lambda_0$, then

$$\sum_{v \in N^+(u) \cap \mathcal{S}} m_v = \sum_{S \in C(u,v)} \min\{\lambda_0, c'(S)\} \geq \lambda_0,$$

which implies

$$x \geq (1 - \epsilon)\lambda_0 > \lambda_0 - 1/2.$$

If $\min\{\lambda_0, \sum_{S \in C(u,v)} c'(S)\} = \sum_{S \in C(u,v)} c'(S)$, then

$$\sum_{v \in N^+(u) \cap \mathcal{S}} m_v = \sum_{S \in C(u,v)} \min\{\lambda_0, c'(S)\} = \sum_{S \in C(u,v)} c'(S),$$

which implies

$$x \in \left[(1-\epsilon) \sum_{S \in C(u,v)} c'(S), (1+\epsilon) \sum_{S \in C(u,v)} c'(S)\right]$$

$$\subseteq \left[\left(\sum_{S \in C(u,v)} c'(S)\right) - \frac{1}{2}, \left(\sum_{S \in C(u,v)} c'(S)\right) + \frac{1}{2}\right].$$

Therefore, u can calculate $\min\{\lambda_0, \sum_{S \in C(u,v)} c'(S)\}$ precisely from x. By Lemma 27, the cost for u to calculate $\min\{\lambda_0, \sum_{S \in C(u,v)} c'(S)\}$ via SR-comm^{apx} is poly log n time.

For each $u \in V_H$, the number of pairs $\{u,v\}$ such that $F^\star(\{u,v\}) = u$ is $O(1)$, so the number of parameters needed to be learned in $\mathcal{I}_1^\diamond(u)$ and $\mathcal{I}_2^\diamond(u)$ by u is $O(1)$. The total number of parameters needed to be learned, for all $u \in V_H$, is at most $|V_H| \cdot O(1) = O(\sqrt{n})$. We fix any ordering of these learning tasks and solve them sequentially. The time and energy cost for learning one parameter is poly log n. Since there are $O(\sqrt{n})$ tasks, the overall time complexity for learning $\mathcal{I}_1^\diamond(u)$ and $\mathcal{I}_2^\diamond(u)$ for all $u \in V_H$ is $O(\sqrt{n}) \cdot \text{poly} \log n = \tilde{O}(\sqrt{n})$.

In view of the above discussion, the minimum cut size of G can be calculated from the following information: (i) $\mathcal{I}_1^\diamond(u)$ and $\mathcal{I}_2^\diamond(u)$ for all $u \in V_H$, (ii) the topology of $G^+[S]$ for each type-3 component S, and (iii) the topology of the subgraph induced by V_H. By replacing $\mathcal{I}_1(u)$ and $\mathcal{I}_2(u)$ with $\mathcal{I}_1^\diamond(u)$ and $\mathcal{I}_2^\diamond(u)$ in the description of the algorithm of Lemma 11, we obtain an algorithm that lets all vertices learn this information using $\tilde{O}(n^{1.5})$ time and $\tilde{O}(\sqrt{n})$ energy. □

The proof of Theorem 3 is similar to that of Theorem 2. The main difference for the setting of s–t minimum cut is that if s or t happens to be within a type-1 or a type-2 component S, then we additionally need to learn the topology of $G^+[S]$. Any type-1 component that does not contain s or t is irrelevant to the s–t minimum cut size.

In the subsequent discussion, we fix s and t to be any two distinct vertices of G. for each $x \in \{s,t\}$, let S_x be the type-1 or type-2 component containing x. In case x is not contained in any type-1 or type-2 component, we let $S_x = \emptyset$. We define $G^\bullet$ as the result of applying the following operations to G.

- Remove all type-1 components, except for S_s and S_t.
- For each pair $\{u,v\}$ of distinct vertices in V_H with $|C(u,v) \setminus \{S_s, S_t\}| > 0$, replace all components in $C(u,v) \setminus \{S_s, S_t\}$ with $\sum_{S \in C(u,v) \setminus \{S_s,S_t\}} c'(S)$ multi-edges between u and v.

Similar to Lemmas 18 and 19, we have the following observation.

Lemma 20. *Both G and $G^\bullet$ have the same minimum s–t cut size.*

Proof. Fix $\mathcal{C} = (X, V \setminus X)$ to be any minimum s–t cut of G, where $s \in X$ and $t \in V \setminus X$. To show that both G and $G^\bullet$ have the same minimum s–t cut size, it suffices to show the following two statements:

- For each type-1 component S that is not S_s and S_t, we must have either $S \subseteq X$ or $S \subseteq V \setminus X$.
- For each pair $\{u, v\}$ of distinct vertices in V_H with $|C(u,v) \setminus \{S_s, S_t\}| > 0$, if u and v belong to different parts of cut $\mathcal{C}$, then the number of cut edges of $\mathcal{C}$ with at least one endpoint in $\bigcup_{S \in C(u,v) \setminus \{S_s, S_t\}} S$ equals $\sum_{S \in C(u,v) \setminus \{S_s, S_t\}} c'(S)$.

The first statement follows from the observation that for each $u \in V_H$, all vertices in $\bigcup_{S \in C(u) \setminus \{S_s, S_t\}} S$ must belong to the part of cut $\mathcal{C}$ that u belongs to, since otherwise $\mathcal{C}$ is not a minimum s–t cut, as moving all vertices in $\bigcup_{S \in C(u) \setminus \{S_s, S_t\}} S$ to the part of cut that u belongs to reduces the number of cut edges.

To show the second statement, consider a pair $\{u, v\}$ of distinct vertices in V_H with $|C(u,v) \setminus \{S_s, S_t\}| > 0$ such that u and v belong to different parts of cut $\mathcal{C}$. Similar to the proof of Lemma 19, for each component $S \in C(u,v) \setminus \{S_s, S_t\}$, we write Z_S to denote the number of cut edges of $\mathcal{C}$ that have at least one endpoint in S. Then we must have $Z_S = c'(S)$, since otherwise $\mathcal{C}$ is not a minimum cut. Therefore, the number of cut edges of $\mathcal{C}$ that have at least one endpoint in $\bigcup_{S \in C(u,v) \setminus \{S_s, S_t\}} S'$ is $\sum_{S \in C(u,v) \setminus \{S_s, S_t\}} c'(S)$. □

We are ready to prove Theorem 3.

Theorem 3. *There is an algorithm that computes an $(1 \pm \epsilon)$-approximate s–t minimum cut size in $\tilde{O}(n^{1.5}) \cdot \epsilon^{-O(1)}$ time and $\tilde{O}(\sqrt{n}) \cdot \epsilon^{-O(1)}$ energy w.h.p. for bounded-genus graphs in No-CD.*

Proof. Let $\tilde{G}^\bullet$ be any graph such that for each pair of vertices $\{u, v\}$, the number of mult-edges in $\tilde{G}^\bullet$ is within a $(1 \pm \epsilon)$ factor of the number of mult-edges in $G^\bullet$. By Lemma 20, the minimum s–t cut size in $\tilde{G}^\bullet$ is a $(1 \pm \epsilon)$-approximation of the minimum s–t cut size of G. Therefore, the task of computing the minimum s–t cut size of G is reduced to computing such a graph $\tilde{G}^\bullet$.

For each $u \in V_H$, we let $\mathcal{I}_2^\bullet(u)$ contain the number $\sum_{S \in C(u,v) \setminus \{S_s, S_t\}} c'(S)$ for all pairs $\{u, v\} \in E(G_H)$ with $F^\star(\{u, v\}) = u$. The same algorithm for learning $\mathcal{I}_2^\Diamond(u)$ presented in the proof of Theorem 2 can be applied here to let all $u \in V_H$ *approximately* learn each number in $\mathcal{I}_2^\bullet(u)$ within a $(1 \pm \epsilon)$ factor, by using SR-comm^{apx} with parameter ϵ. We can tolerate this approximation factor because here our goal is to learn $\tilde{G}^\bullet$.

In view of the above, a $(1 \pm \epsilon)$-approximation of the minimum s–t cut size of G can be calculated from the following information: (i) $\mathcal{I}_2^\bullet(u)$ for all $u \in V_H$, (ii) the topology of $G^+[S]$ for $S = S_s$, $S = S_t$, and each type-3 component S, and (iii) the topology of the subgraph induced by V_H. Same as the proof of Theorem 2, we can obtain an algorithm that lets all vertices learn this information using $\tilde{O}(n^{1.5}) \cdot \epsilon^{-O(1)}$ time and $\tilde{O}(\sqrt{n}) \cdot \epsilon^{-O(1)}$. Here the extra $\epsilon^{-O(1)}$ is due to the use of SR-comm^{apx}, which requires $\epsilon^{-O(1)} \cdot \text{poly} \log n$ time and energy. □

C   Algorithms for Communication Between Two Sets of Vertices

In this section, we present our algorithms for SR-comm and its variants. Recall that SR-comm requires that each vertex $v \in \mathcal{R}$ with $N^+(v) \cap \mathcal{S} \neq \emptyset$ receives a message m_u from *at least one* vertex $u \in N^+(v) \cap \mathcal{S}$ w.h.p.

Lemma 21. (*[4]*) SR-comm *can be solved in time* $O(\log \Delta \log n)$ *and energy* $O(\log \Delta \log n)$.

Proof. By the definition of SR-comm, each vertex $v \in \mathcal{S} \cap \mathcal{R}$ is not required to receive any message from other vertices, as we already have $v \in N^+(v) \cap \mathcal{S}$. Therefore, in the subsequent discussion, we assume that $\mathcal{S} \cap \mathcal{R} = \emptyset$.

The task SR-comm with $\mathcal{S} \cap \mathcal{R} = \emptyset$ can be solved using the well-known *decay* algorithm of [4], which repeats the following routine for $C \log n$ times: For $i = 1$ to $\log \Delta$, let each vertex $u \in \mathcal{S}$ transmit with probability 2^{-i}. Each $v \in \mathcal{R}$ is always listening throughout the procedure. Here $C > 0$ is some large enough constant to be determined.

Consider a vertex $v \in \mathcal{R}$ such that $N(v) \cap \mathcal{S} \neq \emptyset$. Let i^* be the largest integer i such that $2^i \leq 2|N(v) \cap \mathcal{S}|$. Consider a time slot t where each vertex $u \in \mathcal{S}$ transmits with probability 2^{-i^*}. For notational simplicity, we write $n' = |N(v) \cap \mathcal{S}|$ and $p' = 2^{-i^*}$. Our choice of i^* implies that $1/n' \geq p' \geq 1/(2n')$. The probability of the event that exactly one vertex in the set $N(v) \cap \mathcal{S}$ transmits equals $n'p'(1 - p')^{n'-1} \geq 1/(2e)$. The calculation follows from the inequalities $n'p' \geq 1/2$ and $(1 - p')^{n'-1} \geq (1 - 1/n')^{n'-1} \geq 1/e$.

If the above event occurs, then v successfully receives a message m_u from a vertex $u \in N(v) \cap \mathcal{S}$. The probability that v does not receive any message from vertices in $N(v) \cap \mathcal{S}$ throughout the entire algorithm is at most $(1 - 1/(2e))^{C \log n} = n^{-\Omega(C)}$. By setting C to be a large enough constant, the algorithm successfully solves SR-comm w.h.p., and the time and energy complexities of the algorithm are $O(\log \Delta \log n)$. $\qquad \square$

Recall that the goal of SR-comm[all] is to let each vertex $u \in \mathcal{S} \cap N^+(v)$ deliver a message m_u to $v \in \mathcal{R}$, for each $v \in \mathcal{R}$.

Lemma 22. SR-comm[all] *can be solved in time* $O(\Delta' \log n)$ *and energy* $O(\Delta' \log n)$, *where* Δ' *is an upper bound on* $|\mathcal{S} \cap N(v)|$, *for each* $v \in \mathcal{R}$.

Proof. Consider the algorithm which repeats the following routine for $C \cdot \Delta' \log n$ rounds, for some sufficiently large constant $C > 0$. In each round, each vertex $u \in \mathcal{S}$ sends m_u with probability $1/\Delta'$. For each $u \in \mathcal{R}$, if u does not send in this round, then u listens.

Let $e = \{u, v\}$ be any edge with $u \in \mathcal{S}$ and $v \in \mathcal{R}$. In one round of the above algorithm, u successfully sends a message to v if (i) all vertices in $\{v\} \cup (\mathcal{S} \cap N(v)) \setminus \{u\}$ do not send, and (ii) u sends. Therefore, the probability that u successfully sends a message to v is

$$(1 - 1/\Delta')^{|\mathcal{S} \cap N(v)|-1} \cdot (1/\Delta') \geq (1 - 1/\Delta')^{\Delta'-1} \cdot (1/\Delta') \geq 1/(e\Delta')$$

The probability that u does not successfully send a message to v throughout all $C \cdot \Delta' \log n$ rounds is at most $(1 - 1/(e\Delta'))^{C \cdot \Delta' \log n} = n^{-\Omega(C)}$. Selecting a large enough constant C, by a union bound for all $u \in \mathcal{S} \cap N(v)$ and all $v \in \mathcal{R}$, we conclude that the algorithm solves SR-comm[all] w.h.p. The time and energy complexities are $O(\Delta' \log n)$. $\qquad \square$

Recall that the task SR-comm^{multi} requires that each vertex $v \in \mathcal{R}$ receive all distinct messages in $\bigcup_{u \in N^+(v) \cap \mathcal{S}} \mathcal{M}_u$, where is the $\mathcal{M}_u$ is the set of messages hold by u.

Lemma 23. SR-comm^{multi} *can be solved in time* $O(M \log \Delta \log^2 n)$ *and energy* $O(M \log \Delta \log^2 n)$, *where* M *is an upper bound on the number of distinct messages in* $\bigcup_{u \in N^+(v) \cap \mathcal{S}}$, *for each* $v \in \mathcal{R}$.

Proof. Consider the algorithm which repeatedly runs SR-comm for $C \cdot M \log n$ times, where in each iteration, the sets $(\mathcal{S}', \mathcal{R}')$ for SR-comm are chosen randomly as follows. We select $\mathcal{R}'$ as a random subset of $\mathcal{R}$ such that each $v \in \mathcal{R}$ joins $\mathcal{R}'$ with probability $1/2$. We select $\mathcal{S}'$ as a random subset of $\mathcal{S} \setminus \mathcal{R}'$ such that for each message m, all vertices in $\mathcal{S} \setminus \mathcal{R}'$ that hold m join $\mathcal{S}'$ with probability $1/M$, using the shared randomness associated with the message m.

Due to the shared randomness, if $u \in \mathcal{S} \setminus \mathcal{R}'$ joins $\mathcal{S}'$ due to message m, then all vertices in $\mathcal{S} \setminus \mathcal{R}'$ holding the same message m also joins $\mathcal{S}'$. Note that a vertex $u \in \mathcal{S} \setminus \mathcal{R}'$ might hold more than one message in that $|\mathcal{M}_u| > 1$. The probability that $u \in \mathcal{S} \setminus \mathcal{R}'$ joins $\mathcal{S}'$ equals $\Pr[\text{Binomial}(|\mathcal{M}_u|, 1/M) \geq 1]$, because each message $m \in \mathcal{M}_u$ lets u join $\mathcal{S}'$ with probability $1/M$ independently.

To analyze the algorithm, we focus on one vertex $v \in \mathcal{R}$ in one iteration of the above algorithm. Consider any message $m \in \bigcup_{u \in N(v) \cap \mathcal{S}} \mathcal{M}_u \setminus \mathcal{M}_v$. Observe that v receives m if the following three events $\mathcal{E}_1$, $\mathcal{E}_2$, and $\mathcal{E}_3$ occur:

- $\mathcal{E}_1$ is the event that v joins $\mathcal{R}'$.
- $\mathcal{E}_2$ is the event that at least one vertex $u \in N(v) \cap \mathcal{S}$ with $m \in \mathcal{M}_u$ does not join $\mathcal{R}'$.
- $\mathcal{E}_3$ is the event that the subset of vertices of $N(v) \cap \mathcal{S} \setminus \mathcal{R}'$ joining $\mathcal{S}'$ is exactly the set of all vertices $u \in N(v) \cap \mathcal{S} \setminus \mathcal{R}'$ with $m \in \mathcal{M}_u$.

If $\mathcal{E}_1$, $\mathcal{E}_2$, and $\mathcal{E}_3$ occur, then $v \in \mathcal{R}'$, $N(v) \cap \mathcal{S}' \neq \emptyset$, and all vertices $u \in N(v) \cap \mathcal{S}'$ satisfy $m \in \mathcal{M}_u$. Therefore, conditioning on $\mathcal{E}_1$, $\mathcal{E}_2$, and $\mathcal{E}_3$, SR-comm in this iteration allows v to receive message m.

The way $\mathcal{R}'$ is selected implies that $\Pr[\mathcal{E}_1] = 1/2$ and $\Pr[\mathcal{E}_2] \geq 1/2$. Observe that $\mathcal{E}_1$ and $\mathcal{E}_2$ are independent events. The way $\mathcal{S}'$ is selected implies that $\Pr[\mathcal{E}_3 | \mathcal{E}_1 \cap \mathcal{E}_2] \geq \Pr[\text{Binomial}(M, 1/M) = 1] = (1/M) \cdot (1 - 1/M)^{M-1} \geq 1/(eM)$. Therefore, the probability that v receives m in this iteration is at least $1/(4eM)$.

The probability that v does not receive m in all iterations is at most $(1 - 1/(4eM))^{C \cdot M \log n} = n^{-\Omega(C)}$. Selecting a large enough constant C, by a union bound for all $v \in \mathcal{R}$ and all $m \in \bigcup_{u \in N(v) \cap \mathcal{S}} \mathcal{M}_u \setminus \mathcal{M}_v$, we conclude that the algorithm solves SR-comm^{all} w.h.p. The time and energy complexities are $O(M \log \Delta \log^2 n)$, as the number of iterations is $O(M \log n)$ and the time complexity of each iteration is $O(\log \Delta \log n)$ by Lemma 21. $\qquad \square$

Consider the setting where the message m_u sent from each vertex $u \in \mathcal{S}$ contains a key k_u from the key space $[K] = \{1, 2, \ldots, K\}$. Recall that SR-comm^{min} requires that each vertex $v \in \mathcal{R}$ with $N^+(v) \cap \mathcal{S} \neq \emptyset$ receives a message m_u from a vertex $u \in N^+(v) \cap \mathcal{S}$ such that $k_u = \min_{u' \in N^+(v) \cap \mathcal{S}} k_{u'}$.

Lemma 24. *Both* SR-comm^{min} *and* SR-comm^{max} *can be solved in time* $O(K \log \Delta \log n)$ *and energy* $O(\log K \log \Delta \log n)$. *For the special case of* $\mathcal{S} \cap \mathcal{R} = \emptyset$ *and* $|\mathcal{R} \cap N(u)| \leq 1$ *for each* $u \in \mathcal{S}$, *the time complexity can be improved to* $O(\log K \log \Delta \log n)$.

Proof. We only prove the lemma for SR-comm^{min}, as the proof for SR-comm^{max} is the same. The proof presented here is analogous to the analysis of a deterministic version of SR-comm in [10]. Observe that we can do SR-comm once to let each $v \in \mathcal{R}$ test whether or not $N^+(v) \cap \mathcal{S} \neq \emptyset$. If a vertex $v \in \mathcal{R}$ knows that $N^+(v) \cap \mathcal{S} = \emptyset$, then v may remove itself from $\mathcal{R}$. Thus, in the subsequent discussion, we assume $N^+(v) \cap \mathcal{S} \neq \emptyset$ for each $v \in \mathcal{R}$.

Let $v \in \mathcal{R}$, and we define $f_v = \min_{u \in N^+(v) \cap \mathcal{S}} k_u$. The high-level idea of the algorithm is to conduct a binary search to determine all $\log K$ bits of the binary representation of f_v.

General Case. Suppose at some moment each vertex $v \in \mathcal{R}$ already knows the first x bits of f_v. The following procedure allows each $v \in \mathcal{R}$ to learn the $(x+1)$th bit of f_v. For each $(x+1)$-bit binary string s, we do SR-comm with the following choices of $(\mathcal{S}', \mathcal{R}')$:

- $\mathcal{S}'$ is the set of vertices $u \in \mathcal{S}$ such that the first $x+1$ bits of k_u equal s.
- $\mathcal{R}'$ is the set of vertices $v \in \mathcal{R}$ such that the first x bits of f_v equal the first x bits of s.

In this procedure, we perform 2^{x+1} times of SR-comm in total, but each vertex only participates in at most three of them, as each vertex joins $\mathcal{S}'$ at most once and joins $\mathcal{R}'$ at most twice. Thus, the procedure costs $O(2^x \log \Delta \log n)$ time and $O(\log \Delta \log n)$ energy, by Lemma 21. For each $v \in \mathcal{R}$, the messages that v receive during the procedure allows v to determine the $(x+1)$th bit of f_v.

We will run the above procedure for $\log K$ iterations from $x = 0$ to $x = \log K - 1$. Observe that in the last iteration, each vertex $v \in \mathcal{R}$ is guaranteed to receive a message m_u from a vertex $u \in N^+(v) \cap \mathcal{S}$ such that $k_u = f_v = \min_{w \in N^+(v) \cap \mathcal{S}} k_w$, so this algorithm allows us to solve SR-comm^{min}. The overall time complexity of the algorithm is

$$\sum_{x=0}^{\log K - 1} O(2^x \log \Delta \log n) = O(K \log \Delta \log n),$$

and the overall energy complexity of the algorithm is

$$\sum_{x=0}^{\log K - 1} O(\log \Delta \log n) = O(\log K \log \Delta \log n).$$

Special Case. For the rest of the proof, we focus on the special case of $\mathcal{S} \cap \mathcal{R} = \emptyset$ and $|\mathcal{R} \cap N(u)| \leq 1$ for each $u \in \mathcal{S}$. These assumptions imply that the family of sets $(\mathcal{S} \cap N(v)) \cup \{v\}$ for all $v \in \mathcal{R}$ are disjoint. The high-level idea is that for each $v \in \mathcal{R}$, we may let the set of vertices $(\mathcal{S} \cap N(v)) \cup \{v\}$ jointly conduct a binary search to determine all bits of $f_v = \min_{u \in N(v) \cap \mathcal{S}} k_u$, in parallel for all $v \in \mathcal{R}$.

Suppose that for each vertex $v \in \mathcal{R}$, all vertices in the set $(\mathcal{S} \cap N(v)) \cup \{v\}$ already know the first x bits of f_v. We present a more efficient algorithm that let all vertices in the set $(\mathcal{S} \cap N(v)) \cup \{v\}$ learn the $(x+1)$th bit of f_v.

Step 1. Perform SR-comm with the following choices of $(\mathcal{S}', \mathcal{R}')$:
- $\mathcal{R}' = \mathcal{R}$.
- $\mathcal{S}'$ is the subset of $\mathcal{S}$ that contains all vertices $u \in \mathcal{S}$ satisfying the following conditions:
  - The first x bits of k_u equal the first x bits of f_v, where v is the unique vertex in $\mathcal{R} \cap N(u)$.
  - The $(x+1)$th bit of k_u is 0.

This step allows each $v \in \mathcal{R}$ to learn the $(x+1)$th bit of f_v. If $v \in \mathcal{R}$ receives a message in SR-comm, then v knows that the $(x+1)$th bit of f_v is 0. Otherwise, v knows that the $(x+1)$th bit of f_v is 1.

Step 2. Perform SR-comm with the following choices of $(\mathcal{S}', \mathcal{R}')$:
- $\mathcal{R}' = \mathcal{S}$.
- $\mathcal{S}' = \mathcal{R}$.

This step lets each $v \in \mathcal{R}$ send the $(x+1)$th bit of f_v to all vertices in $\mathcal{S} \cap N(v)$.

The time and energy complexities of this algorithm are asymptotically the same as that of SR-comm, which are $O(\log \Delta \log n)$. As discussed earlier, to solve SR-comm^{min}, all we need to do is to run the above algorithm from $x = 0$ to $x = \log K - 1$. The overall time and energy complexities of the algorithm for SR-comm^{min} are $O(\log K \log \Delta \log n)$, as there are $\log K$ iterations.   □

For the rest of the section, we consider the task SR-comm^{apx}, which requires each vertex $v \in \mathcal{R}$ to compute an $(1 \pm \epsilon)$-factor approximation of the summation $\sum_{u \in N^+(v) \cap \mathcal{S}} m_u$. We need the following fact, whose correctness can be verified by means of a simple calculation.

Lemma 25. *There exist three universal constants $0 < \epsilon_0 < 1$, $N_0 \geq 1$, and $c_0 \geq 1$ such that the following statement holds: For any pair of numbers (N, ϵ) such that $N \geq N_0$ and $\epsilon_0 \geq |\epsilon| \geq c_0/\sqrt{N}$,*

$$e^{-1}(1 - 0.51\epsilon^2) \leq (1 + \epsilon)(1 - (1 + \epsilon)/N)^{N-1} \leq e^{-1}(1 - 0.49\epsilon^2).$$

Note that the parameter ϵ in Lemma 25 can be either positive or negative. For the rest of the section, we assume that the message m_u sent from each vertex $u \in \mathcal{S}$ is an integer within the range $[W]$. We first consider the special case of SR-comm^{apx} with $W = 1$. In this case, SR-comm^{apx} is the same as the approximate counting problem whose goal is to let each $v \in \mathcal{R}$ compute $|N^+(v) \cap \mathcal{S}|$, up to a $(1 \pm \epsilon)$-factor error.

Lemma 26. *For $W = 1$, SR-comm^{apx} can be solved in $O((1/\epsilon^5) \log \Delta \log n)$ time and energy.*

Proof. In this proof, we will focus on a slightly different task of estimating $|N(v) \cap S|$ within a $(1 \pm \epsilon)$-factor approximation, for each $v \in \mathcal{R}$. If each $v \in \mathcal{R}$ knows such an estimate of $|N(v) \cap S|$, then v can locally calculate an estimate of $|N^+(v) \cap S|$ within a $(1 \pm \epsilon)$-factor approximation, thereby solving SR-comm^{apx} for the case of $W = 1$.

Basic Setup. Let $C > 0$ be a sufficiently large constant. Let ϵ_0, N_0, and c_0 be the constants in Lemma 25. We assume that $\epsilon \le \epsilon_0$. If this is not the case, then we may reset $\epsilon = \epsilon_0$.

The algorithm consists of two phases. The first phase of the algorithm aims to achieve the following goals: For each $v \in \mathcal{R}$, either (i) v learns the number $|N(v) \cap S|$ exactly or (ii) v detects that $\epsilon \ge 10c_0/\sqrt{|N(v) \cap S|}$. For each vertex $v \in \mathcal{R}$ that calculates the number $|N(v) \cap S|$ exactly in the first phase, we remove v from $\mathcal{R}$. The second phase of the algorithm then solves SR-comm^{apx} for the remaining vertices in $\mathcal{R}$. These vertices $v \in \mathcal{R}$ satisfy $\epsilon \ge 10c_0/\sqrt{|N(v) \cap S|}$.

The First Phase. We define $Z = (10c_0/\epsilon)^2$. The algorithm consists of $C \cdot Z \log n$ rounds, where we do the following in each round:

- Each vertex $u \in S \cup \mathcal{R}$ flips a biased coin that produces head with probability $1/Z$.
- Each $u \in S$ sends $\mathrm{ID}(u)$ if the outcome of its coin flip is head.
- Each vertex $v \in \mathcal{R}$ listens if the outcome of its coin flip is tail.

For each vertex $v \in \mathcal{R}$, there are two cases:

- Suppose that there is a vertex $u \in N(v) \cap S$ such that the number of messages that v receives from is smaller than $0.5 \cdot (C \log n)/e$. Then v decides that $\epsilon \ge 10c_0/\sqrt{|N(v) \cap S|}$ and proceeds to the second phase.
- Suppose that for all vertices $u \in N(v) \cap S$, the number of messages that v receives from is at least $0.5 \cdot (C \log n)/e$. Then v calculate $|N(v) \cap S|$ by the number of distinct IDs that v receives.

The time complexity of the first phase of the algorithm is $C \cdot Z \log n = O((1/\epsilon^2) \log n)$.

Analysis. To analyze the algorithm, let $e = \{u, v\}$ be any edge such that $u \in S$ and $v \in \mathcal{R}$. In one round of the above algorithm, u successfully sends a message to v if and only if (i) the outcome of u's coin flip is head, and (ii) the outcome of the coin flips of all vertices in $(N(v) \cap S) \cup \{v\} \setminus \{u\}$ are all tails. This event occurs with probability $p^* = (1 - 1/Z)^{|N(v) \cap S|} \cdot (1/Z)$. Let X be the number of times v receives a message from u. To prove the correctness of the algorithm, we show the following three concentration bounds:

- If $v \in \mathcal{R}$ satisfies $\epsilon \le 10c_0/\sqrt{|N(v) \cap S|}$, then $\Pr[X \ge 0.8 \cdot (C \log n)/e] = 1 - n^{-\Omega(C)}$.
- If $v \in \mathcal{R}$ satisfies $\epsilon \ge 20c_0/\sqrt{|N(v) \cap S|}$, then $\Pr[X \le 0.2 \cdot (C \log n)/e] = 1 - n^{-\Omega(C)}$.
- If $v \in \mathcal{R}$ satisfies $\epsilon \le 20c_0/\sqrt{|N(v) \cap S|}$, then $\Pr[X \ge 1] = 1 - n^{-\Omega(C)}$.

We show the correctness of the algorithm given these concentration bounds. For the case $\epsilon \geq 20c_0/\sqrt{|N(v) \cap \mathcal{S}|}$, the second bound implies that the number of messages that v receives from u is greater than $0.5 \cdot (C \log n)/e$ w.h.p., so v correctly decides that $\epsilon \geq 10c_0/\sqrt{|N(v) \cap \mathcal{S}|}$ and proceeds to the second phase. For the case $\epsilon \leq 20c_0/\sqrt{|N(v) \cap \mathcal{S}|}$, the third bound implies that v receives at least one message from each vertex in $N(v) \cap \mathcal{S}$ w.h.p., so v can calculate $|N(v) \cap \mathcal{S}|$ precisely. The only remaining thing to show is that when ϵ is at most $10c_0/\sqrt{|N(v) \cap \mathcal{S}|}$, w.h.p. v does not decide that $\epsilon \geq 10c_0/\sqrt{|N(v) \cap \mathcal{S}|}$. This follows from the first bound, which implies that the number of messages that v receives from u is greater than $0.5 \cdot (C \log n)/e$ w.h.p.

We prove the three concentration bounds as follows:

- Suppose that vertex $v \in \mathcal{R}$ satisfies $\epsilon \leq 10c_0/\sqrt{|N(v) \cap \mathcal{S}|}$. We show that in this case the number of messages that v receives from $u \in N(v) \cap \mathcal{S}$ is at least $0.8 \cdot (C \log n)/e$, with probability $1 - n^{-\Omega(C)}$. In this case, we have $Z = (10c_0/\epsilon)^2 \geq |N(v) \cap \mathcal{S}|$, so $p^\star = (1 - 1/Z)^{|N(v) \cap \mathcal{S}|} \cdot (1/Z) \geq (1 - 1/Z)^Z \cdot (1/Z) \geq 0.9/(eZ)$. The expected value μ of X satisfies $\mu = C \cdot Z \log n \cdot p^\star \geq 0.9(C \log n)/e$. By a Chernoff bound, $\Pr[X \leq 0.8 \cdot (C \log n)/e] \leq \exp(-\Omega(C \log n)) = n^{-\Omega(C)}$.
- Suppose that vertex $v \in \mathcal{R}$ satisfies $\epsilon \geq 20c_0/\sqrt{|N(v) \cap \mathcal{S}|}$. We show that in this case the number of messages that v receives from $u \in N(v) \cap \mathcal{S}$ is at most $0.2 \cdot (C \log n)/e$, with probability $1 - n^{-\Omega(C)}$. In this case, we have $Z = (10c_0/\epsilon)^2 \leq |N(v) \cap \mathcal{S}|/4$, so $p^\star = (1 - 1/Z)^{|N(v) \cap \mathcal{S}|} \cdot (1/Z) \leq (1 - 1/Z)^{4Z} \cdot (1/Z) \leq 1/(e^4 Z)$. The expected value μ of X satisfies $\mu = C \cdot Z \log n \cdot p^\star \leq (C \log n)/e^4 < 0.1(C \log n)/e$. By a Chernoff bound, $\Pr[X \geq 0.2 \cdot (C \log n)/e] \leq \exp(-\Omega(C \log n)) = n^{-\Omega(C)}$.
- Suppose that vertex $v \in \mathcal{R}$ satisfies $\epsilon \leq 20c_0/\sqrt{|N(v) \cap \mathcal{S}|}$. We show that in this case the number of messages that v receives from $u \in N(v) \cap \mathcal{S}$ is at least 1, with probability $1 - n^{-\Omega(C)}$. In this case, we have $Z = (10c_0/\epsilon)^2 \geq |N(v) \cap \mathcal{S}|/4$, so $p^\star = (1 - 1/Z)^{|N(v) \cap \mathcal{S}|} \cdot (1/Z) \geq (1 - 1/Z)^{4Z} \cdot (1/Z) \geq 0.9/(e^4 Z)$. We have $\Pr[X < 1] = (1 - p^\star)^{CZ \log n} \leq (1 - 0.9/(e^4 Z))^{CZ \log n} = n^{-\Omega(C)}$.

The Second Phase. For each vertex $v \in \mathcal{R}$ that have already calculated the number $|N(v) \cap \mathcal{S}|$ exactly in the first phase, v removes itself from $\mathcal{R}$. We know that all the remaining vertices in $\mathcal{R}$ satisfy $\epsilon \geq 10c_0/\sqrt{|N(v) \cap \mathcal{S}|}$.

We consider the sequence of sending probabilities: $p_1 = 2/\Delta$, and $p_i = \min\{1, p_{i-1} \cdot (1 + \epsilon)\}$ for $i > 1$. We let $i^\star = O((1/\epsilon) \log \Delta)$ be the smallest index i such that $p_i = 1$.

The second phase of the algorithm consists of $i^\star$ iterations, where the ith iteration repeats the following procedure for $C \cdot (1/\epsilon^4) \log n$ times for all vertices $v \in \mathcal{S} \cup \mathcal{R}$:

- v flips a fair coin.
- If the outcome of the coin flip is head and $v \in \mathcal{S}$, then v sends with probability p_i.
- If the outcome of the coin flip is tail and $v \in \mathcal{R}$, then v listens to the channel.

After finishing the algorithm, each vertex $v \in \mathcal{R}$ finds an index i' such that the number of messages that v successfully receives during the i'th iteration is the highest. Then v decides that $2/p_{i'}$ is an estimate of $|N(v) \cap S|$ within a factor of $(1 \pm \epsilon)$. The time complexity of the second phase of the algorithm is $i^* \cdot C \cdot (1/\epsilon^4) \log n = O((1/\epsilon^5) \log \Delta \log n)$.

Analysis. To show the correctness of the above algorithm, in the subsequent discussion, we focus on a vertex $v \in \mathcal{R}$ in the ith iteration. We say that i is *good* for v if $p_i/2$ is within a $(1 \pm 0.6\epsilon)$-factor of $1/|N(v) \cap S|$, and we say that i is *bad* for v if $p_i/2$ is not within a $(1 \pm \epsilon)$-factor of $1/|N(v) \cap S|$. Our choice of the sequence $(p_1, p_2, \ldots)$ implies that there must be at least one good index i for v.

We write p_i^{suc} to denote the probability that v successfully receives a message in one round of the ith iteration. From the description of the algorithm, we have

$$p_i^{\mathrm{suc}} = (1/2) \cdot |N(v) \cap S| \cdot (p_i/2) \cdot (1 - (p_i/2))^{|N(v) \cap S| - 1}.$$

We define

$$p_{\mathrm{good}} = (1/2) \cdot e^{-1}(1 - 0.51(0.6\epsilon)^2) \quad \text{and} \quad p_{\mathrm{bad}} = (1/2) \cdot e^{-1}(1 - 0.49\epsilon^2).$$

We claim that (i) $p_i^{\mathrm{suc}} \geq p_{\mathrm{good}}$ if i is good for v and (ii) $p_i^{\mathrm{suc}} \leq p_{\mathrm{bad}}$ if i is bad for v.

We first prove this claim for the case that i is good for v. For simplicity, we write $N = |N(v) \cap S|$. Since i is good, $p_i/2 = (1 + \epsilon')/|N(v) \cap S|$ for some $\epsilon' \in [-0.6\epsilon, 0.6\epsilon]$. Using the new notations, we may rewrite p_i^{suc} as

$$p_i^{\mathrm{suc}} = (1/2) \cdot |N(v) \cap S| \cdot (p_i/2) \cdot (1 - (p_i/2))^{|N(v) \cap S| - 1}$$
$$= (1/2) \cdot (1 + \epsilon') \cdot (1 - (1 + \epsilon'))^{N - 1}.$$

By Lemma 25, we infer that $p_i^{\mathrm{suc}} \geq (1/2) \cdot e^{-1}(1 - 0.51(\epsilon')^2) \geq e^{-1}(1 - 0.51(0.6\epsilon)^2) = p_{\mathrm{good}}$.

Now consider the case i is bad for v. Again, we write $N = |N(v) \cap S|$. Since i is bad, $p_i/2 = (1 + \epsilon')/|N(v) \cap S|$ for some $\epsilon' \notin (-\epsilon, \epsilon)$. The above formula for p_i^{suc} still applies to this case, and Lemma 25 implies that $p_i^{\mathrm{suc}} \leq (1/2) \cdot e^{-1}(1 - 0.49(\epsilon')^2) \leq e^{-1}(1 - 0.49\epsilon^2) = p_{\mathrm{bad}}$.

Let X be the number of messages that v receives in the ith iteration of the algorithm. The expected value of X is $\mu = p_i^{\mathrm{suc}} \cdot C \cdot (1/\epsilon^4) \log n$. For the case i is good for v, we have $\mu \geq p_{\mathrm{good}} \cdot C \cdot (1/\epsilon^4) \log n$, so by a Chernoff bound, we have:

$$\Pr[X \leq (1 - 0.01\epsilon^2)p_{\mathrm{good}} \cdot C \cdot (1/\epsilon^4) \log n] = e^{-\Omega(\epsilon^4 \cdot C \cdot (1/\epsilon^4) \log n)} = n^{-\Omega(C)}.$$

For the case i is bad for v, we have $\mu \leq p_{\mathrm{bad}} \cdot C \cdot (1/\epsilon^4) \log n$, so by a Chernoff bound, we have:

$$\Pr[X \geq (1 + 0.01\epsilon^2)p_{\mathrm{bad}} \cdot C \cdot (1/\epsilon^4) \log n] = e^{-\Omega(\epsilon^4 \cdot C \cdot (1/\epsilon^4) \log n)} = n^{-\Omega(C)}.$$

Since $(1 - 0.01\epsilon^2)p_{\mathrm{good}} > (1 + 0.01\epsilon^2)p_{\mathrm{bad}}$, we conclude that w.h.p. the index i' selected by v must be good, which implies that the estimate $2/p_{i'}$ calculated by v is within a $(1 \pm \epsilon)$-factor of $|N(v) \cap S|$, as we know that $p_{i'}/2$ is within a $(1 \pm 0.6\epsilon)$-factor of $1/|N(v) \cap S|$, as i' is good. □

In the following lemma, we extend Lemma 26 to any value of W.

Lemma 27. SR-comm^{apx} *can be solved in* $O((1/\epsilon^6)\log W \log \Delta \log n)$ *time and energy.*

Proof. We let $\epsilon' = \Theta(\epsilon)$ be chosen such that $(1+\epsilon')^2 < 1+\epsilon$ and $(1-\epsilon')^2 > 1-\epsilon$. We consider the following sequence: $w_1 = 1$ and $w_i = \min\{W, (1+\epsilon')w_{i-1}\}$ for $i > 1$. Let $i^\star$ be the smallest index i such that $w_i = W$.

From $i = 1$ to $i^\star$, we run the algorithm of Lemma 26 with the following setting:

- $\mathcal{S}'$ is the vertices $u \in \mathcal{S}$ with $m_u \in (w_{i-1}, w_i]$.
- $\mathcal{R}' = \mathcal{R}$.
- The error parameter is ϵ'.

The algorithm of Lemma 26 lets each $v \in \mathcal{R}'$ compute an $(1 \pm \epsilon')$-factor approximation of $|N^+(v) \cap \mathcal{S}'|$ using $O((1/\epsilon^5)\log \Delta \log n)$ time and energy.

For each $v \in \mathcal{R}$, we write N_i to denote the number of vertices $u \in N^+(v) \cap \mathcal{S}$ such that $m_u \in (w_{i-1}, w_i]$, and we write $\tilde{N}_i$ to denote the estimate of $|N^+(v) \cap \mathcal{S}'|$ computed by v in the ith iteration. We have the following observations:

- $\tilde{N}_i$ is an $(1 \pm \epsilon')$-factor approximation of N_i.
- $\sum_{i=1}^{i^\star} w_i N_i$ is an $(1 \pm \epsilon')$-factor approximation of $\sum_{u \in N^+(v) \cap \mathcal{S}} m_u$.

Thus, $\sum_{i=1}^{i^\star} w_i \tilde{N}_i$, which can be calculated locally at v at the end of the algorithm, is an $(1 \pm \epsilon)$-factor approximation of $\sum_{u \in N^+(v) \cap \mathcal{S}} m_u$, by our choice of ϵ'.

By Lemma 26, the time and energy complexities for each iteration are $O((1/\epsilon^5)\log \Delta \log n)$. The total number of iterations is $i^\star = O((1/\epsilon)\log W)$. Thus, the overall time and energy complexities are $O((1/\epsilon^6)\log W \log \Delta \log n)$. $\qquad\square$

References

1. Akhoondian Amiri, S., Schmid, S., Siebertz, S.: A local constant factor MDS approximation for bounded genus graphs. In: Proceedings of the 2016 ACM Symposium on Principles of Distributed Computing (PODC), pp. 227–233 (2016)
2. Ambühl, C.: An optimal bound for the MST algorithm to compute energy efficient broadcast trees in wireless networks. In: Caires, L., Italiano, G.F., Monteiro, L., Palamidessi, C., Yung, M. (eds.) Automata, Languages and Programming, pp. 1139–1150. Springer, Berlin Heidelberg, Berlin, Heidelberg (2005)
3. Augustine, J., Moses, Jr, W.K., Pandurangan, G.: Distributed MST computation in the sleeping model: awake-optimal algorithms and lower bounds. arXiv preprint arXiv:2204.08385 (2022)
4. Bar-Yehuda, R., Goldreich, O., Itai, A.: On the time-complexity of broadcast in multi-hop radio networks: an exponential gap between determinism and randomization. J. Comput. Syst. Sci. **45**(1), 104–126 (1992)

5. Barenboim, L., Maimon, T.: Deterministic logarithmic completeness in the distributed sleeping model. In: Gilbert, S. (ed.) 35th International Symposium on Distributed Computing (DISC). Leibniz International Proceedings in Informatics (LIPIcs), vol. 209, pp. 10:1–10:19. Schloss Dagstuhl - Leibniz-Zentrum für Informatik, Dagstuhl, Germany (2021). https://doi.org/10.4230/LIPIcs.DISC.2021.10

6. Bender, M.A., Kopelowitz, T., Pettie, S., Young, M.: Contention resolution with log-logstar channel accesses. In: Proceedings of the 48th Annual ACM Symposium on Theory of Computing (STOC), pp. 499–508 (2016). https://doi.org/10.1145/2897518.2897655

7. Berenbrink, P., Cooper, C., Hu, Z.: Energy efficient randomised communication in unknown adhoc networks. Theor. Comput. Sci. **410**(27), 2549–2561 (2009). https://doi.org/10.1016/j.tcs.2009.02.002

8. Bordim, J.L., Jiangtao, C., Hayashi, T., Nakano, K., Olariu, S.: Energy-efficient initialization protocols for ad-hoc radio networks. IEICE Trans. Fundam. Electron. Commun. Comput. Sci. **83**(9), 1796–1803 (2000)

9. Caragiannis, I., Galdi, C., Kaklamanis, C.: Basic computations in wireless networks. In: Deng, X., Du, D.-Z. (eds.) ISAAC 2005. LNCS, vol. 3827, pp. 533–542. Springer, Heidelberg (2005). https://doi.org/10.1007/11602613_54

10. Chang, Y., Dani, V., Hayes, T.P., He, Q., Li, W., Pettie, S.: The energy complexity of broadcast. In: Proceedings of the 2018 ACM Symposium on Principles of Distributed Computing (PODC) (2018)

11. Chang, Y., Kopelowitz, T., Pettie, S., Wang, R., Zhan, W.: Exponential separations in the energy complexity of leader election. In: Proceedings of the 49th Annual ACM SIGACT Symposium on Theory of Computing (STOC), pp. 771–783 (2017)

12. Chang, Y.J., Dani, V., Hayes, T.P., Pettie, S.: The energy complexity of BFS in radio networks. In: Proceedings of the 39th Symposium on Principles of Distributed Computing (PODC), pp. 273–282. ACM (2020). https://doi.org/10.1145/3382734.3405713

13. Chang, Y.J., Duan, R., Jiang, S.: Near-optimal time-energy trade-offs for deterministic leader election. In: Proceedings of the 33th Annual ACM Symposium on Parallelism in Algorithms and Architectures (SPAA). ACM (2021)

14. Chang, Y.J., Jiang, S.: The energy complexity of Las Vegas leader election. In: Proceedings of the 34th ACM Symposium on Parallelism in Algorithms and Architectures (SPAA), pp. 75–86 (2022)

15. Chatterjee, S., Gmyr, R., Pandurangan, G.: Sleeping is efficient: MIS in $O(1)$-rounds node-averaged awake complexity. In: Proceedings of the 39th Symposium on Principles of Distributed Computing (PODC), pp. 99–108. ACM (2020). https://doi.org/10.1145/3382734.3405718

16. Chlamtac, I., Kutten, S.: On broadcasting in radio networks-problem analysis and protocol design. IEEE Trans. Commun. **33**(12), 1240–1246 (1985)

17. Clementi, A.E.F., Crescenzi, P., Penna, P., Rossi, G., Vocca, P.: On the complexity of computing minimum energy consumption broadcast subgraphs. In: Ferreira, A., Reichel, H. (eds.) STACS 2001. LNCS, vol. 2010, pp. 121–131. Springer, Heidelberg (2001). https://doi.org/10.1007/3-540-44693-1_11

18. Czygrinow, A., Hańćkowiak, M., Wawrzyniak, W.: Fast distributed approximations in planar graphs. In: Taubenfeld, G. (ed.) DISC 2008. LNCS, vol. 5218, pp. 78–92. Springer, Heidelberg (2008). https://doi.org/10.1007/978-3-540-87779-0_6

19. Dani, V., Gupta, A., Hayes, T.P., Pettie, S.: Wake up and join me! an energy-efficient algorithm for maximal matching in radio networks. Distrib. Comput. (2022)

20. Dani, V., Hayes, T.P.: How to wake up your neighbors: safe and nearly optimal generic energy conservation in radio networks. arXiv preprint arXiv:2205.12830 (2022)

21. Dufoulon, F., Moses, Jr, W.K., Pandurangan, G.: Sleeping is superefficient: MIS in exponentially better awake complexity. arXiv preprint arXiv:2204.08359 (2022)

22. Ephremides, A., Truong, T.V.: Scheduling broadcasts in multihop radio networks. IEEE Trans. Commun. **38**(4), 456–460 (1990). https://doi.org/10.1109/26.52656

23. Gąsieniec, L., Kantor, E., Kowalski, D.R., Peleg, D., Su, C.: Energy and time efficient broadcasting in known topology radio networks. In: Pelc, A. (ed.) DISC 2007. LNCS, vol. 4731, pp. 253–267. Springer, Heidelberg (2007). https://doi.org/10.1007/978-3-540-75142-7_21

24. Ghaffari, M., Haeupler, B.: Distributed algorithms for planar networks II: low-congestion shortcuts, MST, and min-cut. In: Proceedings of the Twenty-Seventh Annual ACM-SIAM Symposium on Discrete Algorithms (SODA), pp. 202–219. SIAM (2016)

25. Ghaffari, M., Haeupler, B.: Low-congestion shortcuts for graphs excluding dense minors. In: Proceedings of the 2021 ACM Symposium on Principles of Distributed Computing PODC, pp. 213–221 (2021)

26. Ghaffari, M., Parter, M.: Near-optimal distributed DFS in planar graphs. In: 31st International Symposium on Distributed Computing (DISC 2017). Schloss Dagstuhl-Leibniz-Zentrum fuer Informatik (2017)

27. Ghaffari, M., Portmann, J.: Average awake complexity of MIS and matching. In: Proceedings of the 34th ACM Symposium on Parallelism in Algorithms and Architectures (SPAA), pp. 45–55 (2022)

28. Haeupler, B., Li, J., Zuzic, G.: Minor excluded network families admit fast distributed algorithms. In: Proceedings of the 2018 ACM Symposium on Principles of Distributed Computing (PODC), pp. 465–474 (2018)

29. Jurdzinski, T., Kutylowski, M., Zatopianski, J.: Energy-efficient size approximation of radio networks with no collision detection. In: Proceedings of the 8th Annual International Conference on Computing and Combinatorics (COCOON), pp. 279–289 (2002)

30. Jurdzinski, T., Kutylowski, M., Zatopianski, J.: Weak communication in single-hop radio networks: adjusting algorithms to industrial standards. Concurr. Comput. Pract. Exp. **15**(11–12), 1117–1131 (2003)

31. Jurdziński, T., Kutyłowski, M., Zatopiański, J.: Efficient algorithms for leader election in radio networks. In: Proceedings of the 21st Annual ACM Symposium on Principles of Distributed Computing (PODC), pp. 51–57 (2002). https://doi.org/10.1145/571825.571833

32. Kardas, M., Klonowski, M., Pajak, D.: Energy-efficient leader election protocols for single-hop radio networks. In: Proceedings of the 42nd International Conference on Parallel Processing (ICPP), pp. 399–408 (2013)

33. Kirousis, L.M., Kranakis, E., Krizanc, D., Pelc, A.: Power consumption in packet radio networks. Theor. Comput. Sci. **243**(1), 289–305 (2000). https://doi.org/10.1016/S0304-3975(98)00223-0

34. Kutyłowski, M., Rutkowski, W.: Adversary immune leader election in ad hoc radio networks. In: Di Battista, G., Zwick, U. (eds.) ESA 2003. LNCS, vol. 2832, pp. 397–408. Springer, Heidelberg (2003). https://doi.org/10.1007/978-3-540-39658-1_37

35. Lavault, C., Marckert, J.F., Ravelomanana, V.: Quasi-optimal energy-efficient leader election algorithms in radio networks. Inf. Comput. **205**(5), 679–693 (2007)

36. Lenzen, C., Pignolet, Y.A., Wattenhofer, R.: Distributed minimum dominating set approximations in restricted families of graphs. Distrib. Comput. **26**(2), 119–137 (2013)

37. Li, J., Parter, M.: Planar diameter via metric compression. In: Proceedings of the 51st Annual ACM SIGACT Symposium on Theory of Computing (STOC), pp. 152–163 (2019)

38. Miller, G.L., Peng, R., Xu, S.C.: Parallel graph decompositions using random shifts. In: Proceedings of the Twenty-Fifth Annual ACM Symposium on Parallelism in Algorithms and Architectures (SPAA), pp. 196–203. ACM (2013)

39. Nakano, K., Olariu, S.: Randomized leader election protocols in radio networks with no collision detection. In: Goos, G., Hartmanis, J., van Leeuwen, J., Lee, D.T., Teng, S.-H. (eds.) ISAAC 2000. LNCS, vol. 1969, pp. 362–373. Springer, Heidelberg (2000). https://doi.org/10.1007/3-540-40996-3_31

40. Parter, M.: Distributed planar reachability in nearly optimal time. In: 34th International Symposium on Distributed Computing (DISC 2020). Schloss Dagstuhl-Leibniz-Zentrum für Informatik (2020)

41. Roditty, L., Williams, V.V.: Fast approximation algorithms for the diameter and radius of sparse graphs. In: Proceedings 45th ACM Symposium on Theory of Computing (STOC), pp. 515–524 (2013)

42. Sen, A., Huson, M.L.: A new model for scheduling packet radio networks. Wirel. Netw. **3**(1), 71–82 (1997). https://doi.org/10.1023/A:1019128411323

43. Takagi, H., Kleinrock, L.: Optimal transmission ranges for randomly distributed packet radio terminals. IEEE Trans. Commun. **32**(3), 246–257 (1984). https://doi.org/10.1109/TCOM.1984.1096061

44. Wawrzyniak, W.: A strengthened analysis of a local algorithm for the minimum dominating set problem in planar graphs. Inf. Process. Lett. **114**(3), 94–98 (2014)

Search and Rescue on the Line

Jared Coleman[✉]●, Lorand Cheng, and Bhaskar Krishnamachari

University of Southern California, Los Angeles, CA, USA
{jaredcol,lfcheng,bkrishna}@usc.edu

Abstract. We propose and study a problem inspired by a common task in disaster, military, and other emergency scenarios: search and rescue. Suppose an object (victim, message, target, etc.) is at some unknown location on a path. Given one or more mobile agents, also at initially arbitrary locations on the path, the goal is to find and deliver the object to a predefined destination in as little time as possible. We study the problem for the one- and two-agent cases and consider scenarios where the object and agents are arbitrarily (adversarially, even) placed along a path of either known (and finite) or unknown (and potentially infinite) length. We also consider scenarios where the destination is either at the endpoint or in the middle of the path. We provide both deterministic and randomized online algorithms for each of these scenarios and prove bounds on their (expected) competitive ratios.

Keywords: mobile · delivery · search · online · competitive ratio · search and rescue

1 Introduction

In this paper, we study a search and rescue problem where a set of autonomous agents on a one-dimensional path must cooperate to find and deliver an object to its destination (another location on the path) in as little time as possible. Formally, we consider a line with origin 0 onto which n agents with different speeds and an object which must be delivered to 0 are initially located arbitrarily (adversarially, even). We propose algorithms for scenarios where the object and agents are placed on the finite intervals $[0, 1]$ and $[-1, 1]$ but also discuss how slightly modified versions of the algorithms are equally competitive for the infinite intervals $[0, \infty)$ and $(-\infty, \infty)$. Agents can pick up, carry, and give the object to other agents (via physical handover) but can only communicate face-to-face. We assume agents can always move at their maximum speeds and that direction changes and handovers are instantaneous. In the offline setting, where the locations of all other agents and the object are known, this problem is equivalent to the Pony Express Communication Problem, for which an optimal offline algorithm is known [11]. In the online setting, this problem is related to search problems like the cow-path problem but differs in that we must consider the time required to deliver the object after it has been found. A strategy that considers the search and delivery components of the problem separately may not

S. Rajsbaum et al. (Eds.): SIROCCO 2023, LNCS 13892, pp. 297–316, 2023.
https://doi.org/10.1007/978-3-031-32733-9_13

be optimal. For example, a search algorithm that minimizes the time to find the object might force agents into worst-case positions for the subsequent delivery.

Our goal is to find online algorithms with minimal *competitive ratio*. The competitive ratio of an online algorithm A is the maximum over all problem instances I of the ratio between the delivery time by A and the delivery time by an optimal offline algorithm for the same instance. Formally, the competitive ratio of A is

$$\sup_I \frac{T_{A,I}}{T_I^*}$$

where $T_{A,I}$ is the delivery time of algorithm A on instance I and T_I^* is the optimal offline delivery time for instance I. We say an algorithm with a competitive ratio c is c-competitive. For the problem studied in this paper, a c-competitive algorithm guarantees the object is delivered to the origin in at most $c \cdot T^*$ time, where T^* is the optimal (offline) delivery time had the location of the object been known to all agents from the start.

The results of the paper are summarized in the following table.

Table 1. Table of results: lower and upper bounds on the competitive ratios proven for each model studied where $W(x)$ is the product logarithm (Lambert W function [14]) of x and, for the two-agent scenarios, v is the relative speed of the slower agent with respect to the faster agent (i.e. if agents have speeds v_1 and v_2 such that $v_2 \leq v_1$, then $v = v_2/v_1$).

Agents	Destination	Randomized	Lower Bound	Upper Bound	Section
1	endpoint	no	$1 + \sqrt{2}$	$1 + \sqrt{2}$	4.1
		yes	$5/3$	2	4.1
	middle	no	5	5	4.2
		yes	$5/3$	$1 + \frac{1}{2W(1/e)} \approx 2.79556$	4.2
2	endpoint	no	$1 + \sqrt{2}$	$\min\left(1 + \sqrt{2}, \frac{3-v}{1+v}\right)$	5
2 with radios	endpoint	no	$1 + \sqrt{2}$	$\min\left(1 + \sqrt{2}, \frac{3}{1+2v}\right)$	5.1

The layout of the paper is as follows. We survey related work in Sect. 2 and then present some preliminaries on the model and notation in Sect. 3. We begin our study in Sect. 4 by focusing on the problem with a single agent, considering scenarios with the destination at the endpoint (Sect. 4.1) and in the middle (Sect. 4.2), presenting deterministic and randomized algorithms for both scenarios. We present preliminary results for the multi-agent case by studying the problem for agents with no communication ability in Sect. 5 and then consider the case where agents can communicate (i.e. via radio) in Sect. 5.1. Finally, we conclude the paper with a summary of results and a discussion of areas for future work in Sect. 6. All proofs omitted due to space constraints can be found in the appendix.

2   Related Work

Cooperative mobile agents with communication constraints have been used to study search, exploration, rendezvous, message delivery, and other problems related to the search and rescue problem studied in this paper. Cow-path problems, first introduced in 1964 [2], are especially related to the search component of the problem we study. In its simplest form, the cow-path problem involves finding a target on a line with a single agent in as little time as possible. A simple 9-competitive algorithm has been shown to be optimal [2,21]. As a fundamental problem in search theory, many variants of the original cow-path problem have been proposed and solved for different models and using a variety of techniques. For multi-agent systems, it is sometimes framed as an evacuation problem, where the goal is to minimize the time for *every* agent to find and travel to an exit whose location is unknown [3,17]. The Group Search problem, on the other hand, requires any *one* agent to find the target [16]. These problems have been studied for many different topologies including the bounded line [4], the ring [24,28], the disk [15], simple polygons [26], for multiple paths (the original problem is a two-path system - left and right from the starting location of the agent) [25], the plane [20], in graphs [3], and in trees [19]. Competitive algorithms for multi-agent systems have been proposed [3,8,15,17,18,20,24,28], sometimes allowing some of the agents to be faulty [18,24]. A randomized algorithm has also been shown to dramatically improve the competitive ratio by a factor of almost 2 for the original problem (and to a lesser extent for the multi-path variant) [25]. Search for mobile targets has also been studied [6,13,21].

While cow-path problems relate directly with the search component of the problem studied in this paper, they do not consider the rescue component. The subsequent delivery that must occur after the object has been found fundamentally changes the problem. Recently, there has been work in data delivery by systems of mobile agents on the line [7,11], in the plane [10,12], and in graphs by energy-constrained agents [1,5]. The recently proposed Pony Express Communication Problem [11], where agents must cooperate to transmit an object from one endpoint of a line segment to the other, is most similar to the problem we study in this paper. In the offline setting, where the locations of the object and all agents are known to every agent ahead of time, our problem is equivalent to the Pony Express Communication Problem, for which an optimal offline algorithm exists. Essentially, the search and rescue problem we study here can be described as the Pony Express Communication Problem where the initial location of the object is unknown.

We are not aware of any existing work on this problem for the line, though it has been studied on the disk for the one- [23] and two-agent [22] cases. The problem considers agents which start at the center of a unit disk and the object and destination at unknown points on the perimeter of the disk. Algorithms for different communication models (wireless and face-to-face) have been presented and their worst-case delivery times proven. Both algorithms have a constant competitive ratio of $1 + \pi$, but rely on the assumption that the positive arc-distance between the exit and target is known ahead of time. In this paper, we make no assumptions about the distance of the object and also provide algorithms for both wireless and face-to-face communication models for the two-agent case.

Much of the existing work on search and rescue, exploration, and other cooperative tasks for multi-agent systems consider rather complex models and/or environments (obstacles, complex communication networks, communication dropouts, object recognition, urban or disaster environments, etc.). Techniques like queuing theory [9], machine learning [27], and heuristic-based [27] algorithms have been used to great effect. In this paper, we study the problem under a much simpler model in order to provide foundational theoretical guarantees with the hope that they can be used as a basis for future work.

3   Model and Notation

We consider agents that have a constant maximum speed and can start, stop, change directions, and pick up/hand over the object instantaneously. We assume agents can only move finite distances (they cannot move an infinitesimal distance in some direction). Agents may hand over the object to another agent only when they are collocated (face-to-face). In the offline setting, agents know the position of the object and the positions/speeds of all other agents at all times. In the online setting, however, agents do not know the position of the object or the positions/speeds of other agents. Except in Sect. 5.1, agents are assumed to have no ability to wireless communicate with each other. In all other sections, agents can only communicate with each other through face-to-face encounter. In both the face-to-face and wireless communication models, agents may share their entire state with each other instantaneously. We denote the initial position of the object by s and the (unknown) position of the object by y. For the single agent case we assume, without loss of generality, that the agent's speed is 1. For the two agent case, we use v_1 and v_2 to denote the speeds of the two agents such that $v_1 \geq v_2 > 0$. We use $v = v_2/v_1$ to denote the relative speed of the slower agent with respect to the faster agent.

4   A Single Agent

First, we study the problem for the case of a single agent. In this case, we can without any loss of generality assume the agent's speed to be 1. In other words, we simply define a unit of time to be the amount of time it takes for the agent to traverse the unit interval.

4.1   Destination at the Endpoint

In this section, we consider the interval $[0, 1]$ where the destination is at 0 and the agent's initial position is $s \geq 0$. Then, we discuss how the results extend to the unbounded interval $[0, \infty)$.

Deterministic Algorithms. We start by showing a lower bound for any deterministic online single-agent algorithm.

Theorem 4.1. *Any online algorithm for the single-agent case has competitive ratio at least $1 + \sqrt{2}$.*

Proof. Suppose the agent starts at position $\frac{1}{2}$. It is clear that any valid algorithm must eventually reach both endpoints 0 and 1 (otherwise there would exist instances of the problem, with the object at either 0 or 1, where the agent never finds the object). First, consider an algorithm which reaches 1 before 0. By adversarially placing the object at 0, the agent cannot deliver it before time $2 \left(\frac{1}{2} \right) + \frac{1}{2} = \frac{3}{2}$ while an optimal algorithm would have delivered the object in time $\frac{1}{2}$, resulting in a competitive ratio of 3.

Now let's consider algorithms that reach endpoint 0 first. Let x be the largest visited point on the interval $\left[\frac{1}{2}, 1 \right)$. In other words, the agent travels first to x and then to 0. As an adversary, we can choose to place the object either at $y \in (x, 1]$ or at 0. If we choose the former, the delivery time of any algorithm is at least $\left(2x - \frac{1}{2} \right) + 2y$ while the optimal delivery time is $\left(y - \frac{1}{2} \right) + y$. If we choose the latter, however, the delivery time of any algorithm is at least $\left(x - \frac{1}{2} \right) + x$ while the optimal delivery time is $\frac{1}{2}$. Since we have the power to choose whichever is worse for the algorithm, the competitive ratio can be written:

$$\max \left(\sup_{y > x} \left[\frac{2x - \frac{1}{2} + 2y}{2y - \frac{1}{2}} \right], \frac{2x - 1/2}{\frac{1}{2}} \right) = \max \left(\sup_{y > x} \left[1 + \frac{2x}{2y - \frac{1}{2}} \right], 4x - 1 \right) \quad (1)$$

$$= \max \left(\frac{8x - 1}{4x - 1}, 4x - 1 \right) \geq 1 + \sqrt{2} \quad (2)$$

Observe $\sup_{y > x} \left[1 + \frac{2x}{2y - \frac{1}{2}} \right] = \frac{8x - 1}{4x - 1}$ since $\frac{2x}{2y - \frac{1}{2}}$ is decreasing with respect to y. Then, the inequality above follows since $\frac{8x - 1}{4x - 1}$ is decreasing on $\left(\frac{1}{2}, 1 \right)$, $4x - 1$ is increasing on $\left(\frac{1}{2}, 1 \right)$, and they intersect at $x = \frac{1}{2} + \frac{1}{2\sqrt{2}}$. Thus, for any algorithm (which determines a value for x), the competitive ratio is at least $1 + \sqrt{2}$. $\square$

Now, we present Algorithm 1 and prove it to be optimal.

Algorithm 1. Online algorithm for agent starting at $s \in [0, 1]$

1: $x \leftarrow \min \left(1, s \left(1 + \frac{1}{\sqrt{2}} \right) \right)$
2: move along path $s \to x \to 0 \to 1$, returning to 0 with the object once it is found

---

Theorem 4.2. *Algorithm 1 has a competitive ratio of at most $1 + \sqrt{2}$.*

Proof. Essentially, the algorithm involves traveling right toward 1 until reaching a point $x = \min \left(1, s \left(1 + \frac{1}{\sqrt{2}} \right) \right)$, then traveling all the way to 0 (delivering the

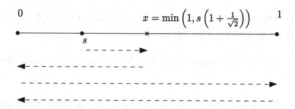

Fig. 1. The dashed line represents movement of agent executing Algorithm 1. The agent travels from its starting position at s to the point x, then to 0, then to 1 and back to 0 again. Once the agent encounters the object along this path it returns to 0 (not drawn).

object if it found it along the way). If the agent still does not have the object, it traverses the entire interval to the object (all the way to 1 if necessary) and back (Fig. 1). Using a similar method as was used in proving Theorem 4.1, there are three interesting cases:

Case 1: $s = 0$. In this case, $x = 0$ and the algorithm clearly performs optimally, since it simply moves right until finding the object and then moves back to 0.

Case 2: $s \geq 2 - \sqrt{2}$. In this case, $x = 1$ and so placing the object at $y < s$ maximizes the competitive ratio (since any $y \geq s$ would result in an optimal delivery time):

$$\frac{(x-s)+x}{s} = \frac{2-s}{s} \leq 1 + \sqrt{2}$$

Case 3: $0 < s < 2 - \sqrt{2}$. In this case, $s < x < 1$ and the maximum competitive ratio is achieved either by placing the object at some position $y < s$ or at some position $y > x$ (since any $s \leq y \leq x$ would result in an optimal delivery time):

$$\max \left(\sup_{y>x} \left[\frac{(2x-s)+2y}{(y-s)+y} \right], \frac{(x-s)+x}{s} \right) = \max \left(\frac{4x-s}{2x-s}, \frac{2x-s}{s} \right)$$

$$= \max \left(1 + \sqrt{2}, 1 + \sqrt{2} \right) \quad (3)$$

$$= 1 + \sqrt{2}$$

where Eq. (3) follows since $x = s \left(1 + \frac{1}{\sqrt{2}} \right)$ in this case. Thus, Algorithm 1 has a competitive ratio of at most $1 + \sqrt{2}$. □

This result is particularly interesting when compared to the search and delivery problems separately. First, observe that for the search problem with no lower bound on the distance of the target to the agent's starting location, there is no competitive online algorithm! The agent must move some distance either left or right to begin with - whatever that distance is, we can place the object an arbitrarily small fraction of the distance in the other direction, making the competitive ratio of the algorithm arbitrarily large. The delivery problem on the line segment, on the other hand, is trivial with one agent - just go to the object and then the destination. A competitive online algorithm for the search and rescue problem, however, *does* exist and is *not* trivial!

It's important to understand that Algorithm 1 does *not* minimize the worst-case delivery time of the object. In fact, a simple algorithm of just going to 1 and then back to 0 terminates in at most time 2 (when the object is at 1) while Algorithm 1 can take up to $2 + \sqrt{2}$ time (when $s = 2 - \sqrt{2} - \epsilon$ for some arbitrarily small $\epsilon > 0$ and the object is at 1). Rather, Algorithm 1 minimizes the delivery time compared to the optimal delivery time if the location of the object were known. The aforementioned simple algorithm, on the other hand, might take time 2 to deliver an object that could have been delivered almost instantaneously! Algorithm 1 guarantees this never happens—an object that can be delivered in time t optimally will be delivered in at most $(1 + \sqrt{2})t$ time. So Algorithm 1 (and any other which minimizes competitive ratio in general) might be described as an algorithm that minimizes the regret that an agent has after discovering the location of the object.

There is another even more important scenario where an algorithm's competitive ratio is more useful than its worst-case runtime: when the worst-case runtime is unbounded. Consider the situation where an agent no longer knows the length of the path ahead of it—only its initial distance to 0. In the extreme case, the object could be anywhere on the interval $[0, \infty)$. In this case, there is no simple exhaustive search algorithm that terminates in bounded time because the path could go on forever (or in a more realistic scenario, for a really, really long time)! Even for this extreme case, Algorithm 2 (a slight modification of Algorithm 1 which simply removes the upper bound on the agent's search to the right of s), will still deliver the object to its destination in $1 + \sqrt{2}$ times the optimal time.

Algorithm 2. Online algorithm for agent starting at $s \in [0, \infty)$

1: $x \leftarrow s \left(1 + \frac{1}{\sqrt{2}}\right)$

2: move along path $s \to x \to 0 \to \infty$, returning to 0 with the object once it is found

---

Corollary 1. *Algorithm 2 has a competitive ratio of* $1 + \sqrt{2}$.

Proof. Follows directly from the proof for Algorithm 1. ☐

Using Randomization. The analysis for Algorithm 1 involved reasoning about how an adversary might place the object at worst-case positions along the line-segment. By using randomization, we can mitigate the damage an adversary can do by not committing to a predictable, deterministic algorithm. First, we prove a lower bound on how well a randomized algorithm can do:

Theorem 4.3. *Every randomized online algorithm for the single-agent case has an expected competitive ratio of at least 5/3.*

Proof. Consider the scenario where the agent is at some position $s < \frac{1}{2}$ and the object is placed on the interval $(s, 2s]$ uniformly at random with probability 2/3 and at 0 with probability 1/3. Observe in the former case, the expected position of the object is $(2s + s)/2 = 3s/2$. Since the object cannot be in the interval $(0, s]$, any optimal online algorithm must involve the agent *either* moving along the path $s \to 0 \to 1$ *or* along the path $s \to x \to 0 \to 1$ (returning to 0 as soon as the object is found, of course) for some $x \in (s, 1]$. If the agent moves to 0 first, then the expected competitive ratio is

$$\frac{1}{3} \cdot 1 + \frac{2}{3} \cdot \left(\frac{2(3s)/2 + s}{2(3s)/2 - s} \right) = \frac{1}{3} + \frac{2}{3} \left(\frac{3s + s}{3s - s} \right) = 5/3$$

If instead, the agent moves to some position $x \in (s, 2s]$, then the probability that the agent finds the object (in optimal time) on $(s, x]$ is $\frac{2}{3} \cdot \frac{x-s}{s}$. On the other hand, the probability that the object is in the interval $(x, 2s]$ is $\frac{2}{3} \cdot \frac{2s-x}{s}$. Given this situation, observe that the expected position of the object in this case is $(2s + x)/2$. Thus the competitive ratio can be written:

$$\frac{1}{3} \cdot \frac{2x - s}{s} + \frac{2}{3} \left(\frac{x - s}{s} \cdot 1 + \frac{2s - x}{s} \cdot \frac{2x - s + 2(x + 2s)/2}{2(x + 2s)/2 - s} \right) = \frac{s^2 + 11sx - 2x^2}{3s^2 + 3sx}$$

which has a maximum value of 5/3 (at both $x = s$ and $x = 2s$). Thus, any deterministic algorithm for this distribution of inputs has an expected competitive ratio of at least 5/3. Finally, by Yao's Minimax Principle [29], every randomized algorithm must have an expected competitive ratio of at least 5/3. $\square$

Now, we present a simple randomized algorithm: with probability $\frac{1}{2}$, the agent will simply execute an algorithm very similar to Algorithm 1, otherwise the agent will move directly to 0 and then, if it still hasn't found the object, move towards 1 until it does and then return to 0.

Algorithm 3. Online randomized algorithm for agent starting at $s \in [0, 1]$

1: Let p be a random bit
2: **if** $p = 0$ **then**
3:     move along path $s \to \min(2s, 1) \to 0 \to 1$, returning to 0 with the object once it is found
4: **else**
5:     move along path $s \to 0 \to 1$, returning to 0 with the object once it is found

---

Theorem 4.4. *Algorithm 3 has an expected competitive ratio of 2.*

Proof. First, we consider the case where $s < 1/2$. Observe the object is either in the interval $(0, s)$, $(s, 2s]$, or $(2s, 1]$. The goal of the adversary now is to place the object in the interval which maximizes the expected competitive ratio. For example, if the object is in the first interval, then the algorithm is optimal with probability $1/2$ (the agent goes towards 0). Otherwise, it has a competitive ratio of $\frac{2(2s)-s}{s} = 3$. The adversary is not required to commit to a deterministic strategy, however. Consider a mixed strategy where the adversary places the object in $(0, s)$ with a probability of q, in $(s, 2s]$ with a probability of r, and in $(2s, 1]$ with a probability of $1 - q - r$. Let CR denote the competitive ratio of Algorithm 3. Then the expected competitive ratio can be written:

$$\mathbb{E}[CR] = \frac{1}{2}\left(q \cdot 1 + r \cdot \frac{2y_2 + s}{2y_2 - s} + (1 - q - r)\frac{2y_3 + s}{2y_3 - s}\right) +$$
$$\frac{1}{2}\left(q \cdot \frac{2(2s - s) + s}{s} + r \cdot 1 + (1 - q - r)\frac{2y_3 + 3s}{2y_3 - s}\right)$$

$$\mathbb{E}[CR] \leq \frac{1}{2}\left(q \cdot 1 + r \cdot 3 + (1 - q - r)\frac{5}{3}\right) + \frac{1}{2}\left(q \cdot 3 + r \cdot 1 + (1 - q - r)\frac{7}{3}\right) = 2$$

where y_2 and y_3 are the expected positions of the object in cases 2 and 3, respectively. The above inequality follows from worst-case values (those which maximize the competitive ratio) y_2 approaches s (from above) and y_3 approaches $2s$ (from above).

Now we must consider the case where $s \geq 1/2$. In this case, there are only two intervals in which the adversary may place the object. Let q' be the probability the object is in $[0, s)$ (case 1) and $1 - q'$ the probability it is in $(s, 1]$ (case 2). Then the competitive ratio can be written:

$$\mathbb{E}[CR] = \frac{1}{2}\left(q' \cdot 1 + (1 - q') \cdot \frac{2y_2' + s}{2y_2' - s}\right) + \frac{1}{2}\left(q' \cdot \frac{2 - s}{s} + (1 - q') \cdot 1\right)$$
$$\mathbb{E}[CR] \leq \frac{1}{2}(q' \cdot 1 + (1 - q') \cdot 3) + \frac{1}{2}(q' \cdot 3 + (1 - q') \cdot 1) = 2$$

where y_2' is the expected position of object in case 2. The above inequality follows from the worst-case value (that which maximizes the competitive ratio) where y_2' approaches s (from above). □

The expected competitive ratio of Algorithm 3 is significantly lower than the competitive ratio of the deterministic Algorithm 1, especially with respect to the lower bound on the competitive ratio of any randomized algorithm.

Remark 1. A slight modification of Algorithm 3 for the unbounded interval $[0, \infty)$ (simply replace 1 in the paths with ∞) has the same expected competitive ratio since the second scenario discussed in the proof for Theorem 4.4 is essentially eliminated and the analysis for the first scenario is equivalent.

4.2 Destination in the Middle

Up until this point, we've only considered scenarios where the destination is at an endpoint. In this section, we consider situations where the object may be on *either* side of the destination.

Deterministic Algorithms. We start by showing the necessity of an additional assumption for this variant of the problem: the agent's initial position cannot be at 0 (the destination).

Lemma 4.1. *For any initial configuration where the agent starts at 0 and the object is on one of two paths emanating from 0, there is no competitive algorithm.*

Proof. Any algorithm must involve the agent doing one of two things at time $t = 0$:

1. wait for some time t'
2. move some distance $d' > 0$ down one of the paths

In the first case, an adversary may simply place the object along any of the paths an arbitrarily small distance $d/2$ away from 0 (so the optimal delivery time is d). The competitive ratio in this case is *at least* t'/d, which is arbitrarily large as $d \to 0$. In the second case, an adversary may place the object an arbitrarily small distance $d/2$ away from 0 along any path *except* the one the agent traveled down. Again, the competitive ratio is at least d'/d, which is arbitrarily large as $d \to 0$ (since $d' > 0$).   □

Thus, we assume, without loss of generality that the agent starts at some position $s > 0$ (all proofs follow for $s < 0$ via a symmetrical argument). We present an online "zig-zag" algorithm (Algorithm 4) which involves the agent searching a distance $2s$ to the left (crossing 0) and returning to s, then a distance $4s$ to the right and returning to s, then a distance $8s$ to left and returning to s, and so on, doubling its search distance in each round (Fig. 2). In other words, the agent follows the trajectory $s \to -s \to 5s \to -7s \to 15s \to \dots$.

Algorithm 4. Online Algorithm for agent starting at $s \neq 0$

1: $i \leftarrow 1$
2: $x \leftarrow -s$
3: **while** object not found **do**
4:    move toward x
5:    **if** arrived at $x \neq s$ or an endpoint **then**
6:       $x \leftarrow s$
7:    **else if** arrived at s **then**
8:       $i \leftarrow i + 1$
9:       $x \leftarrow s + (-1)^i \cdot 2^i s$
10: Return to 0 with object

Now, we will show an upper bound of 5 on its competitive ratio, then prove it is optimal.

Theorem 4.5. *Algorithm 4 has a competitive ratio of 5.*

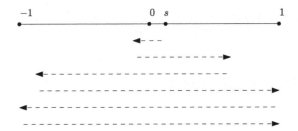

Fig. 2. The dashed line represents movement of agent executing Algorithm 4. The agent travels from its starting position at s to the point $-s$, then to $5s$, and so on. Upon finding the object, the agent returns to 0 (not drawn).

Proof. In Algorithm 4, the agent starts at s and moves left and right in alternating rounds, doubling the distance it travels each round. In round $i = 1, 2, \ldots$, the agent moves a distance $2^i s$ (left in odd rounds and right in even rounds) out and back to s (for a total of 2^{i+1}). In the case an endpoint is reached, the agent would turn around rather than finish travelling the full distance of the round. However, to simplify the analysis it is easier to consider the alternate algorithm A' such that the agent does not turn around early and travels a distance 2^i (out and back) no matter what, traveling beyond the endpoint if necessary. It is clear that our original algorithm cannot perform worse than A' and in cases where an endpoint is never reached before finding the object, the two algorithms are identical. Thus, any upper bound on the competitive ratio of A' is also an upper bound on Algorithm 4. For the following analysis, assume the algorithm we are referring to is A'.

Without loss of generality, suppose $s > 0$ (a symmetric argument follows when $s < 0$). Let y be the position of the object. Then there are three interesting cases: when $-s \leq y \leq s$, when $y > s$, and when $y < -s$. The first case is the simplest to analyze. Since $-s \leq y < s$, the object is found in round 1 and is clearly optimal. For the second case, if the object is found in round 2, then $s < y < 5s$ and the competitive ratio is

$$\frac{3s + 2y}{2y - s} \leq 5 \tag{4}$$

since the left-hand side of Inequality (4) is decreasing with respect to y and $y > s$. Otherwise, $y > 5s$ and so the object must be found in some even round $k > 2$. Also, observe $2^{k-2}s < y - s$ (otherwise the object would have been found in an earlier round), implying $2^k < \frac{4(y-s)}{s}$. The delivery time of A', then, is

$$T_{A'} = \sum_{i=1}^{k-1} 2^{i+1}s + 2y - s = 2s\left(2^k - 2\right) + 2y - s$$

$$T_{A'} < 2s\left(\frac{4(y-s)}{s} - 2\right) + 2y - s = 10y - 13s,$$

while the optimal delivery time is $2y - s$, thus the competitive ratio is at most $(10y - 13s)/(2y - s) \leq 5$. Finally, when $y < -s$, the object must be found in some odd round $k > 2$. Then, we have $2^{k-2}s < s + |y|$ (otherwise the object would have been found in an earlier round just as in the first case), implying $2^k < \frac{4(s+|y|)}{s}$. The delivery time of A' in this case is

$$\sum_{i=1}^{k-1} 2^i s + 2|y| + s = 2s(2^k - 2) + 2|y| + s$$

$$< 2s \left(\frac{4(s + |y|)}{s} - 2 \right) + 2|y| + s = 10|y| + 5s,$$

while the optimal delivery time is $2|y| + s$. Thus the competitive ratio is at most $(10|y| + 5s)/(2|y| + s) \leq 5$. □

Remark 2. Note that algorithm A' is exactly the algorithm for the unbounded case, so the competitive ratio of at most 5 applies to both the bounded and unbounded cases.

Theorem 4.6. *Every online algorithm for the single-agent, line model must have a competitive ratio of at least 5.*

Using Randomization. Again, the analysis of Algorithm 4 involved reasoning about worst-case positions of the object. The following algorithm and subsequent upper bound are very similar to the well-known optimal randomized algorithm for the cow-path search problem [25]. The algorithm is essentially a standard zig-zag algorithm (like Algorithm 4) except that the starting search direction and initial search distance are randomized. In the following Algorithm 5, the random bit p determines the initial search direction and the random number $\epsilon \in (0, 1)$ determines the initial search distance (which is $r^\epsilon s$ for some constant $r > 1$).

Algorithm 5. Online randomized algorithm for agent starting at $s \neq 0$ with expansion rate $r > 1$

1: let p be a random bit
2: sample ϵ uniformly at random from the interval $(0, 1)$
3: $i \leftarrow 1$
4: $x \leftarrow s + (-1)^{i-p} \cdot r^{i+\epsilon} s$
5: **while** object not found **do**
6:     move toward x
7:     **if** arrived at $x \neq s$ or an endpoint **then**
8:         $x \leftarrow s$
9:     **else if** arrived at s **then**
10:         $i \leftarrow i + 1$
11:         $x \leftarrow s + (-1)^{i-p} \cdot r^{i+\epsilon} s$
12: Return to 0 with object

Theorem 4.7. *The expansion rate* $r = \frac{1}{W(1/e)} \approx 3.59112$ *yields an expected competitive ratio of* $1 + \frac{1}{2W(1/e)} \approx 2.79556$ *for Algorithm 5 where* $W(x)$ *is the product logarithm (Lambert W function [14]) of* x.

Remark 3. Note that, again, algorithm A' is exactly the modified version of A that would be used in the unbounded case, so the competitive ratio of at most $1 + \frac{1}{2W(1/e)}$ applies to both the bounded and unbounded cases.

5   Two Agents

In this section, we present results for multi-agent search and rescue, considering the case of two agents initially located at the same point s on the line segment but with different speeds v_1 and v_2. Without loss of generality, we assume $v_1 \geq v_2$. We refer to the agent with speed v_1 as the "first" or "fast" agent and the agent with speed v_2 as the "second" or "slow" agent. The goal is the same, except agents may hand over the object to each other via face-to-face encounter. We denote $v = v_2/v_1$ as the speed of the slower agent relative to the faster agent (observe $0 \leq v \leq 1$). The delivery time of the optimal algorithm, then, is clearly $(2y - s)/v_1$, where the fast agent delivers the object entirely by itself.

Remark 4. By Theorem 4.1, whenever $v_2 = 0$, the lower bound of $1 + \sqrt{2}$ applies to the two-agent case directly.

Now we present Algorithm 6, an online algorithm which involves the slow agent moving toward 1 and the fast agent moving toward 0 only if doing so is better than the fast agent simply executing Algorithm 1 by itself.

Algorithm 6. Online two-agent algorithm for agents starting at $s \in [0, 1]$

1: **if** other agent is faster **then**
2:     $x \leftarrow 1$
3: **else if** $\frac{2-\sqrt{2}}{2+\sqrt{2}} < v < 1$ **then**
4:     $x \leftarrow 0$
5: **else**
6:     $x \leftarrow s(1 + 1/\sqrt{2})$
7: move along path $s \rightarrow x \rightarrow 0 \rightarrow 1$, returning to 0 with the object once it is found and handing it over to any faster agent encountered

---

Theorem 5.1. *For any system with two agents starting at position* $s \in [0, 1]$, *Algorithm 6 has a competitive ratio of* $\min\left(1 + \sqrt{2}, \frac{3-v}{1+v}\right)$ *where* $v = v_2/v_1$.

Proof. First, observe that since both agents start at s, an optimal offline algorithm involves only the first agent moving directly toward the object and then to 0 for delivery. Thus, in the case where $v \leq \frac{2-\sqrt{2}}{2+\sqrt{2}}$ or $v = 1$, the first agent exactly performs Algorithm 1, and so a competitive ratio of $1 + \sqrt{2}$ is achieved. The interesting case, then is when $\frac{2-\sqrt{2}}{2+\sqrt{2}} < v < 1$. Let y be the position of the object along the line segment. Clearly if $y \leq s$, then the algorithm is optimal. If $y > s$, then there are two possible scenarios. If $s + y \leq (y - s)/v$, then the fast agent reaches the object at the same time or before the slow agent, so the competitive ratio is

$$\frac{2y + s}{2y - s} \leq \frac{2y + \frac{1-v}{1+v}y}{2y - \frac{1-v}{1+v}y} = \frac{3 + v}{3v + 1}.$$

If $s + y > (y - s)/v$, then the slow agent reaches the object first, picks it up, and moves toward 0 until encountering the fast agent for a handover. Then the fast agent delivers the object. In this case, the delivery time can be written as the sum of the time it took to meet $(t = \frac{2y}{1+v})$ and the time for the fast agent to carry the object the remaining distance $(y - (tv - (y - s)) = 2y - tv - s)$ for a total time of $\frac{4y}{1+v} - s$. Thus, the competitive ratio is

$$\frac{\frac{4y}{1+v} - s}{2y - s} \leq \frac{3 - v}{1 + v}$$

since the function is decreasing with respect to y in the $y > s$ region and thus obtains its maximum value at $y = s$. Finally, observe the second case dominates the competitive ratio:

$$\frac{3 - v}{1 + v} \geq \frac{3 + v}{1 + 3v} \Rightarrow 3 + 8v - 3v^2 \geq 3 + 4v + v^2 \Rightarrow v(1 - v) \geq 0$$

Of course this condition is always satisfied since $0 \leq v \leq 1$.   □

Observe that since agents start at the same position, Algorithm 1 still has a competitive ratio of $1 + \sqrt{2}$. Algorithm 6 is only better when $v_2 > \frac{2-\sqrt{2}}{2+\sqrt{2}}v_1 \approx 0.1716v_1$. In other words, as long as one agent is not *too* much faster than the other, Algorithm 6 is more competitive.

5.1   Agents with Radios

In Algorithm 6, the fast agent moves toward 0 and only turns around to help the slower agent if it reaches 0 without finding the object. If the slow agent finds the object before the fast agent reaches 0 and the agents can communicate, though, then clearly the fast agent should turn around immediately to acquire the object as quickly as possible.

Theorem 5.2. *Algorithm 7 has a competitive ratio of* $\min\left(1 + \sqrt{2}, \frac{3}{1+2v}\right)$.

Observe that Algorithm 7 has a better competitive ratio than Algorithm 1 whenever the slow agent has speed greater than $\frac{2-\sqrt{2}}{2+2\sqrt{2}} \approx 0.1213$. Recall for the case where agents cannot communicate, the slow agent only helps when its speed is greater than $\frac{2+\sqrt{2}}{2-\sqrt{2}} \approx 0.1716$.

Algorithm 7. Online two-agent algorithm for agent starting at $s \in [0,1]$

1: **if** other agent is faster **then**
2:     $x \leftarrow 1$
3: **else if** $\frac{2-\sqrt{2}}{2+2\sqrt{2}} < v < 1$ **then**
4:     $x \leftarrow 0$
5: **else**
6:     $x \leftarrow s(1 + s/\sqrt{2})$
7: move along path $s \rightarrow x \rightarrow 0 \rightarrow 1$, returning to 0 with the object once it is found and handing it over to any fast agent encountered

6 Conclusion

In this paper, we propose a problem inspired by search and rescue, a task that cooperative robotic systems have long showed promise in assisting with. We provide both deterministic and randomized algorithms for the single-agent case and provide lower and upper bounds on their competitive ratios. We showed the search and rescue problem is fundamentally different than the search and delivery problems considered separately, which are trivial. For the case where the destination is in the middle, however, the optimal (deterministic and randomized) algorithms are essentially the same as the well-known optimal algorithms for search problem, though the resulting competitive ratios are different. For the two-agent case, we essentially provide one algorithm and demonstrate how extra communication ability between the two agents affects its competitive ratio. While the deterministic single-agent algorithms are optimal, it's not clear if the randomized and multi-agent algorithms can be improved. This is an interesting area for future work on this problem. Other areas that deserve more attention are the study of search and rescue for other topologies (i.e. the ring, the plane, and trees/graphs) and for more general multi-agent scenarios (with different starting locations, communication abilities, etc.).

Appendix

A   Proofs From Sect. 4 (A Single Agent)

Theorem 4.6. *Every online algorithm for the single-agent, line model must have a competitive ratio of at least* 5.

Proof. Consider a scenario where the agent is placed at some arbitrarily small distance $s > 0$ away from 0 and the object is at least a distance $2s$ from s. Any algorithm must involve a sequence of positive distances $x_1, x_2, x_3, \ldots$ such that the agent moves left (or right) a distance x_1 and back to s, then right (or left) a distance x_2 and back to s, and so on. Clearly any optimal online algorithm of this form must satisfy $x_{i+2} > x_i$ and any algorithm with a competitive ratio of 5 or better must satisfy $x_i \leq 3x_{i-1}$ where $x_1 \geq 2s$ (since the object at least a distance $2s$ from s). Thus, $x_i < 2s \cdot 3^{i-1}$ for all $i \geq 2$ and so the number of required turning points for an optimal online algorithm can be made arbitrarily large (by setting s to be arbitrarily small).

For some sequence of turning points, suppose the object is found on the kth round (k can be arbitrarily large by the argument above) and observe the competitive ratio can then be written

$$\sup_{k,y,s} \frac{\sum_{i=1}^{k-1} 2x_i + 2y \pm s}{2y \pm s} = \sup_{k} \frac{\sum_{i=1}^{k-1} x_i + x_{k-2}}{x_{k-2}}$$

$$= 1 + \sup_{k} \frac{\sum_{i=1}^{k-1} x_i}{x_{k-2}} \geq 1 + \sup_{k} \frac{\sum_{i=1}^{k-1} r^i}{r^{k-2}} = 1 + \sup_{k} \frac{r^k - r}{(r-1)r^{k-2}}$$

where $r > 1$ (the expansion factor). The above inequality follows from Corollary 7.11 of Sect. 7.2 of [21] (following the method for proving the 9-competitive search on an infinite line in Sect. 8.2.1 of [21]). Then, since $\frac{r^k - r}{(r-1)r^{k-2}}$ is increasing with respect to k (its derivative $\frac{r^{3-k} \ln r}{r-1}$ is greater than 0 for any $r > 1$), we can simplify the competitive ratio to

$$1 + \sup_{k} \frac{r^k - r}{(r-1)r^{k-2}} = 1 + \lim_{k \to \infty} \frac{r^k - r}{(r-1)r^{k-2}} = 1 + \frac{r^2}{r-1}$$

which has a minimum value of 5 for $r = 2$.   □

Theorem 4.7. *The expansion rate* $r = \frac{1}{W(1/e)} \approx 3.59112$ *yields an expected competitive ratio of* $1 + \frac{1}{2W(1/e)} \approx 2.79556$ *for Algorithm 5 where* $W(x)$ *is the product logarithm (Lambert W function [14]) of* x.

Proof. Just as in the proof for Theorem 4.5, we consider the alternate algorithm A' such that the agent does not turn around early upon reaching an endpoint. It is clear that our original algorithm cannot perform worse than A' and in cases where an endpoint is never reached, the two algorithms are identical.

Let $d = s \cdot r^{k+\delta}$ denote the position of the object where $0 \leq \delta < 1$. By executing Algorithm 5, the agent moves a distance $r^{i+\epsilon}$ to the left and right in alternating rounds $i = 1, 2, \ldots$. Consider the round when the agent moves a distance $s \cdot r^k$ for the first time. If the agent moves r^k distance for the first time in the opposite direction of the object, then it will definitely find the object in round $k + 1$. The expected competitive ratio in this case can be written:

$$\mathbb{E}\left[\frac{\sum_{i=1}^{k} s \cdot 2r^{i+\epsilon} + 2d \pm s}{2d \pm s} \right] = \mathbb{E}\left[\frac{\sum_{i=1}^{k} 2r^{i+\epsilon} + 2r^{k+\delta} \pm 1}{2r^{k+\delta} \pm 1} \right]$$

$$= \mathbb{E}\left[1 + \frac{2r^\epsilon \left(r^{k+1} - r \right)}{\left(2r^{k+\delta} \pm 1 \right) \left(r - 1 \right)} \right]$$

$$= 1 + \frac{2 \left(r^{k+1} - r \right)}{\left(2r^{k+\delta} \pm 1 \right) \left(r - 1 \right)} \cdot \mathbb{E}\left[r^\epsilon \right]$$

$$= 1 + \frac{2 \left(r^{k+1} - r \right)}{\left(2r^{k+\delta} \pm 1 \right) \ln r}$$

since $\mathbb{E}[r^\epsilon] = \int_1^r x \frac{1}{x \ln r} dx = \frac{r-1}{\ln r}$.

On the other hand, if the agent moves a distance $s \cdot r^k$ for the first time in the direction of the object, it will find it on round k if $\epsilon \geq \delta$ and on round $k+2$ otherwise. Let B be the event that $\epsilon \geq \delta$, then the expected competitive ratio can be written:

$$\mathbb{E}\left[Pr[B]\left[\frac{\sum_{i=1}^{k-1} s \cdot 2r^{i+\epsilon} + 2d \pm s}{2d \pm s}\right] + (1 - Pr[B])\left[\frac{\sum_{i=1}^{k+1} s \cdot 2r^{i+\epsilon} + 2d \pm s}{2d \pm s}\right]\right]$$

$$= \mathbb{E}\left[Pr[B]\left[1 + \frac{2r^\epsilon(r^k - r)}{(2r^{k+\delta} \pm 1)(r - 1)}\right] + (1 - Pr[B])\left[1 + \frac{2r^\epsilon(r^{k+2} - r)}{(2r^{k+\delta} \pm 1)(r - 1)}\right]\right]$$

$$= Pr[B]\left[1 + \frac{2(r^k - r)}{(2r^{k+\delta} \pm 1)(r - 1)} \cdot \mathbb{E}[r^\epsilon|B]\right]$$

$$+ (1 - Pr[B])\left[1 + \frac{2(r^{k+2} - r)}{(2r^{k+\delta} \pm 1)(r - 1)} \cdot \mathbb{E}[r^\epsilon|\overline{B}]\right]$$

$$= Pr[B]\left[1 + \frac{2(r^k - r)(r - r^\delta)}{(2r^{k+\delta} \pm 1)(r - 1)\ln r Pr[B]}\right]$$

$$+ (1 - Pr[B])\left[1 + \frac{2(r^{k+2} - r)(r^\delta - 1)}{(2r^{k+\delta} \pm 1)(r - 1)\ln r Pr[\overline{B}]}\right]$$

since $\mathbb{E}[r^\epsilon|B] = \int_{r^\delta}^r x \frac{1}{Pr[B] \cdot x \ln r} dx = \frac{r - r^\delta}{\ln r Pr[B]}$ and $\mathbb{E}[r^\epsilon|\overline{B}] = \int_1^{r^\delta} x \frac{1}{Pr[\overline{B}] \cdot x \ln r} dx = \frac{r^\delta - 1}{\ln r Pr[\overline{B}]}$. Then the expression can be further simplified:

$$= 1 + \frac{2(r^k - r)(r - r^\delta)}{(2r^{k+\delta} \pm 1)(r - 1)\ln r} + \frac{2(r^{k+2} - r)(r^\delta - 1)}{(2r^{k+\delta} \pm 1)(r - 1)\ln r}$$

$$= 1 + \frac{2}{(2r^{k+\delta} \pm 1)(r - 1)\ln r}\left((r^k - r)(r - r^\delta) + (r^{k+2} - r)(r^\delta - 1)\right)$$

$$= 1 + \frac{2}{(2r^{k+\delta} \pm 1)(r - 1)\ln r}(r - 1)(r^{\delta+k} + r^{d+k+1} - r^{k+1} - r)$$

$$= 1 + \frac{2(r^{\delta+k} + r^{d+k+1} - r^{k+1} - r)}{(2r^{k+\delta} \pm 1)\ln r}$$

Observe that, since the initial search direction is chosen uniformly randomly, the total expected competitive ratio is

$$\frac{1}{2}\left[1 + \frac{2(r^{k+1} - r)}{(2r^{k+\delta} \pm 1)\ln r}\right] + \frac{1}{2}\left[1 + \frac{2(r^{\delta+k} + r^{d+k+1} - r^{k+1} - r)}{(2r^{k+\delta} \pm 1)\ln r}\right]$$

$$= 1 + \frac{(r^{k+1} - r)}{(2r^{k+\delta} \pm 1)\ln r} + \frac{(r^{\delta+k} + r^{d+k+1} - r^{k+1} - r)}{(2r^{k+\delta} \pm 1)\ln r}$$

$$= 1 + \frac{r^{k+\delta}(1 + r) - 2r}{(2r^{k+\delta} \pm 1)\ln r} \leq 1 + \frac{r^{k+\delta}(1 + r) - (1 + r)}{(2r^{k+\delta} - 1)\ln r - \ln r} = 1 + \frac{(1 + r)(r^{k+\delta} - 1)}{\ln r (2r^{k+\delta} - 2)}$$

$$= 1 + \frac{1 + r}{2 \ln r}.$$

Finally, with an expansion rate of $r = \frac{1}{W(1/e)} \approx 3.59112$, the above upper bound becomes $1 + \frac{1}{2W(1/e)} \approx 2.79556$. □

B Proofs From Sect. 5 (Two Agents)

Theorem 5.2. *Algorithm 7 has a competitive ratio of* $\min\left(1 + \sqrt{2}, \frac{3}{1+2v}\right)$.

Proof. The proof is very similar to that of Theorem 5.1. Without loss of generality, suppose the first agent has a speed of 1 and the second agent a speed of $v \leq 1$. For the case where $v \leq \frac{2-\sqrt{2}}{2+2\sqrt{2}}$ or $v = 1$, the first agent exactly performs Algorithm 1, and so a competitive ratio of $1 + \sqrt{2}$ is achieved. The interesting case, then is when $\frac{2-\sqrt{2}}{2+2\sqrt{2}} < v < 1$. First, if the fast agent still arrives at the object first, the competitive ratio is $\frac{3+v}{1+3v}$ (using the same analysis as used in the proof for Theorem 5.1). Otherwise, if the slow agent arrives at the object before the fast agent reaches 0 (i.e. $s > \frac{y-s}{v} \Rightarrow y < s(v + 1)$), then the competitive ratio is

$$\frac{t + (y - (tv - (y - s)))}{2y - s} = \frac{s(2 - v) - 2y}{v(s - 2y)} \leq \frac{3}{1 + 2v}$$

where $t = \frac{2\left(y - \left(s - \frac{y-s}{v}\right)\right)}{1+v}$ is the time the agents meet for a handover. The first equality follows from substituting this value for t and the final inequality follows since $\frac{s(2-v)-2y}{v(s-2y)}$ is increasing with respect to y (its derivative with respect to y, $\frac{2s(1-v)}{v(s-2y)^2}$, is positive for all $v < 1$) and $y < s(v+1)$. On the other hand, if the slow agent finds the object after the fast agent reaches 0 but still before it catches up ($s < \frac{y-s}{v} < y + s$), then the competitive ratio is

$$\frac{t + (y - (tv - (y - s)))}{2y - s} = \frac{s(1 + v) - 4y}{(1 + v)(s - 2y)} = \frac{s(1 + v) - 4y}{s(1 + v) - 2y(1 + v)} \leq \frac{3}{1 + 2v}$$

where $t = \frac{2y}{1+v}$ is the time the agents meet for a handover. The first equality follows from substituting this value for t and the final inequality follows since $\frac{s(1+v)-4y}{s(1+v)-2y(1+v)}$ is decreasing with respect to y (its derivative with respect to y, $\frac{2s(v-1)}{(1+v)(s-2y)^2}$, is negative for all $v < 1$) and $y > s(1 + v)$. Finally, observe the second and third cases dominate the competitive ratio:

$$\frac{3}{1 + 2v} \geq \frac{3 + v}{1 + 3v} \Rightarrow 3 + 9v \geq 3 + 7v + 2v^2 \Rightarrow v(1 - v) \geq 0$$

Clearly this condition is always satisfied since $0 \leq v \leq 1$. □

References

1. Bärtschi, A., Tschager, T.: Energy-efficient fast delivery by mobile agents. In: Klasing, R., Zeitoun, M. (eds.) FCT 2017. LNCS, vol. 10472, pp. 82–95. Springer, Heidelberg (2017). https://doi.org/10.1007/978-3-662-55751-8_8

2. Beck, A.: On the linear search problem. Israel J. Math. **2**(4), 221–228 (1964)
3. Borowiecki, P., Das, S., Dereniowski, D., Kuszner, Ł: Distributed evacuation in graphs with multiple exits. In: Suomela, J. (ed.) SIROCCO 2016. LNCS, vol. 9988, pp. 228–241. Springer, Cham (2016). https://doi.org/10.1007/978-3-319-48314-6_15
4. Bose, P., De Carufel, J.-L., Durocher, S.: Revisiting the problem of searching on a line. In: Bodlaender, H.L., Italiano, G.F. (eds.) ESA 2013. LNCS, vol. 8125, pp. 205–216. Springer, Heidelberg (2013). https://doi.org/10.1007/978-3-642-40450-4_18
5. Carvalho, I.A., Erlebach, T., Papadopoulos, K.: On the fast delivery problem with one or two packages. J. Comput. Syst. Sci. **115**, 246–263 (2021). https://doi.org/10.1016/j.jcss.2020.09.002
6. Cenek, E.: Chases and escapes by Paul J. Nahin. ACM SIGACT News **40**(3), 48–50 (2009). https://doi.org/10.1145/1620491.1620500
7. Chalopin, J., Jacob, R., Mihalák, M., Widmayer, P.: Data delivery by energy-constrained mobile agents on a line. In: Esparza, J., Fraigniaud, P., Husfeldt, T., Koutsoupias, E. (eds.) ICALP 2014. LNCS, vol. 8573, pp. 423–434. Springer, Heidelberg (2014). https://doi.org/10.1007/978-3-662-43951-7_36
8. Chrobak, M., Gasieniec, L., Gorry, T., Martin, R.: Group search on the line. In: Italiano, G.F., Margaria-Steffen, T., Pokorný, J., Quisquater, J.-J., Wattenhofer, R. (eds.) SOFSEM 2015. LNCS, vol. 8939, pp. 164–176. Springer, Heidelberg (2015). https://doi.org/10.1007/978-3-662-46078-8_14
9. Clark, L., Galante, J., Krishnamachari, B., Psounis, K.: A queue-stabilizing framework for networked multi-robot exploration. IEEE Robot. Autom. Lett. **6**(2), 2091–2098 (2021). https://doi.org/10.1109/LRA.2021.3061304
10. Coleman, J., Kranakis, E., Krizanc, D., Ponce, O.M.: Message delivery in the plane by robots with different speeds. In: Johnen, C., Schiller, E.M., Schmid, S. (eds.) SSS 2021. LNCS, vol. 13046, pp. 305–319. Springer, Cham (2021). https://doi.org/10.1007/978-3-030-91081-5_20
11. Coleman, J., Kranakis, E., Krizanc, D., Morales-Ponce, O.: The pony express communication problem. In: Flocchini, P., Moura, L. (eds.) IWOCA 2021. LNCS, vol. 12757, pp. 208–222. Springer, Cham (2021). https://doi.org/10.1007/978-3-030-79987-8_15
12. Coleman, J., Kranakis, E., Krizanc, D., Morales-Ponce, O.: Delivery to safety with two cooperating robots. CoRR **abs/2210.04080** (2022). https://doi.org/10.48550/arXiv.2210.04080
13. Coleman, J., Kranakis, E., Krizanc, D., Morales-Ponce, O.: Line search for an oblivious moving target. CoRR **abs/2211.03686** (2022). https://doi.org/10.48550/arXiv.2211.03686
14. Corless, R.M., Gonnet, G.H., Hare, D.E.G., Jeffrey, D.J., Knuth, D.E.: On the lambert W function. Adv. Comput. Math. **5**(1), 329–359 (1996). https://doi.org/10.1007/BF02124750
15. Czyzowicz, J., Georgiou, K., Kranakis, E., Narayanan, L., Opatrny, J., Vogtenhuber, B.: Evacuating robots from a disk using face-to-face communication. Discret. Math. Theor. Comput. Sci. **22**(4) (2020). http://dmtcs.episciences.org/6732
16. Czyzowicz, J., Georgiou, K., Kranakis, E.: Group search and evacuation. In: Flocchini, P., Prencipe, G., Santoro, N. (eds.) Distributed Computing by Mobile Entities. LNCS, vol. 11340, pp. 335–370. Springer, Cham (2019). https://doi.org/10.1007/978-3-030-11072-7_14

17. Czyzowicz, J., et al.: Group evacuation on a line by agents with different communication abilities. **212**, 57:1–57:24 (2021). https://doi.org/10.4230/LIPIcs.ISAAC.2021.57

18. Czyzowicz, J., Kranakis, E., Krizanc, D., Narayanan, L., Opatrny, J.: Search on a line with faulty robots. Distrib. Comput. **32**(6), 493–504 (2019). https://doi.org/10.1007/s00446-017-0296-0

19. Devillez, H., Egressy, B., Fritsch, R., Wattenhofer, R.: Two-agent tree evacuation. In: Jurdziński, T., Schmid, S. (eds.) SIROCCO 2021. LNCS, vol. 12810, pp. 204–221. Springer, Cham (2021). https://doi.org/10.1007/978-3-030-79527-6_12

20. Feinerman, O., Korman, A., Lotker, Z., Sereni, J.: Collaborative search on the plane without communication. In: Kowalski, D., Panconesi, A. (eds.) ACM Symposium on Principles of Distributed Computing (PODC 2012), Funchal, Madeira, Portugal, 16–18 July 2012, pp. 77–86. ACM (2012). https://doi.org/10.1145/2332432.2332444

21. Gal, S.: Search games. Wiley Encyclopedia of Operations Research and Management Science (2010)

22. Georgiou, K., Karakostas, G., Kranakis, E.: Search-and-fetch with 2 robots on a disk: wireless and face-to-face communication models. arXiv preprint arXiv:1611.10208 (2016)

23. Georgiou, K., Karakostas, G., Kranakis, E.: Search-and-fetch with 2 robots on a disk - wireless and face-to-face communication models. In: Liberatore, F., Parlier, G.H., Demange, M. (eds.) Proceedings of the 6th International Conference on Operations Research and Enterprise Systems (ICORES 2017), Porto, Portugal, 23–25 February 2017, pp. 15–26. SciTePress (2017). https://doi.org/10.5220/0006091600150026

24. Georgiou, K., Kranakis, E., Leonardos, N., Pagourtzis, A., Papaioannou, I.: Optimal circle search despite the presence of faulty robots. In: Dressler, F., Scheideler, C. (eds.) ALGOSENSORS 2019. LNCS, vol. 11931, pp. 192–205. Springer, Cham (2019). https://doi.org/10.1007/978-3-030-34405-4_11

25. Kao, M., Reif, J.H., Tate, S.R.: Searching in an unknown environment: an optimal randomized algorithm for the cow-path problem. pp. 441–447 (1993). http://dl.acm.org/citation.cfm?id=313559.313848

26. Kleinberg, J.M.: On-line search in a simple polygon. In: Sleator, D.D. (ed.) Proceedings of the Fifth Annual ACM-SIAM Symposium on Discrete Algorithms. 23–25 January 1994, Arlington, Virginia, USA, pp. 8–15. ACM/SIAM (1994). http://dl.acm.org/citation.cfm?id=314464.314473

27. Queralta, J.P., et al.: Collaborative multi-robot search and rescue: planning, coordination, perception, and active vision. IEEE Access **8**, 191617–191643 (2020). https://doi.org/10.1109/ACCESS.2020.3030190

28. Spieser, K., Frazzoli, E.: The cow-path game: a competitive vehicle routing problem. In: Proceedings of the 51th IEEE Conference on Decision and Control (CDC 2012), 10–13 December 2012, Maui, HI, USA, pp. 6513–6520. IEEE (2012). https://doi.org/10.1109/CDC.2012.6426279

29. Yao, A.C.: Probabilistic computations: toward a unified measure of complexity (extended abstract). In: 18th Annual Symposium on Foundations of Computer Science, Providence, Rhode Island, USA, 31 October–1 November 1977, pp. 222–227. IEEE Computer Society (1977). https://doi.org/10.1109/SFCS.1977.24

Routing Schemes for Hybrid Communication Networks

Sam Coy[1], Artur Czumaj[1], Christian Scheideler[2], Philipp Schneider[3(✉)], and Julian Werthmann[2]

[1] University of Warwick, Coventry, UK
{S.Coy,A.Czumaj}@warwick.ac.uk
[2] University of Paderborn, Paderborn, Germany
scheideler@upb.de,jwerth@mail.upb.de
[3] University of Freiburg, Freiburg im Breisgau, Germany
philipp.schneider@cs.uni-freiburg.de

Abstract. We consider the problem of computing routing schemes in the HYBRID model of distributed computing where nodes have access to two fundamentally different communication modes. In this problem nodes have to compute small labels and routing tables that allow for efficient routing of messages in the local network, which typically offers the majority of the throughput. Recent work has shown that using the HYBRID model admits a significant speed-up compared to what would be possible if either communication mode were used in isolation. Nonetheless, if general graphs are used as the input graph the computation of routing schemes still takes polynomial rounds in the HYBRID model.

We bypass this lower bound by restricting the local graph to unit-disc-graphs and solve the problem deterministically with running time $O(|\mathcal{H}|^2 + \log n)$, label size $O(\log n)$, and size of routing tables $O(|\mathcal{H}|^2 \cdot \log n)$ where $|\mathcal{H}|$ is the number of "radio holes" in the network. Our work builds on recent work by Coy et al., who obtain this result in the much simpler setting where the input graph has no radio holes. We develop new techniques to achieve this, including a decomposition of the local graph into path-convex regions, where each region contains a shortest path for any pair of nodes in it.

1 Introduction

The HYBRID model was introduced as a means to study distributed systems which leverage multiple communication modes of different characteristics [3]. Of particular interest are networks that combine a <u>local</u> communication mode, which has a large bandwidth but is restricted to edges of a graph on the nodes

---

This paper takes the form of an extended abstract, summarizing our results, construction, and techniques. A full version is available at https://arxiv.org/abs/2210.05333: the section numbering is the same in both versions.

We would like to thank Martijn Struijs for valuable discussions concerning the geometric insights of the paper. Any errors remain our own.

S. Rajsbaum et al. (Eds.): SIROCCO 2023, LNCS 13892, pp. 317–338, 2023.
https://doi.org/10.1007/978-3-031-32733-9_14

of the network, with a global communication mode where any two nodes may communicate in principle, but the bandwidth is heavily restricted. This concept captures various real distributed systems, notably networks of cellphones that combine high bandwidth but locally restricted wireless communication on a unit-disc-graph with data transmission over the cellular network.

Routing Schemes are one of the most fundamental distributed data structures, most prominently employed in the Internet, and are used to forward packets among connected nodes in a network in order to facilitate data exchange between any pairs of nodes. In the distributed variant of the problem the nodes initially only know their incident neighbors in the network and need to communicate as efficiently as possible using their available means of communication such that subsequently each node knows its label and a routing table with the following properties. Given a packet with the label of the receiver node in the header, any node must be able to forward this packet in the network using the label and its routing table such that the packet eventually reaches the intended receiver. Algorithms for routing schemes in hybrid networks are of increasing importance, as contemporary communication standards support such settings, one prominent example being the 5G standard [2]. Formally we define routing schemes as follows.

Definition 1 (Routing Schemes). *A routing scheme on a connected graph $G = (V, E)$ consists of labels $\lambda(v)$ and routing functions (aka routing table) ρ_v for each $v \in V$. ρ_v maps labels to neighbors of v in G, such that the following holds. Let $s, t \in V$. Let $v_0 = s$ and $v_{i+1} = \rho_{v_i}(\lambda(t))$ for $i \geq 1$. Then there is an $\ell \in \mathbb{N}$, such that $v_\ell = t$. A routing scheme is an approximation with stretch α if $\ell_{st} \leq \alpha \cdot \text{hop}(s, t)$ for all $s, t \in V$, where $\text{hop}(s, t)$ is smallest number of edges of any st-path, and ℓ_{st} is the length of the induced routing path from s to t.[1]*

Since typically large amounts of packets are exchanged between senders and receivers as part of simultaneously ongoing sessions we concentrate on routing schemes for the local network graph, which offers much larger throughput that than what is possible on the global network, since the latter either involves higher costs or is more restricted as infrastructure is shared, like communicating via the cellular network (however, we need very little local communication to actually compute the routing scheme).

Our first goal is to optimize the round complexity of computing such a routing scheme, which is important since frequent changes in the topology of a local network among mobile devices necessitates its fast re-computation. The second goal is to minimize the size of the labels and local routing tables as these must be shared in advance (e.g. via the global network) to initiate a session between two nodes. The third goal is to minimize the *stretch* of the routing path between sender and receiver, minimizing latency and alleviating congestion.

---

[1] Minimizing hop-distance in a unit-disc-graph essentially minimizes the Euclidean distance that the path covers, thus graph weights are not required.

We consider the above problem in the HYBRID model that has received increasing attention during the last few years [1,3,6–8,10–12]. Formally, the HYBRID model builds on the classic principle of synchronous message passing:

Definition 2 (Synchronous message passing, cf. [18]). *We have n computational nodes with some initial state and unique identifiers (IDs) in $[n] := \{1,\ldots,n\}$. Time is slotted into discrete rounds. In each round, nodes receive messages from the previous round; they perform (unlimited) computation based on their internal states and the messages they received so far; and finally, based on those computations, they send messages to other nodes in the network.*

Note that the synchronous message passing model focuses on the analysis of round complexity of a distributed problem (the number of rounds required to solve it). The HYBRID model restricts which nodes may communicate in a given round and to what extent.

Definition 3 (HYBRID model, cf. [3]). *The <u>local</u> communication mode is modeled as a connected graph, in which each node is initially aware of its neighbors and is allowed to send a message of size λ bits to each neighbor in each round. In the <u>global</u> communication mode, each round each node may send or receive γ bits to/from every other node that can be addressed with its ID in $[n]$ in case it is known. If any restrictions are violated in a given round, an arbitrary subset of messages is dropped.*

In this paper, we consider a weak form of the HYBRID model, which sets $\lambda \in O(\log n)$ and $\gamma \in O(\log^2 n)$, which corresponds to the combination of the classic distributed models CONGEST[2] as local mode, and NODE CAPACITATED CLIQUE (NCC)[3] as global mode. Note that while it might appear that more global than local communication is allowed in our model, the local network allows each node to exchange a messages with each neighbor (of which there could be $\Theta(n)$), which is not possible in the global network.

The distributed problem of computing routing schemes on the local communication graph is an excellent fit for the HYBRID model, since the problem is known to require $\widetilde{\Omega}(n)$ rounds of communication (where $\widetilde{O}, \widetilde{\Omega}$ hides polylog n factors) if only <u>either</u> communication via the local mode <u>or</u> the global mode is permitted, see [12]. The lower bound for the local communication mode holds even if the input graph is a path and even for unbounded local communication.

It is natural to wonder if adding a modest amount of global communication on top of a local network significantly improves the required number of communication rounds to establish a routing scheme in the network. This question was recently answered positively by [12], where it was shown that routing schemes with small labels can be computed in $\widetilde{O}(n^{1/3})$ rounds for <u>arbitrary</u> local graphs.

---

[2] Some previous papers that consider hybrid models use $\lambda = \infty$, i.e., the LOCAL model as local mode.

[3] Our methods also work for to the stricter NCC_0 model as the global network, where only incident nodes in the local network and those that have been introduced, can communicate globally.

However, [12] also shows that a polynomial number of rounds is required to solve the problem even approximately (in particular, an exact solution with labels up to size $O(n^{2/3})$ requires $\widetilde{\Omega}(n^{1/3})$ rounds).

To mitigate this lower bound, [8] considers local communication networks that are restricted to certain interesting classes of graphs for which they can compute routing schemes in just $O(\log n)$ rounds. In this article we will continue this line of work and consider local communication graphs that are unit-disk graphs (UDGs). Such a UDG $G = (V, E)$ satisfies the property that nodes are embedded in the Euclidean plane and are connected iff they are at distance at most 1. Note that UDGs have been extensively studied as a model capturing how multiple devices using wireless ad-hoc connections communicate (see, e.g., [4,5,8,13–15], all of which handle routing schemes in such UDGs).

In [8] it was shown that the nodes of a UDG can together simulate a much simpler grid graph structure with constant overhead in round complexity, such that the connectivity, the hole-freeness, and the hop distance up to a constant factor are preserved. Furthermore, [8] shows that a routing scheme on a grid graph can efficiently be transformed into a routing scheme for the underlying UDG, which introduces only a constant overhead on label size and local routing information and takes only a constant number of additional rounds.

This allows them to consider the much simpler grid graphs for the computation of routing schemes to generate good routing schemes for UDGs. However, their actual algorithm for computing a routing scheme for a grid graph comes with a caveat: it works only for grid graphs without holes, which, loosely speaking, are points in the grid without nodes on them that are enclosed by a cycle in the grid graph (more formally in Sect. 2), which implies routing schemes only for UDGs without "radio holes", which roughly correspond to areas enclosed by the UDG that are not covered by nodes (see [8] for the formal definition of radio holes in UDGs).

In this work we extend the solution to UDGs with such radio holes. Note that the transformations given in [8] from UDGs to grid graphs work even if there are holes, which essentially allows us to focus on the computation routing schemes for grid graphs with holes, which gives the same for arbitrary UDGs, which have a set $\mathcal{H}$ of radio holes (formally defined later). Our algorithm is asymptotically as efficient (in computation time, and label and table sizes of the resulting routing scheme) as the algorithm of [8] if the number of holes $|\mathcal{H}|$ is small.

We stress that it is significantly more challenging to compute routing schemes on grid graphs with such holes than on grid graphs without holes. In the paper of Coy et al. [8] the authors heavily exploit the property that any simple st-path can be deformed into any other simple st-path: this is not true in our setting. In particular, it is easy to come up with examples of grid graphs with $|\mathcal{H}|$ holes for which there are at least $2^{|\mathcal{H}|}$ "reasonable" classes of st-paths (intuitively: paths which do not completely encircle or spiral around holes) which cannot be deformed into each other. Worse still, we can make all-but-one of these classes of paths almost arbitrarily long, and so we cannot just consider one arbitrary class of st-paths and obtain an approximate shortest path. We must determine the class in which an exact shortest st-path lies, which seems to require a sparsifying structure that scales in complexity with the number of holes.

1.1 Contributions

Our main contribution is the extension of the result presented in Coy et al., [8], which assumes that no holes are present making the problem much simpler.

Theorem 1 (Main Result for Grid Graphs). *Given a grid graph Γ with a set $\mathcal{H}$ of holes (defined in Sect. 2), we can compute an exact routing scheme for Γ in $O(|\mathcal{H}|^2 + \log n)$ rounds in the HYBRID model (notably, even NCC$_0$ suffices). The labels are of size $O(\log n)$; nodes need to locally store $O(|\mathcal{H}|^2 \log n)$ bits.*

Note that since grid graphs have constant degree, the local network can be simulated using the NCC$_0$ model. Therefore, in the theorem above and in all our subsequent claims about grid graphs, the use of HYBRID can be replaced with the weaker NCC$_0$ model. Although this is interesting, we consider computation of routing schemes solely in the global network as an artificial problem: the fact that we are computing a routing scheme for a local network suggests that this network can be used to help construct it. Furthermore, the local network of the HYBRID model is required by [8] to efficiently transform a unit disc graph into a sparsifying grid graph structure that approximates it well (we briefly summarize this in Sect. 2). This result leads to the following corollary:

Corollary 1 (Main Result for UDGs). *The routing scheme of Theorem 1 can be transformed into a routing scheme which yields constant-stretch shortest paths for unit-disk graphs in the HYBRID model. Round complexity, label size, and local storage are asymptotically the same.*

We believe that several of our technical contributions are of independent interest. Our main technical contribution is a decomposition of a grid graph into simple, path-convex regions which have useful properties for routing. We also provide a small skeleton structure of the UDG called a landmark graph such that shortest paths in the landmark graph are topologically the same (i.e., circumnavigate holes in the same way) as in the original graph, which may be useful when solving shortest paths on grid graphs and UDGs in HYBRID or for similar problems in other models of computation. Furthermore we give an $O(\log n)$ round algorithm for solving SSSP exactly in simple grid graphs and an $O(\log n)$ round algorithm for finding the distance from every node in a simple grid graph to a portal.

1.2 Related Work

Shortest Paths in Hybrid Networks. Previous work in the HYBRID model has mostly focused on shortest path problems [1,3,6,10,11]. In the k-sources shortest path (k-SSP) problem, all nodes must learn their distance in the (weighted) local network to a set of k sources. Particular focus has been given to the all-pairs (APSP, $k = n$) and single-sources (SSSP, $k = 1$) shortest-path problems. Note that solving the APSP problem gives a solution to the routing scheme problem. The complexity of APSP is essentially settled: [3,11] give an algorithm taking $\widetilde{O}(\sqrt{n})$ rounds, and this matches a lower bound of $\widetilde{\Omega}(\sqrt{k})$ to solve k-SSP even for polynomial approximations. This lower bound even matches a deterministic algorithm due to [1], although only with an approximation factor of $O(\frac{\log n}{\log \log n})$.[4] For k-SSP the $\widetilde{\Omega}(\sqrt{k})$ lower bound has been matched by [6] with a constant stretch algorithm, given sufficiently large k (roughly $k \in \Omega(n^{2/3})$) (See Footnote 4). Whether there are any $\widetilde{O}(\sqrt{k})$ round k-SSP algorithms on general graphs for $1 < k < n^{2/3}$ remains open. The state-of-the-art algorithm for exact SSSP is provided by [6] and takes $O(n^{1/3})$ rounds (See Footnote 4). A recent result by [19], which solves SSSP by $\widetilde{O}(1)$ applications of an instruction set called "minor-aggregation" when given access to an oracle that solves the so called Eulerian Orientation problem, can be adapted for a $(1+\varepsilon)$ approxima- tion of SSSP in $\widetilde{O}(1)$ rounds, as was shown in [21] (See Footnote 4). An *exact, deterministic* solution for SSSP in $O(\log n)$ rounds has been achieved on specific classes of graphs (e.g. cactus graphs, which includes trees) by [10]. Another exact SSSP algorithm that takes $\widetilde{O}(\sqrt{SPD})$ rounds (where SPD is the shortest-path- diameter of the local graph) is provided by [3].

Routing Schemes in Distributed Networks. Our work builds on [8], in which they show how to compute a routing scheme for a UDG, by computing a routing scheme for a corresponding grid graph (see Sect. 2). Their approach requires a simplifying assumption: the grid graph needs to be free of holes (for- mally defined in the next section), and this imposes a similar restriction on the underlying UDG. We remove that assumption in this work. In a recent arti- cle, [12] considers computing routing schemes in the HYBRID model on general graphs: they show that in $\widetilde{O}(n^{1/3})$ rounds one can compute exact routing schemes with labels of size $\widetilde{O}(n^{2/3})$ bits, or constant stretch approximations with smaller labels of $O(\log n)$ bits. Interestingly, [12] also gives lower bounds: they show that it takes $\widetilde{\Omega}(n^{1/3})$ rounds to compute exact routing schemes that hold for relabel- ings of size $O(n^{2/3})$ and on unweighted graphs. They also give polynomial lower bounds for constant approximations on weighted graphs, implying that in order to overcome this lower bound in round complexity and label size, the restriction to a class of graphs is necessary! The distributed round complexity of computing routing schemes was also considered in the CONGEST model. In general, it takes $\widetilde{\Omega}(\sqrt{n} + D)$ rounds to solve the problem [20] (where D is the graph diameter). This was nearly matched in a series of algorithmic results [9,16,17], for example, [9] gives a solution with stretch $O(k)$, routing tables of size $\widetilde{O}(n^{1/k})$, labels of size $\widetilde{O}(k)$ in $O(n^{1/2+1/k} + D_G) \cdot n^{o(1)}$ rounds.

---

[4] These results are in the more powerful hybrid combination of LOCAL and NCC.

2   Unit Disk Graphs and Grid Graphs

In this section we mainly introduce concepts, which we require in many of the following sections. We consider the class of Unit Disk Graphs where each node $v \in V$ is associated with a unique point in $\mathbb{R}^2$ and two nodes $u, v \in V$ share an edge $\{u, v\} \in E$ iff $\|u-v\|_2 \leq 1$. Grid graphs can be seen as a sparsifying structure for UDGs which can be easily simulated while preserving certain geometric properties and significantly simplifying the construction of algorithms for the original UDG (cf., [8], more on that further below). A grid graph $\Gamma = (V_\Gamma, E_\Gamma)$ is a graph where the vertices V uniquely correspond to points on a square grid $\mathbb{Z}^2$ and two such vertices are connected by an edge in E_Γ iff their corresponding points on the grid are horizontally or vertically adjacent.

As was shown in [8], one can compute and simulate a grid-graph abstraction $\Gamma = (V_\Gamma, E_\Gamma)$ of any input UDG using only local communication in $O(1)$ rounds. In this simulated grid graph Γ, each grid node is represented with a nearby node in the UDG. Any UDG node represents at most a constant number of grid nodes (and can thus simulate all grid nodes it represents). This means that approximate paths with good stretch on the UDG can be constructed from shortest paths on the grid graph abstraction Γ. In particular, any constant stretch routing scheme on Γ also gives a constant stretch routing scheme on the underlying UDG. Therefore in order to obtain constant stretch routing schemes on UDGs it suffices to consider the problem on the (easier) class of grid graphs.

Theorem 2 (cf. [8]).   *Any algorithm that computes routing schemes with stretch s, labels of at most x and local routing information at most y bits on any grid graph in z rounds (where x, y, z depend on n) implies an algorithm to compute a routing scheme with stretch* $36 \cdot s$ *on any UDG with labels of* $O(x)$ *bits, local routing information of* $O(y)$ *bits in* $O(z)$ *rounds.*

In order to describe the geometric structure of UDGs $G = (V, E)$ and grid graphs $\Gamma = (V_\Gamma, E_\Gamma)$ (which are also UDGs) we require some basic notation. We define a path $\Pi \subseteq E$ as set of edges that form a sequence of incident edges in G. By $|\Pi|$ we denote the number of edges (or hops) of a path Π. The distance between two nodes $u, v \in V$ is defined as $d_(u, v) := \min_{u\text{-}v\text{- path } \Pi} |\Pi|$. We will drop the subscript G when the graph in question is clear from context. Let Π_x be the set of horizontal edges in the grid graph Γ. Then $d_{x, \Gamma}(u, v) := \min_{u\text{-}v\text{- path } \Pi} |\Pi_x|$ is the horizontal distance between u and v. Analogously we define the vertical edges Π_y in Γ and vertical distance $d_{y, \Gamma}$.

To analyze grid graphs we define some geometric structures, starting with portals [8] (see Fig. 1 for a visualization). Let $E_{\Gamma, v}$ be the set of vertical edges of some grid graph Γ. Then the vertical portals are the connected components of the sub graph $(V_\Gamma, E_{\Gamma, v})$. We say that portals p_1 and p_2 are adjacent if p_1 is connected by an (horizontal) edge in E to a node in p_2. Horizontal portals and the corresponding terms are defined analogously.

Next, we (informally) introduce the concept of <u>holes</u> in a grid graph (the formal definition is given in the full version). The area S_Γ that is covered by a grid graph can be described by "filling in" the grid cells (those are unit squares $[a, a+1] \times [b, b+1]$ with $a, b \in \mathbb{Z}$) where all four corner nodes represent nodes in Γ as well as all edges of Γ. The holes are the connected areas in the Euclidean plane which are not "filled" ($\mathbb{R}^2 \setminus S_\Gamma$). The unique unbounded hole is called the outer hole, all others are inner holes, which we describe with the set $\mathcal{H}$. Figure 2 (top left) shows such a grid graph, where S_Γ is given by the white area and holes are shaded gray.

We now give some additional definitions pertaining holes. By the boundary of some $H \in \mathcal{H}$ we describe the subset of nodes of V_Γ that are that are located on the geometric boundary of the set $H \subseteq \mathbb{R}^2$. We say that a portal p is incident to some hole H if p has a non-empty intersection with the boundary of H. Note that [8] gives an efficient algorithm for the special case of computing routing schemes in *simple* grid graphs, i.e., $\mathcal{H} = \emptyset$.

Theorem 3 (cf. [8]). *An exact routing scheme for a simple grid graph Γ using node labels and local space of $O(\log n)$ bits can be computed in $O(\log n)$ rounds in the* HYBRID *model.*

3 Shortest Paths Computations

In this section, we present an $\mathcal{O}(\log n)$ single source shortest paths (SSSP) algorithm for simple grid graphs. As the result of this algorithm's execution, each node learns its distance to some dedicated source node s. Note that nodes can easily infer predecessor pointers comparing their own distances with their neighbors'. We will make use of this algorithm in the later sections as a subroutine. We start by stating our main theorem.

Theorem 4. *SSSP can be solved in a simple grid graph in $\mathcal{O}(\log n)$ rounds.*

To achieve the logarithmic runtime, we split the SSSP problem into two subproblems: Horizontal SSSP, which only considers the horizontal steps a path makes and Vertical SSSP, which only considers the vertical steps a path makes. To be able to efficiently solve these problems, we introduce horizontal and vertical portal graphs, which are depicted in Fig. 1 as well.

Definition 4. *Given a grid graph Γ, we define the <u>vertical portal graph</u> $\mathcal{P}_v$ to be the graph with vertices corresponding to the vertical portals of Γ. Two vertices in $\mathcal{P}_v$ are connected by an edge iff their corresponding portals are connected by a horizontal edge in Γ. The horizontal portal graph $\mathcal{P}_h$ is defined analogously.*

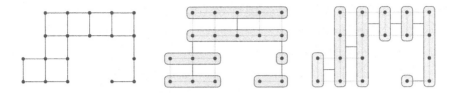

Fig. 1. A simple grid graph (left) and the corresponding horizontal portal graph P_h (center) and vertical portal graph P_v (right).

Note that the P_v graph only retains horizontal distances and the P_h graph only retains vertical distances. As the portal graph of a simple grid graph always is a tree—a fact that we prove in the full version of the paper—we can use the SSSP algorithm for trees presented in [10] to solve Horizontal SSSP for Γ on P_v and Vertical SSSP for Γ on P_h. It remains to combine the resulting distance values, which can be done by each node locally by adding them up. We move the proof of this fact to the full version of the paper as well.

We can expand this algorithm to compute the distance of each node to a dedicated source portal P, i.e., to the closest node on that portal, by picking an arbitrary node of P as the starting node and setting the we weight of all edges on P to 0.

Corollary 2. *Given a simple grid graph Γ and a portal P in Γ, we can compute the distances of every vertex in Γ to P in $\mathcal{O}(\log n)$ rounds.*

4   Pathconvex Region Decomposition

In this section we give our main technical contribution: that grid graphs can be partitioned into comparatively few sets of nodes (with some overlap at the borders) called *regions* $\{R_1, \ldots, R_\ell\}$ with $V = R_1 \cup \cdots \cup R_\ell$ that are simple (i.e., without holes) and path-convex, which is defined as follows:

Definition 5 (Path Convexity). *Let $\Gamma = (V, E)$ be a grid graph and let $R \subseteq V$. Then R is called <u>path-convex</u> if for any pair $u, v \in R$ there is a shortest uv-path contained completely within that region.*

We achieve a simple and convex region decomposition $\mathcal{R}$ by disconnecting parts of Γ and considering each connected component as a separate region. For this, we create copies of nodes that are on the border of at least two regions and only connect each copy to neighbors in <u>one</u> of those regions. The construction breaks down into three main steps. Firstly, we decompose Γ into simple regions. Secondly, we break those regions up further into "tunnels", which overlap in at most two portals (called gates) with their neighboring regions. Finally, we show that such tunnels have crucial properties that allows us to make them path-convex by subdividing them a constant number of times. Figure 2 visualizes the three steps on our example grid graph. Ultimately, we prove the following theorem.

Fig. 2. The top left image shows the grid graph that we will use and develop throughout the sections of this article. In Sect. 4 we split this grid graph into simple regions (top right), tunnels (bottom left) and path-convex regions (bottom right). The color red marks the newly introduced splits in each step. (Color figure online)

Theorem 5. *For any grid graph Γ, a decomposition into $O(|\mathcal{H}|)$ simple, path-convex regions can be computed in $O(|\mathcal{H}| + \log n)$ rounds in the* HYBRID *model.*

Splitting Operations. To achieve the desired region decomposition we split the grid graph at strategic locations (see Fig. 3). The most basic splitting operation is to split a grid graph (or a region) at some portal (in this case vertical), where each node will simulate two copies of itself, a "right copy" which has no left neighbor and a "left copy" which has no right neighbor. This establishes a new grid graph, where nodes that have been split will act only in the role of the nodes they simulate, which blocks paths through the splitting portal and might disconnect Γ.

  If such a splitting portal touches the boundary of a hole, we often further split at a boundary node. In particular, we split the simulated boundary node that is on the "same side" as the hole (say the left copy if the hole is left of the portal) into a "top" and "bottom" copy, which do not have a bottom or top neighbor, respectively. This is where up to three regions may intersect in the resulting region decomposition and can also be used to break up cycles around a hole, thereby making the resulting regions simple. We describe such splitting operations in detail in the full version.

Fig. 3. An example of the "splitting operations" at the blue portal and the blue node. We start by splitting all nodes on the blue portal to make left and right copies, not connected to each other. Then, we split the left copy of the blue node to create upper and lower copies. (Color figure online)

The region decomposition is given implicitly by the connected components in the grid graph formed by the simulated nodes after splitting at a portal or a node. These regions overlap only in node sets that form portals (i.e., vertically connected components). To distinguish those from ordinary portals, we call these gates. Furthermore, we refer to connected segments of the boundary nodes of some region which are not on gates, as walls.

4.1 Decomposition into Simple Regions

The first step is to split our grid graph into regions without holes. We show that a relatively simple procedure, described below, can achieve this goal. The implementation details in our computational model and the proofs are given in the full version.

Splitting Γ into Simple Regions. For each inner hole $H \in \mathcal{H}$ of Γ we repeat the following. Let v_H be the leftmost node on the boundary of H (we make v_H unique by choosing the northernmost among leftmost boundary-nodes). Let P_H be the unique vertical portal with $v_H \in P_H$. We conduct splits at P_H and v_H as described further above. In general, P_H might contain leftmost nodes of boundaries of several different inner holes. In that case we conduct the vertical split at the northernmost node of each such hole (and each such inner hole needs not be considered further, i.e., we only split P_H itself once). Figure 4 shows an example of the procedure.

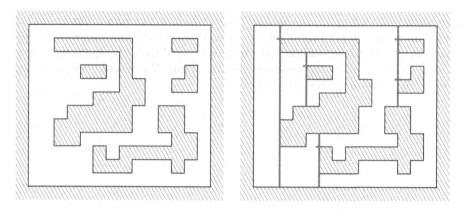

Fig. 4. Decomposition of grid graph into simple regions. Red lines mark portals through a leftmost node of some hole at which a split occurs. Further splits take place at a leftmost node of each hole. In particular the portal on the top right is split at two nodes. (Color figure online)

The connected components that result from the above construction form regions that are simple. The idea is that each such split will make the grid graph at the portal P_H horizontally impassable and at v_H vertically impassable. Roughly speaking, the split at a portal and a node on the portal and hole boundary creates a "thin" hole such that in the resulting graph two holes "merge", and become a single hole. This decreases the overall number of *inner* holes by at least one. By repeating this split for each inner hole we will be left with one hole in the end, which is the outer hole. In the full version we prove that the regions are simple, that there are at most $|\mathcal{H}|+1$ of them and that the construction can be done in $O(\log n)$ rounds in the HYBRID model.

4.2 Decomposition into Tunnel Regions

Our next step is to ensure that each region is a <u>tunnel</u>, which we define as a region that has at most two gates. As a consequence of the previous Sect. 4.1, we start out with a grid graph Γ that is decomposed into simple regions.

In the following, we say that two vertical portals are adjacent to each other, if at least two nodes of either portal are connected by a horizontal grid edge. Recall that we defined walls as connected segments of nodes on the boundary of a some given hole that are not part of gates. It is computationally not very hard for the nodes of the same wall or gate to compute an identifier in $O(\log n)$ rounds (details in the full version) in the HYBRID model.

The notion of walls and gates allows us to define <u>junction portals</u> that we require for the next stage of our decomposition. Informally, a junction portal is a portal at which a simple region "diverges" into at least three "tunnels" (although there are some degenerate cases for such junction portals, where a gate cuts away one of these divergent "tunnels"). Formally we define:

Definition 6 (Junction Portals). *Let R be a simple region. A vertical portal P in R is a junction portal if one of the following conditions holds:*

i. *P has at least 3 adjacent portals each intersecting at least 2 distinct walls.*
ii. *P is a gate, and has at least 2 adjacent portals each intersecting at least 2 distinct walls.*

The idea is to perform portal splitting operations on each junction portal and on specific nodes on the junction portal in order to separate these divergent tunnels. The construction works as follows. Let P be a junction portal according to Definition 6, which implies that there are at least 2 adjacent portals which each intersect multiple distinct walls. Suppose that portals $P_1, P_2, \ldots, P_k$ are such portals with the property of intersecting distinct walls to the left of P (the procedure for those to the right is analogous). We first split the region at the portal P. Then we conduct a node-split at the bottom-most node on P that is adjacent to some node on P_i for $1 \leq i < k$.

In the full version, we prove that if we are given a simple region decomposition, then after splitting at junction portals as described above we obtain a decomposition into a number of tunnel regions that is linear in $|\mathcal{H}|$ and that all required computations can be accomplished in $O(\log n)$ rounds.

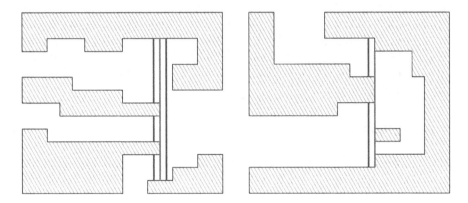

Fig. 5. Examples of Definition 6 case (i) (left) and (ii) (right). Red portals mark the junction portals. Blue portals intersect at least two distinct walls. In case (i) the red portal was not previously a gate but has at least 3 incident blue portals. In case (ii) the red portal is already a gate. Since the area to its right was removed in the previous simplification step it has just two adjacent blue portals. (Color figure online)

4.3 Path-Convex Decomposition

Finally, we make the tunnel regions path-convex (cf. Definition 5), by splitting them at appropriate portals. Essentially, we will separate any pair of nodes in such a tunnel for which there is a shortest path in Γ that travels outside the tunnel. The main question answered in this section is *where* the tunnels should be split in order to make them path-convex. The construction is in fact not very complicated, the main challenge is the proof of correctness, i.e., showing that all "offending" node pairs whose shortest path runs outside of the region are separated. We approach this by first showing the claim for well-behaved special cases and subsequently reduce the general cases to these special ones.

First, we impose the assumption that we have a tunnel region T with gates that are single nodes g and g', see Fig. 6. We show that we can split T into path-convex regions at two portals P_x and P_y. Roughly, if $d_x = d_{x,T}(g, g')$ is the horizontal distance from g to g' in T then P_x is defined by the nodes which are at distance $\frac{d_x}{2}$, which we show forms a vertical portal. The horizontal portal P_y is defined analogously. We split T at P_x and P_y (cf., Fig. 6).

We then remove the assumption that the gates are point-shaped and distinguish two cases, depending on whether there is a horizontal portal connecting the two gates (see Fig. 7). We give a decomposition procedure for each case (visualized in Fig. 7), where the first case (a) is well behaved and the second case (b) has the following property. For any pair of nodes in a region in the "middle part" a path between two nodes in T that goes outside T can also leave and enter T through two fixed nodes g, g' on the gates G, G' without becoming any longer. This essentially allows us to assume that all paths go through g, g and fall back to the proof of correctness for case (b), i.e., we split the "middle part" at P_x, P_y.

In the full version we conclude, first, that each region is path-convex; second, that each tunnel-region is split into a constant number of sub-regions and third, that the whole procedure can be conducted in $O(\log n)$ rounds in the HYBRID model. Combining this with our results from the previous subsections culminates in the proof of Theorem 5.

5   Landmarks

After completing a regionalization of the grid graph, we want to exploit this abstraction to construct a small skeleton graph that captures the topological structure of the graph. We place these landmarks such that for any source node s and target node t, a shortest path can usually be constructed by: (i) routing from s to some landmark in its region; (ii) routing from that landmark to a landmark in t's region; and then (iii) finally routing from this landmark to t. We call a path of this type a landmark path. Sometimes there is no shortest st-path which is a landmark path, but in these cases a shortest st-path still passes through the same regions as a shortest landmark path: we will explain this case at the end of this section.

This result suggests a natural approach. Since s lies in the same region as its closest landmark we can use our SSSP result from Sect. 3 to find a shortest path to it. To support routing between landmarks in different regions, we create "virtual edges" which connect landmarks to each other. Then, by distributing the landmark graph to all nodes, it can be determined locally which landmarks

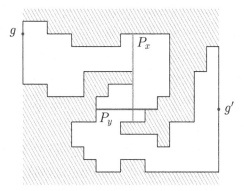

Fig. 6. Splitting tunnels into path-convex regions with point shaped gates.

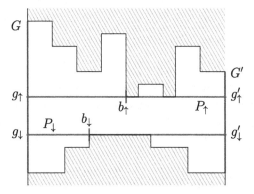

(a) Construction if there is a horizontal portal connecting G, G'.

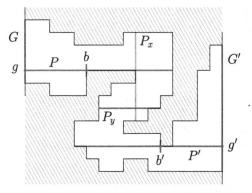

(b) Construction for the case that horizontal "view" between G and G' is obstructed.

Fig. 7. Splitting tunnels into path-convex regions in the two cases.

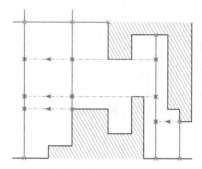 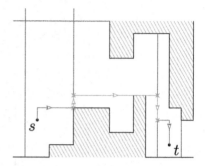

(a) An example of our landmark place-   (b) An example of a landmark path
ment. Type (i) landmarks are in or-   from s to t. The orange edges are edges
ange; type (ii) landmarks are in green;   in E_Λ, and the green edges are com-
and type (iii) landmarks are in red.   posed of multiple edges in E_Γ.

Fig. 8. An overview of our landmark construction.

are on the shortest landmark path, and hence which route a shortest path has to take around the holes $\mathcal{H}$ (covered in more detail in Sect. 6). Distributing this landmark graph efficiently requires that it is small. We choose our landmarks carefully so that there are only $O(|\mathcal{H}|)$ of them in each region (giving $O(|\mathcal{H}|^2)$ in total), and that there are only $O(|\mathcal{H}|^2)$ many "virtual edges" between them in total.

Besides optimizing the number of landmarks, we need to place them in enough locations so that they appear on many shortest paths, and capture all of the different ways in which we could route between and around holes. We place landmarks in the following locations (see also Fig. 8a):

Definition 7. *Suppose $v \in V_\Gamma$ lies on a gate G. Let P be the portal perpendicular to G passing through v, and let R be one of the regions incident to G. Note that v is one of the endpoints of $P \cap R$ (P restricted to R). Then v is a landmark if one of the following holds:*

  *i. v is an **endpoint** of G.*
  *ii. v is an **overhang-induced landmark**. Let u be the other endpoint of $P \cap R$. Then v is a landmark if for some $p \in P \cap R$, p is on a wall W, and u is on either W or a gate which is not G.*
  *iii. v is a **projection landmark** if any node on portal P is a landmark of either of the first two types.*

We also need to select our "virtual edges" carefully to minimize the number of edges and reflect the topological properties of the grid graph Γ. Naively, a virtual edge between every pair of landmarks would give a landmark graph of size $O(|\mathcal{H}|^4)$. Creating virtual edges between all pairs of landmarks in the same region, still results in $O(|\mathcal{H}|^2)$ edges per region, thus $O(|\mathcal{H}|^3)$ in total. We construct an even smaller selection of landmark edges, of size $O(|\mathcal{H}|^2)$ as follows.

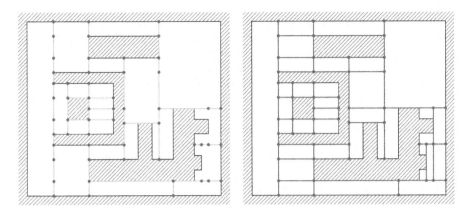

Fig. 9. The landmarks are placed on the gates of the region decomposition according to Definition 7 (left). They are connected with virtual edges according to Definition 8 (right). For clarity, we will omit the virtual edges in future figures.

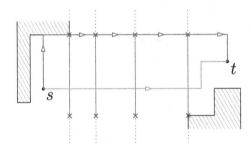

Fig. 10. An example of the case where there are no landmarks on a shortest path from s to t. The red crosses are landmarks; the blue lines are gates; the green path is a shortest landmark path; and the orange path is a shortest st-path. (Color figure online)

Definition 8. *There is a virtual edge between landmarks u and v if: (i) u and v are on the same gate with no landmark between them; or (ii) u and v are on different gates incident to the same region, and v is the closest landmark on its gate to u.*

The process of constructing the landmark graph is depicted in Fig. 9. The fact that each region is bounded by constantly many gates gives us that each landmark has a constant number of incident virtual edges. We remark that nodes can be informed whether they are landmarks, and landmarks can be informed of adjacent landmarks in the landmark graph, in only $O(\log n)$ rounds.

Lemma 1. *The landmark graph has $O(|\mathcal{H}|^2)$ vertices and $O(|\mathcal{H}|^2)$ edges.*

In the full version, we show that the landmark edges that we select suffice to make the resulting landmark graph faithful to the distances in the original graph Γ thus "encoding" shortest paths between any pair of landmarks. This essentially covers the case when a shortest st-path passes through landmarks.

Finally, we give a characterisation of the case where no st-shortest-path is a landmark path. The intuition is that in such a case, it suffices to know which regions to go through: routing greedily between them gives a shortest path. A visual depiction of such a case is given in Fig. 10.

Lemma 2. *If no shortest st-path is a landmark path, then consider the sequence of regions $(R_1 \ldots R_m)$ that are passed by the shortest landmark path from $s \in R_1$ to $t \in R_m$. The path formed by routing from s to the closest point in R_1, then to the closest point in R_2, repeating this way until the closest point in R_m is reached, and finally routing to t, is a shortest st-path.*

6   Routing

Our goal in this section is to define the routing scheme that has the properties described in Theorem 1. Recall Definition 1, where we defined a routing scheme to consist of node labels $\lambda(v)$ and routing tables ρ_v for each $v \in V$. In this section, we will describe how to construct routing tables of size $O(|\mathcal{H}|^2 \cdot \log n)$ and node labels of size $O(\log n)$, and how we can use them to route a packet from a source node to a target node.

Routing Tables. As the purpose of the landmark graph is to capture the structure of the graph, it will form the main part of each node's routing table. In this way, each node learns where the holes are positioned in the graph and how they should be circumnavigated. To enable the nodes to use this knowledge to make routing decisions, we add additional labels to the edges of the landmark graph. Specifically, we label the nodes of the landmark graph with the gate identifiers of the gates they lie on, and the edges of the landmark graph with the region identifiers of the regions they lie on. We do this in order to add information about the region decomposition to the landmark graph, since we need to know which regions the landmarks correspond to.

The landmark graph is of size $O(|\mathcal{H}|^2 \cdot \log n)$ by Lemma 1. Since we only add labels of size $O(\log n)$ to each edge, the size increases by a constant factor, and we can distribute the landmark graph and the labels in $O(|\mathcal{H}|^2 + \log n)$ rounds using well-known techniques for token dissemination.

Lemma 3. *Each node can learn the landmark graph and the landmark graph's labels $O(|\mathcal{H}|^2 + \log n)$ rounds. Each node learns $O(|\mathcal{H}|^2 \cdot \log n)$ bits this way.*

We describe the details of the landmark graph labels' computation, how to distribute them, and the landmark graph itself in the full version.

Node Labels. The landmark graph is part of the routing tables, so the nodes know the structure of the graph. However, to route a packet to a target node, they also need to know where the target node is. This information is encoded in the node label. Specifically, each node's label contains its own identifier, its region identifier, and the distances to close landmarks adjacent to its region. The first two are used to decide whether a given packet is already at its destination or in its destination region. If neither is the case, the third can be used by a node to locally add itself and the target node to the landmark graph, solve SSSP on this augmented landmark graph, and decide in which direction to route a packet. In total, $O(\log n)$ bits per node are required to encode this information.

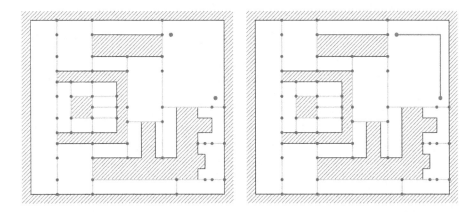

Fig. 11. When source and target of a routing request are in the same region (left), we employ the routing scheme from [8] (right).

Lemma 4. *Each node can learn its region identifier and the distance to the landmarks required to connect it to the landmark graph $O(\log n)$ rounds. Each node learns $O(\log n)$ bits this way.*

The details of the computation of these labels and which landmarks need to be included in a node's label can be found in the full version of the paper.

Routing Scheme. To explain how the nodes can use the routing tables and node labels to make routing decisions, we describe how a node $v = (v.id, v.D, v.rid)$ forwards a packet with destination $t = (t.id, t.D, t.rid)$. id, D, and rid correspond to the nodes' identifiers, their distances to close landmarks and their region identifiers respectively.

If $v.id = t.id$, the packet arrived at its destination and does not need to be forwarded anymore. If the packet is already in the correct region (i.e., $v.rid = t.rid$), the algorithm of [8] can be used, since each region is simple (Fig. 11). Otherwise, v augments its local copy of the landmark graph by adding itself and t to it using $v.D$ and $t.D$. This allows v to locally solve SSSP, learning a shortest

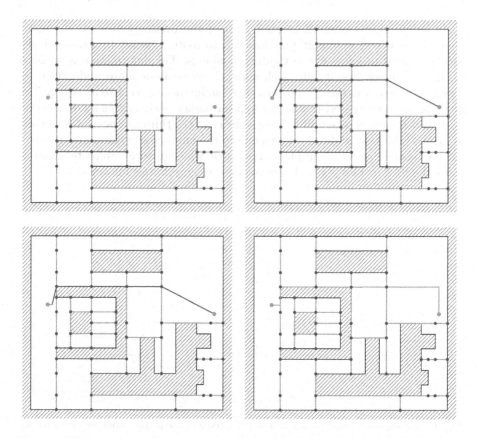

Fig. 12. When source and target node are in different regions (top left), they are added to the landmark graph, enabling each node to locally find a shortest path from the source node to the target node in the landmark graph (top right). After the source node performs this computation, it forwards the packet towards the suggested region boundary (bottom left). This is repeated until the packet the target region, where we it can be forwarded using the algorithm from [8] (bottom right).

path from v to t in the landmark graph. By inspecting the first node of this shortest landmark path, it can decide which gate should be crossed next and infer to which neighbor the packet should be forwarded (Fig. 12). This yields Theorem 1, the details are given in the full version.

7 Conclusion

We believe that there are several interesting directions for interesting follow-up work. Efficiently computing compact routing schemes in more general classes of geometrically interesting graphs (for example planar graphs or visibility graphs) is a natural next step. We suspect that an extension to 3 and higher dimensions might be quite difficult (in particular the geometry required will certainly be more challenging), but the 3-dimensional case could have practical applicability in sensor networks and swarm robotics.

References

1. Anagnostides, I., Gouleakis, T.: Deterministic distributed algorithms and lower bounds in the hybrid model. In: Gilbert, S. (ed.) 35th International Symposium on Distributed Computing (DISC 2021). Leibniz International Proceedings in Informatics (LIPIcs), Dagstuhl, Germany, vol. 209, pp. 5:1–5:19. Schloss Dagstuhl – Leibniz-Zentrum für Informatik (2021). https://doi.org/10.4230/LIPIcs.DISC.2021.5. https://drops.dagstuhl.de/opus/volltexte/2021/14807
2. Asadi, A., Mancuso, V., Gupta, R.: An SDR-based experimental study of outband D2D communications. In: The 35th Annual IEEE International Conference on Computer Communications, IEEE INFOCOM 2016, pp. 1–9 (2016). https://doi.org/10.1109/INFOCOM.2016.7524372
3. Augustine, J., Hinnenthal, K., Kuhn, F., Scheideler, C., Schneider, P.: Shortest paths in a hybrid network model. In: Proceedings of the Thirty-First Annual ACM-SIAM Symposium on Discrete Algorithms, SODA 2020, USA, pp. 1280–1299. Society for Industrial and Applied Mathematics (2020)
4. Bose, P., Morin, P., Stojmenovic, I., Urrutia, J.: Routing with guaranteed delivery in ad hoc wireless networks. Wirel. Netw. **7**(6), 609–616 (2001). https://doi.org/10.1023/A:1012319418150
5. Castenow, J., Kolb, C., Scheideler, C.: A bounding box overlay for competitive routing in hybrid communication networks. In: Proceedings of the 21st International Conference on Distributed Computing and Networking (ICDCN 2020), pp. 14:1–14:10 (2020). https://doi.org/10.1145/3369740.3369777
6. Censor-Hillel, K., Leitersdorf, D., Polosukhin, V.: Distance computations in the hybrid network model via oracle simulations. In: Bläser, M., Monmege, B. (eds.) 38th International Symposium on Theoretical Aspects of Computer Science (STACS 2021). Leibniz International Proceedings in Informatics (LIPIcs), Dagstuhl, Germany, vol. 187, pp. 21:1–21:19. Schloss Dagstuhl – Leibniz-Zentrum für Informatik (2021). https://doi.org/10.4230/LIPIcs.STACS.2021.21. https://drops.dagstuhl.de/opus/volltexte/2021/13666
7. Censor-Hillel, K., Leitersdorf, D., Polosukhin, V.: On sparsity awareness in distributed computations. In: Proceedings of the 33rd ACM Symposium on Parallelism in Algorithms and Architectures, SPAA 2021, pp. 151–161. Association for Computing Machinery, New York (2021). https://doi.org/10.1145/3409964.3461798
8. Coy, S., et al.: Near-shortest path routing in hybrid communication networks. In: Bramas, Q., Gramoli, V., Milani, A. (eds.) 25th International Conference on Principles of Distributed Systems (OPODIS 2021). Leibniz International Proceedings in Informatics (LIPIcs), Dagstuhl, Germany, vol. 217, pp. 11:1–11:23. Schloss Dagstuhl – Leibniz-Zentrum für Informatik (2022). https://doi.org/10.4230/LIPIcs.OPODIS.2021.11. https://drops.dagstuhl.de/opus/volltexte/2022/15786
9. Elkin, M., Neiman, O.: On efficient distributed construction of near optimal routing schemes. In: Proceedings of the 2016 ACM Symposium on Principles of Distributed Computing, pp. 235–244 (2016)

10. Feldmann, M., Hinnenthal, K., Scheideler, C.: Fast hybrid network algorithms for shortest paths in sparse graphs. In: Proceedings of the 24th International Conference on Principles of Distributed Systems (OPODIS 2020), pp. 31:1–31:16 (2020). https://doi.org/10.4230/LIPIcs.OPODIS.2020.31

11. Kuhn, F., Schneider, P.: Computing shortest paths and diameter in the hybrid network model. In: Proceedings of the 39th Symposium on Principles of Distributed Computing, PODC 2020, pp. 109–118. Association for Computing Machinery, New York (2020). https://doi.org/10.1145/3382734.3405719

12. Kuhn, F., Schneider, P.: Routing schemes and distance oracles in the hybrid model. In: International Symposium on Distributed Computing (DISC), vol. 246, pp. 28:1–28:22 (2022)

13. Kuhn, F., Wattenhofer, R, Zhang, Y., Zollinger, A.: Geometric ad-hoc routing: of theory and practice. In: Proceedings of the 22nd ACM Symposium on Principles of Distributed Computing (PODC 2003), pp. 63–72 (2003). https://doi.org/10.1145/872035.872044

14. Kuhn, F., Wattenhofer, R., Zollinger, A.: Asymptotically optimal geometric mobile ad-hoc routing. In: Proceedings of the 6th International Workshop on Discrete Algorithms and Methods for Mobile Computing and Communications (DIAL-M 2002), pp. 24–33 (2002). https://doi.org/10.1145/570810.570814

15. Kuhn, F., Wattenhofer, R., Zollinger, A.: Worst-case optimal and average-case efficient geometric ad-hoc routing. In: Proceedings of the 4th ACM International Symposium on Mobile Ad Hoc Networking and Computing (MobiHoc 2003), pp. 267–278 (2003). https://doi.org/10.1145/778415.778447

16. Lenzen, C., Patt-Shamir, B.: Fast routing table construction using small messages: extended abstract. In: Proceedings of the Forty-Fifth Annual ACM Symposium on Theory of Computing, STOC 2013, pp. 381–390. Association for Computing Machinery, New York (2013). https://doi.org/10.1145/2488608.2488656

17. Lenzen, C., Patt-Shamir, B.: Fast partial distance estimation and applications. In: Proceedings of the 2015 ACM Symposium on Principles of Distributed Computing, pp. 153–162 (2015)

18. Lynch, N.A.: Distributed Algorithms. Elsevier (1996)

19. Rozhoň, V., Grunau, C., Haeupler, B., Zuzic, G., Li, J.: Undirected $(1+\varepsilon)$-shortest paths via minor-aggregates: near-optimal deterministic parallel and distributed algorithms. In: Symposium on Theory of Computing, pp. 478–487 (2022)

20. Sarma, A.D., et al.: Distributed verification and hardness of distributed approximation. SIAM J. Comput. **41**(5), 1235–1265 (2012)

21. Schneider, P.: Power and limitations of Hybrid communication networks. Ph.D. thesis, University of Freiburg (2023). https://doi.org/10.6094/UNIFR/232804

Distributed Half-Integral Matching
and Beyond

Sameep Dahal[iD] and Jukka Suomela[(✉)][iD]

Aalto University, Espoo, Finland
{sameep.dahal,jukka.suomela}@aalto.fi

Abstract. By prior work, it is known that any distributed graph algorithm that finds a *maximal matching* requires $\Omega(\log^* n)$ communication rounds, while it is possible to find a *maximal fractional matching* in $O(1)$ rounds in bounded-degree graphs. However, all prior $O(1)$-round algorithms for maximal fractional matching use arbitrarily fine-grained fractional values. In particular, none of them is able to find a *half-integral* solution, using only values from $\{0, \frac{1}{2}, 1\}$. We show that the use of fine-grained fractional values is necessary, and moreover we give a *complete characterization* on exactly how small values are needed: if we consider maximal fractional matching in graphs of maximum degree $\Delta = 2d$, and any distributed graph algorithm with round complexity $T(\Delta)$ that only depends on Δ and is independent of n, we show that the algorithm has to use fractional values with a denominator at least 2^d. We give a new algorithm that shows that this is also sufficient.

Keywords: maximal matching · fractional matching · half-integral matching · distributed graph algorithms

1 Introduction

By prior work, it is known that there is a distributed graph algorithm that finds a *maximal fractional matching* (see Sect. 1.2) in $O(\Delta)$ rounds in graphs of maximum degree Δ [3]; in particular, the running time is independent of n and only depends on Δ. However, the algorithm uses very fine-grained fractional values; when Δ increases, the denominators grow exponentially fast. In this work we show that this is necessary: any distributed graph algorithm that finds a maximal fractional matching in $T(\Delta)$ rounds, independently of n, has to use fractional values with a denominator at least $2^{\lfloor \Delta/2 \rfloor}$ (and this is tight). In particular, there cannot be a $T(\Delta)$-rounds algorithm for finding a maximal *half-integral* matching.

1.1 Distributed Maximal Matching Is Hard

Maximal matching is one of the classic problems in the field of distributed graph algorithms, studied extensively since the very early days of the field in the 1980s

S. Rajsbaum et al. (Eds.): SIROCCO 2023, LNCS 13892, pp. 339–356, 2023.
https://doi.org/10.1007/978-3-031-32733-9_15

[4,6–8,11,13,17]. In the maximal matching problem, the task is to find a matching (a set of edges without common vertices) that is not a strict subset of another matching. This is something one can trivially find in a centralized setting (pick independent edges greedily until you are stuck), but this is a challenging coordination task in a distributed setting, for two reasons:

1. One has to *break symmetry*. For example, if the input graph is a cycle, one has to select some but not all edges—the input is symmetric, but the output is not. The task is not solvable at all without resorting to, e.g., unique identifiers or randomness, and even then we cannot solve the task in constant number of rounds; maximal matching in cycles requires $\Omega(\log^* n)$ rounds [14,16].
2. One has to solve a *local coordination* task. Even if we have a Δ-regular bipartite graph, with the bipartition given, we still need $\Omega(\Delta)$ rounds to find a maximal matching, at least in sufficiently large graphs [4].

On the positive side, $O(\Delta + \log^* n)$-round distributed algorithms for finding a maximal matching in a graph of maximum degree Δ are known [18]; one can also make different trade-offs between dependency on Δ vs. n [6–8], but it is impossible to achieve a running time of $T(\Delta)$, independent of n [14,16]. All of these results hold in the usual LOCAL model of distributed computing (see Sect. 2.2 for the details).

1.2 Distributed Fractional Matching is Easier

A matching $M \subseteq E$ in a graph $G = (V, E)$ can be interpreted as a function x that assigns value $x(e) = 1$ to each edge $e \in M$. If we let

$$x[v] = \sum_{e \in E : v \in e} x(e)$$

denote the sum of labels on edges incident to node $v \in V$, then we can define that function $x \colon E \to \{0, 1\}$ is a *matching* if $x[v] \le 1$ for all $v \in V$. Moreover, x is a *maximal matching* if for each edge $\{u, v\} \in E$ at least one endpoint is *saturated*, i.e., $x[u] = 1$ or $x[v] = 1$. Finally, x is a *maximum matching* if it maximizes $\sum_e x(e)$.

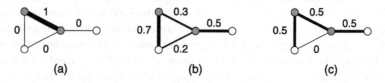

Fig. 1. (a) A maximal matching. (b) A maximal fractional matching. (c) A maximal half-integral matching. The orange nodes are saturated. (Color figure online)

We can now also consider the *fractional relaxation* of this integer program. We say that $x \colon E \to [0, 1]$ is a *fractional matching* if it satisfies $x[v] \le 1$ for each

$v \in V$, it is a *maximal fractional matching* if $x[u] = 1$ or $x[v] = 1$ for each edge $\{u, v\} \in E$, and it is a *maximum fractional matching* if it maximizes $\sum_e x(e)$. See Fig. 1 for illustrations.

Note that any maximal matching is also a maximal fractional matching, but the converse is not necessarily true. However, maximal fractional matchings share many useful properties of maximal matchings. For example, the set of saturated nodes forms a 2-approximation of a minimum vertex cover [5].

When we consider distributed graph algorithms for maximal fractional matchings, one of the obstacles discussed in Sect. 1.1 goes away: *we do not need to break symmetry*. For example, if the graph is a cycle, we can simply label all edges with $1/2$. More generally, if we have a d-regular graph, we can label all edges with $1/d$. The lower bound of $\Omega(\log^* n)$ from [14,16] for symmetry-breaking problems no longer applies.

While the case of non-regular graphs is much more challenging, it is nevertheless possible to design distributed algorithms that find a maximal fractional matching in $O(\Delta)$ rounds, independently of n [3]. It is also known that the local coordination challenge does not disappear; $o(\Delta)$-round algorithms do not exist [10].

1.3 What About Half-Integral Matchings?

The fractional matching polytope is *half-integral* (see e.g. [20, Sect. 30.3]). That is, there exists a maximum fractional matching in which $x(e) \in \{0, \frac{1}{2}, 1\}$ for every edge $e \in E$.

There is also a simple distributed strategy that at first seems to lead to half-integral solutions (see e.g. [2]). First, construct the *bipartite double cover* $G' = (V', E')$ of the graph $G = (V, E)$: for each node $v \in V$ we have two nodes v_1 and v_2 in V', and for each edge $\{u, v\} \in E$ we have two edges $\{u_1, v_2\}$ and $\{u_2, v_1\}$ in E'. Now G' is bipartite, and we know the bipartition, with nodes v_1 on one side and nodes v_2 on the other side. We can now apply any algorithm that finds a matching x' in the bipartite graph G', and this can be mapped into a half-integral matching x by setting

$$x[\{u, v\}] = \frac{x'[\{u_1, v_2\}] + x'[\{u_2, v_1\}]}{2}. \tag{1}$$

Hence, we could use any distributed algorithm designed for bipartite graphs—there is a very simple algorithm that finds a maximal matching in bipartite graphs in $O(\Delta)$ rounds independently of n. Then by applying (1) we could turn it into a fractional matching.

There is, unfortunately, a catch: while (1) will preserve feasibility (given a matching x' it will result in a fractional matching x), it will not preserve maximality: even if x' is a maximal matching, it is not necessarily the case that x is a maximal fractional matching. Could we nevertheless find a half-integral matching efficiently with a distributed algorithm?

If we consider prior distributed algorithms for maximal fractional matching [2,3], they are very far from being able to produce half-integral matchings. For example, [2] uses fractional values with denominators as large as $2^{\Delta-1}$ and [3] is even worse. In this work we show that denominators exponential in Δ are necessary, but we can still do better than prior work.

1.4   Contributions

Our main result is a full characterization of exactly how fine-grained fractional values are necessary:

Theorem 1 (Upper bound). *There is a $T(\Delta)$-round distributed algorithm that finds a maximal fractional matching in graphs of maximum degree $\Delta \leq 2d+1$ using only fractional numbers of the form a/b where $a = 0, 1, \ldots, 2^d$ and $b = 2^d$.*

Theorem 2 (Lower bound). *There is no $T(\Delta)$-round distributed algorithm for any function T that finds a maximal fractional matching in graphs of maximum degree $\Delta \leq 2d + 2$ using only fractional numbers of the form a/b where $a = 0, 1, \ldots, 2^d$ and $b = 1, 2, \ldots, 2^d$.*

We emphasize that the upper bound only uses multiples of $1/2^d$, while the lower bound also excludes the possibility of finding a maximal matching using, e.g., values that are multiples of $1/\Delta$.

As a corollary of these results, we also have a full characterization of the complexity of half-integral matchings:

Corollary 1. *It is possible to find a maximal half-integral matching in graphs of maximum degree $\Delta = 3$ in $O(1)$ rounds.*

Corollary 2. *It is not possible to find a maximal half-integral matching in graphs of maximum degree $\Delta = 4$ in $O(1)$ rounds.*

For larger values of Δ, the range of fractional numbers we use is much smaller than in prior work. In our algorithm, the denominator is upper bounded by $2^{\Delta/2}$, while in prior work [2] it is approximately 2^{Δ}.

1.5   Key New Ideas

While the upper bound of Theorem 1 is a relatively simple adaptation of ideas from prior work, the lower bound of Theorem 2 requires a development of a new proof strategy.

Prior lower-bound techniques in this area tend to fall in one of these categories, each unsuitable for us:

1. The lower-bound construction is a regular graph [4,12]. In Δ-regular graphs we can trivially find a fractional matching using the value $1/\Delta$, which is exponentially far from the lower bound in Theorem 2 that we aim at proving.
2. The lower-bound result aims at establishing that one needs some specific number of rounds, e.g., $\Omega(\Delta)$ rounds [4,10,12]. However, in Theorem 2 we aim at proving that even if the round complexity is, say, exponential in Δ, one cannot avoid using fine-grained fractional values.

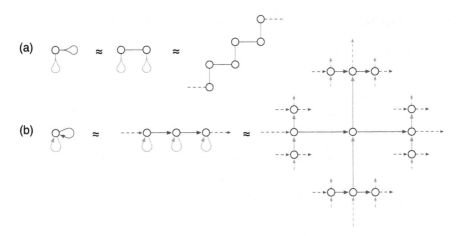

Fig. 2. (a) In prior work [10,12], all the heavy lifting is done in a so-called EC model, in which edges are undirected but colored. Self-loops represent undirected edges. For example, a node with 2 self-loops represents a node in the middle of a 2-regular tree, i.e., a long path. (b) In this work, we work in the PO model. Self-loops represent long directed paths. For example, a node with 2 self-loops represents a node in the middle of a 4-regular tree in which all nodes have indegree 2 and outdegree 2.

Our proof strategy superficially resembles the one used in [10,12] in the sense that we start with one node and k self-loops, which represents the local view of a node in the middle of a regular graph, and then we start unfolding the loops. At each point of the process we see what is the output the algorithm commits to, and then we continue the process until we are left with a concrete lower-bound graph. However, there are major differences; see Fig. 2:

- In [10,12] they start with a *pair of nodes*. The nodes have self-loops, and each self-loop represents an *undirected edge*; the entire argument relies on the fact that an algorithm cannot break symmetry between two ends of such an edge. At each step they unfold a relevant loop, doubling the number of nodes, and then they mix elements from two instances, resulting in another pair of instances. In each iteration they lose one self-loop but force the algorithm to look one step further.
- In this work we start with a *single node*. The node has self-loops, but this time each self-loop represents a *long directed path*; our argument relies on the fact that an algorithm cannot break local symmetry between two nodes near the middle of the path. At each step we unfold a relevant loop, but this will turn one node into a directed path of length $\Theta(T)$. We are interested in the behavior of the algorithm both in the middle of the path and at the endpoints of the path. In each iteration we lose one self-loop but force the algorithm to use at least twice as large denominators.

2   Preliminaries

2.1   Graphs and Self-loops

For a graph $G = (V, E)$, we write $\Delta(G)$ to denote the maximum degree of the graph. We use just Δ when G is clear from the context. For any natural number $d \in \mathbb{N}$, we use $\mathcal{G}_d$ to represent the family of graphs such that $G \in \mathcal{G}_d$ if $\Delta(G) \leq d$. Throughout this work, we will assume that the maximum degree of the input graph G is a globally known constant.

In what follows, we will refer to a self-loop simply as a loop. Each loop counts as one incoming and one outgoing edge (in particular, in $\mathcal{G}_{2d}$ a node can have at most d self-loops). We call a graph *loopy* if each vertex of the graph has at least one loop.

2.2   Model of Computing

Our main results, Theorems 1 and 2, hold in the usual LOCAL model [14,19]. For simplicity, we will focus here on deterministic algorithms (even though the results are not hard to extend to randomized algorithms).

However, to prove the lower bound result, it will be convenient to first prove the lower bound in a weaker model (called PO here, following [10]) and then extend the result from the PO model to the LOCAL model. It will be easiest to define everything we need by starting with the deterministic port-numbering model (PN).

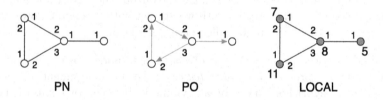

Fig. 3. Models of computing used in this work.

PN Model (Port Numbering) [1,21]. Let $G = (V, E)$ be the input graph. In the PN model, each node $v \in V$ is a computer and each edge $\{u, v\} \in E$ is a communication link between two computers. Initially, each computer is only aware of its degree; nodes of the same degree start with the same initial state.

The endpoints of the edges are labeled with *port numbers*; a node of degree d can refer to its incident edges with the numbers $1, 2, \ldots, d$; see Fig. 3. The port numbering comes from an adversary; a distributed algorithm in the PN model has to work correctly for any given port numbering.

Computation proceeds in *synchronous communication rounds*. In each round, each node can

1. send a message to each neighbor,
2. receive a message from each neighbor, and
3. update its local state based on the current state and the messages it received.

After each round, each node can decide whether it stops and announces its own part of the output—in the case of the maximal fractional problem, the output of a node indicates the fractional value assigned to each incident edge. The *running time* of the algorithm is the number of rounds until all nodes have stopped and announced their local outputs.

PO Model (Port Numbering and Orientation) [10,15]. Algorithms in the PO model behave in exactly the same way as in the PN model. However, there is one additional piece of information available to the algorithm: each edge $\{u, v\} \in E$ is oriented (arbitrarily, by the adversary); see Fig. 3. More precisely, each node knows for each incident edge whether it is "outgoing" or "incoming".

While an arbitrary orientation may not seem particularly useful, note that the PO model is strictly stronger than the PN model. For example, if we have a graph G with two nodes and one edge, it is trivial to find a proper 2-coloring of G in the PO model in 0 rounds, while it is impossible to solve in the PN model in any number of rounds.

LOCAL Model [14,19]. Algorithms in the LOCAL model also behave in exactly the same way as in the PN model, but there is again one additional piece of information available to the algorithm: each node is labeled (arbitrarily, by the adversary) with a *unique identifier* from a polynomially-sized set; see Fig. 3.

Again, the LOCAL model is strictly stronger than the PO model. For example, maximal matching cannot be found in the PO model if the input graph is a cycle that is consistently oriented, while the task is solvable in the LOCAL model in $O(\log^* n)$ rounds.

However, it turns out that *constant-time* algorithms in the LOCAL model are not much stronger than algorithms in the PO model, see e.g. [9,10]. This is the idea we will also make use of in this work: our main goal is to prove a lower bound in the LOCAL model, but it will be convenient to first study the PO model.

2.3   Applying PO Algorithms to Loopy Graphs

To prove the lower-bound result of Theorem 2, we will study the output of a PO algorithm $\mathcal{A}$ in some loopy graph G. However, when we consider distributed graph algorithms, we usually assume that the input graph is loop-free.

However, the output of $\mathcal{A}$ in loopy graphs is nevertheless well-defined. When we refer to the output of $\mathcal{A}$ on some edge e in G, we refer to the result of the following thought experiment: Unfold all loops in G, as shown in Fig. 2b, and hence we arrive at a tree G'. Then apply $\mathcal{A}$ in G' (as the running time of $\mathcal{A}$ is independent of the size of the input graph, this is well-defined). Edge e in G corresponds to infinitely many edges e' in G', but each such edge is symmetric and hence the output of $\mathcal{A}$ on each such edge e' is the same; hence we can take any such edge e' and interpret its label as the output of $\mathcal{A}$ on e.

In particular, if $\mathcal{A}$ finds a maximal fractional matching in any loop-free graph G', it will also produce a maximal fractional matching in the loopy graph G (the label of the loop is counted twice).

3   Lower Bound Result

In this section we prove the lower-bound result, Theorem 2. It turns out that the critical resource is the number of factors of 2 in the denominators. We start by defining sets of rational numbers that will precisely capture how fine-grained values are needed.

3.1   Sets of Rational Numbers

Any natural number $x \geq 1$ can be written as $x = 2^n \cdot m$ where $n \geq 0$ and $m \equiv 1$ mod 2. We refer to $e(x) = 2^n$ as the *even part* of x and $o(x) = m$ as the *odd part* of x. For $x = 0$, we define $e(x) = 0$ and $o(x) = 1$.

We extend this notion to rational numbers as follows. If $x = p/q$ in the reduced form, we define the *even part of the denominator* $\bar{e}(x) = e(q)$ and the *odd part of the denominator* $\bar{o}(x) = o(q)$. For example, $\bar{e}(0/1) = \bar{e}(1/1) = 1$, $\bar{e}(1/3) = 1$ and $\bar{e}(1/4) = 4$.

For each $n \geq 1$, we define

$$R_n = \{x \in \mathbb{Q} : 0 \leq x \leq 1 \text{ and } \bar{e}(x) = 2^n\},$$
$$R_{\leq n} = R_0 \cup R_1 \cup \cdots \cup R_n,$$
$$R_{\geq n} = R_n \cup R_{n+1} \cup \cdots,$$
$$R_{> n} = R_{n+1} \cup R_{n+2} \cup \cdots.$$

For example, we have

$$R_0 = \left\{0, 1, \tfrac{1}{3}, \tfrac{2}{3}, \tfrac{1}{5}, \tfrac{2}{5}, \tfrac{3}{5}, \tfrac{4}{5}, \dots\right\},$$
$$R_1 = \left\{\tfrac{1}{2}, \tfrac{1}{6}, \tfrac{5}{6}, \dots\right\},$$
$$R_2 = \left\{\tfrac{1}{4}, \tfrac{3}{4}, \tfrac{1}{12}, \tfrac{5}{12}, \tfrac{7}{12}, \tfrac{11}{12}, \dots\right\}.$$

We can view R_n as the set of fractional numbers whose denominator has exactly n trailing zeros in its binary representation. Note that for each rational number $x \in [0, 1]$ there exists exactly one n such that $x \in R_n$.

3.2   High-Level Plan

In Sect. 3.3 we prove the following lemma, which essentially shows that we can without loss of generality focus on the PO model:

Lemma 1. *Fix a natural number $\Delta \in \mathbb{N}$. Then, for any natural number $T \in \mathbb{N}$, the following holds: if there exists a T-round algorithm that solves the maximal fractional matching problem using values in a set $\mathscr{R}$ in the LOCAL model on any graph with maximum degree Δ, then there exists a T-round algorithm that solves the maximal fractional matching problem using values in set $\mathscr{R}$ in the PO model for any loopy graph G with maximum degree Δ.*

Then in Sect. 3.4 we prove the following lemma, which captures exactly how fine-grained rational values are needed in the PO model:

Lemma 2. *Fix natural number $d \in \mathbb{N}$. Then, for any natural number $T \in \mathbb{N}$, there does not exist any algorithm in the PO model that uses T rounds and computes a valid solution for the maximal fractional matching problem using the values from $R_{\leq d-1}$ for loopy graphs in graph family $\mathcal{G}_{2d}$.*

By putting together Lemma 1 and Lemma 2, we obtain:

Lemma 3. *Fix a natural number $d \in \mathbb{N}$. Then, for any natural number $T \in \mathbb{N}$, there does not exist any algorithm in the LOCAL model that uses T rounds and computes a valid solution for the maximal fractional matching problem using the values from $R_{\leq d-1}$ for the graph family $\mathcal{G}_{2d}$.*

Now, Theorem 2 directly follows from Lemma 3.

3.3   Proof of Lemma 1

In [10], a similar result is shown with the exception that the edge labels are arbitrary. However, the same proof follows when we add the restriction that the edge labels come from $\mathscr{R}$. This result is a simple extension of [10, Sections 5.3–5.4], where we can see that the simulation argument does not make changes in the value used for the PO model.

3.4   Proof of Lemma 2

Preliminary Observations. We first make a few observations regarding our problem. First recall the way in which we use loops to represent a node in the middle of a directed path (Fig. 2).

Observation 1. *If a node has a loop then it must be saturated.*

Proof. If a node with a loop was not saturated, we would have a directed path of unsaturated nodes and, in particular, edges with unsaturated endpoints.   □

In a saturated node, the labels of incident edges have to sum up to 1. The following observation captures a key property related to how the even parts of the denominators behave when rational numbers sum up to 1.

Observation 2. *Let $n \geq 1$ and $\frac{k}{m \cdot 2^n} \in R_n$. Consider the equation*

$$2\ell_1 + \ldots + 2\ell_r + x_1 + \ldots + x_{r'} + \frac{k}{m \cdot 2^n} = 1,$$

where each ℓ_i and x_i can be any non-negative rational number. Then, either $\ell_i \in R_{>n}$ or $x_i \in R_{\geq n}$ for some i. Put otherwise, either some ℓ_i has the even part of the denominator larger than 2^n or some x_i has the even part of the denominator at least 2^n.

Proof. First consider the equation

$$x_1 + \ldots + x_q + \frac{k}{m \cdot 2^n} = 1$$

in which each x_i can be any non-negative rational number. We show that there exists an index i for which $x_i \in R_{\geq n}$. We can rewrite it as solving the equation

$$x_1 + \ldots + x_q = \frac{m \cdot 2^n - k}{m \cdot 2^n},$$

where $\frac{m \cdot 2^n - k}{m \cdot 2^n} \in R_n$. If each x_i had the even part of the denominator less than 2^n, then $x_1 + \ldots + x_q$ would also have the even part of the denominator less than 2^n. This is because when we add two rationals $\frac{a_1}{b_1}$ and $\frac{a_2}{b_2}$ we get

$$\frac{a_1}{b_1} + \frac{a_2}{b_2} = \frac{a_1 \cdot (\ell/b_1) + a_2 \cdot (\ell/b_2)}{\ell}$$

where $\ell = \mathrm{lcm}(b_1, b_2)$, the least common multiple of b_1 and b_2. The even part of ℓ will be bounded above by the maximum of the even parts of b_1 and b_2. However, if $x_1 + \ldots + x_q$ has the even part of the denominator less than 2^n, then it contradicts the fact that the sum equals $\frac{m \cdot 2^n - k}{m \cdot 2^n}$.

Now, in order to prove the original statement of Observation 2, it is sufficient to replace $x_{r'+i}$ by $2\ell_i$. If $x_{r'+i} \in R_{\geq n}$ then $\ell_i \in R_{>n}$.   □

Assumptions. We now proceed to prove Lemma 2 by contradiction. For the sake of contradiction, we assume that when we fix a nautral number $d \in \mathbb{N}$, there exists a natural number $T \in \mathbb{N}$ such that the following holds: there exists a PO algorithm $\mathcal{A}$ that solves the maximal fractional matching problem in T rounds using values from the set $R_{\leq d-1}$ for graph family $\mathcal{G}_{2d}$.

Properties. Now, our lower bound construction observes the behavior of $\mathcal{A}$ on different kinds of graphs in $\mathcal{G}_{2d}$ to reason about the set of values that is used. We will construct a sequence of loopy graphs $G_0, G_1, \ldots, G_{d-1}$ to argue that the further we go, the more fine-grained value must be used by our algorithm.

For each $i = 0, 1, \ldots d - 1$, we will maintain the following properties:

P1 $G_i \in \mathcal{G}_{2d}$.
P2 Graph G_i without loops forms a tree.
P3 Each node of G_i has at least $d - i$ loops.
P4 There is an integer $j(i) > i$ and a node v_i in G_i such that $\mathcal{A}$ labels at least one loop of v_i with a rational value $x \in R_{j(i)}$.

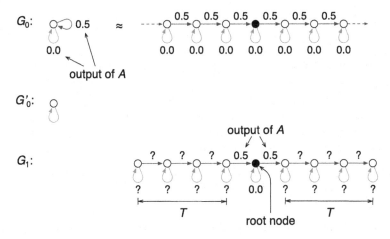

Fig. 4. Construction for $d = 2$ and $T = 3$. Graph G_0 consists of d self-loops. When we apply $\mathcal{A}$ to G_0, at least one of the loops will get labeled by a value in $R_{\geq 1}$; in this example the value was $0.5 \in R_1$. To construct G_1, we remove this loop to arrive at graph G_0', take $2T + 3$ copies of G_0', and connect them with a directed path. The key observation is that given the output of $\mathcal{A}$ in G_0 we also know the output of $\mathcal{A}$ around the node in the middle of G_1—this node is called the *root node* of G_1.

Base Case. Our first graph G_0 consists of a single node v_0 with d oriented self loops (see Fig. 4).

Graph G_0 satisfies properties **P1**, **P2** and **P3** by construction, so we now need to verify only **P4**. Consider that $\mathcal{A}$ assigns values $a_1, \ldots, a_d$ to the loops of v_0. Since v_0 has loops, it must be saturated (recall Observation 1), and hence it must satisfy that $2a_1 + 2a_2 + \ldots + 2a_d = 1$. This is equivalent to solving $a_1 + a_2 + \ldots + a_d = 1/2$ and by Observation 2 we know that there exists an i with $a_i \in R_{\geq 1}$.

Inductive Step. Given G_{i-1}, we construct G_i as follows; see Fig. 4:

S1 Construct the graph G_{i-1}' from G_{i-1} by removing the loop of v_{i-1} for which $\mathcal{A}$ assigned a value in $R_{j(i-1)}$.
S2 Create $2T + 3$ copies of G_{i-1}'.
S3 For each $k = 1, 2, \ldots, 2T + 2$, connect node v_{i-1} in copy number k to node v_{i-1} in copy number $k + 1$; these new edges are called *path edges*.
S4 Node v_{i-1} in copy number $T + 2$ is called the *root node* of G_i.

This way we form a directed path of length $2T + 3$, with the root node in the middle of the path, as shown in Fig. 4. The key observation is that the output of algorithm $\mathcal{A}$ on the root node of G_i is the same as the output of $\mathcal{A}$ for v_{i-1} in G_{i-1}, due to the fact that the radius-T neighborhood of the root node in G_i is isomorphic to the radius-T neighborhoods of v_{i-1} in G_{i-1} (once we conceptually unfold all loops). This property is illustrated in Fig. 4: compare the radius-T neighborhood of the black node in the unfolding of G_0 with the radius-T neighborhoods of the root node of G_1.

Given G_{i-1} satisfies all the properties, we need to show that the same is true for G_i. **P1**, **P2** and **P3** are satisfied by construction. To prove **P4**, consider the root node of G_i. Since its behavior is completely characterized, we know that it will label the incident path edges with values from $R_{j(i-1)}$.

Recall that, by **P2**, the graph G_i without loops forms a tree. We will navigate in this tree, starting from the root node, and moving away from it until we satisfy **P4**. We maintain the following invariant; see Fig. 5:

Definition 1 (path invariant). *If v is the current node, and P is the unique path from v to the root, we have already concluded that A labels each edge of P with a value from $R_{\geq j(i-1)}$.*

To get started, let e be one of the path edges incident to the root node, and let v be the other end of e. As we discussed earlier, we know that e is labeled with a value from $R_{j(i-1)}$.

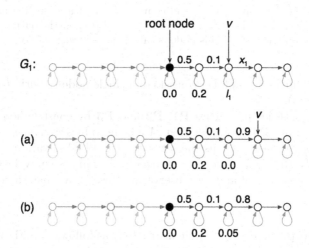

Fig. 5. Inductive step in the proof of Lemma 2 (Sect. 3.4). We have already concluded that all edges in the path between v and the root node are labeled with values from $R_{\geq 1}$. We now ask how algorithm A will label the other edges around v. (a) One possible solution: edge x_1 is labeled with a value $0.9 = \frac{9}{2 \cdot 5} \in R_1$. We did not yet establish property **P4**, but we can extend the R_1-labeled path further away from the root node—eventually we will encounter a leaf node. (b) Another possible solution: we managed to label x_1 with a less fine-grained value $0.8 \in R_0$. However, this means that loop ℓ_1 is labeled with a more fine-grained value $0.05 = \frac{1}{2^2 \cdot 5} \in R_2$. We have established **P4**.

Now assume that we have reached some node v this way. Let P be the path from v to the root, and let e be the first edge of P, let L be the set of loops incident to v, and let X be the set of non-loop edges incident to v that are different from e. That is, we already know the label of edge e, but we do not yet know how $\mathcal{A}$ will label L and X.

Node v is loopy, so it must be saturated. The saturation condition for v is equivalent to solving the equation

$$2\ell_1 + \ldots + 2\ell_r + x_1 + \ldots + x_{r'} + \frac{k}{m \cdot 2^n} = 1,$$

where $n \geq j(i-1)$, values ℓ_i represent the values assigned to the loops in L, values x_i represent the values assigned to the edges in X, and $\frac{k}{m \cdot 2^n}$ refers to the value from $R_{\geq j(i-1)}$ assigned to edge e. With the help of Observation 2, we know that one of the two cases must be true:

1. One of the loops in L has the even part of the denominator $2^{n'}$ for $n' > n$. In this case, we have established **P4**.
2. One of the edges $\{u, v\} \in X$ has the even part of the denominator at least 2^n. We have found another edge labeled with a value from $R_{\geq j(i-1)}$, and we can extend the path P by moving from v to u, still satisfying the path invariant.

Note that this process will eventually terminate, as G_i without loops is a (finite) tree, and hence we will eventually reach a leaf node with $X = \emptyset$. We have established that our construction of graph G_i satisfies properties **P1**–**P4**.

Conclusion. When we take $i = d - 1$, we have a graph $G_{d-1} \in \mathcal{G}_{2d}$ which needs to use even part of the denominator at least 2^d. However, values with denominator 2^d are not present in the set $R_{\leq d-1}$. Thus, we have our desired contradiction.

This concludes the proof of Lemma 2, and hence also the proofs of Lemma 3 and our main lower bound result Theorem 2.

4   Upper Bound Result

Here, we prove the statement of Theorem 1. We will use the notation

$$S(d) = \left\{ \frac{i}{2^d} : i \in \{0, 1, \ldots, 2^d\} \right\}.$$

We need to show that there is a $T(\Delta)$-round, independent of n, distributed algorithm that solves maximal fractional matching in graph family $\mathcal{G}_{2d+1}$ using labels from $S(d)$. We prove the claim by induction, as follows:

- Base case (Lemma 4): $S(1)$ suffices for $\mathcal{G}_2$.
- Odd step (Lemma 5): if $S(d)$ suffices for $\mathcal{G}_{2d}$, then $S(d)$ also suffices for $\mathcal{G}_{2d+1}$.
- Even step (Lemma 6): if $S(d)$ suffices for $\mathcal{G}_{2d+1}$, then $S(d + 1)$ suffices for $\mathcal{G}_{2d+2}$.

We use $T(\Delta)$ to represent the number of rounds taken by our algorithm for graph family $\mathcal{G}_\Delta$. We show that in each of the above steps, $T(\Delta)$ is just a function of Δ and is independent of number of nodes n. We will give a PN algorithm, which implies the existence of a LOCAL algorithm.

Lemma 4. *There is a constant-time PN algorithm that finds a maximal fractional matching in $\mathcal{G}_2$ using values from $S(1)$.*

Proof. In this case, we want to pick $x(e) \in \{0, \frac{1}{2}, 1\}$ for each $e \in E$. We can achieve a simple distributed algorithm with 1 round of communication. Each vertex v, communicates its degree to its neighbors. Any degree 2 vertex can safely assign the value $\frac{1}{2}$ to both of its incident edges. For a degree 1 vertex, it will assign the value $\frac{1}{2}$ to the incident edge if the other endpoint has degree 2 and will assign the value 1, if the other endpoint is 1 as well.

We can see that for each vertex v, the sum of the values assigned to its incident edges is at most 1. By the nature of our algorithm, every degree 2 node is saturated. So, every edge which has a degree 2 endpoint satisfies that one of its endpoints is saturated. The only remaining scenario is when both of the endpoints are degree 1. In this setting, our algorithm assigns the edge with value 1 in which case both of its endpoints are saturated as well. Using 1 round of communication, we have obtained a solution for the maximal fractional matching using values $\{0, \frac{1}{2}, 1\}$ when $\Delta = 2$. This gives us $T(2) = 1$. □

Lemma 5. *Fix $d \in \mathbb{N}$. Assuming that $S(d)$ is sufficient to obtain the solution for $\mathcal{G}_{2d}$, $S(d)$ is sufficient to obtain the solution for $\mathcal{G}_{2d+1}$ as well.*

Proof. Assume that $\mathcal{A}$ is a PN algorithm that computes the solution for $\mathcal{G}_{2d}$ using values in $S(d)$. We now describe PN algorithm $\mathcal{A}'$ that computes the solution for $G \in \mathcal{G}_{2d+1}$ using values in $S(d)$. Algorithm $\mathcal{A}'$ takes the following steps (see Fig. 6 for an illustration):

Step 1: Edge labelling. First, we use the port numbers to define a *label* for each edge. For each edge $e = \{u, v\}$, there exists numbers $i, j \leq \Delta(G)$ such that port i of u is connected to port j of v. We label this edge with the set $\{i, j\}$. Then $L = \{\{i, j\} : 1 \leq i, j \leq \Delta(G)\}$ denotes the set of possible edge labels. We have $|L| = O(\Delta^2)$ different edge labels. For each $\ell \in L$, we define the subgraph G_ℓ of G that contains all the edges labelled ℓ. We write $\deg_{G_\ell}(v)$ for the degree of node v in graph G_ℓ. A key observation is that for each ℓ and v, we have $\deg_{G_\ell}(v) \leq 2$, i.e., each G_ℓ is a collection of paths and cycles.

Step 2: Edge Classification. We classify each edge into two *types*: "Mid" and "End". Consider any edge $e = \{u, v\}$ and say it had label $\ell \in L$. We say that e is of type "Mid" if $\deg_{G_\ell}(u) = 2$ and $\deg_{G_\ell}(v) = 2$. Put otherwise, all edges that are in the middle of the path or part of a cycle in G_ℓ are classified with type "Mid". All other edges are classified as "End". Note that each node can determine the types of its incident edges in two rounds of communication.

Step 3: Solve for "Mid" edges. Consider subgraph G' of G that contains all edges of type "Mid". We argue that $\Delta(G') \leq 2d$. To see this, consider any

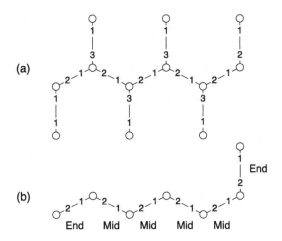

Fig. 6. (a) A graph $G \in \mathcal{G}_3$, with a port numbering. (b) The subgraph G_ℓ for label $\ell = \{1,2\}$, with the edge types "End" and "Mid" indicated.

vertex $v \in V$. If $\deg_G(v) = 2d + 1$, there exists $\ell \in L$ such that $\deg_{G_\ell}(v) = 1$, and therefore at least one edge adjacent to v will receive type "End" and will not be part of G'. Now we have a subgraph G' of G with $\Delta(G') \leq 2d$, and we can simulate $\mathcal{A}$ in G'.

Step 4: Extend for "End" edges. We notice that each edge $e \in G'$ satisfies the maximality condition, i.e., at least one endpoint is saturated. Thus, we now need to ensure the same for edges of type "End". For a label $\ell \in L$, let G_ℓ^{End} be the set of edges labelled ℓ of type "End". We know that edges of type "End" can only be part of paths of length 1 and 2 in G_ℓ^{End}. We proceed to satisfy the maximality condition for edges of type "End" by considering them sequentially on the labels $\ell \in L$. Consider an edge $e = \{u, v\} \in G_\ell^{\text{End}}$. If we assign $x(e) = \min\{1 - x[u], 1 - x[v]\}$ then we can ensure that e satisfies the maximality condition along with ensuring that both u and v satisfy the feasibility condition. The only issue that can arise here is that some other edge adjacent to u or v is trying to update its value in parallel with edge e. Since we are looking at edge of type "End" and proceeding sequentially based on label $\ell \in L$, the above issue can only be caused by paths of length 2. However, the middle vertex of this path can decide the sequential order in which the two edges are considered, after which this issue is avoided.

Step 1 and 2 take a constant number of rounds. Step 3 takes $T(2d)$ rounds to run algorithm $\mathcal{A}$ on graph G'. Step 4 considers $O(\Delta^2)$ labels, and for an individual label ℓ, it takes constant time to assign the values. Overall, the time taken for graph of maximum degree $\Delta = 2d + 1$ is given by the function $T(\Delta) \leq c_1 + c_2\Delta^2 + T(\Delta - 1)$ for some constants c_1 and c_2. Since $T(\Delta - 1)$ is independent of n, $T(\Delta)$ is independent of n as well. Thus, we have obtained a valid solution for the maximal fractional matching problem for graphs in $\mathcal{G}_{2d+1}$ using values from the set $S(d)$. $\qquad\square$

Lemma 6. *Fix $d \in \mathbb{N}$. If $S(d)$ is sufficient to obtain the solution for $\mathcal{G}_{2d+1}$, then $S(d+1)$ is sufficient to obtain the solution for $\mathcal{G}_{2d+2}$.*

Proof. The proof for this theorem uses the same ideas as done in [2]. Consider any graph $G \in \mathcal{G}_{2d+2}$ and let $\mathcal{A}$ be the PN algorithm that uses values in $S(d)$ to compute a valid solution for graphs in $\mathcal{G}_{2d+1}$. We make use of the following definitions from [2]:

Definition 2 (almost-saturating solutions). *A half-integral fractional matching $x\colon E \to \{0, \frac{1}{2}, 1\}$ is almost-saturating if the following conditions hold for each node v:*

- *If $x[v] = 0$, then $x[u] = 1$ for all neighbors u of v.*
- *If $x[v] = 1/2$, then $x[u] = 1$ for at least one neighbor of v.*

Definition 3 (half-saturated edges). *Consider an almost-saturating solution $x\colon E \to \{0, \frac{1}{2}, 1\}$. An edge $e = \{u, v\}$ is:*

- *half-saturated if $x[u] = x[v] = 1/2$,*
- *fully-saturated if $x[u] = 1$ or $x[v] = 1$.*

In [2] there is an algorithm that finds an almost-saturating solution in $O(\Delta)$ rounds. Let $\bar{x}$ denote the almost-saturating solution for G, and we let G' to be the subgraph induced by the half-saturated edges; note that for each node v there has to be at least one incident edge that is not half-saturated. Hence $G' \in \mathcal{G}_{2d+1}$, and we can apply $\mathcal{A}$ to produce a solution x' for G' using values in set $S(d)$. We can then extend domain of x' to E by setting $x'(e) = 0$ for $e \notin G'$. Setting $x(e) = \bar{x}(e) + x'(e)/2$ now gives a maximal fractional matching for the graph G. This is because for any edge $e = \{u, v\}$ in G', we have $\bar{x}[u] = \bar{x}[v] = 1/2$ and $x'[u] = 1$ or $x'[v] = 1$. Moreover, $x(e) \in S(d+1)$. The number of rounds for graphs of degree $\Delta = 2d + 2$ is given by $T(\Delta) \leq c_1 + c_2\Delta + T(\Delta - 1)$ for some constants c_1 and c_2. Since $T(\Delta - 1)$ is independent of n, $T(\Delta)$ is also independent of n. $\qquad\square$

5   Conclusions

Our results give a complete characterization of how fine-grained fractional values are needed in a distributed algorithm that finds a maximal fractional matching in any running time $T(\Delta)$ that only depends on the maximum degree Δ and is independent of n. The main open question is if we can achieve this bound in time $T(\Delta) = O(\Delta)$—this would be optimal by [10].

Acknowledgements. This work was supported in part by the Academy of Finland, Grant 333837. We would like to thank the anonymous reviewers for their helpful feedback, and the members of Aalto Distributed Algorithms group for discussions.

References

1. Angluin, D.: Local and global properties in networks of processors. In: Proceedings of the 12th Annual ACM Symposium on Theory of Computing (STOC 1980), pp. 82–93. ACM Press (1980). https://doi.org/10.1145/800141.804655

2. Åstrand, M., Floréen, P., Polishchuk, V., Rybicki, J., Suomela, J., Uitto, J.: A local 2-approximation algorithm for the vertex cover problem. In: Keidar, I. (ed.) DISC 2009. LNCS, vol. 5805, pp. 191–205. Springer, Heidelberg (2009). https://doi.org/10.1007/978-3-642-04355-0_21

3. Åstrand, M., Suomela, J.: Fast distributed approximation algorithms for vertex cover and set cover in anonymous networks. In: Proceedings of the 22nd ACM Symposium on Parallelism in Algorithms and Architectures (SPAA 2010), pp. 294–302. ACM Press (2010). https://doi.org/10.1145/1810479.1810533

4. Balliu, A., Brandt, S., Hirvonen, J., Olivetti, D., Rabie, M., Suomela, J.: Lower bounds for maximal matchings and maximal independent sets. J. ACM **68**(5), 1–30 (2021). https://doi.org/10.1145/3461458

5. Bar-Yehuda, R., Even, S.: A linear-time approximation algorithm for the weighted vertex cover problem. J. Algorithms **2**(2), 198–203 (1981). https://doi.org/10.1016/0196-6774(81)90020-1

6. Barenboim, L., Elkin, M., Pettie, S., Schneider, J.: The locality of distributed symmetry breaking. In: 2012 IEEE 53rd Annual Symposium on Foundations of Computer Science (FOCS 2012), pp. 321–330 (2012). https://doi.org/10.1109/FOCS.2012.60

7. Barenboim, L., Elkin, M., Pettie, S., Schneider, J.: The locality of distributed symmetry breaking. J. ACM **63**(3), 1–45 (2016). https://doi.org/10.1145/2903137

8. Fischer, M.: Improved deterministic distributed matching via rounding. Distrib. Comput. **33**, 279–291 (2020). https://doi.org/10.1007/s00446-018-0344-4

9. Göös, M., Hirvonen, J., Suomela, J.: Lower bounds for local approximation. J. ACM **60**(5), 1–23 (2013). https://doi.org/10.1145/2528405

10. Göös, M., Hirvonen, J., Suomela, J.: Linear-in-Δ lower bounds in the LOCAL model. Distrib. Comput. **30**(5), 325–338 (2015). https://doi.org/10.1007/s00446-015-0245-8

11. Hanckowiak, M., Karonski, M., Panconesi, A.: On the distributed complexity of computing maximal matchings. SIAM J. Discret. Math. **15**(1), 41–57 (2001). https://doi.org/10.1137/S0895480100373121

12. Hirvonen, J., Suomela, J.: Distributed maximal matching: greedy is optimal. In: Proceedings of the 31st Annual ACM SIGACT-SIGOPS Symposium on Principles of Distributed Computing (PODC 2012), pp. 165–174. ACM Press (2012). https://doi.org/10.1145/2332432.2332464

13. Israeli, A., Itai, A.: A fast and simple randomized parallel algorithm for maximal matching. Inf. Process. Lett. **22**(2), 77–80 (1986). https://doi.org/10.1016/0020-0190(86)90144-4

14. Linial, N.: Locality in distributed graph algorithms. SIAM J. Comput. **21**(1), 193–201 (1992). https://doi.org/10.1137/0221015

15. Mayer, A., Naor, M., Stockmeyer, L.: Local computations on static and dynamic graphs. In: Proceedings of the 3rd Israel Symposium on the Theory of Computing and Systems (ISTCS 1995), pp. 268–278. IEEE (1995). https://doi.org/10.1109/ISTCS.1995.377023

16. Naor, M.: A lower bound on probabilistic algorithms for distributive ring coloring. SIAM J. Discret. Math. **4**(3), 409–412 (1991). https://doi.org/10.1137/0404036

17. Panconesi, A., Rizzi, R.: Some simple distributed algorithms for sparse networks. Distrib. Comput. **14**(2), 97–100 (2001). https://doi.org/10.1007/PL00008932
18. Panconesi, A., Rizzi, R.: Some simple distributed algorithms for sparse networks. Distrib. Comput. **14**, 97–100 (2001). https://doi.org/10.1007/PL00008932
19. Peleg, D.: Distributed Computing: A Locality-Sensitive Approach. Society for Industrial and Applied Mathematics (2000). https://doi.org/10.1137/1. 9780898719772
20. Schrijver, A.: Combinatorial Optimization - Polyhedra and Efficiency. Springer, Heidelberg (2003)
21. Yamashita, M., Kameda, T.: Computing on anonymous networks: part I— characterizing the solvable cases. IEEE Trans. Parallel Distrib. Syst. **7**(1), 69–89 (1996). https://doi.org/10.1109/71.481599

Boundary Sketching with Asymptotically Optimal Distance and Rotation

Varsha Dani[1], Abir Islam[2(✉)] ⓘ, and Jared Saia[2]

[1] Rochester Institute of Technology, Rochester, USA
varsha.dani@rit.edu
[2] University of New Mexico, Albuquerque, USA
{abir,saia}@cs.unm.edu

Abstract. We address the problem of designing a distributed algorithm for two robots that sketches the boundary of an unknown shape. Critically, we assume a certain amount of delay in how quickly our robots can react to external feedback. In particular, when a robot moves, it commits to move along path of length at least λ, *or* turn an amount of radians at least λ for some positive $\lambda \leq 1/2^6$, that is normalized based on a unit diameter shape. Then, our algorithm outputs a polygon that is an ϵ-sketch, for $\epsilon = 8\sqrt{\lambda}$, in the sense that every point on the shape boundary is within distance ϵ of the output polygon. Moreover, our costs are asymptotically optimal in two key criteria for the robots: total distance travelled and total amount of rotation.

Additionally, we implement our algorithm, and illustrate its output on some specific shapes.

Keywords: Boundary Sketch · Robotics · Drones · Distributed Algorithms · Euclidean Plane

1 Introduction

What if a robot cannot react instantaneously? In particular, suppose a robot alternates between (1) analyzing past sensor data in order to plan motion of some minimum amount; and (2) executing that plan and gathering new data. Thus, some small, but finite time elapses between first sensing data; and then planning motion.

Now imagine we want such robots to traverse the boundary of an unknown shape in the Euclidean plane. The robots know nothing about the shape in advance, and can only gather local information as they traverse the shape boundary. If the boundary is a continuous curve, efficiently tracing the exact boundary seems challenging. Instead, our goal is to obtain an *ϵ-sketch*: a traversal curve with the property that every point in the actual boundary is within distance ϵ of some point of the sketch; ϵ will be related to the parameter giving the minimum amount a robot can move.

This work is supported by the National Science Foundation grant IIS 2024520.

Finally, we want to obtain this ϵ-sketch "efficiently". Unfortunately, for most efficiency measures like time or energy usage, cost is a complicated function of the path travelled, since it must account for both angular and linear momentum. This makes it hard to devise an algorithm with provable asymptotic bounds. Instead, prior work generally either provably optimizes at most one parameter related to efficiency, such as amount turned [27], or distance travelled [22].

In this paper, we take a different approach. Our goal is a bicreteria: minimize both (1) distance travelled, and (2) amount turned. Rather serendipitously, we show that using 2 robots it is possible to asymptotically minimize both criteria. This has broad implications for minimizing a large class of efficiency measures. In particular, our algorithm is also asymptotically optimal for any efficiency function that is polynomial in distance traveled and/or amount turned.

Novelty of Result. The novelty of our results is thus three-fold. First, we handle non-zero robot reaction time and also non-instantaneous sensor measurements. Thus, we improve over control-theoretic results which assume instantaneous reaction time, and instantaneous and continual measurements of quantities such as boundary gradient [8,9,15,17], boundary distance [10,16,20,21], or field measurements [26,28,29]. Second, we assume no a priori shape knowledge. Thus we improve over "robotic coverage" results [4,22,27], which assume a priori knowledge of the boundary. Finally, we asymptotically optimize two key criteria: distance travelled and amount turned. Thus, we improve over results [22,27] that provably only optimize only one such criterion.

1.1 Problem Statement

We consider the problem of approximately traversing the boundary of an unknown shape in the Euclidean plane, using two robots.

Problem Parameters. The diameter of the shape is normalized so that it is 1 unit. Our model depends critically on a parameter $\lambda < 1/2^6$, which describes both the "smoothness" of the shape boundary and the reaction time of the robots as described below.

The Robots. We make the following assumptions about the two robots.

- Every time a robot moves, it must commit to travelling a path that has distance of at least λ, *or* turning at least λ radians.
- At any point in time, each robot knows its location and whether it is inside or outside the shape. The robots are both initially located a distance of at most $\sqrt{\lambda}$ from the shape boundary.
- When a robot crosses the shape boundary, it learns the gradient at the crossing point.[1]
- The robots can instantaneously communicate with each other.

---

[1] A robot can consider the last gradient encountered in any path of length λ, so estimation of the gradient at the crossing can be computed efficiently (Details in Sect. A.1).

The Boundary. The boundary of the shape is a *curvilinear polygon*[2], which informally is a closed, non-intersecting loop consisting of a finite number of curves, connected at vertices. Curvilinear polygons include all shapes with boundaries whose gradients are continuous at all but a finite number of points; for example, shapes defined by unions of Gaussians and polygons. They also seem to be the most general shape for which the total rotation of the shape is well-defined.

We make the following additional assumptions about the shape boundary.

- The intersection of the boundary with any ball of a radius $4\sqrt{\lambda}$ centered on a point of the boundary contains *exactly one path component*. (See footnote 2)
- The vertices of the boundary are at least $\sqrt{\lambda}$ distance apart from each other.
- The boundary is twice continuously differentiable except at the vertices.

Our Goal. Our goal is to use the robots to estimate the boundary in the form of an ϵ-sketch, while minimizing both distance travelled and the amount turned by the robots.

Main Result. Our main result is given in the following theorem.

Theorem 1. *For any positive $\lambda < 1/2^6$, there exists an algorithm that uses 2 robots to compute an ϵ-sketch of the boundary, for $\epsilon = 8\sqrt{\lambda}$. Moreover the algorithm requires the robots to travel a total distance and rotate a total amount that are both asymptotically optimal.*

As as a corollary we can use this ϵ-sketch to estimate the area of the shape.

Corollary 1. *Our algorithm can estimate the area of the shape up to an additive error of $O(\ell\sqrt{\lambda})$, where ℓ is the perimeter of the shape.*

1.2   Technical Overview

We now give some intuition behind our algorithm and the proof of Theorem 1.

BOUNDARY-SKETCH *Intuition.* Our algorithm works by trying to ensure a *sandwich invariant*: the robots are traveling in parallel lines on both sides of the boundary. When a robot crosses the boundary, this invariant fails since both robots are now on the same side of the shape. We want the robot that crossed to go back to the other side of the shape in order to reestablish the sandwich invariant. The subroutine CROSS-BOUNDARY performs this function.

The main idea in CROSS-BOUNDARY is to use the boundary gradient learned at the crossing point, to guide the robot back to the other side of the shape and reestablish the sandwich invariant. In CROSS-BOUNDARY, the crossing robot successively takes small steps at a gradually increasing offset from the gradient at the last crossing. The angular offset is in the direction (clockwise or counterclockwise) of the shape boundary. Essentially the robot travels a

---

[2]   These terms are formally defined in Sect. 3.1 in Definition 4 and Definition 6.

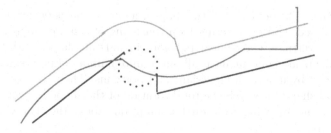

Fig. 1. Illustration of BOUNDARY-SKETCH and CROSS-BOUNDARY. Red curve indicates the shape boundary, blue and green curves indicate the trajectory of the robots that sandwich the boundary. Notice that upon crossing, a dotted blue line (illustrating CROSS-BOUNDARY) indicates a change in direction and step length as discussed in this section. (Color figure online)

regular polygon that approximates a small circle, until it crosses the boundary again. After the crossing, the robot reorients its direction so that both robots are moving in parallel lines the sandwich the boundary. See Fig. 1. By repeatedly re-establishing the sandwich invariant whenever it fails, BOUNDARY-SKETCH progressively computes an ϵ-sketch of the boundary.

BOUNDARY-SKETCH *Analysis.* Our proof of correctness requires tools from real analysis, differential geometry and topology. A main technical challenge is the proof that **BOUNDARY-SKETCH** produces an ϵ sketch, for $\epsilon = 8\sqrt{\lambda}$. Key milestones in this proof include lemmas showing that the sketch exists; it does not self-intersect; and that the sketch and the shape boundary are "close". We use proof by contradiction extensively to show these results. In particular, we repeatedly construct balls of radius $4\sqrt{\lambda}$ that violate the *path component* assumption unless our desired result holds.

Optimality of Distance Traversed. This part of the asymptotic analysis is relatively straightforward. First, we claim when the sandwich invariant fails, the robots at the end of CROSS-BOUNDARY 1) either cover $\Omega(\sqrt{\lambda})$ distance of the shape boundary or 2) traverse a small distance $O(\sqrt{\lambda})$ between successive instances of Case 1 during a number of executions of CROSS-BOUNDARY. This is proven in Lemma 12, which immediately shows that since the former Case 1 occurs at most $O(\ell/\sqrt{\lambda})$ times, the robots traverse $O(\ell)$ distance to restore the sandwich invariant.

Second, the robots take the shortest path when the sandwich invariant holds, since they move in a straight line parallel to each other, they also traverse $O(\ell)$ distance in this case. The optimality of distance traversed follows by combining these two facts.

Optimality of Rotation. To prove bounds on rotation, we need to introduce additional formal definitions in Sect. 3.1, and develop a few helper lemmas in Sect. 3.2. Our first main results is an application of Rolle's Theorem to show the existence, between any points x and y on the shape boundary, of a tangent line somewhere on the boundary between these points that is parallel to the line joining x and y (See Lemma 20). This result has multiple applications including proving two key lemmas, Lemmas 5 and 14. These lemmas were proven via a reduction from the problem for general shapes to shapes that are a polygon. The case of a polygon is one that we can handle easily in the first few lemmas in Sects. 3.2 and 3.4.

Lemma 5 is our first key lemma about our unit-diameter shape. It states that the perimeter of our shape is asymptotically bounded by the total "rotation" in the boundary. In particular, it states that $\ell = O(\phi)$, where ℓ is the shape perimeter and ϕ is the boundary rotation, i.e. the total amount a single robot would rotate if it could follow the shape boundary exactly.

The proof of Lemma 5, requires usage of the property of *uniform continuity* of the curvature (a fact that we prove using *continuity* of the curvature along with some topological properties) to split the curve into a finite number of segments, whose endpoints we define to be vertices of a certain polygon. Next, to compare the perimeter of the shape against the perimeter of this polygon, we borrow a key result in differential geometry from [2] stated as Lemma 2. This lemma from differential geometry compares the path length of a curve with bounded curvature against the length of a line segment connecting two endpoints of that curve, and shows that the former is bounded by a constant times the latter. The other case of unbounded curvature is easy to handle from the definition of total rotation in terms of curvature.

Finally, to compare the total rotation of the polygon against the total rotation of the shape boundary, we recall Lemma 20, which says that the shape has at least some point with a boundary gradient that is parallel to the respective side of the polygon. Thus, the total rotation of the polygon is a lower bound on the total rotation of the shape boundary. Thus, we conclude that the shape boundary rotates at least as much as the constructed polygon boundary.

Lemma 14 is another key lemma for bounding the robot rotation. Lemma 14 bounds the number of times the robots make a turn of $\sqrt{\lambda}$ radians during CROSS-BOUNDARY. Once again, we consider the case where the shape boundary between crossings is a polygon first, and then apply Lemma 20 to derive the asymptotics for the general case. Next, we multiply this bound with the rotation angle $\sqrt{\lambda}$ to bound the overall rotation during all executions of CROSS-BOUNDARY.

Lemma 5 handles an intermediate step where total rotation during CROSS-BOUNDARY include the term ℓ and Lemma 14 handles the rest of the analysis of CROSS-BOUNDARY. Together, these two lemmas prove the optimality of rotation by the robots.

1.3 Related Work

Application Domains. Robot exploration of a shape is a long-standing problem, which has exploded in popularity recently with the advent of drones and other autonomous devices. Application domains are numerous, running the gamut from surveillance of: forest fires [11,13]; harmful algae blooms [18]; mosquito populations [27,30,31]; oil spills [12,23]; radiation leaks [6]; and volcanic emissions [32].

Boundary Search. Our algorithm assumes that the robots are initially located close to the boundary. The *boundary search problem* instead requires the robots to actually find the boundary. Many algorithms for boundary search have been proposed, techniques used include: random walk [6], spiral search [12], gradient following [24], and finite difference approximation based on partial differential equations [7].

Boundary Following. In *boundary following*, the goal is for the robots to traverse the shape boundary, given that they all initially start close to the boundary. This is the problem addressed in our paper. Many control-theoretic algorithms for boundary following offer provable guarantees that their output converges to the exact boundary under certain assumptions on the boundary shape. However, to the best of our knowledge all such results: (1) assume instantaneous and continuous tracking of some quantity such as boundary gradient or distance to boundary; (2) assume infinitesimally accurate control of the robots; and (3) do not give asymptotic bounds on robot travel time or energy expenditure.

Many such prior results use instantaneous and continuous gradient measurements to control the robots tracking the boundary [8,9,15,17]. Some prior results depend on instantaneous and continuous measurements of other quantities; for example, distance from the boundary [10,16,20,21]; or field measurements defining the shape [26,28,29].

Robotic Coverage. In the *robotic coverage problem*, a robot must visit within some given distance of every point in a target shape. Many variants of this problem are known to be NP-Hard, even with a single robot. Thus, many result either use approximation algorithms or heuristics to optimize some criteria such as distance travelled [22] or amount turned [27]. See [4] for a general overview of results. The problem has been extended to multiple robots [5,14]. Our problem is both easier and harder than the typical robotic coverage problem. It is easier in that we only seek to cover a 1-dimensional boundary, and not a 2-dimensional shape. It is harder in that it is online: no information about the shape is known in advance.

2 Our Algorithm: BOUNDARY-SKETCH

Our algorithm BOUNDARY-SKETCH is described in Algorithm 3 (see Fig. 2). It uses a helper algorithm, CROSS-BOUNDARY, that is described in Algorithm 2. We assume an auxiliary function *incomplete* that the robots are capable of, that checks if they have completed a tour around the shape. In addition, by gradient in this algorithm, we mean the direction of the gradient vector.

For simplicity of presentation our algorithms are described in a centralized manner, without explicit communication. To parallelize our algorithms, the robots must sometimes send messages. In particular, if a robot crosses the boundary during some step, it needs to send that information to the other robot.

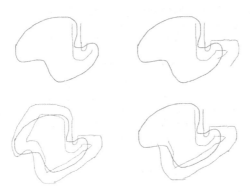

Fig. 2. Figure illustrating a sample execution of BOUNDARY-SKETCH. The red curve indicates the shape boundary and the blue, green curves indicate the path of the robots in progression. (Color figure online)

Algorithm 1. Ensures the robots are at distance $\sqrt{\lambda}$ from each other and are oriented in the same direction. Additional discussion with illustrative diagrams of this synchronization is in Appendix D.

1: **procedure** SYNCHRONIZE(D_1, D_2)
2:     **Path** ← the polyline path of D_2 from last crossing of BOUNDARY-SKETCH with the shape till current position.
3:     ∇ ← the gradient at the last boundary crossing for D_2.
4:     L_1 ← the line in the direction of ∇ through D_1's position.
5:     L_2 ← the line in the direction of ∇ through D_2's position.
6:     **if** L_1 crosses **Path then**
7:         Move D_2 in its current direction until it is $\sqrt{\lambda}$ distance away from L_1. Change direction to ∇ and take a single step of length λ.
8:         Move D_1 along L_1 until it is $\sqrt{\lambda}$ away from D_2.
9:     **else**
10:         Move D_1 in its current direction until it is $\sqrt{\lambda}$ distance away from L_2. Change direction to ∇ and move until the distance from D_2 is $\sqrt{\lambda}$.

Algorithm 2. Reestablishes "Sandwich" Invariant

1: **procedure** CROSS-BOUNDARY(D_1, D_2, α)
2: | $p \leftarrow$ last position of D_1 before crossing
3: | $R \leftarrow$ the vertices of the regular polygon including D_1's position with exterior angle $\sqrt{\lambda}$ and the edge beginning at D_1's position facing the direction of $\nabla + \alpha$.
4: | $P \leftarrow$ the vertices of the convex hull of $R \cup \{p\}$. For all $i : 0 \le i \le |P| - 1$, let P_i be the i-th vertex in this convex hull, ordered such that $P_0 = p$ and $P_1 = D_1$'s current position.
5: | $\nabla \leftarrow$ gradient at the last boundary crossing of D_1
6: | $i \leftarrow 1$.
7: | **while** neither robot has crossed the boundary AND $i + 1 < |P|$ **do**
8: | | D_1 moves to P_{i+1}.
9: | | D_2 moves to closest point from it that is $\sqrt{\lambda}$ distance away from P_i and orthogonal to $\nabla + i\alpha$
10: | | $i \leftarrow i + 1$
11: | **while** neither robot has crossed the boundary **do**
12: | | D_1 moves towards point p taking steps of length λ.
13: | | D_2 moves to closest point from it that is $\sqrt{\lambda}$ distance away from D_1 and orthogonal to D_1's direction.
14: | **if** D_2 crossed the boundary **then**
15: | | SYNCHRONIZE (D_1, D_2)
16: | **else**
17: | | $\nabla \leftarrow$ the current direction of D_1.

Algorithm 3. Initially, robots are $\sqrt{\lambda}$ apart; one inside and one outside

1: **procedure** BOUNDARY-SKETCH(λ)   ▷
2: $D_1, D_2 \leftarrow$ the two robots
3: $\nabla \leftarrow$ boundary gradient at point of crossing with line segment between D_1 and D_2
4: $\alpha \leftarrow \sqrt{\lambda}$
5: | **while** Incomplete(D_1, D_2) **do**
6: | | **if** inside (D_1) XOR inside (D_2) **then**
7: | | | D_1 and D_2 both move λ distance in the direction of ∇
8: | | **if** inside (D_1) = **false** and inside (D_2) = **false then**
9: | | | $\alpha \leftarrow -\sqrt{\lambda}$
10: | | | CROSS-BOUNDARY(D_1, D_2, α)
11: | | **elseif** inside (D_1) = **true** and inside (D_2) = **true**
12: | | | $\alpha \leftarrow \sqrt{\lambda}$
13: | | | CROSS-BOUNDARY (D_2, D_1, α)

3   Analysis

In this section, we give the proof of Theorem 1. We divide the analysis into four sections. First, we formalize the notions of curve, path length and total rotations of a curve in our problem model. In addition, we formally state in the language of topology what *path* and *path component* means. The second section establishes some helper lemmas in computational geometry that will be applied in the later sections. Next, we prove that BOUNDARY-SKETCH terminates and outputs an ϵ-sketch of γ, where $\epsilon = 8\sqrt{\lambda}$. Finally, we provide asymptotic analysis of BOUNDARY-SKETCH.

3.1   Formal Problem Model

The shape is represented by a curve in the Euclidean space. We make use of several definitions, repeated below, about this curve from [19].

Definition 1. *A point $\gamma(t)$ of a parameterized curve γ is called a regular point if $\gamma'(t) \neq 0$; otherwise $\gamma(t)$ is a singular point of γ. A curve is regular if all of its points are regular.*

Definition 2. *A curve $\gamma : [a, b] \rightarrow \mathbb{R}^2$ is called a unit-speed curve if for all $t \in [a, b]$, $|\gamma'(t)| = 1$.*

The next claim which is Proposition 1.3.6 from [19] relates unit-speed parametrization of curves with regular curves.

Lemma 1. *A parametrized curve has a unit-speed reparametrization if and only if it is regular.*

In what follows, we assume γ is regular unless otherwise stated.

Definition 3. *If γ is a unit-speed curve with parameter t, its curvature $\kappa(t)$ at the point $\gamma(t)$ is defined to be $|\gamma''(t)|$.*

Next we generalize the notion of curve by allowing the possibility of *corners*. More precisely, we use the definition 13.2.1 from [19].

Definition 4. *A curvilinear polygon in $\mathbb{R}^2$ is a continuous map $\gamma : \mathbb{R} \rightarrow \mathbb{R}^2$ such that, for some real number T and some values $0 = t_0 < t_1 < ... < t_n = T$:*

1. $\gamma(t) = \gamma(t')$ if and only if $t' - t$ is an integer multiple of T.
2. γ is smooth on each of the open intervals $(t_0, t_1), (t_1, t_2), ..., (t_{n-1}, t_n)$.
3. The one-sided derivatives,

$$\gamma'^-(t_i) = \lim_{t \to t_i^-} \frac{\gamma(t) - \gamma(t_i)}{t - t_i}, \gamma'^+(t_i) = \lim_{t \to t_i^+} \frac{\gamma(t) - \gamma(t_i)}{t - t_i}$$

exist for all $i = 1, ..., n$ and are non-zero and not parallel.

The points $\gamma(t_i)$ are called the vertices of the curvilinear polygon γ, and the segments of it corresponding to the open intervals (t_{i-1}, t_i) are called its edges. Here T is called the period of γ and if the curve has unit-speed i.e. $|\gamma'(t)| = 1$ for all $t \in \mathbb{R}$, then the length of γ, denoted $\ell(\gamma)$ is T, which is the sum of the length of its edges.

Definition 5. *Given a curvilinear polygon γ with vertices at $t_0, t_1, ..., t_n \in [0, T]$ where T is its period , let $\theta_i^{\pm}$ be the angles between $\gamma'^{\pm}(t_i)$ and X-axis. Define $\delta_i = \theta_i^+ - \theta_i^-$ to be the external angle at the vertex $\gamma(t_i)$. The total rotation of γ over the entire period T, denoted ϕ, is defined to be,*

$$\phi = \sum_{i=1}^{n} \delta_i + \int_0^{\ell(\gamma)} |\kappa(t)| dt$$

where we set the speed of γ to be the unit speed.

Next, we state a couple of definitions from the Topology textbook of Munkres [3].

Definition 6. *Given points x and y of a topological space X, a path in X from x to y is a continuous map $f : [a, b] \to X$ of some closed interval in the real line into X, such that $f(a) = x$ and $f(b) = y$. Furthermore, $x, y \in X$ are said to be* path connected *if there is a path from x to y. In addition, define an equivalence relation between pairs $x, y \in X$ if there is a path in X from x to y. The equivalence classes are called the* **path components** *of X.*

Finally, we define ϵ-sketch.

Definition 7. *For $\epsilon > 0$ and a regular curvilinear polygon γ, we say a non self-intersecting polygon P is an ϵ-sketch of γ if every point on γ lies at most an ϵ distance away from P.*

Next we begin the analysis with some helper lemmas, some of which are in the Appendix B.

3.2   Helper Lemmas

We start with a lemma found in the following simplified form (p. 272) in [2].

Lemma 2. *For every real number $K > 0$, any curve in $\mathbb{R}^2$ with curvature at every point not greater than $\chi \in [0, K)$ is not longer than a circular arc of curvature χ whose end points are opposite points of the circumference.*

Lemma 3. *Let $\gamma : [a, b] \to \mathbb{R}^2$ be a regular curve and $\kappa : [a, b] \to \mathbb{R}$ be the curvature function of γ. If $|\kappa(t)| \leq 1/\pi$ for all $t \in [a, b]$, then*

$$\int_{t=a}^{b} |\gamma'(t)| dt \leq \pi |\gamma(b) - \gamma(a)|/2$$

Proof. Let $A = \gamma(a), B = \gamma(b)$ and $\rho = |\gamma(b) - \gamma(a)|$. Now consider the circle drawn from the midpoint of AB with radius $\rho/2$. Since the shape is bounded by a unit square, $\rho \leq \sqrt{2}$. In addition, we have for all $t \in [a, b]$, $|\kappa(t)| \leq 1/\pi \leq 2/\rho$. Setting $\chi = 1/\pi, K = 2/\rho$, we get by Lemma 2,

$$\int_a^b |\gamma'(t)| dt \leq \pi\rho/2 = \pi |\gamma(b) - \gamma(a)|/2$$

This completes the proof.

Next we recall a definition from real analysis.

Definition 8. *A function $f : X \to Y$ with $X \subset \mathbb{R}^n$ and $Y \subset \mathbb{R}^m$ for $n, m \in \mathbb{N}$ is called uniformly continuous on X if for every real number $\epsilon > 0$, there exists a natural number N such that for every $x, y \in X$,*

$$|x - y| < 1/N \implies |f(x) - f(y)| < \epsilon$$

We now state the following lemma that is a simplified form of Theorem 4.19 in [1].

Lemma 4. *Let $f : [a, b] \to \mathbb{R}$ be a continuous mapping with $a, b \in \mathbb{R}$. Then f is uniformly continuous on $[a, b]$.*

Lemma 5. *Let γ be a curvilinear polygon in $\mathbb{R}^2$. Then $\ell(\gamma)$ is $O(\phi)$, where ϕ is the total rotation of γ.*

Proof. **Partition into Segments:**

Let γ be parametrized by its length, then its period $T = \ell$. Suppose γ has m vertices $\gamma(d_1), \gamma(d_2), ..., \gamma(d_m)$ where $d_i \in [0, \ell]$ for all $i = 1, ..., m$. We also set $d_{m+1} = d_1$.

Since $[d_j, d_{j+1}]$ is closed and κ is continuous over $[d_j, d_{j+1}]$, by Lemma 4, κ is uniformly continuous over $[d_j, d_{j+1}]$. That means for all $x, y \in [d_j, d_{j+1}]$ and $j \in [1, m] \cap \mathbb{N}$, there exists $n \in \mathbb{N}$ such that,

$$|x - y| < (d_{j+1} - d_j)/n \implies ||\kappa(x)| - |\kappa(y)|| \leq |\kappa(x) - \kappa(y)| < 1/2\pi \quad (1)$$

We now partition each $[d_j, d_{j+1}]$ into at most n segments of the form $[a_k, b_k]$ where $a_k = (k - 1)\delta_j/n, b_k = k\delta_j/n, \delta_j = d_{j+1} - d_j$ for $k \in [1, n] \cap \mathbb{N}$. Observe that, by inequality 1 for each of these segments $[a_k, b_k]$, either for all $t \in [a_k, b_k]$, $|\kappa(t)| \geq 1/2\pi$ or for all $t \in [a_k, b_k]$, $|\kappa(t)| \leq 1/\pi$. We denote these cases by cases 1, 2 in their respective order.

Finally, over the entire domain $[0, T]$ there are mn segments. Let these segments be indexed by i and let ℓ_i and ϕ_i indicate the perimeter length and the angle turned by the shape in the i-th segment $[a_i, b_i]$.

Case 1:
Since for all $t \in [a_i, b_i]$, $|\kappa(t)| \geq 1/2\pi$ then,

$$\phi_i = \int_{a_i}^{b_i} |\kappa(t)| dt \geq \int_{a_i}^{b_i} 1/2\pi \, dt$$
$$\implies \phi_i \geq 1/2\pi \ell_i$$
$$\implies \ell_i \leq 2\pi\phi_i$$

Case 2:
Next we handle the other case where the segment $[a_i, b_i]$ has the property that for all $t \in [a_i, b_i]$, $|\kappa(t)| \leq 1/\pi$.

Let P be a polygon consisting of vertices equal to the endpoints of each segment of the shape. For a fixed side of this polygon the endpoints are $\gamma(a_i), \gamma(b_i)$. Let ℓ_P be the perimeter length of P.

By Lemma 3, we have $\ell_i \leq \pi|\gamma(b_i) - \gamma(a_i)|/2$. Hence the total length of the shape over all segments covered by these two cases is at most $\pi\ell_P/2$.

By Lemma 20, there exists a value $c \in [a_i, b_i]$ such that $\gamma'(c)$ is parallel to $\gamma(b_i) - \gamma(a_i)$.

Clearly then $\phi \geq \eta$ where η is the sum of the exterior angles of P.

By Lemma 19, $\ell_P \leq \eta$. This means the length of the perimeter of the shape over all the segments covered by this case is at most $\pi\phi/2$.

Conclusion:
Combining both cases gives $\ell(\gamma) \leq 2\pi\phi$ i.e. $\ell(\gamma) = O(\phi)$.

3.3   Correctness of **BOUNDARY-SKETCH**

Let ζ_1, ζ_2 be the parametrized curves for the path of the robots D_1 and D_2 in BOUNDARY-SKETCH. Let $t_i \in [0, \ell(\gamma)]$ such that $\gamma(t_i)$ is i-th point of crossing of either robot with the boundary.

Lemma 6. *The regular polygon constructed in the While loop of Algorithm 2 has diameter at most $2\sqrt{\lambda}$.*

Proof. During Step 10 or Step 13, the regular polygon has diameter at most $\lambda/\sin\sqrt{\lambda}$. By the inequality $\sin(2x) \geq x$ for $x \in [0, \pi/4]$ and since $0 < \sqrt{\lambda} < \pi/4$, we have

$$\lambda/\sin\sqrt{\lambda} \leq 2\sqrt{\lambda}$$

Lemma 7. *Suppose Algorithm 2 is invoked after crossing the shape for the i-th time. Then $\gamma(t_{i+1})$ is at most $3\sqrt{\lambda}$ distance away from the nearest robot for all invocations of Algorithm 2. In addition, $|\gamma(t_i) - \gamma(t_{i+1})| \leq 3\sqrt{\lambda}$ and the nearest robot traverses no more than $3\sqrt{\lambda}$ distance during the execution of Algorithm 2.*

Proof. First we show that $\gamma(t_{i+1})$ is at most $3\sqrt{\lambda}$ distance away for all invocations of Algorithm 2. If $\gamma(t_{i+1})$ is on the boundary of the regular polygon, then it is at most $2\sqrt{\lambda}$ distance away by Lemma 6. Otherwise by triangle inequality it is at most, $2\sqrt{\lambda} + \lambda < 3\sqrt{\lambda}$ distance away, where the first term is the distance from last visited vertex to the starting vertex of the polygon and the second term is the distance from the starting vertex to $\gamma(t_i)$, which are bounded by the diameter of the regular polygon and step length respectively.

Furthermore, following the argument above, $|\gamma(t_i) - \gamma(t_{i+1})| \leq 3\sqrt{\lambda}$ and the nearest robot traverses no more than $3\sqrt{\lambda}$.

Lemma 8. *During the While loop of Algorithm 3, BOUNDARY-SKETCH maintains a distance of at most $\sqrt{\lambda}$ between each of the robots and the shape boundary throughout all executions of Step 7.*

Proof. This is immediate from the assumption that the robots maintain a distance of $\sqrt{\lambda}$ between them and that the shape is sandwiched there.

Lemma 9. BOUNDARY-SKETCH *maintains a distance of at most $8\sqrt{\lambda}$ between the shape boundary and each of the robots.*

Proof. By Lemma 8, throughout all executions of Step 7 inside the While loop in Algorithm 3, every point of γ is at a distance of at most $\sqrt{\lambda}$ from ζ_1 and ζ_2.

Suppose here Algorithm 2 is invoked after crossing the shape for the i-th shape. We will show that over the interval $[t_i, t_{i+1}]$, γ is always at most $7\sqrt{\lambda}$ distance away from the nearest robot.

First by Lemma 7 $\gamma(t_{i+1})$ is at most $3\sqrt{\lambda}$ distance away for all invocations of Algorithm 2. In addition, $|\gamma(t_i) - \gamma(t_{i+1})| \leq 3\sqrt{\lambda}$.

Now define $d(x) = |\gamma(t_{i+1}) - \gamma(x)|$ for all $x \in [t_i, t_{i+1}]$. We claim that $d(x) \leq 4\sqrt{\lambda}$ for all $x \in [t_i, t_{i+1}]$. If not, consider a ball B of radius $4\sqrt{\lambda}$ centered at $\gamma(t_{i+1})$. Observe that the path from $\gamma(t_i)$ to $\gamma(t_{i+1})$ must be contained in B or else we will have two different sections that are disjoint inside this ball. This contradicts our *path component* assumption.

That means we can get to $\gamma(t_{i+1})$ first with at most $3\sqrt{\lambda}$ distance traversal by Lemma 7 and then from $\gamma(t_{i+1})$ to the respective point, which is at most $4\sqrt{\lambda}$ distance away by the above argument. Finally, noting that the robots are apart by at most $\sqrt{\lambda}$ distance and by triangle inequality, the lemma follows. □

Lemma 10. ζ_1, ζ_2 *have finite periods and therefore they intersect the shape finitely many times.*

Proof. Based on our assumption, BOUNDARY-SKETCH selects the direction of the gradient to move away from the region the robots came from. Since the length of γ is finite and the step length of the robots is at least λ, the lemma follows. □

Lemma 11. ζ_1, ζ_2 *do not self-intersect over their respective periods.*

Proof. We will prove this for ζ_1, the proof is identical for ζ_2.

Suppose there exists $u, v \in [0, \ell(\zeta_1)]$ such that $\zeta_1(u) = \zeta_1(v)$ and $u \neq v$.

Observe that, unless crossed γ is always on the same direction (clockwise or counterclockwise) from D_1 and opposite otherwise.

Without loss of generality, assume that γ was on the clockwise direction of D_1 at u. Note that $\zeta_1(u)$ must be inside the shape or else it implies D_1 went back to the direction it came from. Let L be the interval of γ with distance at most $2\sqrt{\lambda}$ from $\zeta_1(u)$ on this direction. By Lemma 8, L is nonempty.

Now consider for v an interval R of γ that is on the counterclockwise direction from D_1 at u and that the distance of every point in R from $\zeta_1(u)$ is at most $2\sqrt{\lambda}$. By Lemma 8 R is nonempty.

Now consider the ball B centered at $\zeta_1(u)$ with radius $2\sqrt{\lambda}$. We now show that the intersection of B with L and R are disjoint. If they are not disjoint, they are *path connected* without crossing themselves, since the latter violates the assumption that γ is a simple i.e. non self-intersecting curve.

If they are *path connected*, BOUNDARY-SKETCH crosses this path since L and R are on different directions of $\zeta_1(u)$. But since D_1 upon crossing the shape, chooses to go away from the direction it came from, it must be that R is on the counterclockwise direction of $\zeta_1(v)$, a contradiction.

Therefore L and R must be disjoint. Finally, we pick any point $c \in L$ and consider a ball B_1 of radius $4\sqrt{\lambda}$ centered at c. Observe that, L and R can only be connected inside this ball in one direction, otherwise it will imply the shape has a bounding box of side length $O(\sqrt{\lambda})$, which contradicts our assumption that the λ is scaled with respect to the diameter of the shape and is at most $1/2^6$.

If L and R connects inside B_1, consider the robot path going in the other direction. We can extend L and R in this direction a distance of at most $8\sqrt{\lambda}$ until we can construct another ball B_2 where L and R do not connect. If this construction is not possible, one of the robots must have crossed the boundary and we can construct this ball B_2 with radius $4\sqrt{\lambda}$ centered at that point of crossing, but this contradicts our assumption on *path component*.

Lemma 12. *For each execution of Algorithm 1, the robots cover $\Omega(\sqrt{\lambda})$ distance of the shape boundary. In addition, the distance traversed by the robots between successive executions of Algorithm 1 and during executions of* CROSS-BOUNDARY *is $O(\sqrt{\lambda})$.*

Proof. The first claim follows immediately since the robot that crosses the boundary changes from D_1 to D_2 and the robots are $\sqrt{\lambda}$ distance apart from each other.

Next, between successive executions of Algorithm 1, robot D_1 may cross the boundary at the end of CROSS-BOUNDARY. The total number of steps robot D_1 can take over this period cannot be more than $2\pi/\sqrt{\lambda}$ since at each step it turns $\sqrt{\lambda}$ and a total turn over a convex path is at most 2π. Since each step is of length λ, the claim follows.

Lemma 13. ζ_1, ζ_2 *are $8\sqrt{\lambda}$-sketches of γ.*

Proof. This follows immediately from Lemmas 9, 11 and Definition 7.

3.4 Asymptotic Analysis

By Lemma 10, BOUNDARY-SKETCH crosses the shape finitely many times. Let m be the number of crossings and for $i \leq m$, ϕ_i be the angle the shape turns over $[t_i, t_{i+1}]$ and define $A = \{1, 2, ..., m\}$ and $f : A \to \mathbb{N}$ such that $f(i)$ is the number of iterations the While loop inside Algorithm 2 executes in between the robots crossing the shape boundary for the i and $i+1$-th time. Some of the lemmas analyzing the case of a polygon are in the Appendix C.

Lemma 14. *If $\sqrt{\lambda} \leq \phi_i \leq \pi/8$ for some positive integer $i \leq m$, $f(i) \leq 8\phi_i/\sqrt{\lambda} + \phi_{i-1}/\sqrt{\lambda} + 1$.*

Proof. The trivial case is where Algorithm 2 is not executed at all i.e. $f(i) = 0$.

Suppose there are j vertices of γ defined over $[t_i, t_{i+1}]$. Let these vertices be indexed $\gamma(a_k)$ where $a_k \in (t_i, t_{i+1})$ for all $1 \le k \le j$. If $j = 0$, select a value $a_1 = (t_{i+1} + t_i)/2$. In addition, let $a_0 = t_i, a_{j+1} = t_{i+1}$.

By Lemma 20, for all $0 \le k \le j$, there exists a $c_k \in (a_k, a_{k+1})$ such that $\gamma'(c_k)$ is parallel to the line segment joining $\gamma(a_k)$ and $\gamma(a_{k+1})$. This means if we consider a polygon P with $j + 1$ vertices being $\gamma(a_k)$ for $0 \le k \le j$, the amount P rotates is at most the amount γ rotates over $[t_i, t_{i+1}]$.

If ϕ_i' is the amount of rotation of P, then by Lemma 23,

$$f(i) \le 8\phi_i'/\sqrt{\lambda} + \phi_{i-1}/\sqrt{\lambda} + 1 \le 8\phi_i/\sqrt{\lambda} + \phi_{i-1}/\sqrt{\lambda} + 1$$

Lemma 15. *If $\pi/8 \ge \phi_i \ge \sqrt{\lambda}$, after resetting ∇ in lines 15 or 17 of Algorithm 2, the robots turn at most $8\phi_i + \phi_{i-1} + \sqrt{\lambda}$ as the algorithm continues to execute Algorithm 3.*

Proof. The angle the robot needs to turn to reorient itself with respect to the boundary just crossed is at most $8\phi_i + \phi_{i-1} + \sqrt{\lambda}$, since the robot orientation itself is no more off than $8\phi_i + \phi_{i-1} + \sqrt{\lambda}$ by Lemma 14.

Lemma 16. *Let I be those indices such that, $\sqrt{\lambda} \le \phi_i \le \pi/8$ for $i \in I$. Then the total radians turned by the algorithm for the ϕ_i values indexed by I is $O(\phi)$.*

Proof. Observe that, $\sum_{i \in I} \phi_i \le \phi$. In addition by Lemma 14 $f(i) \le 8\phi_i/\sqrt{\lambda} + \phi_{i-1}/\sqrt{\lambda} + 1$ and by Lemma 15 the angle turned after resetting ∇ in lines 15 or 17 in Algorithm 2 is at most $8\phi_i + \phi_{i-1} + \sqrt{\lambda}$.

Thus the total radians turned by the algorithm for the ϕ_i values indexed by I is at most:

$$\sum_{i \in I} \sqrt{\lambda} f(i) + 8\phi_i + \phi_{i-1} + \sqrt{\lambda} \le \sum_{i \in I} 16\phi_i + 2\phi_{i-1} + \sqrt{\lambda} = 16\phi + |I|\sqrt{\lambda} = O(\phi)$$

where we note $\phi_i \ge \sqrt{\lambda}$ implies $|I| = O(\phi/\sqrt{\lambda})$.

Theorem 2. *The robots in BOUNDARY-SKETCH traverse a total distance of $O(\ell)$.*

Proof. During Step 7 of Algorithm 3, the robots take the shortest path and thus the distance traversed is bounded by the shape perimeter. Hence overall the total distance traversed during execution of Step 7 is $O(\ell)$.

Now each time Algorithm 1 is executed, the robots cover at least $\Omega(\sqrt{\lambda})$ distance of the shape boundary. This means Algorithm 1 is executed at most $O(\ell/\sqrt{\lambda})$ times. In addition, between successive executions of Algorithm 1 and during executions of CROSS-BOUNDARY the robots traverse a distance at most $O(\sqrt{\lambda})$ by Lemma 12.

Combining we have that $\ell(\zeta_1)$ and $\ell(\zeta_2)$ are both $O(\ell)$.

Theorem 3. *The total rotation by the robots in BOUNDARY-SKETCH is $O(\phi)$.*

Proof. The number of indices $i \leq m$ such that $\phi_i \leq \sqrt{\lambda}$ is $O(\ell/\lambda)$. For each of these indices, the robots rotate $O(\lambda)$ angle in CROSS-BOUNDARY by Lemma 21. Since $\ell = O(\phi)$, for these indices the robots rotate $O(\phi)$.

Now, if $\pi/8 \geq \phi_i \geq \sqrt{\lambda}$, by Lemma 16, the robots rotate a total of $O(\phi)$ radians.

If $\phi_i > \frac{\pi}{8}$ during execution of Algorithm 2, BOUNDARY-SKETCH turns at most 2π. Thus, the radians turned is bounded by $16\phi_i$. In total, across all iterations of Algorithm 2 where $\phi_i > \frac{\pi}{8}$, the robots rotate at most 16ϕ.

In addition, by Lemma 25, the robots asymptotically rotate the same.

Finally, during execution of Step 7 the robots do not rotate at all. Combining all of the above, we have the theorem.

Corollary 2. *The area estimated by BOUNDARY-SKETCH differs from the actual shape area by $O(\ell\sqrt{\lambda})$.*

Proof. BOUNDARY-SKETCH computes the area of the polygon generated by successive positions of one of the robots (say the one that starts from the inside). By Lemma 9, this polygon stays within at most $8\sqrt{\lambda}$ distance from the shape boundary. Hence, the area of the polygon differs from the shape area by at most $O(\ell\sqrt{\lambda})$.

4 Simulations

We present preliminary simulation results to illustrate the precision of ϵ-sketch for various values of λ. To start with, we tested our algorithm for a small number of intersecting gaussians with different variances. Next, we test our algorithm on the boundary of shapes drawn from real world shape data. Both of these simulations demonstrate encouraging convergence of ϵ-sketch to the shape as λ decreases.

Intersecting Gaussians. In our experiments, the shape is an intersection of a few two dimensional gaussians. We discuss experimental details of gradient estimation for this shape in the Appendix A.1.

Definition 9. *In two dimensions, the elliptical gaussian function f for uncorrelated varieties X and Y having a bivariate normal distribution and standard deviations σ_x, σ_y is defined to be,*

$$f(x,y) = \frac{1}{2\pi\sigma_x\sigma_y} e^{-[(x-\sigma_x)^2/(2\sigma_x)^2 + (y-\sigma_y)^2/(2\sigma_y)^2]}$$

Test Shape. We generate a test shape in MATLAB with four gaussians with centers $(1,1), (-2,0), (1,0), (4,0)$ and corresponding standard deviations

$$(0.8, 0.8), (1,1), (1.2, 1.2), (0.7, 0.7).$$

We also ignore the $1/2\pi$ factor in front of the definition.

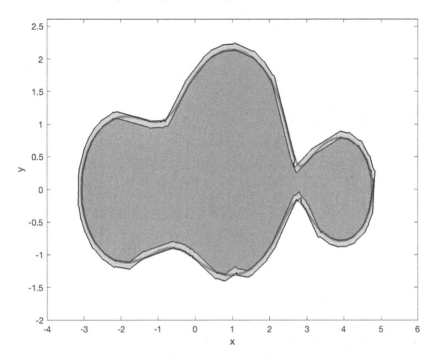

Fig. 3. BOUNDARY-SKETCH for $\lambda = 0.0001$, the green border marks the shape boundary sandwiched by the inner and outer robots. (Color figure online)

BOUNDARY-SKETCH *for different* λ. For the same shape as above, we run BOUNDARY-SKETCH for $\sqrt{\lambda}$ values of $0.0001, 0.000025, 0.000001$. Figure 3 illustrates the output for $\lambda = 0.0001$ and the other two Figs. 5, 6 in Appendix A.1 demonstrate notable improvement in precision as λ gets smaller.

Real World Plume Shapes. We also utilize a real world shape from a volcanic plume in La Palma (see Fig. 7 in Appendix A.2) and use a Python program to generate a polytope approximation of it. The gradient of the boundary then is easily found by the gradient of the corresponding side of the polygon. Next we ran BOUNDARY-SKETCH for different values of λ on this shape. Figure 4 illustrates the output of the algorithm and Figs. 8, 9 in Appendix A.2 demonstrate notable improvement once again in precision as λ gets smaller.

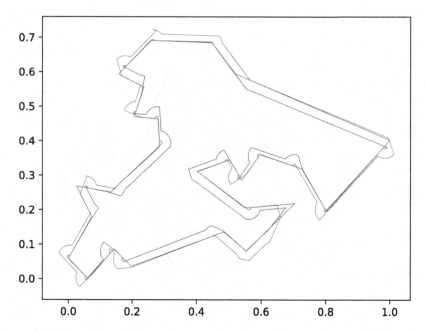

Fig. 4. BOUNDARY-SKETCH for $\lambda = 0.0001$, the green border marks the shape boundary sandwiched by the inner and outer robots (red and blue border). (Color figure online)

5   Conclusion

We have described a distributed algorithm to enable two robots to traverse the perimeter of a curvilinear polygon. Our algorithm is novel in three key ways. First, it does not assume that the robots can respond instantaneously to sensor data, instead assuming a minimum amount of movement or rotation occurs between motion planning events. Second, it does not assume that the robots have access to instantaneous and continual sensor readings. Finally, our algorithm is simultaneously asymptotically optimal in two criteria: both total distance travelled by the robots, and also total amount turned by the robots.

Several open problems remain including the following. First, lower bounds: even though our algorithm is asymptotically optimal in both distance travelled and distance turned, we still would like to know if 2 robots are strictly necessary. Can a single robot achieve the same asymptotic results? We conjecture the answer is no, even if the single robot is attempting to follow a (unknown) function in the Euclidean plane defined from the values $x = 0$ to $x = 1$. Second, Can we reduce the value of ϵ in the ϵ-sketch returned by our algorithm? In particular, is it possible to obtain a value of ϵ that is $o(\sqrt{\lambda})$? Finally, How does our algorithm perform when deployed in the real world? A real-world application domain of particular interest to us is tracing boundaries of volcanic CO2 plumes.

A   Simulation Figures

A.1   Intersecting Gaussians

Gradient Estimation. To estimate the gradient at a point of intersection, we formulate a least squares error (LSQ) problem and solve it using the pseudo-inverse, as described below.

Least Squares Error Formulation. Assume we have three points a, b, c that are close together, and that we have the values $f(a)$, $f(b)$, $f(c)$. For the sake of the BOUNDARY-SKETCH, these three points can be said to be successive points of a robot's path with the middle one being where it crosses the shape boundary. Then we estimate the gradient $\nabla f(b) = (x', y')$ as a vector that minimizes the following sum:

$$(f(a) - f(b) - \nabla f(b) \cdot (a - b))^2 + (f(c) - f(b) - \nabla f(b) \cdot (c - b))^2$$

Setting the rows of a 2×2 matrix A to be $a - b$, $c - b$ respectively and the entries of vector β to be $f(a) - f(b), f(c) - f(b)$ respectively, this problem is formally solving for the vector v such that, $|Av - \beta|^2$ is smallest.

Finally, we utilize the pseudo-inverse method for solving LSQ as discussed in numerous sources e.g. in Sect. 4.5 of [25].

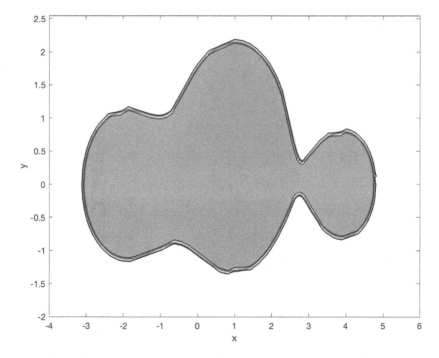

Fig. 5. BOUNDARY-SKETCH for $\lambda = 0.000025$, the green border marks the shape boundary sandwiched by the inner and outer robots. (Color figure online)

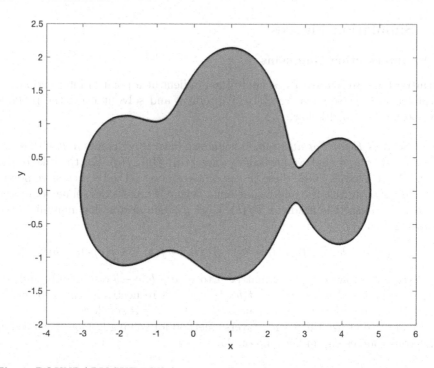

Fig. 6. BOUNDARY-SKETCH for $\lambda = 0.000001$, the green border marks the shape boundary sandwiched by the inner and outer robots. (Color figure online)

A.2   Real World Plume Shapes

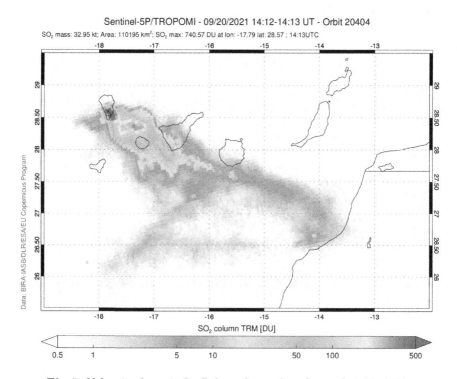

Fig. 7. Volcanic plume in La Palma observed on September 20, 2021.

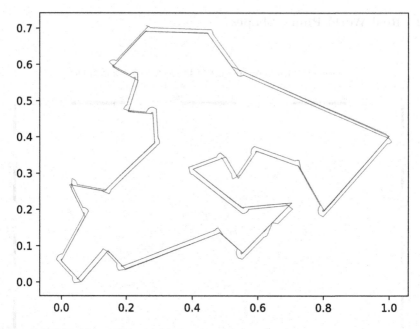

Fig. 8. BOUNDARY-SKETCH for $\lambda = 0.000025$, the green border marks the shape boundary sandwiched by the inner and outer robots (red and blue border). (Color figure online)

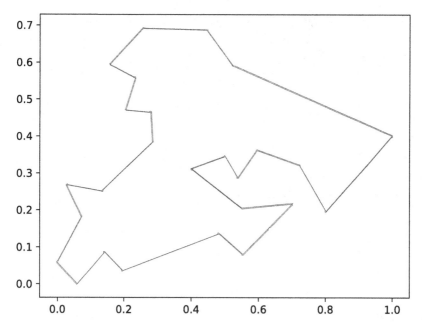

Fig. 9. BOUNDARY-SKETCH for $\lambda = 0.000001$, the green border marks the shape boundary sandwiched by the inner and outer robots (red and blue border). (Color figure online)

B Helper Lemmas

Lemma 17. ℓ *is* $\Omega(1)$

Proof. Since the diameter of the shape is scaled to be 1, $\ell \geq 1$ and the lemma follows.

Lemma 18. *The number of vertices in* γ *is at most* $\ell/\sqrt{\lambda}$.

Proof. This is immediate from the assumption that each of the vertices are at least $\sqrt{\lambda}$ distance apart.

Lemma 19. *If a curve* γ *is a polygon and* ϕ *is the total rotation of this polygon, then* $\ell(\gamma) \leq \phi$.

Proof. Suppose the vertices of the polygon P are given by a list of n points $v_1, v_2, ..., v_n$. Then $\ell = \sum_{i=1}^{n} |v_i - v_{i+1}|$ where we set $v_{n+1} = v_1, v_{n+2} = v_2$.

Now fix three successive vertices, v_i, v_{i+1}, v_{i+2} for $1 \leq i \leq n$ and denote them A, B, C respectively. In addition, let $a = |AB|, b = |BC|, c = |CA|, \alpha = \angle BAC, \beta = \angle ABC, \omega = \angle BCA$.

By the law of sines,

$$\frac{a}{\sin \alpha} = \frac{b}{\sin \beta} = \frac{c}{\sin \omega} = 2\rho$$

where ρ is the radius of the circumcircle of the triangle ABC.

Since the polygon is bounded by a unit square, $\rho \leq 1$. By the inequality $\sin x \leq x$ for all $x \in \mathbb{R}$, we have,

$$a \leq 2\alpha$$

$$b \leq 2\beta$$

Hence,

$$a + b \leq 2(\alpha + \beta) = 2(\pi - w) = 2\phi_i$$

where ϕ_i is the i-th exterior angle of P.

Summing over all $i \in [1, n]$ we get,

$$2\ell(\gamma) = \sum_{i=1}^{n} |v_i - v_{i+1}| + |v_{i+1} - v_{i+2}| \leq 2 \sum_{i=1}^{n} \phi_i = 2\phi$$

This implies,

$$\ell(\gamma) \leq \phi$$

Lemma 20. *Let $\gamma : [0, L] \to \mathbb{R}^2$ be a regular curve parametrized by its arc length L such that $\gamma(0) \neq \gamma(L)$. Then there exists $c \in (0, L)$ such that $\gamma'(c)$ is parallel to the line segment joining $\gamma(0)$ and $\gamma(L)$.*

Proof. Let $\gamma(t) = (x(t), y(t))$ for all $t \in [0, L]$ where x, y are differentiable single valued real functions defined over $[0, L]$.

Let u be the vector from $\gamma(0)$ to $\gamma(L)$. That is,

$$u = \gamma(0) - \gamma(0) = (x(L) - x(0), y(L) - y(0))$$

Let v be a vector perpendicular to $\gamma(L) - \gamma(0)$. Since, $\langle u, v \rangle = 0$, we can write,

$$v = (y(0) - y(L), -x(0) + x(L))$$

Now consider the function f defined as follows over $[0, L]$,

$$f(t) = \gamma(t) \cdot v = x(t)(y(0) - y(L)) + y(t)(x(L) - x(0))$$

Observe that, $f(0) = f(L)$. Hence by Rolle's theorem there exists $c \in (0, L)$ such that, $f'(c) = 0$. Since, for all $t \in (0, L)$, $f'(t) = \langle \gamma'(t), v \rangle + \langle \gamma(t), v' \rangle = \langle \gamma'(t), v \rangle$, we conclude, $\langle \gamma'(c), v \rangle = 0$.

This means $\gamma'(c)$ is perpendicular to v. Since u is perpendicular to v as well, we conclude that, $\gamma'(c)$ is parallel to u.

C   Asymptotic Analysis

Lemma 21. *Let $j > 1$ be an index such that, after crossing the boundary at point D (Fig. 10), D_1, D_2 are both outside or D_1, D_2 are both inside. Then the number of times the While loop in Step 7 of Algorithm 2 executes before the robots cross the line DB is at most $\phi_{j-1}/\sqrt{\lambda} + 1$.*

Proof. Define ψ_j to be the change of gradient in radians between $\gamma(t_{j-1})$ and $\gamma(t_j)$. Clearly, $\phi_{j-1} \geq \psi_j$.

The vertical distance from the robot at the beginning of the execution of Algorithm 2 to the line DB is at most $\lambda \sin(\psi_j + \sqrt{\lambda})$.

Thus after $\psi_j/\sqrt{\lambda} + 1 \leq \phi_{j-1}/\sqrt{\lambda} + 1$ steps, the robots will cross DB, which concludes the proof.

Lemma 22. *Suppose γ defines a polygon. Given an instance of crossing the shape at a point D, let β be the first exterior angle of the shape continuing from D. If β is the only exterior angle of γ from D to the next point of crossing and $\sqrt{\lambda} \leq \beta \leq \pi/8$, then the While loop in Step 7 of Algorithm 2 executes at most $8\beta/\sqrt{\lambda}$ times to cover the distance from B to C.*

Proof. Figure 10 illustrates this lemma where $\angle BAC = \beta$. Note that the radius of the circumcircle of the triangle $\triangle ABC$ is at most the radius ρ of the circumcircle of the regular polygon in the diagram indicated by dotted lines. Given the exterior angle $\sqrt{\lambda}$ and side length λ of the regular polygon, the radius of the circumcircle is $\rho = \lambda/2\sin(\sqrt{\lambda})$. By Lemma 6, $\rho \leq \sqrt{\lambda}$.

Now by the law of sines, $|BC|/\sin\beta = 2\rho$ and this implies $|BC| \leq 2\beta\sqrt{\lambda}$.

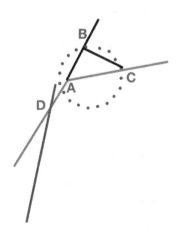

Fig. 10. Figure for Lemmas 21 and 22, the red color indicates the shape boundary, blue color indicates path of the nearest robot, black colored segment is a construction for the proof.

Next, the angle β' formed by the chord BC with the center of the regular polygon is at most,

$$2\arcsin\left(\frac{2\beta\sin(\sqrt{\lambda})}{\sqrt{\lambda}}\right) \leq 2\arcsin(2\beta) \leq \frac{4\beta}{\sqrt{1-(2\beta)^2}} \leq \frac{4\beta}{\sqrt{1-\pi^2/16}} \leq 8\beta$$

Next we multiply this angle with the radius to get the arc length between B and C,

$$s = \rho\beta' \leq 8\beta\sqrt{\lambda}$$

Finally, noting that the arc length covered by every step of the robots is at least λ, we get that after $8\beta/\sqrt{\lambda}$ steps from B the robots will cross AC.

Lemma 23. *If γ is a polygon and if for some $i \leq m$, $\sqrt{\lambda} \leq \phi_i \leq \pi/8$, then $f(i) \leq 8\phi_i/\sqrt{\lambda} + \phi_{i-1}/\sqrt{\lambda} + 1$.*

Proof. The trivial case is where Algorithm 2 is not executed at all i.e. $f(i) = 0$.

Observe that the motion of the robots during the execution of Algorithm 2 forms part of the perimeter of a convex polygon with side lengths and exterior angles being λ and $\sqrt{\lambda}$ respectively (except for the first and last sides).

Now suppose there are j vertices of γ defined over $[t_i, t_{i+1}]$. Let these vertices be indexed $\gamma(a_k)$ where $a_k \in [t_i, t_{i+1}]$ for all $k \in [1, j]$. Finally, let β_k be the exterior angle at $\gamma(a_k)$. By Lemma 21, there will be at most $\phi_{i-1}/\sqrt{\lambda}+1$ before the robot crosses the nearest side of the exterior angle at $\gamma(a_1)$.

Next, observe that by Lemma 22 the nearest robot to the shape will cross one of the sides of the exterior angle at $\gamma(a_k)$ by at most $8\beta_k/\sqrt{\lambda}$ iterations of the While loop in Algorithm 2.

In addition, the shape boundary turns either in convex or concave manner. If the turn at an index changes from convex to concave or concave to convex, it may actually move the sides of γ closer for the robot and hence the amount the angle β_k contributes to the overall iterations run inside the While loop of Algorithm 2 is at most $8\beta_k/\sqrt{\lambda}$. Therefore,

$$f(i) \leq \sum_{k=1}^{j} 8\beta_k/\sqrt{\lambda} + \phi_{i-1}/\sqrt{\lambda} + 1 = 8\phi_i/\sqrt{\lambda} + \phi_{i-1}/\sqrt{\lambda} + 1$$

D   Synchronization Details

Our final two lemmas show that the other robot do not rotate or traverse asymptotically more than the robot nearest to the boundary. In addition, we discuss briefly the synchronization steps in Algorithm 1.

Fig. 11. Figure for synchronization steps in CROSS-BOUNDARY, green and blue curves indicate robot path, red curve indicates the shape boundary. (Color figure online)

Observe that in Fig. 11, after reorientating itself to the curve gradient direction, the green and blue curves are "closer" to each other. But this can easily be handled by letting the blue or green curve based robot traverse a little longer, in particular greater than $\sqrt{\lambda} - \lambda > \lambda$ distance and then turn. This synchronization that maintains the distance of $\sqrt{\lambda}$ between the two robots guarantees that between successive executions of Algorithm 1, the robots will cover $\Omega(\sqrt{\lambda})$ distance of the shape boundary, in turn we are able to prove Lemma 12.

Lemma 24. *For each execution of Algorithm 2, D_2 traverses a distance of $O(\sqrt{\lambda})$.*

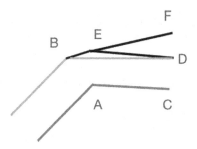

Fig. 12. Figure for Lemmas 24 and 25

Proof. In Fig. 12, the line segment AC describes the path of D_1 and D_2 can move either directly from B to D or B to D via E. Now, by Algorithm 2 description, $|AC| = |DE| = \lambda$. Note that, $\angle BAE = \sqrt{\lambda}$ and $|BA| = |AE| = \sqrt{\lambda}$. Thus by the law of sine, $|EB|/\sin \sqrt{\lambda} = \sqrt{\lambda}/\sin((\pi - \sqrt{\lambda})/2)$ Hence,

$$|EB| \leq \lambda/\cos(\sqrt{\lambda}/2) \leq 2\lambda.$$

Thus, the path length of D_2 at each step is at most 3λ.

In the last step independent of which robot encountered the shape boundary and whether they are both inside or outside or one inside and one outside, D_2

will traverse at most a constant factor of $\sqrt{\lambda}$ by the above discussion based on Fig. 11. Finally, noting that D_1 traverses $O(\sqrt{\lambda})$ distance according to Lemma 7, D_2 traverses $O(\sqrt{\lambda})$ during each execution of Algorithm 2.

Lemma 25. *For each execution of Algorithm 2, D_2 does not rotate asymptotically more than D_1.*

Proof. We only need to check that the rotation of the green path towards BE and then ED is asymptotically bounded by D_1's rotation. For brevity, we simply note this follows by elementary euclidean geometric comparison of various angles, beginning from the rotation of the red path to AC.

Finally, in the last step, both robots turn towards a common direction i.e. gradient at a point of crossing the boundary.

References

1. Rudin, W.: Principles of Mathematical Analysis. McGraw-Hill Education, New York (1976)
2. Dekster, B.V.: The length of a curve in space with curvature at most K. Am. Math. Soc. **79**(2), 271–278 (1980)
3. Munkres, J.: Topology. Prentice-Hall Pearson, Hoboken (2000)
4. Choset, H.: Coverage for robotics-a survey of recent results. Ann. Math. Artif. Intell. **31**, 113–126 (2001)
5. Koenig, S., Szymanski, B., Liu, Y.: Efficient and inefficient ant coverage methods. Ann. Math. Artif. Intell. **31**, 41–76 (2001)
6. Bruemmer, D.J., et al.: A robotic swarm for spill finding and perimeter formation. Technical report. Idaho National Engineering and Environmental Lab Idaho Falls (2002)
7. Marthaler, D., Bertozzi, A.L.: Collective motion algorithms for determining environmental boundaries. In: SIAM Conference on Applications of Dynamical Systems, pp. 1–15 (2003)
8. Kemp, M., Bertozzi, A.L., Marthaler, D.: Multi-UUV perimeter surveillance. In: Proceedings of IEEE/OES Autonomous Underwater Vehicles, pp. 102–107 (2004)
9. Marthaler, D., Bertozzi, A.L.: Tracking environmental level sets with autonomous vehicles". In: Butenko, S., Murphey, R., Pardalos, P.M. (eds.) Recent developments in cooperative control and optimization. Cooperative Systems, vol. 3, pp. 317–332. Springer, Boston (2004). https://doi.org/10.1007/978-1-4613-0219-3_17
10. Zhang, F., Justh, E.W., Krishnaprasad, P.S.: Boundary following using gyroscopic control. In: 2004 43rd IEEE Conference on Decision and Control (CDC) (IEEE Cat. No. 04CH37601), vol. 5, pp. 5204–5209. IEEE (2004)
11. Casbeer, DW., et al.: Forest fire monitoring with multiple small UAVs. In: 2005 Proceedings of the American Control Conference, pp. 3530–3535. IEEE (2005)
12. Clark, J., Fierro, R.: Cooperative hybrid control of robotic sensors for perimeter detection and tracking. In: 2005 Proceedings of the American Control Conference, pp. 3500–3505. IEEE (2005)
13. Casbeer, D.W., et al.: Cooperative forest fire surveillance using a team of small unmanned air vehicles. Int. J. Syst. Sci. **37**(6), 351–360 (2006)
14. Spears, D., Kerr, W., Spears, W.: Physics-based robot swarms for coverage problems. Int. J. Intell. Control Syst. **11**(3), 11–23 (2006)

15. Hsieh, M.A., Kumar, V., Chaimowicz, L.: Decentralized controllers for shape generation with robotic swarms. Robotica **26**(5), 691–701 (2008)
16. Zhang, F., Haq, S.: Boundary following by robot formations without GPS. In: 2008 IEEE International Conference on Robotics and Automation, pp. 152–157. IEEE (2008)
17. Zhang, F., Leonard, N.E.: Cooperative filters and control for cooperative exploration. IEEE Tran. Autom. Control **55**(3), 650–663 (2010)
18. Pettersson, L.H., Pozdnyakov, D.: Monitoring of Harmful Algal Blooms. Springer, Heidelberg (2013). https://doi.org/10.1007/978-3-540-68209-7
19. Pressley, A.: Elementary Differential Geometry. Springer, London (2012). https://doi.org/10.1007/978-1-84882-891-9
20. Hoy, M.: A method of boundary following by a wheeled mobile robot based on sampled range information. J. Intell. Robot. Syst. **72**(3–4), 463–482 (2013)
21. Matveev, A.S., Hoy, M.C., Savkin, A.V.: The problem of boundary following by a unicycle-like robot with rigidly mounted sensors. Robot. Auton. Syst. **61**(3), 312–327 (2013)
22. Tokekar, P., et al.: Tracking aquatic invaders: autonomous robots for monitoring invasive fish. IEEE Robot. Autom. Mag. **20**(3), 33–41 (2013)
23. Fahad, M., et al.: Robotic simulation of dynamic plume tracking by unmanned surface vessels. In: 2015 IEEE International Conference on Robotics and Automation (ICRA), pp. 2654–2659. IEEE (2015)
24. Saldana, D., Assunção, R., Campos, Campos, M.F.: A distributed multi-robot approach for the detection and tracking of multiple dynamic anomalies. In: 2015 IEEE International Conference on Robotics and Automation (ICRA), pp. 1262–1267. IEEE (2015)
25. Goodfellow, I., Bengio, Y., Courville, A.: Deep Learning. The MIT Press, Cambridge (2016)
26. Chatterjee, S., Wu, W.: Cooperative curve tracking in two dimensions without explicit estimation of the field gradient. In: 2017 4th International Conference on Control, Decision and Information Technologies (CoDIT), pp. 0167–0172. IEEE (2017)
27. Nguyen, A., et al.: Using a UAV for destructive surveys of mosquito population. In: 2018 IEEE International Conference on Robotics and Automation (ICRA), pp. 7812–7819. IEEE (2018)
28. Chatterjee, S., Wu, W.: A modular approach to level curve tracking with two nonholonomic mobile robots. In: International Design Engineering Technical Conferences and Computers and Information in Engineering Conference, vol. 59292, p. V009T12A051. American Society of Mechanical Engineers (2019)
29. Al-Abri, S., Zhang, F.: A distributed active perception strategy for source seeking and level curve tracking. IEEE Trans. Autom. Control **67**(5), 2459–2465 (2021)
30. Stanton, M.C., et al.: The application of drones for mosquito larval habitat identification in rural environments: a practical approach for malaria control? Malaria J. **20**(1), 1–17 (2021)
31. Carrasco-Escobar, G., et al.: The use of drones for mosquito surveillance and control. Parasites Vectors **15**(1), 473 (2022)
32. Ericksen, J., et al.: Aerial survey robotics in extreme environments: mapping volcanic CO_2 emissions with flocking UAVs. Front. Control Eng. **3**, 7 (2022)

Cops & Robber on Periodic Temporal Graphs: Characterization and Improved Bounds

Jean-Lou De Carufel[1]([⊠]), Paola Flocchini[1], Nicola Santoro[2], and Frédéric Simard[1]

[1] School of Electrical Engineering and Computer Science, University of Ottawa, Ottawa, Canada
jdecaruf@uottawa.ca
[2] School of Computer Science, Carleton University, Ottawa, Canada

Abstract. We study the classical *Cops and Robber* game when the cops and the robber move on an infinite periodic sequence $\mathcal{G} = (G_0, \ldots, G_{p-1})^*$ of graphs on the same set V of n vertices: in round t, the topology of $\mathcal{G}$ is $G_i = (V, E_i)$ where $i \equiv t \pmod{p}$. As in the traditional case of *static graphs*, the main concern is on the characterization of the class of *periodic temporal graphs* where k cops can capture the robber. Concentrating on the case of a *single* cop, we provide a characterization of copwin periodic temporal graphs. Based on this characterization, we design an algorithm for determining if a periodic temporal graph is copwin with time complexity $O(p\, n^2 + n\, m)$, where $m = \sum_{i \in \mathbb{Z}_p} |E_i|$, improving the existing $O(p\, n^3)$ bound. Let us stress that, when $p = 1$ (i.e., in the *static* case), the complexity becomes $O(n\, m)$, improving the best existing $O(n^3)$ bound.

1 Introduction

Cops & Robber Games. *Cops & Robber* (*C&R*) is a pursuit-evasion game played in rounds on a finite graph G between a set of $k \geq 1$ cops and a single robber. Before starting the game, an initial position on the vertices of G is chosen first by the cops, then by the robber. Then, in each round, first the cops, then the robber, move to a neighbouring vertex or (if allowed by the variant of the game) stay in their current location. The game ends if at least one cop moves to the vertex currently occupied by the robber, in which case the cops *capture* the robber and win. The robber wins by forever avoiding capture. In the original version [30,33], the graph G is connected and undirected, there is a single cop and, in each round, the players are allowed to move to a neighbouring vertex or not to move. Moreover, the cops and the robber have perfect information. It has been then extended to permit multiple cops [2]. This version, which we shall call *standard*, is the most commonly investigated.

Supported by the Natural Sciences and Engineering Research Council of Canada.

Among the many variants of this game (for a partial list, see [4,5]), two are of particular interest to us. The first is the (much less investigated) natural generalization when the graph G is a strongly connected directed graph [25,27]; we shall refer to this version as *directed*. Also of interest is the variant, called *fully active* (or *restless*), in which the players must move in every round; proposed for the standard game [19], it can obviously be extended to the directed version.

In the extensive research (see [5] for a review), the main focus is on characterizing the class of *k-copwin* graphs; i.e., those graphs where there exists a strategy allowing k cops to capture the robber regardless of the latter's decisions. Related questions are to determine the minimum number of cops capable of winning in G, called the *copnumber of G*, or just to decide whether k cops suffice. Currently, the most efficient algorithm for deciding whether or not a graph is k-copwin in the standard game is $O(k\, n^{k+2})$ [32], which yields $O(n^3)$ for the case $k = 1$.

In the existing literature on the *C&R* game, with only a couple of recent exceptions, all results are based on the assumption that the graph on which the game is played is *static*; that is, its link structure is the same in every round. The question naturally arises: what happens if the link structure of the graph changes in time, possibly in every round? This question is particularly relevant in view of the intense research on time-varying graphs in the last two decades.

Temporal Graphs. The extensive investigations on computational aspects of time-varying graphs have been motivated by the development and increasing importance of highly dynamic networks, where the topology is continuously changing. Such systems occur in a variety of settings, ranging from wireless ad-hoc networks to social networks. Various formal models have been advanced to describe the dynamics of these networks (e.g., [8,21,34]).

When time is *discrete*, as in the *C&R* game, the dynamics of these networks is usually described as an infinite sequence $\mathcal{G} = (G_0, G_1, \dots)$, called *temporal graph* (or *evolving graph*), of static graphs $G_i = (V, E_i)$ on the same set V of vertices; the graph G_i is called *snapshot* (of $\mathcal{G}$ at time i), and the aggregate graph $G = (V, \cup_i E_i)$ is called the *footprint* (or *underlying*) graph. This model, originally suggested in [15,20], has become the de-facto standard in the ensuing investigations.

All the studies are being carried out under some assumptions restricting the arbitrariness of the changes. Some of these assumptions are on the "connectivity" of the graphs G_i in the sequence; they range from the (strong) *1-interval connectivity* requiring every G_i to be connected (e.g., [22,26,31]), to the weaker *temporal connectivity* allowing each G_i to be disconnected but requiring the sequence to be *connected over time* (e.g., [7,17]). Another class of assumptions is on the "frequency" of the existence of the links in the sequence. An important assumption in this class is *periodicity*: there exists a positive integer p such that $G_i = G_{i+p}$ for all $i \in \mathbb{Z}$ (e.g., [16,23,24]).

A large number of studies has focused on *mobile entities* operating on temporal graphs, under different combinations of the above (and other) restrictive assumptions. Among them, computations include *graph exploration*, *dispersion*, and *gathering* (e.g., [1,6,10–13,17,18]; for a recent survey see [9]). Until very recently, none of these studies considered *C&R*.

Conceptually, the extension of *C&R* to a temporal graph $\mathcal{G} = (G_0, G_1, \dots)$ is quite natural. Initially, first the cops, then the robber, choose a starting position on the vertices of G_0. At the beginning of round $t \geq 0$, the players are in G_t and, after making their decisions and moves (according to the rules of the game), they find themselves in G_{t+1} in the next round. The game ends if and only if a cop moves to the vertex currently occupied by the robber; in this case the cops have won. The robber wins by forever preventing the cops from winning.

Existing Results. This extension has been first investigated by Erlebach and Spooner [14]. They considered the standard game with a single cop under the *periodic* frequency restriction; they presented an algorithm to determine if a periodic temporal graph is copwin, and mentioned that it can be extended to $k > 1$ cops. In this pioneering study, the results are obtained by reformulating the problem in terms of a reachability problem and solving the latter; this, unfortunately, does not provide insights on the temporal nature of the game.

Using the same reduction to reachability games, and thus with the same drawbacks as [14], Balev et al. [3] studied the standard game in temporal graphs under the *1-interval connectivity* restriction. They showed how to determine whether a single cop can capture the robber in a fixed temporal window, and indicated how their algorithm can be extended to the case of $k > 1$ cops. They also considered an "on-line" version of the problem, i.e. where the sequence of graphs is a priori unknown; these results however are not relevant for the "full-disclosure" problem studied here.

Finally, if the temporal graph is not given explicitly (i.e., as the sequence of snapshots), but only implicitly by means of the Boolean *edge-presence* function (e.g., [8]), the problem of deciding whether a single cop has a winning strategy in the standard game on a periodic temporal graph has been shown to be NP-hard [28,29], answering a question raised in [14].

Contributions. We focus on the *C&R* game in *periodic* temporal graphs, concentrating on the case of a *single* cop. We study the *unified* version of the game defined as follow: in every round $i \geq 0$, G_i is directed and the players are restless. Observe that the standard and the directed versions, both in the original and restless variant, can be expressed as a restless game played on (appropriately chosen) directed graphs.

For the unified game, we provide a complete characterization of copwin periodic temporal graphs, establishing several basic properties on the nature of a copwin game in such graphs. We do so by employing a compact representation of periodic temporal graphs, introducing the novel notion of augmented arenas, and using these structures to extend to the temporal domain classical concepts such as *corners* and *covers*.

These characterization results are *general*, in the sense that they do not rely on any assumption on properties such as connectivity, symmetry, reflexivity held (or not held) by the individual snapshot graphs in the sequence. The only requirement, for the game to be defined, and thus *playable*, is that every node in the graph must have an outgoing edge.

Based on these results, we design an algorithm that determines if a periodic temporal graph is copwin in time $O(p\,n^2 + n\,m)$, where $m = \sum_{i \in \mathbb{Z}_p} |E_i|$, improving on the existing $O(p\,n^3)$ bound established by [14]. Let us stress that, in the static case studied in the literature, the complexity becomes $O(n\,m)$, improving the best existing $O(n^3)$ bound [32]; in particular our bound becomes $O(n^2)$ for sparse graphs.

All our results are established for the unified version of the game. Therefore, all the characterization properties and algorithmic results hold for the standard and for the directed games studied in the literature, both when the players are restless and when they are not. They hold also for all those settings, not considered in the literature, where there is a mix of nodes: those where the players must leave and those where the players can wait; furthermore such a mix might be time-varying (i.e., different in every round). Due to space constraints, some proofs are missing.

2   Definitions and Terminology

2.1   Graphs and Time

Static Graphs. We denote by $G = (V, E)$, or sometimes by $G = (V(G), E(G))$, the *directed graph* with set of vertices V and set of edges $E \subseteq V \times V$. A self-loop is an edge of the form (u, u); if $(u, u) \in E$ for all $u \in V$, then we will say that G is *reflexive*. If $(v, u) \in E$ whenever $(u, v) \in E$, we will say that G is *symmetric* (or *undirected*). Given a graph G', if $V(G') \subseteq V(G)$ and $E(G') \subseteq E(G)$, then we say G' is a *subgraph* of G and write $G' \subseteq G$. A subgraph $G' \subseteq G$ is *proper*, written $G' \subset G$, if $G' \neq G$. For reasons apparent later, we shall refer to a graph G so defined as a *static graph*, and say it is *playable* if every node has at least one outgoing edge.

Temporal Graphs. A *time-varying graph* $\mathcal{G}$ is a graph whose set of edges changes in time[1]. A *temporal graph* is a time-varying graph where time is assumed to be discrete and to have a start; i.e., *time* is the set $\mathbb{Z}^+$ of positive integers including 0. A temporal graph $\mathcal{G}$ is represented as an infinite sequence $\mathcal{G} = (G_0, G_1, \dots)$ of static graphs $G_i = (V, E_i)$ on the same set of vertices V; we shall denote by $n = |V|$ the number of vertices. The graph G_i is called the *snapshot* of $\mathcal{G}$ at time $i \in \mathbb{Z}^+$, and the aggregate graph $G = (V, \bigcup_i E_i)$ is called the *footprint* of $\mathcal{G}$. A temporal graph $\mathcal{G}$ is said to be *reflexive* if all its snapshots are reflexive, *symmetric* if all its snapshots are symmetric.

Given two nodes $x, y \in V$, a strict *journey* (or *temporal walk*), from x to y starting at time t is any finite sequence $\pi(x, y) = \langle (z_0, z_1), (z_1, z_2), \dots, (z_{k-1}, z_k) \rangle$ where $z_0 = x$, $z_k = y$, and $(z_i, z_{i+1}) \in E_{t+i}$ for $0 \le i < k$. In the following, for simplicity, we will omit the adjective "strict". A temporal graph $\mathcal{G}$ is *temporally connected* if for any $u, v \in V$ and any time $t \in \mathbb{Z}^+$ there is a journey from u to v that starts at time t. Observe that, if $\mathcal{G}$ is temporally connected, then its

---

[1] The terminology in this section is from [8].

footprint is strongly connected even when all its snapshots are disconnected. A temporal graph $\mathcal{G}$ is said to be *always connected* (or 1-*interval connected*) if all its snapshots are strongly connected.

A temporal graph $\mathcal{G}$ is *periodic* if there exists a positive integer p such that for all $i \in \mathbb{Z}^+$, $G_i = G_{i+p}$. If p is the smallest such integer, then p is called the *period* of $\mathcal{G}$ and $\mathcal{G}$ is said to be p-*periodic*. We shall represent a p-periodic temporal graph $\mathcal{G}$ as $\mathcal{G} = (G_0, \ldots, G_{p-1})^*$; all operations on the indices will be taken modulo p. An example of a temporal periodic graph $\mathcal{G}$ with $p = 4$ is shown in Fig. 1; observe that $\mathcal{G}$ is temporally connected, however most of its snapshots are disconnected directed graphs, and none of them is strongly connected.

Let $\mathcal{G} = (G_0, G_1, \ldots, G_{p-1})^*$ and $\mathcal{H} = (H_0, H_1, \ldots, H_{p-1})^*$ be two temporal periodic graphs with the same period on the same set V of vertices; we say $\mathcal{H}$ is a *periodic subgraph* of $\mathcal{G}$, written $\mathcal{H} \subseteq \mathcal{G}$, if $H_i \subseteq G_i$ for every $i \in \mathbb{Z}_p = \{0, 1, \ldots, p-1\}$. We shall denote by $\mathcal{H} \subset \mathcal{G}$ the fact that $\mathcal{H}$ is a *proper* subgraph of $\mathcal{G}$; i.e., $\mathcal{H} \subseteq \mathcal{G}$ but $\mathcal{H} \neq \mathcal{G}$. Let us point out the obvious but useful fact that static graphs are temporal periodic graphs with period $p = 1$. In this paper we focus on *C&R* games in *periodic* temporal graphs, henceforth referred to simply as *periodic graphs*, concentrating on the case of a *single* cop.

Consider the following class of directed static graphs, we shall call *arenas*.

Definition 1 (Arena). *Let $k \geq 1$ be an integer and W be a non-empty finite set. An* arena *of length k on W is any static directed graph $\mathcal{M} = (\mathbb{Z}_k \times W, E(\mathcal{M}))$ where $E(\mathcal{M}) \subseteq \{((i, w), ([i+1]_k, w')) | i \in \mathbb{Z}_k$ and $w, w' \in W\}$, and $[i]_k$ denotes i modulo k.*

A periodic graph $\mathcal{G} = (G_0, \ldots, G_{p-1})^*$ with period p and set of nodes V has a unique correspondence with the arena $\mathcal{D} = (\mathbb{Z}_p \times V, E(\mathcal{D}))$ where, for all $i \in \mathbb{Z}_p$, $((i, u), ([i+1]_p, v)) \in E(\mathcal{D})$ if and only if $(u, v) \in E_i$, called the *arena of $\mathcal{G}$*. In particular, the arena $\mathcal{D}$ of $\mathcal{G}$ explicitly preserves the snapshot structure of $\mathcal{G}$: for all $i \in \mathbb{Z}_p$, there is an obvious one-to-one correspondence between the snapshot G_i of $\mathcal{G}$ and the subgraph S_i of $\mathcal{D}$, called *slice* (or *stage*), where $V(S_i) = \{(i, v), v \in V\}$ and $E(S_i) = \{((i, u), ([i+1]_p, v)) | (u, v) \in E_i)\}$. An example of a periodic graph $\mathcal{G}$ and its arena $\mathcal{D}$ is shown in Fig. 1. In the following, when no ambiguity arises, $\mathcal{D}$ shall indicate the arena of $\mathcal{G}$.

The vertices of an arena $\mathcal{D}$ will be called *temporal nodes*. Given a temporal node $(i, u) \in V(S_i)$ we shall denote by $N_i(u, \mathcal{D})$ the set of its outneighbours, and by $\Gamma_i(u, \mathcal{D}) = \{v \in V | ([i+1]_p, v) \in N_i(u, \mathcal{D})\}$ the corresponding set of nodes in G_i. We define $\Gamma_i^{in}(u, \mathcal{D})$ similarly for the inneighbours. A temporal node $(i, u) \in V(S_i)$ is said to be a *star* if $\Gamma_i(u, \mathcal{D}) = V$. It is said to be *anchored* if there exists a journey from some node $(0, v) \in V(S_0)$ to (i, u). A *subarena* of $\mathcal{D} = (\mathbb{Z}_p \times V, E(\mathcal{D}))$ is any arena $\mathcal{D}' = (\mathbb{Z}_p \times V, E(\mathcal{D}'))$ where $E(\mathcal{D}') \subseteq E(\mathcal{D})$; we shall denote by $\mathcal{D}' \subset \mathcal{D}$ the fact that $\mathcal{D}'$ is a subarena of $\mathcal{D}$ with $E(\mathcal{D}') \subset E(\mathcal{D})$.

2.2 Cop & Robber Game in Periodic Graphs

The extension of the game from static to temporal graphs is quite natural. Initially, first the cop, then the robber, chooses a starting position on the vertices

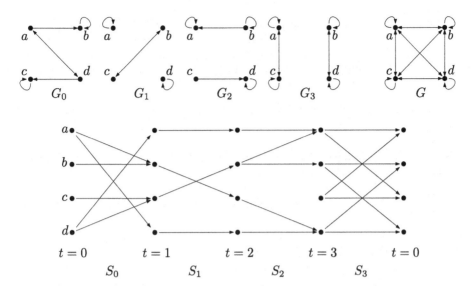

Fig. 1. A periodic graph $\mathcal{G} = (G_0, G_1, G_2, G_3)^*$, its footprint G, and the corresponding arena.

of G_0. Then, at each time $t \in \mathbb{Z}^+$, first the cop, then the robber, moves to a vertex adjacent to its current position in G_i, where $i = [t]_p$. Thus, in round t, the players are in $G_{[t]_p}$ and, after making their decisions and moves, they find themselves in $G_{[t+1]_p}$ in the next round. The game ends if and only if the cop moves to the vertex currently occupied by the robber; in this case the cop has won. The robber wins by forever preventing the cop from winning. Moreover, the cops and the robber have perfect information.

A play on the arena $\mathcal{D}$ of $\mathcal{G}$ follows the play on $\mathcal{G}$ in a direct obvious way: at each time $t \in \mathbb{Z}^+$, first the cop, then the robber, chooses a new node in the outneighbourhood of its current position and moves there. The cop wins and the game ends if it manages to move to a temporal node $([t+1]_p, u)$ while the robber is on $([t]_p, u)$. The robber wins by forever escaping capture from the cop, in which case the game never ends.

We consider the version of the game where all players are *restless*, i.e., they all move to a different node in each round. In this version, the only requirement on $\mathcal{G}$ is that it is *playable*: in each snapshot, every node must have an outgoing edge. In what follows we only consider playable periodic graphs. No other requirement such as connectivity, symmetry or reflexivity is imposed on $\mathcal{G}$.

We call this version of the game *unified*. Observe that the standard version, both in the original or restless variant, as well as the non-restless directed version can actually be redefined as a restless game played in this unified version: a pair of directed edges between a pair of nodes corresponds to an unidirected link between them, and the presence of a self-loop at a node allows the players currently there not to move to a different node in the current round.

A *configuration* is a triple $(t, c, r) \in \mathbb{Z}^+ \times V \times V$, denoting the position $c \in V$ of the cop and $r \in V$ of the robber at the beginning of round $t \in \mathbb{Z}^+$. Let $\mathcal{CG} = (V(\mathcal{CG}), E(\mathcal{CG}))$ be the infinite directed graph, called *configuration graph* of $\mathcal{D}$, describing all the possible configurations (t, u, v) together with the following subset of their temporal connections in $\mathcal{D}$:

$$V(\mathcal{CG}) = \{(t, u, v) \mid t \in \mathbb{Z}^+; ([t]_p, u), ([t]_p, v) \in V(\mathcal{D})\}$$

$$E(\mathcal{CG}) = \{((t, u, v), (t+1, u', v')) \mid t \in \mathbb{Z}^+; u \neq v; u' \in \Gamma_{[t]_p}(u, \mathcal{D}), v' \in \Gamma_{[t]_p}(v, \mathcal{D})\}.$$

Observe that $\mathcal{CG}$ is acyclic; the *source* nodes (i.e., the nodes with no in-edges) are those with $t = 0$, the *sink* nodes (i.e., the nodes with no out-edges) are those with $u = v$. A playing *strategy for the cop* is any function $\sigma_c : V(\mathcal{CG}) \rightarrow V$ where, for every $(t, u, v) \in V(\mathcal{CG})$, $\sigma_c(t, u, v) \in \Gamma_{[t]_p}(u, \mathcal{D})$, and $\sigma_c(t, u, v) = u$ if $u = v$; it specifies where the cop should move in round t if the cop is at $([t]_p, u)$, the robber is at $([t]_p, v)$, and it is the cop's turn to move. A playing strategy σ_r for the robber is defined in a similar way.

A configuration (t, u, v) is said to be *copwin* if there exists a strategy σ_c such that, starting from (t, u, v), the cop wins the game regardless of the strategy σ_r of the robber; such a strategy σ_c will be said to be *copwin for* (t, u, v). A strategy σ_c is said to be *copwin* if there exists a temporal node $(0, u)$ such that σ_c is winning for all $(0, u, v)$, $v \in V$. If a copwin strategy exists, then $\mathcal{G}$ and its arena $\mathcal{D}$ are said to be *copwin*, else they are *robberwin*.

3   Copwin Periodic Graphs

3.1   Preliminary

In the analysis of the standard game played in a static graph, an important role is played by the notions of *corner* node and its *cover*. The usual meaning is that if the robber is on the corner, after the cop has moved to the cover, no matter where the robber plays, the robber gets captured by the cop *in the next round*.

In an arena $\mathcal{D}$, the same meaning is provided directly by the notions of "temporal corner" and "temporal cover".

Definition 2 *(Temporal Corner and Temporal Cover). A temporal node (t, u) in an arena $\mathcal{D}$ is said to be a* temporal corner *of temporal node $(t+1, v)$ if $u \neq v$ and $\Gamma_t(u, \mathcal{D}) \subseteq \Gamma_{t+1}(v, \mathcal{D})$. The temporal node $(t+1, v)$ is said to be a* temporal cover *of (t, u).*

Lemma 1. *Every copwin arena contains a temporal corner.*

This necessary condition, although important, provides only limited indications on how to solve the characterization problem.

3.2   Augmented Arenas and Characterization

The crucial element in the characterization of copwin periodic graphs is the notion of *augmented arena*.

Definition 3 *(Augmented Arena). Let $\mathcal{D}$ be the arena of $\mathcal{G}$. An augmented arena $\mathcal{A}$ of $\mathcal{D}$ is an arena of length p such that $\mathcal{D} \subseteq \mathcal{A}$ and, for each edge $((t, x), (t + 1, y)) \in E(\mathcal{A})$, the configuration (t, x, y) is winning for the cop in $\mathcal{D}$.*

We shall refer to the edges of the augmented arena $\mathcal{A}$ of $\mathcal{D}$ as *shadow edges*. Observe that, by definition, all edges of $\mathcal{D}$ are shadow edges of $\mathcal{A}$. Let $\mathbb{A}(\mathcal{D})$ denote the set of augmented arenas of $\mathcal{D}$. Observe that, by definition of $\mathcal{D}$, for each edge $((t, x), (t + 1, y)) \in E(\mathcal{D})$, the configuration (t, x, y) is winning for the cop in $\mathcal{D}$. Therefore, $\mathcal{D} \in \mathbb{A}(\mathcal{D})$. Further observe the following:

Property 1. The partial order $(\mathbb{A}(\mathcal{D}), \subset)$ induced by edge-set inclusion on $\mathbb{A}(\mathcal{D})$ is a complete lattice. Hence $(\mathbb{A}(\mathcal{D}), \subset)$ has a maximum which we denote by $\mathcal{A}^*$.

We have now the elements for the characterization of copwin periodic graphs.

Theorem 1 (Characterization Property). *An arena $\mathcal{D}$ is copwin if and only $\mathcal{A}^*$ contains an anchored star.*

Proof (**only if**). Let $\mathcal{A}^*$ contain an anchored star (t, u), $t \in \mathbb{Z}_p$. By definition of star, $\Gamma_t(u, \mathcal{A}^*) = V$; thus, by definition of augmented arena, for every $v \in V$ the configuration (t, u, v) is copwin, i.e. there is a copwin strategy σ_c from (t, u, v).

Since (t, u) is anchored, there exists a journey $\pi((0, x), (t, u))$, starting at time 0 and ending at time t, to (t, u) from some temporal node $(0, x)$. Consider now the cop strategy σ_c' of: (1) initially positioning itself on the temporal node $(0, x)$, (2) then moving according to the journey $\pi((0, x), (t, u))$ and, once on (t, u), (3) following the copwin strategy σ_c from (t, u, w), where w is the position of the robber at the beginning of round t. This strategy σ_c' is winning for all $(0, x, v)$, $v \in V$; hence $\mathcal{D}$ is copwin.

(**if**) Let $\mathcal{D}$ be copwin. We then show that there must exist an augmented arena $\mathcal{A}$ of $\mathcal{D}$ that contains an anchored star. Since $\mathcal{D}$ is copwin, by definition, there must exist some starting position $(0, c)$ for the cop such that, for all positions $(0, r)$ initially chosen by the robber, the cop eventually captures the robber. In other words, all the configurations $(0, c, v)$ with $v \in V$ are copwin; thus the arena $\mathcal{A}$ obtained by adding to $E(\mathcal{D})$ the set of edges $\{((0, c), (1, v)) | v \in V\}$ is an augmented arena of $\mathcal{D}$ and $(0, c)$ is an anchored star. By Property 1, $E(\mathcal{A}) \subseteq E(\mathcal{A}^*)$ and the theorem follows. $\square$

The characterization of copwin periodic graphs provided by Theorem 1 indicates that, to determine whether or not an arena $\mathcal{D}$ is copwin, it suffices to check whether $\mathcal{A}^*$ contains an anchored star. To be able to transform this fact into an effective solution procedure, some additional concepts need to be introduced and properties established.

3.3   Shadow Corners and Augmentation

Other crucial elements in the analysis of copwin periodic graphs are the concepts of corner and cover, introduced in Sect. 3.1 for arenas, now in the context of augmented arenas.

Definition 4 *(Shadow Corner and Shadow Cover). Let $\mathcal{A}$ be an augmented arena of $\mathcal{D}$. A temporal node (t, u) is a* shadow corner *of a temporal node $(t+1, v)$, with $v \neq u$, if $\Gamma_t(u, \mathcal{D}) \subseteq \Gamma_{t+1}(v, \mathcal{A})$. The temporal node $(t + 1, v)$ will then be called the* shadow cover *of (t, u).*

By definition, any temporal corner is a shadow corner, and its temporal covers are shadow covers. An example is shown in Fig. 2; the red links indicate the neighbours of node (t, u) in $\mathcal{D}$, while in green are indicated the edges to the neighbours of $(t + 1, v)$ that exists in $\mathcal{A}$ but not in $\mathcal{D}$.

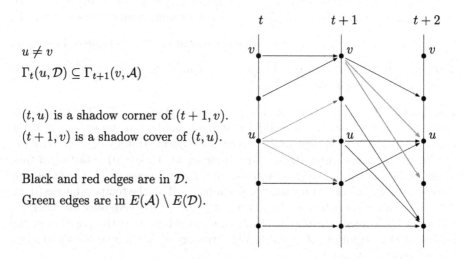

$u \neq v$

$\Gamma_t(u, \mathcal{D}) \subseteq \Gamma_{t+1}(v, \mathcal{A})$

(t, u) is a shadow corner of $(t + 1, v)$.

$(t + 1, v)$ is a shadow cover of (t, u).

Black and red edges are in $\mathcal{D}$.

Green edges are in $E(\mathcal{A}) \setminus E(\mathcal{D})$.

Fig. 2. Node (t, u) is a shadow corner of $(t + 1, v)$. (Color figure online)

The role that shadow corners play with regards to the set $\mathbb{A}(\mathcal{D})$ of augmented arenas of $\mathcal{D}$ is expressed by the following.

Theorem 2 (Augmentation Property). *Let $\mathcal{A} \in \mathbb{A}(\mathcal{D})$; $(t, x), (t, y) \in V(\mathcal{D})$; and $z \in \Gamma_t(x, \mathcal{D})$. If (t, y) is a shadow corner of $(t + 1, z)$, then the arena $\mathcal{A}' = \mathcal{A} \cup \{((t, x), (t+1, y))\}$ is an augmented arena of $\mathcal{D}$.*

Proof. Let $\mathcal{A}$ be an augmented arena of $\mathcal{D}$ and let $(t, x), (t, y), (t + 1, z) \in V(\mathcal{D})$ where $z \in \Gamma_t(x, \mathcal{D})$ and (t, y) is a shadow corner of $(t + 1, z)$. The theorem follows if $((t, x), (t + 1, y))$ is already an edge of $\mathcal{A}$. Consider the case where $((t, x), (t + 1, y)) \notin E(\mathcal{A})$. Since (t, y) is a shadow corner of $(t + 1, z)$, then for every $w \in \Gamma_t(y, \mathcal{D})$ we have that $((t+1, z), (t+2, w)) \in E(\mathcal{A})$; i.e., $(t+1, z, w)$ is winning for the cop. Since $z \in \Gamma_t(x, \mathcal{D})$, if the cop moves from (t, x) to $(t + 1, z)$ when the robber is on (t, y), then regardless of the robber's move, the resulting configuration would be winning for the cop. In other words, (t, x, y) is a winning configuration for the cop. It follows that $\mathcal{A}' = \mathcal{A} \cup \{((t, x), (t + 1, y))\}$ is an augmented arena of $\mathcal{D}$. □

In other words, given an augmented arena, by identifying a (still unconsidered) shadow corner and its covers, new shadow edges may be determined and added to form a denser augmented arena.

3.4   Determining $\mathcal{A}^*$

The properties expressed by Theorem 2, in conjunction with that of Theorem 1, provide an algorithmic strategy to construct $\mathcal{A}^*$: start from an augmented arena; determine new shadow edges; add them to the set of shadow edges, creating a denser augmented arena; repeat this process until the current augmented arena $\mathcal{A}$ either contains an anchored star or is $\mathcal{A}^*$.

To be able to employ the above strategy, a condition is needed to determine if the current augmented arena of $\mathcal{D}$ is indeed $\mathcal{A}^*$. This is provided by the following.

Theorem 3 (Maximality Property). *Let $\mathcal{A} \in \mathbb{A}(\mathcal{D})$. Then $\mathcal{A} = \mathcal{A}^*$ if and only if, for every edge $((t,x),(t+1,y)) \notin E(\mathcal{A})$, there exists no $z \in \Gamma_t(x,\mathcal{D})$ such that $\Gamma_t(y,\mathcal{D}) \subseteq \Gamma_{t+1}(z,\mathcal{A})$.*

Proof (**only if**). By contradiction, let $\mathcal{A} = \mathcal{A}^*$ but there exists an edge $((t,x),(t+1,y)) \notin E(\mathcal{A})$ and a temporal node $z \in \Gamma_t(x,\mathcal{D})$ such that $\Gamma_t(y,\mathcal{D}) \subseteq \Gamma_{t+1}(z,\mathcal{A})$. This means that (t,y) is a shadow corner of $(t+1,z)$. By Theorem 2, $\mathcal{A}' = \mathcal{A} \cup \{((t,x),(t+1,y))\}$ is an augmented arena of $\mathcal{D}$; however, $E(\mathcal{A}')$ contains one more edge than $E(\mathcal{A})$, contradicting the assumption that $\mathcal{A}$ is maximum.

(**if**) Let $\mathcal{A} \neq \mathcal{A}^*$; that is, there exists $((t,x),(t+1,y)) \in E(\mathcal{A}^*) \setminus E(\mathcal{A})$. By definition, the configuration (t,x,y) is copwin; let σ_c be a cop winning strategy for the configuration (t,x,y); i.e., starting from (t,x,y), the cop wins the game regardless of the strategy σ_r of the robber.

Let $\mathcal{C} = (V(\mathcal{C}),E(\mathcal{C})) \subseteq \mathcal{CG}$ be the directed acyclic graph of configurations induced by σ_c starting from (t,x,y), and defined as follows: (1) $(t,x,y) \in V(\mathcal{C})$; (2) if $(t',u,v) \in V(\mathcal{C})$ with $t' \geq t$ and $u \neq v$, then, for all $w \in \Gamma_{t'}(v,\mathcal{D})$, $(t'+1,\sigma_c(t'+1,u,v),w) \in V(\mathcal{C})$ and $((t',u,v),(t'+1,\sigma_c(t'+1,u,v),w))) \in E(\mathcal{C})$.

Observe that in $\mathcal{C}$ there is only one source (or root) node, (t,x,y), and every $(t',w,w) \in V(\mathcal{C})$ is a sink (or terminal) node. Since σ_c is a winning strategy for the root, every node in $\mathcal{C}$ is a copwin configuration, and every path from the root terminates in a sink node (see Fig. 3).

Partition $V(\mathcal{C})$ into two sets, U and W where $U = \{(i,u,v)|((i,u),(i+1,v)) \in E(\mathcal{A})\}$ and $W = V(\mathcal{C}) \setminus U$. Observe that every sink of $V(\mathcal{C})$ belongs to U; on the other hand, since $((t,x),(t+1,y)) \notin E(\mathcal{A})$ by assumption, the root belongs to W (see Fig. 4). Given a node $\kappa = (i,u,v) \in V(\mathcal{C})$, let $\mathcal{C}[\kappa]$ denote the subgraph of $\mathcal{C}$ rooted in κ.

Claim. *There exists $\kappa \in V(\mathcal{C})$ such that all nodes of $\mathcal{C}[\kappa]$ except the root belong to U.*

Proof of Claim. Let P_0 be the set of sinks of $\mathcal{C}$. Starting from $k = 0$, consider the set P_{k+1} of all in-neighbours of any node of P_k; if P_{k+1} does not contains an

σ_c winning strategy for (t, x, y) (every path terminates)

● sink

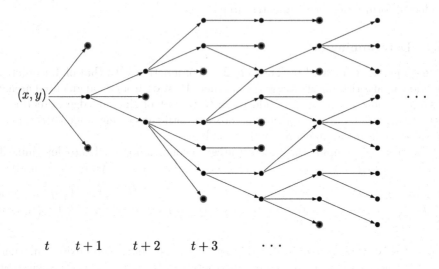

Fig. 3. The directed acyclic graph $\mathcal{C}$ of configurations induced by σ_c starting from (t, x, y). (Color figure online)

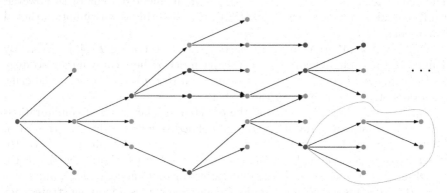

W: Shadow edges are missing.   U: Shadow edges are not missing.

Fig. 4. The sets U (green) and W (purple). (Color figure online)

element of W, then increase k and repeat the process. Since $(t, x, y) \in W$, this process terminates for some $j \geq 1$, and the Claim holds for every $\kappa \in P_j$. $\square$

Let (t', x', y') be a node of $V(\mathcal{C})$ satisfying the above Claim (see Fig. 5). Thus $((t', x'), (t' + 1, y')) \notin E(\mathcal{A})$ but, since (t', x', y') is copwin, $((t', x'), (t' + 1, y')) \in \mathcal{A}^*$. By the Claim, all other nodes of $\mathcal{C}[(t', x', y')]$ belong to U, in particular the set of nodes $\{(t' + 1, w, z) | w = \sigma_c(t', x', y'), z \in \Gamma_{t'}(y', \mathcal{D})\}$. This means that,

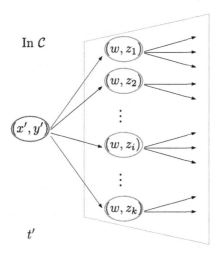

No shadow edges missing

Fig. 5. (t', x', y') satisfies the Claim. (Color figure online)

for every $z \in \Gamma_{t'}(y', \mathcal{D})$, $(t' + 1, w, z) \in E(\mathcal{A})$. In other words, $\Gamma_{t'}(y', \mathcal{D}) \subseteq \Gamma_{t'+1}(w, \mathcal{A})$; that is, (t', y') is a shadow corner of $(t' + 1, w)$ (see Fig. 6).

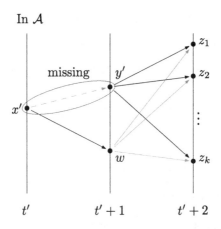

Fig. 6. (t', y') is a shadow corner of $(t' + 1, w)$. (Color figure online)

Summarizing: by assumption $\mathcal{A} \neq \mathcal{A}^*$; as shown, $((t', x'), (t' + 1, y')) \notin E(\mathcal{A}^*) \setminus E(\mathcal{A})$, and $w \in \Gamma_{t'}(x', \mathcal{D})$ is a shadow cover of (t', y'); that is, $\Gamma_{t'}(y', \mathcal{D}) \subseteq \Gamma_{t'+1}(w, \mathcal{A})$, concluding the proof of the **if** part of the theorem. $\square$

4   Algorithmic Determination

In this section we show that the results established in the previous sections provide all the tools necessary to design an algorithm to determine whether or not a periodic graph $\mathcal{G}$ is copwin. Furthermore, if $\mathcal{G}$ is copwin, the algorithm can actually provide a winning cop strategy σ_c.

4.1   Solution Algorithm

General Strategy. Given a periodic graph $\mathcal{G}$, or equivalently its arena $\mathcal{D}$, to determine whether or not it is copwin, by Theorem 1, it is sufficient to determine whether or not its maximal augmented arena $\mathcal{A}^*$ contains an anchored star. Hence, informally, a basic solution approach is to start from $\mathcal{A} = \mathcal{D}$, repeatedly determine a "new" shadow edge (i.e., in $E(\mathcal{A}^*) \setminus E(\mathcal{A})$), using Theorem 2, and consider the new augmented arena obtained by adding such an edge. This process is repeated until either the current augmented arena $\mathcal{A}$ contains an anchored star, or no other "missing" shadow edge exists. In the former case, by Theorem 1, $\mathcal{D}$ is copwin; in the latter case, by Theorem 3, the current augmented arena is $\mathcal{A}^*$ and, if it does not contain an anchored star, $\mathcal{D}$ is robberwin.

A general strategy based on this approach operates in a sequence of iterations, each composed of two operations: the examination of a shadow edge, and the examination of new shadow corners (if any) determined in the first operation. More precisely, in each iteration: (i) A "new" (i.e., not yet examined) shadow edge $e = ((t, x), (t + 1, y))$ is examined to determine if its presence transforms some nodes into new shadow corners of (t, x). (ii) Each of these new shadow corners is examined, determining if its presence generates new shadow edges. By the end of the iteration, the shadow edge e and the new shadow corners of (t, x) examined in this iteration are removed from consideration. This iterative process continues until there are no new shadow edges to be examined (i.e. $\mathcal{A} = \mathcal{A}^*$) or there is an anchored star in $\mathcal{A}$.

GENERAL STRATEGY

1. While there is a still unexamined shadow edge $e = ((t, x), (t + 1, y))$ in $\mathcal{A}$ do:
2.   If there are still unexamined shadow corners covered by (t, x) then:
3.   For each such shadow corner $(t - 1, z)$ do:
4.   If there are new shadow edges due to $(t - 1, z)$ then:
5.   Add them to $\mathcal{A}$ to be examined.
6.   Remove $(t - 1, z)$ from consideration as a shadow corners of (t, x).
                                (i.e., mark it as examined).
7.   Remove e from consideration (i.e., mark it as examined).
8. If there is an anchored star in $\mathcal{A}$, then $\mathcal{D}$ is copwin else it is robberwin.

Fig. 7. Outline of general strategy where the iterative process terminates when $\mathcal{A} = \mathcal{A}^*$.

An outline of the strategy, where the iterative process is made to terminate when $\mathcal{A} = \mathcal{A}^*$, is shown in Fig. 7.

Algorithm Description. Let us present the proposed algorithm, CopRob-berPeriodic, which follows directly the general strategy described above to determine whether or not an arena $\mathcal{D} = ((Z_p \times V), E(\mathcal{D}))$ is copwin, where $V = \{v_1, \ldots, v_n\}$.

We denote by $\mathcal{A}$ the current augmented arena of $\mathcal{D}$, by A its adjacency matrix, and by A_t the adjacency matrix of slice S_t of $\mathcal{A}$. Auxiliary structures used by the algorithm include the queue $\mathcal{SE}$, of the known shadow edges that have not been examined yet; a $n \times n$ Boolean matrix SE_t for each t, initialized to A_t, used to indicate shadow edges already known; a $n \times n$ Boolean matrix SC_t for each t, initialized to zero and used to indicate the detected shadow corners; more precisely, $SC_t[x, y] = 1$ indicates that (t, x) has been determined to be a shadow corner of $(t + 1, y)$,

The algorithm is composed of two phases: *Initialization*, in which all the necessary structures are set up and preliminary computations are performed; and *Iteration*, a repetitive process where the two basic operations of the general strategy (described in Sect. 4.1) are performed in each iteration: examination of a "new" shadow edge (to determine "new" shadow corners generated by that edge) and examination of the "new" shadow corners (to determine "new" shadow edges generated by that corner).

The structure used to determine new shadow corners is the set $\{\mathrm{DIF}(t, x, y) : t \in Z_p, x, y \in V\}$ of $n^2 p$ Boolean arrays of dimension n. For all $x, v \in V$ and $t \in Z_p$, the value of the cell $\mathrm{DIF}(t, x, y)[i]$ indicates whether $v_i \in \Gamma_t(x, \mathcal{D}) \setminus \Gamma_{t+1}(y, \mathcal{A})$ (in which case $\mathrm{DIF}(t, x, y)[i] = 1$) or $v_i \in \Gamma_t(x, \mathcal{D}) \cap \Gamma_{t+1}(y, \mathcal{A})$ (in which case $\mathrm{DIF}(t, x, y)[i] = 0$). Note that, if $v_i \notin \Gamma_t(x, \mathcal{D})$, the value of $\mathrm{DIF}(t, x, y)[i]$ is left undefined; indeed, the algorithm only initializes and uses the $|\Gamma_t(x, \mathcal{D})|$ cells corresponding to the elements of $\Gamma_t(x, \mathcal{D})$; we shall call those cells the *core* of $\mathrm{DIF}(t, x, y)$.

The algorithm also maintains a variable $\phi(\mathrm{DIF}(t, x, y))$ indicating the current number of core cells with value "1" in array $\mathrm{DIF}(t, x, y)$; this variable is initialized to $|\Gamma_t(x, \mathcal{D})|$. Observe that, by definition of $\mathrm{DIF}(t, x, y)$, $\phi(\mathrm{DIF}(t, x, y)) = 0$ iff (t, x) is a shadow corner of $(t + 1, y)$.

In each iteration of the *Iteration* phase, a new shadow edge is taken from $\mathcal{SE}$, added to the augmented arena $\mathcal{A}$, and examined. The examination of a shadow edge $((t, x), (t + 1, y))$ involves (i) the update of $\mathrm{DIF}(t - 1, z, x)[y]$ for any in-neighbour $(t - 1, z)$, in $\mathcal{D}$, of (t, y) and, for any such in-neighbour, (ii) the test to see if the presence of the edge $((t, x), (t + 1, y))$ in the augmented arena has created new shadow corners among such in-neighbours[2]. If new shadow corners exist, they may in turn have created new shadow edges originating from the in-neighbours, in $\mathcal{D}$, of (t, x). In fact, any in-neighbour $(t - 1, w)$ of (t, x) such that $((t - 1, w), (t, z))$ is not already in the augmented arena is a new shadow

---

[2] Such would be any $(t - 1, z)$ for which the update has resulted in an array $\mathrm{DIF}(t - 1, z, x)$ that contains only zero entries.

edge: a move of the cop from $(t-1, w)$ to (t, x) is fatal for the robber wherever it goes; in such a case, the algorithm then adds $((t-1, w), (t, z))$ to $\mathcal{SE}$.

The pseudo code of the algorithm is shown in Algorithm 1. Not shown are several very low level (rather trivial) implementation details. These include, for example, the fact that the core cells of $\mathrm{DIF}(t, x, y)$ are connected through a doubly linked list, and that, for efficiency reasons, we also maintain two additional doubly linked lists: one going through the core cells of the array containing "1", the other linking the core cells containing "0".

4.2   Analysis

Correctness. Let us prove the correctness of Algorithm CopRobberPeriodic. Let $\mathcal{D} = (\mathbb{Z}_p \times V, E(\mathcal{D}))$ be the arena of a p-periodic graph with $n = |V|$ and $m = |E(\mathcal{D})|$.

Lemma 2. *Algorithm* CopRobberPeriodic *terminates after at most* $|E(\mathcal{A}^*)|$ $- |E(\mathcal{D})|$ *iterations.*

Given an augmented arena $\mathcal{A}$ and a shadow edge $e = ((t, x), (t+1, y)) \in E(\mathcal{A}^*) \setminus E(\mathcal{A})$, we shall say that e is an *implicit* shadow edge of $\mathcal{A}$ if there exists $z \in \Gamma_t(x, \mathcal{D})$ such that (t, y) is a shadow corner of $(t+1, z)$ in $\mathcal{A}$.

Lemma 3. *At the end of the* Initialization *phase: (i) for all and only the temporal corners* (t, x) *of* $(t+1, y)$ *in* $\mathcal{D}$, $SC_t[x, y] = 1$ *and* $\phi(\mathrm{DIF}(t, x, y)) = 0$; *(ii) all implicit shadow edges of* $\mathcal{D}$ *are in* $\mathcal{SE}$; *furthermore, the entry in SE of all edges of* $\mathcal{D}$ *and implicit shadow edges of* $\mathcal{D}$, *is 1.*

Let us consider the *Initialization* phase as iteration 0 of the *Iteration* phase; hence, the entire algorithm can be viewed as a sequence of iterations. Denote by $\mathcal{A}_j$ the augmented arena at the beginning of the j-th iteration, with $\mathcal{A}_0 = \mathcal{D}$. We now show that, at the beginning of iteration j, all shadow corners of $\mathcal{A}_{j-1}$ have been examined and all implicit shadow edges of $\mathcal{A}_{j-1}$ are in $\mathcal{SE}$.

Lemma 4. *At the beginning of iteration* $j > 0$:

(a) $\phi(\mathrm{DIF}(t, x, y)) = 0$ *if and only if* (t, x) *is a shadow corner of* $(t+1, y)$ *in* $\mathcal{A}_{j-1}$; *furthermore, in such a case,* $SC_t[x, y] = 1$.

(b) $\mathcal{SE}$ *contains all the implicit shadow edges of* $\mathcal{A}_{j-1}$; *furthermore, in SE, the entry of the edges of* $\mathcal{A}_{j-1}$ *and of the implicit shadow edges of* $\mathcal{A}_{j-1}$ *is 1.*

Theorem 4. *Algorithm* CopRobberPeriodic *correctly determines whether or not an arena* $\mathcal{D}$ *is copwin.*

Complexity. Let us analyze the cost of Algorithm CopRobberPeriodic. Given $\mathcal{D} = (\mathbb{Z}_p \times V, E(\mathcal{D}))$, let m_i denote the number of edges of slice S_i of $\mathcal{D}$, $i \in \mathbb{Z}_p$, and $m = |E(\mathcal{D})| = \sum_{i=0}^{p-1} m_i$ the total number of edges of $\mathcal{D}$. As usual, $n = |V|$.

Algorithm 1: COPROBBERPERIODIC

Input: Arena $\mathcal{D} = (\mathbb{Z}_p \times V, E(\mathcal{D}))$, with $V = \{v_1, \ldots, v_n\}$

1 *Initialization*
2 $\mathcal{A} := \mathcal{D}$
3 $SE := A$
4 $\mathcal{SE} := \emptyset$
5 $SC := Zero$ /* a table of p zero matrices, each of size $n \times n$ */
6 **foreach** $t \in \mathbb{Z}_p$, $u, v \in V$ **do**
7     $\phi(\text{DIF}(t, u, v)) := |\Gamma_t(u, \mathcal{D})|$
8   **foreach** $w \in \Gamma_t(u, \mathcal{D})$ **do**
9   **if** $A_{t+1}[v, w] = 1$ **then**
10           $\text{DIF}(t, u, v)[w] := 0$
11           $\phi(\text{DIF}(t, u, v)) := \phi(\text{DIF}(t, u, v)) - 1$
12   **if** $\phi(\text{DIF}(t, u, v)) = 0$ **and** $SC_t[u, v] = 0$ **then**
13             $SC_t[u, v] := 1$
14   **foreach** $z \in \Gamma_{t+1}^{\text{in}}(v, \mathcal{D})$ **do**
15   **if** $SE_t[z, u] = 0$ **then**
16                 $SE_t[z, u] := 1$
17                 $\mathcal{SE} \leftarrow ((t, z), (t + 1, u))$

18   **else**
19           $\text{DIF}(t, u, v)[w] := 1$

20 *Iteration*
21 **while** $\mathcal{SE} \neq \emptyset$ **do**
22     $((t, x), (t + 1, y)) \leftarrow \mathcal{SE}$
23     $A_t(x, y) := 1$
24   **foreach** $z \in \Gamma_t^{\text{in}}(y, \mathcal{D})$ **do**
25   **if** $\text{DIF}(t - 1, z, x)[y] = 1$ **then**
26           $\text{DIF}(t - 1, z, x)[y] := 0$
27           $\phi(\text{DIF}(t - 1, z, x)) := \phi(\text{DIF}(t - 1, z, x)) - 1$
28   **if** $\phi(\text{DIF}(t - 1, z, x)) = 0$ **and** $SC_{t-1}[z, x] = 0$ **then**
29           $SC_{t-1}[z, x] := 1$
30   **foreach** $w \in \Gamma_t^{\text{in}}(x, \mathcal{D})$ **do**
31   **if** $SE_{t-1}[w, z] = 0$ **then**
32               $SE_{t-1}[w, z] := 1$
33               $\mathcal{SE} \leftarrow ((t - 1, w), (t, z))$

34 **if** $\mathcal{A}$ contains an anchored star **then** $\mathcal{D}$ is copwin
35 **else** $\mathcal{D}$ is robberwin

Theorem 5. *Algorithm* COPROBBERPERIODIC *determines in time* $O(n^2 p + nm)$ *whether or not $\mathcal{D}$ is copwin.*

Proof. We first derive the cost of the *Initialization* phase. Observe that the initialization of $\mathcal{A}$, SE, SC (Lines 2–4) can be performed with $O(n^2 p)$ operations.

Then, Line 7 is executed n^2p times. The cost of the initialization of DIF and of $\phi(\text{DIF})$ (Lines 6–13, 18–19), which includes the update of some entries of SC, plus the cost of the initialization of $\mathcal{SE}$ (Lines 14–17), which includes the update of some entries of SE, require at most

$$n^2p + \sum_{i \in \mathbb{Z}_p, u \in V} O(|\Gamma_i(u, \mathcal{D})|) + \sum_{i \in \mathbb{Z}_p, v \in V} O(|\Gamma_i^{in}(v, \mathcal{D})|) = \sum_{i=0}^{p-1} O(n(m_i + m_{i-1})),$$

which sums up to $O(n^2p + nm)$ operations for the *Initialization* phase.

Let us consider now the *Iteration* phase. The while loop will be repeated until in the current augmented arena $\mathcal{A}$ there are no more shadow edges to be examined (i.e. $\mathcal{A} = \mathcal{A}^*$). By Lemma 2, the total number of iterations is $|E(\mathcal{A}^*)| - |E(\mathcal{D})| \leq n^2p - m$. Further observe that every operation performed during an iteration requires constant time.

In each iteration, two processes are being carried out. The first process (Lines 24–27) is the determination of all new shadow corners (if any) of (t, x) created by (the addition of) the shadow edge $((t, x), (t + 1, y))$ being examined. The total cost of this process in this iteration is at most two operations for each in-neighbour of (t, y), i.e., at most $2c_1|\Gamma_t^{in}(y, \mathcal{D})|$, where $c_1 \in O(1)$ is the constant cost of performing a single operation in this process.

This process is repeated in all iterations, each time with a different shadow edge being examined. Thus, the cost of $2c_1|\Gamma_t^{in}(y, \mathcal{D})|$ will be incurred for all $((t, x), (t + 1, y)) \in E(\mathcal{A}^*)$; that is, at most n times. Summarizing, for each $y \in V, t \in \mathbb{Z}_p$ this process costs $2c_1n|\Gamma_t^{in}(y, \mathcal{D})|$. Hence the total cost of this process over all iterations is

$$\sum_{y \in V, t \in \mathbb{Z}_p} 2\,c_1\,n\,|\Gamma_t^{in}(y, \mathcal{D})| = 4\,c_1\,n\,\sum_{t=0}^{p-1} m_t = O(nm).$$

The second process, to be performed only if new shadow corners of (t, x) have been found in the first process, is the determination (Lines 28–33) of all the new shadow edges (if any) created by the found new shadow corners, and their addition to $\mathcal{SE}$. The cost of this process for a new shadow corner in this iteration is $c_2|\Gamma_t^{in}(x, \mathcal{D})|$, where $c_2 \in O(1)$ is the constant cost of performing a single operation in this process. Observe that, if a new shadow corner of (t, x) is found in this iteration, it will not be considered in any subsequent iteration (Lines 28–29). Hence, the cost $c_2|\Gamma_t^{in}(x, \mathcal{D})|$ will be incurred at most once for each shadow corner of (t, x); that is, at most n times. Summarizing, for each $x \in V, t \in \mathbb{Z}_p$ this process costs at most $2c_2n|\Gamma_t^{in}(x, \mathcal{D})|$. Hence the total cost of this process over all iterations is

$$\sum_{x \in V, t \in \mathbb{Z}_p} 2\,c_2\,n\,|\Gamma_t^{in}(x, \mathcal{D})| = 4\,c_2\,n\,\sum_{t=0}^{p-1} m_t = O(nm).$$

Consider now the last step of the algorithm, of determining if the constructed $\mathcal{A}$ contains an anchored star. To determine all the stars (if any) in $\mathcal{A}^*$ can be done by checking the degree of each temporal node in $\mathcal{A}^*$, i.e., in $O(np)$ time. To determine if at least one of them is anchored can be done by a DFS traversal of $\mathcal{A}^*$ starting from each root node $(0, x)$, for a total of at most $O(n^2 + nm)$ operations. It follows that the total cost of the algorithm is $O(pn^2 + nm)$ as claimed.   □

The bound established by Theorem 5 improves on the existing $O(p\, n^3)$ bound [14]; in particular, in periodic graphs with sparse snapshots the proposed algorithm terminates in $O(p\, n^2)$ time. Furthermore, since a static graph is a periodic graph with $p = 1$, the bound of Theorem 5 becomes $O(n\, m)$, improving the existing $O(n^3)$ bound [32]; in particular our bound becomes $O(n^2)$ for sparse graphs.

4.3   Extensions

Determining a Copwin Strategy. The algorithm, as described, determines whether or not the arena $\mathcal{D}$ (and, thus, the corresponding temporal graph $\mathcal{G}$) is copwin. Simple additions to the algorithm would allow it to easily determine a copwin strategy σ_c if $\mathcal{D}$ is copwin. For any shadow edge $e = ((t, x), (t + 1, y))$, let $\rho(t, x, y)$ be defined as follows. If $e = ((t, x), (t + 1, y)) \in E(\mathcal{D})$, then $\rho(t, x, y) = y$. If $e = ((t, x), (t + 1, y)) \in E(\mathcal{A}^*) \setminus E(\mathcal{D})$, when e is inserted in $\mathcal{SE}$, either during the Initialization or the Iteration phase, then $\rho(t, x, y) = z$ where $(t + 1, z)$ is the shadow cover of (t, y) determined in the corresponding phase of the algorithm (Line 12 if Initialization, Line 28 if Iteration).

Recall that, if $\mathcal{D}$ is copwin, $\mathcal{A}^*$ must contain an anchored star, say (t, x). Since (t, x) is a star, if the cop is located on (t, x) and the robber is located on (t, y), by moving according to ρ (starting with $\rho(t, x, y)$) the cop will eventually capture the robber. Since (t, x) is anchored, it is reachable from some node in G_0, say $(0, v)$; that is, there is a journey $\pi((0, v), (t, x))$ from $(0, v)$ to (T, x), where $[T]_p = t$. Consider now the following strategy σ_c for the cop: (1) choose as initial location $(0, v)$; (2) follow $\pi((0, v), (t, x))$; (3) follow ρ. Using this strategy, the cop will eventually capture the robber.

More Cops & One Robber. The framework presented so far can be generalized to the case when there are $k > 1$ cops. By shifting from a representation in terms of directed graphs to one in terms of directed multi-hypergraphs, it is possible to extend all the basic concepts introduced for $k = 1$. Indeed, all the fundamental properties of augmented arenas continue to hold in this extended setting, and the same strategy can be used to determine if a periodic graph is k-copwin. The strategy can be implemented by a direct extension of the solution algorithm for $k = 1$.

References

1. Agarwalla, A., Augustine, J., Moses, W., Madhav, S., Sridhar, A.K.: Deterministic dispersion of mobile robots in dynamic rings. In: 19th International Conference on Distributed Computing and Networking, pp. 19:1–19:4 (2018)

2. Aigner, M., Fromme, M.: A game of cops and robbers. Discret. Appl. Math. **8**, 1–12 (1984)
3. Balev, S., Laredo, J.L.J., Lamprou, I., Pigné, Y., Sanlaville, E.: Cops and robbers on dynamic graphs: offline and online case. In: 27th International Colloquium on Structural Information and Communication Complexity (SIROCCO) (2020)
4. Bonato, A., MacGillivray, G.: Characterizations and algorithms for generalized Cops and Robbers games. Contrib. Discret. Math. **12**(1) (2017)
5. Bonato, A., Nowakowski, R.: The Game of Cops and Robbers on Graphs. American Mathematical Society (2011)
6. Bournat, M., Dubois, S., Petit, F.: Computability of perpetual exploration in highly dynamic rings. In: 37th IEEE International Conference on Distributed Computing Systems, pp. 794–804 (2017)
7. Casteigts, A., Flocchini, P., Mans, B., Santoro, N.: Measuring temporal lags in delay-tolerant networks. IEEE Trans. Comput. **63**(2), 397–410 (2014)
8. Casteigts, A., Flocchini, P., Quattrociocchi, W., Santoro, N.: Time-varying graphs and dynamic networks. Int. J. Parallel Emergent Distrib. Syst. **27**(5), 387–408 (2012)
9. Di Luna, G.A.: Mobile agents on dynamic graphs. In: Flocchini, P., Prencipe, G., Santoro, N. (eds.) Distributed Computing by Mobile Entities, Current Research in Moving and Computing. LNCS, vol 11340, pp. 549–584. Springer, Cham (2019). https://doi.org/10.1007/978-3-030-11072-7_20. https://dblp.org/rec/series/lncs/Luna19.bib
10. Di Luna, G.A., Dobrev, S., Flocchini, P., Santoro, N.: Distributed exploration of dynamic rings. Distrib. Comput. **33**, 41–67 (2020)
11. Di Luna, G.A., Flocchini, P., Pagli, L., Prencipe, G., Santoro, N., Viglietta, G.: Gathering in dynamic rings. Theor. Comput. Sci. **811**, 79–98 (2020)
12. Erlebach, T., Hoffmann, M., Kammer, F.: On temporal graph exploration. In: 42nd International Colloquium on Automata, Languages, and Programming, pp. 444–455 (2015)
13. Erlebach, T., Spooner, J.T.: Faster exploration of degree-bounded temporal graphs. In: Proceedings of 43rd International Symposium on Mathematical Foundations of Computer Science (MFCS), pp. 1–13 (2018)
14. Erlebach, T., Spooner, J.T.: A game of cops and robbers on graphs with periodic edge-connectivity. In: Chatzigeorgiou, A., et al. (eds.) SOFSEM 2020. LNCS, vol. 12011, pp. 64–75. Springer, Cham (2020). https://doi.org/10.1007/978-3-030-38919-2_6
15. Ferreira, A.: Building a reference combinatorial model for MANETs. IEEE Netw. **18**(5), 24–29 (2004)
16. Flocchini, P., Mans, B., Santoro, N.: On the exploration of time-varying networks. Theor. Comput. Sci. **469**, 53–68 (2013)
17. Gotoh, T., Flocchini, P., Masuzawa, T., Santoro, N.: Exploration of dynamic networks: tight bounds on the number of agents. J. Comput. Syst. Sci. **122**, 1–18 (2021)
18. Gotoh, T., Sudo, Y., Ooshita, F., Kakugawa, H., Masuzawa, T.: Group exploration of dynamic tori. In: Proceedings of the IEEE 38th International Conference on Distributed Computing Systems (ICDCS), pp. 775–785 (2018)
19. Gromovikov, I., Kinnersley, W., Seamone, B.: Fully active cops and robbers. Australas. J. Comb. **76**(2), 248–265 (2020)
20. Harary, F., Gupta, G.: Dynamic graph models. Math. Comput. Model. **25**(7), 79–88 (1997)

21. Holme, P., Saramäki, J.: Temporal networks. Phys. Rep. **519**(3), 97–125 (2012)
22. Ilcinkas, D., Klasing, R., Wade, A.M.: Exploration of constantly connected dynamic graphs based on cactuses. In: Halldórsson, M.M. (ed.) SIROCCO 2014. LNCS, vol. 8576, pp. 250–262. Springer, Cham (2014). https://doi.org/10.1007/978-3-319-09620-9_20
23. Ilcinkas, D., Wade, A.M.: On the power of waiting when exploring public transportation systems. In: Fernàndez Anta, A., Lipari, G., Roy, M. (eds.) OPODIS 2011. LNCS, vol. 7109, pp. 451–464. Springer, Heidelberg (2011). https://doi.org/10.1007/978-3-642-25873-2_31
24. Jathar, R., Yadav, V., Gupta, A.: Using periodic contacts for efficient routing in delay tolerant networks. Ad Hoc Sens. Wirel. Netw. **22**(1,2), 283–308 (2014)
25. Khatri, D., et al.: A study of cops and robbers in oriented graphs, pp. 1–23 (2018)
26. Kuhn, F., Lynch, N., Oshman, R.: Distributed computation in dynamic networks. In: 42nd ACM Symposium on Theory of Computing, pp. 513–522 (2010)
27. Loh, P.-S., Oh, S.: Cops and robbers on planar-directed graphs. J. Graph Theory **86**(3), 329–340 (2017)
28. Morawietz, N., Rehs, C., Weller, M.: A timecop's work is harder than you think. In: 45th International Symposium on Mathematical Foundations of Computer Science (MFCS), pp. 71:1–14 (2020)
29. Morawietz, N., Wolf, P.: A timecop's chase around the table. In: 46th International Symposium on Mathematical Foundations of Computer Science (MFCS), pp. 77:1–18 (2021)
30. Nowakowski, R., Winkler, P.: Vertex-to-vertex pursuit in a graph. Discret. Math. **43**(2–3), 235–239 (1983)
31. O'Dell, R., Wattenhofer, R.: Information dissemination in highly dynamic graphs. In: Joint Workshop on Foundations of Mobile Computing, pp. 104–110 (2005)
32. Petr, J., Portier, J., Versteegen, L.: A faster algorithm for cops and robbers. Discret. Appl. Math. **320**, 11–14 (2022)
33. Quilliot, A.: Jeux et points fixes sur les graphs. Ph.D. thesis. Université de Paris VI (1978)
34. Wehmuth, K., Ziviani, A., Fleury, E.: A unifying model for representing time-varying graphs. In: 2nd IEEE International Conference on Data Science and Advanced Analytics, (DSAA), pp. 1–10 (2015)

Minimum Cost Flow in the CONGEST Model

Tijn de Vos[✉][iD]

University of Salzburg, Salzburg, Austria
tdevos@cs.sbg.ac.at

Abstract. We consider the CONGEST model on a network with n nodes, m edges, diameter D, and integer costs and capacities bounded by poly n. In this paper, we show how to find an exact solution to the minimum cost flow problem in $n^{1/2+o(1)}(\sqrt{n} + D)$ rounds, improving the state of the art algorithm with running time $m^{3/7+o(1)}(\sqrt{n}D^{1/4} + D)$ [16], which only holds for the special case of unit capacity graphs. For certain graphs, we achieve even better results. In particular, for planar graphs, expander graphs, $n^{o(1)}$-genus graphs, $n^{o(1)}$-treewidth graphs, and excluded-minor graphs our algorithm takes $n^{1/2+o(1)}D$ rounds. We obtain this result by combining recent results on Laplacian solvers in the CONGEST model [3,16] with a CONGEST implementation of the LP solver of Lee and Sidford [29], and finally show that we can round the approximate solution to an exact solution. Our algorithm solves certain linear programs, that generalize minimum cost flow, up to additive error ϵ in $n^{1/2+o(1)}(\sqrt{n} + D)\log^3(1/\epsilon)$ rounds.

Keywords: CONGEST model · Minimum Cost Flow · LP solver

1 Introduction

The CONGEST model [38] is one of the most widely studied distributed models. It consists of a network of n nodes that communicate in synchronous rounds, where each node can exchange a message of size $O(\log n)$ with each of its neighbors. The minimum cost flow problem is considered one of the harder problems in the CONGEST model. Although the highest lower bound is $\tilde{\Omega}(\sqrt{n}+D)$, which is the same as for 'easier' problems such as shortest path, minimum spanning trees, bipartiteness, s-t connectivity [15,39,43], it was only recently that the first distributed algorithm was presented [16]. For the approximate version there exists some further, also quite recent, results [6,19]. These results use the powerful *Laplacian paradigm* to obtain their results.

The Laplacian paradigm encompasses a series of algorithms that combine numerical and combinatorial techniques. The *Laplacian matrix* of a weighted

This work is supported by the Austrian Science Fund (FWF): P 32863-N. This project has received funding from the European Research Council (ERC) under the European Union's Horizon 2020 research and innovation programme (grant agreement No 947702).

S. Rajsbaum et al. (Eds.): SIROCCO 2023, LNCS 13892, pp. 406–426, 2023.
https://doi.org/10.1007/978-3-031-32733-9_18

graph G is defined as $L(G) := \mathrm{Deg}(G) - A(G)$, where $\mathrm{Deg}(G)$ is the diagonal weighted degree matrix: $\mathrm{Deg}(G)_{uu} := \sum_{(u,v) \in E} w(u,v)$ and $\mathrm{Deg}(G)_{uv} := 0$ for $u \neq v$, and $A(G)$ is the adjacency matrix: $A(G)_{uv} := w(u,v)$. This line of research was initiated by Spielman and Teng [46], who showed that linear equations in the Laplacian matrix of a graph can be solved in near-linear time. More efficient sequential and parallel Laplacian solvers have been presented since [10, 23, 25–28, 41]. The Laplacian paradigm has booked many successes, including but not limited to flow problems [5, 12, 22, 33, 34, 36, 37, 40, 44], bipartite matching [7], and (parallel) shortest paths [4, 31].

Recently, these developments have also made their way to the distributed world [3, 6, 16, 17, 19]. In particular, Forster, Goranci, Liu, Peng, Sun, and Ye [16] provide a Laplacian solver that takes $n^{o(1)}(\sqrt{n} + D)$ rounds, which is near-optimal: they provide a $\tilde{\Omega}(\sqrt{n} + D)$ lower bound. Furtermore, they show that their Laplacian solver leads to an implementation of (minimum cost) maximum flow algorithms [12, 37] in the CONGEST model. In this paper, we significantly improve the round complexity of the algorithms solving the exact variants of these flow problems.

1.1   Our Results

Our main result is an algorithm that solves the minimum cost flow problem, so in particular also the maximum flow problem.

Theorem 1. *There exists an algorithm that, given a directed graph $G = (V, E, w)$ with integer costs $q \in \mathbb{Z}_{>0}^m$ and capacities $c \in \mathbb{Z}_{>0}^m$ satisfying $||q||_\infty, ||c||_\infty \leq M$, computes a minimum cost maximum s-t flow in $\tilde{O}(\sqrt{n} T_{\mathrm{Laplacian}}(G) \log^3 M)$ rounds in the CONGEST model, where $T_{\mathrm{Laplacian}}(G)$ is the number of rounds needed to solve a Laplacian system on G.*

We know that $T_{\mathrm{Laplacian}}(G) = n^{o(1)}(\sqrt{n} + D)$ for general graphs [16], which is near-optimal. However, for certain graphs we can get better results. This is based on the concept of *universally optimal* algorithms, which takes the topology of the input graph into account. The details regarding this can be found in Sect. 1.2. In particular, we have $T_{\mathrm{Laplacian}}(G) = n^{o(1)}D$ for planar graphs, expander graphs, $n^{o(1)}$-genus graphs, $n^{o(1)}$-treewidth graphs, and excluded-minor graphs.

Further we remark that Cohen, Mądry, Sankowski, and Vladu [12] show that the negative weight single source shortest path problem can be reduced to minimum cost flow and a non-negative weight shortest path computation. Using [8] for the latter in $\tilde{O}(\sqrt{n}D^{1/4} + D)$ rounds, we obtain the following corollary.

Corollary 1. *There exists an algorithm that, given a directed graph $G = (V, E, w)$ with integer weights $w \in \mathbb{Z}^m$ satisfying $||w||_\infty \leq M$, and source $s \in V$, computes shortest paths from s in $\tilde{O}(\sqrt{n} T_{\mathrm{Laplacian}}(G) \log^3 M)$ rounds in the CONGEST model, where $T_{\mathrm{Laplacian}}(G)$ is the number of rounds needed to solve a Laplacian system on G.*

We obtain Theorem 1 by writing the problem as an LP, solving this LP up to high precision and rounding the result. Hereto we present an LP solver for certain linear programs in the CONGEST model.

Formally, the setting is as follows. Let $A \in \mathbb{R}^{m \times n}$, $b \in \mathbb{R}^n$, $c \in \mathbb{R}^m$, $l_i \in \mathbb{R} \cup \{-\infty\}$, and $u_i \in \mathbb{R} \cup \{+\infty\}$ for all $i \in [m]$, where we assume $l_i \neq -\infty$ or $u_i \neq +\infty$. The linear program we want to solve is as follows

$$\text{OPT} := \min_{\substack{x \in \mathbb{R}^m : A^T x = b \\ \forall i \in [m] : l_i \leq x_i \leq u_i}} c^T x.$$

We assume that the set of feasible solutions to the LP $\Omega^\circ := \{x \in \mathbb{R}^m : A^T x = b, \ l_i \leq x_i \leq u_i\}$ is non-empty.

Theorem 2. *Let $A \in \mathbb{R}^{m \times n}$ be a constraint matrix with $\text{rank}(A) = n$, let $b \in \mathbb{R}^n$ be a demand vector, and let $c \in \mathbb{R}^m$ be a cost vector. Moreover, let $x_0 \in \Omega^\circ$ be a given initial point. Suppose a CONGEST network consists of n nodes, where each node i knows both every entire j-th row of A for which $A_{ji} \neq 0$ and knows $(x_0)_j$ if $A_{ji} \neq 0$. Moreover, suppose that for every $y \in \mathbb{R}^n$ and positive diagonal $W \in \mathbb{R}^{m \times m}$ we can compute $(A^T W A)^{-1} y$ up to precision $\text{poly}(1/m)$ in $T_{\text{Laplacian}}(G)$ rounds. Let $U := \max\{\|1/(u - x_0)\|_\infty, \|1/(x_0 - l)\|_\infty, \|u - l\|_\infty, \|c\|_\infty\}$. Then with high probability the CONGEST algorithm* `LPSolve` *outputs a vector $x \in \Omega^\circ$ with $c^T x \leq \text{OPT} + \epsilon$ in $\tilde{O}(\sqrt{n} \log^3(U/\epsilon) T_{\text{Laplacian}}(G))$ rounds.*

Intuitively, U is a bound on the size of any constants or variables appearing in the LP. For graph problems, this is usually bounded by a polynomial in n and M.

The formal statement of this theorem might seem somewhat convoluted; essentially it means that we can solve linear programs whose constraint matrix can be expressed in terms of the adjacency matrix, where each node knows the entries in the constraint matrix corresponding to its incident edges. Analogously, a node has to output the variables corresponding to its incident edges. This includes flow problems, see Sect. A. It also includes approximate fractional maximal matching. However, here the running time does not come close to the $O(\log(nM)/\epsilon^2)$ running time of Ahmadi, Kuhn, and Oshman [2] (at least for $\epsilon = \Omega(1/n^{1/4})$).

1.2   Related Work

Distributed Flow Algorithms. Our main point of reference is Forster, Goranci, Liu, Peng, Sun, and Ye [16]. They provide the previous best minimum cost flow solver, which takes $m^{3/7+o(1)}(\sqrt{n}D^{1/4} + D)$ rounds[1]. Their approach uses the framework from Cohen, Mądry, Sankowski, and Vladu [12], which uses $\tilde{O}(m^{3/7})$ iterations of another interior point method to solve a (different) LP representing the problem. We bring this number of iterations down to $\tilde{O}(n^{1/2})$. Moreover, this approach leads to an approximate solution that has to be made into an exact solution by running $\tilde{O}(m^{3/7})$ shortest path computations. Currently, the state of the art for algorithm for shortest path computations takes $\tilde{O}(\sqrt{n}D^{1/4} + D)$

---

[1] For simplicity, we restrict ourselves to graphs with weights bounded by $\text{poly}(m)$ when discussing related work.

rounds [8], which is already (slightly) worse than the global (near optimal) round complexity for solving Laplacian systems. Moreover, this means that their set-up cannot benefit from the recent progress of (almost) universally optimal Laplacian solvers. Our approach solves the LP up to a higher precision, such that an internal rounding procedure gives the exact solution, and no further shortest path computations are necessary. A further improvement is that [16] only solves minimum cost flow in graphs with *unit capacities*, where we solve it for arbitrary capacities. Further, [16] provides a maximum flow algorithm for graphs with arbitrary capacities, which takes $\tilde{O}(m^{3/7}U^{1/7}(n^{o(1)}(\sqrt{n} + D) + \sqrt{n}D^{1/4}) + \sqrt{m})$ rounds. This is the previous best result for maximum flow in the CONGEST model.

For approximate versions, there exist some further results that only hold for undirected graphs. Ghaffari et al. [19] give a $(1 + \epsilon)$-approximate maximum flow in weighted undirected graphs in $n^{o(1)}(\sqrt{n} + D)/\epsilon^3$ rounds. Further, Becker et al. [6] gave a $(1 + \epsilon)$-approximation to unit capacity minimum cost flow in undirected graphs in $\tilde{O}(n/\epsilon^2)$ rounds.

Interior Point Methods for Flow Problems. The line of work giving solutions for flow problems through interior point methods is initiated by Daitch and Spielman [13], who leverage the Laplacian solver of Spielman and Teng [46] in an $\tilde{O}(m^{3/2})$ time algorithm. The most recent development is the near-linear time algorithm of Chen et al. [9]. However, their algorithm uses $\Omega(m)$ iterations, which seems to render it hard to implement it efficiently in a distributed setting, as any intuitive implementation uses at least one round per iteration. The algorithms with lowest iteration counts have either $\Theta(m^{3/7})$ iterations [12,37], or $\Theta(\sqrt{n})$ iterations [29]. In our work, we show how to implement the latter efficiently in the CONGEST model.

Distributed Laplacian Solvers and Shortcut Quality. Forster et al. [16] provide a CONGEST model algorithm with $T_{\text{Laplacian}}(G) = n^{o(1)}(\sqrt{n} + D)$, and show that this is existentially optimal. For any graph, we know that $T_{\text{Laplacian}}(G) = \Omega(D)$, however it turns out that the $\sqrt{n}$-term is not necessary for every instance. To make this precise, we define the *shortcut quality* of a graph, as introduced by Ghaffari and Haeupler [18]. Intuitively, the shortcut quality tells us how easy it is, given some partition of the nodes, to compute some simple function (e.g., a minimum over the values held by nodes) on each part separately. Since distributed algorithm design often has such functions at its core, the shortcut quality can be used both for better upper and lower bounds.

Definition 1. *Let $G = (V, E)$ be an undirected graph whose node set V is partitioned into k disjoint subsets $V = P_1 \cup P_2 \cup \cdots \cup P_k$, such that each induced subgraph $G[P_i]$ is connected. A collection of k subgraphs $H_1, \cdots, H_k$ is called a shortcut of G with congestion c and dilation d if*

1. *the (hop) diameter of $G[P_i] \cup H_i$ is at most d;*
2. *every edge is included in at most c graphs H_i.*

The quality *of the shortcut is defined as $c+d$. The* shortcut quality *of G, denoted by $SQ(G)$, is defined as the smallest shortcut quality of the worst-case partition of V into connected parts.*

Anagnostides et al. [3] provide efficient algorithms for Laplacian solving in terms of the shortcut quality. Moreover, they provide an $\tilde{\Omega}(SQ(G))$ lower bound.

Theorem 3 ([3]). *There exists a Laplacian solver with error $\epsilon > 0$ in the CONGEST model that, given a graph G, takes $n^{o(1)} \text{poly}(SQ(G)) \log(1/\epsilon)$ rounds. In graphs with minor density δ and hop-diameter D, it takes $n^{o(1)} \delta D \log(1/\epsilon)$.*

Note that on graphs with minor density $n^{o(1)}$ the algorithm takes $n^{o(1)} D \log(1/\epsilon)$ rounds, matching the lower bound up to $n^{o(1)}$ factors. This includes planar graphs, $n^{o(1)}$-genus graphs, $n^{o(1)}$-treewidth graphs, and excluded-minor graphs.

Further note that in particular on graphs G with $SQ(G) = n^{o(1)}$ the algorithm takes $n^{o(1)} \log(1/\epsilon)$ rounds. This includes expanders, hop-constrained expanders, and the classes mentioned above with restricted diameter $D = n^{o(1)}$.

2   Overview and Techniques

2.1   LP Solver

For our LP solver, we give an implementation of Lee and Sidford's [29] LP solver in the CONGEST model. For correctness, we refer to [29]. In a similar fashion, Forster and de Vos [17] gave an implementation of this algorithm in the Broadcast Congested Clique. In this distributed model, each round each node can send *the same* $O(\log n)$-bit message to every other node in the network. This is in contrast to the CONGEST model, where nodes can send different messages, but only to its neighbors. Forster and de Vos essentially show that this LP framework uses $\sqrt{n}$ iterations, where each iteration involves:

(1) Matrix-vector multiplication involving some matrix with entries corresponding to edges;
(2) Approximately solving a Laplacian system;
(3) Computing leverage scores;
(4) Projecting on a mixed norm ball.

For this section (and the formulation of Theorem 2), we assume that matrix-vector multiplication can be done efficiently. In practice, this means that we have to be able to write the constraint matrix in terms of the adjacency matrix. In other words, there should only be constraints that correspond to edges.

For (2), we can use [3,16], as mentioned in Sect. 1.2.

Concerning (3); the leverage score of a matrix M is defined by $\sigma(M) := \text{diag}(M(M^T M)^{-1} M^T)$, where $\text{diag}(\cdot)$ returns the diagonal vector. We remark (similar to [17,29]) that, using the Johnson-Lindenstrauss Lemma, we can compute a sufficient approximation by some local sampling and a small number of matrix-vector multiplications and Laplacian solves. Details can be found in Sect. 3.

It remains to show that we can 'project on a mixed norm ball'. The objective here is to project a vector $a \in \mathbb{R}^m$ onto a ball of mixed norm. In particular, given $l \in \mathbb{R}^m$, we consider the ball of mixed norm 1: $\mathcal{B} := \{x : ||x||_2 + ||l^{-1}x||_\infty \le 1\}$. Now we need to compute $x \in \mathcal{B}$ closest to a, more formally we need to compute

$$\arg \max_{||x||_2 + ||l^{-1}x||_\infty \le 1} a^T x.$$

We do this by borrowing ideas from [17,29]. Details can be found in Sect. 3.

2.2   Minimum Cost Flow

Let $G = (V, E)$ be a directed graph, with integer capacities $c \in \mathbb{Z}_{\ge 0}^m$, integer costs $q \in \mathbb{Z}_{\ge 0}^m$, and source and target nodes s and t respectively. The *minimum cost (maximum) flow* problem is to find an s-t flow of minimum cost, among all such flows of maximum value. More formally, we say that $f \in \mathbb{R}_{\ge 0}^E$ is a *s-t flow* if $f_e \le c_e$ for all $e \in E$ and $\sum_{e \in E : v \in e} f_e = 0$. The *value* of the flow is $\sum_{v \in V : (s,v) \in E} f_{(s,v)}$. The *maximum flow*, is the flow of maximum value and the *minimum cost (maximum) flow* is the flow of minimum cost $\sum_{e \in E} f_e q_e$ among all flows of maximum value.

The minimum cost flow problem has a natural corresponding linear program. However, the state-of-the-art LP solvers only provide an approximate solution. To turn this efficiently into an exact solution, we do not consider the textbook LP formulation, but one that is closely related. After solving this up to high precision (additive error $\epsilon = O(1/\operatorname{poly}(m))$), we use the well-known fact that a minimum cost flow problem with integer input admits an optimal solution with integer values [24], and we (internally) round the approximate fractional solution to an optimal integer solution.

The technical contribution is to show that this particular LP formulation satisfies the demands of Theorem 2. This is done in Sect. A.

3   A Distributed LP Solver

In this section, we present our algorithm to solve a linear program, given a Laplacian solver. First, we reiterate the formal description of the problem. Let $A \in \mathbb{R}^{m \times n}$, $b \in \mathbb{R}^n$, $c \in \mathbb{R}^m$, $l_i \in \mathbb{R} \cup \{-\infty\}$, and $u_i \in \mathbb{R} \cup \{+\infty\}$ for all $i \in [m]$, where we assume $l_i \neq -\infty$ or $u_i \neq +\infty$. The linear program we try to solve is as follows

$$\text{OPT} := \min_{\substack{x \in \mathbb{R}^m : A^T x = b \\ \forall i \in [m] : l_i \le x_i \le u_i}} c^T x.$$

We assume that the set of feasible solutions to the LP $\Omega^\circ := \{x \in \mathbb{R}^m : A^T x = b,\ l_i \le x_i \le u_i\}$ is non-empty.

Theorem 2. *Let $A \in \mathbb{R}^{m \times n}$ be a constraint matrix with $\operatorname{rank}(A) = n$, let $b \in \mathbb{R}^n$ be a demand vector, and let $c \in \mathbb{R}^m$ be a cost vector. Moreover, let $x_0 \in \Omega^\circ$ be a given initial point. Suppose a CONGEST network consists of n nodes, where each node i knows both every entire j-th row of A for which $A_{ji} \neq 0$ and knows $(x_0)_j$ if*

$A_{ji} \neq 0$. *Moreover, suppose that for every $y \in \mathbb{R}^n$ and positive diagonal $W \in \mathbb{R}^{m \times m}$ we can compute $(A^T W A)^{-1} y$ up to precision $\text{poly}(1/m)$ in $T_{\text{Laplacian}}(G)$ rounds. Let $U := \max\{||1/(u - x_0)||_\infty, ||1/(x_0 - l)||_\infty, ||u - l||_\infty, ||c||_\infty\}$. Then with high probability the CONGEST algorithm* LPSolve *(outputs a vector) $x \in \Omega^\circ$ with $c^T x \leq \text{OPT} + \epsilon$ in $\tilde{O}(\sqrt{n} \log^3(U/\epsilon) T_{\text{Laplacian}}(G))$ rounds.*

The algorithm we provide in this section is an implementation of Lee and Sidford's [29] in the CONGEST model. We refer to them for the proof of correctness. For the two subroutines that we change, computing leverage scores and projecting on a mixed norm ball, we provide a correctness analysis. The remainder of this section consists of presenting the algorithm and proving the bound on the number of rounds. In both we follow notation of Forster and de Vos [17], who provide the equivalent for the Broadcast Congested Clique.

Lee and Sidford [30][2] show that it is sufficient to solve equations involving $A^T D A$ up to precision $\text{poly}(1/m)$. We use this fact for our running time, and simplify our presentation by writing as if we solve such equations exactly. Similarly, we need to perform matrix-vector multiplication with the adjacency matrix and diagonal matrices only up to precision $\text{poly}(\epsilon/(mU))$. Further we can assume all values are upper bounded by $\text{poly}(mU/\epsilon)$. Due to the bandwidth constraint of the CONGEST model, these multiplications take $\tilde{O}(\log(U/\epsilon))$ rounds. At the end of the computation both incident nodes to an edge know its value.

Further, we use throughout in runtime bounds that $D = \tilde{O}(T_{\text{Laplacian}}(G))$, which holds since $T_{\text{Laplacian}}(G) = \tilde{\Omega}(SQ(G)) = \tilde{\Omega}(D)$ [3].

Definitions and Set-Up. On a high level, we perform a weighted path following interior point method. This means that throughout a number of iterations, given a current point $x^{(i)} \in \Omega^\circ$, we find a point $x^{(i+1)} \in \Omega^\circ$ closer to the optimal solution. To control that a point $x^{(i)}$ stays away from the boundary, we need to control $l_j \leq x_j^{(i)} \leq u_j$ for $j \in [n]$. This is done using a *barrier function* $\phi_i(x_i)$, which goes to ∞ when x_i goes to the boundary, i.e., to l_j or u_j. The path then looks as follows:

$$x^{(i)} = \arg \min_{A^T x = b} \left(i \cdot c^T x + \sum_{j \in [m]} \phi_j(x_j) \right). \tag{1}$$

To make this work, we need some more properties on ϕ, leading to the definition of a self-concordant barrier function.

Definition 2. *A convex, thrice continuously differentiable function $\phi \colon K \to \mathbb{R}^n$ is a ν-self-concordant barrier function for open convex set $K \subseteq \mathbb{R}^n$ if the following three conditions are satisfied*

1. $\lim_{i \to \infty} \phi(x_i) = \infty$ *for all sequences $(x_i)_{i \in \mathbb{N}}$ with $x_i \in K$ converging to the boundary of K.*
2. $|\phi'''(x)[h, h, h]| \leq 2|\phi''(x)[h, h]|^{3/2}$ *for all $x \in K$ and $h \in \mathbb{R}^n$.*

---

[2] In this section, we refer to the arXiv version [30] rather than the conference version [29], whenever the technical details can only be found there.

3. $|\phi'(x)[h]| \leq \sqrt{\nu} |\phi''(x)[h,h]|^{1/2}$ for all $x \in K$ and $h \in \mathbb{R}^n$.

In our case, we choose ϕ as follows.

- If l_i is finite and $u_i = +\infty$, we use a log barrier: $\phi_i(x) := -\log(x - l_i)$.
- If $l_i = -\infty$ and u_i is finite, we use a log barrier: $\phi_i(x) := -\log(u_i - x)$.
- If l_i and u_i are finite, we use a trigonometric barrier: $\phi_i(x) := -\log\cos(a_i x + b_i)$, where $a_i := \frac{\pi}{u_i - l_i}$ and $b_i := -\frac{\pi}{2}\frac{u_i + l_i}{u_i - l_i}$.

This ϕ is a 1-concordant barrier function [30]. It can be computed internally in the CONGEST model, since we only require local knowledge of the constraints. By using this function in Eq. 1, we obtain a $\tilde{O}(\sqrt{m}\log(1/\epsilon))$ iteration method [42].

We can generalize Eq. 1 to

$$x^{(i)} = \arg\min_{A^T x = b}\left(i \cdot c^T x + \sum_{j \in [m]} g_j(x)\phi_j(x_j)\right), \tag{2}$$

for some *weight functions* $g_j : \Omega^\circ \to \mathbb{R}^m_{\geq 0}$. Lee and Sidford [29] show that using *regularized Lewis weights* we only need $\tilde{O}(\sqrt{n}\log(1/\epsilon))$ iterations.

To give the formal definition of the regularized Lewis weight function, we first introduce some general notation.

- For any matrix $M \in \mathbb{R}^{n \times n}$, we let $\mathrm{diag}(M) \in \mathbb{R}^n$ denote the diagonal of M, i.e., $\mathrm{diag}(M)_i := M_{ii}$.
- For any vector $x \in \mathbb{R}^n$, we write upper case $X \in \mathbb{R}^{n \times n}$ for the diagonal matrix associated to x, i.e., $X_{ii} := x_i$ and $X_{ij} := 0$ if $i \neq j$.
- For $x \in \Omega^\circ$, we write $A_x := (\Phi''(x))^{-1/2}A$.
- For $h: \mathbb{R}^n \to \mathbb{R}^m$ and $x \in \mathbb{R}^n$, we write $J_h(x) \in \mathbb{R}^{m \times n}$ for the Jaccobian of h at x, i.e., $[J_h(x)]_{ij} := \frac{\partial}{\partial x_j}h(x)_i$.
- For positive $w \in \mathbb{R}^n_{>0}$, we let $||\cdot||_w$ the norm defined by $||x||^2_w = \sum_{i \in [n]} w_i x_i^2$, and we let $||\cdot||_{w+\infty}$ the *mixed norm* defined by $||x||_{w+\infty} = ||x||_\infty + C_{\mathrm{norm}}||x||_w$ for some constant $C_{\mathrm{norm}} > 0$ to be defined later.
- Whenever we apply scalar operation to vectors, these operations are applied coordinate-wise, e.g., for $x, y \in \mathbb{R}^n$ we have $[x/y]_i := x_i/y_i$, and $[x^{-1}]_i := x_i^{-1}$.

Definition 3. *A differentiable function* $g: \Omega^\circ \to \mathbb{R}^m_{>0}$ *is a* (c_1, c_s, c_k)-*weight function if the following bounds holds for all* $x \in \Omega^\circ$ *and* $i \in [m]$:

- *size bound:* $\max\{1, ||g(x)||_1\} \leq c_1$;
- *sensitivity bound:* $e_i^T G(x)^{-1} A_x (A_x^T G(x)^{-1} A_x)^{-1} A_x^T G(x)^{-1} e_i \leq c_s$;
- *consistency bound:* $||G(x)^{-1} J_g(x)(\Phi''(x))^{-1/2}||_{g(x)+\infty} \leq 1 - c_k < 1$.

We denote $C_{\mathrm{norm}} := 24\sqrt{c_s}c_k$.

Following Lee and Sidford [29], we use the *regularized Lewis weights*. The ℓ_p-Lewis weights generalize a ℓ_2 measure of row importance called leverage scores. They are a key tool in approximating matrix ℓ_p-norms.

Definition 4. *For* $M \in \mathbb{R}^{m \times n}$ *with* $\text{rank}(M) = n$, *we let*

$$\sigma(M) := \text{diag}(M(M^T M)^{-1} M^T)$$

denote the leverage scores *of* M. *For all* $p > 0$, *we define the* ℓ_p-*Lewis weights* $w_p(M)$ *as the unique vector* $w \in R_{>0}^m$ *such that* $w = \sigma(W^{\frac{1}{2} - \frac{1}{p}} M)$, *where* $w = \text{diag}(W)$. *We define the* regularized Lewis weights *as* $g(x) := w_p(M_x) + c_0$, *for* $p = 1 - \frac{1}{\log(4m)}$ *and* $c_0 := \frac{n}{2m}$.

The regularized Lewis weight function g is a (c_1, c_s, c_k)-weight function with $c_1 \le \frac{3}{2}n$, $c_s \le 4$, and $c_k \le 2\log(4m)$ [30].

As said before, Lee and Sidford show that using Eq. 2 with this weight function for $\tilde{O}(\sqrt{n}\log(1/\epsilon))$ iterations gives an ϵ-approximate solution to our LP.

Computing Leverage Scores. Existing techniques for computing regularized Lewis weights compute leverage scores $\sigma(M) := \text{diag}(M(M^T M)^{-1} M^T)$ as an intermediate step. As shown later, by repeatedly computing leverage scores, we can approximate the Lewis weights. Unfortunately, there are no known efficient algorithms for computing leverage scores exactly. However, obtaining a sufficiently close approximation is feasible [11,14,32,35,45,47]. We observe that $\sigma(M)_i = ||M(M^T M)^{-1} M^T e_i||_2^2$, and note that by the Johnson-Lindenstrauss lemma [21] this norm can be approximately preserved under projections onto a low dimensional subspace. In particular Achlioptas [1] gives an explicit (randomized) construction.

Theorem 4 ([1]). *Let* $m > 0$ *be an integer, let* $\eta, \beta > 0$ *be parameters, let* $k = \Omega(\beta \log m / \eta^2)$ *be an integer, and let* $R \in \mathbb{R}^{k \times m}$ *be a random matrix, where* $R_{ij} = \pm 1/\sqrt{k}$, *each with probability* $1/2$. *Then for any* $x \in \mathbb{R}^m$ *we have*

$$\mathbb{P}[(1 - \eta)||x||_2 \le ||Rx||_2 \le (1 + \eta)||x||_2] \ge 1 - m^{-\beta}.$$

Now we are ready to given an algorithm to compute $\sigma^{(\text{apx})}$ such that $(1 - \eta)\sigma(M)_i \le \sigma_i^{(\text{apx})} \le (1 + \eta)\sigma(M)_i$, for all $i \in [m]$.

Algorithm 1: COMPUTELEVERAGESCORES(M, η)

---

1 Set $k = \Theta(\log(m)/\eta^2)$.
2 Let $R \in \mathbb{R}^{k \times m}$ be a matrix where $R_{ij} = \pm 1/\sqrt{k}$, each with probability $1/2$.
3 Compute $p^{(j)} = M(M^T M)^{-1} M^T R^{(j)}$ for $j \in [k]$.
4 **return** $\sum_{j=1}^{k} \left(p^{(j)}\right)^2$.

Lemma 1. *For any* $\eta > 0$, *with probability at least* $1 - 1/m^{O(1)}$ *the CONGEST model algorithm* ComputeLeverageScores(M, η) *computes* $\sigma^{\text{apx}}(M)$ *such that*

$$(1 - \eta)\sigma(M)_i \le \sigma^{\text{apx}}(M)_i \le (1 + \eta)\sigma(M)_i$$

for all $i \in [m]$. *If* $M = WA$, *for some diagonal* $W \in \mathbb{R}^{m \times m}$, *then it terminates in* $\tilde{O}((\log(U/\epsilon) + T_{\text{Laplacian}}(G))/\eta^2)$ *rounds.*

Proof. $\texttt{ComputeLeverageScores}(M, \eta)$ returns $\sigma^{\mathrm{apx}}(M)_i$. Using that the matrix $M(M^T M)^{-1}M^T$ is symmetric, we obtain

$$\sigma^{\mathrm{apx}}(M)_i := \sum_{j=1}^{k}(M(M^T M)^{-1}M^T R^{(j)})_i^2$$

$$= \sum_{j=1}^{k}(RM(M^T M)^{-1}M^T)_{ji}^2$$

$$= ||RM(M^T M)^{-1}M^T e_i||_2^2.$$

Since we also have $\sigma(M)_i = ||M(M^T M)^{-1}M^T e_i||_2^2$, Theorem 4 gives us that

$$(1 - \eta)\sigma(M)_i \le \sigma^{\mathrm{apx}}(M)_i \le (1 + \eta)\sigma(M)_i,$$

with probability at least $1 - 1/m^{O(1)}$. Using a union bound, we can get the same guarantee for all $i \in [m]$ simultaneously.

In the CONGEST model, we construct the required random matrix R as follows. For each edge, the node with higher ID flips k coins to determine the values $\pm 1/\sqrt{k}$ and sends the result over the edge. This takes $O(k/\log n) = O(1/\eta^2)$ rounds.

For the computation in line 3, we note that we can view this as k times

- a matrix-vector multiplication $M^T R^{(j)}$, followed by
- a Laplacian system solve $(M^T M)^{-1}M^T R^{(j)}$, as $M^T M = A^T W^2 A$, followed by
- a matrix-vector multiplication $M(M^T M)^{-1}M^T R^{(j)}$.

The first and last step can be done in $\tilde{O}(\log(U/\epsilon))$ rounds, and the Laplacian solve can be done in $T_{\mathrm{Laplacian}}(G)$ rounds. Finally, line 4 can be done internally. Hence we have total running time $\tilde{O}((\log(U/\epsilon) + T_{\mathrm{Laplacian}}(G))/\eta^2)$.

Computing the Weight Function. We continue by providing the algorithms for computing the initial weights, and for updating the weights throughout the path finding algorithm. As we use the latter for the former, we give the latter first.

Algorithm 2: COMPUTEAPXWEIGHTS$(M, p, w^{(0)}, \eta)$

---

1 $L = \max\{4, \frac{8}{p}\}$, $r = \frac{p^2(4-p)}{2^{20}}$, and $\delta = \frac{(4-p)\eta}{256}$.

2 $T = \left\lceil 80\left(\frac{p}{2} + \frac{2}{p}\right)\log\left(\frac{pn}{32\eta}\right) \right\rceil$.

3 **for** $j = 1, \ldots, T - 1$ **do**

4     $\sigma^{(j)} = \texttt{ComputeLeverageScores}(W_{(j)}^{\frac{1}{2}-\frac{1}{p}}M, \delta/2)$.

5   **for** $i \in [m]$ **do**

6   Let $w_i^{(j+1)}$ be the median of $(1 - r)w_i^{(0)}$, $w_i^{(j)} - \frac{1}{L}\left(w_i^{(0)} - \frac{w_i^{(0)}}{w_i^{(j)}}\sigma_i^{(j)}\right)$,

and $(1 + r)w_i^{(0)}$.

7 **return** $w^{(T)}$.

Lemma 2. *Let $W \in R^{m \times m}$ be some diagonal matrix, let $w^{(0)} \in \mathbb{R}_{>0}^m$ be a vector, and let $\eta \in (0,1]$ and $p \in [1 - 1/\log(4m), 2]$ be parameters. Set $M = WA$. Then* COMPUTEAPXWEIGHTS$(M, p, w^{(0)}, \eta)$ *returns approximate weights in*

$$\tilde{O}(\tfrac{\log(1/\eta)}{\eta^2}(\log(U/\epsilon) + T_{\text{Laplacian}(G)}))$$

rounds.

Proof. The algorithm consists of $T = \tilde{O}((p + \frac{1}{p})\log(p/\eta))$ iterations. Using the assumption that $p \in [1 - 1/\log(4m), 2]$, we get $T = \tilde{O}(\log(1/\eta))$. In each iteration, we call ComputeLeverageScores$(W_{(j)}^{\frac{1}{2} - \frac{1}{p}} M, \delta/2)$ and compute some medians, the latter of which can be done internally. The call to ComputeLeverageScores takes $\tilde{O}((\log(U/\epsilon) + T_{\text{Laplacian}}(G))/(\delta/2)^2) = \tilde{O}((\log(U/\epsilon) + T_{\text{Laplacian}}(G))/\eta^2)$ rounds, giving us the total running time as stated.

For the properties and correctness of the approximate weights we refer to [30]. Using the following algorithm, we compute the initial weights. We do this by iteratively bringing the all-ones vector closer to the initial weight vector.

Algorithm 3: COMPUTEINITIALWEIGHTS$(A, p_{\text{target}}, \eta)$

---

1 $p = 2$.
2 $w = 12c_k \mathbf{1}$.
3 **while** $p \neq p_{\text{target}}$ **do**
4      $h = \dfrac{\min\{2, p\}}{\sqrt{n}\log\frac{me^2}{n}} \cdot r$.
5   Let $p^{(\text{new})}$ be the median of $p - h$, p_{target}, and $p + h$.
6      $w = $ComputeApxWeights$(A, p^{(\text{new})}, w^{p^{(\text{new})}}/p, \frac{p^2(4-p)}{2^{22}})$.
7      $p = p^{(\text{new})}$.
8 **return** ComputeApxWeights$(A, p_{\text{target}}, w, \eta)$.

Lemma 3. *Let $\eta \in (0,1]$ and $p_{\text{target}} \in [1 - 1/\log(4m), 2]$ be parameters, then the CONGEST model algorithm* COMPUTEINITIALWEIGHTS$(A, p_{\text{target}}, \eta)$ *returns initial weights in* $\tilde{O}((\sqrt{n} + \frac{\log(1/\eta)}{\eta^2}(\log(U/\epsilon) + T_{\text{Laplacian}(G)}))$ *rounds.*

Proof. The while loop of line 3 finishes in $O(\sqrt{n}(p_{\text{target}} + \frac{1}{p_{\text{target}}})\log(m/n))$ iterations. Using that $p_{\text{target}} \in [1 - 1/\log(4m), 2]$, this simplifies to $\tilde{O}(\sqrt{n})$ iterations. Each iteration consists of internally computing h and some medians, and a call to ComputeApxWeights. This call requires precision $\frac{p^2(4-p)}{2^{22}}$, which is $\Omega(1)$ for our range of p. So the while loop takes $\tilde{O}(\sqrt{n}(\log(U/\epsilon) + T_{\text{Laplacian}}(G)))$ rounds in total.

Then in line 8 we call ComputeApxWeights with precision η, which takes $\tilde{O}(\frac{\log(1/\eta)}{\eta^2}(\log(U/\epsilon) + T_{\text{Laplacian}(G)}))$ rounds. Together this gives the stated running time.

For the properties and correctness of the initial weights we refer to [30].

Algorithm. In this section, we give the formal algorithm for solving the LP, together with a series of lemmas proving the running time of each subroutine.

Algorithm 4: LPSOLVE(x_0, ϵ)

Input: an initial point x_0 such that $A^T x_0 = b$.

1 w =ComputeInitialWeights$(A, 1 - 1/\log(4m), \frac{1}{2^{16}\log^3 m}) + \frac{n}{2m}$, $d = -w\phi'(x_0)$.

2 $t_1 = (2^{27}m^{3/2}U^2 \log^4 m)^{-1}$, $t_2 = \frac{2m}{\eta}$, $\eta_1 = \frac{1}{2^{18}\log^3 m}$, and $\eta_2 = \frac{\epsilon}{8U^2}$.

3 $(x^{(\text{new})}, w^{(\text{new})})$=PathFollowing$(x_0, w, 1, t_1, \eta_1, d)$.

4 $(x^{(\text{final})}, w^{(\text{final})})$=PathFollowing$(x^{(\text{new})}, w^{(\text{new})}, t_1, t_2, \eta_2, c)$.

5 **return** $x^{(\text{final})}$.

After computing the initial weights, this algorithm calls PathFollowing twice, first to move the given initial point towards a central starting point with respect to the cost vector c, and second to move the path along from there. The algorithm PathFollowing is as follows.

Algorithm 5: PATHFOLLOWING$(x, w, t_{\text{start}}, t_{\text{end}}, \eta, c)$

---

1 $t = t_{\text{start}}$, $R = \frac{1}{768c_k^2 \log(36c_1 c_s c_k m)}$, and $\alpha = \frac{R}{1600\sqrt{n}\log^2 m}$.

2 **while** $t \neq t_{\text{end}}$ **do**

3 $\quad$ (x, w) =CenteringInexact(x, w, t, c).

4 $\quad$ Let t be the median of $(1 - \alpha)t$, t_{end}, and $(1 + \alpha)t$.

5 **for** $i = 1, \ldots, 4c_k \log(\frac{1}{\eta})$ **do**

6 $\quad$ (x, w)=CenteringInexact$(x, w, t_{\text{end}}, c)$.

7 **return** (x, w).

The progress steps in PathFollowing are made by CenteringInexact, which is as follows.

Algorithm 6: CENTERINGINEXACT(x, w, t, c)

---

1 $R = \frac{1}{768c_k^2 \log(36c_1 c_s c_k m)}$, and $\eta = \frac{1}{2c_k}$.

2 $\delta = \left\| P_{x,w}\left(\frac{tc + w\phi'(x)}{w\sqrt{\phi''(x)}}\right) \right\|_{w+\infty}$ // where
$$P_{x,w} := I - W^{-1}A_x(A_x^T W^{-1} A_x)^{-1}A_x^T.$$

3 $x^{(\text{new})} = x - \frac{1}{\sqrt{\phi''(x)}}P_{x,w}\left(\frac{tc - w\phi'(x)}{w\sqrt{\phi''(x)}}\right)$.

4 $z = \log\left(\text{ComputeApxWeights}(A_{x^{(\text{new})}}, 1 - 1/\log(4m), w, e^R - 1)\right)$.

5 $u = \left(1 - \frac{6}{7c_k}\right)\delta\cdot\text{ProjectMixedBall}(-\nabla\Phi_{\frac{\eta}{12R}}(z - \log(w)), C_{\text{norm}}\sqrt{w})$.

6 $w^{(\text{new})} = \exp(\log(w) + u)$.

7 **return** $(x^{(\text{new})}, w^{(\text{new})})$.

We present the subroutine ProjectMixedBall in Sect. 4. We prove the running times of these three algorithms in reverse order.

Lemma 4. *The CONGEST model algorithm* CenteringInexact(x, w, t, c) *terminates in* $\tilde{O}(\log^2(U/\epsilon) T_{\text{Laplacian}}(G))$ *rounds.*

Proof. Computing $P_{x,w}\left(\frac{tc + w\phi'(x)}{w\sqrt{\phi''(x)}}\right)$ takes $\tilde{O}(T_{\text{Laplacian}}(G))$ rounds, using internal computation for multiplying with diagonal matrices and a Laplacian solve. To compute δ and make it known to every node, we use $\tilde{O}(D \log(U/\epsilon))$ rounds.

Next, we call ComputeApxWeights with precision $\eta = \tilde{\Omega}(1)$, so this takes $\tilde{O}(\log(U/\epsilon) + T_{\text{Laplacian}}(G))$ rounds by Lemma 2. Finally we call the algorithm ProjectMixedBall, which takes

$$\tilde{O}(D \log^2(U/\epsilon)) = \tilde{O}(\log^2(U/\epsilon) T_{\text{Laplacian}}(G))$$

rounds by Lemma 7.

We use this result to prove the running time of PathFollowing.

Lemma 5. *Let* $t_{\text{start}}, t_{\text{end}} \geq 1$, *and* $\eta \in (0, 1]$ *be parameters. The CONGEST model algorithm* PathFollowing$(x, w, t_{\text{start}}, t_{\text{end}}, \eta, c)$ *terminates in*

$$\tilde{O}(\sqrt{n}(|\log(t_{\text{end}}/t_{\text{start}})| + \log(1/\eta)) \log^2(U/\epsilon) T_{\text{Laplacian}}(G))$$

rounds.

Proof. First, we note that the while loop of line 2 uses $\tilde{O}(\sqrt{n}(|\log(t_{\text{end}}/t_{\text{start}})| + \log(1/\eta)))$ iterations [30]. Each such iteration consists of a call to CenteringInexact and internal computations. Then the for loop of line 5 takes $O(c_k \log(1/\eta))$ iterations, each consisting of a call to CenteringInexact. Clearly this is dominated by the running time of the while loop.

Since CenteringInexact takes $\tilde{O}(\log^2(U/\epsilon) T_{\text{Laplacian}}(G))$ rounds by Lemma 4, we obtain the stated running time.

Finally, we give the running time of the complete algorithm.

Lemma 6. *Given* $\epsilon > 0$, *the CONGEST model algorithm* LPSolve(x_0, ϵ) *terminates in* $\tilde{O}(\sqrt{n} \log^3(U/\epsilon) T_{\text{Laplacian}}(G))$ *rounds.*

Proof. Apart from some internal computation, this algorithm consists of three different parts: computing initial weight and two calls to PathFollowing with different parameters.

The call to ComputeInitialWeights takes

$$\tilde{O}(\sqrt{n}(\log(U/\epsilon) + T_{\text{Laplacian}(G)}))$$

rounds, since we call it with precision $\frac{1}{2^{16} \log^3 m}$.

The execution of PathFollowing$(x_0, w, 1, t_1, \eta_1, d)$ takes

$$\tilde{O}(\sqrt{n} \log(U) \log^2(U/\epsilon) T_{\text{Laplacian}(G)})$$

rounds, by Lemma 5 and plugging in t_1 and η_1.

The execution of PathFollowing$(x^{(\text{new})}, w^{(\text{new})}, t_1, t_2, \eta_2, c)$ takes

$$\tilde{O}(\sqrt{n} \log^3(U/\epsilon) T_{\text{Laplacian}}(G))$$

rounds, by Lemma 5 and plugging in t_1, t_2 and η_2.

The last running time dominates the first two and gives the stated result.

4   Projecting on a Mixed Norm Ball

In this section, we present a CONGEST algorithm for projecting on a mixed norm ball. This problem is defined as follows. Given $a, l \in \mathbb{R}^m$, find

$$\arg\max_{||x||_2 + ||l^{-1}x||_\infty \le 1} a^T x.$$

In the original work, Lee and Sidford [30] initially sort m values and precompute m functions on a and l. In the CONGEST model, these are expensive routines. We borrow ideas from Forster and de Vos [17], who overcame the same problem for the Broadcast Congested Clique. The rough idea is to only sort implicitly, and perform a binary search to reduce the number of functions that we have to compute to a manageable amount. We provide pseudocode in Algorithm 7, with more details in the proof of Lemma 7. The pseudocode has a rather complicated binary search and some daunting equations in it. Both are probably best understood by examining the proof.

Note that this problem has little to do with the graph structure in the CONGEST model, and as expected the algorithm actually does not make use of the graph structure other than establishing a shortest path tree for communication.

Algorithm 7: PROJECTMIXEDBALL(a, l)

1 Determine the minimum value, maximum value, and step size of $\{|a_i|/l_i : i \in [m]\}$, denote this space of possible values S.

2 For $s \in S$, let i be the index of the value $|a_i|/l_i$ closest to s.

3 Perform a binary search on S w.r.t. g_i:

4   Compute $\sum_{k \in [j]} |a_k||l_k|$, $\sum_{k \in [j]} a_k^2$, and $\sum_{k \in [j]} l_k^2$ for $j \in \{i - 1, i\}$.

5   Internally compute

$$g_i := \max_{t: i_t = i} t \sum_{k \in [i]} |a_k||l_k| + \sqrt{(1-t)^2 - t^2 \sum_{k \in [i]} l_k^2} \sqrt{||a||_2^2 - \sum_{k \in [i]} a_k^2}.$$

6 Let t be the index corresponding to the maximal g_i.

7 $x_j^i := \begin{cases} \frac{t}{1-t} \operatorname{sign}(a_j) l_j & \text{if } j \in [i] \\ \sqrt{\frac{1 - \left(\frac{t}{1-t}\right)^2 \sum_{k \in [i]} l_k^2}{||a||_2^2 - \sum_{k \in [i]} a_k^2}} a_j & \text{otherwise.} \end{cases}$

8 return x

Lemma 7. *Suppose the vectors $a, l \in \mathbb{R}^m$ are distributed over the network such that: 1) for each $i \in [m]$, a_i and l_i are known by exactly one node, 2) a node knows a_i if and only if it knows l_i. Moreover, suppose that $||a||_\infty, ||l||_\infty \le O(\operatorname{poly}(m)U)$. Then the algorithm ProjectMixedBall(a, l) finds*

$$\arg\max_{||x||_2 + ||l^{-1}x||_\infty \le 1} a^T x$$

up to precision $O(1/(\operatorname{poly}(mU/\epsilon))$ in $\tilde{O}(D \log^2(U/\epsilon))$ rounds in the CONGEST model.

Proof. We rewrite the problem into maximizing over some concave function, which has a unique maximum that can be found using a binary search over the domain. We start by parameterizing the ℓ_2-norm:

$$\max_{||x||_2+||l^{-1}x||_\infty \leq 1} a^T x = \max_{0 \leq t \leq 1} \left[\max_{||x||_2 \leq 1-t, \ -tl_i \leq x_i \leq tl_i} a^T x \right]$$

$$= \max_{0 \leq t \leq 1} (1-t) \left[\max_{||x||_2 \leq 1, \ -\frac{t}{1-t} l_i \leq x_i \leq \frac{t}{1-t} l_i} a^T x \right]$$

$$= \max_{0 \leq t \leq 1} g(t),$$

where we define $g(t)$ as

$$g(t) := (1-t) \left[\max_{||x||_2 \leq 1, \ -\frac{t}{1-t} l_i \leq x_i \leq \frac{t}{1-t} l_i} a^T x \right].$$

We *conceptually* sort the values of a and l with $|a_i|/l_i$ monotonically decreasing, i.e., we only sort them for this notation in the proof, the algorithm does not sort the values. Next, we write i_t for the first coordinate $i \in [m]$ such that

$$\frac{1-\left(\frac{t}{1-t}\right)^2 \sum_{k \in [i_t]} l_k^2}{||a||_2^2 - \sum_{k \in [i_t]} a_k^2} \leq \frac{\left(\frac{t}{1-t}\right)^2 l_i^2}{a_i^2}.$$

Now it can be shown (see e.g. [30]) that the vector that attains the maximum in $g(t)$ is $x^{i_t} \in \mathbb{R}^m$, defined by

$$x_j^{i_t} := \begin{cases} \frac{t}{1-t} \operatorname{sign}(a_j) l_j & \text{if } j \in [i_t] \\ \sqrt{\frac{1-\left(\frac{t}{1-t}\right)^2 \sum_{k \in [i_t]} l_k^2}{||a||_2^2 - \sum_{k \in [i_t]} a_k^2}} a_j & \text{otherwise.} \end{cases}$$

We substitute this into the definition of $g(t)$:

$$g(t) = t \sum_{k \in [i_t]} |a_k| |l_k| + \sqrt{(1-t)^2 - t^2 \sum_{k \in [i_t]} l_k^2} \sqrt{||a||_2^2 - \sum_{k \in [i_t]} a_k^2}.$$

We note that $g(t)$ is a concave function (its second derivative is non-positive), hence it has a unique maximum. We find this maximum by searching over the domain. To do this, we rewrite g in terms of the index i_t:

$$\max_{0 \leq t \leq 1} g(t) = \max_{0 \leq t \leq 1} \max_{i \in [m]} g_i(t)$$

$$= \max_{i \in [m]} \max_{t: i_t = i} g_i(t),$$

where

$$g_i(t) := t \sum_{k \in [i]} |a_k| |l_k| + \sqrt{(1-t)^2 - t^2 \sum_{k \in [i]} l_k^2} \sqrt{||a||_2^2 - \sum_{k \in [i]} a_k^2}.$$

Now fix a index i, and suppose a node knows $\sum_{k\in[j]}|a_k||l_k|$, $\sum_{k\in[j]}a_k^2$, and $\sum_{k\in[j]}l_k^2$ for $j \in \{i-1,i\}$. Then we can internally compute $g_i := \max_{t:i_t=i}g_i(t)$, because we can internally find the range of t where $i_t = i$, since we have that $i_t \geq i_s$ if $t \leq s$, hence the set of t such that $i_t = j$ is an interval.

Next, we describe how to compute the sums $\sum_{k\in[j]}$. We do this by constructing a shortest path tree of diameter D from the node holding the values a_j, l_j. Along the tree, we aggregate the values of $|a_k||l_k|$, a_k^2, or l_k^2 respectively, for all $k \leq j$. The result can be broadcasted to all nodes without incurring extra costs. Note that if the indices are not known explicitly, the node holding a_j and l_j, can first broadcast $|a_j|/l_j$, and then other nodes only add their values $|a_i||l_i|$ (and others respectively) if $|a_i|/l_i \leq |a_j|/l_j$. Since the values need to be maintained with precision $\text{poly}(mU/\epsilon)$, sending one message needs at most $\tilde{O}(\log(U/\epsilon))$ rounds, so the whole procedure takes at most $\tilde{O}(D\log(U/\epsilon))$ rounds.

Naively, we would now be done by a simple binary search over $i \in [m]$, however we have the complication that we have only conceptually sorted the indices and hence nodes do not know which indices belong to the values they are holding. Instead we do a binary search over the possible values of $|a_i|/l_i$. Again using a communication tree from an arbitrary leader, we can find the global minimum, global maximum, and step size (least common multiple of denominators l_i) for the $|a_i|/l_i$ values. As not all values in the search space appear, we take the closest appearing value for a given value in the binary search. This gives a total search space of size $O(\text{poly}(mU/\epsilon))$, so we need $\tilde{O}(\log(U/\epsilon))$ iterations, each taking $\tilde{O}(D\log(U/\epsilon))$ rounds.

A Minimum Cost Flow

In this section, we prove Theorem 1 by applying Theorem 2 to a suitable linear program and rounding the result to an exact solution accordingly. This particular LP formulation of minimum cost flow has first been presented by Daitch and Spielman [13], and is used by Lee and Sidford [30], and Forster and de Vos [17]. As opposed to the formulation of Theorem 1, we use $|V|$ and $|E|$ in this section to indicate the size of the node and edge set. We reserve n and m for the dimensions of the linear program, in line with Sect. 3. We write M for the maximal edge capacity and cost.

Let $B \in \mathbb{R}^{(|V|-1)\times|E|}$ be the edge-node incidence matrix with the row for the source s omitted. The variables of the LP consist of $x \in \mathbb{R}^{|E|}$, $y, z \in \mathbb{R}^{|V|}$ and $F \in \mathbb{R}$. The linear program is defined as follows.

$$\min \tilde{q}^T x + \lambda(1^T y + 1^T z) - 2n\tilde{M}F$$
$$\text{subject to } Bx + y - z = Fe_t,$$
$$0 \le x_i \le c_i,$$
$$0 \le y_i \le 4V|M,$$
$$0 \le z_i \le 4V|M,$$
$$0 \le F \le 2V|M,$$

where $\tilde{M} := 8|E|^2 M^3$, $\lambda := 28160|E|^8 2M^9$, and $\tilde{q} = c + r$, where for each edge r_e is a uniformly random number from $\left\{ \frac{1}{4|E|^2 M^2}, \frac{2}{4|E|^2 M^2}, \ldots, \frac{2|E|M}{4|E|^2 M^2} \right\}$. Daitch and Spielman [13] show that with probability at least $1/2$ this problem has a unique solution, which is also a valid solution to the original problem. After applying this reduction we (conceptually) scale everything by $4|E|^2 M^2$ to ensure the cost vector is integral again.

We set the variables as follows to obtain an initial interior point: $F = |V|M$, $x = \frac{c}{2}$, $y = 2|V|M1 - (B\frac{c}{2})^- + Fe_t$, $z = 2|V|M1 + (B\frac{c}{2})^+$, where we denote a^+ and a^- for the vectors defined by

$$(a^+)_i := \begin{cases} a_i & \text{if } a_i \ge 0; \\ 0 & \text{else.} \end{cases} \quad \text{and} \quad (a^-)_i := \begin{cases} a_i & \text{if } a_i \le 0 \\ 0 & \text{else.} \end{cases}$$

respectively.

Next, we describe how we transform an ϵ-approximate solution x to this LP into an exact solution for the minimum cost flow problem. By introducing extra variables y and z, we might have overshot the flow by at most $1^T y + 1^T z \le \epsilon$. To correct for this we set $\tilde{x} = (1 - \epsilon)x$. We set $\epsilon := \frac{1}{320|E|^4 M^5}$, and then the error with respect to the unique solution is at most $1/6$ [30], so we have we can simply round the flow on each edge to the closest integer. Clearly both these steps can be done internally in the CONGEST model.

To solve the above LP, we use Theorem 2 with $A = [B \; I \; -I \; -e_t]^T$. Actually, this does not use the entire network, but only $n = |V| - 1$ nodes, since the source does not need to participate in the computation. Since the knowledge of the node-incident matrix B is distributed as required, the knowledge of A is distributed as required. The last step is to show that we can solve equations in $A^T W A$ in $T_{\text{Laplacian}}$. This follows from [20] and is made explicit in [17,23], who show that $A^T W A$ is symmetric and diagonally dominant, hence equations in $A^T W A$ can be solved by solving two Laplacian equations. We get $U/\epsilon = M \text{poly}(|V|)$, so $\log^3(U/\epsilon) = \tilde{O}(\log^3 M)$.

References

1. Achlioptas, D.: Database-friendly random projections: Johnson-lindenstrauss with binary coins. J. Comput. Syst. Sci. **66**(4), 671–687 (2003). https://doi.org/10.1016/S0022-0000(03)00025-4, announced at PODS 2001

2. Ahmadi, M., Kuhn, F., Oshman, R.: Distributed approximate maximum matching in the CONGEST model. In: Proceedings of the 32nd International Symposium on Distributed Computing, DISC 2018. LIPIcs, vol. 121, pp. 6:1–6:17. Schloss Dagstuhl - Leibniz-Zentrum für Informatik (2018). https://doi.org/10.4230/LIPIcs.DISC.2018.6

3. Anagnostides, I., Lenzen, C., Haeupler, B., Zuzic, G., Gouleakis, T.: Almost universally optimal distributed laplacian solvers via low-congestion shortcuts. In: Proceedings of the 36th International Symposium on Distributed Computing, DISC 2022. LIPIcs, vol. 246, pp. 6:1–6:20. Schloss Dagstuhl - Leibniz-Zentrum für Informatik (2022). https://doi.org/10.4230/LIPIcs.DISC.2022.6, announced at PODC 2022

4. Andoni, A., Stein, C., Zhong, P.: Parallel approximate undirected shortest paths via low hop emulators. In: Proceedings of the 52nd Annual ACM SIGACT Symposium on Theory of Computing, STOC 2020, pp. 322–335. ACM (2020)

5. Axiotis, K., Mądry, A., Vladu, A.: Circulation control for faster minimum cost flow in unit-capacity graphs. In: Proceedings of the 61st IEEE Annual Symposium on Foundations of Computer Science, FOCS 2020, pp. 93–104. IEEE (2020). https://doi.org/10.1109/FOCS46700.2020.00018

6. Becker, R., Forster, S., Karrenbauer, A., Lenzen, C.: Near-optimal approximate shortest paths and transshipment in distributed and streaming models. SIAM J. Comput. $\mathbf{50}$(3), 815–856 (2021). https://doi.org/10.1137/19M1286955, announced at DISC 2017

7. van den Brand, J., et al.: Bipartite matching in nearly-linear time on moderately dense graphs. In: Proceedings of the 61st IEEE Annual Symposium on Foundations of Computer Science, FOCS 2020, pp. 919–930. IEEE (2020). https://doi.org/10.1109/FOCS46700.2020.00090

8. Chechik, S., Mukhtar, D.: Single-source shortest paths in the CONGEST model with improved bounds. Distributed Comput. $\mathbf{35}$(4), 357–374 (2022). https://doi.org/10.1007/s00446-021-00412-8, announced at PODC 2020

9. Chen, L., Kyng, R., Liu, Y.P., Peng, R., Gutenberg, M.P., Sachdeva, S.: Maximum flow and minimum-cost flow in almost-linear time. CoRR abs/2203.00671 (2022)

10. Cohen, M.B., et al.: Solving SDD linear systems in nearly $\text{mlog}^{1/2}\text{n}$ time. In: Proceedings of the Symposium on Theory of Computing, STOC 2014, pp. 343–352. ACM (2014). https://doi.org/10.1145/2591796.2591833

11. Cohen, M.B., Lee, Y.T., Musco, C., Musco, C., Peng, R., Sidford, A.: Uniform sampling for matrix approximation. In: Proceedings of the Conference on Innovations in Theoretical Computer Science (ITCS 2015), pp. 181–190 (2015). https://doi.org/10.1145/2688073.2688113

12. Cohen, M.B., Mądry, A., Sankowski, P., Vladu, A.: Negative-weight shortest paths and unit capacity minimum cost flow in $\tilde{o}(m^{10/7} \log w)$ time (extended abstract). In: Proceedings of the Twenty-Eighth Annual ACM-SIAM Symposium on Discrete Algorithms, SODA 2017, pp. 752–771. SIAM (2017). https://doi.org/10.1137/1.9781611974782.48

13. Daitch, S.I., Spielman, D.A.: Faster approximate lossy generalized flow via interior point algorithms. In: Proceedings of the 40th Annual ACM Symposium on Theory of Computing (STOC 2008), pp. 451–460 (2008). https://doi.org/10.1145/1374376.1374441

14. Drineas, P., Magdon-Ismail, M., Mahoney, M.W., Woodruff, D.P.: Fast approximation of matrix coherence and statistical leverage. J. Mach. Learn. Res. $\mathbf{13}$(1), 3475–3506 (2012). announced at ICML 2012

15. Elkin, M.: An unconditional lower bound on the time-approximation trade-off for the distributed minimum spanning tree problem. SIAM J. Comput. **36**(2), 433–456 (2006). https://doi.org/10.1137/S0097539704441058

16. Forster, S., Goranci, G., Liu, Y.P., Peng, R., Sun, X., Ye, M.: Minor sparsifiers and the distributed Laplacian paradigm. In: Proceedings of the 62nd IEEE Annual Symposium on Foundations of Computer Science, FOCS 2021, pp. 989–999. IEEE (2021). https://doi.org/10.1109/FOCS52979.2021.00099, https://doi.org/10.1109/FOCS52979.2021.00099

17. Forster, S., de Vos, T.: The Laplacian paradigm in the broadcast congested clique. In: Proceedings of the ACM Symposium on Principles of Distributed Computing, PODC 2022, pp. 335–344. ACM (2022). https://doi.org/10.1145/3519270.3538436, https://doi.org/10.1145/3519270.3538436

18. Ghaffari, M., Haeupler, B.: Distributed algorithms for planar networks II: low-congestion shortcuts, MST, and min-cut. In: Krauthgamer, R. (ed.) Proceedings of the Twenty-Seventh Annual ACM-SIAM Symposium on Discrete Algorithms, SODA 2016, pp. 202–219. SIAM (2016). https://doi.org/10.1137/1.9781611974331.ch16

19. Ghaffari, M., Karrenbauer, A., Kuhn, F., Lenzen, C., Patt-Shamir, B.: Near-optimal distributed maximum flow. SIAM J. Comput. **47**(6), 2078–2117 (2018). https://doi.org/10.1137/17M113277X, announced at PODC 2015

20. Gremban, K.D.: Combinatorial preconditioners for sparse, symmetric, diagonally dominant linear systems. Ph.D. thesis, Carnegie Mellon University, Pittsburgh (1996)

21. Johnson, W.B., Lindenstrauss, J.: Extensions of Lipschitz mappings into a Hilbert space. Contemp. Math. **26**, 189–206 (1984)

22. Kelner, J.A., Lee, Y.T., Orecchia, L., Sidford, A.: An almost-linear-time algorithm for approximate max flow in undirected graphs, and its multicommodity generalizations. In: Proceedings of the Twenty-Fifth Annual ACM-SIAM Symposium on Discrete Algorithms, SODA 2014, pp. 217–226. SIAM (2014). https://doi.org/10.1137/1.9781611973402.16

23. Kelner, J.A., Orecchia, L., Sidford, A., Zhu, Z.A.: A simple, combinatorial algorithm for solving SDD systems in nearly-linear time. In: Proceedings of the 45th Annual ACM Symposium on Theory of Computing (STOC 2013), pp. 911–920 (2013). https://doi.org/10.1145/2488608.2488724

24. Kleinberg, J.M., Tardos, É.: Algorithm Design. Addison-Wesley, Boston (2006)

25. Koutis, I., Miller, G.L., Peng, R.: A nearly-m log n time solver for SDD linear systems. In: Proceedings of the IEEE 52nd Annual Symposium on Foundations of Computer Science, FOCS 2011, pp. 590–598. IEEE Computer Society (2011). https://doi.org/10.1109/FOCS.2011.85

26. Koutis, I., Miller, G.L., Peng, R.: Approaching optimality for solving SDD linear systems. SIAM J. Comput. **43**(1), 337–354 (2014). https://doi.org/10.1137/110845914, announced at FOCS 2010

27. Kyng, R., Lee, Y.T., Peng, R., Sachdeva, S., Spielman, D.A.: Sparsified cholesky and multigrid solvers for connection Laplacians. In: Proceedings of the 48th Annual ACM SIGACT Symposium on Theory of Computing, STOC 2016, pp. 842–850. ACM (2016). https://doi.org/10.1145/2897518.2897640

28. Kyng, R., Sachdeva, S.: Approximate gaussian elimination for Laplacians - fast, sparse, and simple. In: Proceedings of the IEEE 57th Annual Symposium on Foundations of Computer Science, FOCS 2016, pp. 573–582. IEEE Computer Society (2016). https://doi.org/10.1109/FOCS.2016.68

29. Lee, Y.T., Sidford, A.: Path finding methods for linear programming: solving linear programs in $\tilde{O}(\sqrt{\text{rank}})$ iterations and faster algorithms for maximum flow. In: Proceedings of the 55th IEEE Annual Symposium on Foundations of Computer Science (FOCS 2014), pp. 424–433. IEEE Computer Society (2014). https://doi.org/10.1109/FOCS.2014.52

30. Lee, Y.T., Sidford, A.: Solving linear programs with $\tilde{O}(\sqrt{\text{rank}})$ linear system solves. CoRR abs/1910.08033 (2019)

31. Li, J.: Faster parallel algorithm for approximate shortest path. In: Proceedings of the 52nd Annual ACM SIGACT Symposium on Theory of Computing, STOC 2020, pp. 308–321. ACM (2020). https://doi.org/10.1145/3357713.3384268

32. Li, M., Miller, G.L., Peng, R.: Iterative row sampling. In: Proceedings of the 54th Annual IEEE Symposium on Foundations of Computer Science (FOCS 2013), pp. 127–136 (2013). https://doi.org/10.1109/FOCS.2013.22

33. Liu, Y.P., Sidford, A.: Faster divergence maximization for faster maximum flow. CoRR abs/2003.08929 (2020). https://arxiv.org/abs/2003.08929

34. Liu, Y.P., Sidford, A.: Faster energy maximization for faster maximum flow. In: Proceedings of the 52nd Annual ACM SIGACT Symposium on Theory of Computing, STOC 2020, pp. 803–814. ACM (2020). https://doi.org/10.1145/3357713.3384247

35. Mahoney, M.W.: Randomized algorithms for matrices and data. Found. Trends® Mach. Learn. **3**(2), 123–224 (2011). https://doi.org/10.1561/2200000035

36. Mądry, A.: Navigating central path with electrical flows: from flows to matchings, and back. In: Proceedings of the 54th Annual IEEE Symposium on Foundations of Computer Science, FOCS 2013, pp. 253–262. IEEE Computer Society (2013). https://doi.org/10.1109/FOCS.2013.35

37. Mądry, A.: Computing maximum flow with augmenting electrical flows. In: Proceedings of the IEEE 57th Annual Symposium on Foundations of Computer Science, FOCS 2016, pp. 593–602. IEEE Computer Society (2016). https://doi.org/10.1109/FOCS.2016.70

38. Peleg, D.: Distributed computing: a locality-sensitive approach. SIAM (2000)

39. Peleg, D., Rubinovich, V.: A near-tight lower bound on the time complexity of distributed minimum-weight spanning tree construction. SIAM J. Comput. **30**(5), 1427–1442 (2000). https://doi.org/10.1137/S0097539700369740, announced at FOCS 1999

40. Peng, R.: Approximate undirected maximum flows in O(mpolylog(n)) time. In: Proceedings of the Twenty-Seventh Annual ACM-SIAM Symposium on Discrete Algorithms, SODA 2016, pp. 1862–1867. SIAM (2016). https://doi.org/10.1137/1.9781611974331.ch130

41. Peng, R., Spielman, D.A.: An efficient parallel solver for SDD linear systems. In: Proceedings of the Symposium on Theory of Computing, STOC 2014, pp. 333–342. ACM (2014). https://doi.org/10.1145/2591796.2591832

42. Renegar, J.: A polynomial-time algorithm, based on newton's method, for linear programming. Math. Program. **40**(1–3), 59–93 (1988)

43. Sarma, A.D., et al.: Distributed verification and hardness of distributed approximation. SIAM J. Comput. **41**(5), 1235–1265 (2012). https://doi.org/10.1137/11085178X, announced at STOC 2011

44. Sherman, J.: Nearly maximum flows in nearly linear time. In: Proceedings of the 54th Annual IEEE Symposium on Foundations of Computer Science, FOCS 2013, pp. 263–269. IEEE Computer Society (2013). https://doi.org/10.1109/FOCS.2013.36

45. Spielman, D.A., Srivastava, N.: Graph sparsification by effective resistances. SIAM J. Comput. **40**(6), 1913–1926 (2011). https://doi.org/10.1137/080734029, announced at STOC 2008
46. Spielman, D.A., Teng, S.: Nearly-linear time algorithms for graph partitioning, graph sparsification, and solving linear systems. In: Proceedings of the 36th Annual ACM Symposium on Theory of Computing (STOC 2004), pp. 81–90. ACM (2004). https://doi.org/10.1145/1007352.1007372
47. Woodruff, D.P.: Sketching as a tool for numerical linear algebra. Found. Trends® Theor. Comput. Sci. **10**(1–2), 1–157 (2014). https://doi.org/10.1561/0400000060

The Communication Complexity of Functions with Large Outputs

Lila Fontes[1]![ORCID], Sophie Laplante[2]![ORCID], Mathieu Laurière[3]![ORCID],
and Alexandre Nolin[4(✉)]![ORCID]

[1] Swarthmore College, Swarthmore, USA
`fontes@cs.swarthmore.edu`
[2] IRIF, Université Paris Cité, Paris, France
`laplante@irif.fr`
[3] NYU-ECNU Institute of Mathematical Sciences, NYU Shanghai, Shanghai, China
`ml5197@nyu.edu`
[4] CISPA Helmholtz Center for Information Security, Saarbrücken, Germany
`alexandre.nolin@cispa.de`

Abstract. We study the two-party communication complexity of functions with large outputs, and show that the communication complexity can greatly vary depending on what output model is considered. We study a variety of output models, ranging from the *open model*, in which an external observer can compute the outcome, to the *XOR model*, in which the outcome of the protocol should be the bitwise XOR of the players' local outputs. This model is inspired by XOR games, which are widely studied two-player quantum games.

We focus on the question of error-reduction in these new output models. For functions of output size k, applying standard error reduction techniques in the XOR model would introduce an additional cost linear in k. We show that no dependency on k is necessary. Similarly, standard randomness removal techniques, incur a multiplicative cost of 2^k in the XOR model. We show how to reduce this factor to $O(k)$.

In addition, we prove analogous error reduction and randomness removal results in the other models, separate all models from each other, and show that some natural problems – including Set Intersection and Find the First Difference – separate the models when the Hamming weights of their inputs is bounded.

Keywords: Communication complexity · error reduction · non-Boolean functions

1 Introduction

Most of the literature on the topic of communication complexity has focused on Boolean functions. The usual definition stipulates that at the end of the protocol, one of the players knows the value of the function. In the rectangle based lower bounds, the assumption is slightly stronger: at the end of the protocol, the transcript of the protocol determines a combinatorial rectangle of

S. Rajsbaum et al. (Eds.): SIROCCO 2023, LNCS 13892, pp. 427–458, 2023.
https://doi.org/10.1007/978-3-031-32733-9_19

inputs that all evaluate to the same outcome. This means that given the transcript (together with the public coins, in the randomized public-coin setting), an external observer can determine the output. In the case of Boolean functions, this assumption makes no significant difference since the player who knows the value of the function can send it in the last message of the protocol, at an additional cost of at most one bit. When the function has large outputs, however, sending the value of the function as part of the transcript could cost more than all the prior communication. When this happens, then what should be considered the "true" communication complexity of the problem?

When studying functions with large outputs, several fundamental questions and issues emerge. What lower bound techniques extend to non-Boolean functions? When composing protocols with large outputs, it may not be useful for both players to know the values of the intermediate functions, and the aggregated cost of relaying the outcome at each intermediate step could exceed the complexity of the composed problem. These issues are also applicable to information complexity, where the cost is measured in information theoretic terms instead of in number of bits of communication. Requiring protocols to reveal the outcome as part of the transcript could be an obstacle to finding very low information protocols. It also raises the following issue: how does one amplify success when outputs are large? Amplification schemes typically involve repeating a protocol and taking a majority outcome, but finding said majority outcome naïvely incurs a cost that depends on the length of the output. We explore these issues, and give new models and amplification schemes.

Well-studied examples of functions with large outputs include asymmetric games, like the NBA problem [34,35] (see also [32, Example 4.53, p. 64]), and many problems where the output is essentially of the same size as the input (e.g., computing the intersection of two sets [14,15]). A decision version of a large-output function may or may not have a similar communication complexity (e.g., Set Disjointness [3,29,39], as opposed to deciding if the parties' numbers sum to something greater than a given constant [33,43]). Large output functions also appear when studying whether multiple instances of the same function exhibit economies of scale, known as direct sum problems, along with their variants such as agreement and elimination [1,2,6]. In these and other problems, computing one bit of the output can be just as hard or significantly easier than computing the full output, depending on the function and on the model. Finally, simulation protocols, whose output are transcripts of another protocol, have played a key role in compression [9–11,30,40] as well as structural results [13,24,25,37]. The Find the First Difference problem has been instrumental in compression protocols. Better protocols are known when weaker output conditions are required [4,5].

1.1 Output Models

We put forward several natural alternatives to the model where the transcript and public randomness reveal (possibly without containing it explicitly) the value of the function (we call this the *open* model). In the *local* model, both players

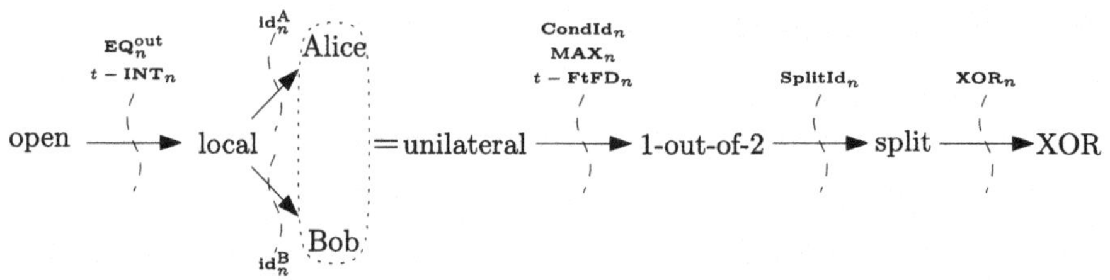

Fig. 1. The various models of communication and problems separating them. An arrow from A to B indicates that a communication protocol for a task of type A is also a communication protocol for a task of type B. Details of the stronger separations are provided in Appendix A.

can determine the value of the function locally (but an external observer might not be able to do so – unlike in the open model). In the *unilateral* model, one player always learns the answer. In the *one-out-of-two* model, the player who knows the answer can vary. In the *split* model, the bits of the output are split between the players in an arbitrary way known to both players. Finally, in the *XOR model*, each player outputs a string and the result is the bitwise XOR of these outputs. The models form a hierarchy, shown in Fig. 1. We defer formal definitions to Appendix A.

In the context of protocols, we make a distinction between what the players output and what the protocol computes. For example, in the XOR model, players output strings a and b but the result of the protocol is $a \oplus b$. We will use the word "output" to designate what the players output at the end of the protocol, and "result" or "outcome" to be the outcome of the protocol (which should be – either probably or certainly – the value or output of the function). Similarly, we will use the term "protocol" to designate the full mechanism for producing the result, and "communication protocol" for the interactive part of the protocol where the players exchange messages, not including the output mechanism.

Among all the models we propose, the XOR model is perhaps the most interesting. This model was partly inspired by (quantum) XOR games, where the players do not exchange any messages (for example [7,16,36]). One interesting property of the XOR model is that it could be the case, for example, that the output distribution of each player, taken individually, is uniform[1], revealing nothing about either input or even the value of the function, when run as a black box.

Moreover, it is common in communication complexity to consider the complexity of Boolean functions composed with some "gadget" applied to the inputs. For example, for a Boolean function f, one can ask what is the communication complexity of $F(u, v) = f(u \oplus v)$, where bitwise XOR is applied as a gadget on the inputs. The XOR model can be seen as applying the XOR gadget to the outputs instead of the inputs: the players output (a, b), and we require $F(u, v) = a \oplus b$ for the computation to be correct.

---

[1] Any protocol in this model can be converted into a protocol of same complexity with this property: the players pick a shared random string r of the same length as the output, and output $a \oplus r$ ($b \oplus r$), where a, b were the outputs of the original protocol.

1.2   Our Contributions

We focus on the XOR model where the players each output a string and the outcome of the protocol is the bitwise XOR of these strings.

Error Reduction. We consider the question of error reduction in Sect. 5. Error reduction is usually a simple task: repeat a computation enough times, and take the majority outcome. However, in the XOR model, neither of the players knows any of the outcomes, so neither can compute the majority outcome without additional communication. Sending over all the outcomes so one of the players can compute the majority would add a prohibitive $\Theta(k)$ term, where k is the length of the output. Removing this dependency on k is possible, however, and doing so requires quite elaborate protocols that highlight, and circumvent, the inherent limitations of the XOR model (Theorem 2).

We further improve the dependency on the error parameter ϵ for direct sum problems (Theorem 4), by combining protocols for amortized Equality [20] and Find the First Difference [21], as well as Gap Hamming Distance [18,27,41,42].

Deterministic Versus Randomized Complexity. In Sect. 6, we revisit the classical result that states that for any Boolean function, the deterministic communication complexity is at most exponential in the private coin randomized complexity. Once again, if the size of the output is k, then applying existing schemes naively to our weaker models adds a multiplicative cost of 2^k. We show that a dependency of a factor of k suffices (Theorem 6).

Gap Majority Composed with XOR. To prove our results for the XOR model, we consider the non-Boolean *Gap Majority* problem composed with an XOR gadget. In the standard majority problem, the input is a set of elements and the goal is to find the element which appears most often. The gap majority problem adds the promise that the majority element should appear at least some a fixed fraction (more than half) of the time. Composition with an XOR gadget turns the problem into a communication complexity problem (see Sect. 5). We show that the communication complexity of this problem is closely related to the problems of reducing error and removing randomness in the XOR model.

Other Models and Separations. We define several communication models and give problems that maximally separate them (Appendix A).

We provide additional results in the full version of this paper [23]. We revisit error reduction and randomness removal in other models. Notably, we reduce the dependency on k to a factor of $\log(k)$ in the one-out-of-two model, and remove this dependency entirely when the error parameter ϵ is bounded by $1/3$. We study a few additional problems which exhibit gaps between our various communication models. In particular, several common problems exhibit a gap when the Hamming weights of their inputs are bounded. We also show how lower bound techniques can be adapted to our weak output models by revisiting the notion of monochromatic rectangles associated with the leaves of a protocol tree.

It is important to note that our results mostly do not apply to large-output *relations* (such as the variants of direct sum, elimination and agreement), as many of our proofs crucially rely on the fact that there is a single correct answer.

2   Related Work

Previous works have addressed the question of the output model for large output functions. Braverman et al. [12] make a distinction between "simulation" and "strong simulation" of a protocol. In a strong simulation, an external observer can determine the result without any knowledge of the inputs. In their paper on compression to internal information [5], Bauer et al. stress the importance, when compressing to internal information, that the compression itself need not reveal information to an external observer. They consider two output models which they call internal and external computation. In external computation (which we call the open model), an external observer can determine the result of the protocol, whereas in internal computation (which we call the local model), the players both determine the result at the end of the protocol.[2] They observe that in the deterministic setting, for total functions, the two models coincide, but they can differ in the distributional setting. They consider a key problem of finding the first bit where two strings differ, when each player has one of the two strings. This problem is used in reconciliation protocols to find the first place where transcripts differ. Feige et al. [21] externally (openly) solve Find the First Difference in $O(\log(\frac{n}{\epsilon}))$, which was shown to be tight by Viola [43]. Bauer et al. [5] give an internal (local) protocol with a better complexity, where the improvement depends on the entropy of the input distribution.

3   Preliminaries

An introduction to communication complexity can be found in Kushilevitz and Nisan's [32], and Rao and Yehudayoff's [38] textbooks.

We denote by $\mathcal{X}$ (resp. $\mathcal{Y}$) the set of inputs of Alice (resp. Bob), $\mathcal{R}_A$ her private randomness ($\mathcal{R}_B$ for Bob), and $\mathcal{R}^{\text{pub}}$ the public randomness accessible to both players. When $|\mathcal{X}| = |\mathcal{Y}|$, we denote by n the size of the input (so that $n = \lceil \log(|\mathcal{X}|) \rceil$). When computing a function, we denote by k the length of the output, $\mathcal{Z}$ the *image* of the function and $k = \lceil \log(|\mathcal{Z}|) \rceil$. We sometimes consider an additional output symbol $\top$.

We define a *full protocol* as the combination of a *communication protocol* and an *output mechanism* (this is discussed in Appendix A). We define a (two-player) communication protocol Π as a full binary tree where each non-leaf node v is assigned a player $\mathcal{P}^v$ amongst A(lice) and B(ob), and a mapping $\mathcal{N}^v$ into $\{0,1\}$ whose input space depends on which player the node was assigned to. When $\mathcal{P}^v = A$ (resp. B) then $\mathcal{N}^v$'s input space is $\mathcal{X} \times \mathcal{R}_A \times \mathcal{R}^{\text{pub}}$ (resp. $\mathcal{Y} \times \mathcal{R}_B \times \mathcal{R}^{\text{pub}}$).

---

[2] We prefer the terms *open* and *local* to avoid any confusion between the notions of *internal* and *external* computation, and *internal* and *external* information.

Note that the tree and each node's owner are fixed and do not depend on the input. In an execution of a communication protocol, the two players walk down the tree together, starting from the root, until they reach a leaf. Each step down the tree is done by letting the player who owns the current node v apply its corresponding mapping $\mathcal{N}^v$, and sending the result to the other player. If it is 0, the players replace the current node by its left child, and otherwise by its right child. The *communication cost* $\mathrm{CC}(\Pi)$ of a protocol Π is the total number of bits exchanged for the worst case inputs.

Since an execution of a communication protocol Π is entirely defined by the players' inputs $((x, y) \in \mathcal{X} \times \mathcal{Y})$ and the randomness (the players' private randomness $r_A \in \mathcal{R}_A$ and $r_B \in \mathcal{R}_B$ as well as the public randomness $r \in \mathcal{R}^{\mathrm{pub}}$), we also view the communication protocol as a function $\Pi : \mathcal{X} \times \mathcal{Y} \times \mathcal{R}_A \times \mathcal{R}_B \times \mathcal{R}^{\mathrm{pub}} \to \{0, 1\}^*$ whose values we call *transcripts* of Π. For the purposes of this paper, we do not include the public randomness as part of the transcript. For a given protocol Π, we denote by $T_\pi = \Pi(X, Y, R_A, R_B, R)$ the random variable over transcripts of the protocol that naturally arises from X, Y, R_A, R_B, and R, taken as random variables. We denote by $\mathcal{T}_\pi$ the support of the distribution T_π. We denote by $x, y, z, r_A, r_B, r, t_\pi$ elements of the sets $\mathcal{X}, \mathcal{Y}, \mathcal{Z}, \mathcal{R}_A, \mathcal{R}_B, \mathcal{R}^{\mathrm{pub}}, \mathcal{T}_\pi$, respectively, which in turn are the supports of the random variables $X, Y, Z, R_A, R_B, R, T_\pi$.

We recall definitions and known bounds of functions that will be used in this paper. For all of these problems, note that the communication complexity is of the same order of magnitude whether we require that both players know the output or only one of them, since the size of the output is no larger than the communication required for one player to know the output. In the remainder of this section, we denote by $R_\epsilon(f)$ the minimal communication cost of a randomized protocol computing function f with error at most ϵ when, say, Bob outputs. $D(f) = R_0(f)$ denotes the deterministic communication complexity. Unless otherwise specified, our protocols use both private and public coins. We use the 'priv' superscript to indicate when only private randomness is used.

Definition 1 (Find the First Difference problem). $\mathbf{FtFD}_n : \{0, 1\}^n \times \{0, 1\}^n \to \{0, \ldots, n\}$ *is defined as* $\mathbf{FtFD}_n(x, y) = \min(\{i : x_i \neq y_i\} \cup \{n\})$.

Proposition 1. *For any* $0 < \epsilon < \frac{1}{2}$, $R_\epsilon(\mathbf{FtFD}_n) \in \Theta(\log(n) + \log(1/\epsilon))$ [21,43].

The upper bound uses a walk on a tree where steps are taken according to results from hash functions. The lower bound is from a lower bound on the Greater Than function $\mathbf{GT}_n$, which reduces to $\mathbf{FtFD}_n$. For a good exposition of the upper bound, see Appendix C in [4].

Definition 2 (Gap Hamming Distance problem). *Let* n, L, U *be integers such that* $0 \leq L < U \leq n$. $\mathbf{GHD}_n^{L,U} : \{0, 1\}^n \times \{0, 1\}^n \to \{0, 1\}$ *is a promise problem where the input satisfies the promise that the Hamming distance between inputs* x, y *is either* $\geq U$ *or* $\leq L$. *Then* $\mathbf{GHD}_n^{L,U}(x, y) = 1$ *in the first case and* 0 *in the second case.*

The bounds on Gap Hamming Distance vary depending on the parameters. In this paper we use a linear upper bound which is essentially tight in the regime we require. Many other bounds are known for other regimes [17,18,31,41,42,44].

Definition 3 (Equality problem). $\mathbf{EQ}_n : \{0,1\}^n \times \{0,1\}^n \to \{0,1\}$ *is defined as* $\mathbf{EQ}_n(x,y) = \mathbf{1}_{x=y}$. *The k-fold Equality problem is* $\mathbf{EQ}_n^{\otimes k}((x_1,\ldots,x_k),$ $(y_1,\ldots,y_k)) = (\mathbf{EQ}_n(x_1,y_1),\ldots,\mathbf{EQ}_n(x_k,y_k))$, *where* $(x_i, y_i) \in \{0,1\}^n$ *for all i.*

Proposition 2. *For* $0 < \epsilon < \frac{1}{2}$, $R_\epsilon(\mathbf{EQ}_n^{\otimes k}) \in \Theta(k + \log(1/\epsilon))$.

The algorithm from [26] which achieves optimal communication uses hashing just like the algorithm for a single instance. It saves on communication compared to k successive uses of a protocol for equality with error ϵ/k by having players hash all k instances simultaneously, exchange results, and repeat this process, exploiting that they have less and less to communicate about. Intuitively, the number of unequal instances to discover should decrease as the algorithm runs. Once it has been determined for an instance (x_i, y_i) that $x_i \neq y_i$ through unequal hashes, the players do not need to speak further about this instance. An unequal instance is unlikely to survive many tests, which means that late in the algorithm the players can exchange their hashes using that most of them should agree. The idea was also present in previous algorithms [20] which improved on the trivial algorithm. The lower bound is just from $\Omega(k)$ bits of communication being necessary to send k bits worth of information, even with ϵ error.

4   The *XOR* model

In the XOR model, each player outputs a string and the value of the function is the bitwise XOR of the two outputs (Definition 4). This model is inspired by XOR games which have been widely studied in the context of quantum nonlocality as well as unique games.

Definition 4 (XOR computation). *Consider a function f whose output set is* $\mathcal{Z} = \{0,1\}^k$. *A protocol* Π *is said to* XOR*-compute f with* ϵ *error if there exist two mappings* $\mathcal{O}_A$ *and* $\mathcal{O}_B$ *with* $\mathcal{O}_A : \mathcal{T}_\pi \times \mathcal{R}^{\mathsf{pub}} \times \mathcal{R}_A \times \mathcal{X} \to \{0,1\}^k$ *and similarly* $\mathcal{O}_B : \mathcal{T}_\pi \times \mathcal{R}^{\mathsf{pub}} \times \mathcal{R}_B \times \mathcal{Y} \to \{0,1\}^k$ *such that for all* $(x,y) \in \mathcal{X} \times \mathcal{Y}$, $\Pr_{r,r_A,r_B}[\mathcal{O}_A(t_\pi,r,r_A,x) \oplus \mathcal{O}_B(t_\pi,r,r_B,y) = f(x,y)] \geq 1 - \epsilon$.

We define $D^{\mathsf{xor}}(f)$ (resp. $R_\epsilon^{\mathsf{xor}}(f)$) as the best communication cost of any protocol that computes f in the XOR model with error $\epsilon = 0$ (resp. with error at most ϵ, for $0 < \epsilon < \frac{1}{2}$). (Notations are defined similarly for our other models with superscripts open, loc, A, B, uni, 1of2, spl.)

5   Error Reduction and the Gap Majority Problem

We study the cost of reducing the error of communication protocols in our weaker models of communication where the outcome of the protocol is not known to both of the players. We focus on the more interesting case of the XOR model here, and results for the other models are in the full version of this paper [23].

Standard error reduction schemes work by repeating a protocol many times in order to compute and output the most frequently occurring value among all the executions. Repeating the protocol enough times ensures that with high probability, the output that appears the most is correct. One can derive an upper bound on the number of iterations needed from Hoeffding's inequality.

Lemma 1 (Hoeffding's inequality). *Consider N independent Bernoulli trials $(V_i)_{i \in [N]}$ of expected value p. We have*

$$\Pr\left[\left|\tfrac{1}{N}\sum_{i=1}^{N} V_i - p\right| \geq \delta\right] \leq 2 \cdot \exp\left(-\frac{\delta^2 N}{2p(1-p)}\right).$$

The following holds in the setting where Bob outputs the value of the function at the end of the protocol.

Theorem 1 *(Folklore, see [32]). Let $0 < \epsilon' < \epsilon < \tfrac{1}{2}$, and $C_{\epsilon,\epsilon'} = \frac{2\epsilon(1-\epsilon)}{(\tfrac{1}{2}-\epsilon)^2}\ln\left(\tfrac{2}{\epsilon'}\right)$. For all functions $f : \mathcal{X} \times \mathcal{Y} \to \mathcal{Z}$, $R^{\mathsf{B}}_{\epsilon'}(f) \leq C_{\epsilon,\epsilon'} \cdot R^{\mathsf{B}}_{\epsilon}(f)$.*

Note that it is important here that f is a function, not a relation, so that there is a unique correct output and the player(s) can compute the majority.

In the XOR model, finding the majority result among some number T of runs is much more difficult than in the standard model, since neither of the players can identify reasonable candidates as the majority answer. Exchanging all of the T k-bit outputs would result in a bound of $R^{\mathsf{xor}}_{\epsilon'}(f) \leq C_{\epsilon,\epsilon'}(R^{\mathsf{xor}}_{\epsilon}(f) + k)$. We show that this dependence on k is unnecessary.

Theorem 2. *Let $0 < \epsilon' < \epsilon < \tfrac{1}{2}$, $C_{\epsilon,\epsilon'} = 8\epsilon(\tfrac{1}{2} - \epsilon)^{-2}\ln\left(\tfrac{8}{\epsilon'}\right)$. For all $f : \mathcal{X} \times \mathcal{Y} \to \{0,1\}^k$, $R^{\mathsf{xor}}_{\epsilon'}(f) \leq C_{\epsilon,\epsilon'} \cdot R^{\mathsf{xor}}_{\epsilon}(f) + O(C_{\epsilon,\epsilon'})$.*

In order to prove this result, we introduce the Gap Majority (**GapMAJ**) problem, show how Theorem 2 reduces to solving **GapMAJ∘XOR** (Lemma 2), then give an upper bound on solving **GapMAJ∘XOR** (Theorem 3).

The partial function **GapMAJ**$_{N,k,\epsilon,\mu}$ has N strings of length k as input and the promise is that there is a string z of length k that appears with μ weight at least $(1 - \epsilon)$ among the N strings, where μ is a distribution over indices in $[N]$.

Definition 5 (Gap Majority). *In the Gap Majority problem* **GapMAJ**$_{N,k,\epsilon,\mu}$: $\left(\{0,1\}^k\right)^N \to \{0,1\}^k$ *the input is $(Z_1,\ldots,Z_N)$, and μ is a fixed distribution over the indices $[N]$. When unspecified, μ is understood to be the uniform distribution. The promise is that $\exists z \in \{0,1\}^k$ such that $\mu(\{i \in [N] : Z_i = z\}) \geq (1 - \epsilon)$. Then*

$$\mathbf{GapMAJ}_{N,k,\epsilon,\mu}((Z_i)_{i\in[N]}) = z \quad s.t. \quad \mu(\{i : Z_i = z\}) \geq (1 - \epsilon).$$

In **GapMAJ∘XOR**, the players are given N strings of length k and their goal is to compute **GapMAJ** on the bitwise XOR of their inputs whenever the **GapMAJ** promise is satisfied. (Notice that when $k = 1$, this is equivalent to the Gap Hamming Distance problem (Definition 2) with parameters $L = \epsilon N$, $U = (1 - \epsilon)N$.)

For inputs $(X_1, \ldots, X_N), (Y_1, \ldots, Y_N)$ to $\mathbf{GapMAJ}_{N,k,\epsilon,\mu} \circ \mathbf{XOR}$, we will refer to a pair (X_i, Y_i) as a *row*, and we call X_i Alice's ith row, and Y_i Bob's ith row. As a warm-up exercise, we show that error reduction reduces to solving an instance of $\mathbf{GapMAJ} \circ \mathbf{XOR}$.

Lemma 2. *Let* $0 < \epsilon' < \epsilon < \frac{1}{2}$ *and* $C_{\epsilon,\epsilon'} = 2\epsilon \left(\frac{1}{2} - \epsilon \right)^{-2} \ln \left(\frac{4}{\epsilon'} \right)$. *For every* $f :$ $\mathcal{X} \times \mathcal{Y} \to \{0,1\}^k$, $R_{\epsilon'}^{\mathsf{xor}}(f) \le C_{\epsilon,\epsilon'} \cdot R_{\epsilon}^{\mathsf{xor}}(f) + R_{\epsilon'/2}^{\mathsf{xor}} \left(\mathbf{GapMAJ}_{C_{\epsilon,\epsilon'},k,\frac{1}{4}+\frac{\epsilon}{2}} \circ \mathbf{XOR} \right)$.

Proof (Proof of Lemma 2). Let π be a protocol which XOR-computes $f(x,y)$ with ϵ-error and π' be a protocol which computes $\mathbf{GapMAJ}_{C_{\epsilon,\epsilon'},k,\frac{1}{4}+\frac{\epsilon}{2}} \circ \mathbf{XOR}$ in the XOR model, with error $\epsilon'/2$. We consider the following protocol, which we denote by $\hat{\pi}$: first, run π $C_{\epsilon,\epsilon'}$ times; then, use the outputs produced by this computation as inputs for π', run the latter protocol, and output the result. We analyze the new protocol $\hat{\pi}$ as follows. The outputs produced in the first step are strings $X_1, \cdots, X_{C_{\epsilon,\epsilon'}}$ on Alice's side, and $Y_1, \cdots, Y_{C_{\epsilon,\epsilon'}}$ for Bob. A run of π is correct iff $X_i \oplus Y_i = f(x,y)$. By Hoeffding's bound (Lemma 1), applied with $N = C_{\epsilon,\epsilon'}$, $V_i = 1$ if $X_i \oplus Y_i \ne f(x,y)$ and $V_i = 0$ otherwise for $i = 1, \ldots, N$, $p = \mathbb{E}[V_i] \le \epsilon$, and $\delta = \frac{1}{2} (\frac{1}{2} - \epsilon)$, we get that with probability at least $1 - 2e^{-\delta^2 N/(2p(1-p))} \ge 1 - \epsilon'/2$, a fraction $p + \delta \le (\frac{1}{2} + \epsilon)/2$ of the N computations err. In other words, with probability at most $\epsilon'/2$, the above strings fail to satisfy the promise in the definition of $\mathbf{GapMAJ}_{C_{\epsilon,\epsilon'},k,\frac{1}{4}+\frac{\epsilon}{2}} \circ \mathbf{XOR}$. Conditioned on this not happening (i.e., on the promise being met), π' (hence $\hat{\pi}$) errs with probability at most $\epsilon'/2$. The overall error is at most ϵ'. $\quad\square$

To derive a general upper bound on error reduction using Lemma 2, it would suffice to have an upper bound on $R_{\epsilon'}^{\mathsf{xor}}(\mathbf{GapMAJ}_{N,k,\epsilon} \circ \mathbf{XOR})$. When the error parameter is large ($\epsilon \le \epsilon'$), $\mathbf{GapMAJ} \circ \mathbf{XOR}$ in the XOR model is trivial: the players just need to sample a common row and output according to that row. However, Lemma 2 requires solving a $\mathbf{GapMAJ} \circ \mathbf{XOR}$ instance with small error $\epsilon'/2$, which takes us back to square one: finding an error reduction scheme that we can apply to $\mathbf{GapMAJ} \circ \mathbf{XOR}$.

In the remainder of the section, we give a protocol for $\mathbf{GapMAJ} \circ \mathbf{XOR}$ (Sect. 5.1) followed by an error reduction scheme for direct sum functions (Sect. 5.2). In both cases, we use the structure of the XOR function and a protocol for Equality on pairs of rows to find a majority outcome. The error reduction scheme for direct sum functions is a refinement of Lemma 2 and is useful in cases where the starting error is very close to $\frac{1}{2}$ and where computing one bit of the output is significantly less costly than computing the full output.

5.1 Solving GapMAJ∘XOR

Given an instance of $\mathbf{GapMAJ}_{N,k,\epsilon} \circ \mathbf{XOR}$, if Alice and Bob pick a row and output what they have on this row, they get the correct output with probability $\ge 1 - \epsilon$. Recall that we would like to achieve error $\epsilon' < \epsilon$ without incurring a dependence on parameter k, which in our application to error reduction corresponds to the length of the output. We show that this is possible.

Theorem 3. *Let $0 < \epsilon' < \epsilon < \frac{1}{2}$. Then*

$$R_{\epsilon'}^{\text{xor}}(\textbf{GapMAJ}_{N,k,\epsilon}\circ\textbf{XOR}) \leq O\big(N + \log\big(\tfrac{1}{\epsilon'}\big)\big).$$

Proof Idea. We use the fact that $a \oplus b = a' \oplus b'$ iff $a \oplus a' = b \oplus b'$. Therefore, the players can identify rows that XOR to a same string by solving instances of Equality. This idea alone is enough to obtain a protocol for **GapMAJ**$_{N,k,\epsilon}\circ$**XOR** of complexity $O\big(N^2 + \log\big(\tfrac{1}{\epsilon'}\big)\big)$ by computing Equality for all $\binom{N}{2}$ pairs of rows to identify the majority outcome. We improve on this by reducing the number of computed Equality instances using Erdős-Rényi random graphs (Lemma 3).

Lemma 3 (Variation of eq. (9.18) in [19]). *Let $G(n, p(n))$ be the distribution over graphs of n vertices where each edge is sampled with independent probability $p(n)$. Let $L_1(G)$ be the size of the largest connected component of G. Then:*

$$\forall \alpha \in [0,1], c \in \mathbb{R}^+, \qquad \Pr[L_1(G(n,c/n)) < (1-\alpha)n] \leq e^{\left(\ln(2) - \frac{\alpha}{2}(1-\frac{\alpha}{2})c\right)n}.$$

In particular this probability goes to 0 as n goes to infinity when $\alpha c > 4\ln(2)$.

For completeness, the proof is given in Appendix C.1.

Proof (Proof of Theorem 3). Consider the **GapMAJ∘XOR** instance as a $N \times k$ matrix such that $(X_i)_{i\in[N]}$ are the rows of Alice and $(Y_i)_{i\in[N]}$ are the rows of Bob. By the promise of the **GapMAJ∘XOR** problem, we know there exists a $z \in \{0,1\}^k$ such that $\{i : X_i \oplus Y_i = z\} \geq (1-\epsilon)N$. The goal is now for Alice and Bob to identify a row belonging to this large set of rows that XOR to the same k-bit string.

Let i and j be the indices of two rows. The event that the two rows XOR to the same string is expressed as $X_i \oplus Y_i = X_j \oplus Y_j$, which is equivalent to $X_i \oplus X_j = Y_i \oplus Y_j$. This means that we can test whether any two rows XOR to the same bit string with a protocol for Equality.

The protocol goes through the following steps:

1. The players pick rows randomly, enough rows so that with high probability, a constant fraction of the rows XOR to the majority element z.
2. The players solve instances of Equality to find large sets of rows that XOR to the same string. In each such large set of rows, they pick a single row. This leaves them with a constant number of candidate rows that might XOR to the majority element z.
3. The players decide between those candidates by comparing them with all the rows. There is one candidate row that XORs to the same string as most rows; this row XORs to the majority element z.

Step 1. Using public randomness, Alice and Bob now pick a multiset S of all their rows of size $|S| = T_{\epsilon'} = 50\ln\big(\tfrac{10}{\epsilon'}\big)$. Each element of S is picked uniformly and independently. Using Hoeffding's inequality (Lemma 1), with probability $\geq 1 - \frac{\epsilon'}{5}$ more than $\frac{2}{5}$ of those executions XOR to the majority element z.

Step 2. We now consider S as the vertices V of a random graph $G = G(V, E)$, in which each edge is picked with a probability $\frac{c}{|V|}$ with $c > 0$. Consider the subgraph G' of G induced on the vertices $V' \subseteq V$ that correspond to executions that XOR to the majority element z. From the previous step, we know that $|V'| \geq \frac{2}{5}T_{\epsilon'} = 20\ln\left(\frac{10}{\epsilon'}\right)$. The subgraph G' is a random graph where each edge was picked with the same probability $\frac{c}{|V|} = \frac{c'}{|V'|}$ where $c' = c\frac{|V'|}{|V|} \geq \frac{2}{5}c$. By Lemma 3, this subgraph G' contains a connected component of size $\geq (1 - \frac{1}{12})|V'| \geq \frac{11}{30}|V|$ with probability $\geq 1 - 2^{-|V'|} \geq 1 - \frac{\epsilon'}{5}$ for $c \geq \frac{720}{143}\ln(2) \approx 3.49$ as $|V'| \geq 20\ln\left(\frac{10}{\epsilon'}\right) \geq \log\left(\frac{5}{\epsilon'}\right)$.

At this point, Alice (resp. Bob) computes the bitwise XOR of all pairs of executions that correspond to an edge in G: $(X_i \oplus X_j)_{(i,j)\in E, i<j}$ (resp. $(Y_i \oplus Y_j)_{(i,j)\in E, i<j}$). For ϵ' small enough, with high probability ($\geq 1 - \frac{\epsilon'}{5}$), the set of edges of G is smaller than $2c \cdot T_{\epsilon'}$ by Hoeffding's inequality (the players can abort the protocol otherwise). Then, Alice and Bob solve $\leq 2c \cdot T_{\epsilon'}$ instances of Equality with (total) error $\leq \frac{\epsilon'}{5}$ to discover a large set of rows that XOR to a same bit string. We now have groups of rows that we know XOR to the same bit string, at least one of which represents more than $\frac{11}{30}$ of S's rows because of the Hoeffding argument combined with the random graph lemma. Now for each submultiset of rows of S that XOR to the same bit string and represents more than $\frac{11}{30}$ of all of S's rows, pick an arbitrary row in the submultiset. If there is only one such submultiset, Alice and Bob can end the protocol here, outputting the content of the row selected in this submultiset. If there were two such submultisets, then let i_1 and i_2 be the indices picked in each submultiset.

Step 3. To decide between their two candidates, Alice and Bob solve N Equality instances between $X_{i_1} \oplus X_j$ and $Y_{i_1} \oplus Y_j$ for all $j \in [N]$ with error $\leq \frac{\epsilon'}{5}$. If more than half of the N rows XOR to the same string as the i_1^{th} row, Alice and Bob output their i_1^{th} row. Otherwise, they output the other candidate row i_2.

The complexity of computing $\mathbf{GapMAJ}_{N,k,\epsilon}\circ\mathbf{XOR}$ with error $\epsilon' < \epsilon$ satisfies

$$R_{\epsilon'}^{\mathsf{xor}}\left(\mathbf{GapMAJ}_{N,k,\epsilon}\circ\mathbf{XOR}\right) \leq R_{\epsilon'/5}\left(\mathbf{EQ}_k^{\otimes 2cT_{\epsilon'}}\right) + R_{\epsilon'/5}\left(\mathbf{EQ}_k^N\right).$$

To conclude, we apply an amortized protocol for Equality (Proposition 2).

Combining Lemma 2 and Theorem 3 concludes the proof of Theorem 2. We give additional results on the $\mathbf{GapMAJ}\circ\mathbf{XOR}$ problem in the full version of the paper [23].

5.2   XOR Error Reduction for Direct Sum Functions

The protocol of Theorem 2 first generates a full instance of $\mathbf{GapMAJ}\circ\mathbf{XOR}$, then solves this instance. The generation of this instance might create an implicit dependency on the output length k of f, which in the regime where ϵ is very close to $1/2$ can be prohibitive. We give a different protocol in which the players are not required to fully generate these intermediate results.

For large output functions, generating one bit of the output can be much less costly than generating all k, for example, when f is a direct sum of k instances of a function g. We state our stronger amplification theorem for the case of direct sum problems of Boolean functions, but we note that the protocol could be used for other problems where computing one bit of the output is less costly than computing the entire output.

Theorem 4. *Let* $0 < \epsilon' < \epsilon < \frac{1}{2}$ *and* $C_{\epsilon,\epsilon'} = 8\epsilon\left(\frac{1}{2} - \epsilon\right)^{-2}\ln\left(\frac{12}{\epsilon'}\right)$. *For any* $g : \mathcal{X} \times \mathcal{Y} \to \{0,1\}$ *and* $f = g^{\otimes k}$,

$$R_{\epsilon'}^{\text{xor}}(f) \le 50\ln\left(\frac{12}{\epsilon'}\right) \cdot R_{\epsilon}^{\text{xor}}(f) + C_{\epsilon,\epsilon'} \cdot R_{\epsilon}^{\text{xor}}(g) + O(C_{\epsilon,\epsilon'} + \log(k)) \ .$$

Notice that the $C_{\epsilon,\epsilon'}$ factor – which scales with $\left(\frac{1}{2} - \epsilon\right)^{-1}$ – applies to the complexity of g, not of f.

Proof Idea. Instead of iterating the basic protocol $C_{\epsilon,\epsilon'}$ times, we will start by iterating it a smaller number of times which does not depend on ϵ, but only on $\log(\frac{1}{\epsilon'})$. This number of iterations suffices to guarantee that the most frequent outcome represents more than a $1/3$ fraction of the rows. If no other outcome represents a large fraction of the rows, we output according to a row from this large fraction. Otherwise, still, at most two outcomes can represent more than a $1/3$ fraction of the rows. We identify a "critical index" of the output function, one that will help us identify the majority result among the two candidate outcomes. We do so by solving a Gap Hamming Distance instance on the critical index. In these remaining $C_{\epsilon,\epsilon'}$ runs, we only need one of the k bits of the output.

Details of the proof are given in the full version of the paper [23].

6   Deterministic Versus Randomized Complexity

We now turn to removing randomness from private coin protocols.

The standard scheme to derive a deterministic protocol from a private coin protocol[3] proceeds as follows [32, Lemma 3.8, page 31]. The players exchange messages to estimate the probability of each transcript. They use the fact that the probability of a transcript can be factored into two parts, each of which can be computed by one of the two players. One of the players sends all of its factors to the other, up to some precision, and the second player can then estimate the probability of each transcript. Each transcript determines an output, therefore from the estimate for the transcripts' probabilities, this player can derive an estimate for the probability of each output, and output the majority answer.

---

[3] For public coins, the exponential upper bounds do not hold, for example in the case of the Equality function, which has an $O(1)$ public coin randomized protocol, but requires n bits of communication to solve deterministically.

Theorem 5 (Lemma 3.8 in [32], page 31). *For any function* $f : \mathcal{X} \times \mathcal{Y} \to \mathcal{Z}$ *and* $0 < \epsilon < \frac{1}{2}$, *let* $R = R_\epsilon^{\text{priv}}(f)$. *Then* $D(f) \leq 2^R \left(R + \log \left(\frac{1}{\frac{1}{2} - \epsilon} \right) + 1 \right)$.

Using this well-known result for our output models (first adding k bits of communication to the original protocol of cost R to obtain a protocol that works in the unilateral model) would add $2^R R \cdot 2^k$ bits to the complexity. For the XOR model, we reduce the dependency to a $O(2^R k)$ term. In the full version of the paper [23], we show some lower dependencies on k in our other models.

We formalize the problem which we call Transcript Distribution Estimation. Let $\Delta(\mu, \nu) = \frac{1}{2} \sum_{u \in \mathcal{U}} |\mu(u) - \nu(u)|$ be the total variation distance between two probability distributions μ and ν over a universe $\mathcal{U}$. For a protocol Π, let $\mathcal{T}_\pi$ be the set of transcripts of Π, and for $(x, y) \in \mathcal{X} \times \mathcal{Y}$, let us denote by $T_\pi^{x,y}$ the distribution over $\mathcal{T}_\pi$ witnessed when running Π on (x, y).

The key step of the proof of Theorem 5 is a protocol (in the standard model) for the following problem.

Definition 6 (Transcript Distribution Estimation problem). *For any protocol* Π *and* $\delta < \frac{1}{2}$, *we say that a protocol* $\widetilde{\Pi}$ *solves* **TDE**$_{\Pi,\delta}$ *in model* $\mathcal{M}$ *if, for each input* (x, y), $\widetilde{\Pi}$ *computes in the sense of model* $\mathcal{M}$ *a distribution* $\widetilde{T}_\pi^{x,y}$ *such that* $\Delta(\widetilde{T}_\pi^{x,y}, T_\pi^{x,y}) \leq \delta$.

Lemma 4 (Implicit in [32], page 31). *Let* Π *be a private coin communication protocol and* $\mathcal{T}_\pi$ *its set of possible transcripts. For any* $0 < \delta < \frac{1}{2}$, $D(\textbf{TDE}_{\Pi,\delta}) \leq |\mathcal{T}_\pi| \cdot \lceil \log \left(\frac{|\mathcal{T}_\pi|}{\delta} \right) \rceil$.

In their proof, Kushilevitz and Nisan [32] require only one of the players to learn an estimate of the probability of each leaf. Here we require both players to learn the same estimate, which can be achieved with a factor of two in the communication. Details are given in the full version of the paper [23].

In the XOR model, however, sharing such an estimate is not sufficient to remove randomness. At each leaf, each player outputs values with some probability (depending on their private randomness), so there can be as many as $|\mathcal{Z}|$ outputs per leaf by each player, making identifying the majority outcome impossible. We prove the following bound on deterministic communication in the XOR model.

Theorem 6. *Let* $0 < \epsilon < 1/2$ *and* $f : \mathcal{X} \times \mathcal{Y} \to \mathcal{Z} = \{0, 1\}^k$. *Let* $R = R_\epsilon^{\text{xor,priv}}(f)$, $M = 16 \cdot \left(\frac{1}{2} - \epsilon \right)^{-2} \cdot 2^R$, *and* $\epsilon' = \frac{5}{8} - \frac{\epsilon}{4}$. *Then*

$$D^{\text{xor}}(f) \leq D^{\text{loc}}(\textbf{TDE}_{\Pi_f, \epsilon' - \frac{1}{2}}) + D^{\text{xor}}(\textbf{GapMAJ}_{M,k,\epsilon',\mu} \circ \textbf{XOR})$$

$$\leq \left(2^{R+1} \right) \cdot \left(R + \log \left(\frac{8}{\frac{1}{2} - \epsilon} \right) + 1 \right) + k \cdot \left(\frac{5 - 2\epsilon}{4} M + 1 \right).$$

where μ *is an unspecified distribution over* $[M]$ *known to both players.*

Proof Idea. We reduce the problem of finding the majority outcome to a much smaller instance of **GapMAJ∘XOR** by discretizing the probabilities of the outputs. This lets us reduce the dependence on the size of the output to just a factor of $k = \log(|\mathcal{Z}|)$ (instead of a factor of $2^{2k} = |\mathcal{Z}|^2$).

Proof (Proof of Theorem 6). Let Π be an optimal private coin XOR protocol for f. The players start running a protocol for $\mathbf{TDE}_{\Pi,\delta}$ in the local model (an adaptation of Lemma 4; full details are in the complete version of this paper [23]) with $\delta = \frac{1}{4}(\frac{1}{2} - \epsilon)$, thus learning within statistical distance δ the probability distribution over leaves that results from the protocol.

Let $o_\mathsf{A}(.\mid w, x)$ and $o_\mathsf{B}(.\mid w, y)$ be the two independent probability distributions over $\{0,1\}^k$ according to which Alice and Bob output, conditioned on reaching leaf w, having received inputs x and y. To reduce the problem to $\mathbf{GapMAJ}\circ\mathbf{XOR}$, they discretize o_A and o_B into $\lceil\delta^{-1}\rceil$ events. Let $\dot{o}_\mathsf{A}$ denote the discretization of o_A with following properties for Alice (Similarly for $\dot{o}_\mathsf{B}$):

$$\forall z, w: \quad \dot{o}_\mathsf{A}(z\mid w, x) \cdot \lceil\delta^{-1}\rceil \in \mathbb{N} \quad \text{and} \quad |o_\mathsf{A}(z\mid w, x) - \dot{o}_\mathsf{A}(z\mid w, x)| \le \frac{1}{\lceil\delta^{-1}\rceil}.$$

A simple greedy approach to discretization goes like this:

1. Replace all $o_\mathsf{A}(z\mid w, x)$ by $\dot{o}_\mathsf{A}(z\mid w, x) = \frac{1}{\lceil\delta^{-1}\rceil}\lfloor\lceil\delta^{-1}\rceil o_\mathsf{A}(z\mid w, x)\rfloor$.
2. While the probabilities of $\dot{o}_\mathsf{A}$ sum to less than 1, pick a z s.t. $o_\mathsf{A}(z\mid w, x) - \dot{o}_\mathsf{A}(z\mid w, x)$ is maximal. For that z, set $\dot{o}_\mathsf{A}(z\mid w, x) = \frac{1}{\lceil\delta^{-1}\rceil}\lceil\lceil\delta^{-1}\rceil o_\mathsf{A}(z\mid w, x)\rceil$.

The players then construct a distributional $\mathbf{GapMAJ}\circ\mathbf{XOR}$ instance with M rows where $M = \lceil\delta^{-1}\rceil^2|\mathcal{T}_\pi|$ in the following way:

- For each leaf w the players define $\lceil\delta^{-1}\rceil^2$ rows. Rows are indexed by $(i, j) \in [[\lceil\delta^{-1}\rceil]] \times [[\lceil\delta^{-1}\rceil]]$ and are such that:
  - For each z, there are exactly $\lceil\delta^{-1}\rceil\dot{o}_\mathsf{A}(z\mid w, x)$ indices $i_z \in [[\lceil\delta^{-1}\rceil]]$ such that Alice outputs z on all rows of the form $(i_z, j), \forall j$.
  - For each z, there are exactly $\lceil\delta^{-1}\rceil\dot{o}_\mathsf{B}(z\mid w, y)$ indices $j_z \in [[\lceil\delta^{-1}\rceil]]$ such that Bob outputs z on all rows of the form $(i, j_z), \forall i$.
- The probability of the row (i, j) associated to the leaf w under the distribution μ is taken to be $p^{\mathsf{lf}}(w\mid x, y) \cdot \lceil\delta^{-1}\rceil^{-2}$, where $p^{\mathsf{lf}}(w\mid x, y)$ is the probability of ending in a leaf w in the original protocol Π. (μ is the unspecified distribution over $[M]$ in the statement of Theorem 6.)

The players then solve the $\mathbf{GapMAJ}\circ\mathbf{XOR}$ instance and output the result. Clearly, the above procedure has the previously claimed communication complexity. It remains to show that the players built a valid $\mathbf{GapMAJ}\circ\mathbf{XOR}$ instance whose result is $f(x, y)$, that is, picking a random row according to μ from this $\mathbf{GapMAJ}\circ\mathbf{XOR}$ instance gives outputs z_A and z_B on Alice and Bob's sides such that $z_\mathsf{A} \oplus z_\mathsf{B} = f(x, y)$ with probability $> \frac{1}{2}$.

1. In the original protocol Π, let $p^{\mathsf{out}}(z\mid x, y)$ be the probability of computing z (after the XOR), $p^{\mathsf{out}}(z\mid w, x, y)$ that same probability conditioned on the protocol ending in leaf w, and for all w let $o_\mathsf{A}(.\mid w, x)$ (resp. $o_\mathsf{B}(.\mid w, y)$) be

the distribution according to which Alice (resp. Bob) outputs once in leaf w. Then $p^{\text{out}}(z \mid x, y)$ can be expressed as:

$$p^{\text{out}}(z \mid x, y) = \sum_w p^{\text{lf}}(w \mid x, y) \cdot p^{\text{out}}(z \mid w, x, y)$$

$$= \sum_w p^{\text{lf}}(w \mid x, y) \cdot \sum_{\substack{z_A, z_B \\ z_A \oplus z_B = z}} o_A(z_A \mid w, x) \cdot o_B(z_B \mid w, x).$$

By correctness of the protocol, $p^{\text{out}}(f(x, y) \mid x, y) \geq 1 - \epsilon$.

2. Consider $p'^{\text{lf}}(. \mid x, y)$, $p'^{\text{out}}(. \mid x, y)$, $p'^{\text{out}}(. \mid w, x, y)$, $\dot{o}_A(. \mid w, x)$ and $\dot{o}_B(. \mid w, y)$ the approximations of the above quantities encountered when building our instance of **GapMAJ∘XOR**. The probability $p'^{\text{out}}(z \mid x, y)$ that a random row of our weighted **GapMAJ∘XOR** instance corresponds to a given z is:

$$p'^{\text{out}}(z \mid x, y) = \sum_w p'^{\text{lf}}(w \mid x, y) \cdot \sum_{\substack{z_A, z_B \\ z_A \oplus z_B = z}} \dot{o}_A(z_A \mid w, x) \cdot \dot{o}_B(z_B \mid w, x).$$

3. $p'^{\text{lf}}(. \mid x, y)$ is δ-close to $p^{\text{lf}}(. \mid x, y)$ in statistical distance. $\dot{o}_A(. \mid w, x)$ is point-wise δ-close to $o_A(. \mid w, x)$ (and similarly for $\dot{o}_B$ and o_B).

Consider $o_A \cdot o_B$ the distribution over $z \in \{0, 1\}^k$ defined by $o_A \cdot o_B(z) = \sum_{z'} o_A(z' \mid w, x) \cdot o_B(z \oplus z' \mid w, y)$. Similarly define $o_A \cdot \dot{o}_B$ and $\dot{o}_A \cdot \dot{o}_B$. Point 3 above implies that $\dot{o}_A \cdot \dot{o}_B$ is point-wise δ-close to $o_A \cdot \dot{o}_B$, which is itself point-wise δ-close to $o_A \cdot o_B$. One can check that $\dot{o}_A \cdot \dot{o}_B$ is point-wise 2δ-close to $o_A \cdot o_B$.

Using Lemma 5 (Appendix C.2) with $V \sim p^{\text{out}}$, $V' \sim p'^{\text{out}}$, $U \sim p^{\text{lf}}$, $U' \sim p'^{\text{lf}}$, $V_u \sim o_A \cdot o_B$ and $V'_u \sim \dot{o}_A \cdot \dot{o}_B$, we get that p and p' are point-wise 3δ-close. Since δ was taken to be $\frac{1}{4}\left(\frac{1}{2} - \epsilon\right)$, the probability that the random row of the **GapMAJ∘XOR** instance corresponds to $f(x, y)$ is: $p'^{\text{out}}(f(x, y)) \geq p^{\text{out}}(f(x, y)) - 3\delta \geq (1 - \epsilon) - \frac{3}{4}\left(\frac{1}{2} - \epsilon\right) = \frac{1}{2} + \frac{1}{4}\left(\frac{1}{2} - \epsilon\right) > \frac{1}{2}$.

7 Conclusion and Open Questions

We have presented output models that are tailored for non-Boolean functions. We hope that these will find many applications, including extensions to information complexity, a better understanding of direct sum problems, simulation protocols, new lower bounds for these models, to name just a few.

The Gap Majority composed with XOR problem (Definition 5) is closely related to the Gap Hamming Distance, extended to a large alphabet but with an additional promise, so lower bounds for **GHD** do not apply. We conjecture that its deterministic communication complexity is $\Omega(\epsilon N k)$, matching the trivial upper bound. If true, this would indicate that our randomness removal scheme (Theorem 6) is close to tight.

Acknowledgments. We thank Jérémie Roland and Sagnik Mukhopadhyay for helpful conversations and the anonymous referees for their numerous suggestions to improve the paper's presentation. This work was funded in part by the ANR grant FLITTLA ANR-21-CE48-0023.

A   Models for Large-Output Functions

One standard definition of communication complexity requires that at the end of the communication protocol, the output of the computation can be determined from the transcript of the communication and the public randomness (it is the model used in rectangle bounds). It is easy to find examples where such a definition makes it necessary to exchange much more communication than seems natural. For example,

Example 1. Consider the function $f : \{0,1\}^n \times \{0,1\}^n \to \{0,1\}^n, f(x,y) = x$, and assume we want to compute it with the promise $x = y$.

A protocol for f requires n bits of communication if the result of the protocol has to be apparent from the communication and the public randomness, even though both players know $f(x,y)$ right from the start.

In this section, we formally define the output models and prove separation results. The most interesting models are arguably the weakest ones: the one-out-of-two (Definition 12), the split (Definition 14), and the XOR models (Definition 4).

A.1   The Open Model

We start with the formal definition of our model which reveals the most information regarding the outcome of the computation. We call it the *open* model.

This is the model for which the partition bounds [28], in the form in which they appear in the literature, give lower bounds.

Definition 7 (Open computation). *A protocol Π is said to* openly *compute f with ϵ error if there exists a mapping $\mathcal{O} : \mathcal{T}_\pi \times \mathcal{R}^{\mathsf{pub}} \to \mathcal{Z}$ such that: for all $(x,y) \in \mathcal{X} \times \mathcal{Y}$,*

$$\Pr_{r,r_A,r_B} [\mathcal{O}(t_\pi, r) = f(x,y)] \geq 1 - \epsilon.$$

A.2   The Local Model

In the previous model, protocols are *revealing*, in the sense that the result of the computation can not be a secret only known to the players. In the *local* model, we only require that both players, at the end of the protocol, can output the value of the function (or the same valid output, in the case of a relation).

Definition 8 (Local computation). *A protocol Π is said to* locally *compute f with ϵ error if there exist two mappings $\mathcal{O}_A$ and $\mathcal{O}_B$ with $\mathcal{O}_A : \mathcal{T}_\pi \times \mathcal{R}^{\mathsf{pub}} \times \mathcal{R}_A \times \mathcal{X} \to \mathcal{Z}$ and similarly $\mathcal{O}_B : \mathcal{T}_\pi \times \mathcal{R}^{\mathsf{pub}} \times \mathcal{R}_B \times \mathcal{Y} \to \mathcal{Z}$ such that: for all $(x,y) \in \mathcal{X} \times \mathcal{Y}$,*

$$\Pr_{r,r_A,r_B} [\mathcal{O}_A(t_\pi, r, r_A, x) = \mathcal{O}_B(t_\pi, r, r_B, y) = f(x,y)] \geq 1 - \epsilon.$$

Bauer et al. [5] remarked that for total functions and relations, the deterministic open and local communication complexities are the same. Example 1 shows a separation between the deterministic complexities of computing a function with a promise.

For randomized communication, the local model is separated from the open model by the following total function, as seen in Theorem 7 (Fig. 2):

Definition 9 (Equality with output problem). $\mathbf{EQ}_n^{\mathsf{out}} : \{0,1\}^n \times \{0,1\}^n \to \{0,1\}^n \cup \{\top\}$ *is defined as*

$$\mathbf{EQ}_n^{\mathsf{out}}(x,y) = \begin{cases} x & if\, x = y \\ \top & otherwise \end{cases}$$

0	⊤	⊤	⊤	⊤	⊤	⊤	⊤
⊤	1	⊤	⊤	⊤	⊤	⊤	⊤
⊤	⊤	2	⊤	⊤	⊤	⊤	⊤
⊤	⊤	⊤	3	⊤	⊤	⊤	⊤
⊤	⊤	⊤	⊤	4	⊤	⊤	⊤
⊤	⊤	⊤	⊤	⊤	5	⊤	⊤
⊤	⊤	⊤	⊤	⊤	⊤	6	⊤
⊤	⊤	⊤	⊤	⊤	⊤	⊤	7

Fig. 2. The communication matrix of $\mathbf{EQ}_3^{\mathsf{out}}$

Theorem 7. $\forall f : \mathcal{X} \times \mathcal{Y} \to \mathcal{Z}$ *with* $k = \lceil \log|\mathcal{Z}| \rceil$ *and* $\epsilon > 0$,

$$R_\epsilon^{\mathsf{loc}}(f) \le R_\epsilon^{\mathsf{open}}(f) \le R_\epsilon^{\mathsf{loc}}(f) + k, \quad and$$
$$R_{1/4}^{\mathsf{loc}}(\mathbf{EQ}_n^{\mathsf{out}}) \le 4, \quad R_{1/4}^{\mathsf{open}}(\mathbf{EQ}_n^{\mathsf{out}}) \in \Omega(n).$$

We provide a full proof of this theorem, but because all the results of the form $R_\epsilon^{\mathcal{M}_1}(f) \le R_\epsilon^{\mathcal{M}_2}(f)$ or $R_\epsilon^{\mathcal{M}_1}(f) \le R_\epsilon^{\mathcal{M}_2}(f) + k$ for two models $\mathcal{M}_1$ and $\mathcal{M}_2$ can be proved by essentially the same proof, we will omit them in proofs of later similar theorems, only proving the separation result.

Proof (Proof of Theorem 7).

Proof of $R_\epsilon^{\mathsf{loc}}(f) \le R_\epsilon^{\mathsf{open}}(f)$**:** An open protocol for a function f is also a local protocol for f, as the players can take as mappings $\mathcal{O}_A$ and $\mathcal{O}_B$ the mapping $\mathcal{O}$ of the open protocol (ignoring both players' randomness and input).

Proof of $R_\epsilon^{\mathsf{open}}(f) \le R_\epsilon^{\mathsf{loc}}(f) + k$: Let Π be a local protocol for computing f with error at most ϵ. Consider Π', the protocol that consists of first running the protocol Π, and then Alice sends $\mathcal{O}_\mathsf{A}(t_\pi, r, r_\mathsf{A}, x)$ – what she would output at the end of Π to locally compute f – over the communication channel. This only requires k additional bits of communication. Now Π' is an open protocol, since an external observer can use the last k bits of the transcript as probable $f(x, y)$.

Both the lower bound and the upper bound on **EQ**$^\mathsf{out}$ directly follow from propositions and theorems previously seen in this manuscript.

Local model upper bound: The players apply the standard protocol for **EQ** (Proposition 2). If the strings are different, they output $\top$, otherwise Alice outputs x and Bob outputs y.

Open model lower bound: Consider the mapping $\mathcal{O}$ of the open protocol Π and notice that for all x, $\Pr_r[\mathcal{O}(\Pi(x, x, r), r) = x] \ge 3/4$. Consider that the players have a public randomness source $\mathcal{R}^\mathsf{pub}$ that is the uniformly random distribution over $\{0, 1\}^k$. Then the above statement implies $|\mathcal{O}^{-1}(x)| \ge \frac{3}{4} \cdot 2^k$. Since $\cup_x \mathcal{O}^{-1}(x) \subseteq \mathcal{T}_\pi \times \{0, 1\}^k$, we have that $\frac{3}{4} \cdot 2^k \cdot 2^n \le 2^{\mathsf{CC}(\Pi)} \cdot 2^k$ hence $\mathsf{CC}(\Pi) \ge n + \log\left(\frac{3}{4}\right) \in \Omega(n)$. This is also true when the source of public randomness is not a uniform distribution over $\{0, 1\}^k$ because of the fact that any non-uniform source of randomness can be simulated with arbitrary precision by a uniform source of randomness.

In the full version of the paper, we generalize this to show that any open protocol for a problem requires $\Omega(k)$ communication. This result follows from analyzing a lower bound known as the weak partition bound [22].

A.3   The Unilateral Models

In this section, we consider models of communication complexity where we require that at the end of the protocol, one player can output the value of the function (or a valid output, in the case of a relation). One-way problems are usually stated in this model.

Definition 10 (Unilateral computation).   *A protocol Π is said to Alice-compute f with ϵ error if there exists a mapping $\mathcal{O}_\mathsf{A} : \mathcal{T}_\pi \times \mathcal{R}^\mathsf{pub} \times \mathcal{R}_\mathsf{A} \times \mathcal{X} \to \mathcal{Z}$ such that: for all $(x, y) \in \mathcal{X} \times \mathcal{Y}$,*

$$\Pr_{r, r_\mathsf{A}, r_\mathsf{B}}[\mathcal{O}_\mathsf{A}(t_\pi, r, r_\mathsf{A}, x) = f(x, y)] \ge 1 - \epsilon.$$

Bob-computation is defined in a similar manner.

A protocol is said to unilaterally *compute f if it Alice-computes or Bob-computes f.*

Our definition of the unilateral model corresponds to a minimum of two models, each assigned to a player. The unilateral models are separated from each other and the local model by the following functions, where a given player possesses all the information about the output (Fig. 3).

Definition 11 (Unilateral identity problems). $\mathrm{id}_n^A : \{0,1\}^n \times \{0,1\}^n \to \{0,1\}^n$ *is defined as*

$$\mathrm{id}_n^A(x,y) = x$$

id_n^B *is defined similarly, with opposite roles for Alice and Bob.*

0	0	0	0	0	0	0	0
1	1	1	1	1	1	1	1
2	2	2	2	2	2	2	2
3	3	3	3	3	3	3	3
4	4	4	4	4	4	4	4
5	5	5	5	5	5	5	5
6	6	6	6	6	6	6	6
7	7	7	7	7	7	7	7

0	1	2	3	4	5	6	7
0	1	2	3	4	5	6	7
0	1	2	3	4	5	6	7
0	1	2	3	4	5	6	7
0	1	2	3	4	5	6	7
0	1	2	3	4	5	6	7
0	1	2	3	4	5	6	7
0	1	2	3	4	5	6	7

Fig. 3. The communication matrix of id_3^A and id_3^B

Theorem 8. $\forall f : \mathcal{X} \times \mathcal{Y} \to \mathcal{Z}$ *with* $k = \lceil \log|\mathcal{Z}| \rceil$, $\lambda \in [0,1]$ *and* $\epsilon > 0$

$$R_\epsilon^{\mathrm{uni}}(f) \le R_\epsilon^{\mathrm{loc}}(f) \le R_\epsilon^{\mathrm{open}}(f) \le R_\epsilon^{\mathrm{uni}}(f) + k,$$

$$D^{\mathrm{loc}}(f) \le D^A(f) + D^B(f), \qquad R_\epsilon^{\mathrm{loc}}(f) \le R_{\lambda\epsilon}^A(f) + R_{(1-\lambda)\epsilon}^B(f), \quad and$$

$$D^{\mathrm{uni}}(\mathrm{id}_n^A) = D^A(\mathrm{id}_n^A) = D^B(\mathrm{id}_n^B) = 0, \qquad R_{1/4}^{\mathrm{loc}}(\mathrm{id}_n^A) = R_{1/4}^{\mathrm{loc}}(\mathrm{id}_n^B) \in \Omega(n).$$

The first line also holds for relations, but the second line does not: consider as counterexample the relation $f : \{0,1\}^n \times \{0,1\}^n \to 2^{\{0,1\}^n}$, $f(x,y) = \{x,y\}$. This problem does not require any communication in both unilateral models ($D^A(f) = D^B(f) = 0$), but in the local model, the fact that the players need to agree on a single output makes the communication of order $\Omega(n)$ in both the deterministic and the randomized setting ($D^{\mathrm{loc}}(f) \ge R_\epsilon^{\mathrm{loc}}(f) \in \Omega(n)$).

Proof (Proof of Theorem 8). We omit the proof of the first two lines, that are only based on using the same protocol with the different proper mappings, or sending what one would output in a lower model over the communication channel.

We prove a slightly stronger result for the separation: that $R_{1/4}^B(\mathrm{id}_n^A) \in \Omega(n)$.

Alice model upper bound: Alice outputs her x, which requires no communication.

Bob model lower bound: Let us consider $D_{1/4}^B(\mathrm{id}_n^A, \mu)$ where μ is the uniform distribution. Bob has to output one of 2^n equiprobable answers. With communication C, Bob can only have 2^C different answers, so Bob is wrong with probability $\ge 1 - 2^{C-n}$. Since Bob is supposed to make less than $\frac{1}{4}$ error, we have: $C \ge n + \log\left(\frac{3}{4}\right)$, so $R_{1/4}^B(\mathrm{id}_n^A) \in \Omega(n)$.

A.4   The One-Out-of-Two Model

In the unilateral models, the player that outputs the result at the end of the protocol is fixed. In particular, it does not depend on the inputs. In the one-out-of-two model, we relax this condition: correctly computing a function in the one-out-of-two model corresponds to an execution such that at the end of the protocol:

- one player outputs a special symbol $\top \notin \mathcal{Z}$ (which corresponds to silence)
- the other players outputs $f(x,y)$.

Intuitively, we not only require that one of the players outputs the correct answer, but also that she knows that her output is probably correct, while the other knows that other player has a good answer to output. If we were only requiring that one player gives the correct answer, then all Boolean functions would be solved with zero communication in this model. In contrast, our model does not trivialize the communication complexity of Boolean functions.

Definition 12 (One-out-of-two computation). *A protocol Π is said to one-out-of-two* compute f *with ϵ error if there exist two mappings $\mathcal{O}_A$ and $\mathcal{O}_B$ with $\mathcal{O}_A : \mathcal{T}_\pi \times \mathcal{R}^{\text{pub}} \times \mathcal{R}_A \times \mathcal{X} \to \mathcal{Z} \cup \{\top\}$ and similarly $\mathcal{O}_B : \mathcal{T}_\pi \times \mathcal{R}^{\text{pub}} \times \mathcal{R}_B \times \mathcal{Y} \to \mathcal{Z} \cup \{\top\}$ such that: for all $(x,y) \in \mathcal{X} \times \mathcal{Y}$,*

$$\Pr_{r,r_A,r_B} [(\mathcal{O}_A(t_\pi, r, r_A, x), \mathcal{O}_B(t_\pi, r, r_B, y)) \in \{(f(x,y), \top), (\top, f(x,y))\}] \geq 1 - \epsilon.$$

The next proposition shows that any one-out-of-two protocol can be transformed into another one-out-of-two protocol of lesser or equal error and using only one additional bit of communication, such that at the end of the protocol it is always the case that exactly one player outputs a value in $\mathcal{Z}$ and the other stays silent (outputs $\top$).

Proposition 3. *Consider a function $f : \mathcal{X} \times \mathcal{Y} \to \mathcal{Z}$ and Π a one-out-of-two protocol for f with error $\epsilon > 0$ of communication cost C. Then there exists a one-out-of-two protocol Π' of communication cost $(C+1)$ that computes f with the same error but with mappings such that it is always the case that only one of them speaks at the end:*

$$\forall x, y, r_A, r_B, r, t_{\pi'} = \Pi'(x, y, r_A, r_B, r) :$$
$$(\mathcal{O}'_A(t_{\pi'}, r, r_A, x), \mathcal{O}'_B(t_{\pi'}, r, r_B, y)) \in (\mathcal{Z} \times \{\top\}) \cup (\{\top\} \times \mathcal{Z}).$$

Proof (Proof of Proposition 3). Let Π be a one-out-of-two protocol for f and $\mathcal{O}_A, \mathcal{O}_B$ the associated mappings. We define the protocol Π' to be a protocol that first behaves as Π (getting a transcript t_π) and when we hit a leaf in the protocol for Π, Alice sends a bit of communication to Bob following this rule:

- If $\mathcal{O}_A(t_\pi, r, r_A, x) = \top$, Alice sends 0 to Bob.
- Otherwise Alice sends 1 to Bob.

Let c_A be this control bit, sent by Alice in the last round of the new protocol Π'. Then, Alice keeps the same mapping $\mathcal{O}_A$ whereas Bob's new mapping $\mathcal{O}'_B$ is such that:

$$\mathcal{O}'_B(t_{\pi'}, r, r_B, y) = \begin{cases} \top & \text{if } c_a = 1, \\ \mathcal{O}_B(t_\pi, r, r_B, y) & \text{if } c_a = 0 \text{ and } \mathcal{O}_B(t_\pi, r, r_B, y) \neq \top, \\ z & \text{picked u.a.r. in } \mathcal{Z}, \text{ otherwise.} \end{cases}$$

Intuitively, Alice tells Bob whether to speak or not, and he obeys. Since the only cases where this changes what the players output is when they were going to both speak or both stay silent, the error does not increase in the process. We separate the one-out-of-two model from the unilateral models with the following function, where the first bit essentially determines which player possesses the output of the function (Fig. 4).

Definition 13 (Conditional identity problem). *The function* $\mathbf{CondId}_n :$ $\{0,1\}^n \times \{0,1\}^n \rightarrow \{0,1\}^n$ *is defined as*

$$\mathbf{CondId}_n(x, y) = \begin{cases} x & \text{if } x_0 = y_0, \\ y & \text{otherwise,} \end{cases}$$

where x_0 is the fist bit of x, similarly for y.

0	0	0	0	4	5	6	7
1	1	1	1	4	5	6	7
2	2	2	2	4	5	6	7
3	3	3	3	4	5	6	7
0	1	2	3	4	4	4	4
0	1	2	3	5	5	5	5
0	1	2	3	6	6	6	6
0	1	2	3	7	7	7	7

Fig. 4. The communication matrix of $\mathbf{CondId}_3$

Theorem 9. $\forall f : \mathcal{X} \times \mathcal{Y} \rightarrow \mathcal{Z}$ *with* $k = \lceil \log|\mathcal{Z}| \rceil$ *and* $\epsilon > 0$

$$R_\epsilon^{\text{1of2}}(f) \leq R_\epsilon^{\text{uni}}(f) \leq R_\epsilon^{\text{loc}}(f) \leq R_\epsilon^{\text{open}}(f) \leq R_\epsilon^{\text{1of2}}(f) + k + 1, \quad \text{and}$$
$$D^{\text{1of2}}(\mathbf{CondId}_n) \in O(1), \quad R_\epsilon^{\text{uni}}(\mathbf{CondId}_n) \in \Omega(n).$$

Proof (Proof of Theorem 9). Again, we focus on the separation result.

One-out-of-two model upper bound: Alice and Bob send each other x_0 and y_0. If $x_0 = y_0$, Alice outputs x, otherwise Bob outputs y. This only takes 2 bits of communication.

Unilateral model lower bound: Let us consider $D_{1/4}^{\mathsf{B}}(\mathbf{CondId}_n, \mu)$ where μ is the uniform distribution over (x, y) such that $x_0 = y_0$. Having received any given x, Bob has to output one of 2^{n-1} equiprobable answers. With communication C, Bob can only have 2^C different answers, so Bob is wrong with probability $\geq 1 - 2^{C-n+1}$. Since Bob is supposed to make less than $\frac{1}{4}$ error, we have: $C \geq n - 1 + \log\left(\frac{3}{4}\right)$, so $R_{1/4}^{\mathsf{B}}(\mathbf{CondId}_n) \in \Omega(n)$. By symmetry, we also have $R_{1/4}^{\mathsf{A}}(\mathbf{CondId}_n) \in \Omega(n)$, so $R_{1/4}^{\mathsf{uni}}(\mathbf{CondId}_n) \in \Omega(n)$.

A.5   The Split Model

In our next model, we allow the answer to be split between the two players. In the one-out-of-two model, one of the player had to output the full output, while the other stayed fully silent. In contrast, in the split model we allow both players to output part of the result. We only require that any given bit is output by exactly one player (the other player stays silent on this particular bit). In a valid split computation, it may be that the first bit of $f(x, y)$ is output by Alice, while the second one is output by Bob.

Definition 14 (Split computation). *A protocol Π is said to* split *compute f with ϵ error if there exist two mappings $\mathcal{O}_{\mathsf{A}}$ and $\mathcal{O}_{\mathsf{B}}$ with $\mathcal{O}_{\mathsf{A}} : \mathcal{T}_\pi \times \mathcal{R}^{\mathsf{pub}} \times \mathcal{R}_{\mathsf{A}} \times \mathcal{X} \to \{0, 1, *\}$ and similarly $\mathcal{O}_{\mathsf{B}} : \mathcal{T}_\pi \times \mathcal{R}^{\mathsf{pub}} \times \mathcal{R}_{\mathsf{B}} \times \mathcal{Y} \to \{0, 1, *\}$ such that: for all $(x, y) \in \mathcal{X} \times \mathcal{Y}$,*

$$\Pr_{r, r_{\mathsf{A}}, r_{\mathsf{B}}} [\mathcal{O}_{\mathsf{A}}(t_\pi, r, r_{\mathsf{A}}, x) \rtimes\kern-0.6em\ltimes \mathcal{O}_{\mathsf{B}}(t_\pi, r, r_{\mathsf{B}}, y) = f(x, y)] \geq 1 - \epsilon.$$

$$\text{where } (a \rtimes\kern-0.6em\ltimes b)_i \begin{cases} a_i & \text{if } b_i = *, \\ b_i & \text{if } a_i = *, \\ * & \text{otherwise.} \end{cases}$$

We call *weave* the binary operator $\rtimes\kern-0.6em\ltimes : \{0, 1, *\}^k \times \{0, 1, *\}^k \to \{0, 1, *\}^k$ described at the end of Definition 14, that recombines the parts split among the players.

To separate this model from the one-out-of-two model, we introduce a problem where the information about the output is naturally split between the two players (Fig. 5). We do so in a manner which makes computing this problem in the split model trivial, while the fact that one of the players must aggregate complete information about the output in the one-out-of-two model leads to a large amount of communication.

Definition 15 (Split identity problem). $\mathbf{SplitId}_n : \{0, 1\}^n \times \{0, 1\}^n \to \{0, 1\}^n$ *is defined as*

$$\mathbf{SplitId}_n(x, y)_i = \begin{cases} x_i & \text{if } i = 0 \mod 2, \\ y_i & \text{otherwise.} \end{cases}$$

0	0	2	2	0	0	2	2
1	1	3	3	1	1	3	3
0	0	2	2	0	0	2	2
1	1	3	3	1	1	3	3
4	4	6	6	4	4	6	6
5	5	7	7	5	5	7	7
4	4	6	6	4	4	6	6
5	5	7	7	5	5	7	7

Fig. 5. The communication matrix of $\mathbf{SplitId}_3$

Theorem 10. $\forall f : \mathcal{X} \times \mathcal{Y} \to \mathcal{Z}$ with $k = \lceil \log |\mathcal{Z}| \rceil$ and $\epsilon > 0$

$$R_\epsilon^{\mathsf{spl}}(f) \le R_\epsilon^{\mathsf{1of2}}(f) \le R_\epsilon^{\mathsf{spl}}(f) + \lfloor k/2 \rfloor + 1, \quad \text{and}$$
$$D^{\mathsf{spl}}(\mathbf{SplitId}_n) \in O(1), \qquad R_\epsilon^{\mathsf{1of2}}(\mathbf{SplitId}_n) \in \Omega(n).$$

Proof (Proof of Theorem 10). There is a small subtlety here, that the players may make the error of having too many or too few $*$ symbols at the end of the split protocol. Our proof that $R_\epsilon^{\mathsf{1of2}}(f) \le R_\epsilon^{\mathsf{spl}}(f) + \lfloor k/2 \rfloor + 1$ must not rely on this assumption: we can not, for instance, say "the player with fewer $*$ symbols speaks first", as this could result in an ambiguous protocol.

Proof of $R_\epsilon^{\mathsf{1of2}}(f) \le R_\epsilon^{\mathsf{spl}}(f) + \lfloor k/2 \rfloor + 1$: Let Π be an optimal split protocol. At the end of Π, Alice counts how many $*$ symbols she would output in the split protocol. She sends a 1 bit if that number is greater than $\lfloor k/2 \rfloor$, 0 otherwise. If she sent a 0, she then sends $\lfloor k/2 \rfloor$ bits, the first of which are, in order, the non-$*$ symbols she would have output, in order, in the split protocol. If she sent a 1, it is Bob that sends the first $\lfloor k/2 \rfloor$ non-$*$ bits that he would have output in the split protocol. In both cases, if there are not enough bits to send, the players append 0's as needed to reach $\lfloor k/2 \rfloor$ bits.
If it is Alice that is sending the non-$*$ symbols of her split output, then Bob will replace the $*$ symbols in his split output by the bits sent by Alice before outputting it as final step of the one-out-of-two protocol. The situation is symmetric if Bob is sending his non-$*$ bits. If there are too many or not enough bits to replace the $*$ symbols, the bits are discarded or we just put 0. This protocol is unambiguous (it does not rely on Alice and Bob not having exactly k stars together) and is correct in the one-out-of-two model whenever the original protocol was correct in the split model.

The separation result again bounds the size of rectangles that do not make too many errors.

Split model upper bound: Alice replaces odd positions in x by $*$, Bob replaces even positions of y by $*$. They then each output their resulting string, which computes $\mathbf{SplitId}_n(x, y)$ in the split model. This requires no communication.

One-out-of-two model lower bound: Consider $D^{1of2}_{1/4}(\mathbf{SplitId}_n, \mu)$, where μ is the uniform distribution over (x, y) such that $x_i = 0$ for odd i and $y_i = 0$ for even i, and consider the communication matrix $\widetilde{M}_{\mathbf{SplitId}_n}$ of this reduced (but still total) problem. This reduces the number of inputs to 2^n. Let Π be an optimal deterministic one-out-of-two protocol of communication $C = D^{1of2}_{1/4}(\mathbf{SplitId}_n, \mu)$.

Π partitions the communication matrix $\widetilde{M}_{\mathbf{SplitId}_n}$ with striped rectangles: in any given rectangle, the output of the one-out-of-two protocol can depend on either the row or on the column, but not both. But for our problem, every cell of the communication matrix has a different output, so any rectangle of width and height both at least 2 makes an error in at least half its cells.

A rectangle of width or height at most 1 contains at most $2^{n/2}$ elements, therefore at most $2^{C+n/2}$ elements are covered by a rectangle that makes less than half error on its elements. Therefore at least $2^n - 2^{C+n/2}$ inputs are covered by rectangles with at least $1/2$ error, so Π makes error at least $2^{-n} \cdot \frac{1}{2}(2^n - 2^{C+n/2})$. This error has to be less than $\frac{1}{4}$, so:

$$\frac{1}{4} \geq 2^{-n} \cdot \frac{1}{2}\left(2^n - 2^{C+n/2}\right) \Rightarrow C \geq n/2 - 1$$

Which completes our proof that $R^{1of2}_{1/4}(\mathbf{SplitId}_n) \geq D^{1of2}_{1/4}(\mathbf{SplitId}_n, \mu) \in \Omega(n)$.

The XOR Model. In our final model, the players both output a k bit string at the end of the protocol. A computation correctly computes the value of $f(x, y)$ when the bit-wise XOR of the two strings is equal to $f(x, y)$.

Definition 4 (XOR computation). *Consider a function f whose output set is $\mathcal{Z} = \{0, 1\}^k$. A protocol Π is said to XOR-compute f with ϵ error if there exist two mappings $\mathcal{O}_A$ and $\mathcal{O}_B$ with $\mathcal{O}_A : \mathcal{T}_\pi \times \mathcal{R}^{pub} \times \mathcal{R}_A \times \mathcal{X} \to \{0, 1\}^k$ and similarly $\mathcal{O}_B : \mathcal{T}_\pi \times \mathcal{R}^{pub} \times \mathcal{R}_B \times \mathcal{Y} \to \{0, 1\}^k$ such that for all $(x, y) \in \mathcal{X} \times \mathcal{Y}$, $\Pr_{r, r_A, r_B}[\mathcal{O}_A(t_\pi, r, r_A, x) \oplus \mathcal{O}_B(t_\pi, r, r_B, y) = f(x, y)] \geq 1 - \epsilon$.*

The XOR model is separated from the one-out-of-two model by the following function (Fig. 6):

Definition 16. XOR$_n$: $\{0, 1\}^n \times \{0, 1\}^n \to \{0, 1\}^n$ *is defined by* $\mathbf{XOR}_n(x, y) = (x_i \oplus y_i)_{i \in [n]}$

Theorem 11. $\forall f : \mathcal{X} \times \mathcal{Y} \to \mathcal{Z}$ *with* $k = \lceil \log|\mathcal{Z}| \rceil$ *and* $\epsilon > 0$,

$$R^{xor}_\epsilon(f) \leq R^{spl}_\epsilon(f) \leq R^{1of2}_\epsilon(f) \leq R^{uni}_\epsilon(f) \leq R^{xor}_\epsilon(f) + k, \quad and$$
$$D^{xor}(\mathbf{XOR}_n) = 0, \quad R^{spl}_\epsilon(\mathbf{XOR}_n) \in \Omega(n).$$

0	1	2	3	4	5	6	7
1	0	3	2	5	4	7	6
2	3	0	1	6	7	4	5
3	2	1	0	7	6	5	4
4	5	6	7	0	1	2	3
5	4	7	6	1	0	3	2
6	7	4	5	2	3	0	1
7	6	5	4	3	2	1	0

Fig. 6. The communication matrix of $\mathbf{XOR}_3$

Proof (Proof of Theorem 11).

XOR model upper bound: Alice and Bob can just each output their input, which requires no communication.

Split model lower bound: Let us consider $D^{\mathsf{spl}}_{1/4}(\mathbf{XOR}_n, \mu)$ where μ is the uniform distribution. Let Π be an optimal deterministic one-out-of-two protocol of communication $C = D^{\mathsf{spl}}_{1/4}(\mathbf{XOR}_n, \mu)$.

Π partitions the communication matrix $M_{\mathbf{XOR}_n}$ into 2^C rectangles. Let us first assume that in each rectangle, each bit of the output is output by a fixed player. We will see later that our argument still holds without this assumption.

In each of the 2^C rectangles, one of the players has to output less than $n/2$ bits of the output. Let us consider a rectangle where Bob outputs at most half the bits of the output. Then, on a given row of this rectangle, there can be at most $2^{n/2}$ different outputs. But the $\mathbf{XOR}_n$ problem is such that on a given row, all cells have a different output. We will argue that this bounds the size of the rectangles that do not make a lot of error.

Let a rectangle contain at least $2^{3n/2+1}$ elements. Since a row or column contains at most 2^n elements, such a rectangle contains at least $2^{n/2+1}$ rows and columns. Therefore, the player that outputs at most half the bits of the output in the split model will output at most $2^{n/2}$ different strings on a given row or column that contains more than $2^{n/2+1}$ different values, so the rectangle has error on at least half of its elements.

If the players do not always split the outputs bits in the same way, consider the largest set of rows such that Alice outputs a given subset of the output bits, and the largest set of columns such that Bob outputs a given subset of the output bits. If the sets of output bits that Alice and Bob output on those rows and columns are not the complement of each other, the rectangle is in error on at least half of its elements. If the sets correctly partition the output bits, we do the same argument as before: let us assume that Bob outputs at most half the bits in the subrectangle we defined. Then no more than 2^n

cells can be correct in any row of this subrectangle, and rows outside of the subrectangle are also mostly error, therefore the rectangle has error on at least half of its elements.

At most $2^{C+3n/2+1}$ elements are in rectangles with error strictly less than half, so the error made by the protocol is at least $\frac{1}{2} \cdot 2^{-2n}\left(2^{2n} - 2^{C+3n/2+1}\right)$. The error has to be less than $\frac{1}{4}$, so:

$$C \geq n/2 - 2$$

Which completes our proof that $R_{1/4}^{\mathsf{spl}}(\mathbf{XOR}_n) \geq D_{1/4}^{\mathsf{spl}}(\mathbf{XOR}_n, \mu) \in \Omega(n)$.

A.6   Relations Between Models

The next proposition summarizes the relations between models in Theorems 7 to 11.

Proposition 4. $\forall f : \mathcal{X} \times \mathcal{Y} \to \mathcal{Z}$ with $k = \lceil \log|\mathcal{Z}| \rceil$ and $\epsilon > 0$ we have:

$$R_\epsilon^{\mathsf{open}}(f) \geq R_\epsilon^{\mathsf{loc}}(f) \geq \max\left(R_\epsilon^{\mathsf{A}}(f), R_\epsilon^{\mathsf{B}}(f)\right)$$
$$\geq \min\left(R_\epsilon^{\mathsf{A}}(f), R_\epsilon^{\mathsf{B}}(f)\right) = R_\epsilon^{\mathsf{uni}}(f)$$
$$\geq R_\epsilon^{\mathsf{1of2}}(f) \geq R_\epsilon^{\mathsf{spl}}(f) \geq R_\epsilon^{\mathsf{xor}}(f) \qquad (1)$$
$$R_{2\epsilon}^{\mathsf{loc}}(f) \leq R_\epsilon^{\mathsf{A}}(f) + R_\epsilon^{\mathsf{B}}(f) \qquad (2)$$
$$R_\epsilon^{\mathsf{open}}(f) \leq R_\epsilon^{\mathsf{uni}}(f) + k \qquad (3)$$
$$R_\epsilon^{\mathsf{open}}(f) \leq R_\epsilon^{\mathsf{1of2}}(f) + k + 1 \qquad (4)$$
$$R_\epsilon^{\mathsf{1of2}}(f) \leq R_\epsilon^{\mathsf{spl}}(f) + \lceil k/2 \rceil + 1. \qquad (5)$$
$$R_\epsilon^{\mathsf{uni}}(f) \leq R_\epsilon^{\mathsf{xor}}(f) + k. \qquad (6)$$

The same statements hold for deterministic communication and communication with private randomness only. All statements except subproposition 2 also hold for relations and nondeterministic communication.

Proposition 4 shows that the models form a natural hierarchy and can be ordered from most to least communication intensive. We also summarize this hierarchy in Fig. 1, in the main text. This figure also displays separating problems other than those in this section, in Appendix.

B   Summary of Our Results

In this section, we summarize the results in this paper. Table 1 summarizes the problems we have studied which show gaps between the different output models. Table 2 summarizes the bounds on **GapMAJ∘XOR** in various models. Table 3 summarizes error reduction bounds and derandomization.

The upper bounds on the *Gap Majority* problem, are summarized in Table 2. We conjecture a matching lower bound to our stated deterministic $O(\epsilon Nk)$ upper

Table 1. Summary of the communication complexities of our separating problems in all models. The definitions of the problems and the proofs are in Appendix A and the full version of this paper [23]. In this table, n is the input length, k is the output length, $\mathcal{M}$ is an output model, $\mathcal{M} \in \{\text{open}, \text{loc}, \text{A}, \text{B}, \text{uni}, \text{1of2}, \text{xor}\}$, and t is the Hamming weight of an instance.

	open		local	unilateral	1-out-of-2	XOR
$\mathbf{EQ}_n^{\text{out}}$	$R_{1/3}^{\mathcal{M}} \in \Theta(n)$		$R_{1/3}^{\mathcal{M}} \in \Theta(1)$			
$t\mathbf{-INT}_n$	$R_{1/3}^{\mathcal{M}} \in \Theta(t \cdot \log(n))$		$R_{1/3}^{\mathcal{M}} \in \Theta(t)$			
$\mathbf{id}_n^{\text{A}}$	$R_{1/3}^{\mathcal{M}} \in \Theta(n)$			$D^{\mathcal{M}} = 0$		
$\mathbf{CondId}_n$	$R_{1/3}^{\mathcal{M}} \in \Theta(n)$				$D^{\mathcal{M}} = 2$	
$\mathbf{MAX}_n$	$R_{1/3}^{\mathcal{M}} \in \Theta(n)$				$R_{1/3}^{\mathcal{M}} \in \Theta(\log(n))$	
$t\mathbf{-FtFD}_n$	$R_{1/3}^{\mathcal{M}} \in \Theta(\log(n))$				$R_{1/3}^{\mathcal{M}} \in \Theta(\log(t) + \log\log(n))$	
$\mathbf{XOR}_n$	$R_{1/3}^{\mathcal{M}} \in \Theta(n)$					$D^{\mathcal{M}} = 0$
$\mathbf{GapMAJ}_{N,k,1/3}\circ\mathbf{XOR}$	$R_{1/3}^{\mathcal{M}} \in \Theta(k)$					$R_{1/3}^{\mathcal{M}} = 0$
$\mathbf{GapMAJ}_{N,k,2/5}\circ\mathbf{XOR}$	$R_{1/3}^{\mathcal{M}} \in \Theta(k)$					$R_{1/3}^{\mathcal{M}} \in O(1)$

Table 2. Upper bounds on $\mathbf{GapMAJ}\circ\mathbf{XOR}$. In this table, N, k, ϵ are the parameters of the Gap Majority problem, and ϵ' is the error parameter. Proofs in the full version of this paper [23].

		Upper bounds
$\epsilon' \geq \epsilon$	$R_{\epsilon'}^{\text{xor}}$	0
	$R_{\epsilon'}^{\text{xor,priv}}$	$\log(N)$
	$R_{\epsilon'}^{\text{open}}$	$2k$
	$R_{\epsilon'}^{\text{open,priv}}$	$2k + \log(N)$
$0 < \epsilon' < \epsilon$	$R_{\epsilon'}^{\text{xor}}$	$O\left(\min\left(C_{\epsilon,\epsilon'}, N + \log\left(\frac{1}{\epsilon'}\right)\right)\right)$
$\epsilon' = 0$	D^{uni}	$(2\epsilon N + 1)k$

bound. Studying the communication complexity of this problem is of theoretical interest, as we have seen in this paper that fundamental results in communication complexity, namely error reduction and derandomization, are related to the $\mathbf{GapMAJ}\circ\mathbf{XOR}$ problem in the XOR model. Improving the deterministic upper bound on $\mathbf{GapMAJ}\circ\mathbf{XOR}$ would yield a better derandomization result through Theorem 6. Similarly, improving the randomized upper bounds could improve error reduction through Lemma 2. Conversely, considering that we have an upper bound of $\log(N)$ on the private coin XOR communication complexity of $\mathbf{GapMAJ}\circ\mathbf{XOR}$, proving a $\Omega(Nk)$ lower bound on its deterministic communication complexity would indicate that our derandomization theorem in the XOR model (Theorem 6) is close to tight.

Table 3. Summary of our error reduction and derandomization schemes. In all statements above, f is a function whose output length is k, ϵ is the starting error parameter, ϵ' is the target error parameter, $R = R_\epsilon^\mathcal{M}(f)$, $C_{\epsilon,\epsilon'} \in O\left(\epsilon(\frac{1}{2} - \epsilon)^{-2}\log\left(\frac{1}{\epsilon'}\right)\right)$ and $C'_{\epsilon,\epsilon'} \in O\left(\log\left(\frac{1}{\epsilon'}\right) + \log\left(\frac{1}{\frac{1}{2}-\epsilon}\right)\right)$.

Error reduction		
model	Upper bounds	(condition)
open local unilateral	$R_{\epsilon'}(f) \leq C_{\epsilon,\epsilon'} \cdot R_\epsilon(f)$	
1-out-of-2	$R_{\epsilon'}(f) \leq C_{\epsilon,\epsilon'}(R_\epsilon(f) + 1) + C'_{\epsilon,\epsilon'}$	
split	$R_{\epsilon'}(f) \leq C_{\epsilon,\epsilon'}R_\epsilon(f) + O(C_{\epsilon,\epsilon'})$	
XOR	$R_{\epsilon'}(f) \leq C_{\epsilon,\epsilon'}R_\epsilon(f) + O(C_{\epsilon,\epsilon'})$	
	$R_{\epsilon'}(f) \leq 50\ln\left(\frac{12}{\epsilon'}\right)R_\epsilon(f) + C_{\epsilon,\epsilon'}R_\epsilon(g) + O(C_{\epsilon,\epsilon'} + \log(k))$	$(f = g^{\otimes k})$

Derandomization		
model	Upper bounds	(condition)
open local	$D(f) \in O\left(2^R\left(R + \log(\frac{1}{\frac{1}{2}-\epsilon})\right)\right)$	
unilateral	$D(f) \in O\left(2^R\left(R + \log(\frac{1}{\frac{1}{2}-\epsilon})\right)\right)$	
1-out-of-2	$D(f) \in O\left(2^R\left(R + \log(\frac{1}{\frac{1}{2}-\epsilon})\right)\right)$	
	$D(f) \in O\left(2^R\left(R + \log(\frac{1}{\frac{1}{2}-\epsilon})\right) + \log(k)\right)$	
split	$D(f) \in O\left(2^R\left(R + \log(\frac{1}{\frac{1}{2}-\epsilon})\right) + k\right)$	
	$D(f) \in O\left(2^R\left(R + \log(\frac{1}{\frac{1}{2}-\epsilon})\right) + 2^R\left(\frac{1}{2} - \epsilon\right)^{-2}k\right)$	
XOR	$D(f) \in O\left(2^R\left(R + \log(\frac{1}{\frac{1}{2}-\epsilon})\right) + 2^R\left(\frac{1}{2} - \epsilon\right)^{-2}k\right)$	

C   Technical Lemmas

C.1   Proof of the Random Graph Lemma

The proof of the random graph lemma stated in Sect. 5.1 and used to solve **GapMAJ∘XOR** is a simple variation of a result of Erdős and Rényi [19]. The result they proved is in a model of random graphs where a fixed number of edges are picked randomly from the set of all possible edges, while we are interested in a model of random graphs where each edge is picked with a fixed probability p independently of other edges. The two models are known to have essentially similar asymptotic behaviours. Readers interested in the theory of random graphs might refer to [8].

Proof (Proof of Lemma 3). We observe as in [19] that if no connected component of more than $(1-\alpha)n$ vertices exists, then we can partition the vertices into two disconnected sets of size n_0 and n_1 such that $\frac{\alpha}{2}n \leq n_0 \leq n_1 \leq (1-\frac{\alpha}{2})n$.

Given a partition of the vertices into sets of size n_0 and n_1, the probability that those two sets are disconnected is $(1-p(n))^{n_0 n_1}$. With $p(n) = \frac{c}{n}$, and since there are less than 2^n possible partitions, the probability that there is no connected component of more than $(1-\alpha)n$ vertices is bounded by:

$$2^n\left(1-\frac{c}{n}\right)^{n_0 n_1} \leq 2^n e^{-c\frac{n_0 n_1}{n}} \leq 2^n e^{-c\frac{\alpha}{2}\left(1-\frac{\alpha}{2}\right)n} = e^{\left(\ln(2)-\frac{\alpha}{2}\left(1-\frac{\alpha}{2}\right)c\right)n}$$

C.2   Distribution Distance Lemma

The following lemma is used in Sect. 6.

Lemma 5. *Let U and V be random variables over their respective domain $\mathcal{U}$ and $\mathcal{V}$. For all $u \in \mathcal{U}$, let us consider $V_{U=u}$ the random variable V conditioned on the event $[U = u]$. Assume there exists two constants δ_U and δ_V and two random variables U' and V' over the same domains as U and V such that:*

$$\Delta(U, U') \leq \delta_U \qquad \forall u \in \mathcal{U} : d_\infty(V_{U=u}, V'_{U'=u}) \leq \delta_V.$$

Then:

$$d_\infty(V, V') \leq \delta_U + \delta_V.$$

Proof (Proof of Lemma 5). Let us show that $\forall v \in \mathcal{V}, |\Pr[V = v] - \Pr[V' = v]| \leq \delta_U + \delta_V$. Fix an arbitrary $v \in \mathcal{V}$, then the probabilities $\Pr[V = v]$ and $\Pr[V' = v]$ can be written as:

- $\Pr[V = v] = \sum_{u \in \mathcal{U}} \Pr[U = u] \cdot \Pr[V = v \mid U = u],$
- $\Pr[V' = v] = \sum_{u \in \mathcal{U}} \Pr[U' = u] \cdot \Pr[V' = v \mid U' = u].$

Hence using our two hypotheses above we get:

$$\Pr[V = v] - \Pr[V' = v]$$
$$= \sum_{u \in \mathcal{U}} (\Pr[U = u] \cdot \Pr[V = v \mid U = u] - \Pr[U' = u] \cdot \Pr[V' = v \mid U' = u])$$
$$\leq \sum_{u \in \mathcal{U}} ((\Pr[U = u] - \Pr[U' = u]) \Pr[V = v \mid U = u] + \delta_V \Pr[U' = u])$$
$$\leq \sum_{u \in \mathcal{U}: \Pr[U=u] > \Pr[U'=u]} (\Pr[U = u] - \Pr[U' = u]) + \delta_V$$
$$\leq \delta_U + \delta_V.$$

We can prove $\Pr[V = v] - \Pr[V' = v] \geq -(\delta_U + \delta_V)$ following the same proof method, and combining the two we get the desired result:

$$\forall v \in \mathcal{V} : |\Pr[V = v] - \Pr[V' = v]| \leq \delta_U + \delta_V.$$

References

1. Aaronson, S.: The complexity of agreement. In: Proceedings of the Thirty-Seventh Annual ACM Symposium on Theory of Computing, pp. 634–643. ACM (2005). https://doi.org/10.1145/1060590.1060686
2. Ambainis, A., Buhrman, H., Gasarch, W., Kalyanasundaram, B., Torenvliet, L.: The communication complexity of enumeration, elimination, and selection. J. Comput. Syst. Sci. **63**(2), 148–185 (2001). https://doi.org/10.1006/jcss.2001.1761
3. Bar-Yossef, Z., Jayram, T.S., Kumar, R., Sivakumar, D.: An information statistics approach to data stream and communication complexity. J. Comput. Syst. Sci. **68**(4), 702–732 (2004). https://doi.org/10.1016/j.jcss.2003.11.006
4. Barak, B., Braverman, M., Chen, X., Rao, A.: How to compress interactive communication. SIAM J. Comput. **42**(3), 1327–1363 (2013). https://doi.org/10.1137/100811969
5. Bauer, B., Moran, S., Yehudayoff, A.: Internal compression of protocols to entropy. In: Approximation, Randomization, and Combinatorial Optimization. Algorithms and Techniques, APPROX/RANDOM 2015, pp. 481–496 (2015). https://doi.org/10.4230/LIPIcs.APPROX-RANDOM.2015.481
6. Beimel, A., Ben Daniel, S., Kushilevitz, E., Weinreb, E.: Choosing, agreeing, and eliminating in communication complexity. Comput. Complex. **23**(1), 1–42 (2013). https://doi.org/10.1007/s00037-013-0075-7
7. Bell, J.S.: On the Einstein Podolsky Rosen paradox. Physics **1**, 195–200 (1964). https://doi.org/10.1103/PhysicsPhysiqueFizika.1.195
8. Bollobás, B.: Random Graphs. Cambridge Studies in Advanced Mathematics, 2nd edn. Cambridge University Press, Cambridge (2001). https://doi.org/10.1017/CBO9780511814068
9. Braverman, M.: Interactive information complexity. SIAM J. Comput. **44**(6), 1698–1739 (2015). https://doi.org/10.1137/130938517
10. Braverman, M., Kol, G.: Interactive compression to external information. In: Proceedings of the 50th Annual ACM SIGACT Symposium on Theory of Computing, STOC, pp. 964–977 (2018). https://doi.org/10.1145/3188745.3188956
11. Braverman, M., Rao, A.: Information equals amortized communication. IEEE Trans. Inf. Theory **60**(10), 6058–6069 (2014). https://doi.org/10.1109/TIT.2014.2347282
12. Braverman, M., Rao, A., Weinstein, O., Yehudayoff, A.: Direct products in communication complexity. In: 54th Annual IEEE Symposium on Foundations of Computer Science, FOCS, pp. 746–755 (2013). https://doi.org/10.1109/FOCS.2013.85
13. Brody, J., Buhrman, H., Koucký, M., Loff, B., Speelman, F., Vereshchagin, N.: Towards a reverse Newman's theorem in interactive information complexity. Algorithmica **76**(3), 749–781 (2016). https://doi.org/10.1007/s00453-015-0112-9
14. Brody, J., Chakrabarti, A., Kondapally, R., Woodruff, D.P., Yaroslavtsev, G.: Beyond set disjointness: the communication complexity of finding the intersection. In: ACM Symposium on Principles of Distributed Computing, PODC 2014, pp. 106–113 (2014). https://doi.org/10.1145/2611462.2611501
15. Brody, J., Chakrabarti, A., Kondapally, R., Woodruff, D.P., Yaroslavtsev, G.: Certifying equality with limited interaction. Algorithmica **76**(3), 796–845 (2016). https://doi.org/10.1007/s00453-016-0163-6
16. Buhrman, H., Cleve, R., Massar, S., de Wolf, R.: Nonlocality and communication complexity. Rev. Mod. Phys. **82**, 665–698 (2010). https://doi.org/10.1103/RevModPhys.82.665

17. Buhrman, H., Cleve, R., Wigderson, A.: Quantum vs. classical communication and computation. In: Proceedings of the Thirtieth Annual ACM Symposium on Theory of Computing, pp. 63–68. ACM (1998). https://doi.org/10.1145/276698.276713

18. Chakrabarti, A., Regev, O.: An optimal lower bound on the communication complexity of gap-hamming-distance. SIAM J. Comput. **41**(5), 1299–1317 (2012). https://doi.org/10.1137/120861072

19. Erdős, P., Rényi, A.: On the evolution of random graphs. In: Publication of the Mathematical Institute of the Hungarian Academy of Sciences, pp. 17–61 (1960)

20. Feder, T., Kushilevitz, E., Naor, M., Nisan, N.: Amortized communication complexity. SIAM J. Comput. **24**(4), 736–750 (1995). https://doi.org/10.1137/S0097539792235864

21. Feige, U., Raghavan, P., Peleg, D., Upfal, E.: Computing with noisy information. SIAM J. Comput. **23**(5), 1001–1018 (1994). https://doi.org/10.1137/S0097539791195877

22. Fontes, L., Jain, R., Kerenidis, I., Laplante, S., Laurière, M., Roland, J.: Relative discrepancy does not separate information and communication complexity. ACM Trans. Comput. Theory **9**(1), 4:1–4:15 (2016). https://doi.org/10.1145/2967605

23. Fontes, L., Laplante, S., Laurière, M., Nolin, A.: Communication complexity of functions with large outputs. Technical report 2304.00391. arXiv (2023). https://arxiv.org/abs/2304.00391

24. Ganor, A., Kol, G., Raz, R.: Exponential separation of information and communication for Boolean functions. J. ACM **63**(5), 46:1–46:31 (2016). https://doi.org/10.1145/2907939

25. Ganor, A., Kol, G., Raz, R.: Exponential separation of communication and external information. SIAM J. Comput. **50**(3) (2021). https://doi.org/10.1137/16M1096293

26. Huang, D., Pettie, S., Zhang, Y., Zhang, Z.: The communication complexity of set intersection and multiple equality testing. SIAM J. Comput. **50**(2), 674–717 (2021). https://doi.org/10.1137/20M1326040

27. Indyk, P., Woodruff, D.P.: Tight lower bounds for the distinct elements problem. In: Proceedings 44th Symposium on Foundations of Computer Science (FOCS 2003), pp. 283–288 (2003). https://doi.org/10.1109/SFCS.2003.1238202

28. Jain, R., Klauck, H.: The partition bound for classical communication complexity and query complexity. In: Proceedings of the 25th Annual IEEE Conference on Computational Complexity, CCC 2010, pp. 247–258 (2010). https://doi.org/10.1109/CCC.2010.31

29. Kalyanasundaram, B., Schnitger, G.: The probabilistic communication complexity of set intersection. SIAM J. Discrete Math. **5**(4), 545–557 (1992). https://doi.org/10.1137/0405044

30. Kol, G.: Interactive compression for product distributions. In: Proceedings of the 48th Annual ACM SIGACT Symposium on Theory of Computing (STOC), pp. 987–998. ACM (2016). https://doi.org/10.1145/2897518.2897537

31. Kozachinskiy, A.: Some bounds on communication complexity of gap hamming distance. Technical Report 1511.08854, arXiv (2015). https://doi.org/10.48550/arXiv.1511.08854

32. Kushilevitz, E., Nisan, N.: Communication Complexity. Cambridge University Press, New York (1997). https://doi.org/10.1017/CBO9780511574948

33. Nisan, N.: The communication complexity of threshold gates. In: Combinatorics, Paul Erdős is eighty, Bolyai Society Mathematical Studies, vol. 1, pp. 301–315. János Bolyai Mathematical Society (1993)

34. Orlitsky, A.: Worst-case interactive communication I: two messages are almost optimal. IEEE Trans. Inform. Theory **36**(5), 1111–1126 (1990). https://doi.org/10.1109/18.57210

35. Orlitsky, A.: Worst-case interactive communication - II: two messages are not optimal. IEEE Trans. Inform. Theory **37**(4), 995–1005 (1991). https://doi.org/10.1109/18.86993

36. Palazuelos, C., Vidick, T.: Survey on nonlocal games and operator space theory. J. Math. Phys. **57**(1), 015220 (2016). https://doi.org/10.1063/1.4938052

37. Rao, A., Sinha, M.: Simplified separation of information and communication. Theory Comput. **14**(1), 1–29 (2018). https://doi.org/10.4086/toc.2018.v014a020

38. Rao, A., Yehudayoff, A.: Communication Complexity. Cambridge University Press, Cambridge (2020). https://doi.org/10.1017/9781108671644

39. Razborov, A.A.: On the distributional complexity of disjointness. Theor. Comput. Sci. **106**(2), 385–390 (1992). https://doi.org/10.1016/0304-3975(92)90260-M

40. Sherstov, A.: Compressing interactive communication under product distributions. SIAM J. Comput. **47**(2), 367–419 (2018). https://doi.org/10.1137/16M109380X

41. Sherstov, A.A.: The communication complexity of gap hamming distance. Theory Comput. **8**(1), 197–208 (2012). https://doi.org/10.4086/toc.2012.v008a008

42. Vidick, T.: A concentration inequality for the overlap of a vector on a large set, with application to the communication complexity of the gap-hamming-distance problem. Chicago J. Theor. Comput. Sci. **18**, 1–12 (2012). https://doi.org/10.4086/cjtcs.2012.001

43. Viola, E.: The communication complexity of addition. Combinatorica **35**(6), 703–747 (2015). https://doi.org/10.1007/s00493-014-3078-3

44. Watson, T.: Communication complexity with small advantage. In: 33rd Computational Complexity Conference, CCC, pp. 9:1–9:17 (2018). https://doi.org/10.4230/LIPIcs.CCC.2018.9

On the Power of Threshold-Based Algorithms for Detecting Cycles in the CONGEST Model

Pierre Fraigniaud[1], Maël Luce[1(✉)], and Ioan Todinca[2]

[1] Institut de Recherche en Informatique Fondamentale (IRIF),
CNRS and Université Paris Cité, Paris, France
`luce@irif.fr`
[2] Laboratoire d'Informatique Fondamentale d'Orléans (LIFO),
Université d'Orléans, Orléans, France

Abstract. It is known that, for every $k \geq 2$, C_{2k}-freeness can be decided by a generic Monte-Carlo algorithm running in $n^{1-1/\Theta(k^2)}$ rounds in the CONGEST model. For $2 \leq k \leq 5$, faster Monte-Carlo algorithms do exist, running in $O(n^{1-1/k})$ rounds, based on upper bounding the number of messages to be forwarded, and aborting search sub-routines for which this number exceeds certain thresholds. We investigate the possible extension of these *threshold-based* algorithms, for the detection of larger cycles. We first show that, for every $k \geq 6$, there exists an infinite family of graphs containing a $2k$-cycle for which any threshold-based algorithm fails to detect that cycle. Hence, in particular, neither C_{12}-freeness nor C_{14}-freeness can be decided by threshold-based algorithms. Nevertheless, we show that $\{C_{12}, C_{14}\}$-freeness can still be decided by a threshold-based algorithm, running in $O(n^{1-1/7}) = O(n^{0.857\cdots})$ rounds, which is faster than using the generic algorithm, which would run in $O(n^{1-1/22}) \simeq O(n^{0.954\cdots})$ rounds. Moreover, we exhibit an infinite collection of families of cycles such that threshold-based algorithms can decide $\mathcal{F}$-freeness for every $\mathcal{F}$ in this collection.

Keywords: Cycle-Freeness · Distributed Computing · CONGEST model

1 Introduction

1.1 Objective

Graphs excluding a fixed family $\mathcal{F}$ of graphs, whether it be as subgraphs, induced subgraphs, topological subgraphs, or minors, play a huge role in theoretical computer science, especially in graph theory as well as in algorithm design and complexity, from standard and parametrized complexity, to the design of approximation and exact algorithms. Famous examples in structural graph theory are

P. Fraigniaud—Additional support from the ANR Project DUCAT (ref. ANR-20-CE48-0006).

M. Luce—Additional support from the ANR Project QuDATA (ref. ANR-18-CE47-0010).

S. Rajsbaum et al. (Eds.): SIROCCO 2023, LNCS 13892, pp. 459–481, 2023.
https://doi.org/10.1007/978-3-031-32733-9_20

Wagner's theorem stating that a finite graph is planar if and only if it does not have K_5 or $K_{3,3}$ as a minor, and the *forbidden subgraph problem* which looks for the maximum number of edges in any n-vertex graph excluding a given graph G as induced subgraph. In the algorithm and complexity framework, it is known that the vertex coloring problem is NP-hard in triangle-free (i.e., C_3-free) graphs, but many families $\mathcal{F}$ have been identified, for which computing the chromatic number of graphs excluding every graph in $\mathcal{F}$ as induced subgraphs can be done in polynomial time. For instance, it is known that, for a graph H of at most six vertices, vertex coloring for $\{C_3, H\}$-free graphs is polynomial-time solvable if H is a forest not isomorphic to $K_{1,5}$, and NP-hard otherwise [2]. Another recent illustration of the importance of $\mathcal{F}$-free graphs, is graph isomorphism, which can be tested in time $n^{\text{polylog}(k)}$ on all n-node graphs excluding an arbitrary k-node graph as a topological subgraph [13].

In the context of distributed computing for networks however, still very little is known about $\mathcal{F}$-free graphs, even for the most basic case where the graphs in $\mathcal{F}$ must be excluded as mere subgraphs (not necessarily induced). In fact, up to our knowledge, most of the work in this domain has focused on the standard CONGEST model, and its variants. Recall that the CONGEST model is a distributed computing model for networks where the nodes of a graph execute the same algorithm, as a sequence of synchronous rounds, during which every node is bounded to exchange messages of $O(\log n)$ bits with each of its neighbors (see [14]). Also recall that a distributed algorithm $\mathcal{A}$ *decides* a graph property P if, for every input graph G, the following holds: G satisfies P if and only if $\mathcal{A}$ accepts at every node of G. Deciding H-freeness is a fruitful playground for inventing new techniques for the design of efficient CONGEST algorithms. Indeed, the problem itself is *local*, yet the limited bandwidth of the links imposes severe limitations on the ability of every node to gather information about nodes at distance more than one from it.

An important case is checking the absence of a cycle of given size as a subgraph. On the negative side, for every $k \geq 2$, deciding C_{2k+1}-freeness requires $\tilde{\Omega}(n)$ rounds in CONGEST, even for randomized algorithms [6]. However, for every $k \geq 2$, C_{2k}-freeness can be solved by Monte-Carlo algorithms performing in a *sub-linear* number of rounds. For instance C_4-freeness can be decided (deterministically) in $O(\sqrt{n})$ rounds [6], and, for every $k \geq 3$, the round-complexity of C_{2k}-freeness is at most $O(n^{1-2/(k^2-2k+4)})$ if k is even, and $O(n^{1-2/(k^2-k+2)})$ if k is odd (see [9]).

The round-complexity of deciding C_{2k}-freeness has been recently improved (see [4]), for small values of k, by an elegant algorithm which, for every $2 \leq k \leq 5$, runs in $O(n^{1-1/k})$ rounds. For $k = 2$ the (randomized) algorithms in [4,9] runs with the same asymptotic complexity as the (deterministic) algorithm in [6], i.e., in $O(\sqrt{n})$ rounds, and this cannot be improved, up to a logarithmic multiplicative factor [6]. However, for $k \in \{3, 4, 5\}$, the current best-known upper bound on the round-complexity of deciding C_{2k}-freeness is $O(n^{1-1/k})$. Interestingly, the algorithm in [4] also allows to decide whether the girth of a network is at most g, in $\tilde{O}(n^{1-2/g})$ rounds. In other words, the algorithm decides $\{C_k, 3 \leq k \leq g\}$-freeness for any given g.

In a nutshell, the algorithm in [4] is based on the notion of *light* and *heavy* nodes, where a node is light if its degree is at most $n^{1/k}$, and heavy otherwise. Cycles of length $2k$ composed of light nodes only can be found in at most $\sum_{i=0}^{k-1} n^{i/k} = \Theta(n^{1-1/k})$ rounds, by brute-force search, using color-coding [1]. For finding cycles containing at least one heavy node, it is noticed that, by picking a node s uniformly at random, the probability that s is neighbor of a heavy node is at least $n^{1/k}/n$, and thus, by repeating the experiment $\Theta(n^{1-1/k})$ times, a neighbor of a heavy node belonging to some $2k$-cycle will be found with constant probability, if it exists. The node s chosen at a given time of the algorithm initiates brute-force searches from all its heavy neighbors in parallel, each one searching for a cycle containing it, using color coding. The main point in the algorithm is the following. It is proved that, for every $k \in \{2, 3, 4, 5\}$, and every $i \in \{1, \ldots, k-1\}$, there is a *constant* threshold $T_k(i)$ such that, if a node colored i or $2k - i$ has to forward more than $T_k(i)$ searches initiated from the heavy neighbors of s, then that node can safely abort the search, without preventing the algorithm from eventually detecting a $2k$-cycle, if it exists. It follows that the parallel searches initiated by the random source s run in $O(1)$-rounds, and thus the "threshold-based" algorithm in [4] runs in $O(n^{1-1/k})$ rounds overall.

The objective of this paper is to determine under which condition, and for which graph family $\mathcal{F}$, threshold-based algorithms can be used for deciding $\mathcal{F}$-freeness.

1.2   Our Results

Our first contribution is a negative result. For every $k \geq 6$, we exhibit an infinite family of graphs in which any threshold-based algorithm fails to decide C_{2k}-freeness. That is, we show that, for $k \geq 6$, a threshold-based algorithm must forward a non-constant amount of messages at some step to guarantee that the parallel searches initiated by the random source s detect a $2k$-cycle. More specifically, we show the following.

Theorem 1. *For every $k \geq 6$, there exists an infinite family $\mathcal{G}$ of graphs containing a unique $2k$-cycle $C = (u_0, u_1, \ldots, u_{2k-1})$ such that, for every $T \in o(n^{1/6}/\log n)$, the threshold-based algorithm fails to detect C in at least one n-node graph in $\mathcal{G}$ if the thresholds are set to T.*

In other words, Theorem 1 says that, for every $k \geq 6$, there are no efficient threshold-based algorithms capable to decide C_{2k}-freeness. In particular, neither C_{12}-freeness nor C_{14}-freeness can be decided by a threshold-based algorithm. Nevertheless, our second contribution states that this is not the case of determining whether a graph is free of both C_{12} and C_{14}.

Theorem 2. *$\{C_{12}, C_{14}\}$-freeness can be decided by a threshold-based algorithm running in $O(n^{1-\frac{1}{7}})$ rounds.*

Note that the generic algorithm from [9] would run in $O(n^{1-\frac{1}{22}}) = O(n^{0.954\ldots})$ rounds for deciding $\{C_{12}, C_{14}\}$-freeness by checking separately whether the graph contains a C_{12}, and whether the graph contains a C_{14}. Instead, our algorithm performs in $O(n^{1-1/7}) = O(n^{0.857\cdots})$ rounds. Note that establishing that $\{C_{10}, C_{12}\}$-freeness can be decided by a threshold-based algorithm running in $O(n^{1-\frac{1}{6}})$ rounds is rather easy because C_{10}-freeness can be decided by such an algorithm. The point is that, again, thanks to Theorem 1, neither C_{12}-freeness nor C_{14}-freeness can be decided by a threshold-based algorithm.

Finally, note that, by construction, threshold-based algorithms can decide $\mathcal{F}_k$-freeness, for every $k \geq 2$, where $\mathcal{F}_k = \{C_{2\ell} \mid 2 \leq \ell \leq k\}$. This raises the question of identifying infinite collections of smaller families $\mathcal{F}$ of cycles for which threshold-based algorithms succeed to decide $\mathcal{F}$-freeness. We identify two such families.

Theorem 3. *Let $k \geq 2$, $\mathcal{F}'_k = \{C_{4\ell} \mid 1 \leq \ell \leq k\}$, and $\mathcal{F}''_k = \{C_{4\ell+2} \mid 1 \leq \ell \leq k\}$. Both $\mathcal{F}'_k$-freeness and $\mathcal{F}''_k$-freeness can be decided by threshold-based algorithms running in $\tilde{O}(n^{1-1/2k})$ rounds, and $\tilde{O}(n^{1-1/(2k+1)})$ rounds, respectively.*

Due to lack of space, the proof of this latter theorem in placed in Appendix F.

1.3 Related Work

Deciding H-freeness for a given graph H has been considered in [7], which describes an algorithm running in $\tilde{O}(n^{2-2/(3k+1)+o(1)})$ rounds for k-node graphs H. This round complexity is nearly matching the general lower bound $\tilde{\Omega}(n^{2-\Theta(1/k)})$ established in [9]. This latter bound can be overcome for specific graphs H, and typically when H is a cycle.

For every $k \geq 3$, C_k-freeness can be decided in $O(n)$ rounds (see, e.g., [11]). However, the exact round-complexity of deciding C_k-freeness varies a lot depending on whether k is even or odd. It was proved in [6] that deciding C_{2k+1}-freeness requires $\tilde{\Omega}(n)$ rounds for $k \geq 2$. Nevertheless, sub-linear algorithms are known for even cycles. In particular, the round-complexity of deciding C_4-freeness was established as $\tilde{\Theta}(n^{1/2})$ in [6]. The lower bound $\tilde{\Omega}(n^{1/2})$ rounds is also known to hold for deciding C_{2k}-freeness, for every $k \geq 3$ [11]. The best generic upper bound for deciding C_{2k}-freeness is $\tilde{O}(n^{1-\Theta(1/k^2)})$ rounds [9]. Faster algorithms are known, but for specific values of k only. Specifically, for every $k \in \{3,4,5\}$, C_{2k}-freeness can be decided in $O(n^{1-1/k})$ rounds [4]. The special case of triangle detection, i.e., deciding C_3-freeness is widely open.

It may also be worth mentioning the study of cycle-detection in the context of a model stronger than CONGEST, namely in the CONGESTED CLIQUE model. In this model, efficient algorithms have been designed. In particular, it was shown in [5] that C_3-freeness can be decided in $O(n^{0.158})$ rounds, C_4-freeness can be decided in $O(1)$ rounds, and C_k-freeness can be decided in $O(n^{0.158})$ rounds for any $k \geq 5$.

The interested reader is referred to [3] for a recent survey on subgraph detection, and related problems, in CONGEST or similar models.

2 Preliminaries

In this section, we recall the main techniques used for deciding whether the graph contain a cycle of a given length as a subgraph, and we summarize the threshold-based algorithms defined in [4].

2.1   Subgraph Detection

Recall that a graph H is a *subgraph* of a graph G if $V(H) \subseteq V(G)$ and $E(H) \subseteq E(G)$. Given a graph H, a deterministic distributed algorithm $\mathcal{A}$ for the CONGEST model decides H-freeness in $R(\cdot)$ rounds if, for every n-node graph G, whenever $\mathcal{A}$ runs in G, each node outputs "accept" or "reject" after $R(n)$ rounds, and

$$G \text{ contains } H \text{ as a subgraph} \iff \text{at least one node of } G \text{ rejects.}$$

A randomized Monte-Carlo algorithm decides H-freeness if, for every n-node graph G,

$$\begin{cases} G \text{ contains } H \text{ as a subgraph} \implies \Pr[\text{at least one node of } G \text{ rejects}] \geq 2/3. \\ G \text{ does not contain } H \text{ as a subgraph} \implies \Pr[\text{all nodes accept}] \geq 2/3. \end{cases}$$

In fact, most algorithms for deciding H-freeness are 1-sided, i.e., they alway accept H-free graphs, and may err only by failing to detect an existing copy of H in G. By repeating the execution of 1-sided error algorithms for sufficiently many times, one can make the error probability as small as desired.

In the case $H = C_{2k}$, which is the framework of this paper, one standard technique, called *color-coding* [1], plays a crucial role and was used in many algorithms for detecting cycles in various contexts (see, e.g., [4,8,10]).

Color Coding. Let $G = (V, E)$, and $W \subseteq V$. For deciding whether there is a $2k$-cycle including one node in W, let every node of G pick a color in $\{0, \ldots, 2k-1\}$ uniformly at random. Then every node $w \in W$ colored 0 launches a search, called color-BFS(k, w), by sending its identifier to all its neighbors colored 1 and $2k-1$. Every node colored 1 receiving an identifier from a node colored 0 forwards that identifier to all its neighbors colored 2, while every node colored $2k-1$ receiving an identifier from a node colored 0 forwards it to all its neighbors colored $2k-2$. More generally, for every $i = 2, \ldots, k-1$, every node colored i receiving an identifier from a node colored $i-1$ forwards it to all its neighbors colored $i+1$, and, for every $i = 2k-2, \ldots, k+1$, every node colored i receiving an identifier from a node colored $i+1$ forwards it to all its neighbors colored $i-1$. If a node colored k receives a same identifier from a neighbor colored $k-1$, and from a neighbor colored $k+1$, then it rejects.

The number of rounds required by color-BFS(k, W) is at most $k|W|$. Also, if the nodes in the graphs have maximum degree Δ, then the number of rounds is at most $O(\Delta^{k-1})$. Overall, we have

$$\#\text{rounds color} - \text{BFS}(k, W) = O(\min\{k|W|, \Delta^{k-1}\}). \tag{1}$$

If there is a $2k$-cycle passing through a node in W, then the probability that this cycle is colored appropriately is at least $\rho = 1/(2k)^{2k}$, and therefore the cycle is found with probability at least ρ. By repeating the procedure a constant number of times proportional to $(2k)^{2k}$, the cycle is found with probability at least $2/3$.

2.2 Threshold-Based Algorithms

We denote the algorithm defined in [4] by $\mathcal{A}^\star$. This algorithm, summarized in Algorithm 1, heavily uses color-coding. A node u of G is called *light* if $\deg(u) \leq n^{1/k}$, and *heavy* otherwise. A $2k$-cycle C containing only light nodes is called *light cycle*, and is heavy otherwise.

Detecting Light Cycles. Detecting whether there is a light $2k$-cycle is easy by applying color-coding in the subgraph $G[U]$ of G induced by light nodes U (i.e., only light nodes participate). By Eq. (1) with $\Delta = n^{1/k}$, we get that the detection of light cycles takes $O(n^{1-1/k})$ rounds. If a light $2k$-cycle exists in the graph, some light node rejects with constant probability, and we are done.

Detecting Heavy Cycles. For detecting heavy cycles, $\mathcal{A}^\star$ picks a node s uniformly at random in the graph[1]. The idea is that if there is a heavy $2k$-cycle in the graph, say $C = (u_0, u_1, ..., u_{2k-1})$ where u_0 is heavy, then the probability that a neighbor s of u_0 is picked is at least $n^{-(1-1/k)}$ since $\deg(u_0) \geq n^{1/k}$. Therefore, by repeating $\Theta(n^{1-1/k})$ times the choice of s, a neighbor of u_0 will be picked with constant probability. For each choice of s, the goal is to proceed with searching a $2k$-cycle in a constant number of rounds.

   The chosen node s launches color-BFS(k, s) for figuring out whether there is a $2k$-cycle passing through s. By Eq. (1) with $|W| = 1$, this takes $O(1)$ rounds. If a $2k$-cycle is detected, some node rejects, and we are done.

   We therefore assume from now that s does not belong to a $2k$-cycle. The source node s then sends a message to all its heavy neighbors W, and each of these neighbors w launches color-BFS(k, w), in parallel. At this point, one cannot simply rely on Eq. (1) with $|W| \leq \deg(s)$ to bound the round-complexity of color-BFS(k, W) because s may have non-constant degree. The central trick used in [4] consists to provide each node with a threshold for the number of messages the node can forward at a given step of a color-BFS. In case the number of messages to be transmitted exceeds the threshold, then the node aborts, i.e., it stops participating to the current color-BFS. It is shown that such threshold-based approach may prevent the nodes to detect $2k$-cycles, but not too often, and that a $2k$-cycle will be detected with constant probability anyway, if it exists. This is summarized by the following lemma.

Lemma 1 ([4]). *Let $C = (u_0, u_1, ..., u_{2k-1})$ be a $2k$-cycle in G, with u_0 heavy, and of maximum degree among the nodes in C. For every $k \in \{2, 3, 4, 5\}$, there exists a constant $\alpha_k > 0$, and there exist constant thresholds $T_k(i)$, $i = 1, ..., k-1$, such that, even if nodes colored i or $2k - i$ abort the search launched from the set W of heavy neighbors of $s \in N_G(u_0)$ at the i-th step of color $-$ BFS(k, W) whenever they generate a congestion larger than $T_k(i)$, still, for a fraction at least α_k of the neighbors s of u_0, the cycle C will be found, unless s itself belongs to a $2k$-cycle.*

---

[1] This can be done by letting each node choosing an integer value in $\{1, ..., n^3\}$ uniformly at random; The node s with smallest value is the chosen node. With high probability, this node is unique.

That is, $\mathcal{A}^\star$ sets thresholds (depending on k), and if the volume of communication generated by the color-BFS(k, W) launched in parallel by all the heavy neighbors W of a random source s exceeds these thresholds, then the search aborts. Yet, it is proved in [4] that this does not prevent a $2k$-cycle to be found, if it exists.

Algorithm 1. Deciding C_{2k}-Freeness by the Threshold Algorithm $\mathcal{A}^\star$ from [4]

1: color-BFS(k, U) in $G[U]$ ▷ $U = \{u \in V(G) \mid \deg(u) \leq n^{1/k}\}$
2: **for** $i = 1$ **to** $\Theta(n^{1-1/k})$ **do**
3:     $s \leftarrow$ random node in G ▷ $W = \{v \in N_G(s) \mid \deg(v) > n^{1/k}\}$
4:     color-BFS(k, s)
5:     color-BFS(k, W) with threshold $T_k(i), i \in \{1, \ldots, k-1\}$
6: **end for**

The algorithm $\mathcal{A}^\star$ is summarized in Algorithm 1. Each color-BFS includes $\sim (2k)^{2k}$ executions of color-coding to guarantee $2k$-cycle detection with probability at least $2/3$. Instruction 1 performs in $\Theta(n^{1-1/k})$ rounds because $G[U]$ has maximum degree $n^{1/k}$. It finds a light $2k$-cycle, if it exists, with probability $2/3$. Instruction 4 performs in $O(1)$ rounds for each constant $k \leq 5$, and, if s belongs to a $2k$-cycle, it finds such a cycle with probability $2/3$. Instruction 5 also performs in $O(1)$ rounds as well, thanks to the thresholds specified in Lemma 1. If there is a heavy $2k$-cycle, and if s does not belong to a $2k$-cycle, then that heavy $2k$-cycle is found with probability $2/3$. Overall, $\mathcal{A}^\star$ performs in $O(n^{1-1/k})$ rounds, and succeeds with probability $2/3$.

In the next section, we shall show that thresholds $T_k(i)$, $i = 1, \ldots, k-1$, such as the ones specified in Lemma 1 cannot be set for $k \geq 6$.

3 Limits of the Threshold-Based Algorithms

This section is entirely dedicated to the proof of Theorem 1, which is essentially based on proving the impossibility of setting a constant $T_k(k-3)$ for $k \geq 6$. For this purpose, we exhibit a class of graph $\{G_k \mid k \geq 6\}$ such that each G_k does not contain any light cycle C_{2k}, and contains exactly one heavy cycle C_{2k}. The construction of G_k for $k \geq 6$ is split in two cases: a generic construction, which works for all $k \geq 7$, and a specific construction for G_6. We begin the proof by the generic case.

Let $k \geq 7$. The graph G_k is composed of the following nodes (see Fig. 1), for $N \geq 1$:

- The $2k$ nodes of the unique $2k$-cycle $C^\star = (u_0, u_1, \ldots, u_{2k-1})$;
- The set $S = \{s^p, p \in \{1, \ldots, N\}\}$ of N neighbors of u_0;
- The set $W = \{w_{k-4}^q, q \in \{1, \ldots, N\}\}$ of N neighbors of u_{k-3};
- For $(p, q) \in \{1, \ldots, N\}^2$, the set $\{w_j^{p,q}, j \in \{0, \ldots, k-5\}\}$ of the nodes on a path from node s^p to node in w_{k-4}^q;

- For $(p, q) \in \{1, \ldots, N\}^2$, the set $\{v_0^{p,q,r}, \ r \in \{1, \ldots, N\}\}$ of private neighbors of node $w_0^{p,q}$ (these nodes are added in order to ensure that $w_0^{p,q}$ is heavy);
- For $(p, q) \in \{1, \ldots, N\}^2$, the set $\{v_{k-5}^{p,q,r} : r \in \{1, \ldots, N\}\}$ of private neighbors of node $w_{k-5}^{p,q}$ (as above, this makes node $w_{k-5}^{p,q}$ heavy).

The number of nodes in G_k is $n = \Theta(N^3)$.

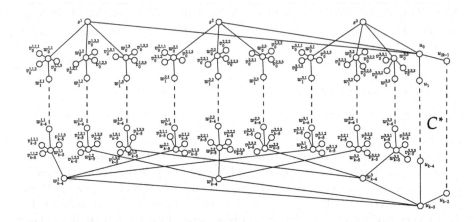

Fig. 1. The graph G_k for $k \geq 7$ and $N = 3$.

The proof of the following result can be found in Appendix A.

Lemma 2. *For every $k \geq 7$, $C^\star$ is the unique $2k$-cycle in G_k, and is a heavy cycle.*

As a consequence of Lemma 2, a $2k$-cycle in G_k can only be detected if the algorithm picks the random source s in $N_{G_k}(C^\star)$, i.e., it must pick $s \in S \cup W \cup C^\star$. Also, if $s \in S \cup W$, then s does not belong to a $2k$-cycle, and thus s will initiate the search for C_{2k} from each of its heavy neighbors.

Lemma 3. *Let $T \in o(n^{1/3}/\log n)$, and let us set $T_k(k-3) = T$ in the threshold-based algorithm. If $s \in S$ (resp., $s \in W$), then the probability that u_{k-3} (resp., u_0) forwards at most T messages during a search phase from heavy nodes is $\exp(-\Theta(n^{1/3}))$.*

Proof. By the symmetry of G_k, the roles of $S \cup \{u_0\}$ and $W \cup \{u_{k-3}\}$ are identical. We shall thus prove the lemma only for $s \in S$, i.e., $s = s^p$ for some $p \in \{1, \ldots, N\}$. For every $q \in \{1, \ldots, N\}$, let X_q be the following Bernoulli random variable, assuming each node picks a color in $\{0, \ldots, 2k-1\}$ uniformly at random. We say that the path $w_0^{p,q}, \ldots, w_{k-5}^{p,q}, w_{k-4}^q$ is well-colored if, for every $i \in \{0, \ldots, k-5\}$, $w_i^{p,q}$ is colored i, and w_{k-4}^q is colored $k-4$. We define

$$X_q = \begin{cases} 1 \text{ if the path } (w_0^{p,q}, \ldots, w_{k-5}^{p,q}, w_{k-4}^q) \text{ is well-colored;} \\ 0 \text{ otherwise.} \end{cases}$$

Let then $X = \sum_{q=1}^{N} X_q$ be the random variable that counts the number of identifiers different from $\mathrm{id}(u_0)$ that u_{k-3} has to forward, that is, the identifiers of all the nodes $w_0^{p,q}$ satisfying $X_q = 1$. X follows a Binomial law of parameters N and $r = (\frac{1}{2k})^{k-3}$, so its expectation is $E(X) = Nr$. Since every node $w_0^{p,q}$ has a unique path with length $k-3$ to node u_{k-3}, we get that

$$\Pr[X \leq T] = \sum_{t=0}^{T} \Pr[X = t] = \sum_{t=0}^{T} \binom{N}{t} r^t (1-r)^{N-t}$$

$$\leq \sum_{t=0}^{T} N^t\, r^t e^{(N-t)\ln(1-r)} \leq (T+1)\, N^T\, r^T\, e^{N\ln(1-r)}$$

$$= N^T e^{-\Theta(N)}.$$

Therefore, the probability that u_{k-3} has to forward at most T messages is $O(N^T e^{-\Theta(N)})$. If $T = o(N/\log N) \simeq o(n^{1/3}/\log n)$, then this probability is asymptotically equal to $\exp(-\Theta(n^{1/3}))$.   □

To conclude the proof of Theorem 1 for $k \geq 7$ note that, even by fixing all thresholds to $T \in o(n^{1/3}/\log n)$, Algorithm $\mathcal{A}^\star$ fails to detect the unique (heavy) cycle $C^\star$ almost surely. Indeed when picking vertex s in S (or, symmetrically, s in W), the algorithm succeeds with probability $\exp(-\Theta(n^{1/3}))$, since vertex u_{k-3} aborts almost surely. The other possibility of detecting the cycle is when the algorithm picks s directly on $C^\star$, which only contains $2k$ vertices, so the success possibility is $O(1/n)$. Hence, although the algorithm makes $\tilde{O}(n^{1-1/k})$ independent random choices of s, the probability of success is only $\tilde{O}(n^{-1/k})$.

The specific case $k = 6$ is treated in Appendix B, which completes the proof of Theorem 1.

4   Deciding $\{C_{12}, C_{14}\}$-Freeness

This section is entirely dedicated to the proof of Theorem 2. We rely mostly on the threshold algorithm as such, with the following slight modification, for simplifying the analysis.

Remark. For exhibiting the thresholds $T_{2k}(i)$, $1 \leq i \leq 2k-1$, it is convenient to assume that, instead of repeating $O(n^{1-1/k})$ random choices of s, and then, for each chosen s, repeating $\sim (2k)^{2k}$ random choices of colors (for the color-BFSs), the algorithm proceeds as follows: The outer loop repeats $\sim (2k)^{2k}$ random assignments of colors, and the inner loop repeats $O(n^{1-1/k})$ random choices of s (for each of the $\sim (2k)^{2k}$ color-assignments). In fact, it simplifies the presentation even further by assuming that the random colors are in the range $\{-1, 0, \ldots, 2k-1\}$. The extra color -1 is used only by s, and s launches color-BFS(W) only under the condition that s has random color -1. None of these changes affect the performances of the algorithm, up to a constant factor in the round-complexity.

The algorithm starts by checking the existence of a light 12-cycle or a light 14-cycle. This is achieved in $O(n^{6/7})$ rounds, by parallel color-BFSs running on the light nodes only (see Sect. 2.2). For detecting heavy cycles, the algorithm proceeds as the threshold algorithm, by repeating the choice of a random node s. For each choice, the chosen node s checks whether it belongs to a 12-cycle or to a 14-cycle, by performing two series of color-BFS(s), one for detecting a possible 12-cycle passing through s, and one for detecting a possible 14-cycle. If no such cycles are detected, then s proceeds as follows.

Looking for 14-cycles. Node s launches color-BFS(W), from the set W of all its heavy neighbors, with appropriate thresholds $T_7(i)$, $1 \le i \le 6$, that will be specified later. The crucial point here is that if the algorithm proceeds by checking the existence of 12-cycles *and* of 14-cycles. By checking both lengths, we will be able to establish a result similar to Lemma 1, that is, if there is a 14-cycle in G, say $C = (u_0, \ldots, u_{13})$, with u_0 heavy, and of maximum degree among the nodes in C, then there exists a constant $\alpha > 0$, such that, even if nodes colored i or $14 - i$ abort the search launched from the set W at the i-th step of color $-$ BFS(W) whenever they generate a congestion larger than $T_7(i)$, still, for a fraction at least α of the neighbors s of u_0, the cycle C will be found with probability at least $2/3$. In other words, if a node rejects during this phase, it is because there is a 12-cycle, or there is a 14-cycle. On the other hand, the fact that all nodes accept during this phase only provides a (statistical) guarantee on the absence of 14-cycles, but provides little information on the absence of 12-cycles.

Looking for 12-cycles. Again, node s launches color-BFS(W), but for 12-cycles now, with the mere thresholds $T_6(i) = 1$ for all $i = 1, \ldots, 5$. The crucial point here is that, assuming that the graph is C_{14}-free, then a threshold of 1 suffices. There will only ever be one message crossing an edge in a well-colored heavy 12-cycle. This latter fact is easy to establish, so most of the proof consists in proving the existence of the thresholds when looking for 14-cycles.

To set the values $T_7(i)$ for $1 \le i \le 6$, let us define $T_7(0) = 1$. Our construction is then inductive, and, for $i > 0$, we shall set $T_7(i) = f(i) \cdot T_7(i - 1)$ for appropriate constants $f(i)$. Let us assume that the graph contains a 14-cycle, denoted by $C^\star = (u_0, u_1, \ldots, u_{13})$, where u_0 is of maximum degree in $C^\star$, and, for every $i = 0, \ldots, 13$, node u_i is colored i. From now on, we will work only on the nodes $u_0, u_1, \ldots, u_7$. By symmetry, the same arguments will apply to nodes $u_0, u_{13}, \ldots, u_7$. Before further defining the setting of the proof, recall that, as underlined before, the $O(n^{1-1/k})$ drawings of nodes s are performed on a given coloring of the graph with colors in $\{-1, 0, \ldots, 13\}$, and that only a picked node s colored -1 invokes color-BFS(W).

The lemma below is generic, as it applies to all $k \ge 2$. Recall that a path $s, w_0, \ldots, w_{i-1}, u_i$ from node s to node u_i is *well-colored* if s is colored -1, u_i is colored i, and, for every $j = 0, \ldots, i - 1$, node w_j is colored j by color-coding.

Lemma 4. *Let $k \geq 2$ be an integer. For every $i \in \{1, \ldots, k-1\}$, let ρ be the maximum number of node-disjoint well-colored paths from s to u_i. If s launches color-BFS(W) from all its heavy neighbors colored 0, then u_i cannot receive more than $\rho \cdot T_k(i-1)$ identifiers from nodes colored $i-1$.*

Proof. Let S be a set of ρ node-disjoint, well-colored paths from s to u_i. Let $w_0 \in W$ be a heavy neighbor of s colored 0, and let us assume that id(w_0) has reached u_i. It follows that there is a well-colored path P of length i from w_0 to u_i. The path P must intersect some path $P' = \{w'_0, \ldots, w'_{i-1}\}$ in S (perhaps even $P = P'$). As a consequence, id(w_0) is included in the at most $T_k(i-1)$ identifiers that node w'_{i-1} may forward to u_i. Therefore, the number of identifiers received by u_i during color-BFS(W) does not exceed $\rho \cdot T_k(i-1)$. □

Given $f : \{1, \ldots, 6\} \to \mathbb{N}$ to be fixed later, we define, for every $i \in \{1, \ldots, 6\}$, the set of nodes

$$B(i) = \{s \in N_G(u_0) \mid (\text{color}(s) = -1) \wedge (s \notin C_{12} \vee C_{14}) \wedge (\rho(s) > f(i))\},$$

where $\rho(s)$ denotes the maximum number of node-disjoint well-colored paths from s to node u_i in the graph. Thanks to Lemma 4, we have that a neighbor of u_0 colored -1 and not in any 12- or 14-cycle, causing u_i to receive more than $T_7(i)$ identifiers, is in $B(i)$. This set of nodes thus represents the *bad* neighbors of u_0, those that will prevent us from detecting any cycle whenever any such neighbor is picked.

The rest of this section will prove that the bad nodes represent only a fraction of the neighbors of u_0. It follows that, by performing sufficiently many choices of s, the probability to select a *good* neighbour s of u_0, which will not cause congestion, and will thus allow detecting the cycle, is still $\Omega(n^{-6/7})$. The parameter $f(i)$ makes the connection between the parameter $T_7(i)$ used by the algorithm, and the set of nodes we do not want to pick as the source s. Formally we are aiming at showing the following result.

Proposition 1. *Let us set $f(1) = 60$, $f(2) = f(3) = 10$, $f(4) = f(5) = 5$, and $f(6) = 6$. With this setting, we get $\left| \bigcup_{i=1}^{6} B(i) \right| \leq \frac{35}{72} \deg(u_0) + 3$.*

The thresholds yielded by the function f defined in Proposition 1 are:

$$T_7(1) = 60 \qquad T_7(2) = 600 \qquad T_7(3) = 6\,000$$
$$T_7(4) = 30\,000 \quad T_7(5) = 150\,000 \quad T_7(6) = 900\,000$$

To prove Proposition 1, our strategy is to bound each $|B(i)|$ separately by a fraction of the degree of u_0 (for $i = 1, 2, 3$), or by a constant (for $i = 4, 5, 6$). Let us now consider the values of $i = 1, \ldots, 6$ successively.

Case $i = 1$. Our goal is to bound $|B(1)|$, given $f(1) = 60$. To achieve that, we will show that the nodes colored 0 whose identifiers can reach u_1 when a node $s \in B(1)$ is picked have to be sufficiently many compared to $B(1)$ itself. Otherwise a 12-cycle that involves nodes from $B(1)$ would appear, which contradicts the definition of the set of bad nodes.

Lemma 5. *If $f(1) \geq 60$ then $|B(1)| \leq \frac{1}{4} \deg(u_0)$.*

Proof. Let W_0 denote the set of nodes $x \neq u_0$ colored 0 such that x is a heavy neighbor of a node in $B(1)$, and a neighbor of u_1. This means that for any node $s \in B(1)$ that is picked, the identifiers that u_1 receives are those of u_0 and of nodes in W_0. Let us then consider the bipartite graph H formed by nodes of $B(1)$ and W_0, and the edges between $B(1)$ and W_0. Let H' be the subgraph of H obtained by iteratively deleting all nodes of degree at most 11. If H' is not empty, then, since all its vertices have degree at least 12, we can construct a path of length 11 starting from any vertex of H'. Thanks to the fact that H' is bipartite, this path has either both endpoints in $B(1)$, or both in W_0, meaning that they are linked to u_0 or u_1, creating a 12-cycle with the path. This cannot be true as it would mean that some nodes in $B(1)$ are in a 12-cycle. It follows that H' is empty. As a consequence,

$$60 \cdot |B(1)| \leq f(1) \cdot |B(1)| \leq |E(H)| < 12 \left(|B(1)| + |W_0| \right),$$

where the second inequality comes the fact that any node in $B(1)$ has a degree larger than $f(1)$ in H, and the third inequality comes from the fact that our iterative removing of nodes of degree at most 11 in H has removed all of the nodes. This yields $|W_0| > 4|B(1)|$. Under our assumption that u_0 has maximum degree in C^*, we then get $\deg(u_0) \geq \deg(u_1) \geq |W_0| \geq 4\,|B(1)|$. $\quad\square$

Case $i = 2$. To prove upper bounds on the number of bad nodes for $i = 2$, as well as for $i > 2$, we use the following lemma that allows us to assume the existence of node-disjoint well-colored paths from different nodes in $B(i)$ to u_i.

Lemma 6. *Let $b \geq 1$, and let U be a set of nodes. If $f(i) \geq (b-1)i + |U|$ then either $|B(i)| < b$, or, for any $c \in \{0, \dots, b-1\}$, any nodes $s^1, \dots, s^c \in B(i)$, any collection C of c node-disjoint well-colored paths from $B(i)$ to u_i that do not intersect U, and any $s \in B(i) \smallsetminus \{s^1, \dots, s^c\}$, there exists well-colored path P that does not intersect U nor any path in C.*

Proof. Let $C = \left\{ P^j = (s^j, w_0^j, \dots, w_{i-1}^j) \mid j = 1, \dots, c \right\}$. A well-colored path from any node $s \in B(i)$ to u_i cannot go through any other node in $B(i)$ as the bad nodes are colored -1. Such a path may however contain some nodes in U, or in the paths in C. There are less than $ci + |U|$ such nodes in total. Since $f(i) \geq ci + |U|$, any node $s \in B(i) \smallsetminus \{s^1, \dots, s^c\}$ has at least $ci + |U| + 1$ node-disjoint well-colored paths to u_i. Therefore, there is a well-colored path from s to u_i that does not contain any node in U nor any node in the paths in C, as claimed. $\quad\square$

To find upper bounds on the sizes of $B(2)$ and $B(3)$, the strategy is similar to the case $i = 1$. The novelty is to consider each color of the node-disjoint paths from nodes in $B(i)$ to u_i, and to show that the paths cannot *merge* at nodes of the considered color without making the nodes of $B(i)$ appear in a 12- or 14-cycle.

Lemma 7. *If* $f(2) \geq 10$ *then* $|B(2)| \leq \frac{1}{9} \deg(u_0) + 2$.

Proof. We say that two well-colored paths from $B(i)$ to u_i *merge at color* j if (1) they are node-disjoint before color j, and (2) they use the same nodes of color j. We first consider paths merging at color 1.

Claim 1. *If* $f(2) \geq 6$, $|B(2)| \geq 3$, *and there exist two distinct nodes* $s^1, s^2 \in B(2)$ *with paths merging at color 1, then one of those paths goes through* u_0.

Proof of claim. For the purpose of contraposition, let us assume that there exist two distinct nodes s^1, s^2 in $B(2)$ that have well-colored paths $P^1 = (s^1, w_0^1, x_1)$, and $P_z = (s^2, z_0, x_1)$ to u_2 merging at color 1, with $w_0^1 \neq u_0$ and $z_0 \neq u_0$. Then, by applying Lemma 6 with $b = 3$ and $U_1 = \{u_0, z_0\}$, we get that, if $|B(2)| \geq 3$, then (1) there exists a well-colored path $P^2 = (s^2, w_0^2, w_1^2)$ to u_2 that does not intersect $U_1 \cup P^1$, and (2) for all $s^3 \in B(2) \smallsetminus \{s^1, s^2\}$, there is a well-colored path $P^3 = (s^3, w_0^3, w_1^3)$ to u_2 that does not intersect $U_1 \cup P^1 \cup P^2$. As a consequence,

$$(u_0, s^1, w_0^1, x_1, z_0, s^2, w_0^2, w_1^2, u_2, w_1^3, w_0^3, s^3)$$

is a 12-cycle (see Fig. 2-right), which contradicts $s^1, s^2, s^3 \in B(2)$. This means that nodes in $B(2)$ cannot have paths that merge at color 1 except if one of these paths goes through u_0.   ◇

We now consider paths merging at color 0.

Claim 2. *Let* $s^1, s^2 \in B(2)$ *with* $s^1 \neq s^2$, *and let* $x_0 \neq u_0$ *such that* s^1 *has a well-colored path* $P^1 = (s^1, x_0, w_1^1)$ *to* u_2, *and* $s^2 \in N(x_0)$. *If* $f(2) \geq 8$ *then there are no two distinct nodes* $s^3, s^4 \in B(2) \smallsetminus \{s^1, s^2\}$ *having paths merging at a node colored 0 different from* u_0 *and* x_0.

Proof of claim. Let us assume, for the purpose of contradiction, that there exist $s^3, s^4 \in B(2) \smallsetminus \{s^1, s^2\}$ that have well-colored paths $P^3 = (s^3, y_0, w_1^3)$ and $P_y = \{s^4, y_0, w_1^3\}$ to u_2, merging at color 0 with $y_0 \neq u_0, x_0$. Applying Lemma 6 with $b = 4$ and $U = \{u_0, x_0, y_0\}$, we get that if $|B(2)| \geq 4$, then there is a well-colored path $P^4 = (s^4, w_0^4, w_1^4)$ to u_2 that does not intersect $U \cup P^1 \cup P^2 \cup P^3$. It follows that

$$(u_0, s^1, x_0, s^2, w_0^2, w_1^2, u_2, w_1^4, w_0^4, s^4, y_0, s^3)$$

is a 12-cycle (see Fig. 2-left), which contradicts the fact that $s^1, s^2, s^3, s^4 \in B(2)$. This means that nodes in $B(2) \smallsetminus \{s^1, s^2\}$ cannot have well-colored paths to u_2 merging at any node colored 0 different than u_0 or x_0.   ◇

In the end, by combining the impossibility results of Claims 1 and 2, two situations can occur. The first scenario is that the nodes s^1 and s^2 defined in Claim 2 do not exist, and any two nodes $s, s' \in B(2)$ cannot merge their well-colored paths to u_2, except in u_0. As every node in $B(2)$ has at least $f(2) + 1$ node-disjoint well-colored paths to u_2. By discarding (if it exists) the one going through u_0, we still have $f(2)$ paths not merging with any other well-colored path from $B(2)$ to u_2. It follows that $f(2) \cdot |B(2)| \leq \deg(u_2)$.

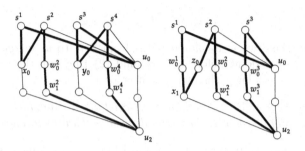

Fig. 2. Bold cycles are 12-cycles appearing whenever nodes in B(2) have merged paths.

The other scenario is that the nodes s^1 and s^2 as in Claim 2 do exist. In this case, any other two nodes $s, s' \in$ B(2) cannot merge their well-colored paths to u_2, except in u_0 or x_0. Discarding paths going through those two nodes, any node $s \in$ B(2) $\smallsetminus \{s^1, s^2\}$ still has at least $f(2) - 1$ paths not merging with any other well-colored path from B(2) $\smallsetminus \{s^1, s^2\}$ to u_2. It follows that

$$(f(2) - 1) \cdot (|B(2)| - 2) \leq \deg(u_2).$$

Therefore, in all cases, we have $(|B(2)| - 2)(f(2) - 1) \leq \deg(u_2) \leq \deg(u_0)$, which proves Lemma 7. $\qquad\qquad\square$

Case $i = 3$. We show that the nodes in B(3) cannot have their paths merging before reaching u_3 (see proof in Appendix C).

Lemma 8. *If $f(3) \geq 10$ then $|B(3)| \leq \frac{1}{8} \deg(u_0)$.*

Case $i \in \{4,5,6\}$. For $i = 4, 5, 6$, with values of $f(i)$ satisfying the inequality in the statement of Lemma 6, the mere existence of one or two nodes in B(i) is impossible, as such nodes would appear in a 12- or 14-cycle. This is shown below (see proof in Appendix D).

Lemma 9. *The following holds:*

- *If $f(4) \geq 5$ then $|B(4)| \leq 1$.*
- *If $f(5) \geq 5$ then $B(5) = \varnothing$.*
- *If $f(6) \geq 6$ then $B(6) = \varnothing$.*

Proposition 1 directly follows from Lemmas 5, and 7–9. Since, for every $i \in \{1, \ldots, 6\}$, $T_7(14 - i) = T_7(i)$ induces the same upper bound for $|B(14 - i)|$ as for $|B(i)|$, we get that $\left| \bigcup_{i \in \{1,\ldots,13\} \smallsetminus \{7\}} B(i) \right| \leq \frac{35}{36} \deg(u_0) + 6$. It follows that $\left| N(u_0) \smallsetminus \bigcup_{i \in \{1,\ldots,13\} \smallsetminus \{7\}} B(i) \right| \geq \frac{1}{36} \deg(u_0) - 6$. As a consequence, the number of *good* neighbours s of u_0 (not belonging to any $B(i)$ is at least a constant fraction of $\deg(u_0)$. This means that after $\Theta(n^{6/7})$ repetitions of the choice of s, a node in $N(u_0) \smallsetminus \bigcup_{i \in \{1,\ldots,13\} \smallsetminus \{7\}} B(i)$ that is colored -1 will be picked with probability at least $2/3$. By the previous Lemmas, this node will lead to a rejection, detecting a 12- or 14-cycle.

Looking for a C_{12}. At this point, we can assume that there are no 14-cycle in the graph because, if there were, then the algorithm would have rejected before. Let us then assume the existence of a 12-cycle $C^\star = \{u_0, \ldots, u_{11}\}$, that is well colored, where u_0 a heavy node. Then, by fixing $T_6(i) = 1$ for every $i = 1, \ldots, 5$, u_6 will reject. Indeed, recall that there are $\Omega(n^{1/6})$ neighbors of u_0 colored -1. Therefore, by performing $O(n^{6/7})$ iterations, the probability of picking $s \in N(u_0)$ colored -1 is at least $2/3$. Whenever such a node s is picked, u_0 sends its identifier. Suppose that u_i is the first node in $C^\star$ to receive 2 identifiers. Then, by Lemma 4, one of these identifiers comes from a path $w_0, w_1, \ldots, w_{i-1}$ that is node-disjoint from $\{u_0, \ldots, u_{i-1}\}$. Therefore, $(s, w_0, w_1, \ldots, w_{i-1}, u_i, u_{i+1}, \ldots, u_{11}, u_0)$ is a 14-cycle, a contradiction, which completes the proof of Theorem 2.

Remark. A threshold algorithm deciding $\{C_{10}, C_{12}\}$-freeness in $O(n^{1-1/6})$ rounds is given in Appendix E.

5 Conclusion

The threshold-based approach, as used in [4], is appealing for the design of efficient CONGEST algorithms deciding C_{2k}-freeness, for arbitrary $k \geq 2$. It was successfully applied to $k \in \{2, \ldots, 5\}$, resulting in algorithms deciding C_{2k}-freeness in $O(n^{1-1/k})$ rounds. We have shown that it is hopeless to use the threshold-based approach as such for $k \geq 6$.

Nevertheless, we have also shown that, despite this limit, the threshold-based approach can be used to design algorithms for deciding $\{C_{12}, C_{14}\}$-freeness in $n^{1-\frac{1}{7}}$ rounds, even if neither C_{12}-freeness nor C_{14}-freeness can be decided in the same round-complexity by threshold-based algorithms. We do not know whether this is just a specific case, or whether there is an infinite collection of pairs (k, k') with $k > k'$ such that $\{C_{2k'}, C_{2k}\}$-freeness can be decided in $O(n^{1-1/k})$ rounds by a threshold-based algorithm.

So far, the best known generic algorithm, i.e., an algorithm applying to all $k \geq 2$, decides C_{2k}-freeness in $n^{1-1/\Theta(k^2)}$ rounds [9], and it is open whether one can do better in general. It may actually be the case that the round-complexity of deciding C_{2k}-freeness is precisely $\Theta(n^{1-1/k})$ for all $k \geq 2$, but this is just a guess.

Acknowledgements. The authors are thankful to Pedro Montealegre and Ivan Rapaport for fruitful initial discussions about the power and limit of the threshold-based algorithms from [4].

Appendix

A   Proof of Lemma 2

We can view G_k as a weighted graph $\widehat{G}_k$, with the following edge-weights (see Fig. 3):

- For every $p \in \{1, \ldots, N\}$, the edge between u_0 and s^p has weight 1; Similarly, for every $q \in \{1, \ldots, N\}$, the edge between u_{k-3} and w_{k-4}^q has weight 1;
- All the part of G_k including the nodes $w_j^{p,q}$ for $(p, q) \in \{1, \ldots, N\}^2$ and $j \in \{0, \ldots, k-5\}$ is replaced by a complete bipartite graph $K_{N,N}$ with partitions S and W, with edge-weights $k-3$;
- The path $u_0, \ldots, u_{k-3}$ is replaced by two parallel edges e_1 and e_2 with respective weights $k-3$ and $k+3$;
- The private neighbors of $w_0^{p,q}$ and $w_{k-5}^{p,q}$, $(p, q) \in \{1, \ldots, \alpha\}^2$, are simply discarded.

Note that there is a one-to-one correspondence between the cycles in G_k and the cycles in $\widehat{G}_k$. Moreover, if the lengths of the cycles in $\widehat{G}_k$ take into account the edge-weights, then this correspondence also preserve the lengths.

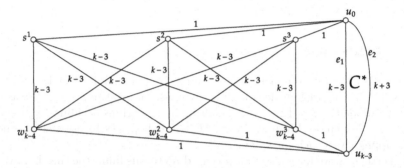

Fig. 3. The weighted graph $\widehat{G}_k$ for $k \geq 7$ and $N = 3$.

Any cycle in $\widehat{G}_k$ different from $C^\star = (e_1, e_2)$, but using e_2 has length larger than $2k$. Any cycle in $\widehat{G}_k$ different from $C^\star = (e_1, e_2)$, but using e_1 must contain an odd number $2x + 1$ of edges from the complete bipartite subgraph $K_{N,N}$. Therefore it has length $k - 3 + 1 + (2x + 1)(k - 3) + 1$ for some integer $x \geq 0$. For $x = 0$, this length is $2k - 4$, and, for $x \geq 1$, this length is at least $4k - 10 > 2k$ for every $k \geq 6$. For similar reasons, every cycle passing through u_0 or u_{k-3}, but not both, is also of length $k - 3 + 1 + (2x + 1)(k - 3) + 1$ for some integer $x \geq 0$, and the same analysis holds. Every cycle passing through u_0 and u_{k-3} but not using e_1 nor e_2 contains an even number of edges from the $K_{N,N}$. Thus is has a length of the form $2(k - 3) + 4 + 2x(k - 3)$ for some integer $x \geq 0$. For $x = 0$, this length is $2k - 2$, and, for $x \geq 1$, this length is at least $4k - 8 > 2k$ for any $k \geq 6$. Finally, every cycle fully included in the complete bipartite graph $K_{N,N}$ has length of the form $2(x + 2)(k - 3)$ for some integer $x \geq 0$. This length is at least $4k - 12 > 2k$, for every $k \geq 7$. □

B   The Specific Construction for $k = 6$

Lemma 2 does not hold for $k = 6$. Indeed, any 4-tuple $(s, s', w_2, w_2') \in S^2 \times W^2$ induces a 12-cycle in the complete bipartite graph $K_{N,N}$ in $\widehat{G}_6$. Nevertheless, there are no C_{12} in G_6 passing through u_0 or u_3, other than $C^\star$. We slightly modify G_6 to extend Lemma 2 to $k = 6$. Specifically, we replace the complete bipartite graph in $\widehat{G_6}$ by a dense bipartite graph with no 4-cycle, using the following lemma.

Lemma 10 ([12]). *There exists an infinite family of C_4-free graphs $\{G_d \mid d \text{ prime}\}$ such that, for every prime number d, G_d is a d-regular bipartite graph in which each partition has size d^2, and $|E(G_d)| = d^3 + o(d^3)$.*

Let $N = d^2$. In $\widehat{G_6}$, we replace the complete bipartite graph $K_{N,N}$ by G_d. With this modification, Lemma 2 holds. We now revisit Lemma 3. X becomes a random variable following a Binomial law with parameters $\sqrt{N}$ and $(\frac{1}{12})^3$. As a consequence,

$$\Pr[X \leq T] = O(N^T e^{-\Theta(\sqrt{N})}).$$

Therefore, the probability that u_3 has to forward at most T messages is $O(N^T e^{-\Theta(\sqrt{N})})$. If $T = o(\sqrt{N}/\log N) \simeq o(n^{1/6}/\log n)$, then this probability is asymptotically equal to $\exp(-\Theta(n^{1/6}))$. So for $k = 6$ we obtain a result similar to Lemma 3, by simply replacing the value of threshold T by $o(\sqrt{N}/\log N) \simeq o(n^{1/6}/\log n)$. The proof of Theorem 1 follows as for $k \geq 7$. □

C   Proof of Lemma 8

We first consider paths merging at color 2.

Claim 3. *If $f(3) \geq 8$, and if there exist two distinct nodes $s^1, s^2 \in B(3)$ that have well-colored paths merging at a node of color 2 different from u_2, then one of these paths goes through u_0 or u_1.*

Proof of claim. Suppose that there exist $s^1, s^2 \in B(3)$ that have well-colored paths $P^1 = (s^1, w_0^1, w_1^1, x_2)$ and $P_z = (s^2, z_0, z_1, x_2)$ to u_3 merging at color 2 with $u_0 \notin \{z_0, w_0^1\}$, $u_1 \notin \{z_1, w_1^1\}$, and $x_2 \neq u_2$. Then, applying Lemma 6 with $b = 2$ and $U = \{u_0, u_1, u_2, z_0, z_1\}$, we get that if $|B(3)| \geq 2$, then there is a well-colored path $P^2 = (s^2, w_0^2, w_1^2, w_2^2)$ to u_3 that does not intersect $U \cup P^1$. Therefore

$$(u_0, s^1, w_0^1, w_1^1, x_2, z_1, z_0, s^2, w_0^2, w_1^2, w_2^2, u_3, u_2, u_1)$$

is a 14-cycle (see Fig. 4-right), which contradicts $s^1, s^2 \in B(3)$. This means that nodes in $B(3)$ cannot have well-colored paths to u_3 merging at any node colored 2 (except if they go through u_0 or u_1, or merge at u_2).   ◇

We now consider paths merging at color 1.

Claim 4. *If $f(3) \geq 7$, and if there exist two distinct nodes $s^1, s^2 \in B(2)$ that have well-colored paths merging at a node of color 1 different from u_1, then one of these paths goes through u_0, or they both go through u_2.*

Proof of claim. Suppose that there exist $s^1, s^2 \in B(3)$ that have well-colored paths $P^1 = (s^1, w_0^1, x_1, w_2^1)$ and $P_z = (s^2, z_0, x_1)$ to u_3 merging at color 1, with $u_0 \notin \{w_0^1, z_0\}$ and $w_2^1 \neq u_2$. Applying Lemma 6 with $b = 2$ and $U = \{u_0, u_1, u_2, z_0\}$, we get that if $|B(3)| \geq 2$, then there is a well-colored path $P^2 = (s^2, w_0^2, w_1^2, w_2^2)$ to u_3 that does not intersect $U \cup P^1$. This implies that

$$(u_0, s^1, w_0^1, x_1, z_0, s^2, w_0^2, w_1^2, w_2^2, u_3, u_2, u_1)$$

is a 12-cycle (see Fig. 4-center), which contradicts $s^1, s^2 \in B(3)$. This means that nodes in $B(3)$ cannot have well-colored paths to u_3 merging at any node colored 1 other than u_1, except if it goes through u_0 or u_2.   ◇

We finally consider paths merging at color 0.

Claim 5. *If $f(3) \geq 8$ and $|B(3)| \geq 3$, then there are no two distinct nodes $s^1, s^2 \in B(2)$ that have paths merging at a node of color 0 different from u_0.*

Proof of claim. Suppose that there exists $s^1, s^2 \in B(3)$ that have well-colored paths $P^1 = (s^1, x_0, w_1^1, w_2^1)$ and $P_x = \{s^2, x_0, w_1^1, w_1^1\}$ to u_3 merging at color 0, with $x_0 \neq u_0$. Applying Lemma 6 with $b = 3$ and $U = \{u_0, x_0\}$, we get that if $|B(3)| \geq 3$, then (1) there is a well-colored path $P^2 = (s^2, w_0^2, w_1^2, w_2^2)$ to u_3 that does not intersect $U \cup P^1$, and (2) for all $s^3 \in B(3) \smallsetminus \{s^1, s^2\}$, there is a well-colored path $P^3 = (s^3, w_0^3, w_1^3, w_2^3)$ to u_3 that does not intersect $U \cup P^1 \cup P^2$. Therefore,

$$(u_0, s^1, x_0, s^2, w_0^2, w_1^2, w_2^2, u_3, w_2^3, w_1^3, w_0^3, s^3)$$

is a 12-cycle (see Fig. 4-left), which contradicts $s^1, s^2, s^3 \in B(3)$. This means that nodes in $B(3)$ cannot have well-colored paths to u_3 merging at color 0 (except u_0).   ◇

By combining the impossibility results of Claims 3, 4 and 5, it follows that paths from $B(3)$ to u_3 can only merge if they go through u_0, u_1 or u_2. Let us discard paths passing through those nodes. As each node in $B(3)$ has at least $f(3)+1$ node-disjoint paths to u_3, at least $f(3)-2$ of these paths do not intersect any other path from $B(3)$. Therefore, $(f(3) - 2) \cdot |B(3)| \leq \deg(u_3) \leq \deg(u_0)$, which proves Lemma 8.   □

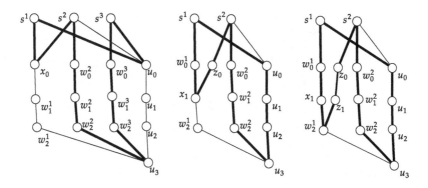

Fig. 4. Bold cycles are 12- and 14-cycles appearing whenever nodes in B(3) have merged paths.

D Proof of Lemma 9

We treat each $i = 4, \ldots, 6$ sequentially.

Claim 6. *If $f(4) \geq 5$ then $|B(4)| \leq 1$.*

Proof of claim. Let $U = \{u_0\}$. According to Lemma 6 applied with $b = 2$, we have that if $|B(4)| \geq 2$ then (1) for all $s^1 \in B(4)$, there is a well-colored path $P^1 = (s^1, w_0^1, w_1^1, w_2^1, w_3^1)$ to u_4 that does not intersect U, and (2) for all $s^2 \in B(3) \smallsetminus \{s^1\}$, there is a well-colored path $P^2 = (s^2, w_0^2, w_1^2, w_2^2, w_3^2)$ to u_4 that does not intersect $U \cup P^1$. It follows that

$$(u_0, s^1, w_0^1, w_1^1, w_2^1, w_3^1, u_4, w_3^2, w_2^2, w_1^2, w_0^2, s^2)$$

is a 12-cycle (see Fig. 5-left), which contradicts $s^1, s^2 \in B(4)$. ◇

The arguments for $i = 5, 6$ are a reformulation of Observation 2 in [4], we give them for the sake of completeness.

Claim 7. *If $f(5) \geq 5$ then $B(5) = \emptyset$.*

Proof of claim. Let $U = \{u_0, u_1, u_2, u_3, u_4\}$. By Lemma 6 applied with $b = 1$, if $B(5)$ is not empty, then for every $s^1 \in B(5)$, there is a well-colored path $P^1 = (s^1, w_0^1, w_1^1, w_2^1, w_3^1, w_4^1)$ to u_5 that does not intersect U. Therefore,

$$(u_0, s^1, w_0^1, w_1^1, w_2^1, w_3^1, w_4^1, u_5, u_4, u_3, u_2, u_1)$$

is a 12-cycle (see Fig. 5-center), which contradicts $s^1 \in B(5)$. ◇

Claim 8. *If $f(6) \geq 6$ then $B(6) = \emptyset$.*

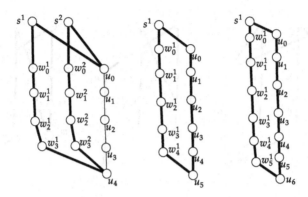

Fig. 5. Bold cycles are 12- and 14-cycles appearing whenever there are two nodes in B(4), or one node in B(5), B(6).

Proof of claim. Let $U = \{u_0, u_1, u_2, u_3, u_4, u_5\}$. Applying Lemma 6 with $b = 1$, we get that if B(6) is not empty, then, for every $s^1 \in$ B(6), there is a well-colored path $P^1 = (s^1, w_0^1, w_1^1, w_2^1, w_3^1, w_4^1, w_5^1)$ to u_6 that does not intersect U. As a consequence,

$$(u_0, s^1, w_0^1, w_1^1, w_2^1, w_3^1, w_4^1, w_5^1, u_6, u_5, u_4, u_3, u_2, u_1)$$

is a 14-cycle (see Fig. 5-right), which contradicts $s^1 \in$ B(6).   ◇

The lemma directly follows from Claims 6–8.   □

E   Deciding $\{C_{10}, C_{12}\}$-Freeness

We begin by deciding C_{10}-freeness on its own. This is doable using the threshold approach, with the threshold $T_5(i)$, $i = 1, \ldots, 4$, given in [4]. If a heavy 10-cycle $C = (u_0, \ldots, u_9)$ exists, then, by repeating $O(n^{4/5})$ times the choice of s, the probability that C is detected is at least $9/10$. If no node has rejected during the search for a 10-cycle, then we can assume that the graph is C_{10}-free. The search for a 12-cycle is still performed by the threshold algorithm. Note that we have proved that deciding C_{12}-freeness cannot be done by a threshold algorithm. However, we are now working under the assumption that the graph is C_{10}-free. Searching for light 12-cycles can be done in $O(n^{5/6})$ rounds. Let us assume that there exists a heavy 12-cycle $C^\star = \{u_0, \ldots, u_{11}\}$ where u_0 has maximum degree in the cycle, and, for every $i = 0, \ldots, 11$, node u_i has picked color i. Let us fix $T_6(i) = T_5(i)$ for $i = 1, \ldots, 4$. Then there is at least a constant fraction of neighbors s of u_0 such that, if s is chosen by the algorithm, then u_i and u_{12-i} receive at most $T_6(i)$ identifiers, for all $1 \le i \le 4$.

Finally let $T_6(5) = T_6(4)$. This is sufficient because, by Lemma 4, if u_5 receives more than $T_6(4)$ identifiers, then there exist two node-disjoint well-colored paths from s to u_5. The combination of these two paths is a 12-cycle. Therefore a node s picked by the algorithm is either in a 12-cycle of his own or will not cause u_5 to receive more than $T_6(4)$ identifiers.

F   Proof of Theorem 3

This section is dedicated to the proof of Theorem 3. In essence, the algorithm deciding $\{C_{4\ell} \mid 1 \leq \ell \leq k\}$-freeness performs by successively deciding $C_{4\ell}$-freeness, for $\ell = 1, \ldots, k$, using threshold-based algorithms. This sequence of algorithms runs in $O(n^{1-1/2k})$ rounds. The same holds for deciding $\{C_{4\ell+2} \mid 1 \leq \ell \leq k\}$-freeness, with round-complexity $O(n^{1-1/(2k+1)})$. The proof mostly consists in showing that thresholds can be defined with the guarantee that, if the graph contains a 4ℓ-cycle (or a $(4\ell + 2)$-cycle), then this cycle is detected with constant probability. Let us start with $\mathcal{F}_k = \{C_{4\ell} \mid 1 \leq \ell \leq k\}$, $k \geq 1$.

The base case is $\ell = 1$, i.e., deciding C_4-freeness. It was shown in [4] that, in this case, a threshold $T_2(1) = 1$ suffices. Let us now assume that we have set up appropriate thresholds for deciding $\mathcal{F}_{k-1}$-freeness. For deciding $\mathcal{F}_k$-freeness, it is sufficient to decide $\mathcal{F}_{k-1}$-freeness first, and then deciding C_{4k}-freeness. Therefore it is sufficient that the algorithm deciding C_{4k}-freeness succeeds whenever the graph is $\mathcal{F}_{k-1}$-free. So, from this point on, we assume that the graph has no 4ℓ-cycles, for all $\ell = 1, \ldots, k-1$. To decide C_{4k}-freeness under this latter hypothesis, we set the thresholds as $T_{2k}(0) = 1$, and, for $i \geq 1$,

$$T_{2k}(i) = \begin{cases} T_{2k}(i-1) & \text{if } i \text{ is odd,} \\ (i+1) \cdot T_{2k}(i-1) & \text{if } i \text{ is even.} \end{cases}$$

To prove that the thresholds work, let $C = (u_0, \ldots, u_{4k-1})$ be a $4k$-cycle in the graph, where u_0 is the heavy node with largest degree in C. Let us assume that, for every $i = 0, \ldots, 4k - 1$, node u_i is colored i. Note that this occurs with probability at least $1/(4k+1)^{4k}$. Let s be a neighbor of u_0 not belonging to any $4k$-cycle, and let us assume that s is colored -1. We define a well-colored path from s to a node u_i as a path of the form $s, w_0, \ldots, w_{i-1}, u_i$ where, for every $j = 0, \ldots, i - 1$, node w_j is colored j.

To prove that our thresholds are sufficient, we now consider separately the odd and even indices.

Odd Indices. Let i be an odd index with $1 \leq i \leq 2k - 1$, and let ρ be the maximum number of node-disjoint, well-colored paths from s to u_i. Let us prove by contradiction that $\rho = 1$ (see Fig. 6(left) for an illustration of the case $\rho \geq 2$). Indeed, $\rho \geq 1$ as $\{u_0, \ldots, u_{i-1}\}$ is a well-colored path from s to u_i. Suppose now that there exist two node-disjoint well-colored paths from s to u_i, denoted by P and P'. Then $\{s\} \cup P \cup \{u_i\} \cup P'$ is a $(2i + 2)$-cycle. As i is odd, we get that $2i + 2$ is a multiple of 4. If $i \leq 2k - 3$, this contradicts our assumption that the graph is $\mathcal{F}_{k-1}$-free. If $i = 2k - 1$, then s is in a $4k$-cycle that the algorithm finds when it performs color-BFS(s). Since $\rho = 1$, by Lemma 4, node u_i receives at most $T_{2k}(i - 1)$ identifiers.

Even Indices. Let i be an even index with $1 \leq i \leq 2k - 2$, and let ρ be the maximum number of node-disjoint well-colored paths from s to u_i. Let us assume that s is such that u_i receives more than $(i + 1) \cdot T_{2k}(i - 1)$ identifiers from nodes colored $i - 1$. By Lemma 4, $\rho \geq i + 2$. Therefore, there exists a well-colored path from s to u_i that does not go through u_0. Let us denote this path $P = \{w_0, \ldots, w_{i-1}\}$. Let s' be the next neighbor of u_0 colored -1 that the algorithm picks. Let us define ρ' as the maximum number of node-disjoint well-colored paths from s' to u_i. If for the source node s', node u_i also receives more than $(i + 1) \cdot T_{2k}(i - 1)$ identifiers from nodes colored $i - 1$, then, thanks to Lemma 4, $\rho' \geq i+2$. It follows that there exists a well-colored path from s' to u_i that does not go through any of the $i + 1$ nodes $u_0, w_0, \ldots, w_{i-1}$. Let us denote this path by P'. As a consequence, $\{s, u_0, s'\} \cup P' \cup \{u_i\} \cup P$ is a $(2i + 4)$-cycle (see Fig. 6(right)). Since i is even, $2i + 4$ is a multiple of 4. If $i \leq 2k - 4$, this contradicts the assumption that the graph is $\mathcal{F}_{k-1}$-free. If $i = 2k - 2$, then s is in a $4k$-cycle, which is detected by s. Therefore, if the algorithm picks a neighbor s of u_0 colored -1 that causes u_i to receive more than $T_{2k}(i)$ identifiers, then no other neighbor s' of u_0 can also cause u_i to receive more than $T_{2k}(i)$ identifiers, as s would then be in a cycle that would be detected by s.

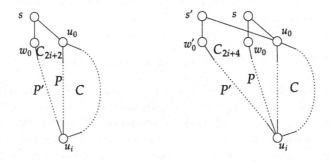

Fig. 6. Cycles of $\mathcal{F}_{k-1}$ or $4k$-cycles appearing whenever neighbors of u_0 have node-disjoint well-colored paths to u_i. On the left, i is odd; On the right, i is even.

Wrap Up. It results from our analyses for odd and even indices that at most $k - 1$ neighbors s of u_0 colored -1 may prevent the detection of the $4k$-cycle $(u_0, \ldots, u_{4k-1})$, namely at most one source s for each node u_{2i}, for $1 \leq i \leq k-1$. Since there are $\Omega(n^{1/2k})$ neighbors of u_0 colored -1, the algorithm will randomly choose at least k different neighbors of u_0 whenever $\Theta(n^{1-1/2k})$ random choices are performed. This completes the proof for $\{C_{4\ell} \mid 1 \leq \ell \leq k\}$-freeness, $k \geq 1$.

$\{C_{4\ell+2} \mid 1 \leq \ell \leq k\}$-*freeness.* The proof for $\{C_{4\ell+2} \mid 2 \leq \ell \leq k\}$-freeness can be adapted with very little changes from the proof for $\{C_{4\ell} \mid 2 \leq \ell \leq k\}$-freeness by inverting the roles of odd and even values of i to compute $T_{2k+1}(i)$.

Indeed, let i be an even index. Similarly to the case of odd indices for $\{C_{4\ell} \mid 2 \leq \ell \leq k\}$-freeness, if there are two node-disjoint well-colored paths from a picked s to u_i, then s is in a $(2i + 2)$-cycle (see Fig. 6(left)). With i being even, the cycle belongs to $\{C_{4\ell+2} \mid 2 \leq \ell \leq k\}$.

On the other hand, let i be an odd index. Similarly to the case of even indices for $\{C_{4\ell} \mid 2 \leq \ell \leq k\}$-freeness, if u_i receives more than $(i+1)T_{2k}(i-1)$ identifiers for two different source nodes s and s', then s is in a $(2i+4)$-cycle (see Fig. 6(right)). With i being odd, the cycle belongs to $\{C_{4l+2} \mid 2 \leq l \leq k\}$.

Consequently, the algorithm works by fixing $T_{2k}(0) = 1$ and, for $i \geq 1$,

$$T_{2k}(i) = \begin{cases} (i+1) \cdot T_{2k}(i-1) & \text{if } i \text{ is odd}, \\ T_{2k}(i-1) & \text{if } i \text{ is even}. \end{cases}$$

This completes the proof of Theorem 3. □

References

1. Alon, N., Yuster, R., Zwick, U.: Color-coding. J. ACM **42**(4), 844–856 (1995)
2. Broersma, H., Golovach, P.A., Paulusma, D., Song, J.: Determining the chromatic number of triangle-free 2P3-free graphs in polynomial time. Theor. Comput. Sci. **423**, 1–10 (2012)
3. Censor-Hillel, K.: Distributed subgraph finding: progress and challenges. arXiv preprint arXiv:2203.06597 (2022)
4. Censor-Hillel, K., Fischer, O., Gonen, T., Gall, F.L., Leitersdorf, D., Oshman, R.: Fast distributed algorithms for girth, cycles and small subgraphs. In: 34th International Symposium on Distributed Computing (DISC). LIPIcs, vol. 179, pp. 33:1–33:17. Schloss Dagstuhl - Leibniz-Zentrum für Informatik (2020)
5. Censor-Hillel, K., Kaski, P., Korhonen, J.H., Lenzen, C., Paz, A., Suomela, J.: Algebraic methods in the congested clique. In: 34th ACM Symposium on Principles of Distributed Computing (PODC), pp. 143–152 (2015)
6. Drucker, A., Kuhn, F., Oshman, R.: On the power of the congested clique model. In: 33rd ACM Symposium on Principles of Distributed Computing (PODC), pp. 367–376 (2014)
7. Eden, T., Fiat, N., Fischer, O., Kuhn, F., Oshman, R.: Sublinear-time distributed algorithms for detecting small cliques and even cycles. Distrib. Comput. **35**(3), 207–234 (2022)
8. Even, G., et al.: Three notes on distributed property testing. In: 31st International Symposium on Distributed Computing (DISC). LIPIcs, vol. 91, pp. 15:1–15:30. Schloss Dagstuhl - Leibniz-Zentrum für Informatik (2017)
9. Fischer, O., Gonen, T., Kuhn, F., Oshman, R.: Possibilities and impossibilities for distributed subgraph detection. In: 30th on Symposium on Parallelism in Algorithms and Architectures (SPAA), pp. 153–162 (2018)
10. Fraigniaud, P., Olivetti, D.: Distributed detection of cycles. ACM Trans. Parallel Comput. **6**(3), 1–20 (2019)
11. Korhonen, J.H., Rybicki, J.: Deterministic subgraph detection in broadcast congest. arXiv preprint arXiv:1705.10195 (2017)
12. Kővári, P., T Sós, V., Turán, P.: On a problem of Zarankiewicz. In: Colloquium Mathematicum, vol. 3, pp. 50–57. Polska Akademia Nauk (1954)
13. Neuen, D.: Isomorphism testing for graphs excluding small topological subgraphs. In: 33rd ACM Symposium on Discrete Algorithms (SODA), pp. 1411–1434 (2022)
14. Peleg, D.: Distributed computing: a locality-sensitive approach. SIAM (2000)

Energy-Efficient Distributed Algorithms for Synchronous Networks

Pierre Fraigniaud[1], Pedro Montealegre[2](✉), Ivan Rapaport[3], and Ioan Todinca[4]

[1] Institut de Recherche en Informatique Fondamentale (IRIF),
CNRS and Université Paris Cité, Paris, France
`pierre.fraigniaud@irif.fr`

[2] Facultad de Ingeniería y Ciencias, Universidad Adolfo Ibáñez,
Santiago, Chile
`p.montealegre@uai.cl`

[3] Departamento de Ingeniería Matemática - Centro de Modelamiento Matemático
(UMI 2807 CNRS), Universidad de Chile, Santiago, Chile
`rapaport@dim.uchile.cl`

[4] Laboratoire d'informatique fondamentale d'Orléans (LIFO),
Université d'Orléans, Orléans, France
`Ioan.Todinca@univ-orleans.fr`

Abstract. We study the design of energy-efficient algorithms for the LOCAL and CONGEST models. Specifically, as a measure of complexity, we consider the maximum, taken over all the edges, or over all the nodes, of the number of rounds at which an edge, or a node, is active in the algorithm. We first show that every Turing-computable problem has a CONGEST algorithm with constant node-activation complexity, and therefore constant edge-activation complexity as well. That is, every node (resp., edge) is active in sending (resp., transmitting) messages for only $O(1)$ rounds during the whole execution of the algorithm. In other words, every Turing-computable problem can be solved by an algorithm consuming the least possible energy. In the LOCAL model, the same holds obviously, but with the additional feature that the algorithm runs in $O(\mathrm{poly}(n))$ rounds in n-node networks. However, we show that insisting on algorithms running in $O(\mathrm{poly}(n))$ rounds in the CONGEST model comes with a severe cost in terms of energy. Namely, there are problems requiring $\Omega(\mathrm{poly}(n))$ edge-activations (and thus $\Omega(\mathrm{poly}(n))$ node-activations as well) in the CONGEST model whenever solved by algorithms bounded to run in $O(\mathrm{poly}(n))$ rounds. Finally, we demonstrate the existence of a sharp separation between the edge-activation complexity and the node-activation complexity in the CONGEST model, for algorithms bounded to run in $O(\mathrm{poly}(n))$ rounds. Specifically, under this

This work was performed during the visit of the first and last authors to Universidad de Chile, and to Universidad Adolfo Ibañez, Chile.

P. Fraigniaud—Additional support from ANR project DUCAT (ref. ANR-20-CE48-0006).

P. Montealegre and I. Rapaport—Additional support from ANID via PIA/Apoyo a Centros Científicos y Tecnológicos de Excelencia AFB 170001, Fondecyt 1220142 and Fondecyt 1230599.

S. Rajsbaum et al. (Eds.): SIROCCO 2023, LNCS 13892, pp. 482–501, 2023.
https://doi.org/10.1007/978-3-031-32733-9_21

constraint, there is a problem with $O(1)$ edge-activation complexity but $\tilde{\Omega}(n^{1/4})$ node-activation complexity.

Keywords: Synchronous distributed algorithms · LOCAL and CONGEST models · Energy efficiency

1   Introduction

1.1   Objective

Designing computing environments consuming a limited amount of energy while achieving computationally complex tasks is an objective of utmost importance, especially in distributed systems involving a large number of computing entities. In this paper, we aim at designing energy-efficient algorithms for the standard LOCAL and CONGEST models of distributed computing in networks [11]. Both models assume a network modeled as an n-node graph $G = (V, E)$, where each node is provided with an identifier, i.e., an integer that is unique in the network, which can be stored on $O(\log n)$ bits. All nodes are assumed to run the same algorithm, and computation proceeds as a series of synchronous rounds (all nodes start simultaneously at round 1). During a round, every node sends a message to each of its neighbors, receives the messages sent by its neighbors, and performs some individual computation. The two models LOCAL and CONGEST differ only in the amount of information that can be exchanged between nodes at each round.

The LOCAL model does not bound the size of the messages, whereas the CONGEST model allows only messages of size $O(\log n)$ bits. Initially, every node $v \in V$ knows solely its identifier $\mathrm{id}(v)$, an upper bound of the number n of nodes, which is assumed to be polynomial in n and to be the same for all nodes, plus possibly some input bit-string $x(v)$ depending on the task to be solved by the nodes. In this paper, we denote by N the maximum between the largest identifier and the upper bound on n given to all nodes. Hence $N = O(\mathrm{poly}(n))$, and is supposed to be known by all nodes. After a certain number of rounds, every node outputs a bit-string $y(v)$, where the correctness of the collection of outputs $y = \{y(v) : v \in V\}$ is defined with respect to the specification of the task to be solved, and may depend on the collection of inputs $x = \{x(v) : v \in V\}$ given to the nodes, as well as on the graph G (but not on the identifiers assigned to the nodes, nor on the upper bound N).

Activation Complexity. We measure the energy consumption of an algorithm A by counting how many times each node and each edge is activated during the execution of the algorithm. More specifically, a node v (resp., an edge e) is said to be *active* at a given round r if v is sending a message to at least one of its neighbors at round r (resp., if a message traverses e at round r). The *node-activation* and the *edge-activation* of an algorithm A running in a graph $G = (V, E)$ are respectively defined as

$$\mathsf{nact}(A) := \max_{v \in V} \#\mathrm{activation}(v), \quad \text{and} \quad \mathsf{eact}(A) := \max_{e \in E} \#\mathrm{activation}(e),$$

where #activation(v) (resp., #activation(e)) denotes the number of rounds during which node v (resp., edge e) is active along the execution of the algorithm A. By definition, we have that, in any graph of maximum degree Δ,

$$\mathsf{eact}(A) \leq 2 \cdot \mathsf{nact}(A), \quad \text{and} \quad \mathsf{nact}(A) \leq \Delta \cdot \mathsf{eact}(A). \tag{1}$$

Objective. Our goal is to design *frugal* algorithms, that is, algorithms with *constant* node-activation, or to the least *constant* edge-activation, independent of the number n of nodes and of the number m of edges. Indeed, such algorithms can be viewed as consuming the least possible energy for solving a given task. Moreover, even if the energy requirement for solving the task naturally grows with the number of components (nodes or edges) of the network, it grows *linearly* with this number whenever using frugal algorithms. We refer to *node-frugality* or *edge-frugality* depending on whether we focus on node-activation or edge-activation, respectively.

1.2   Our Results

We first show that every Turing-computable problem[1] can thus be solved by a node-frugal algorithm in the LOCAL model as well as in the CONGEST model. It follows from Eq. 1 that every Turing-computable problem can be solved by an edge-frugal algorithm in both models. In other words, every problem can be solved by an energy-efficient distributed algorithm. One important question remains: what is the round complexity of frugal algorithms?

In the LOCAL model, our node-frugal algorithms run in $O(\mathrm{poly}(n))$ rounds. However, they may run in exponentially many rounds in the CONGEST model. We show that this cannot be avoided. Indeed, even if many symmetry-breaking problems such as computing a maximal-independent set (MIS) and computing a $(\Delta + 1)$-coloring can be solved by a node-frugal algorithm performing in $O(\mathrm{poly}(n))$ rounds, we show that there exist problems (e.g., deciding C_4-freeness or deciding the presence of symmetries in the graph) that cannot be solved in $O(\mathrm{poly}(n))$ rounds in the CONGEST model by any edge-frugal algorithm.

Finally, we discuss the relation between node-activation complexity and edge-activation complexity. We show that the bounds given by Eq. 1 are essentially the best that can be achieved in general. Precisely, we identify a problem, namely DEPTH FIRST POINTER CHASING (DFPC), which has edge-activation complexity $O(1)$ for all graphs with an algorithm running in $O(\mathrm{poly}(n))$ rounds in the CONGEST model, but satisfying that, for every $\Delta = O\left(\left(\frac{n}{\log n}\right)^{1/4}\right)$, its node-activation complexity in graphs with maximum degree Δ is $\Omega(\Delta)$ whenever solved by an algorithm bounded to run in $O(\mathrm{poly}(n))$ rounds in the CONGEST model.

Our main results are summarized in Table 1.

---

[1] A problem is Turing-computable if there exists a Turing machine that, given any graph with identifiers and inputs assigned to the nodes, computes the output of each node in the graph.

Table 1. *Summary of our results where, for a problem Π, $\Pi \in O(f(n))$ means that the corresponding complexity of Π is $O(f(n))$ (same shortcut for Ω).*

	Awakeness	Node-Activation	Edge-Activation
LOCAL	• $\forall \Pi, \Pi \in O(\log n)$ with $O(\text{poly}(n))$ rounds [2] • ST $\in \Omega(\log n)$ [2]	• $\forall \Pi, \Pi \in O(1)$ with $O(\text{poly}(n))$ rounds	• $\forall \Pi, \Pi \in O(1)$ with $O(\text{poly}(n))$ rounds
CONGEST	• MIS $\in O(\text{polyloglog}(n))$ with $O(\text{polylog}(n))$ rounds [6] (randomized) • MST $\in O(\log n)$ with $O(\text{poly}(n))$ rounds [1]	• $\forall \Pi, \Pi \in O(1)$ • poly(n) rounds $\Rightarrow \exists \Pi \in \Omega(\text{poly}(n))$ • poly(n) rounds $\Rightarrow$ DFPC $\in \tilde{\Omega}(n^{1/4})$	• $\forall \Pi, \Pi \in O(1)$ • poly(n) rounds $\Rightarrow \exists \Pi \in \Omega(\text{poly}(n))$ • DFPC $\in O(1)$ with $O(\text{poly}(n))$ rounds • $\Pi \in$ FO and $\Delta = O(1)$ $\Rightarrow \Pi \in O(1)$ with $O(\text{poly}(n))$ rounds [8]

Our Techniques. Our upper bounds are mostly based on similar types of upper bounds techniques used in the sleeping model [2,4] (cf. Sect. 1.3), based on constructing spanning trees along with gathered and broadcasted information. However, the models considered in this paper do not suffer from the same limitations as the sleeping model (cf. Sect. 2), and thus one can achieve activation complexity $O(1)$ in scenarios where the sleeping model limits the awake complexity to $\Omega(\log n)$.

Our lower bounds for CONGEST are based on reductions from 2-party communication complexity. However, as opposed to the standard CONGEST model in which the simulation of a distributed algorithm by two players is straightforward (each player performs the rounds sequentially, one by one, and exchanges the messages sent across the cut between the two subsets of nodes handled by the players at each round), the simulation of distributed algorithms in which only subsets of nodes are active at various rounds requires more care. This is especially the case when the simulation must not only control the amount of information exchanged between these players, but also the number of communication steps performed by the two players. Indeed, there are 2-party communication complexity problems that are hard for k steps, but trivial for $k + 1$ steps [10], and some of our lower bounds rely on this fact.

1.3 Related Work

The study of frugal algorithms has been initiated in [8], which focuses on the edge-frugality in the CONGEST model. It is shown that for *bounded-degree graphs*, any problem expressible in first-order logic (e.g., C_4-freeness) can be solved by an edge-frugal algorithm running in $O(\text{poly}(n))$ rounds in the CONGEST model. This also holds for planar graphs with no bounds on the maximum degree, whenever the nodes are provided with their local combinatorial embedding. Our results show that these statements cannot be extended to arbitrary graphs as we prove that any algorithm solving C_4-freeness in $O(\text{poly}(n))$ rounds in the CONGEST model has edge-activation $\tilde{\Omega}(\sqrt{n})$.

More generally, the study of energy-efficient algorithms in the context of distributed computing in networks has been previously considered in the framework of the *sleeping* model, introduced in [4]. This model assumes that nodes can be in two states: *awake* and *asleep*. A node in the awake state performs as in the LOCAL and CONGEST models, but may also decide to fall asleep, for a prescribed amount of rounds, controlled by each node, and depending on the algorithm executed at the nodes. A sleeping node is totally inactive in the sense that it does not send messages, it cannot receive messages (i.e., if a message is sent to a sleeping node by an awake neighbor, then the message is lost), and it is computationally idle (apart from counting rounds). The main measure of interest in the sleeping model is the *awake complexity*, defined as the maximum, taken over all nodes, of the number of rounds at which each node is awake during the execution of the algorithm.

In the LOCAL model, it is known [2] that all problems have awake complexity $O(\log n)$, using algorithms running in $O(\text{poly}(n))$ rounds. This bound is tight in the sense that there are problems (e.g., spanning tree construction) with awake complexity $\Omega(\log n)$ [2,3].

In the CONGEST model, It was first shown [4] that MIS has constant *average* awake complexity, thanks to a *randomized* algorithm running in $O(\text{polylog}(n))$ rounds. The round complexity was improved in [7] with a *randomized* algorithm running in $O(\log n)$ rounds. The (worst-case) awake complexity of MIS was proved to be $O(\log \log n)$ using a *randomized* Monte-Carlo algorithm running in $O(\text{poly}(n))$ rounds [6]. This (randomized) round complexity can even be reduced to $O(\log^3 n \cdot \log \log n \cdot \log^\star n)$, at the cost of slightly increasing the awake complexity to $O(\log \log n \cdot \log^\star n)$. MST has also been considered, and it was proved [1] that its (worst-case) awake complexity is $O(\log n)$ thanks to a (deterministic) algorithm running in $O(\text{poly}(n))$ rounds. The upper bound on the awake complexity of MST is tight, thank to the lower bound for spanning tree (ST) in [2].

2   Preliminaries

In this section, we illustrate the difference between the standard LOCAL and CONGEST models, their sleeping variants, and our node- and edge-activation variants. Figure 1(a) displays the automaton corresponding to the behavior of a node in the standard models. A node is either *active* (A) or *terminated* (T). At each clock tick (i.e., round) a node is subject to message events corresponding to sending and receiving messages to/from neighbors. A node remains active until it terminates.

Figure 1(b) displays the automaton corresponding to the behavior of a node in the sleeping variant. In this variant, a node can also be in a *passive* (P) state. In this state, the clock event can either leave the node passive, or awake the node, which then moves back to the active state.

Finally, Fig. 1(c) displays the automaton corresponding to the behavior of a node in our activation variants. It differs from the sleeping variant in that a passive node is also subject to message events, which can leave the node passive, but may also move the node to the active state. In particular, a node does not need to be active for receiving messages, and incoming messages may not trigger an immediate response from the node (e.g., forwarding information). Instead, a node can remain passive while collecting information from each of its neighbors, and eventually react by becoming active.

Example 1: Broadcast. Assume that one node of the n-node cycle C_n has a token to be broadcast to all the nodes. Initially, all nodes are active. However, all nodes but the one with the token become immediately passive when the clock ticks for entering the second round. The node with the token sends the token to one of its neighbors, and becomes passive at the next clock tick. Upon reception of the token, a passive node becomes active, forwards the token, and terminates. When the source node receives the token back, it becomes active, and terminates. The node-activation complexity of broadcast is therefore $O(1)$, whereas it is known that broadcasting has awake complexity $\Omega(\log n)$ in the sleeping model [2].

Example 2: At-Least-One-Leader. Assume that each node of the cycle C_n has an input-bit specifying whether the node is leader or not, and the nodes must collectively check that there is at least one leader. Every leader broadcasts a token, outputs accept, and terminates. Non-leader nodes become passive immediately after the beginning of the algorithm, and start waiting for N rounds (recall that N is an upper bound on the number n of nodes). Whenever the "sleep" of a (passive) non-leader is interrupted by the reception of a token, it becomes active, forwards the token, outputs accept, and terminates. After N rounds, a passive node that has not been "awaken" by a token becomes active, outputs reject, and terminates. This guarantees that there is at least one leader if and only if all nodes accept. The node-activation complexity of this algorithm is $O(1)$, while the awake complexity of at-least-one-leader is $\Omega(\log n)$ in the sleeping model, by reduction to broadcast.

The following observation holds for LOCAL and CONGEST, by noticing that every algorithm for the sleeping model can be implemented with no overheads in terms of node-activation.

Observation 1. *In n-node graphs, every algorithm with awake complexity $a(n)$ and round complexity $r(n)$ has node-activation complexity at most $a(n)$ and round complexity at most $r(n)$.*

It follows from Observation 1 that all upper bound results for the awake complexity directly transfer to the node-activation complexity. However, as we shall show in this paper, in contrast to the sleeping model in which some problems (e.g., spanning tree) have awake complexity $\Omega(\log n)$, even in the LOCAL model, all problems admit a frugal algorithm in the CONGEST model, i.e., an algorithm with node-activation $O(1)$.

Definition 1. *A LOCAL or CONGEST algorithm is* node-frugal *(resp., edge-frugal) if the activation of every node (resp., edge) is upper-bounded by a constant independent of the graph, and of the identifiers and inputs given to the nodes.*

3   Universality of Frugal Algorithms

In this section we show that every Turing-computable problem can be solved by frugal algorithms, both in the LOCAL and CONGEST models. Thanks to Eq. 1, it is sufficient to prove that this holds for node-frugality.

Lemma 1. *There exists a CONGEST algorithm electing a leader, and constructing a BFS tree rooted at the leader, with node-activation complexity $O(1)$, and performing in $O(N^2) = O(\mathrm{poly}(n))$ rounds.*

Proof. The algorithm elects as leader the node with smallest identifier, and initiates a breadth-first search from that node. At every node v, the protocol performs as follows.

- If v has received no messages until round $\mathrm{id}(v) \cdot N$, then v elects itself as leader, and starts a BFS by sending message $(\mathrm{id}(v), 0)$ to all its neighbors. Locally, v sets its parent in the BFS tree to $\bot$, and the distance to the root to 0.
- Otherwise, let r be the first round at which vertex v receives a message. Such a message is of type $(\mathrm{id}(u), d)$ where u is the neighbor of v which sent the message to v, and d is the distance from u to the leader in the graph. Node v sets its parent in the BFS tree to $\mathrm{id}(u)$, its distance to the root to $d+1$, and, at round $r+1$, it sends the message $(\mathrm{id}(v), d+1)$ to all its neighbors. (If v receives several messages at round r, from different neighbors, then v selects the messages coming from the neighbors with smallest identifier).

The node v with smallest identifier is indeed the node initiating the BFS, as the whole BFS is constructed between rounds $\mathrm{id}(v) \cdot N$ and $\mathrm{id}(v) \cdot N + N - 1$, and $N \geq n$. The algorithm terminates at round at most $O(N^2)$.   □

An instance of a problem is a triple (G, id, x) where $G = (V, E)$ is an n-node graph, $\mathrm{id} : V \to [1, N]$ with $N = O(\mathrm{poly}(n))$, and $x : V \to [1, \nu]$ is the input assignment to the nodes. Note that the input range ν may depend on n, and even be exponential in n, even for classical problems, e.g., whenever weights assigned to the edges are part of the input. A solution to a graph problem is an output assignment $y : V \to [1, \mu]$, and the correctness of y depends on G and x only, with respect to the specification of the problem. We assume that μ and ν are initially known to the nodes, as it is the case for, e.g., MST, in which the weights of the edges can be encoded on $O(\log n)$ bits.

Theorem 1. *Every Turing-computable problem has a LOCAL algorithm with $O(1)$ node-activation complexity, and running in $O(N^2) = O(\mathrm{poly}(n))$ rounds.*

Proof. Once the BFS tree T of Lemma 1 is constructed, the root can (1) gather the whole instance (G, id, x), (2) compute a solution y, and (3) broadcast y to all nodes. Specifically, every leaf v of T sends the set

$$E(v) = \big\{\{(\mathsf{id}(v), x(v)), (\mathsf{id}(w), x(w))\} : w \in N(v)\big\}$$

to its parent in T. An internal node v waits for receiving a set of edges $S(u)$ from each of its children u in T, and then forwards the set

$$S(v) = E(v) \cup (\cup_{u \in \mathrm{child}(v)} S(u))$$

to its parent. This set can be encoded in $O(N^2)$ bits by the adjacency matrix of the subgraph induced by the edges in $S(v)$. Each node of T is activated once during this phase, and thus the node-activation complexity of gathering is 1. Broadcasting the solution y from the leader to all the nodes is achieved along the edges of T, again with node-activation 1. □

The algorithm used in the proof of Theorem 1 cannot be implemented in CONGEST due to the size of the messages, which may require each node to be activated more than a constant number of times. To keep the node-activation constant, we increased the round complexity of the algorithm.

Lemma 2. *Every node-frugal algorithm $\mathcal{A}$ performing in R rounds in the LOCAL model with messages of size at most M bits[2] can be implemented by a node-frugal algorithm $\mathcal{B}$ performing in $R\,2^M$ rounds in the CONGEST model.*

Proof. Let v be a node sending a message m through an incident edge e at round r of $\mathcal{A}$. Then, in $\mathcal{B}$, v sends one "beep" through edge e at round $r\,2^M + t$ where t is lexicographic rank of m among the at most 2^M messages generated by $\mathcal{A}$. □

Theorem 2. *Every Turing-computable problem has a CONGEST algorithm with $O(1)$ node-activation complexity, and running in $2^{\mathrm{poly}(n)(1+\log(\nu\mu))}$ rounds for inputs in the range $[1, \nu]$ and outputs in the range $[1, \mu]$.*

Proof. The algorithm used in the proof of Theorem 1 used messages of size at most $N^2 + N \log \nu$ bits during the gathering phase, and of size at most $N \log \mu$ bits during the broadcast phase. The result follows from Lemma 2. □

Of course, there are many problems that can be solved in the CONGEST model by a frugal algorithm much faster than the bound from Theorem 2. This is typically the case of all problems that can be solved by a sequential greedy algorithm visiting the nodes in arbitrary order, and producing a solution at the currently visited node based only on the partial solution in the neighborhood of the node. Examples of such problems are MIS, $\Delta + 1$-coloring, etc. We call such problem *sequential-greedy.*

---

[2] Without loss of generality, in $\mathcal{A}$ each node sends the same message to all its neighbors at each round when it is active. Otherwise, the different messages can be merged into one, by adding the identifiers of the neighbors.

Theorem 3. *Every sequential-greedy problem whose solution at every node can be encoded on $O(\log n)$ bits has a node-frugal CONGEST algorithm running in $O(N) = O(\mathrm{poly}(n))$ rounds.*

Proof. Every node $v \in V$ generates its output at round $\mathrm{id}(v)$ according to its current knowledge about its neighborhood, and sends this output to all its neighbors.   □

4 Limits of CONGEST Algorithms with Polynomially Many Rounds

Given a graph $G = (V, E)$ such that V is partitioned in two sets V_A, V_B, the set of edges with one endpoint in V_A and the other in V_B is called the *cut*. We denote by $e(V_A, V_B)$ the number of edges in the cut, and by $n(V_A, V_B)$ the number of nodes incident to an edge of the cut. Consider the situation where there are two players, namely Alice and Bob. We say that a player controls a node v if it knows all its incident edges and its input. For a CONGEST algorithm $\mathcal{A}$, we denote $\mathcal{A}(\mathcal{I})$ the output of $\mathcal{A}$ on input $\mathcal{I} = (G, \mathrm{id}, x)$. We denote $R_{\mathcal{A}}(n)$ the round complexity of $\mathcal{A}$ on inputs of size n.

Lemma 3 (Simulation lemma). *Let $\mathcal{A}$ be an algorithm in the CONGEST model, let $\mathcal{I} = (G, \mathrm{id}, x)$ be an input for $\mathcal{A}$, and let V_A, V_B be a partition of $V(G)$. Suppose that Alice controls all the nodes in V_A, and Bob controls all the nodes in V_B. Then, there exists a communication protocol $\mathcal{P}$ between Alice and Bob with at most $2 \cdot \min(n(V_A, V_B) \cdot \mathsf{nact}(\mathcal{A}), e(V_A, V_B) \cdot \mathsf{eact}(\mathcal{A}))$ rounds and using total communication $\mathcal{O}(\min(n(V_A, V_B) \cdot \mathsf{nact}(\mathcal{A}), e(V_A, V_B) \cdot \mathsf{eact}(\mathcal{A})) \cdot (\log n + \log R_{\mathcal{A}}(n))$, such that each player computes the value of $\mathcal{A}(\mathcal{I})$ at all nodes he or she controls.*

Proof. In protocol $\mathcal{P}$, Alice and Bob simulate the rounds of algorithm $\mathcal{A}$ in all the nodes they control. The simulation run in phases. Each phase is used to simulate up to a certain number of rounds t of algorithm $\mathcal{A}$, and takes two rounds of protocol $\mathcal{P}$ (one round for Alice, and one round for Bob). By simulating $\mathcal{A}$ up to t rounds, we mean that Alice and Bob know all the states of all the nodes they control, on every round up to round t.

In the first phase, players start simulating $\mathcal{A}$ from the initial state. Let us suppose that both Alice and Bob have already executed $p \geq 0$ phases, meaning that they had correctly simulated $\mathcal{A}$ up to round $t = t(p) \geq 0$. Let us explain phase $p + 1$ (see also Fig. 2).

Starting from round t, Alice runs an *oblivious simulation* of algorithm $\mathcal{A}$ over all nodes that she controls. By oblivious, we mean that Alice assumes that no node of V_B communicates a message to a node in V_A in any round at least t. The oblivious simulation of Alice stops in one of the following two possible scenarios:

(1) All nodes that she controls either terminate or enter into a passive state that quits only on an incoming message from V_B.

(2) The simulation reaches a round r_a where a message is sent from a node in V_A to a node in V_B.

At the same time, Bob runs and oblivious simulation of $\mathcal{A}$ starting from round t (i.e. assuming that no node of V_A sends a message to a node in V_B in any round at least t). The oblivious simulation of Bob stops in one of the same two scenarios analogous to the ones above. In this case, we call r_b the round reached by Bob in his version of scenario (2).

At the beginning of a phase, it is the turn of Alice to speak. Once the oblivious simulation of Alice stops, she is ready to send a message to Bob. If the simulation stops in the scenario (1), Alice sends a message "*scenario 1*" to Bob. Otherwise, Alice sends r_a together with all the messages sent from nodes in V_A to nodes in V_B at round r_a, to Bob. When Bob receives the message from Alice, one of the following situations holds:

Case 1: the oblivious simulation of both Alice and Bob stopped in the first scenario. In this case, since $\mathcal{A}$ is correct, there are no deadlocks. Therefore, all vertices of G reached a terminal state, meaning that the oblivious simulation of both players was in fact a real simulation of $\mathcal{A}$, and the obtained states are the output states. Therefore, Bob sends a message to Alice indicating that the simulation is finished, and indeed Alice and Bob have correctly computed the output of $\mathcal{A}$ for all the nodes they control.

Case 2: the oblivious simulation of Alice stopped in scenario (1), and the one of Bob stopped in the scenario (2). In this case, Bob infers that his oblivious simulation was correct. He sends r_b and all the messages communicated in round r_b through the cut to Alice. When Alice receives the message of Bob, she updates the state of the nodes she controls up to round r_b. It follows that both players have correctly simulated algorithm $\mathcal{A}$ up to round $r_b > t$.

Case 3: the oblivious simulation of Alice stopped in scenario (2), and the one of Bob stopped in scenario (1). In this case, Bob infres that the simulation of Alice was correct up to round r_a. He sends a message to Alice indicating that she has correctly simulated $\mathcal{A}$ up to round r_a, and he updates the states of all the nodes he controls up to round r_a. It follows that both players have correctly simulated $\mathcal{A}$ up to round $r_a > t$.

Case 4: the oblivious simulation of both players stopped in scenario (2), and $r_a > r_b$. Bob infers that his oblivious simulation was correct up to r_b, and that the one of Alice was not correct after round r_b. Then, the players act in the same way as described in Case 2. Thus, both players have correctly simulated $\mathcal{A}$ up to round r_b.

Case 5: the oblivious simulation of both players stopped in scenario (2), and $r_b > r_a$. Bob infers that his oblivious simulation was incorrect after round r_a, and that the one of Alice was correct up to round r_a. Then, the players act in the same way as described in Case 3. Thus, both players have correctly simulated $\mathcal{A}$ up to round r_a.

Case 6: the oblivious simulation of both players stopped in scenario (2), and $r_b = r_a$. Bob assumes that both oblivious simulations were correct. He sends r_b together with all the messages communicated from his nodes at round r_b through the cut. Then, he updates the states of all the nodes he controls up to round r_b. When Alice receives the message from Bob, she updates the states of the nodes she controls up to round r_b. It follows that both players have correctly simulated $\mathcal{A}$ up to round $r_b > t$.

Observe that, except when the algorithm terminates, on each phase of the protocol, at least one node controlled by Alice or Bob is activated. Since the number of rounds of $\mathcal{P}$ is twice the number of phases, we deduce that the total number of rounds is at most

$$2 \cdot \min(n(V_A, V_B) \cdot \mathsf{nact}(\mathcal{A}), e(V_A, V_B) \cdot \mathsf{eact}(\mathcal{A})).$$

Moreover, on each round of $\mathcal{P}$, the players communicate $O((\log(R_{\mathcal{A}}(n)) + \log n) \cdot e(V_A, V_B))$ bits. As a consequence, the total communication cost of $\mathcal{P}$ is

$$O((\log(R_{\mathcal{A}}(n)) + \log n) \cdot e(V_A, V_B)) \cdot \min(n(V_A, V_B) \cdot \mathsf{nact}(\mathcal{A}), e(V_A, V_B) \cdot \mathsf{eact}(\mathcal{A}))),$$

which completes the proof.   □

We use the simulation lemma to show that there are problems that cannot be solved by a frugal algorithm in a polynomial number of rounds. In problem C4-FREENESS, all nodes of the input graph G must accept if G has no cycle of 4 vertices, and at least one node must reject if such a cycle exists. Observe that this problem is expressible in first-order logic, in particular it has en edge-frugal algorithm with a polynomial number of rounds in graphs of bounded degree [8]. We show that, in graphs of unbounded degree, this does not hold anymore. We shall also consider problem SYMMETRY, where the input is a graph G with $2n$ nodes indexed from 1 to $2n$, and with a unique edge $\{1, n+1\}$ between $G_A = G[\{1, \ldots, n\}]$ and $G_B = G[\{n+1, \ldots, 2n\}]$. Our lower bounds holds even if every node is identified by its index. All nodes must output *accept* if the function $f : \{1, \ldots, n\} \to \{n+1, \ldots, 2n\}$ defined by $f(x) = x + n$ is an isomorphism from G_A to G_B, otherwise at least one node must output *reject*.

The proof of the following theorem is based on classic reductions from communication complexity problems EQUALITY and SET DISJOINTNESS (see, e.g., [9]), combined with Lemma 3. It can be found in Appendix A.

Theorem 4. *Any CONGEST algorithm solving* SYMMETRY *(resp.,* C4-FREE-NESS*) in polynomially many rounds has node-activation and edge-activation at least* $\Omega\left(\frac{n^2}{\log n}\right)$ *(resp.,* $\Omega\left(\frac{\sqrt{n}}{\log n}\right)$*).*

5   Node Versus Edge Activation

In this section we exhibit a problem that admits an edge-frugal CONGEST algorithm running in a polynomial number of rounds, for which any algorithm running in a polynomial number of rounds has large node-activation complexity.

We proceed by reduction from a two-party communication complexity problem. However, unlike the previous section, we are now also interested in the number of rounds of the two-party protocols. We consider protocols in which the two players Alice and Bob do not communicate simultaneously. For such a protocol $\mathcal{P}$, a *round* is defined as a maximal contiguous sequence of messages emitted by a same player. We denote by $R(\mathcal{P})$ the number of rounds of $\mathcal{P}$.

Let G be a graph, and S be a subset of nodes of G. We denote by ∂S the number of vertices in S with a neighbor in $V \setminus S$.

Lemma 4 (Round-Efficient Simulation lemma). *Let $\mathcal{A}$ be an algorithm in the CONGEST model, let $\mathcal{I} = (G, \mathrm{id}, x)$ be an input for $\mathcal{A}$, and let V_A, V_B be a partition of $V(G)$. Let us assume that Alice controls all the nodes in V_A, and Bob controls all the nodes in V_B, and both players know the value of $\mathsf{nact}(\mathcal{A})$. Then, there exists a communication protocol $\mathcal{P}$ between Alice and Bob such that, in at most $\min(\partial V_A, \partial V_B) \cdot \mathsf{nact}(\mathcal{A})$ rounds, and using total communication $O\Big(\big((\partial(V_A) + \partial(V_B)) \cdot \mathsf{nact}(\mathcal{A})\big)^2 \cdot (\log n + \log R_{\mathcal{A}}(n))\Big)$ bits, each player computes the value of $\mathcal{A}(\mathcal{I})$ at all the nodes he or she controls.*

Proof. In protocol $\mathcal{P}$, Alice and Bob simulate the rounds of algorithm $\mathcal{A}$ at all the nodes each player controls. Without loss of generality, we assume that algorithm $\mathcal{A}$ satisfies that the nodes send messages at different rounds, by merely multiplying by N the number of rounds.

Initially, Alice runs an oblivious simulation of $\mathcal{A}$ that stops in one of the following three cases:

1. Every node in V_A has terminated;
2. Every node in V_A entered into the passive state that it may leave only after having received a message from a node in V_B (this corresponds to what we called the "first scenario" in the proof of Lemma 3);
3. The number of rounds $R_{\mathcal{A}}(n)$ is reached.

Then, Alice sends to Bob the integer $t_1 = 0$, and the set M_A^1 of all messages sent from nodes in V_A to nodes in V_B in the communication rounds that she simulated, together with their corresponding timestamps. If the number of messages communicated by Alice exceeds $\mathsf{nact}(\mathcal{A}) \cdot \partial A$, we trim the list up to this threshold.

Let us suppose that the protocol $\mathcal{P}$ has run for p rounds, and let us assume that it is the turn of Bob to speak at round $p + 1$—the case where Alice speaks at round $p + 1$ can be treated in the same way. Moreover, we assume that $\mathcal{P}$ satisfies the following two conditions:

1. At round p, Alice sents an integer $t_p \geq 0$, and a list of timestamped messages M_A^p corresponding to messages sent from nodes in V_A to nodes in V_B in an oblivious simulation of $\mathcal{A}$ starting from a round t_p.
2. Bob had correctly simulated $\mathcal{A}$ at all the nodes he controls, up to round t_p.

We now describe round $p + 1$ (see also Fig. 3). Bob initiates a simulation of $\mathcal{A}$ at all the nodes he controls. However, this simulation is *not* oblivious. Specifically,

Bob simulates $\mathcal{A}$ from round t_p taking into account all the messages sent from nodes in V_A to nodes in V_B, as listed in the messages M_A^p. The simulation stops when Bob reaches a round $t_{p+1} > t_p$ at which a node in V_B sends a message to a node in V_A. Observe that, up to round t_{p+1}, the oblivious simulation of Alice was correct. At this point, Bob initiates an oblivious simulation of $\mathcal{A}$ at all the nodes he controls, starting from t_{p+1}. Finally, Bob sends to Alice t_{p+1}, and the list M_B^{p+1} of all timestamped messages sent from nodes in V_B to nodes in V_A resulting from the oblivious simulation of the nodes he controls during rounds at least t_{p+1}. Using this information, Alice infers that her simulation was correct up to round t_{p+1}, and she starts the next round for protocol $\mathcal{P}$.

The simulation carries on until one of the two players runs an oblivious simulation in which all the nodes he or she controls terminate, and no messages were sent through the cut in at any intermediate round. In this case, this player sends a message "*finish*" to the other player, and both infer that their current simulations are correct. As a consequence, each player has correctly computed the output of $\mathcal{A}$ at all the nodes he or she controls.

At every communication round during which Alice speaks, at least one vertex of V_A which has a neighbor in V_B is activated. Therefore, the number of rounds of Alice is at most $\partial V_A \cdot \mathsf{nact}(\mathcal{A})$. By the same argument, we have that the number of rounds of Bob is at most $\partial V_B \cdot \mathsf{nact}(\mathcal{A})$. It follows that

$$R(\mathcal{P}) = \min(\partial V_A, \partial V_B) \cdot \mathsf{nact}(\mathcal{A}).$$

At each communication round, Alice sends at most $\partial(V_A) \cdot \mathsf{nact}(\mathcal{A})$ timestamped messages, which can be encoded using $O\big(\partial(V_A) \cdot \mathsf{nact}(\mathcal{A}) \cdot (\log n + \log R_{\mathcal{A}}(n))\big)$ bits. Similarly, Bob sends $O\big(\partial(V_B) \cdot \mathsf{nact}(\mathcal{A}) \cdot (\log n + \log R_{\mathcal{A}}(n))\big)$ bits. It follows that

$$C(\mathcal{P}) = O\Big(\big((\partial(V_A) + \partial(V_B)) \cdot \mathsf{nact}(\mathcal{A})\big)^2 \cdot (\log n + \log R_{\mathcal{A}}(n))\Big),$$

which completes the proof.   □

In order to separate the node-activation complexity from the edge-activation complexity, we consider a problem called DEPTH FIRST POINTER CHASING, and we show that this problem can be solved by an edge-frugal CONGEST algorithm running in $O(\mathrm{poly}(n))$ rounds, whereas the node-activation complexity of any algorithm running in $O(\mathrm{poly}(n))$ rounds for this problem is $\Omega(\Delta)$, for any $\Delta \in O\big((n/\log n)^{1/4}\big)$. The lower bound is proved thanks to the Round-Efficient Simulation Lemma (Lemma 3), by reduction from the two-party communication complexity problem POINTER CHASING, for which too few rounds imply large communication complexity [10].

In the DEPTH FIRST POINTER CHASING, each node v of the graph is given as input its index $\mathrm{DFS}(v) \in [n]$ in a depth-first search ordering (as usual we denote $[n] = \{1, \ldots, n\}$). Moreover the vertex indexed i is given a function $f_i : [n] \to [n]$, and the root (i.e., the node indexed 1) is given a value $x \in [n]$ as part of its input. The goal is to compute the value of $f_n \circ f_{n-1} \circ \cdots \circ f_1(x)$ at the root.

Lemma 5. *There exists an edge-frugal CONGEST algorithm for problem* DEPTH FIRST POINTER CHASING, *with polynomial number of rounds.*

The lemma is established using an algorithm that essentially traverses the DFS tree encoded by the indices of the nodes, and performs the due partial computation of the function at every node, that is, the node with index i computes $f_i \circ f_{i-1} \ldots f_1(x)$, and forwards the result to the node with index $i+1$. The detailed proof can be found in Appendix B.

Let us recall the POINTER CHASING problem as defined in [10]. Alice is given a function $f_A : [n] \to [n]$, and a number $x_0 \in [n]$. Bob is given function $f_B : [n] \to [n]$. Both players have a parameter $k \in [n]$. Note that the size of the input given to each player is $\Theta(n \log n)$ bits. The goal is to compute $(f_A \circ f_B)^k(x_0)$, i.e., k successive iterations of $f_A \circ f_B$ applied to x_0. We give a slightly simplified version of the result in [10].

Lemma 6 (Nissan and Wigderson [10]). *Any two-party protocol for* POINTER CHASING *using less than $2k$ rounds has communication complexity* $\Omega(n - k \log n)$.

We have now all ingredients for proving the main result of this section.

Theorem 5. *For every* $\Delta \in O\left((n/\log n)^{1/4}\right)$, *every CONGEST algorithm solving* DEPTH FIRST POINTER CHASING *in graphs of maximum degree Δ with polynomialy many rounds has node-activation complexity* $\Omega(\Delta)$.

Proof. Let k be the parameter of POINTER CHASING that will be fixed later. The lower bound is established for this specific parameter k. Let us consider an arbitrary instance of POINTER CHASING $f_A, f_B : [n] \to [n]$, and $x_0 \in [n]$, with parameter k. We reduce that instance to a particular instance of DEPTH FIRST POINTER CHASING (see Fig. 4).

The graph is a tree T on n vertices, composed of a path $(v_1, \ldots, v_{n-2k})$, and $2k$ leaves $v_{n-2k+1}, \ldots, v_n$, all adjacent to v_{n-2k}. Node v_1 is called the root, and node v_{n-2k} is said central. Note that the ordering obtained by taking $\text{DFS}(v_i) = i$ is a depth-first search of T, rooted at v_1. The root v_1 is given value x_0 as input. If $i \leq n - 2k$, then function f_i is merely the identity function f (i.e., $f(x) = x$ for all x). For every $j \in [k]$, let $a_j = v_{n-k+2j-1}$, and $b_j = v_{n-k+2j}$. All nodes b_j get as input the function f_B, and all nodes a_j get the function f_A. Observe that the output of DEPTH FIRST POINTER CHASING on this instance is precisely the same as the output of the initial instance of POINTER CHASING. Indeed, $f_{n-2k} \circ f_{n-2k-1} \circ \cdots \circ f_1$ is the identity function, and the sequence $f_n \circ f_{n-1} \circ \cdots \circ f_{n-2k+2} \circ f_{n-2k+1}$ alternates nodes of "type" a_j with nodes of "type" b_j, for decreasing values of $j \in [k]$, and thus corresponds to $f_A \circ f_B \circ \cdots \circ f_A \circ f_B$, where the pair $f_A \circ f_B$ is repeated k times, exactly as in problem POINTER CHASING.

We can now apply Round-Efficient Simulation Lemma. Let Alice control all vertices a_j, for all $j \in [k]$, and vertices $v_1, \ldots, v_{n-2k}$. Let Bob control vertices b_j, for all $j \in [k]$. See Fig. 4. Note that Alice and Bob can construct the subgraph that they control, based only on their input in the considered POINTER CHASING instance, and they both now value k.

Claim. If there exists a CONGEST algorithm $\mathcal{A}$ for DEPTH FIRST POINTER CHASING on n-node graphs performing in $R_{\mathcal{A}}$ rounds with node-activation smaller than $2k$, then POINTER CHASING can be solved by a two-party protocol $\mathcal{P}$ in less than $2k$ rounds, with communication complexity $O(k^4(\log n + \log R_{\mathcal{A}}))$ bits.

The claim directly follows from Lemma 4. Indeed, by construction, $\partial V_A = 1$ and $\partial V_B = k$. Since we assumed $\mathsf{nact}(\mathcal{A}) < 2k$, the two-way protocol $\mathcal{P}$ provided by Lemma 4 solves the POINTER CHASING instance in less than $2k$ rounds, and uses $O(k^4(\log n + \log R_{\mathcal{A}}))$ bits.

By Lemma 6, we must have $k^4(\log n + \log R_{\mathcal{A}}) \in \Omega(n - k \log n)$. Therefore, if our CONGEST algorithm $\mathcal{A}$ has polynomially many rounds, we must have $k \in \Omega\left(\left(\frac{n}{\log n}\right)^{1/4}\right)$. Since our graph has maximum degree $\Delta = 2k + 1$, the conclusion follows. $\qquad\square$

6   Conclusion

In this paper, we have mostly focused on the round complexity of (deterministic) *frugal* algorithms solving general graph problems in the LOCAL or CONGEST model. It might be interesting to consider specific classical problems. As far as "local problems" are concerned, i.e., for locally checkable labeling (LCL) problems, we have shown that MIS and $(\Delta+1)$-coloring admit frugal algorithms with polynomial round complexities. It is easy to see, using the same arguments, that problems such as maximal matching share the same properties. It is however not clear that the same holds for $(2\Delta - 1)$-edge coloring.

Open Problem 1. *Is there a (node or edge) frugal algorithm solving $(2\Delta - 1)$-edge-coloring with round complexity $O(\mathrm{poly}(n))$ in the CONGEST model?*

In fact, it would be desirable to design frugal algorithms with sub-polynomial round complexities for LCL problems in general. In particular:

Open Problem 2. *Is there a (node or edge) frugal algorithm solving MIS or $(\Delta + 1)$-coloring with round complexity $O(\mathrm{polylog}(n))$ in the LOCAL model?*

The same type of questions can be asked for global problems. In particular, it is known that MST has no "awake frugal" algorithms, as MST has awake complexity $\Omega(\log n)$, even in the LOCAL model. In contrast, frugal algorithms for MST do exist as far as node-activation complexity is concerned. The issue is about the round complexities of such algorithms.

Open Problem 3. *Is there a (node or edge) frugal algorithm solving MST with round complexity $O(\mathrm{poly}(n))$ in the CONGEST model?*

Another intriguing global problem is depth-first search (DFS), say starting from an identified node. This can be performed by an edge-frugal algorithm performing in a linear number of rounds in CONGEST. However, it is not clear whether the same can be achieved by a node-frugal algorithm.

Open Problem 4. *Is there a node-frugal algorithm solving* DFS *with round complexity* $O(\text{poly}(n))$ *in the CONGEST model?*

Finally, we have restricted our analysis to *deterministic* algorithms, and it might obviously be worth considering *randomized* frugal algorithms as well.

Acknowledgements. The authors are thankful to Benjamin Jauregui for helpful discussions about the sleeping model.

Appendix

A   Proof of Theorem 4

In problem EQUALITY, two players Alice and Bob have a boolean vector of size k, x_A for Alice and x_B for Bob. Their goal is to answer *true* if $x_A = x_B$, and *false* otherwise. The communication complexity of this problem is known to be $\Theta(k)$ [9]. Let $k = n^2$. We can interpret x_A and x_B as the adjacency matrix of two graphs G_A and G_B in an instance of SYMMETRY. It is a mere technicality to "shift" G_B as if its vertices were indexed from 1 to n, such that SYMMETRY is true for G iff $x_A = x_B$. Moreover, Alice can construct G_A from its input x_A, and Bob can construct G_B from x_B. Both can simulate the unique edge joining the two graphs in G. Therefore, by Lemma 3 applied to G, if Alice controls the vertices of G_A, and Bob controls the vertices of G_B, then any CONGEST algorithm $\mathcal{A}$ solving SYMMETRY in polynomially many rounds yields a two-party protocol for EQUALITY on n^2 bits. Since graphs G_A and G_B are linked by a unique edge, the total communication of the protocol is $O(\text{eact}(\mathcal{A}) \cdot \log n)$ and $O(\text{nact}(\mathcal{A}) \cdot \log n)$. The result follows.

In SET DISJOINTNESS, each of the two players Alice and Bob has a Boolean vector of size k, x_A for Alice, and x_B for Bob. Their goal is to answer *true* if there is no index $i \in [k]$ such that both $x_A[i]$ and $x_B[i]$ are true (in which case, x_A and x_B are disjoint), and *false* otherwise. The communication complexity of this problem is known to be $\Theta(k)$ [9]. We use the technique in [5] to construct an instance G for C_4 freeness, with a small cut, from two Boolean vectors x_A, x_B of size $k = \Theta(n^{3/2})$. Consider a C_4-free n-vertex graph H with a maximum number of edges. Such a graph has $k = \Theta(n^{3/2})$ edges, as recalled in [5]. We can consider the edges $E(H)$ as indexed from 1 to k, and $V(H)$ as $[n]$. Let now x_A and x_B be two Boolean vectors of size k. These vectors can be interpreted as edge subsets $E(x_A)$ and $E(x_B)$ of H, in the sense that the edge indexed i in $E(H)$ appears in $E(x_A)$ (resp. $E(x_B)$) iff $x_A[i]$ (resp. $x_B[i]$) is true. Graph G is constructed to have $2n$ vertices, formed by two sub-graphs $G_A = G[\{1, \ldots, n\}]$ and $G_B = G[\{n+1, \ldots, 2n\}]$. The edges of $E(G_A)$ are exactly the ones of $E(x_A)$.

Similarly, the edges of $E(G_B)$ correspond to $E(x_A)$, modulo the fact that the vertex indexes are shifted by n, i.e., for each edge $\{u, v\} \in E(x_B)$, we add edge $\{u+n, v+n\}$ to G_B. Moreover we add a perfect matching to G, between $V(G_A)$ and $V(G_B)$, by adding all edges $\{i, i+n\}$, for all $i \in [n]$. Note that G is C_4-free if and only if vectors x_A and x_B are disjoint. Indeed, since G_A, G_B are isomorphic to sub-graphs of H, they are C_4-free. Thus any C_4 of G must contain two vertices in G_A and two in G_B, in which case the corresponding edges in G_A and G_B designate the same bit of x_A and x_B respectively. Moreover Alice and Bob can construct G_A and G_B, as well as the edges in the matching, from their respective inputs x_A and x_B. Therefore, thanks to Lemma 3, a CONGEST algorithm $\mathcal{A}$ for C4-FREENESS running in a polynomial number of rounds can be used to design a protocol $\mathcal{P}$ solving SET DISJOINTNESS on $k = \Theta(n^{3/2})$ bits, where Alice controls $V(G_A)$ and Bob controls $V(G_B)$. The communication complexity of the protocol is $O(\mathsf{eact}(\mathcal{A}) \cdot n \cdot \log n)$, and $O(\mathsf{nact}(\mathcal{A}) \cdot n \cdot \log n)$, since the cut between G_A and G_B is a matching. The result follows.   □

B   Proof of Lemma 5

At round 1, each node v transmits its depth-first search index $\mathrm{DFS}(v)$ to its neighbors. Therefore, after this round, every node knows its parent, and its children in the DFS tree. Then the algorithm merely forwards messages of type $m(i) = f_i \circ f_{i-1} \ldots f_1(x)$, corresponding to iterated computations for increasing values i, along the DFS tree, using the DFS ordering. That is, for any node v, let $\mathrm{MaxDFS}(v)$ denote the maximum DFS index appearing in the subtree of the DFS tree rooted at v. We will not explicitly compute this quantity but it will ease the notations. At some round, vertex v of DFS index i will receive a message $m(i-1)$ from its parent (of index $i-1$). Then node v will be in charge of computing message $m(\mathrm{MaxDFS}(v))$, by "calling" its children in the tree, and sending this message back to its parent. In this process, each edge in the subtree rooted at v is activated twice.

The vertex of DFS index 1 initiates the process at round 2, sending $f_1(x)$ to its child of DFS index 2. Any other node v waits until it receives a message from its parent, at a round that we denote $r(v)$. This message is precisely $m(i-1) = f_{i-1} \circ f_{i-2} \ldots f_1(x)$, for $i = \mathrm{DFS}(v)$. Then v computes message $m(i) = f_i \circ f_{i-1} \ldots f_1(x)$ using its local function f_i. If it has no children, then it sends this message $m(i)$ to its parent at round $r(v) + 1$. Assume now that v has j children in the DFS tree, denoted $u_1, u_2, \ldots, u_j$, sorted by increasing DFS index. Observe that, by definition of DFS trees, $\mathrm{DFS}(u_k) = \mathrm{MaxDFS}(u_{k-1}) + 1$ for each $k \in \{2, \ldots, j\}$. Node v will be activated j times, once for each edge $\{v, u_k\}$, $1 \leq k \leq j$, as follows. At round $r(v) + 1$ (right after receiving the message from its parent), v sends message $m(i)$ to its child u_1, then it awaits until round $r^1(v)$ when it gets back a message from u_1.

The process is repeated for $k = 2, \ldots, j$: at round $r^{k-1}(v) + 1$, node v sends the message $m(\mathrm{DFS}(u_k) - 1)$ received from u_{k-1} to u_k, and waits until it gets back a message from u_k, at round $r^k(v)$. Note that if $k < j$ then this message is

$m(\text{DFS}(u_{k+1}) - 1)$, and if $k = j$ then this message is $m(\text{MaxDFS}(v))$. At round $r^j(v) + 1$, after having received messages from all its children, v backtracks message $m(\text{MaxDFS}(v))$ to its parent. If v is the root, then the process stops.

The process terminates in $O(n)$ rounds, and, except for the first round, every edge of the DFS tree is activated twice: first, going downwards, from the root towards the leaves, and, second, going upwards. At the end, the root obtains the requested message $m(n) = f_n \circ f_{n-1} \ldots f_1(x)$. □

C   Figures

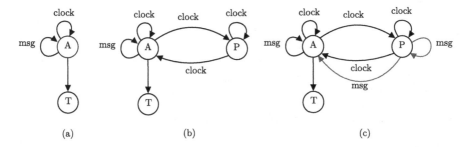

Fig. 1. (a) Classical model (b) Sleeping model, (c) Activation model.

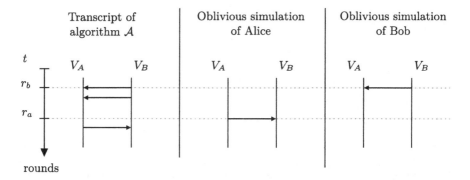

Fig. 2. Illustration of one phase of the simulation protocol. Assuming that the players agree on the simulation of algorithm $\mathcal{A}$ up to round t, each player runs an oblivious simulation at the nodes they control. In the example of the figure, the next message corresponds to a node controlled by Bob, who sends a message to a node in V_A at round r_b. The oblivious simulation of Alice is not aware of this message, and incorrectly considers that a message is sent from V_A to V_B at round $r_a > r_b$. Using the communication rounds in this phase, the players agree that the message of Bob was correct. Thus the simulation is correct up to round r_b, for both players.

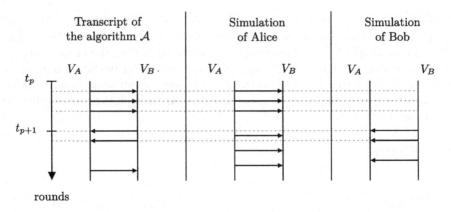

Fig. 3. Illustration of the round-efficient simulation protocol for algorithm $\mathcal{A}$. After round p, Alice has correctly simulated the algorithm up to round t_p. It is the turn of Bob to speak in round $p + 1$. In round p, Alice sent to Bob the set of messages M_A^p, obtained from an oblivious simulation of $\mathcal{A}$ starting from t_p. Only the first three messages are correct, since at round t_{p+1} Bob communicates a message to Alice. Then, Bob runs an oblivious simulation of $\mathcal{A}$ starting from t_{p+1}, and communicates all the messages sent from nodes V_B to nodes in V_A. In this case the two first messages are correct.

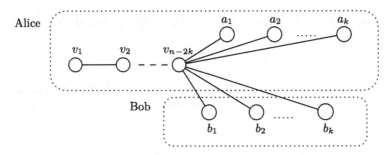

Fig. 4. Reduction from POINTER CHASING to DEPTH FIRST POINTER CHASING.

References

1. Augustine, J., Moses, W.K., Pandurangan, G.: Brief announcement: distributed MST computation in the sleeping model: awake-optimal algorithms and lower bounds. In: 41st ACM Symposium on Principles of Distributed Computing (PODC), pp. 51–53 (2022). https://doi.org/10.1145/3519270.3538459
2. Barenboim, L., Maimon, T.: Deterministic logarithmic completeness in the distributed sleeping model. In: 35th International Symposium on Distributed Computing (DISC). LIPIcs, vol. 209, pp. 10:1–10:19. Schloss Dagstuhl - Leibniz-Zentrum für Informatik (2021). https://doi.org/10.4230/LIPIcs.DISC.2021.10
3. Chang, Y., Dani, V., Hayes, T.P., He, Q., Li, W., Pettie, S.: The energy complexity of broadcast. In: 37th ACM Symposium on Principles of Distributed Computing (PODC), pp. 95–104 (2018). https://doi.org/10.1145/3212734.3212774

4. Chatterjee, S., Gmyr, R., Pandurangan, G.: Sleeping is efficient: MIS in $O(1)$-rounds node-averaged awake complexity. In: 39th ACM Symposium on Principles of Distributed Computing (PODC), pp. 99–108 (2020). https://doi.org/10.1145/3382734.3405718

5. Drucker, A., Kuhn, F., Oshman, R.: On the power of the congested clique model. In: Proceedings of the 2014 ACM Symposium on Principles of Distributed Computing, PODC 2014, pp. 367–376. Association for Computing Machinery, New York (2014). https://doi.org/10.1145/2611462.2611493, https://doi.org/10.1145/2611462.2611493

6. Dufoulon, F., Moses, W.K., Pandurangan, G.: Sleeping is superefficient: MIS in exponentially better awake complexity (2022). https://doi.org/10.48550/ARXIV.2204.08359

7. Ghaffari, M., Portmann, J.: Average awake complexity of MIS and matching. In: 34th ACM Symposium on Parallelism in Algorithms and Architectures (SPAA), pp. 45–55 (2022). https://doi.org/10.1145/3490148.3538566

8. Grumbach, S., Wu, Z.: Logical locality entails frugal distributed computation over graphs (extended abstract). In: Paul, C., Habib, M. (eds.) WG 2009. LNCS, vol. 5911, pp. 154–165. Springer, Heidelberg (2010). https://doi.org/10.1007/978-3-642-11409-0_14

9. Kushilevitz, E., Nisan, N.: Communication Complexity. Cambridge University Press, Cambridge (1997)

10. Nisan, N., Wigderson, A.: Rounds in communication complexity revisited. SIAM J. Comput. **22**(1), 211–219 (1993). https://doi.org/10.1137/0222016

11. Peleg, D.: Distributed Computing: A Locality-Sensitive Approach. SIAM (2000)

Spanning Trees with Few Branch Vertices in Graphs of Bounded Neighborhood Diversity

Luisa Gargano[ID] and Adele A. Rescigno[✉][ID]

Department of Computer Science, University of Salerno, Fisciano, Italy
{lgargano,arescigno}@unisa.it

Abstract. A branch vertex in a tree is a vertex of degree at least three. We study the NP-hard problem of constructing spanning trees with as few branch vertices as possible. This problem generalizes the famous Hamiltonian Path problem which corresponds to the case of no vertices having degree three or more. It has been extensively studied in the literature and has important applications in network design and optimization. In this paper, we study the problem of finding a spanning tree with the minimum number of branch vertices in graphs of bounded neighborhood diversity. Neighborhood diversity, a generalization of vertex cover to dense graphs, plays an important role in the design of algorithms for such graphs.

Keywords: Spanning tree · Fixed parameterized algorithms · Neighborhood diversity

1 Introduction

A *branch vertex* of a tree is a vertex having degree at least three. Let $G = (V, E)$ be a connected graph. The MINIMUM BRANCH VERTICES spanning tree problem asks to find a spanning tree of G having the smallest number of branch vertices among all the spanning trees of G. We notice that a spanning tree without branch vertices is a Hamilton path of G, hence a hamiltonian path can be regarded as a spanning tree with no branch vertices.

The problem of determining a spanning tree with a bounded number of branch vertices, is of practical importance in scenarios where it is desirable to minimize the number of vertices that need to be considered or processed in some way, such as in network design and optimization. The MINIMUM BRANCH VERTICES was first studied in relation to a problem in wavelength-division multiplexing (WDM) technology in optical networks, where one wants to minimize the number of lightsplitting switches in a light-tree; the interested reader is referred to [24] for more details.

In general, switching between different service providers incurs into switching costs. Cognitive radio networks (CRN) operate across a wide frequency range

© The Author(s), under exclusive license to Springer Nature Switzerland AG 2023
S. Rajsbaum et al. (Eds.): SIROCCO 2023, LNCS 13892, pp. 502–519, 2023.
https://doi.org/10.1007/978-3-031-32733-9_22

in the spectrum and frequently require frequency switching; therefore, bounding the number of switches has high importance both in terms of delay and energy consumption [29,39]. Notice that the energy consumption aspect of this switching cost is especially important in the recently active research area of green networks [8,10]. Furthermore, operating with a wide range of frequencies is a property not only of CRNs but also of other 5G technologies.

1.1 Notation and Definitions

Given an undirected graph $G = (V, E)$, where V is the set of vertices and E is the set of edges, we use n and m to denote the number of vertices and edges in the graph, respectively. The neighborhood of a vertex v is denoted by $\Gamma_G(v) = \{u \in V \mid \{u, v\} \in E\}$. In general, the neighborhood of a set $U \subseteq V$ is denoted by $\Gamma_G(U) = \{v \in V - U \mid \{u, v\} \in E, \ u \in U\}$. The degree of a vertex $v \in V(G)$ is the number of edges incident on it, $d_G(v) = |\Gamma_G(v)|^1$.

A *branch vertex* is a vertex having degree at least three. If G is a connected graph, we let $b(G)$ denote the smallest number of branch vertices in any spanning tree of G. Since a spanning tree without branch vertices is a Hamiltonian path of G, we have $b(G) = 0$ if and only if G admits a Hamiltonian path. We study the following problem:

MINIMUM BRANCH VERTICES (MBV)
Instance: A connected graph $G = (V, E)$.
Goal: Find a spanning tree of G having $b(G)$ branch vertices.

1.2 Previous Works

Since its proposal, MBV has been a widely studied problem, both from the algorithmic and the graph-theoretic point of view. Most of the previous work on this problem has yielded upper bounds on the number of branch vertices in the resulting tree, but these bounds were not tight. Gargano *et al.* [25] proved that it is NP-complete to decide whether, given a graph G and an integer k, G admits a spanning tree with at most k branch vertices, even in cubic graphs. In the same paper, the authors give an algorithm that finds a spanning tree with 1 branch vertex if each set of 3 independent vertices of the input graph G has degree sum at least equal to $|V(G)| - 1$. Results for the MBV problem have been given in [38]. The author proves the existence of an algorithm that finds a spanning tree with $O(\log |V(G)|)$ branch vertices whenever the degree of each vertex of the input graph is $\Omega(n)$; moreover, an approximation factor better than $O(\log |V(G)|)$ would imply that $NP \subseteq DTIME(n^{O(\log \log n)})$. An algorithm to construct spanning trees with few branch vertices in claw-free graphs has been presented in [35]. Other results including mathematical formulations, heuristics and approximation results can be found in [4–6,9,11–13,27,32–34,37,40–42].

In this paper we study MBV from a parameterized complexity point of view.

---

[1] In the following we use omit the graph name G (e.g. we use $d(v)$, instead of $d_G(v)$) whenever the graph is clear from the context.

1.3 Parameterized Complexity

Parameterized complexity is a refinement to classical complexity theory in which one takes into account not only the input size, but also other aspects of the problem given by a parameter p. A problem with input size n and parameter p is called *fixed parameter tractable (FPT)* if it can be solved in time $f(p) \cdot n^c$, where f is a computable function only depending on p and c is a constant.

Neighborhood Diversity. Graphs of bounded neighborhood diversity can be seen as the simplest of dense graphs and thus neighborhood diversity plays an important role in the design of algorithms for such graphs.

Given a graph $G = (V, E)$, two vertices $u, v \in V$ are said to have the same *type* if $\Gamma(v) - \{u\} = \Gamma(u) - \{v\}$. The *neighborhood diversity* $\mathsf{nd}(G)$ of a graph G is the minimum number nd of sets in a partition $V_1, V_2, \ldots, V_{\mathsf{nd}}$, of the vertex set V, such that all the vertices in V_i have the same type, for $i \in \{1, \ldots, \mathsf{nd}\}$.

The neighborhood diversity parameter, was first introduced by Lampis in [31]. It has then received much attention [1,2,7,14–16,23,26,28,43], also due to the fact that, contrary to other popular parameters, it is computable in linear time.

The family $\{V_1, V_2, \ldots, V_{\mathsf{nd}}\}$ is called the *type partition* of G. Notice that each V_i induces either a clique or an independent set in G. Moreover, for each V_i and V_j in the type partition, either each vertex in V_i is a neighbor of each vertex in V_j or no vertex in V_i has a neighbor in V_j. Hence, between each pair V_i and V_j there is either a complete bipartite graph or no edge at all.

Starting from a graph G and its type partition $\mathcal{V} = \{V_1, \ldots, V_{\mathsf{nd}}\}$, we can see each element of $\mathcal{V}$ as a vertex of a new graph H, called the *type graph* of G, with

- $V(H) = \{1, 2, \cdots, \mathsf{nd}\}$
- $E(H) = \{\{x, y\} \mid x \neq y$ and for each $u \in V_x,\ v \in V_y$ it holds that $\{u, v\} \in E\ \}$.

For sake of clearness, we will refer to the vertices of H as *types* and reserve the term vertex to those in G.

Example 1. Figure 1 shows a graph G and its type graph H corresponding to the type partition $\mathcal{V} = \{V_1, \ldots, V_{10}\}$ where $V_1 = \{a\}, V_2 = \{c, d\}, V_3 = \{b\}, V_4 = \{e\}, V_5 = \{f, g\}, V_6 = \{h\}, V_7 = \{i\}, V_8 = \{l\}, V_9 = \{m\}, V_{10} = \{n\}$. Note that G cannot contain a hamiltonian path; hence, $b(G) \geq 1$.

Other Graph Parameters. It was proved in [3] that MBV is FPT with respect to treewidth. The treewidth parameter, which represents a way to describe the distance between a graph and a tree, was introduced by Robertson and Seymour [36] and has been widely used in parameterized complexity of graph optimization problems [18,19]. We recall that graphs of small treewidth are necessarily sparse and that neighborhood diversity, which is a good parameters in case of dense graphs, is incomparable to the treewidth.

Cliquewidth, defined in [17], covers a larger family of graphs with respect both to treewidth and neigborhood diversity, including many dense graphs. However, several natural problems become W[1]-hard when parameterized by this measure.

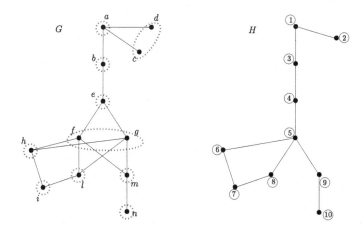

Fig. 1. A graph G (on the left) and its type graph H (on the right). Dashed circles group vertices having the same type.

In particular, this is the case with the MBV problem. Indeed, it was proven in [21] that the (MBV special case) HAMILTONIAN PATH problem is W[1]-Hard when parameterized by cliquewidth.

When the parameter is the modular-width [22], the HAMILTONIAN PATH problem becomes fixed parameter tractable. We notice that, modular-width is stronger than neighborhood diversity in the sense that graphs of bounded neighborhood diversity have bounded modular-width, while the converse may not be true.

We present a FPT algorithm for the problem of determining the minimum number of branch vertices in any spanning tree of G, parameterized by neighborhood diversity, leaving open the question to asses the paremterized complexity of MBV with respect to modular-width.

A summary of the relations which hold between some popular graph parameters is given in Fig. 2. We refer to [20] for the formal definitions of the parameters.

2   The Algorithm

This rest of the paper is devoted to prove the following theorem.

Theorem 1. *The* MINIMUM BRANCH VERTICES *spanning tree problem is FPT when parameterized by neighborhood diversity.*

We first give a characterization of the spanning trees of a graph G with $b(G)$ branch vertices in terms of neighborhood diversity.

Lemma 1. *Let $G = (V, E)$ be a connected graph with type partition $\mathcal{V} = \{V_1, V_2, \ldots, V_{\mathsf{nd}}\}$. Any spanning tree of G with $b(G)$ branch vertices has at most one branch vertex belonging to any set of the type partition $\mathcal{V}$. Hence, $b(G) \leq \mathsf{nd}$.*

Proof. Let T be a spanning tree of G with $b(G)$ branch vertices and assume, by contradiction, that there exists a set $V_i \in \mathcal{V}$ containing at least two branch

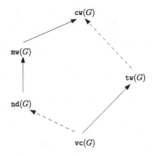

Fig. 2. A summary of the relations holding among some popular parameters. We use $\mathsf{mw}(G)$, $\mathsf{tw}(G)$, $\mathsf{cw}(G)$ and $\mathsf{vc}(G)$ to denote modular-width, treewidth, cliquewidth and minimum vertex cover of a graph G, respectively. Solid arrows denote generalization, e.g., modular-width generalizes neighborhood diversity. Dashed arrows denote that the generalization may exponentially increase the parameter.

vertices in T. Using the fact that all the vertices in V_i share the same neighborhood outside V_i we can construct a new spanning tree T' of G with exactly one branch vertex in V_i. In particular, rooting T in any vertex and choosing any of the branch vertices in V_i, let say u, we can move the children of all the other branch vertices in V_i so that they become children of u.   □

Let $G = (V, E)$ be a connected graph with type partition $\mathcal{V} = \{V_1, V_2, \ldots, V_{\mathsf{nd}}\}$ and let $H = (\{1, \ldots, \mathsf{nd}\}, E(H))$ be the corresponding type graph. By exploiting Lemma 1, the algorithm proceeds by considering all the subsets $B_H \subseteq \{1, \ldots, \mathsf{nd}\}$, ordered by size, and verifying if it there exists a spanning tree in G with $|B_H|$ branch vertices each chosen from a different type set V_i with $i \in B_H$.

We notice that, since modular-width generalizes neighborhood diversity, the algorithm in [22] also gives a FPT algorithm for HAMILTONIAN PATH parameterized by neighborhood diversity. As a consequence, the algorithm in [22] can be used for the case $B_H = \emptyset$. Hence, in the following we assume $|B_H| \geq 1$.

The identification of the spanning tree of G goes through the solution of an Integer Linear Program that uses the properties of the type partition $\mathcal{V} = \{V_1, V_2, \ldots, V_{\mathsf{nd}}\}$ of G. Namely, if the set B_H is such that the ILP does not admit a solution, then the set is discarded; if for B_H the ILP admits a solution, we will show how to obtain a spanning tree of G having exactly $|B_H|$ branch vertices (recall that the sets B_H are considered by increasing size). The optimal spanning tree will be indeed shown to correspond to the smallest B_H for which the ILP admits a solution. We now give a description of our algorithm.

For each set $B_H \neq \emptyset$, we select an arbitrary $r \in B_H$ and construct a digraph

$$H_{B_H} = (\{1, \ldots, \mathsf{nd}\} \cup \{s\}, A_{B_H}),$$

where $s \notin V(H)$ is an additional vertex that will be called the source type. H_{B_H} is obtained from the type graph H by replacing every edge $\{i, j\} \in E(H)$ by the two directed arcs (i, j) and (j, i), and then adding the directed arc (s, r). Formally, $A_{B_H} = \{(s, r)\} \cup \{(i, j), (j, i) \mid \{i, j\} \in E(H)\}$.

We use the solution of the following Integer Linear Programming (ILP) to select arcs of H_{B_H} that will help to construct the desired spanning tree in G.

For each arc $(i, j) \in A_{B_H}$, the non negative decision variable x_{ij} represents the load to be put on (i, j). The load of the arc (s, r) is set to 1.

$$x_{sr} = 1 \tag{1}$$

$$\sum_{j:(j,i) \in A_{B_H}} x_{ji} \leq |V_i| \qquad \forall i \in \{1, \ldots, \mathsf{nd}\} \text{ s.t. } V_i \text{ is a clique} \tag{2}$$

$$\sum_{j:(j,i) \in A_{B_H}} x_{ji} = |V_i| \qquad \forall i \in \{1, \ldots, \mathsf{nd}\} \text{ s.t. } V_i \text{ is an ind. set} \tag{3}$$

$$\sum_{\ell:(i,\ell) \in A_{B_H}} x_{i\ell} - \sum_{j:(j,i) \in A_{B_H}} x_{ji} \leq 0 \qquad \forall i \in \{1, \ldots, \mathsf{nd}\} - B_H \tag{4}$$

$$y_{sr} = \mathsf{nd} \tag{5}$$

$$\sum_{j:(j,i) \in A_{B_H}} y_{ji} - \sum_{\ell:(i,\ell) \in A_{B_H}} y_{i\ell} = 1 \qquad \forall i \in \{1, \ldots, \mathsf{nd}\} \tag{6}$$

$$y_{ij} \leq \mathsf{nd}\, x_{ij} \qquad \forall (i, j) \in A_{B_H} \tag{7}$$

$$y_{ij}, x_{ij} \in \mathbb{N} \qquad \forall (i, j) \in A_{B_H} \tag{8}$$

The total incoming load at type $i \in \{1, \ldots, \mathsf{nd}\}$ has to be at most $|V_i|$ in case V_i is a clique and exactly $|V_i|$ in case V_i is an independent set. Constraints (2) and (3) correspond to this requirement. Constraint (4) binds the relation between the total incoming and outgoing loads at any type $i \notin B_H$, namely i must have an outgoing load upper bounded by its incoming load.

Constraints (5) and (6) use a single commodity flow in which s is used as the source and the other types are demand vertices. For each arc $(i, j) \in A_{B_H}$, the non negative decision variable y_{ij} represents the quantity of flow from type i to type j.

Each type $i \in \{1, \ldots, \mathsf{nd}\}$ has demand of one unit; therefore, the difference between the inflow and the outflow must be exactly one. Meanwhile, the supply quantity at the source s has to be exactly nd to reach each of the types in $\{1, \ldots, \mathsf{nd}\}$.

Constraint (7) stresses variable $y_{ij} = 0$ whenever $x_{ij} = 0$; thus if no load is put on (i, j) then type j is not to be reached trough type i.

3   The Spanning Tree Construction

Given an integer solution (y, x), if any, to the above ILP, the values of variables y imply that each type $i \in \{1, \ldots, \mathsf{nd}\} - \{r\}$ is reached from the source s. Then, by the construction of digraph H_{B_H}, each type $i \in \{1, \ldots, \mathsf{nd}\}$ is reached from type r. Furthermore, by the relation between variables x and y (constraint (7)), we know that each type $i \in \{1, \ldots, \mathsf{nd}\}$ obtains incoming load from at least one of its neighbors.

Claim 1. *The subgraph H_x of H_{B_H} with vertex set $\{1, \ldots, \text{nd}\}$ and arc set $\{(i,j) \mid x_{ij} \geq 1\}$ contains a directed path from r to any other type.*

Example 1 (cont.). Figure 3 shows the digraph H_x subgraph of H_{B_H} obtained from H, given in Fig. 1, for $B_H = \{1\}$ and the solution (y, x) of ILP. The pair on each arc (i,j) corresponds to (y_{ij}, x_{ij}).

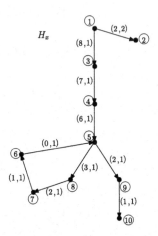

Fig. 3. Digraph H_x subgraph of H_{B_H} obtained from H, given in Fig. 1, for $B_H = \{1\}$ and the solution (y, x) of ILP. The pair on each arc (i,j) corresponds to (y_{ij}, x_{ij}).

Our algorithm uses the values of variables x to obtain a spanning tree T of G with $k = |B_H|$ branch vertices, one in each of the type sets V_i with $i \in B_H$.

To describe the construction of T we first introduce some useful notation.

For each $i \in \{1, \ldots, \text{nd}\}$, we denote by $In(i)$ the set of the types from which there exist arcs in H_x toward i, that is, $In(i) = \{j \mid x_{ji} \geq 1\}$;. Moreover, we define

$$\alpha_i = \sum_{j:j \in In(i)} x_{ji} \tag{9}$$

that represents the number of vertices of V_i whose parent in T is a vertex outside V_i, and

$$\beta_i = \begin{cases} \sum_{\ell:i \in In(\ell)} x_{i\ell} & \text{if } i \notin B_H \\ 1 & \text{if } i \in B_H \end{cases} \tag{10}$$

that represents the number of vertices of V_i, that will be the parent of some vertex in $\cup_{\ell:i \in In(\ell)} V_\ell$. In particular, if $i \notin B_H$ then $x_{i\ell}$ vertices of V_i will be chosen to be each the parent of exactly one vertex in V_ℓ, while if $i \in B_H$ then exactly one vertex of V_i will be the parent of all the $\sum_{\ell:i \in In(\ell)} x_{i\ell}$ vertices in V_ℓ. Notice that by Claim 1 ($\alpha_i \geq 1$) and constraint (4), it follows that for each $i = 1, \ldots, \text{nd}$

$$\alpha_i \geq \beta_i.$$

Finally, we set $s_i = |V_i| - \alpha_i$. We notice that if V_i is an independent set then $s_i = 0$ since the constraint (3) of ILP imposes $\sum_{j:(j,i)\in A_{B_H}} x_{ji} = |V_i|$. Hence, if $s_i > 0$ then any selection of s_i vertices of V_i induces a clique; moreover, no branch vertex of T will be among them.

3.1   The Case $s_i = 0$, for Each Type i

For sake of simplicity, we first describe the proposed algorithm in the case $s_i = 0$ for each $i \in \{1, \ldots, \mathsf{nd}\}$. Next, we show how a simple modification of the algorithm allows to cover the general case in which $s_i > 0$ for some type i.

The algorithm TREE constructs a spanning tree of G iteratively by exploring unexplored vertices of G, until possible, and maintains a main subtree T and a forest whose roots are progressively connected to T to assemble the spanning tree. The process stops when all the vertices of G are explored.

The algorithm uses a queue Q to enqueue the explored vertices and maintains a set R of the roots of trees of explored vertices that wait to be connected to the main tree T. The forest structure is described through the parent function π.

At the beginning the set R is empty. Chosen any vertex $u_r \in V_r$ (recall that by construction $r \in B_H$), the procedure EXPLORE(u_r) carries out the construction of the main tree T rooted at u_r and marks as explored all the reached vertices (adding them to the set Ex). Clearly, for each explored vertex v there is a path in T joining u_r to v.

Algorithm 1. TREE($G, \mathcal{V}, B_H$)

1: $R = \emptyset$, $B = \emptyset$, $Ex = \emptyset$
2: $\pi(u) = nhil$ for each $u \in V(G)$
3: Let $u_r \in B_H$
4: EXPLORE(u_r)
5: **while** $V - Ex \neq \emptyset$ **do**
6:     - Let V_j be any set s.t. $(V_j - Ex \neq \emptyset \neq V_j \cap Ex)$ and $\beta_j \geq 1$
7:     - Let $w \in V_j \cap Ex$ and $u \in (V_j - Ex) - R$
8:     - Set $\pi(u) = \pi(w)$, $Ex = Ex - \{w\}$, $R = R \cup \{w\}$
9:     - EXPLORE(u)
10: **end while**
11: **return** π, B

However, it can occur that some of the vertices have not been explored (i.e., $V - Ex \neq \emptyset$). In such a case an explored vertex w is chosen so that it belongs to a type set V_j, which also contains at least a never explored vertex u able to explore at least one unexplored neighbour (that is, $w \in V_j \cap Ex$ and $u \in (V_j - Ex) - R$ and $\beta_j \geq 1$; the existence of such a set V_j is assured by Lemma 4). By using the properties of the neighborhood diversity, we know that since u and w belong to the same type set V_j, they have the same neighborhood. Hence, the algorithm lets that:

Algorithm 2. EXPLORE(u)

1: Let Q be an empty queue
2: $Q.enqueue(u)$, $Ex = Ex \cup \{u\}$
3: **while** $Q \neq \emptyset$ **do**
4:     $v = Q.dequeue$
5:     Let $v \in V_i$
6:     **if** $i \notin B_H$ and $\beta_i \geq 1$ **then**
7:         - Let $v' \in V_\ell - Ex$ for some ℓ s.t. $i \in In(\ell)$
8:         - $\pi(v') = v$, $Ex = Ex \cup \{v'\}$
9:         **if** $v' \notin R$ **then** $Q.enqueue(v')$
10:         **else** $R = R - \{v'\}$
11:         **end if**
12:         - $\alpha_\ell = \alpha_\ell - 1$, $\beta_i = \beta_i - 1$
13:     **else if** $i \in B_H$ and $\beta_i = 1$ **then**
14:         - $B = B \cup \{v\}$
15:         **for each** ℓ s.t. $i \in In(\ell)$ **do**
16:             - Let $A_{i\ell} \subseteq V_\ell - Ex$ s.t. $|A_{i\ell}| = x_{i\ell}$
17:             - $\alpha_\ell = \alpha_\ell - x_{i\ell}$,
18:             **for each** $v' \in A_{i\ell}$ **do**
19:                 - $\pi(v') = v$ $Ex = Ex \cup \{v'\}$
20:                 **if** $v' \notin R$ **then** $Q.enqueue(v')$
21:                 **else** $R = R - \{v'\}$
22:                 **end if**
23:             **end for**
24:         **end for**
25:         - $\beta_i = \beta_i - 1$
26:     **end if**
27: **end while**

- the parent of w (recall that w is explored) becomes the parent of u, and
- w (the root of a subtree of explored vertices) is added to R and removed from Ex (this will allow w to be later explored and added, together with its subtree, to the main tree T), and
- EXPLORE(u) is called to start a new exploration from u.

Notice that the algorithm modifies the forest by assigning to u the parent of w and only later (after adding u and some descendants of u) adding again the subtree rooted in w to the main tree T. This allows connecting new vertices in $V - Ex$ to the main tree T; the particular choice of u and w will be shown to avoid the possibility that the algorithm fails, due to the fact that no arc can be added to T without forming a cycle or creating an extra branch vertex (see Fig. 4 for an example). The process is iterated as long as there are unexplored vertices, i.e. $V(G) - Ex \neq \emptyset$.

The procedure EXPLORE(u) starts with Q containing only the vertex u and pads the main tree T (recall that (unless $u = u_r$) the parent of u is a vertex already in T thus constructing a subtree rooted at u spanning on all the newly explored vertices). EXPLORE(u) uses for each type i the values of α_i and β_i that are initially defined as in (9) and (10). The value $\alpha_i = \sum_{j:j \in In(i)} x_{ji}$ counts the number of vertices of V_i that must be assigned a parent outside V_i; in particular, x_{ji} vertices of V_i have to be explored by vertices in V_j. The value β_i counts the number of vertices of V_i on which the EXPLORE procedure must be called; in particular,

- if $i \in B_H$ then exactly 1 vertex in V_i becomes a branch vertex in T: it is set as the parent of $x_{i\ell}$ unexplored vertices in V_ℓ for each ℓ such that $x_{i\ell} \geq 1$ (i.e., $i \in In(\ell)$), and
- if $i \notin B_H$ then $\sum_{\ell:i \in In(\ell)} x_{i\ell}$ vertices in V_i are chosen and each one becomes the parent of one unexplored vertex in V_ℓ.

Recall that, by the ILP constraints, we know that $\alpha_i \geq \beta_i$ and that we assume $s_i = 0$. As a consequence, we have $\alpha_i = |V_i|$ for each $i \in \{1, \ldots, \mathsf{nd}\}$. When a vertex $v \in V_i$ is dequeued from Q in EXPLORE(u) then the value of β_i is decreased by one if v explores (i.e., if $\beta_i \geq 1$); furthermore, the value α_ℓ is decreased by the number of vertices in V_ℓ that v explores, for $i \in In(\ell)$. Hence, at the beginning of each iteration of the while loop in EXPLORE(u) the values of α_i represents the number of vertices in V_i that remain to be explored while β_i is the number of vertices in V_i on which EXPLORE must still be called. Note that when a vertex $v \in V_i$ is dequeued from Q in EXPLORE(u), with $u \neq u_r$, and $\beta_i \geq 1$, the algorithm checks if the neighbour v', that v explores, is in R (i.e., v' is a root of a tree in the forest). In this case v' is not enqueued in Q and v', with the tree rooted at it, is connected to the main tree T.

Example 1 (cont.). Figure 4 shows the construction of the spanning tree of G made by algorithm TREE($G, \mathcal{V}, B_H = \{1\}$), where $a \in V_1$ is the only branch vertex and is the root of the spanning tree. The red edges in Fig. 4(a) show the main tree obtained at the end of EXPLORE(a) when $Ex = \{a, b, c, d, e, g, m, n\}$. Since $V(G) - Ex \neq \emptyset$, the algorithm TREE finds $V_5 = \{f, g\}$ having $g \in V_5 \cap Ex$ and $f \in V_5 - Ex$ and $\beta_5 = 1$, so obtaining the forest shown by the red edges in Fig. 4(b), with two trees one rooted at a and the other one rooted at g where $R = \{g\}$. EXPLORE(f) explores the remaining unexplored vertices of G and reconnect g with its descendants to the main tree, so obtaining the spanning tree of G with branch vertex a, as shown in Fig. 4(c).

Lemma 2. *At the end of EXPLORE(u_r) the function π describes a tree, rooted at u_r, spanning the set $Ex \subset V$ of explored vertices. The vertices in $B \cap Ex$ are the branch vertices.*

Proof. When EXPLORE(u_r) is called, vertex u_r is the first explored vertex (i.e. it is added to the set Ex) and is enqueued in Q. After that, each time a vertex $v \in V_i$ is dequeued from Q (recall, $v \in Ex$ is an explored vertex), the algorithm can either stop its exploration (i.e., v is a leaf in T and this occurs when $\beta_i = 0$)

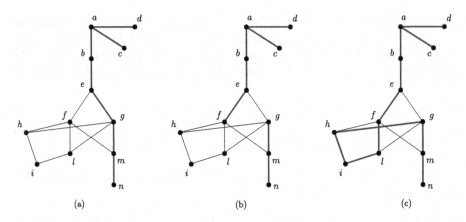

Fig. 4. Construction by TREE($G, \mathcal{V}, B_H = \{1\}$) of the spanning tree with $B = \{a\}$. In order to have exactly one branch vertex, the initial choice of g as child of e is incorrect (see (a)). Hence, the algorithm fixes such an error by choosing f as child of e, while keeping the already explored subtree rooted at g (see (b)). Finally, the exploration from f allows to reach g and have the desired spanning tree (see (c)).

or explore one o more unexplored neighbors of v; since $R = \emptyset$ (i.e., no tree is in the forest), such neighbors are enqueued in Q and become children of v in T, through the function π. Hence, any explored vertex has u_r has ancestor, i.e., the function π describes a path joining any explored vertex to u_r. Noticing that no vertex can be enqueued twice in Q (since any enqueued vertex is also marked as explored), we have that the function π does not create cycles.

Now, we prove that when a vertex v is dequeued from Q, it has the needed number of unexplored neighbors. If $\beta_i = 0$ then v is e leaf in T; henca, we only have to consider the case $\beta_i \geq 1$. If $v \in V_i$ with $i \notin B_H$ then v has $\beta_i \geq 1$ unexplored neighbor and one of them can be added to T as child of v. If $i \in B_H$ then v is the first vertex of V_i to be explored and all the $x_{i\ell}$ vertices in V_ℓ are unexplored and can be added to T as children of v, for each ℓ such that $x_{i\ell} \geq 1$. Hence v becomes a branch vertex in T and is put in B.   □

Fix any iteration of the while loop in algorithm TREE and define $\hat{H}_x$ as the subgraph of H_x containing the arc (i, j) if, at the beginning of the while loop, less than x_{ij} vertices in V_j have been assigned a parent in V_i.

Lemma 3. *Let j be any type in $\hat{H}_x$. The following properties hold for $\hat{H}_x$:*

(a) If $V_j \subseteq Ex$ then the type j is isolated in $\hat{H}_x$.

(b) The type r s.t. V_r contains the root u_r of T has no outgoing arcs in $\hat{H}_x$.

(c) $V_j \not\subseteq Ex$ if and only if j has at least one incoming arc in $\hat{H}_x$.

(d) If $V_j \cap Ex = \emptyset$ then j keeps in $\hat{H}_x$ all the incoming and outgoing arcs it has in H_x.

(e) If $V_j - Ex \subseteq R$ then j has no outgoing arcs in $\hat{H}_x$.

Proof. Consider property (a). If each vertex in V_j is explored, then it has a parent in T and, recalling that $\sum_{t:t\in In(j)} x_{tj} = |V_j|$ (note that $s_j = 0$ and then constraints (2) and (3) in ILP are satisfied in equality) it follows that j has no incoming arc in $\hat{H}_x$. Moreover, procedure EXPLORE implies that once a vertex in V_j is explored (i.e. it has assigned a parent) then it will be assigned at least one child as long as there exist vertices to be explored from V_j (i.e., if $\beta_j \geq 1$). In particular, if $j \notin B_H$ then since $|V_j| \geq \sum_{t:j\in In(t)} x_{jt}$ (see constraint (4) in ILP), it follows that x_{jt} vertices in V_t have a parent in V_j, for each t such that $j \in In(t)$. Hence j has no outgoing arc surviving in $\hat{H}_x$. If, otherwise, $j \in B_H$ then the first explored vertex in V_j has $x_{jt} > 0$ children in V_t for each t such that $j \in In(t)$. Hence, also in this case j has no outgoing arc surviving in $\hat{H}_x$. Property (b) follows by noticing that, by construction, u_r is a branch vertex of T and has x_{rt} children in each V_t such that $x_{rt} > 0$, (i.e., $r \in In(t)$).

Property (c) follows by noticing that $V_j \not\subseteq Ex$ is equivalent to say that

$$\sum_{t:t\in In(j)} x_{tj} = |V_j| > |V_j \cap Ex|.$$

Property (d) follows by noticing that $V_j \cap Ex = \emptyset$ implies that V_j still has an incoming neighbour for each t such that $x_{tj} > 0$ and an outgoing neighbor for each t such that $x_{jt} > 0$.

Property (e) follows by noticing that when the algorithm TREE disconnects a vertex w and adds it in R, vertex w has already been assigned its child/children. Hence, if V_j does not contain any unexplored vertex outside R then β_j has been decreased to 0 meaning that all the x_{tj} arcs from a vertex in V_t to one in V_j have been added to the forest, for each $j = 1, \ldots, \mathrm{nd}$.   □

Lemma 4. *Let Ex be the set of explored vertices at the beginning of any iteration of the while loop in algorithm TREE. If $V - Ex \neq \emptyset$ then there exists a type j such that $V_j - Ex \neq \emptyset \neq V_j \cap Ex$ and $\beta_j \geq 1$.*

Proof. By (a) and (c) of Lemma 3, we know that each type j in $\hat{H}_x$ either is isolated or has at least an incoming arc. Hence, focus on the subset of non-isolated types. Knowing that each of them has an incoming arc, we have that $\hat{H}_x$ contains a cycle. Then, each type j on such a cycle has an outgoing arc and satisfies $\beta_j \geq 1$.

We show now that at least one type j on the cycle has $V_j \cap Ex \neq \emptyset$. Point (b) of Lemma 3 implies that $\hat{H}_x$ does not contain any path from r to any type on the cycle. If we suppose that for each type j in the cycle $V_j \cap Ex = \emptyset$, then (d) of Lemma 3 implies that also H_x does not contain a path from r to j, thus contradicting Claim 1.   □

Lemma 5. *After each call of EXPLORE(u) the function π describes a forest spanning the vertices in $Ex \cup R$ of explored vertices and consisting of $|R| + 1$ trees respectively rooted at u_r and at the vertices in R. The vertices in B are the only branch vertices in the forest.*

Proof. When EXPLORE(u) is called, the function π describes a forest, spanning the current set $Ex \cup R$, whose roots are the vertices in $\{u_r\} \cup R$ and where $R \subset V(G) - Ex$.

We notice that by Lemma 2, this is true the first time EXPLORE is called, that is, after the call to EXPLORE(u_r) (at that time $R = \emptyset$). We prove that the claim is also true at the end of each call to EXPLORE(u).

When EXPLORE(u) is called, Q is empty; vertex u is explored (i.e. it is added to Ex) and enqueued in Q. Then EXPLORE(u) proceeds, exactly as in EXPLORE(u_r), dequeueing vertices from Q and exploring their unexplored neighbors, so constructing a subtree of the main tree T rooted at u described by function π.

The only difference with EXPLORE(u_r) is when one of the vertices explored is $v' \in R$. Vertex $v' \in R$ is removed from R and connected to the main tree T through the function π and marked as explored (see lines 8, 19) exactly as any other explored vertex. However v' is not enqueued in Q since it has already explored its neighbors; hence, v' is connected to T together with its subtree of explored vertices. □

We are now able to prove the following result.

Lemma 6. *The algorithm TREE returns a spanning tree of G, described by function π, with branch vertex set B.*

Proof. By Lemma 2 we know that algorithm TREE constructs through procedure EXPLORE(u_r) a main tree T, described by π. In case T does not span all the vertices in $V(G)$ then, Lemma 4 and (e) in Lemma 3 assure that the algorithm finds a type set V_j with an explored vertex $w \in V_j \cap Ex$ and an unexplored vertex $u \in (V_j - Ex) - R$ that allows disconnecting w (with its subtree) from the main tree T, so that it becomes one of the roots of trees in R, to use the parent of w in T to connect u to T and to start a new exploration from u (since $u \notin R$ and $\beta_j \geq 1$) calling EXPLORE(u). By Lemma 5, this allows padding T with the subtree rooted a u of new explored vertices and/or some of the trees rooted at vertices in R. The iteration of the above procedure until no unexplored vertex exists in $V(G)$ gives the lemma. □

3.2 The General Case

In this section we present a simple modification to the algorithm given in the previous section to cover the case in which $s_i \geq 1$ for some type $i \in \{1, \ldots, \text{nd}\}$.

First of all recall that $s_i = |V_i| - \sum_{j:j \in In(i)} x_{ji}$ and that by constraint (2) and (3) in ILP, it can occur that $s_i \geq 1$ only if V_i is a clique.

Our algorithm proceeds first selecting a set $S_i \subseteq V_i$ such that $|S_i| = s_i$ for each i with $s_i \geq 1$. Then, we consider the subgraph G' of G induced by the vertex set $V(G) - \cup_{i:s_i \geq 1} S_i$, and the type partition $\mathcal{V}' = \{V_1', \cdots, V_{\text{nd}}'\}$ of G' where $V_i' = V_i - S_i$, and we call algorithm TREE($G', \mathcal{V}, B_H$). Let T' be the spanning tree of G' described by the function π returned by the algorithm TREE, and let B be the branch vertex set of T'. Now, we pad T' with $\cup_{i:s_i \geq 1} S_i$ so that it becomes a spanning tree of G and keeps the branch vertex set B unchanged.

- For each i such that $s_i \geq 1$ and $i \in B_H$, we make the branch vertex $u \in V_i' \cap B$ the parent of all the vertices in S_i; formally, $\pi(v) = u$ for each $v \in S_i$.
- For each i such that $s_i \geq 1$ and $i \notin B_H$, we choose any vertex $u \in V_i' = V_i - S_i$ and substitute the arc to u from its parent $\pi(u)$ by a path P_i going from $\pi(u)$ to u through all the vertices in S_i; formally, let $S_i = \{v_1, \ldots, v_{s_i}\}$, we set $\pi(v_1) = \pi(u)$, $\pi(v_{j+1}) = v_j$ for $j = 1, \ldots, s_i - 1$, and $\pi(u) = v_{s_i}$.

4   The Algorithm Complexity

Summarizing, the proposed method:
For each fixed set $B_H \subseteq \{1, \ldots, \mathsf{nd}\}$, ordered by size, the algorithm

- solves the corresponding ILP
- if a solution exists for the current set B_H, it uses algorithm TREE to construct a spanning tree of G with $|B_H|$ branch vertices.

Jansen and Rohwedderb [30] have recently showed that the time needed to find a feasible solution of an ILP with p integer variables and q constraints is $O(\sqrt{q}\Delta)^{(1+o(1))q} + O(qp)$, where Δ is the biggest absolute value of any coefficient in the ILP. As our ILP has at most $q = \mathsf{nd}^2 + 3\mathsf{nd} + 2$ constraints, at most $p = 2(\mathsf{nd}^2 + 1)$ variables and $\Delta = \mathsf{nd}$, the time needed to solve it is $O(\mathsf{nd}^2)^{(1+o(1))(\mathsf{nd}^2+3\mathsf{nd}+2)} + O(\mathsf{nd}^4)$. Using the solution (y, x) of the ILP, the algorithm TREE returns the spanning tree of G in time $O(n^2)$. Overall, the algorithm requires time $2^{\mathsf{nd}}(O(\mathsf{nd}^2)^{(1+o(1))(\mathsf{nd}^2+3\mathsf{nd}+2)} + O(\mathsf{nd}^4)) + O(n^2)$.

5   Optimality

In this section we show that if no set $B_H \subseteq \{1, \ldots, \mathsf{nd}\}$, with $|B_H| = k$, exists for which the ILP admits a solution then does not exist a spanning tree in G with k branch vertices, that is $b(G) \geq k + 1$.

This will allow to say that the optimal spanning tree in G corresponds to the smallest set $B_H \subseteq \{1, \ldots, \mathsf{nd}\}$ for which the ILP admits a solution, if any.

Lemma 7. *If there exists a spanning tree in G with $k \geq 1$ branch vertices then there exists a set $B_H \subseteq \{1, \ldots, \mathsf{nd}\}$ with $|B_H| = k$, and a solution (x, y) of the corresponding ILP.*

Proof. Let T be a spanning tree in G with branch vertex set B such that $|B| = k$. We show how to obtain from T and B an assignment of values to the variables in x and y that satisfy the constraint (1)–(8) of ILP. Let $B_H = \{i \mid B \cap V_i \neq \emptyset, \ i = 1, \ldots, \mathsf{nd}\}$. By Lemma 1 we have $|B_H| = |B|$. Choose any $r \in B_H$ and a vertex $u_r \in B \cap V_r$. Root T at u_r and direct each edge in T so that there is a path of directed arcs from u_r to any vertex $u \in V(G) - \{u_r\}$. Let A_T be the set of all the arcs in T.

We set $x_{sr} = 1$ (satisfying constraint (1) of ILP), and for $i, j \in \{1, \ldots, nd\}$,

$$x_{ij} = |\{(u, v) \mid (u, v) \in A_T, \ u \in V_i, \ v \in V_j\}|.$$

Let $In(i) = \{j \mid x_{ji} \geq 1\}$, for $i = 1, \ldots, nd$. Since each vertex $u \in V_i$ has a parent in T, we have that if V_i is an independent set then the parent of each $u \in V_i$ is a vertex in some V_j with $j \in In(i)$, while if V_i is a clique then the parent of $u \in V_i$ can be either a vertex in V_i or a vertex in V_j with $j \in In(i)$. This implies that

$$\sum_j x_{ji} = |V_i| \text{ if } V_i \text{ is an independent set}, \qquad \sum_j x_{ji} \leq |V_i| \text{ if } V_i \text{ is a clique},$$

satisfying constraints (2) and (3) of ILP.

If $i \notin B_H$ then V_i does not contain branch vertices. Hence, each vertex $u \in V_i$ can be the parent of at most one vertex. Hence,

$$\sum_{\ell:(i,\ell)} x_{i\ell} \leq \sum_{j:(j,i)} x_{ji}$$

satisfying constraint (4) of ILP.

To assign values to the variables y, we introduce the digraph H^x having vertex set $\{1, \ldots, nd\}$ and arc set $\{(i, j) \mid x_{ij} \geq 1\}$. Let T_x be the tree rooted at r obtained by a BFS visit of H^x. For each $i \in \{1, \ldots, nd\} - \{r\}$, let $p(i)$ the parent of i in T_x. Pad T_x, adding arc (s, r) (i.e., $p(r) = s$). We set $y_{sr} = nd$ (satisfying constraint (5) of ILP) and for $i \in \{1, \ldots, nd\} - \{r\}$ we set

$$y_{ji} = \begin{cases} \text{the number of vertices in the subtree of } T_x \text{ rooted at } i & \text{if } j = p(i) \\ 0 & \text{if } j \neq p(i) \end{cases}$$

Hence,

$$\sum_{j:(j,i)} y_{ji} = y_{p(i)i} = 1 + \sum_{\ell:p(\ell)=i} y_{i\ell} = 1 + \sum_{\ell:(\ell,i)} y_{i\ell}$$

satisfying constraint (6) of ILP.

We notice that the number of vertices in the subtree of T_x rooted at i is at most nd, for each $i \in \{1, \ldots, nd\}$. Moreover, recalling that $x_{p(i)i} \geq 1$, we know that H^x contains $(p(i), i)$. Therefore, we get

$$y_{p(i)i} \leq nd \leq nd \, x_{p(i)i}$$

satisfying constraint (7) of ILP since $y_{ji} = 0$ for each $j \neq p(i)$. ☐

6 Conclusion and Open Problems

In this paper, we have studied the parameterized complexity of finding a spanning tree with the minimum number of branch vertices. We have shown that the problem is fixed-parameter tractable when parameterized by the neighborhood diversity of the input graph. We have provided an exact algorithm for finding such a spanning tree, which is based on a ILP approach.

Future work can investigate the complexity of the problem when other graph properties are used as parameter. In particular, it would be interesting to asses the parameterized complexity of MBT with respect to modular-width.

References

1. Agrawal, A., et al.: Parameterized complexity of happy coloring problems. Theoret. Comput. Sci. **835**, 58–81 (2020)
2. Araujo, J., Ducoffe, G., Nisse, N., Suchan, K.: On interval number in cycle convexity. Discrete Math. Theoret. Comput. Sci. **20**(1), 1–28 (2018)
3. Watel, D., Baste, J.: An FPT algorithm for node-disjoint subtrees problems parameterized by treewidth. Available at SSRN: https://ssrn.com/abstract=4197048 or https://dx.doi.org/10.2139/ssrn.4197048
4. Bermond, J.-C., Gargano, L., Rescigno, A.A.: Gathering with minimum delay in tree sensor networks. In: Shvartsman, A.A., Felber, P. (eds.) SIROCCO 2008. LNCS, vol. 5058, pp. 262–276. Springer, Heidelberg (2008). https://doi.org/10.1007/978-3-540-69355-0_22
5. Bermond, J.-C., Gargano, L., Rescigno, A.A.: Gathering with minimum completion time in sensor tree networks. J. Interconnect. Netw. **11**, 1–33 (2010)
6. Bermond, J.-C., Gargano, L., Peénnes, S., Rescigno, A.A., Vaccaro, U.: Optimal time data gathering in wireless networks with multidirectional antennas. Theoret. Comput. Sci. **509**, 122–139 (2013)
7. Bhyravarapu, S., Reddy, I.V.: On structural parameterizations of star coloring. arXiv preprint: arXiv:2211.12226 (2022)
8. Bianzino, A.P., Chaudet, C., Rossi, D., Rougier, J.L.: A survey of green networking research. IEEE Commun. Surv. Tutorials **14**, 3–20 (2012)
9. Carrabs, F., Cerulli, R., Gaudioso, M., Gentili, M.: Lower and upper bounds for the spanning tree with minimum branch vertices. Comput. Optim. Appl. **56**(2), 405–438 (2013). https://doi.org/10.1007/s10589-013-9556-5
10. Celik, A., Kamal, A.E.: Green cooperative spectrum sensing and scheduling in heterogeneous cognitive radio networks. IEEE Trans. Cognitive Commun. Networking **2**, 238–248 (2016)
11. Cerrone, C., Cerulli, R., Raiconi, A.: Relations, models and a memetic approach for three degree-dependent spanning tree problems. Eur. J. Oper. Res. **232**(3), 442–453 (2014)
12. Cerulli, R., Gentili, M., Iossa, A.: Bounded-degree spanning tree problems: models and new algorithms. Comput. Optim. Appl. **42**(3), 353–370 (2009)
13. Chimani, V.M., Spoerhase, J.: Approximating spanning trees with few branches. Theory Comput. Syst. **56**(1), 181–196 (2015)
14. Cordasco, G., Gargano, L., Rescigno, A.A., Vaccaro, U.: Evangelism in social networks: algorithms and complexity. Networks **71**(4), 346–357 (2018)
15. Cordasco, G., Gargano, L., Rescigno, A.A.: Iterated type partitions. In: Gąsieniec, L., Klasing, R., Radzik, T. (eds.) IWOCA 2020. LNCS, vol. 12126, pp. 195–210. Springer, Cham (2020). https://doi.org/10.1007/978-3-030-48966-3_15

16. Cordasco, G., Gargano, L., Rescigno, A.A.: Parameterized complexity of immunization in the threshold model. In: Mutzel, P., Rahman, M.S., Slamin (eds.) WALCOM: Algorithms and Computation. WALCOM 2022. Lecture Notes in Computer Science(), vol. 13174. Springer, Cham (2022). https://doi.org/10.1007/978-3-030-96731-4_23

17. Courcelle, B., Olariu, S.: Upper bounds to the clique width of graphs. Discret. Appl. Math. **101**(1–3), 77–144 (2000)

18. Cygan, M., et al.: Lower bounds for kernelization. In: Parameterized Algorithms, pp. 523–555. Springer, Cham (2015). https://doi.org/10.1007/978-3-319-21275-3_15

19. Downey, R.G., Fellows, M.R.: Parameterized Complexity. Springer-Verlag, New York (1999)

20. Fomin, F.V., Golovach, P.A., Lokshtanov, D., Saurabh, S.: Clique-width: on the price of generality. In: Proceedings of SODA 2009, pp. 825–834 (2009)

21. Fomin, F.V., Golovach, P.A., Lokshtanov, D., Saurabh, S.: Intractability of clique-width parameterizations. SIAM J. Comput. **39**(5), 1941–1956 (2010)

22. Gajarsky, J., Lampis, M., Ordyniak, S.: Parameterized algorithms for modular-width. In: Gutin, G., Szeider, S. (eds.) Parameterized and Exact Computation. IPEC 2013. Lecture Notes in Computer Science, vol. 8246. Springer, Cham (2013). https://doi.org/10.1007/978-3-319-03898-8_15

23. Ganian, R.: Using neighborhood diversity to solve hard problems (2012). arXiv:1201.3091

24. Gargano, L., Hammar, M., Hell, P., Stacho, L., Vaccaro, U.: Spanning spiders and light splitting switches. Discret. Math. **285**(1), 83–95 (2004)

25. Gargano, L., Hell, P., Stacho, L., Vaccaro, U.: Spanning trees with bounded number of branch vertices. In: Widmayer, P., Eidenbenz, S., Triguero, F., Morales, R., Conejo, R., Hennessy, M. (eds.) ICALP 2002. LNCS, vol. 2380, pp. 355–365. Springer, Heidelberg (2002). https://doi.org/10.1007/3-540-45465-9_31

26. Gargano, L., Rescigno, A.A.: Complexity of conflict-free colorings of graphs. Theoret. Comput. Sci. **566**, 39–49 (2015)

27. Gargano, L., Rescigno, A.A.: Collision-free path coloring with application to minimum-delay gathering in sensor networks. Discret. Appl. Math. **157**(8), 1858–1872 (2009)

28. Gavenciak, T., Koutecký, M., Knop, D.: Integer programming in parameterized complexity: five miniatures. Discrete Optim. **44**, 100596 (2022)

29. Gozupek, D., Buhari, S., Alagoz, F.: A spectrum switching delay-aware scheduling algorithm for centralized cognitive radio networks. IEEE Trans. Mobile Comp. **12**, 1270–1280 (2013)

30. Jansen, K., Rohwedderb, L.: On integer programming, discrepancy, and convolution. Math. Operat. Res., 1–15 (2023)

31. Lampis, M.: Algorithmic meta-theorems for restrictions of treewidth. Algorithmica **64**, 19–37 (2012)

32. Landete, M., Marín, A., Sainz-Pardo, J.L.: Decomposition methods based on articulation vertices for degree-dependent spanning tree problems. Comput. Optim. Appl. **68**(3), 749–773 (2017). https://doi.org/10.1007/s10589-017-9924-7

33. Marin, A.: Exact and heuristic solutions for the minimum number of branch vertices spanning tree problem. Eur. J. Oper. Res. **245**(3), 680–689 (2015)

34. Melo, R.A., Samer, P., Urrutia, S.: An effective decomposition approach and heuristics to generate spanning trees with a small number of branch vertices. Comput. Optim. Appl. **65**(3), 821–844 (2016). https://doi.org/10.1007/s10589-016-9850-0

35. Matsuda, H., Ozeki, K., Yamashita, T.: Spanning trees with a bounded number of branch vertices in a claw-free graph. Graphs and Combinatorics **30**, 429–437 (2014)
36. Robertson, N., Seymour, P.D.: Graph minors II. algorithmic aspects of tree-width. J. Algorithms **7**, 309–322 (1986)
37. Rossi, A., Singh, A., Shyam, S.: Cutting-plane-based algorithms for two branch vertices related spanning tree problems. Optim. Eng. **15**, 855–887 (2014)
38. Salamon, G.: Spanning tree optimization problems with degree-based objective functions. In: 4th Japanese-Hungarian Symposium on Discrete Mathematics and Its Applications, pp. 309–315 (2005)
39. Shami, N., Rasti, M.: A joint multi-channel assignment and power control scheme for energy efficiency in cognitive radio networks. In: Proceedings IEEE Wireless Communications and Networking Conference, WCNC 2016, pp. 1–6 (2016)
40. Silva, R.M.A., Silva, D.M., Resende, M.G.C., Mateus, G.R., Goncalves, J.F., Festa, P.: An edge-swap heuristic for generating spanning trees with minimum number of branch vertices. Optim. Lett. **8**(4), 1225–1243 (2014)
41. Silvestri, S., Laporte, G., Cerulli, R.: A branch-and-cut algorithm for the minimum branch vertices spanning tree problem. Comput. Oper. Res. **81**, 322–332 (2017)
42. Sundar, S., Singh, A., Rossi, A.: New heuristics for two bounded-degree spanning tree problems. Inf. Sci. **195**, 226–240 (2012)
43. Toufar, T., Masařík, T., Koutecký, M., Knop, D.: Simplified algorithmic metatheorems beyond MSO: treewidth and neighborhood diversity. Log. Methods Comput. Sci. **15** (2019)

Overcoming Probabilistic Faults in Disoriented Linear Search

Konstantinos Georgiou[1]([✉]), Nikos Giachoudis[1], and Evangelos Kranakis[2][ID]

[1] Department of Mathematics, Toronto Metropolitan University, Toronto, ON, Canada
konstantinos@torontomu.ca
[2] School of Computer Science, Carleton University, Ottawa, ON, Canada
kranakis@scs.carleton.ca

Abstract. We consider search by mobile agents for a hidden, idle target, placed on the infinite line. Feasible solutions are agent trajectories in which all agents reach the target sooner or later. A special feature of our problem is that the agents are p-faulty, meaning that every attempt to change direction is an independent Bernoulli trial with known probability p, where p is the probability that a turn fails. We are looking for agent trajectories that minimize the worst-case expected termination time, relative to the distance of the hidden target to the origin (competitive analysis). Hence, searching with one 0-faulty agent is the celebrated linear search (cow-path) problem that admits optimal 9 and 4.59112 competitive ratios, with deterministic and randomized algorithms, respectively.

First, we study linear search with one deterministic p-faulty agent, i.e., with no access to random oracles, $p \in (0, 1/2)$. For this problem, we provide trajectories that leverage the probabilistic faults into an algorithmic advantage. Our strongest result pertains to a search algorithm (deterministic, aside from the adversarial probabilistic faults) which, as $p \to 0$, has optimal performance $4.59112 + \epsilon$, up to the additive term ϵ that can be arbitrarily small. Additionally, it has performance less than 9 for $p \le 0.390388$. When $p \to 1/2$, our algorithm has performance $\Theta(1/(1 - 2p))$, which we also show is optimal up to a constant factor.

Second, we consider linear search with two p-faulty agents, $p \in (0, 1/2)$, for which we provide three algorithms of different advantages, all with a bounded competitive ratio even as $p \to 1/2$. Indeed, for this problem, we show how the agents can simulate the trajectory of any 0-faulty agent (deterministic or randomized), independently of the underlying communication model. As a result, searching with two agents allows for a solution with a competitive ratio of $9 + \epsilon$ (which we show can be achieved with arbitrarily high concentration) or a competitive ratio of $4.59112 + \epsilon$. Our final contribution is a novel algorithm for searching with two p-faulty agents that achieves a competitive ratio $3 + 4\sqrt{p(1 - p)}$, with arbitrarily high concentration.

Keywords: Linear Search · Probabilistic Faults · Mobile Agents

Research supported in part by NSERC, and by the Fields Institute for Research in Mathematical Sciences

Extended Abstract—The full version of this paper appears on arXiv [17].

S. Rajsbaum et al. (Eds.): SIROCCO 2023, LNCS 13892, pp. 520–535, 2023.
https://doi.org/10.1007/978-3-031-32733-9_23

1 Introduction

Linear search refers to the problem of searching for a point target which has been placed at an unknown location on the real line. The searcher is a mobile agent that can move with maximum speed 1 and is starting the search at the origin of the real line. The goal is to find the target in minimum time. This search problem provides a paradigm for understanding the limits of exploring the real line and has significant applications in mathematics and theoretical computer science.

In the present paper we are interested in linear search under a faulty agent which is disoriented in that when it attempts to change direction not only it may fail to do so but also cannot recognize that the direction of movement has changed. More precisely, for some $0 \leq p \leq 1$, a successful turn occurs with probability $1-p$ but the agent will not be able to recognize this until it has visited an anchor, a known, preassigned point, placed on the real line. Despite this faulty behaviour of the agent it is rather surprising that it is possible to design algorithms which outperform the well-known zig-zag algorithm whose competitive ratio is 9.

1.1 Related Work

Search by a single agent on the real line was initiated independently by Bellman [9] and Beck [6–8] almost 50 years ago; the authors prove the well known result that a single searcher whose max speed is 1 cannot find a hidden target placed at an initial distance d from the searcher in time less than $9d$. These papers gave rise to numerous variants of linear search. Baeza-Yates et. al. [3,4] study search problems by agents in other environments, e.g. in the plane or starting at the origin of w concurrent rays (also known as the "Lost Cow" problem). Group search was initiated in [11] where evacuation (a problem similar to search but one minimizing the time it takes for the last agent to reach the target) by multiple agents that can communicate face-to-face was studied. An extension to the problem, where one tries to minimize the weighted average of the evacuation times was studied in [18]. There is extensive literature on this topic and [12] provides a brief survey of more recent topics on search.

Linear search with multiple agents some of which may be faulty, Crash or Byzantine, was initiated in the work of [15] and [13], respectively. In this theme, one uses the power of communication in order to overcome the presence of faults. For three agents one of which is Byzantine, [22] shows that the proportional schedule presented in [15] can be analyzed to achieve an upper bound of 8.653055. Recently, [14] gives a new class of algorithms for n agents when the number of Byzantine faulty among them is near majority, and the best known upper bound of 7.437011 on an infinite line for three agents one of which is Byzantine.

The present paper focuses on probabilistic search. The work of Bellman [9] and Beck [6–8], also mentioned above, has probabilistic focus. In addition numerous themes on probabilistic models of linear search can be found in the book [2] of search games and rendezvous, as well as in [1,21].

Search which takes into account the agent's turning cost is the focus of [2][Section 8.4] as well as the paper [16]. Search with uncertain detection is studied in [2][Section 8.6]. According to this model the searcher is not sure to find the target when reaching it; instead it is assumed that the probability the searcher will find it on its k-th visit is p_k, where $\sum_{k \geq 0} p_k = 1$. A particular case of this is search with geometric detection probability [2][Section 8.6.2] in which the probability of finding the target in the k-th visit is $(1 - p)^{k-1}p$. [20] investigates searching for a one-dimensional random walker and [5] is concerned with rendezvous search when marks are left at the starting points. In another result pertaining to different kind of probabilistic faults, [10] studies the problem on the half-line (or 1-ray), where detecting the target exhibits faults, i.e. every visitation is an independent Bernoulli trial with a known probability of success p. Back to searching the infinite line, a randomized algorithm with competitive ratio 4.59112 for the cow path problem can be found in [19] and is also shown to be optimal. In a strong sense, the results in this work are direct extensions of the optimal solutions for deterministic search in [3] and for randomized search in [19]. To the best of our knowledge the linear search problem considered in our paper has never been investigated before. We formally define our problem in Sect. 2. Then in Sect. 3 we elaborate further on the relevance of our results to [3] and [19].

2   Model and Problem Definition (p-PDLS)

We introduce and study the so-called Probabilistically Disoriented Linear Search problem (or p-PDLS, for short), associated with some probability p. We generalize the well studied linear search problem (also known as cow-path) where the searcher's trajectory decisions exhibit probabilistic faults. The value p will quantify a notion of probabilistic failure (disorientation).

In p-PDLS, an agent (searcher) can move at unit speed on an infinite line, where any change of direction does not incur extra cost. On the line there are two points, *distinguishable* from any other point. Those points are the *origin*, i.e. the agent's starting location, and the *target*, which is what the agent is searching for and which can be detected when the agent walks over it.

The agents have a faulty behaviour. If the agent tries to change direction (even after stopping), then with known probability p the agent will fail and she will still move towards the same direction. Consequent attempts to change direction are independent Bernoulli trials with probability of success $1 - p$. Moreover the agent is *oblivious* to the result of each Bernoulli trial, i.e. the agent is not aware if it manages to change direction. We think of this probabilistic behaviour as a co-routine of the agent that fails to be executed with probability p. An agent which satisfies this property for a given p is called p-faulty. Moreover we assume, for the sake of simplicity, that at the very beginning the p-faulty agent starts moving to a specific direction, without fault.

The agent's faulty behaviour is compensated by that it can utilize the origin and the target to recover its perception of orientation. Indeed, suppose that the agent passes over the origin and after time 1 it decides to change direction. In additional time 1, the agent has either reached the origin, in which case it realizes it turned successfully, or it does not see the origin, in which case it detects that it failed to turn. We elaborate more on this idea later.

A solution to p-PDLS is given by the agent's trajectory (or agents' trajectories), i.e. instructions for the agent(s) to turn at specific times which may depend on previous observations (e.g. visitations of the origin and when they occurred). A *feasible trajectory* is a trajectory in which every point on the infinite line is visited (sooner or later) with probability 1 (hence a 1-faulty agent admits no feasible trajectory).

For a point $x \in \mathbb{R}$ (target) on the line, we define the termination cost of a feasible trajectory in terms of competitive analysis. Indeed, if $E(x)$ denotes the expected time that target x is reached by the last agent (so by the only agent, if searching with one agent), where the expectation is over the probabilistic faults or even over algorithmic randomized choices, then the termination cost for target (input) x is defined as $E(x)/|x|$. The competitive ratio of the feasible trajectory is defined then as $\lim \sup_x E(x)/|x|$.

For the sake of simplicity, our definition deviates from the standard definition of the competitive ratio for linear search in which the performance is defined as the supremum over x with absolute value bounded by a constant d, usually $d = 1$. However it can be easily seen that the two measures differ by at most ϵ, for any $\epsilon > 0$ using a standard re-scaling trick (see for example [18]) that shows why the value of d is not important, rather what is important is that d is known to the algorithm.

Specifications When Searching with Two Agents: When searching with two p-faulty agents, the value of p is common to both, as well as the probabilistic faults they exhibit are assumed to be independent Bernoulli trials. The search by two p-faulty agents can be done either in the wireless or the face-to-face model. In the former model, we assume that agents are able to exchange messages instantaneously, whereas in the face-to-face model messages can be exchanged only when the agents are co-located.

In either communication model, we assume that the two agents can detect that (and when) they meet. As a result, we naturally assume that upon a meeting, a p-faulty agent can also act as a *distinguished point* (same as the origin), hence helping the other agent to turn. Later, we will call an agent who facilitates the turn *Follower*, and the agent who performs the turn *Leader*. As long as the agents have a way to resolve the two roles (which will be built in to our algorithms), we also assume that the Leader moving in any direction can "pick up" another faulty agent she meets so that the two continue moving in that direction. This means that two agents meeting at a point can continue moving to the direction of the leader (with probability 1) even if the non-leader is idle. This property is motivated by that, effectively, the leader does not change direction, and hence there is no risk to make a mistake. Finally we note that two p-faulty agents can move at speed at most 1, independently of each other, as well as any of them can slow down or even stay put, complying still with the faulty turn specifications.

2.1 Notes on Algorithmic and Adversarial Randomness

The algorithms (feasible trajectories) that we consider are either deterministic or randomized, independently of the randomness induced by the faultiness. In particular, the efficiency measure is defined in the same way, where any expectations are calculated over the underlying probability space (induced by the combination of probabilistic faults and the possible randomized algorithmic choices). Moreover, if the algorithm is randomized then an additional random mechanism is used that is independent of the faulty behaviour. For example, a randomized agent (algorithm) can choose a number between 0 and 1 uniformly at random. A p-faulty agent that has access to a random oracle will be called *randomized*, and *deterministic* otherwise (but both exhibit probabilistically failed turns).

It follows by our definitions that 0-PDLS with one deterministic agent is the celebrated linear search problem (cow-path) which admits a provably optimal trajectory of competitive ratio 9 [3]. In the other extreme, 1-PDLS does not admit a feasible solution, since the agent moves indefinitely along one direction. In a similar spirit we show next in Lemma 1 that the problem is meaningful only when $p < 1/2$.

Lemma 1. *No trajectory for p-PDLS has bounded competitive ratio when $p \geq 1/2$.*

It is essential to note that in our model, the probabilistic faulty turns of a agent do hinder the control of the trajectory, but also introduce uncertainty of the algorithmic strategy for the adversary. As a result, the probabilistic movement of the agent (as long as $p > 0$), even though it is not controlled by the algorithm, it can be interpreted as an algorithmic choice that is set to stone. Therefore, the negative result of [19] implies the following lower bound for our problem.

Corollary 1. *For any $p \in (0, 1/2)$, no solution for p-PDLS with one agent (deterministic or randomized) has competitive ratio lower than 4.59112.*

3   Contributions' Outline and Some Preliminary Results

Our main results in this work pertain to upper bounds for the competitive ratio that (one or two) faulty agents can achieve for p-PDLS.

3.1   Results Outline for Searching with One Faulty Agent

This is an extended abstract. Any omitted proofs can be found in the full version of the paper [17]. Here, we start our exposition with search algorithms for one p-faulty agent. In Sect. 4 we analyze the performance of a deterministic search algorithm, whose performance is summarized in Theorem 2 on page 8. The section serves as a warm-up for the calculations we need for our main result when searching with one randomized p-faulty agent. Indeed, in Sect. 5 we present a *randomized* algorithm whose performance is summarized in Theorem 3 on page 10. The reader can see the involved formulas in the formal statements of the theorems, so here we summarize our results graphically in Fig. 1. Some important observations are in place.

Comments on the expansion factors: We emphasize that both our algorithms are adaptations of the standard zig-zag algorithms of [3] and [19] which are optimal for the deterministic and randomized model, respectively, when searching with a 0-faulty agent. The zig-zag algorithms are parameterized by the so-called *expansion factor g* that quantifies the rate by which the searched space is increased in each iteration that the searcher changes direction. In our case however, we are searching with p-faulty agents, where $p > 0$, and as a result, the algorithmic choice for how the searched space expands cannot be fully controlled (since turns are subject to probabilistic faults). This is the reason that, in our case, the analyses of these algorithms are highly technical, which is also one of our contributions.

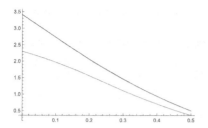

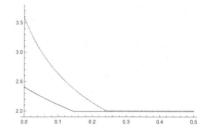

(a) Competitive ratio comparison between the deterministic algorithm of Theorem 2 in blue and the randomized algorithm of Theorem 3 in yellow. The competitive ratios are scaled by $(1/2 - p)$.

(b) Intended expansion factors of the zig-
zag algorithms of the deterministic algorithm of Theorem 2 in blue and of the randomized algorithm of Theorem 3 in yellow.

Fig. 1. Graphical summary of the positive results pertaining to searching with one p-faulty agent, $p \in (0, 1/2)$.

On a relevant note, the optimal expansion factor for the optimal deterministic 0-faulty agent is 2, while the expansion factors $g = g(p)$ we use for the deterministic p-faulty agent, $p \in (0, 1/2)$, are decreasing in p. As $p \to 0$, we use expansion factor $1 + \sqrt{2}$, and the expansion factor drops to 2, for all $p \geq 0.146447$. When it comes to our randomized algorithm, the chosen expansion factor is again decreasing in p, starting from the same choice as for the optimal randomized 0-faulty agent of [19], and being equal to 2 for all $p \geq 0.241516$.[1] The expansion factors are depicted in Fig. 1b.

Comments on the established competitive ratios: By the proof of Lemma 1 it follows that as $p \to 1/2$, the optimal competitive ratio for p-PDLS is of order $\Omega(1/(1 - 2p))$. Hence for the sake of better exposition, we depict in Fig. 1a the established competitive ratios scaled by $1/2 - p$. Moreover, the results are optimal up to a constant factor when $p \to 1/2$.

It is also interesting to note that for small enough values of p, the established competitive ratios are better than the celebrated optimal competitive ratio 9 for 0-PDLS. This is because our algorithms leverage the probabilistic faults to their advantage, making the adversarial choices weaker. Indeed, algebraic calculations show that the competitive ratios of the deterministic and the randomized algorithms are less than 9 when $p \leq 0.390388$ and when $p \leq 0.436185$, respectively. Moreover, when searching with one p-faulty agent and $p \to 0$, the competitive ratio of our deterministic algorithm tends to $6.82843 < 9$ and of our randomized algorithm to the provably optimal competitive ratio of $\frac{1}{W(\frac{1}{e})} + 1 \approx 4.59112$, where $W(\cdot)$ is the Lambert W-Function. It is important to also note that for deterministic algorithms only, there is no continuity at $p = 0$, since when the agent exhibits no faults, the adversary has certainty over the chosen trajectory and hence is strictly more powerful.

---

[1] For simplicity, we only give numerical bounds on p. All mentioned bounds of p in this section have explicit algebraic representations that will be discussed later.

Values of p close to $1/2$ give rise to interesting observations too. Indeed, it is easy to see in Fig. 1a that the derived comparative ratios are of order $\Theta\left(1/(1-2p)\right)$. More interestingly, the difference of the established competitive ratios of the deterministic and the randomized algorithms is $\Theta\left(1/(1-2p)\right)$ too, when $p \to 1/2$. Hence the improvement when utilizing controlled (algorithmic) randomness is significant. We ask the following critical question:

> *"Does access to a random oracle provide an advantage for p-PDLS with one agent?"*

Somehow surprisingly, we answer the question in the *negative*! More specifically, we show, for all $p \in (0, 1/2)$ and using the probabilistic faults in our advantage, how a deterministic p-faulty agent (deterministic algorithm) can simulate a randomized p-faulty agent (randomized algorithm). Hence our improved upper bound of Theorem 3 which is achieved by a randomized algorithm can be actually simulated by a deterministic p-faulty agent. The proof of this claim relies on that our randomized algorithm assumes access to a randomized oracle that samples only from uniform distributions a finite number of times (in fact only 2). The main idea is that a deterministic agent can stay arbitrarily close to the origin, sampling arbitrarily many random bits using it's faulty turns, allowing her to simulate queries to a random oracle.

Theorem 1. *For any $p \in (0, 1/2)$, let c be the competitive ratio achieved by a randomized faulty agent for p-PDLS, having access (finite many times) to an oracle sampling from the uniform distribution. Then for every $\epsilon > 0$, there is a deterministic p-faulty agent for the same problem with competitive ratio at most $c + \epsilon$.*

3.2 Results Outline for Searching with Two Faulty Agents

We conclude our contributions in Sect. 6 where we study p-PDLS with two agents. First we show how two (deterministic) p-faulty agents (independently of the underlying communication model) can simulate the trajectory of any (one) 0-faulty agent. As an immediate corollary, we derive in Theorem 4 on page 13 a method for finding the target with two p-faulty agents that has competitive ratio $9 + \epsilon$, for every $\epsilon > 0$. Most importantly as long as $p > 0$, the result holds not only in expectation, but with arbitrarily large concentration.

Motivated by similar ideas, we show in Theorem 5 on page 14 how two deterministic p-faulty agents can simulate the celebrated optimal randomized algorithm for one agent for 0-PDLS, achieving competitive ratio arbitrarily close to 4.59112. The result holds again regardless of the communication model. However, in this case we cannot guarantee a similar concentration property as before.

Finally, we study the problem of searching with two *wireless* p-faulty agents. Here we are able to show in Theorem 6 on page 14 how both agents can reach any target with competitive ratio $3 + 4\sqrt{p(1-p)}$, in expectation. The performance is increasing in $p < 1/2$, ranging from 3 (the optimal competitive ratio when searching with two wireless 0-fault agents) to 5. However, as before we can control the concentration of the performance, making it arbitrarily close to $3 + 4\sqrt{p(1-p)}$ with arbitrary confidence. Hence, this gives an advantage over Theorem 5 for small values of p, i.e. when the derived competitive ratio is smaller than 4.59112. In this direction, we show that when

$p > 0.197063$ the competitive ratio exceeds 4.59112, and hence each of the results described above are powerful in their own right.

4   Searching with One Deterministic p-Faulty Agent

We start by describing a deterministic algorithm for searching with a p-faulty agent, where $p \in (0, 1/2)$. Our main result in this section reads as follows.

Theorem 2. p-PDLS *with one agent admits a deterministic algorithm with competitive ratio equal to* $2(\sqrt{2} + 2)$ *when* $0 < p \leq \frac{1}{4}(2 - \sqrt{2})$, *and equal to* $6 - 4p + \frac{1}{1-2p}$ *when* $\frac{1}{4}(2 - \sqrt{2}) < p < 1/2$.

We emphasize that having $p > 0$ will be essential in our algorithm. This is because the probabilistic faults introduce uncertainty for the adversary. For this reason, it is interesting but not surprising that we can in fact have competitive ratio $2(\sqrt{2} + 2) \approx 6.82843 < 9$ for all $p < \frac{1}{8}\left(\sqrt{17} - 1\right) \approx 0.390388$.

First we give a verbose description of our algorithm, that takes as input $p \in (0, 1)$, and chooses parameter $g = g(p)$ which will be the intended expansion rate of the searched space. In each iteration of the algorithm, the agent will be passing from the origin with the intention to expand up to g^i in a certain direction. When distance g^i is covered, the agent attempts to return to the origin. After additional time g^i the agent knows if the origin is reached, in which case she expands in the opposite direction with intended expansion g^{i+1}. If not, the agent knows she did not manage to turn, and proceeds up to point g^{i+1} (this is why we require that $g \geq 2$). Then, she makes another attempt to turn. This continues up to the event that the agent succeeds in turning at g^j, for some $j > i$, and then the agent attempts to expand in the opposite direction with intended expansion length g^{j+1}. The algorithm starts by searching towards an arbitrary direction, say right, with intended expansion g^0. later on, it will become clear that the termination time of the algorithm converges only if $g < 1/p$.

In order to simplify the exposition (and avoid repetitions), we introduce first subroutine Algorithm 1 which is the baseline of both algorithms we present for searching with one agent. Here, this is followed by Algorithm 2 which is our first search algorithm.

As we explain momentarily, Theorem 2 follows directly by the following technical lemma.

Lemma 2. *Fix* $p \in (0, 1/2)$ *and* $g \in [2, 1/p)$. *If the target is placed in* $(g^t, g^{t+1}]$, *then the competitive ratio of Algorithm 2 is at most* $\frac{(1-p)g}{(1-gp)}\left(\frac{1+(1-2p)g}{g-1} + (1 - 2p)^{t+1}\right) + 1$.

Taking the limit when $t \to \infty$ of the expression of Lemma 2 shows that the competitive ratio of Algorithm 2 is

$$f_p^{\text{DET}}(g) := \frac{1 - g(2\, g(p - 2)p + g + 2)}{(1 - g)(1 - gp)}$$

for all p, g complying with the premise.[2]

---

[2] If one wants to use the original definition of the competitive ratio, then by properly re-scaling the searched space (just by scaling the intended turning points), one can achieve competitive ratio which is additively off by at most ϵ from the achieved value, for any $\epsilon > 0$.

Algorithm 1. Baseline Search

Require: g, p, i, d

 1: **repeat**
 2:     **repeat**
 3:         Move towards point dg^i with speed 1
 4:     **until** The target is found $\lor$ g^i time has passed
 5:     **while** The origin is ¬found $\land$ the target is ¬found **do**
 6:         Set $d \leftarrow -d$ {This changes the direction but it can fail with probability p}
 7:         Set $i \leftarrow i + 1$
 8:         **repeat**
 9:             Move towards point dg^i with speed 1
10:         **until** Either the target is found or $g^i - g^{i-1}$ time has passed or the origin is found
11:     **end while**
12: **until** The target is found

Algorithm 2. Faulty Deterministic Linear Search

Require: $2 \leq g \leq 1/p$ and $0 < p < 1/2$

 1: Set $i \leftarrow 0$ and $d \leftarrow 1$ {$d \in \{1, -1\}$ represents direction (1 going right and -1 going left)}
 2: Run Algorithm 1 with parameters g, p, i, d.

Next we optimize function $f_p^{\mathrm{DET}}(g)$. When $p \leq \frac{1}{4}\left(2 - \sqrt{2}\right)$, the optimal expansion factor (optimizer of $f_p^{\mathrm{DET}}(g)$) is $g_0(p) := \frac{-(\sqrt{2}+2)p+\sqrt{2}+1}{1-2p^2}$. It is easy to see that $2 \leq g_0(p) < 1/p$, for all $p \leq \frac{1}{4}\left(2 - \sqrt{2}\right)$ (in fact the strict inequality holds for all $0 < p < 1$). In this case the induced competitive ratio becomes $2\left(\sqrt{2} + 2\right) \approx 6.82843$.

When $p \geq \frac{1}{4}\left(2 - \sqrt{2}\right)$, the optimal expansion factor (at least 2) is $g = 2$, in which case the competitive ratio becomes $6 - 4p + \frac{1}{1-2p}$. Interestingly, the competitive ratio becomes at least 9 for $p \geq \frac{1}{8}\left(\sqrt{17} - 1\right) \approx 0.390388$. In other words, 0.390388 is a threshold for the probability associated with the agent's faultiness for which, at least for the proposed algorithm, that probabilistic faultiness is useful anymore towards beating the provably optimal deterministic bound of 9. On a relevant note, recall that by Lemma 1, there is a threshold probability p' such that any algorithm (even randomized) has competitive ratio at least 9 when searching with a p-faulty agent, when $p \geq p'$.

5   Searching with One Randomized (Improved Deterministic) p-Faulty Agent

In this section we equip the p-faulty agent with the power of randomness, and we show the next positive result. Note that due to Theorem 1, the results can be simulated by a deterministic p-faulty agent too, up to any precision.

Theorem 3. *Let* $p_0 = \frac{1}{8}\left(5 - \sqrt{1 + \ln^2(2) + 6\ln(2)} - \ln(2)\right) \approx 0.241516$. *For each* $p < p_0$, *let also* g_p *denote the unique root, no less than 2, of*

$$h_p(g) := (1 - gp)(1 + g(1 - 2p)) - g(1 - p)\ln g. \tag{1}$$

Then for each $p \in (0, 1/2)$, p-PDLS *with one agent admits a randomized algorithm with competitive ratio at most*

$$\begin{cases} g_p\left(\frac{1-p}{1-g_p p}\right)^2 + 1, & p \in (0, p_0] \\ (1 - p)\frac{1+2(1-2p)}{(1-2p)\ln(2)} + 1, & p \in (p_0, 1/2). \end{cases}$$

Elementary calculations show that the competitive ratio of Theorem 3 remains at most 9, as long as $p \leq \frac{1}{8}\left(7 + \sqrt{1 + 256\ln^2(2) - 96\ln(2)} - 16\ln(2)\right) \approx 0.436185$.

First we argue that the premise regarding g_p of Theorem 3 is well defined. Indeed, consider $h_p(g)$ as in (1). We have that $\partial h_p(g)/\partial p = 4g^2 p - g^2 - 3g + g\ln g \leq -3g + g\ln g < 0$, for all $p \leq 1/4$. Therefore, $h_p(2) = 8p^2 - 10p + 2p\ln(2) + 3 - 2$ is decreasing in p, and hence $h_p(2) > h_{p_0}(2) = 0$, by the definition of p_0. Also $h_p(4) = 32p^2 - 28p + 4p\ln(4) + 5 - 4\ln 4$ is decreasing in p too, and therefore $h_p(4) \leq h_0(4) = 5 - 4\ln 4 \approx -0.545177 < 0$. This means that $h_p(g)$ has indeed a root, with respect to g, in $[2, 4)$, for all $p \in (0, p_0]$. Next we show that this root is unique. Indeed, we have $\partial h_p(g)/\partial g = 4gp^2 - 2gp + p\ln g - \ln g - 2p \leq -\ln g < 0$, for $p \in [0, 1/2]$. Therefore, for all $p \leq p_0 < 1/2$, function $h_p(g)$ is decreasing in g, and hence any root is unique, that is g_p is indeed well-defined and lies in the interval $[2, 1/p)$, for all $p \in (0, 1/2)$.

Next, we observe that as $p \to 0$, the competitive ratio promised by Theorem 3 when searching with a p-faulty agent (i.e. a nearly non-faulty agent) is given by g_0 being the root of $g - g\ln g + 1$, i.e. $g_0 = \frac{1}{W(\frac{1}{e})} \approx 3.59112$, where $W(\cdot)$ is the Lambert W-Function.[3] Moreover, the induced competitive ratio is $1 + g_0 \approx 4.59112$, which is exactly the competitive ratio of the optimal randomized algorithm for the original linear search problem due to Kao et al [19]. We view this also as a sanity check regarding the correctness of our calculations.

Not surprisingly, our algorithm that proves Theorem 3 is an adaptation of the celebrated optimal randomized algorithm for linear search with a 0-faulty agent of [19]. As before, the search algorithm, see Algorithm 3 below, is determined by some expansion factor g, that represents the (intended, in our case, due to faults) factor by which the searched space is extended to, each time the direction of movement changes. The randomized algorithm makes two queries to a random oracle. First it chooses a random bit, representing an initial random direction to follow. Second, the algorithm samples random variable ϵ that takes a value in $[0, 1]$, uniformly at random. Variable ϵ quantifies a random scale to the intended turning points. It is interesting to note that setting $\epsilon = 0$ in Algorithm 3 deterministically, and removing the initial random direction choice, gives rise to the previous deterministic Algorithm 2.

---

[3] The Lambert W-Function is the inverse function of $L(x) = xe^x$.

Algorithm 3. Faulty Randomized Linear Search

Require: $g \geq 2$ and $0 < p < 1/2$
1: Choose ϵ from $[0, 1)$ uniformly at random
2: Set $i \leftarrow \epsilon$ and $d \leftarrow 1$ {$d \in \{1, -1\}$ represents direction (1 going right and -1 going left)}
3: Run Algorithm 1 with parameters g, p, i, d.

We show next how Theorem 3 follows by the following technical lemma.

Lemma 3. *For any $p \in (0, 1/2)$ and any $g \in [2, 1/p)$, if the target is placed in the interval $(g^t, g^{t+1}]$, then the competitive ratio of Algorithm 3 is at most*

$$1 + \frac{1}{(1-gp)\ln g} \left(\begin{array}{c} (1-p)(1-g(-1+2p))(1+(-1+2p)^t) \\ + \frac{2(1-p)}{g^t} + 2\,g(1-p)(p+g^t(-1+p)(-1+2p)^t) \end{array} \right).$$

Recall that $p \in (0, 1/2)$, so taking the limit of the expression of Lemma 3 when $t \to \infty$, and after simplifying algebraically the expression, shows that the competitive ratio of Algorithm 3 is at most

$$f_p^{\mathrm{RAND}}(g) := 1 + \frac{(1-p)(1+g(1-2p))}{(1-gp)\ln g},$$

for all p, g complying with the premise. Next we optimize $f_p^{\mathrm{RAND}}(g)$ for all parameters $p \in (0, 1/2)$, under the constraint that $2 \leq g < 1/p$. In particular, we show that the optimizers of $f_p^{\mathrm{RAND}}(g)$ are g_p, if $p \leq p_0$, and $g = 2$ otherwise resulting in the competitive ratios as described in Theorem 3.

First we compute $\frac{\partial}{\partial g} f_p^{\mathrm{RAND}}(g) = \frac{1-p}{g(1-gp)^2 \ln^2 g} h_p(g)$, where $h_p(g)$ is the same as (1). As already proven below the statement of Theorem 3, we have that $h_p(g)$ has a unique root in the interval $[2, 4)$. Next we show that $f_p^{\mathrm{RAND}}(g)$ is convex.

Indeed, we have that $\frac{\partial^2}{\partial g^2} f_p^{\mathrm{RAND}}(g) = \frac{1-p}{g^2(1-gp)^3 \ln^3(g)} s_p(g)$, where $s_p(g) = \sum_{i=0}^{3} \alpha_i p^i$ is a degree 3 polynomial in p with coefficients $\alpha_3 = 2\,g + g(-\ln g) + \ln(g) + 2$, $\alpha_2 = 2g\left(-2g + g\ln^2 g - \ln g - 4\right)$, $\alpha_1 = g^2\left(2g - 2\ln^2 g + g\ln g + 3\ln g + 10\right)$, and $\alpha_0 = -2g^3(\ln g + 2)$. As a result, it is easy to verify that $s_p(g)$ remains positive for all $p \in (0, 1/2)$, condition on that $g \in [2, 4]$ (in fact the optimizers as described in Theorem 3 do satisfy this property). We conclude that $f_p^{\mathrm{RAND}}(g)$ is convex in g.

Together with our previous observation, this means that, under constraint $g \geq 2$, function $f_p^{\mathrm{RAND}}(g)$ is minimized at the unique root of $\frac{\partial}{\partial g} f_p^{\mathrm{RAND}}(g)$ when $p \leq p_0$, and at $g = 2$ when $p \in [p_0, 1/2)$. These are exactly the optimizers described in Theorem 3, where in particular the competitive ratio $f_p^{\mathrm{RAND}}(g)$ is simplified taking into consideration that for the chosen value of g we have that $s_p(g) = 0$, for all $p \leq p_0$.

Lastly, it remains to argue that all optimizers of $f_p^{\mathrm{RAND}}(g)$ are indeed at most $1/p$. For this, it is enough to show that the unique root g_p of $h_p(g)$ is at most $1/p$, for all $p \leq p_0$ (since for larger values of p we use expansion factor $g = 2$). For this, and since $g \geq 2$, we have $h_p(g) \leq 2\,g^2 p^2 - g^2 p - 3\,gp + gp\ln 2 + g - g\ln 2 + 1$. The latest expression is a polynomial in p of degree 2, which has only one of its roots positive, namely

$$\frac{-\sqrt{(-3p + p \ln 2 + 1 - \ln 2)^2 - 4(2p^2 - p)} + 3p + p(-\ln 2) - 1 + \ln 2}{2p(2p - 1)}.$$

Simple calculations then can show that the latter expression is at most $1/p$ for all $p \in (0, 1/2)$. In fact one can show that the expression above, multiplied by p, is strictly increasing in p and at $p = 1/4 > p_0$ becomes $\frac{1}{4}\left(1 - \ln 8 + \sqrt{9 + (\ln 8 - 2) \ln 8}\right) \approx 0.486991 < 1/2$. That shows that for each $p < p_0$ we have that for the unique root g_p of $h_p(g)$ the inequality $g_p < 1/2p < 1/p$ is valid, as desired.

6   Searching with Two p-Faulty Agents

In this section we present algorithms for p-PDLS for two faulty agents, for all $p \in (0, 1/2)$. Central to our initial results is the following subroutine that, at a high level, will be used by a p-faulty agent, the *Leader*, in order to make a "forced" turn, with the help of a *Follower*, which can be either a distinguished immobile point, e.g. the target or the origin, or another p-faulty agent. In this process the p-faulty agent who undertakes the role of the Follower may need to either slow down or even halt for some time, still complying with the probabilistic faulty turns (once halted, she can continue moving in the previous direction, but changing it is subject to a fault).

Algorithm 4. Force Change Direction (Instructions for a Leader)

Require: γ small real number, Follower either mobile or immobile.

1: **repeat**
2:    Change direction {This fails with probability p}
3:    Move for time γ if Follower is mobile, and $\gamma \leftarrow 2\gamma$ if Follower is immobile.
4: **until** You meet with follower
5: Communicate to Follower the Leader's direction

It will be evident, in the proof of Lemma 4 below, that Algorithm 4 will allow an agent to change direction arbitrarily close to an intended turning point, and with arbitrary concentration (both controlled by parameter γ).

The next lemma refers to a task that two p-faulty agents can accomplish independently of the underlying communication model. At a high level, the lemma establishes that two p-faulty agents can bypass the probabilistic faults at the expense of giving up the independence of the searchers' moves.

Lemma 4. *For every $p \in [0, 1/2)$, two p-faulty agents can simulate the trajectory of a deterministic 0-faulty agent within any precision (and any probability concentration).*

Proof. Consider two p-faulty agents that are initially collocated at the origin. We show how the agents can simulate (at any precision) a deterministic trajectory. For this we need to show how the two agents can successfully make a turn at any point without deviating (in expectation) from that point.

Indeed, consider a scheduled turning point and consider the two p-faulty agents approaching that point. For some $\gamma > 0$ small enough, at time 2γ before the agents arrive at the point, the two agents undertake two distinguished roles, that of a Leader and that of a Follower. The roles can remain invariant throughout the execution of the algorithm. The Follower instantaneously slows down so that when the distance of the Leader and the Follower becomes 2γ, the Follower is $\gamma/(1-p)$ before the turning point (which is strictly more than γ and strictly less than 2γ), and as a result the Follower has passed the turning point by $\frac{1-2p}{1-p}\gamma$. At this moment, the Follower resumes full speed, and both agents move towards the same direction as before. Then, the Leader runs Algorithm 4 with mobile Follower being the other p-faulty agent.

Note that if at any moment the two agents meet, it is because after a successful turn the two have moved towards each other for time γ, whereas if a turn is unsuccessful the two preserve their relative distance γ. Therefore, since the moment of the first turning attempt, the two agents meet in expected time $\sum_{i=0}^{\infty}(i + 1)\gamma(1 - p)p^i = \frac{\gamma}{1-p}$. We conclude that the expected meeting (and turning) point is the original turning point. Most importantly, the probability that the resulting turning point is away from the given turning point drops exponentially with p, and is also proportional to γ, which can be independently chosen to be arbitrarily small.   □

Lemma 4 is quite powerful, since it shows how to simulate deterministic turns with arbitrarily small deviation from the actual turning points. More importantly, that deviation can be chosen to drop arbitrarily fast, dynamically, hence we can achieve smaller expected deviation later in the execution of the algorithm, compensating this way for the passed time. Therefore, we obtain the following theorem.

Theorem 4. *For all $p \in [0, 1/2)$, two deterministic faulty agents can solve p-PDLS with competitive ratio $9 + \epsilon$, for every $\epsilon > 0$, independently of the underlying communication model. Also, the performance is concentrated arbitrarily close to $9 + \epsilon$.*

Is it worthwhile noticing that agents' movements, in the underlying algorithm of Theorem 4 is still probabilistic, due to the probabilistic faulty turns. However, choosing appropriate parameters every time Algorithm 4 is invoked, one can achieve arbitrary concentration in the expected performance of the algorithm, hence the bound of $9 + \epsilon$ can be practically treated as deterministic.

In contrast, using the same trick, we can achieve a much better competitive ratio, but only in expectation equal to the one of [19] (with uncontrolled concentration). To see how, note that by the proof of Theorem 1, the two p-faulty agents can stay together in order to collect sufficiently many random bits and simulate any finite number of queries to a random oracle. Then using Lemma 4, the agents simulate the optimal randomized algorithm of [19] with performance 4.59112, designed originally for one randomized 0-faulty agent that makes only 2 queries to the uniform distribution. In other words, the two deterministic p-faulty agents can overcome their faulty turns using Lemma 4 and the lack of random oracle by invoking Theorem 1. To conclude, we have the following theorem which requires that $p > 0$.

Theorem 5. *Two deterministic faulty agents can solve p-PDLS with competitive ratio* $4.59112 + \epsilon$, *for every* $\epsilon > 0$ *and for every* $p \in (0, 1/2)$, *independently of the underlying communication model.*

In our final main result we show that two p-faulty agents operating in the wireless model can do better than 9 for all $p < 1/2$, as well as better than 4.59112 for a large spectrum of p values. Note that the achieved competitive ratio is at least 3, which is the optimal competitive ratio for searching with two 0-faulty agents in the wireless model, and that our result matches this known bound when $p \to 0$.

Theorem 6. *Two deterministic p-faulty agents in the wireless model can solve p-PDLS with competitive ratio* $3 + 4\sqrt{p(1-p)} + \epsilon$, *for every* $\epsilon > 0$ *and for every* $p \in [0, 1/2)$.

Algorithm 5. Search with two wireless p-faulty agents

Require: p-faulty agents with distinct roles of Leader and Follower, $s < 1$ and $\gamma > 0$.
 1: Agents search in opposite direction until target is found and reported.
 2: Target finder becomes Leader, and non-finder becomes Follower.
 3: Non-finder changes speed to s, attempts a turn (that fails with probability), and continues moving until she meets with the finder.
 4: Finder moves in same direction for $\gamma > 0$ and runs Algorithm 4 (target plays role of Follower), until the target is reached again.
 5: Finder (Leader) continues until she meets the non-finder.
 6: Non-finder (Follower) stays put until met by the Finder (Leader) again.
 7: Leader continues moving in the same direction (away from the target and the Follower) for time γ and then runs Algorithm 4 with the Follower being the immobile agent, in order to turn.
 8: When the Leader turns successfully, she picks up the Follower, and continuing in the same direction, together, they move to the target.

Proof (sketch of Theorem 6). The proof is given by the performance analysis of Algorithm 5 for a proper choice of speed $s < 1$. In this simplified (sketch of) proof, we make the assumption that the target finder (using the target) as well as the two agents when walking together can make a deterministic turn. Indeed, using Algorithm 4 one can show, we show how the actual probabilistically faulty turns have minimal impact in the competitive ratio.

We assume that the target is reported by the finder at time 1, when the distance of the two agents is 2. As for the non-finder, she turns successfully when she receives the wireless message with probability $1 - p$. Since the finder moves towards her, their relative speed $1 + s$. This means that they meet in additional time $2/(1 + s)$, during which time the non-finder has moved closer to the target by $2s/(1 + s)$. Hence, when the two agents meet, they are at distance $2 - 2s/(1 + s)$ from the target.

On the other hand with probability p the non-finder fails to turn, and the two agents continue to move towards the same direction, only that the non-finder's new speed is s. So, their relative speed in this case is $1 - s$. This means that they meet in additional time $2/(1 - s)$, during which time the non-finder has moved further from the target by $2s/(1 - s)$. Hence, when the two agents meet, they are at distance $2 + sW + s(2 + sW)/(1 - s)$ from the target. Also recall that when they meet, they can make together a forced turn (that affects minimally the termination time), inducing this way total termination time (and competitive ratio, since the target was at distance 1) $3 + p \left(\frac{2}{1-s} + \frac{2s}{1-s} \right) + (1 - p) \left(\frac{2}{1+s} - \frac{2s}{1+s} \right) = \frac{5-s(s+4-8p)}{1-s^2}$. We choose $s = s(p) = \frac{1-2\sqrt{p-p^2}}{1-2p}$, the minimizer of the latter expression, that can be easily seen to attain values in $(0, 1)$ for all $p \in (0, 1/2)$, hence it is a valid choice for a speed. Now we substitute back to the formula of the competitive ratio, and after we simplify algebraically, the expression becomes $3 + 4\sqrt{(1 - p)p}$. □

It is worthwhile noticing that the upper bound 4.59112 of Theorem 5 holds in expectation, without being able to control the deviation. However, the upper bound of Theorem 6 holds again in expectation, but the resulting performance can be concentrated around the expectation with arbitrary precision. Moreover, the derived competitive ratio is strictly increasing in $p < 1/2$, and ranges from 3 to 5. Hence, the drawback of Theorem 6 is that for high enough values of p ($p > 0.197063$), the induced competitive ratio exceeds 4.59112. Therefore, we have the incentive to choose either the algorithm of Theorem 6, when $p \leq 0.197063$, and the algorithm of Theorem 5 otherwise. It would be interesting to investigate whether a hybrid algorithm, combining the two ideas, could accomplish an improved result

7   Conclusion

In this paper we studied a new mobile agent search problem whereby an agent's ability to navigate in the search space exhibits probabilistic faults in that every attempt by the agent to change direction is an independent Bernoulli trial (the agent fails to turn with probability $p < 1/2$). When searching with one agent, our best performing algorithm has optimal performance 4.59112 as $p \to 0$, performance less that 9 for $p \leq 0.436185$, and optimal performance up to constant factor and unbounded as $p \to 1/2$. When searching with two faulty agents, we provide 3 algorithms with different attributes. One algorithm has (expected) performance 9 with arbitrary concentration, the other has performance 4.59112, and finally one has performance $3 + 4\sqrt{p(1 - p)}$ (ranging between 3 and 5) again with arbitrary concentration.

It is rather surprising that even in this probabilistic setting with one searcher, we can design algorithms that outperform the well-known zig-zag algorithm for linear search whose competitive ratio is 9, as well as that the problem with two searchers admits bounded competitive ratio for all $p \in (0, 1/2)$, and unlike the one search problem. Interesting questions for further research could definitely arise in the study of similar, related "probabilistic navigation" faults either for their own sake or in conjunction with "communication" faults in more general search domains (e.g., in the plane or more general cow path with w rays) and for multiple (possibly collaborating) agents.

References

1. Ahlswede, R., Wegener, I.: Search Problems. Wiley, Hoboken (1987)
2. Alpern, S., Gal, S.: The Theory of Search Games and Rendezvous, vol. 55. Springer, Cham (2006)
3. Baeza-Yates, R., Culberson, J., Rawlins, G.: Searching in the plane. Inf. Comput. **106**(2), 234–252 (1993)
4. Baeza-Yates, R., Schott, R.: Parallel searching in the plane. Comput. Geom. **5**(3), 143–154 (1995)
5. Baston, V., Gal, S.: Rendezvous search when marks are left at the starting points. Naval Res. Logistics (NRL) **48**(8), 722–731 (2001)
6. Beck, A.: On the linear search problem. Israel J. Math. **2**(4), 221–228 (1964)
7. Beck, A.: More on the linear search problem. Israel J. Math. **3**(2), 61–70 (1965)
8. Beck, A., Newman, D.: Yet more on the linear search problem. Israel J. Math. **8**(4), 419–429 (1970)
9. Bellman, R.: An optimal search. SIAM Rev. **5**(3), 274–274 (1963)
10. Bonato, A., Georgiou, K., MacRury, C., Prałat, P.: Algorithms for p-faulty search on a half-line. Algorithmica (2022)
11. Chrobak, M., Gąsieniec, L., Gorry, T., Martin, R.: Group search on the line. In: Italiano, G.F., Margaria-Steffen, T., Pokorný, J., Quisquater, J.-J., Wattenhofer, R. (eds.) SOFSEM 2015. LNCS, vol. 8939, pp. 164–176. Springer, Heidelberg (2015). https://doi.org/10.1007/978-3-662-46078-8_14
12. Czyzowicz, J., Georgiou, K., Kranakis, E.: Group search and evacuation. In: Flocchini, P., Prencipe, G., Santoro, N. (eds.) Distributed Computing by Mobile Entities, Lecture Notes in Computer Science(), vol. 11340, pp. 335–370. Springer, Cham (2019). https://doi.org/10.1007/978-3-030-11072-7_14
13. Czyzowicz, J., et al. : Search on a line by byzantine robots. Int. J. Found. Comput. Sci., 1–19 (2021)
14. Czyzowicz, J., Killick, R., Kranakis, E., Stachowiak, G.: Search and evacuation with a near majority of faulty agents. In: SIAM Conference on Applied and Computational Discrete Algorithms (ACDA 2021), pp. 217–227. SIAM (2021)
15. Czyzowicz, J., Kranakis, E., Krizanc, D., Narayanan, L., Opatrny, J.: Search on a line with faulty robots. Distrib. Comput. **32**(6), 493–504 (2019)
16. Demaine, E., Fekete, S., Gal, S.: Online searching with turn cost. Theoret. Comput. Sci. **361**(2), 342–355 (2006)
17. Georgiou, K., Giachoudis, N., Kranakis, E.: Overcoming probabilistic faults in disoriented linear search. CoRR, abs/2303.15608 (2023)
18. Georgiou, K., Lucier, J.: Weighted group search on a line & implications to the priority evacuation problem. Theoret. Comput. Sci. **939**, 1–17 (2023)
19. Kao, M.-Y., Reif, J.H., Tate, S.R.: Searching in an unknown environment: an optimal randomized algorithm for the cow-path problem. Inf. Comput. **131**(1), 63–79 (1996)
20. McCabe, B.J.: Searching for a one-dimensional random walker. J. Appl. Probab. **11**(1), 86–93 (1974)
21. Stone, D.L.: Theory of optimal search. Academic Press, Cambridge (1975). Incorporated
22. Sun, X., Sun, Y., Zhang, J.: Better upper bounds for searching on a line with byzantine robots. In: Du, D.-Z., Wang, J. (eds.) Complexity and Approximation. LNCS, vol. 12000, pp. 151–171. Springer, Cham (2020). https://doi.org/10.1007/978-3-030-41672-0_9

Packet Forwarding with Swaps

Cameron Matsui and Will Rosenbaum$^{(\boxtimes)}$ [ID]

Amherst College, Amherst, MA, USA
{cmatsui22,wrosenbaum}@amherst.edu

Abstract. We consider packet forwarding in the adversarial queueing theory (AQT) model introduced by Borodin et al. In this context, a series of recent works have established optimal bounds for buffer space usage of $O(\log n)$ for simple network topologies, where n is the size of the network. Optimal buffer space usage, however, comes at a cost: any protocol that achieves $o(n)$ buffer space usage cannot guarantee bounded packet latency.

In this paper, we introduce a generalization of the AQT model that allows for packet swaps in addition to regular forwarding operations. We show that in this model, it is possible to simultaneously achieve both optimal buffer space usage and packet latency when the network is a path of length n. To this end, we introduce an analytic tool we call the *smoothed configuration* of the network. We employ the smoothed configuration to reason about packet latency for a large family of local forwarding protocols, whereby we derive our main result. We also employ the smoothed configuration to analyze the total buffer space usage of forwarding protocols under stochastic packet arrivals. We show that the total network load is n in its steady state, but that the system takes exponential time in expectation to reach a total load of n.

Keywords: adversarial queueing theory · packet forwarding · packet swaps · buffer space usage · packet latency

1 Introduction

Packet forwarding is a fundamental problem that arises throughout computer science, electrical engineering, and operations research. In store-and-forward networks indivisible items—referred to as **packets**—arrive spontaneously in a network and must be routed to their respective destinations. Packet movement is restricted by the network topology and link capacity constraints. As packets move through the network, they are stored in buffers in intermediate locations. Buffers are limited in the number of packets they can store, so efficient buffer management is essential.

In a seminal work, Borodin et al. [2] introduced model of queueing networks known as Adversarial Queueing Theory (AQT). In AQT, packets are injected into a network with a prescribed path from their source to destination. AQT parameterizes injection patterns in terms of long-term average edge utilization

© The Author(s), under exclusive license to Springer Nature Switzerland AG 2023
S. Rajsbaum et al. (Eds.): SIROCCO 2023, LNCS 13892, pp. 536–557, 2023.
https://doi.org/10.1007/978-3-031-32733-9_24

as well as short-term "burstiness." The goal is to determine scheduling protocols that are *stable* in the sense that for all feasible injection patterns, the buffers remain bounded. AQT also focused primarily on buffer space usage, and not other measures of efficiency such as packet latency.

Stability gives a qualitative measure of buffer space usage, in that it only distinguishes bounded and unbounded buffer requirements. Adler and Rosén [1], however, gave the first quantitative bounds for AQT. Specifically, they showed that for DAG networks, using "longest-in-system" (LIS) results in buffer space requirements that are at most linear in the network size.

A recent series of work initiated by Miller and Patt-Shamir [4] analyzes the buffer space requirement for non-greedy forwarding protocols on simple network topologies such as information gathering networks. Miller and Patt-Shamir showed that when the network is a tree with n nodes and all packets share the same destination, a centralized (non-greedy) protocol achieves buffer space $O(1)$, whereas any greedy protocol requires space $\Omega(n)$. Subsequent work analyzed trade-offs between buffer space requirement and protocol locality [3,7,8,10], while other works consider tradeoffs between buffer space usage and injection rate [5,9].

In [3,7], Dobrev et al. and Patt-Shamir and Rosenbaum independently introduced the same local forwarding protocol, *Odd-Even Downhill (OED)* forwarding, which achieves optimal buffer space usage of $\Theta(\log n)$ for single-destination paths of length n. Patt-Shamir and Rosenbaum, however, showed that this optimal buffer space usage comes at a cost: any protocol that guarantees $o(n)$ buffer space usage cannot simultaneously achieve bounded packet latency. Their argument crucially relies on the assumption that packet motion is one-directional (a standard assumption in AQT).

1.1   Our Contributions

In this paper, we introduce an augmented model for AQT which allows for adjacent buffers to swap packets in addition to normal forwarding operations. In this model, we show that the buffer space/latency dichotomy demonstrated in [7] no longer holds for the single destination path network. More generally, we introduce a family of "1/2-local" forwarding protocols. Informally, these are protocols for which the decision of whether a packet is forwarded across an edge is determined only by the states of the edge's endpoints. (Greedy and OED forwarding are both examples of 1/2-local protocols.) We analyze a large subclass of 1/2-local protocols, which we call "solid" protocols, that are particularly well-behaved. We show that a natural packet swapping strategy ensures that all solid protocols deliver packets in the same order given the same arrival order. As a result, we show that a modified OED algorithm with swaps simultaneously achieves (asymptotically) optimal buffer space usage and maximum latency.

Theorem 1. *Suppose G is a path network of length n, all packets share a common destination, and packets arrive according to a (ρ, σ)-bounded injection pattern with $\rho \leq 1$. Then OED with swaps incurs a worst-case maximum buffer* *space usage of $\Theta(\log n + \sigma)$, and maximum packet latency of at most $n + \sigma$. This buffer space usage is asymptotically optimal for local forwarding protocols, and the latency is optimal for all protocols.*

To prove Theorem 1, we introduce an analytic tool, which we call the *smoothed configuration*. The smoothed configuration essentially describes a linear ordering of packets in the network by the order in which they will be delivered. The dynamics of individual packets in an arbitrary solid protocol may be complex, but the dynamics of the corresponding smoothed configuration are always simple: every packet in the smoothed configuration always progress a single step towards the destination each round, and these dynamics are the same for all solid protocols with swaps.

We then employ the machinery of the smoothed configurations to reason about stochastic, rather than adversarial, injection patterns. We introduce a "random injection model" in which a single packet is injected into a uniformly random buffer each round. We analyze the total network load of solid protocols in this model. We show that for a path of length n, the total load grows to n almost surely. Interestingly, once the total load becomes sufficiently large (αn for some $\alpha < 1$), it increases exponentially slowly. The expected times between steps in which the total load increases is exponential in n.

The remainder of the paper is organized as follows: In Sect. 2, we provide basic definitions state related results. In Sect. 3, we state the main lemmas and prove Theorem 1. Section 4 contains our analysis of the random injection model, and we conclude with open questions in Sect. 5. Complete proofs of lemmas appear in the appendices.

2   Background and Prerequisites

2.1   Network and Protocols

We consider the adversarial queueing theory (AQT) model introduced by Borodin et al. [2]. In this model, the network consists of a directed graph $G = (V, E)$. Each edge $e = (u, v) \in E$ has an associated buffer (also denoted e) that stores packets as they wait to cross the edge e from u to v. Each edge e has an associated **capacity**, denoted $C(e)$ that specifies the number of packets that can be forwarded across e in a single step. In this paper, we restrict attention to networks with unit-capacity edges—i.e., $C(e) = 1$ for all e.

A **packet** p is a pair $p = (t_p, P_p)$ where t_p is the round in which p is injected, and $P = (v_0, v_1, \ldots, v_\ell)$ is a directed path in G. We call the buffer $e_1 = (v_0, v_1)$ p's **source** and v_ℓ is p's **destination**. The interpretation is that $p = (t, P)$ is **injected** into buffer $e_1 = (v_0, v_1)$ in round t. The packet must be forwarded along buffers in P to its destination, at which point we say p is **absorbed**. An **adversary** or **injection pattern** is a multiset of packets.

An execution proceeds in synchronous rounds, $t = 1, 2, \ldots$. Each round t consists of two steps: an **injection step** in which new packets arrive in the network, and a **forwarding step** in which buffers forward packets across their corresponding edges. In each round, the injection step precedes the forwarding step. When referring to the state of the network, we will use t to represent round t *after* the injection step and *before* the forwarding step. We use t' to represent the time after *both* the injection and forwarding steps of round t.

A *configuration* $\mathbf{C}(G)$ specifies the contents of—i.e., the multiset of packets contained in—each buffer in the network. We use $\mathbf{C}^t(G)$ and $\mathbf{C}^{t'}(G)$ to denote the configurations of the network at times t and t' in an execution (with t and t' as specified in the previous paragraph). Similarly, we use $\mathbf{C}(e)$ and $\mathbf{C}(S)$ to refer to the configuration of a single buffer or set $S \subseteq E$ of buffers. For a given configuration, the *load* of a buffer, $\mathbf{L}(e)$, is the number of packets stored in e; i.e., $\mathbf{L}(e) = |\mathbf{C}(e)|$. As with $\mathbf{C}$, we use a superscript t or t' to specify the load of a buffer at a particular time step.

A *forwarding protocol*, Π, is a rule that specifies which packets are forwarded across each link in the network. That is, a forwarding protocol determines for each configuration $\mathbf{C}(G)$ of a network the updated configuration $\mathbf{C}'(G)$ after packets are forwarded.

In this paper, we consider the restricted setting in which the network G consists of a path of $n + 1$ nodes, labeled $1, 2, \ldots, n + 1$, and n edges/buffers $E = (1, 2), (2, 3), \ldots, (n, n + 1)$. We assume that all edges have unit capacity ($C(e) = 1$) and all packets share the common destination $n + 1$. We refer to this restricted network as the *single destination path* of size n. In this setting, we adopt simplified notation where we identify each node $x \in \{1, 2, \ldots, n\}$ with its unique outgoing buffer. For example, $\mathbf{C}(x)$ and $\mathbf{L}(x)$ specify a configuration and load of the buffer $(x, x + 1)$. We also use the notation $[a, b]$ where $a \leq b \in \{1, 2, \ldots, n\}$ to denote the set of buffers with labels $\{a, a + 1, \ldots, b\}$. For an interval A, we write $\mathbf{L}(A) = \sum_{x \in A} \mathbf{L}(x)$.

Parameterized Injection Patterns. AQT considers (adversarial) injection patterns that parameterized by their long-term average rate (ρ) as well as their burstiness (σ). In the case of the single-destination path, such a parameterization simply bounds the total number of packets injected into the network in every time interval.

Definition 1. *Let $\rho, \sigma \geq 0$ be parameters, and let A be an injection pattern. We say that A is (ρ, σ)-bounded if for every T consecutive rounds, the total number of packets injected by A during those rounds is at most $\rho T + \sigma$. We denote the family of all (ρ, σ)-bounded adversaries by $\mathcal{A}(\rho, \sigma)$.*

Since we assume that network edge capacities are all 1, we restrict attention to (ρ, σ)-bounded adversaries with $\rho \leq 1$. With this assumption, we always have $\rho = O(1)$, so ρ does not appear our asymptotic expressions for buffer space usage.

Recently, Rosenbaum [10] introduced a refinement of the (ρ, σ) model that parameterizes adversaries jointly by injection round and location.

Definition 2 ([10])**.** *Let* $\rho, \sigma, B \geq 0$ *be parameters and let* A *be an injection pattern. Then we say that* A *is* **locally** (ρ, σ, B)**-bounded** *if for every* T *consecutive rounds and every subset* S *of buffers, the total number packets injected by* A *into buffers in* S *during the* T *rounds is at most* $\rho T + B |S| + \sigma$. *We denote the family of all locally* (ρ, σ, B)*-bounded adversaries by* $\mathcal{A}(\rho, \sigma, B)$.

We note that every locally (ρ, σ, B)-bounded adversary is also $(\rho, \sigma + nB)$-bounded, but the converse does not hold. That is, $\mathcal{A}(\rho, \sigma, B) \subset \mathcal{A}(\rho, \sigma + nB)$ (and the opposite inclusion does not hold).

1/2-Local and Solid Protocols. We consider stateless forwarding protocols for which the decision of whether or not each buffer x forwards depends only on the configurations of x and $x + 1$. We refer to such protocols as *1/2-local forwarding protocols*.

When we are not concerned with latency, we represent a 1/2-local protocol Π for a unit-capacity network as a function $\Pi : \mathbb{N}^+ \times \mathbb{N} \rightarrow \{0, 1\}$ where the first argument to the function is the load of a buffer x, the second argument is the load of the following buffer $x + 1$, and $\Pi(\mathbf{L}(x), \mathbf{L}(x + 1)) = 1$ if and only if Π stipulates that x forwards a packet to $x + 1$.

Example 1. The following protocols are 1/2-local forwarding protocols:

Greedy For greedy forwarding, $\Pi(a, b) = 1$ for all $a > 0$ and for all b. That is, under greedy forwarding, every non-empty buffer forwards.

Downhill The downhill forwarding protocol is defined by $\Pi(a, b) = 1 \iff a > b$. That is, a buffer forwards precisely when its load is strictly larger than the next buffer's load.

OED The odd-even downhill (OED) protocol of [3,7] has $\Pi(a, b) = 1 \iff a > b$ or $a = b$ and a is odd.

We will show that a certain class of 1/2-local protocols, which we call *solid protocols*, behave similarly with respect to packet dynamics.

Definition 3. *A 1/2-local protocol* Π *is a* **solid forwarding protocol** *if*

1. $\Pi(1, 1) = 1$, *and*
2. *for all* $n, m \in \mathbb{N}$ *such that* $n > m$, $\Pi(n, m) = 1$.

For example, the greedy and odd-even downhill forwarding protocols are solid protocols, but downhill forwarding is not a solid protocol (as for downhill forwarding, $\Pi(1, 1) = 0$).

As defined above, 1/2-local and solid forwarding protocols define whether or not a buffer should forward, but do not specify *which* packet should be forwarded. For the remainder of the paper, we assume that packets are forwarded according to *longest in system* (**LIS**) scheduling. A buffer always forwards its oldest packet (if any). To this end we assume packets are ordered according to lexicographical *latency-ordering*, and that each buffer acts as a priority queue with respect to this ordering. In the case of a tie (i.e., multiple packets are injected in the same round), ties are broken in an arbitrary, but consistent manner (e.g., according to injection buffer).

In order to simplify our bookkeeping going forward, we assume that each packet p has a **unique identifier** $\mathbf{id}(p) \in \mathbb{N}$, and that ids are assigned sequentially. That is, the "first" injected packet has $\mathbf{id}(p) = 1$, the second has id 2, and so on. If multiple packets are injected in the same round, they receive distinct consecutive ids. We say that p is **older** than q if $\mathbf{id}(p) < \mathbf{id}(q)$.

For a given buffer x, we will use $h(x)$ to refer to the **highest-priority** or **oldest** packet in x in $\mathbf{C}(G)$, and we will use $l(x)$ to refer to the **lowest-priority** or **youngest** packet in x.

Buffer Space Usage and Latency. Two primary measures of the quality of a forwarding protocol Π are its buffer space requirement and maximum latency.

Definition 4. *Let Π be a protocol and A an injection pattern. Then the **buffer space requirement** of Π against A is $\sup_{t,i} \mathbf{L}^t(i)$. That is, the buffer space requirement is the supremumum of buffer sizes in an execution of Π with injection pattern A. More generally, given any family $\mathcal{F}$ of injection patterns, we define the buffer space requirement of Π against $\mathcal{F}$ to be $\sup_{A \in \mathcal{F}} \sup_{t,i} \mathbf{L}^t(i)$.*

In independent works, Dobrev et al. [3] and Patt-Shamir and Rosenbaum [7] showed that OED forwarding (Example 1) achieves asymptotically optimal buffer space usage among *local* forwarding protocols against (ρ, σ)-bounded adversaries. Recently, Rosenbaum [10] showed that OED has optimal buffer space usage against locally (ρ, σ, B)-bounded adversaries, even when compared to centralized protocols.

Theorem 2 (Rosenbaum[10], cf.[3,7]). *Let $\mathcal{F} = \mathcal{A}(\rho, \sigma, B)$ be the family of locally (ρ, σ, B)-bounded adversaries on a single destination path of size n. Then OED has buffer space requirement $\Theta(B \log n + \sigma)$. This buffer space requirement is asymptotically optimal, as every (centralized, randomized) protocol has buffer space requirement $\Omega(B \log n + \sigma)$. In particular, for a standard (ρ, σ)-bounded adversary, OED's buffer space requirement is $\Theta(\log n + \sigma)$.*[1]

Packet latency measures how many rounds elapse between when a packet is injected and when the packet is absorbed at its destination.

Definition 5. *Fix a protocol Π and injection pattern A and packet $p \in A$. Let t_p denote the time p is injected and let s_p denote the time p is absorbed at its destination, with the convention $s_p = \infty$ if p is never absorbed. The **latency** of*

---

[1] For the standard (ρ, σ)-bounded injection model, OED's buffer space requirement is only optimal among *local* forwarding protocols—i.e., protocols where each buffer's decision to forward depends only on the state of that buffer's distance $O(1)$ neighborhood. Indeed, there are centralized protocols whose buffer space usage is $O(\rho + \sigma)$ [4,5] in the standard injection model. In locally (ρ, σ, B)-bounded model, however, OED's buffer space usage is optimal even when compared to centralized protocols. See [10].

p is $s_p - t_p$. The **maximum latency** of Π against A is $\sup_{p \in A} s_p - t_p$. Given a family $\mathcal{F}$ of injection patterns, the maximum latency of Π is the supremum of the maximum latencies of Π against $A \in \mathcal{F}$.

Greedy forwarding with LIS scheduling achieves optimal latency against (ρ, σ)-bounded adversaries.

Theorem 3 (cf.[1]). *Let G be a single destination path of size n. Then the maximum latency of greedy forwarding with LIS priority against locally (ρ, σ, B)-bounded adversaries is $\Theta(n+\sigma)$. Moreover, the maximum latency of any protocol is $\Omega(n + \sigma)$.*

The buffer-space optimal OED protocol, however, does not guarantee finite latency. In fact, Patt-Shamir and Rosenbaum showed a strict dichotomy between buffer space usage and latency.

Theorem 4 ([7]). *Let G be a single destination path with unit capacity edges, and let $\mathcal{F}$ be the family of $(1,0)$-bounded adversaries on G. Then for any protocol Π, if Π has maximum latency $< \infty$, then the buffer space usage of Π is $\Omega(n)$.*

The argument of Theorem 4 crucially relies on the assumption that packet movement in the single destination path is one-directional. We next introduce a model that allows for packet swaps between adjacent buffers. Or main result shows that the conclusion of Theorem 4 fails in the augmented model.

Protocols with Swaps. In order to achieve linear latency on the path, we consider an extended model which additionally allows each buffer x to send one packet backwards to buffer $x - 1$. In this model, we can augment solid forwarding protocols to define **solid forwarding protocols with swaps**.

Definition 6. *Let Π be a solid forwarding protocol. Π's associated **solid forwarding protocol with swaps** Π_S is a protocol which, given a buffer x such that $\mathbf{L}(x) > 0$:*

1. *If $\Pi(\mathbf{L}(x), \mathbf{L}(x + 1)) = 1$, x forwards its oldest packet to $x + 1$, and*
2. *If $\Pi(\mathbf{L}(x), \mathbf{L}(x + 1)) = 0$ and x's oldest packet is older than $x + 1$'s youngest packet, then x forwards its oldest packet to $x + 1$ and $x + 1$ sends its youngest packet back to x.*

The following lemma asserts that for single destination paths, a protocol Π and its associated protocol with swaps Π_S have the same buffer space usage.

Lemma 1. *Let A be any injection pattern on a single-destination path. Let Π be any solid forwarding protocol, and Π_S the associated protocol with swaps. Then for all times t, all buffers x have the same load under Π_S as Π.*

Proof. We argue by induction on t. The base case $t = 0$ is trivial, as all buffers are initially empty. For the inductive step, suppose the lemma holds a time t. To see that both protocols incur the same loads at time t' (i.e., after forwarding), observe that (1) if buffer x forwards according to Π, then so does Π_S, and (2) if x does not forward according to Π, then either x does not forward in Π_S or x and $x + 1$ swap. In the case of a packet swap, x's net load is unchanged, hence x has the same load under Π and Π_S.

2.2   Smoothed Dynamics

The primary tool we use to derive our results is the *smoothed configuration*, which is derived from a packet configuration on a path network. For a configuration $\mathbf{C}(G)$ of a path G of length n, the corresponding smoothed configuration $\mathbf{S}(\mathbf{C}(G))$ is the result of applying Algorithm 1 to $\mathbf{C}(G)$. For convenience, we write $\mathbf{S}(G) = \mathbf{S}(\mathbf{C}(G))$ (Fig. 1).

Algorithm 1. Smoothing $\mathbf{C}(G)$

$x \leftarrow n$
$Q \leftarrow$ an empty priority queue, with priority by packet age
$\mathbf{S}(G) \leftarrow$ an empty configuration with buffers $\ldots, -2, -1, 0, 1, 2, \ldots, n$
while $x \geq 1$ or $|Q| > 0$ **do**
    for $p \in \mathbf{C}(x)$ **do**
        Q.enqueue(p)
    end for
    if $|Q| > 0$ **then**
        $\mathbf{S}(x) \leftarrow Q$.dequeue()
    end if
    $x \leftarrow x - 1$
end while
return $\mathbf{S}(G)$

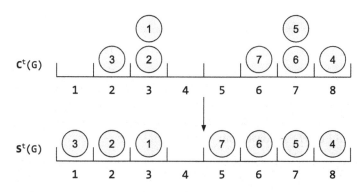

Fig. 1. The result of applying Algorithm 1 to the packet configuration in the top panel, $\mathbf{C}^t(G)$, is the corresponding *smoothed configuration* $\mathbf{S}^t(G)$ in the bottom panel.

In order to create the smoothed configuration $\mathbf{S}(G)$ from $\mathbf{C}(G)$, the algorithm moves left from the rightmost buffer in $\mathbf{C}(G)$, picking up all of the packets in each buffer, placing them into the queue Q, and then dropping one packet into $\mathbf{S}(x)$ if the queue is non-empty. Note that for any $x \in (-\infty, n]$ we have $|\mathbf{S}(x)| \in \{0, 1\}$. We refer to Q as the **smoothing queue** for $\mathbf{S}(G)$, and denote the set of packets in Q over a buffer x in Algorithm 1 after enqueueing the packets in x and before dequeueing a packet by $\mathbf{Q}(x)$. From Algorithm 1, we observe that

$$|\mathbf{Q}(x)| = \begin{cases} \mathbf{L}(x) & x = n, \\ \mathbf{L}(x) + \max\{0, |\mathbf{Q}(x+1)| - 1\} & 1 \le x < n, \\ \max\{0, |\mathbf{Q}(x+1)| - 1\} & x < 1. \end{cases}$$

In our arguments, we will also use $s(x)$ when $\mathbf{SL}(x) = 1$ to refer to the packet in buffer x in the smoothed configuration.

To characterize the packet dynamics of the smoothed configuration, we break a smoothed configuration $\mathbf{S}(G)$ down into its component **smoothed plateaus**. A smoothed plateau is a maximal interval I of buffers in a smoothed configuration $\mathbf{S}(G)$ such that for each buffer x in I, $\mathbf{SL}(x) = 1$.

3   Latency Analysis

In this section, we prove two primary results for the latency of solid protocols with swaps. First, Theorem 5 states that any solid forwarding protocol on a path of length n has a maximum packet latency of $n + \sigma - 1$ against any $(1, \sigma)$ adversary. Second, Theorem 6 states that any protocol in the packet-swapping model on a path of n buffers must have worst-case packet latency of $n + \sigma - 1$.

Theorem 5. *Let G be a single destination path of length n, let Π_S be a solid forwarding protocol with swaps, and let $A \in \mathcal{A}(1, \sigma)$. Then the maximum latency of Π_S against A is at most $n + \sigma - 1$.*

Theorem 6. *For any protocol Π in the packet-swapping model on a single-destination path network G of length n, there exists a $(1, \sigma)$ adversary $\mathcal{A}$ such that the maximum latency of packets on G is at least $n + \sigma - 1$.*

Theorem 1 follows from Theorem 5 applied to OED forwarding together with Theorem 2.

Proof (Theorem 1). Let Π be OED forwarding and Π_S OED with swaps. By Theorem 2, the buffer space usage of Π is $\Theta(\log n + \sigma)$, which is optimal. By Lemma 1, Π_S also has (optimal) buffer space requirement $\Theta(\log n + \sigma)$. The latency bound of Π_S follows from Theorem 5.

The idea of our proof of Theorem 5 is to connect the dynamics of solid protocols with swaps with the dynamics of smoothed configurations and their component smoothed plateaus. The main technical piece is Lemma 4, which states that each packet in a smoothed configuration moves one buffer forward after a round's forwarding step. More formally, Lemma 4 implies the following corollary:

Corollary 1. *Let G be a path network of length n, Π a 1/2-local forwarding protocol with swaps, and $\mathbf{C}(G)$ a configuration of packets on G. Let $\Pi(\mathbf{C}(G))$ denote the configuration that results from forwarding the packets in a configuration of G according to Π. Then $\mathbf{S} \circ \Pi(\mathbf{C}(G)) = \Pi \circ \mathbf{S}(\mathbf{C}(G))$; the operations of smoothing and forwarding under Π commute (Fig. 2).*

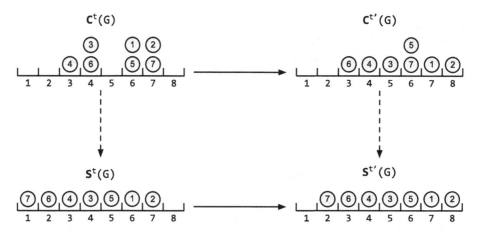

Fig. 2. An illustration of Corollary 1. Starting from $\mathbf{C}^{t}(G)$, *smoothing* ($\mathbf{S}^{t}(G)$) and then *forwarding* ($\mathbf{S}^{t'}(G)$) is equivalent to *forwarding* ($\mathbf{C}^{t'}(G)$) and then *smoothing* ($\mathbf{S}^{t'}(G)$).

Corollary 1 is valuable because the dynamics of *smoothed* configurations under all solid protocols are the same: since every buffer in a smoothed configuration contains either 0 or 1 packets, every buffer in the smoothed configuration forwards. Thus, in the smoothed configuration, every packet progresses one step towards the destination during forwarding. Together with Lemma 5, which states that a preexisting packet's location in the smoothed configuration is unaffected by injection, Lemma 4 implies that $s^{t}(x) = s^{t+1}(x+1)$ for all rounds t. Lemma 6 allows us to conclude that the left-most non-empty buffer in a smoothed configuration in any given round t before forwarding is $1 - \sigma$, and thus the maximum latency of any solid protocol with swaps is $n + \sigma - 1$.

For a path network G of length n, we assume **there are two extra buffers** $n + 1$ and $n + 2$ which are cleared after the forwarding step and before the injection step of each round in order to simplify the arguments that follow.

The first component of our argument for Lemma 4, informally speaking, is to show that smoothed plateaus move forward one buffer during any round's forwarding step. In order to do so, we break each smoothed plateau down into **islands**, and characterize how these islands move during forwarding in Lemma 2. The proofs of Lemmas 2, 3, 4, and 5 can be found in the appendix.

Definition 7. *Let G be a path network of length n with configuration $\mathbf{C}(G)$, and let $I = [a, b]$ be an interval of buffers such that:*

1. For all $x \in I$, $\mathbf{L}(x) > 0$,
2. Either $a = 1$ or $\mathbf{L}(a - 1) = 0$, and
3. $\mathbf{L}(b + 1) = 0$.

*Then I is an **island** in $\mathbf{C}(G)$.*

Lemma 2. *Let G be a path network of length n with associated solid forwarding protocol Π. If $I = [a, b]$ is an island in $\mathbf{C}^t(G)$ before forwarding, then the packets that were in I occupy the interval I' in $\mathbf{C}^{t'}(G)$ after forwarding, where I' is either $[a+1, b+1]$ if $\mathbf{L}^{t'}(a) = 0$ or $[a, b+1]$ if $\mathbf{L}^{t'}(a) > 0$. Furthermore, in either case, for all x in I', $\mathbf{L}^{t'}(x) > 0$.*

Lemma 3. *Let H be a path network with an associated solid forwarding protocol Π, and suppose that $M = [a, b]$ is a smoothed plateau in $\mathbf{C}^t(H)$. Then $M' = [a + 1, b + 1]$ is a smoothed plateau in $\mathbf{C}^{t'}(H)$.*

To show that smoothed plateaus move forward one buffer during forwarding, we consider a smoothed plateau as composed of islands and **gaps**, intervals of empty buffers in the packet configuration. We use induction in combination with Lemma 2 to show that if these gaps were filled before forwarding, then they are filled after forwarding.

Next, we show that each individual packet in a smoothed plateau moves forward one buffer. Since a smoothed configuration is made up of only smoothed plateaus, Lemma 4 implies that each packet in any smoothed configuration moves forward one buffer during forwarding.

Lemma 4. *Let G be a path network of length n with associated solid forwarding protocol with swaps Π_S, and suppose there is a smoothed plateau at $M = [a, b]$ in $\mathbf{S}^t(G)$ before forwarding. Then in $\mathbf{S}^{t'}(G)$, there is a smoothed plateau at $M' = [a + 1, b + 1]$ such that for all $x \in M$, $s^t(x) = s^{t'}(x + 1)$.*

In order to prove Lemma 4, we consider the states of the smoothing queue over a buffer x before forwarding and over the next buffer $x + 1$ after forwarding. Specifically, we show that for each buffer x in the smoothed configuration, $\mathbf{Q}^{t'}(x + 1) \subseteq \mathbf{Q}^t(x)$ and $s^t(x) \in \mathbf{Q}^{t'}(x + 1)$, which together imply that $s^t(x) = s^{t'}(x + 1)$.

For the final technical piece of Theorem 7, we show that any preexisting packet in the smoothed configuration remains in the same location after injection.

Lemma 5. *Let G be a path network of length n, Π a $1/2$-local forwarding protocol with swaps, and $\mathbf{S}^{(t-1)'}(G)$ a smoothed configuration of packets for G in round $t - 1$ before injection for round t but after forwarding for round $t - 1$. Then for all buffers x in $(-\infty, n]$ such that $\mathbf{SL}^{(t-1)'}(x) > 0$, we have $s^{(t-1)'}(x) = s^t(x)$.*

To prove Lemma 5, we show that for each buffer x in the smoothed configuration ($x \in (-\infty, n]$), $\mathbf{Q}^{(t-1)'}(x) \subseteq \mathbf{Q}^t(x)$ and $\mathbf{Q}^t(x) \setminus \mathbf{Q}^{(t-1)'}(x)$ contains only packets injected in round t. Theorem 7 then follows almost immediately from Lemmas 4 and 5.

Theorem 7. *Let G be a path network of length n, and let x be a buffer in $\mathbf{S}^t(G)$, i.e., $x \in (-\infty, n]$. Let $p = s^t(x)$. Then p is absorbed in round $t + (n - x)$.*

Proof (Theorem 7). We first show that for any round $t + s$ where $0 \leq s \leq n - x$, before forwarding for round $t + s$, $s^{t+s}(x + s) = p$. We induct on s.

<u>Base Case</u> ($s = 0$): Clear.

<u>Inductive Step</u> ($0 < s \leq n - x$): Suppose that $s^{t+s-1}(x + s - 1) = p$. After forwarding for round $t + s - 1$, since $x + s - 1$ is in a smoothed plateau before forwarding, $s^{t+s-1}(x + s - 1) = s^{(t+s-1)'}(x + s)$ by Lemma 4. Next, after injection for round $t + s$, by Lemma 5, we still have $s^{t+s}(x + s) = s^{(t+s-1)'}(x + s) = p$ and the claim holds.

Finally, Lemma 6 allows us to conclude that the minimum non-empty buffer in a smoothed configuration is $1 - \sigma$, and thus that the maximum packet latency of any solid forwarding protocol with swaps is $n + \sigma - 1$.

Lemma 6. *Suppose A is a $(1, \sigma)$-bounded injection pattern and Π is a solid protocol. Then for all times t and buffers $x \in [n]$, we have $\mathbf{L}^t([1, x]) \leq x + \sigma$.*

We are now ready to prove Theorem 5.

Proof (Theorem 5). We claim that, for any round $t \in \mathbb{N}$ before forwarding, the minimum smoothed configuration buffer $x \in (-\infty, n]$ such that $\mathbf{SL}^t(x) > 0$ is $1 - \sigma$. We will utilize Lemma 6, which states that before forwarding for any $y \in [1, n]$, there are at most $y + \sigma$ packets in $[1, y]$. As in previous arguments, we also assume there is an extra buffer $n + 1$ which is cleared after forwarding.

For a configuration at timestep t, let y be the minimum buffer in $[1, n]$ such that $|\mathbf{Q}^t(y)| = 0$, and let $z = y - 1$. Then for all $x \in [1, z]$, $\mathbf{SL}^t(x) > 0$ and $\mathbf{L}^t([1, z]) \leq z + \sigma$. Thus, at most $z + \sigma$ packets are enqueued and z packets are dequeued over $[1, z]$ and $|\mathbf{Q}^t(0)| \leq \sigma$, and the minimum non-empty buffer in the smoothed configuration is $1 - \sigma$ as claimed. From Theorem 7, this packet will be absorbed in $n - (1 - \sigma) = n + \sigma - 1$ rounds.

We also provide a lower bound on protocol latency in the packet-swapping model. Specifically, any protocol in this model has worst-case packet latency of $n + \sigma - 1$, and thus, all solid forwarding protocols with swaps have optimal worst-case latency.

Proof (Theorem 6). Let $\mathcal{A}$ be any $(1, \sigma)$ adversary which injects $1 + \sigma$ packets into buffer $x = 1$ in round 1, and let $\mathcal{A}^1$ denote the set of packets injected in round 1 by $\mathcal{A}$. Since there is no buffer behind buffer 1, $x = 1$ can only lose at most one packet per round. Thus, at timestep $t = 1 + \sigma$ before forwarding, at least one packet in $\mathcal{A}^1$ remains in buffer 1, i.e. $|\mathcal{A}^1 \cap \mathbf{C}^t(1)| > 0$. Let $p \in \mathcal{A}^t \cap \mathbf{C}^t(1)$. Since p can move forward at most one buffer per round, at the earliest, p will be absorbed in round $t + (n - 1) = \sigma + n$. Thus, p's latency is at best $\sigma + n - 1$.

4   Total Load Analysis

In this section, we turn our attention to stochastic, rather than adversarial, packet injections. Again, we consider the single-destination path. Specifically, we analyze the following stochastic process: each round, (the index of) a buffer $x \in [1, n]$ is chosen independently and uniformly at random, and a single packet is injected into buffer x. We identify such an injection pattern with an infinite sequence of iid random variables $x_1, x_2, \ldots$ where each x_i is chosen uniformly from $[1, n]$. We refer to this process as the ***random injection model***.

In this section, we analyze the *total (network) load* in the random injection model when packets are forwarded according to a solid protocol. Here the ***total (network) load*** at time t is the total number of packets in the network at time t. We denote the total load by $\mathbf{L}^t$.

We first observe that $\mathbf{L}^t$ is a non-decreasing function of t. To see this, note that the random injection model injects precisely one packet into the network per round, while the final buffer forwards at most one packet per round. For an injection pattern A and (solid) forwarding protocol Π, we define A's ***increasing sequence*** $t_1 < t_2 < t_3 < \cdots$ to be the sequence of times at which $\mathbf{L}^t$ increases. More formally, we define

$$t_i = \inf \left\{ t \,\middle|\, \mathbf{L}^t \geq i \right\}.$$

Our main result in this section concerns the expected growth of the sequence (t_i). Specifically, we show that under the random injection model, any solid forwarding protocol total network load reaches, but does not exceed, n almost surely. On the other hand, $\mathbf{L}^t$ grows slowly: once $\mathbf{L}^t$ exceeds a certain threshold (αt for some constant $\alpha < 1$), the expected time until $\mathbf{L}^t$ increases is exponential in n.

Theorem 8. *Let Π be a solid forwarding protocol, and suppose packets are injected into a single destination path of size n according to the random injection model. Let $t_1, t_2, \ldots$ denote the increasing sequence for the process. Then the following hold:*

1. *$t_n < \infty$ almost surely (with probability 1),*
2. *$t_k = \infty$ for all $k > n$, and*
3. *there exists a constant $\alpha < 1$ such that for all $k \geq \alpha n$, $E(t_{k+1} - t_k) = 2^{\Omega(n)}$.*

The first claim of Theorem 8 follows from the observation that if A makes n consecutive injections into buffer 1, then the total load will be (at least) n for any forwarding protocol. Indeed, after $n-1$ forwarding steps, none of the packets injected into buffer 1 can have been absorbed, hence in round n there are (at least) n packets in the network. In the random injection model, the probability that any n consecutive injections are made to buffer 1 is $n^{-n} > 0$ Thus, in an infinite execution the total load reaches n almost surely. The second claim of Theorem 8 follows from Lemma 6, and the observation that every injection pattern in the random injection model is $(1, 0)$-bounded.

The focus of this section is to establish Claim 3 of Theorem 8. To this end, we employ the machinery of smoothed configurations developed in the previous sections. We state the following consequence of (the proof of) Theorem 5:

Corollary 2. *Let G be a the single destination path of size n. Suppose A is $(1, \sigma)$-bounded, Π is a solid forwarding protocol, and $\mathbf{C}$ is a configuration of G during some round of an execution (after injection and before forwarding). Let $\mathbf{S}$ be the associated smoothed configuration. Then for all $i < 1 - \sigma$, $\mathbf{S}(i) = 0$. That is, in the smoothed configuration, the left-most packet is in buffer at most $1 - \sigma$.*

In the case of the random injection model, Corollary 2 implies that in a smoothed configuration $\mathbf{S}$, only buffers in $[n]$ are non-empty. Now let $\mathbf{S}'$ denote the smoothed configuration after forwarding and before the next injection step. By Lemma 3, $\mathbf{S}'(i) = \mathbf{S}(i - 1)$. In particular, $\mathbf{S}'(1) = 0$.

Definition 8. *Let $\mathbf{C}$ be a configuration of a single destination path and $\mathbf{S}$ its associated smoothed configuration. We say that buffer i is a **gap** in $\mathbf{S}$ if $\mathbf{S}(i) = 0$.*

In accordance with Lemma 3, with think of gaps as moving forward (along with packets) during each forwarding step for the smoothed configuration. A packet injection may, however, **destroy** a gap in $\mathbf{S}$. The following lemma shows that the total load of the network increases during an injection round if and only if a gap remains at buffer n after injection. If a gap persists from its creation in some round t' to buffer n in round $t + n$, we say that the gap **survives**.

Lemma 7. *Consider the random injection model with solid protocol Π. Let $\mathbf{C}^t$ be a configuration at time t, and $\mathbf{S}^t$ its associated smoothed configuration. Suppose the total load of the network is $k = \mathbf{L}^t$, so that $t_k \leq t$. Then $\mathbf{L}^{t+1} = \mathbf{L}^t + 1$ (equivalently $t_{k+1} = t + 1$) if and only if $\mathbf{S}^t$ has a gap in buffer n.*

From the discussion above, we make the following observations. In each forwarding step of a round t, a gap is created in buffer 1. During the injection step of round $t + 1$, the injected packet destroys a gap in the corresponding smoothed configuration. If the injection occurs at buffer i, then the right-most gap $j \leq i$ is the gap that is destroyed as the result of the injection. All other gaps remain after injection. During a forwarding step, Lemma 3 implies that all gaps progress forward a single step. Observe that if the gap survives then, $\mathbf{S}^{t+n}$ contains a gap in buffer n (namely, the gap created in round t). Applying Lemma 7, we obtain the following result.

Lemma 8. *Consider an execution of a solid protocol Π in the random injection model. Then for any time $t \geq 1$, the total network load increases in round $t + n$ if and only if the gap created during the forwarding step of round t survives.*

Lemma 8 illuminates our path towards proving claim 3 of Theorem 8: in order to show that it is unlikely that the total load increases in round $t + n$, it suffices to show that the gap created in round t is unlikely to survive. The following definition and lemma provide necessary conditions for a gap to survive.

Definition 9. *Suppose a gap g is created in round t so that the gap occupies buffers $1, 2, \ldots$ in rounds $t, t + 1, \ldots$ until it either survives or is destroyed. We say that the injection in round $t + k$ is **ahead** of g if the injection is made to a buffer $j \geq k$.*

Lemma 9. *Suppose g is a gap created in round t. Then if g survives, there were at most $n - \mathbf{L}^t$ injections ahead of g in rounds $t + 1, t + 2, \ldots, t + n$.*

The following corollary (proven in the appendix) is obtain by applying a suitable Chernoff bound to bound the probability that few packets are injected ahead of a gap.

Corollary 3. *Let δ be any constant satisfying $0 < \delta < \frac{1}{2}$. Suppose $\mathbf{L}^t \geq (\frac{1}{2} + \delta)n$. Then the probability that the gap g formed in round t survives is at most $e^{-n\delta^2/4}$.*

Corollary 3 implies that if the total load satisfies $\mathbf{L}^t \geq (\frac{1}{2} + \delta)n$, then the total load is unlikely to increase in round $t + n$. Since $\mathbf{L}^t$ is non-decreasing, the same bound applies to the probability that the total load increases in rounds $t + n + 1, t + n + 2, \ldots$. Thus, a simple union bound implies that probability the load increases at any round in the range $t + n, t + n + 1, \cdots, t + n + C$ is at most $Ce^{-\Omega(n)}$. However, this argument does not give an upper bound $\mathbf{L}^{t+n}$. Nonetheless, assuming $\mathbf{L}^t = (\frac{1}{2} + \delta)n$, we obtain the following bound on the probability that $\mathbf{L}^{t+n}$ is large.

Lemma 10. *Suppose the total network load at time t is $\mathbf{L}^t = (\frac{1}{2} + \delta)n$. Then for $\alpha = 1 - (\frac{1}{2} - \delta)^2/16$, we have $\Pr(\mathbf{L}^{t+n} \geq \alpha n) = e^{-\Omega(n)}$.*

We are now ready to complete our proof of Theorem 8.

Proof (Claim 3 of Theorem 8). Fix δ to be any value satisfying $0 < \delta < \frac{1}{2}$, and let t denote the first round in which $\mathbf{L}^t = (\frac{1}{2} + \delta)n$. Let $k = \alpha n$, where α is the value in Lemma 10. By Lemma 10, $t_k \geq t + n$ with probability $1 - e^{\Omega(n)}$. Conditioned on $t_k \geq t + n$, Corollary 3 implies that for any round $t' > t_k$, the probability that $t' = t_{k+1}$ is $e^{-\Omega(n)}$. Applying a union bound, we find that for any $c = e^{-o(n)}$, the probability that $t_{k+1} \leq t_k + c$ is $e^{-\Omega(n)}$, from which Claim 3 of Theorem 8 follows.

5   Conclusions and Open Questions

Our main result from Sect. 3 establishes that optimal buffer space usage and optimal latency are simultaneously achievable for the single destination path network in a model that allows for packets to be moved backwards (or swapped). From previous results, this additional capability is necessary. We suggest several lines of future work here.

1. To what extent can our results and techniques be generalized to networks beyond the single-destination path? It seems plausible that our techniques might be generalized to single destination trees (the context of [3,7]). Can our techniques be further generalized to paths with multiple destinations (for example in the model of [5]), or to more general topologies?
2. Our results focus on the case of a network with unit capacity edges. Can our results be extended to higher capacities as well? A bundling strategy analyzed in [10] shows that optimal buffer space usage can be achieved on a general capacity path, but the strategy poses issues with respect to latency because some packets are ignored by the general capacity variant of OED. Nonetheless, it seems plausible that a variant of forwarding with swaps might generalize to general capacity setting.

To our knowledge, the stochastic process described in Sect. 4 has not been studied previously. We think it would be interesting to get a better understanding of the dynamics of different forwarding protocols in the random injection model.

1. What is a typical *maximum* load in the steady state for the random injection model for greedy forwarding? For OED forwarding? Preliminary experiments indicate that typical loads for OED are quite small (possibly $O(1)$), while greedy forwarding incurs large loads in the final buffer (e.g., $\Omega(n)$). Can these observations be explained analytically?
2. Can the analysis of the total load in the random injection model be tightened? Our simulations seem to indicate that the total load grows exponentially slowly even for a total load of $(1/2 + \delta)n$, but our techniques only give this result for total loads much closer to n.

Appendix: Latency of Solid Forwarding Protocols with Swaps

Proof (Lemma 2). First, it is clear that any packet in I at t can only occupy a buffer in $[a, b + 1]$ at t'. Furthermore, since I is an island in $\mathbf{C}^t(G)$, either $\mathbf{L}^t(a - 1) = 0$ or $a = 1$, and $\mathbf{L}^t(b + 1) = 0$, so no other packets occupy any buffer in $[a, b+1]$ after forwarding. We next induct on the suffixes of I', $[b + 1 - i, b + 1]$ where $0 \leq i < b + 1 - a$, to show that each buffer in $[a + 1, b + 1]$ is non-empty in $\mathbf{C}^{t'}(G)$.

<u>Base Case</u> $(i = 0)$: Since I is an island in $\mathbf{C}^t(G)$, $\mathbf{L}^t(b + 1) = 0$ so b forwards a packet to $b + 1 \in I'$ and $\mathbf{L}^{t'}(b + 1) > 0$.

<u>Inductive Step</u> $(0 < i < b+1-a)$: Assume that the lemma holds for the interval $[b + 1 - (i - 1), b + 1]$ where $0 \leq i < b + 1 - a$. Since $i < b + 1 - a$, $\mathbf{L}^t(b - i) > 0$. If either $\mathbf{L}^t(b + 1 - i) > 1$, or $\mathbf{L}^t(b + 1 - i) = 1$ but $b + 1 - i$ does not forward during round t, then we know that $\mathbf{L}^{t'}(b + 1 - i) \geq 1$. On the other hand if $\mathbf{L}^t(b+1-i) = 1$ and $b+1-i$ forwards during round t, then since $\mathbf{L}^t(b-i) > 0$, by Definition 3, $b + 1 - i$ receives a packet during forwarding from $b - i \in I$ and thus $\mathbf{L}^{t'}(b + 1 - i) = 1$. Combining this with the inductive hypothesis, we get that for each buffer $x \in [b + 1 - i, b + 1]$, $\mathbf{L}^{t'}(x) > 0$.

Thus, the statement holds for all $i < b + 1 = a$. If $\mathbf{L}^t(a) > 1$ or $\mathbf{L}^t(a) = 1$ but a does not forward during round t, then the lemma holds for $I' = [a, b+1]$. Otherwise, it holds for $I' = [a+1, b+1]$.

Proof (Lemma 3). Let $M_1 = [a_1, b_1], M_2 = [a_2, b_2,], \ldots, M_k = [a_k, b_k] \subseteq M$ be the islands contained in M (from right to left) in $\mathbf{C}^t(H)$. We first claim that for each buffer $x \in M'$, $\mathbf{SL}^{t'}(x) = 1$. Notice that we must have $b = b_1$.

Define $M'_1 = [a_1+1, b_1+1], M'_2 = [a_2+1, b_2+1], \ldots, M'_k = [a_k+1, b_k+1], G_1 = [b_2+1, a_1-1], G_2 = [b_3+1, a_2-1], \ldots, G_{k-1} = [b_k+1, a_{k-1}-1], G_k = [a, a_k-1]$, and $G'_1 = [b_2+2, a_1], G'_2 = [b_3+2, a_2], \ldots, G'_{k-1} = [b_k+2, a_{k-1}], G'_k = [a+1, a_k]$.

The G_is represent the gaps between the islands contained in M in $\mathbf{C}^t(H)$. Note that the intervals M_i, G_i are disjoint and that M'_i, G'_i are also disjoint. Furthermore, note that $\cup_{i=1}^k (M_i \cup G_i) = M$ and that $\cup_{i=1}^k (M'_i \cup G'_i) = M'$. Lemma 2 implies that for all $x \in \cup_{i=1}^k M'_i$, $\mathbf{SL}^{t'}(x) = 1$.

We will make frequent use of the following observations, where $I = [a, b]$ is any interval of buffers, in our arguments:

1. If $|\mathbf{Q}(b)| \geq |I|$, then $\forall x \in I$, $\mathbf{SL}(x) = 1$. Furthermore, if each buffer x in I is empty but has $\mathbf{SL}(x) = 1$, then $|\mathbf{Q}(b)| \geq |I|$.
2. $|\mathbf{Q}(a - 1)| \geq \mathbf{L}(I) - |I|$.

To prove our first claim, we show that for all i such that $1 \leq i \leq k$, that for all $x \in G'_i$, $\mathbf{SL}^{t'}(x) = 1$.

<u>Base Case</u> $(i = 1)$: Since M is a smoothed plateau in $\mathbf{C}^t(H)$, we know that $|\mathbf{Q}^t(b_1 + 1)| = 0$, and since $\forall x \in G_1 \cup M_1$ we have $\mathbf{SL}^t(x) = 1$, we know that $|\mathbf{Q}^t(a_1 - 1)| = \mathbf{L}^t(M_1) - |M_1| \geq |G_1|$. There are two cases to consider:

1. $\mathbf{L}^{t'}(a_1) > 0$. From Lemma 2, since M_1 is an island in $\mathbf{C}^t(H)$,

$$\mathbf{L}^t(M_1) = \mathbf{L}^{t'}(M'_1) + \mathbf{L}^{t'}(a_1) \implies \mathbf{L}^{t'}(M'_1) + \mathbf{L}^{t'}(a_1) - |M_1| \geq |G_1|,$$

and thus

$$\mathbf{L}^{t'}(M'_1) + \mathbf{L}^{t'}(a_1) - (|M_1| + 1) \geq |G'_1| - 1 \implies \left|\mathbf{Q}^{t'}(a_1 - 1)\right| \geq |G'_1| - 1.$$

Combining this with the assumption that $\mathbf{L}^{t'}(a_1) > 0 \implies \mathbf{SL}^{t'}(a_1) = 1$, we get that $\forall x \in G'_1$, $\mathbf{SL}^{t'}(x) = 1$ since all buffers in G'_1 except a_1 are empty.

2. $\mathbf{L}^{t'}(a_1) = 0$. In this case, $\mathbf{L}^{t'}(M'_1) = \mathbf{L}^t(M_1)$ and $|M_1| = |M'_1|$, so $\mathbf{L}^{t'}(M'_1) - |M'_1| \geq |G'_1|$, and therefore $\left|\mathbf{Q}^{t'}(a_1)\right| \geq |G'_1|$ where all buffers in G'_1 are empty in $\mathbf{C}^{t'}(H)$. Thus, $\forall x \in G'_1, \mathbf{SL}^{t'}(x) = 1$.

<u>Inductive Step</u> $(k \geq i > 1)$: Suppose that for all $j < i$, for all $x \in G'_j$, we have $\mathbf{SL}^{t'}(x) = 1$. Let $N_i = [a_i, b]$. Since M is a smoothed plateau in $\mathbf{C}^t(H)$ and $\mathbf{L}^t(a_i - 1) = 0$, we know that $|\mathbf{Q}^t(a_i - 1)| = \mathbf{L}^t(N_i) - |N_i|$ and that $|\mathbf{Q}^t(a_i - 1)| \geq |G_i|$. There are two cases to consider:

1. $\mathbf{L}^{t'}(a_i) = 0$. Let $N_i' = [a_i + 1, b + 1]$. First, because $\mathbf{L}^{t'}(a_i) = 0$, by Lemma 2 we know that $\mathbf{L}^{t'}(N_i') = \mathbf{L}^t(N_i)$, and since $|N_i'| = |N_i|$, we have that $\mathbf{L}^{t'}(N_i) - |N_i'| = \mathbf{L}^t(N_i) - |N_i|$, and thus by observation 2

$$\left|\mathbf{Q}^{t'}(a_i)\right| \geq \mathbf{L}^t(N_i) - |N_i| = \left|\mathbf{Q}^t(a_i - 1)\right| \geq |G_i| = |G_i'|.$$

Thus, $\left|\mathbf{Q}^{t'}(a_i)\right| \geq |G_i'|$ where all buffers in G_i' are empty in $\mathbf{C}^{t'}(H)$, so $\forall x \in G_i', \mathbf{SL}^{t'}(x) = 1$.

2. $\mathbf{L}^{t'}(a_i) > 0$. Let $N_i' = [a_i, b + 1]$. In this case we know that $\mathbf{L}^t(N_i) - |N_i| - 1 = \mathbf{L}^{t'}(N_i') - |N_i'|$ since $\mathbf{L}^t(N_i) = \mathbf{L}^{t'}(N_i')$ and $|N_i| = |N_i'| - 1$. By the same reasoning as case 1,

$$\left|\mathbf{Q}^{t'}(a_i - 1)\right| \geq \mathbf{L}^{t'}(N_i') - |N_i'| = \mathbf{L}^t(N_i) - |N_i| - 1$$
$$= \left|\mathbf{Q}^t(a_i - 1)\right| - 1 \geq |G_i| - 1 = |G_i'| - 1.$$

Since by assumption $\mathbf{L}^{t'}(a_i) > 0 \implies \mathbf{SL}^{t'}(a_i) = 1$, and since $\left|\mathbf{Q}^{t'}(a_i - 1)\right| \geq |G_i'| - 1$, where all buffers in G_i' except a_i are empty in $\mathbf{C}^{t'}(H)$, we have that for all $x \in G_i', \mathbf{SL}^{t'}(x) = 1$.

Combining the arguments above, we see that for all $x \in M', \mathbf{SL}^{t'}(x) > 0$. Finally, we show that $\mathbf{SL}^{t'}(a) = \mathbf{SL}^{t'}(b + 2) = 0$. M is one of possibly many smoothed plateaus in $\mathbf{C}^t(H)$. Label the $l \geq 1$ smoothed plateaus in $\mathbf{C}^t(H)$ (from right to left) $R_1 = [c_1, d_1], \ldots, R_l = [c_l, d_l]$.

We show for all $1 \leq i \leq l$ that $\mathbf{SL}^{t'}(c_i) = \mathbf{SL}^{t'}(d_i + 2) = 0$ by induction. We will use the following observation: for a suffix $I = [a, n + 2]$ of buffers, we must have $\mathbf{L}^t(I) \geq \mathbf{SL}^t(I)$, and in particular, if $|\mathbf{Q}^t(a - 1)| = 0$, i.e., $\mathbf{SL}^t(a - 1) = 0$, then $\mathbf{L}^t(I) = \mathbf{SL}^t(I)$.

<u>Base Case</u> ($i = 1$): Since $\mathbf{L}^t([d_1 + 1, n]) = 0$, we have $\mathbf{L}^{t'}([d_1 + 2, n]) = 0$ so $\mathbf{SL}^{t'}(d_1 + 2) = 0$. Let $I = [c_1, n + 2]$. For c_1, since $\mathbf{SL}^t(c_1 - 1) = 0$ by assumption, c_1 receives no packets during forwarding so $\mathbf{L}^t(I) = \mathbf{L}^{t'}(I)$, and furthermore $\mathbf{L}^t(I) = \mathbf{SL}^t(I)$. If $\mathbf{SL}^{t'}(c_1) = 1$, then $\mathbf{SL}^{t'}(I) = \mathbf{SL}^t(I) + 1$ since by our previous results $\mathbf{SL}^t(R_1) = \mathbf{SL}^t(R_1')$ where $R_1' = [c_1 + 1, d_1 + 1]$. Thus, $\mathbf{SL}^{t'}(I) > \mathbf{SL}^t(I)$ which, since $\mathbf{L}^{t'}(I) \geq \mathbf{SL}^{t'}(I)$ implies $\mathbf{L}^{t'}(I) > \mathbf{L}^t(I)$, a contradiction. Thus, $\mathbf{SL}^{t'}(c_1) = 0$.

<u>Inductive Step</u> ($1 < i \leq l$): Assume that for all $j < i, \mathbf{SL}^{t'}(c_j) = \mathbf{SL}^{t'}(d_j + 2) = 0$. For $d_i + 2$, by assumption $d_i + 2 \leq c_{i-1}$, and all buffers x in $[d_i + 1, c_{i-1} - 1]$ have $\mathbf{SL}^t(x) = 0$. If $d_i + 2 < c_{i-1}$, then $\mathbf{SL}^t(d_i + 2) = 0$ and since $\mathbf{SL}^t(d_i + 1) = 0$, we still have $\mathbf{SL}^{t'}(d_i + 2) = 0$ after forwarding. On the other hand if $c_{i-1} = d_i + 2$, then by inductive hypothesis $\mathbf{SL}^{t'}(c_{i-1}) = 0 \iff \mathbf{SL}^{t'}(d_i + 2) = 0$. For c_i, let $I = [c_i, n + 2]$. By assumption, c_i receives no packets during forwarding and $\mathbf{SL}^t(c_i - 1) = 0$ so $\mathbf{L}^t(I) = \mathbf{L}^{t'}(I) = \mathbf{SL}^t(I)$. If $\mathbf{SL}^{t'}(c_i) = 1$,

then $\mathbf{SL}^{t'}(I) = \mathbf{SL}^t(I) + 1$ since each R_m for $m \le i$ moves forward one buffer during forwarding. As in the base case, this is a contradiction and thus $\mathbf{SL}^{t'}(c_i) = 0$.

Proof (Lemma 4). First, because the buffer loads of each $x \in [1, n]$ are the same under Π_S as under Π, Lemma 3 implies that there will be a smoothed plateau at $M' = [a+1, b+1]$ in $\mathbf{S}^{t'}(G)$. What is left to show is that for all i such that $0 \le i \le b - a$, $s^t(b-i) = s^{t'}(b-i+1)$. We prove the claim using induction.

<u>Base Case</u> ($i = 0$): Since M is a smoothed plateau in $\mathbf{C}^t(G)$, $|\mathbf{Q}^t(b+1)| = 0$ but $|\mathbf{Q}^t(b)| > 0$, so $\mathbf{L}^t(b) > 0$. Thus, b forwards $h^t(b) = s^t(b)$ to the empty buffer $b+1$. At t', $h^t(b)$ is the only packet in $b+1$ and from Lemma 3, we will have $\left|\mathbf{Q}^{t'}(b+2)\right| = 0$, so $s^t(b) = s^{t'}(b+1)$.

<u>Inductive Step</u> ($1 \le i \le b - a$): Suppose for all $0 \le j < i$ that $s^t(b-j) = s^{t'}(b-j+1)$. Let $x = b-i$, and $p = s^t(x)$. From Algorithm 1, p is the oldest packet in $\mathbf{Q}^t(x)$, which is the set of packets in $\mathbf{C}^t([x,b])$ and not in $\{s^t(y) \mid y \in [x+1, b]\}$. Furthermore, $s^{t'}(x+1)$ is the oldest packet in $\mathbf{Q}^{t'}(x+1)$, which is the set of packets in $\mathbf{C}^{t'}([x+1, b+1])$ and not in $\left\{s^{t'}(y) \mid y \in [x+2, b+1]\right\}$. We show 1. that $\mathbf{Q}^{t'}(x+1) \subseteq \mathbf{Q}^t(x)$, and 2. that $p \in \mathbf{Q}^{t'}(x+1)$, which together imply that $p = s^{t'}(x+1)$.

1. Let r be any packet in $\mathbf{Q}^{t'}(x+1)$. At timestep t, r could have been in any buffer in the interval $[x, b+2]$. Since $\mathbf{L}^t(b+1) = 0$, r was not in $b+1$ at t. Nor was r in $b+2$ at t since under Π_S, no buffer will send a packet backwards to an empty buffer, so r was in $[x, b]$ at t. Furthermore, from Algorithm 1 and by assumption that $r \in \mathbf{Q}^{t'}(x+1)$,

$$r \notin \left\{s^{t'}(y) \mid y \in [x+2, b+1]\right\}.$$

   By the inductive hypothesis, this set is the same as $\{s^t(y) \mid y \in [x+1, b]\}$ so $r \in \mathbf{Q}^t(x)$ and $\mathbf{Q}^{t'}(x+1) \subseteq \mathbf{Q}^t(x)$.

2. First, since $p \in \mathbf{Q}^t(x)$,

$$p \notin \left\{s^t(y) \mid y \in [x+1, b]\right\} = \left\{s^{t'}(y) \mid y \in [x+2, b+1]\right\}.$$

   If p was in a buffer in $[x+2, b]$ at t, then at t', p is in a buffer in $[x+1, b+1]$ since packets can only move forward or backward one buffer during forwarding. Thus, in this case $p \in \mathbf{Q}^{t'}(x+1)$.

   If p was in $x+1$ and p was sent backward during forwarding, then there was an older packet in x at t than p, so $p \ne s^t(x)$, a contradiction.

   Finally, if p was in x at timestep t, there are two cases. Notice that $p = h^t(x)$. First, if p is in x at t', then $l^t(x+1) > p$ and $\mathbf{L}^t(x+1) \ge 2$, so $l^t(x+1)$ was older than p and in $\mathbf{Q}^t(x)$, so $p \ne s^t(x)$. Second, if p is in $x-1$ at t', then $\mathbf{L}^t(x) \ge 2$ and $p = l^t(x)$, so $p \ne s^t(x)$. Thus, if p was in x at timestep t, then p must be in $x+1$ at t' and so $p \in \mathbf{Q}^{t'}(x+1)$.

Proof (Lemma 5). We prove the following claim, which implies the lemma: for all $x \in (-\infty, n]$, $\mathbf{Q}^{(t-1)'}(x) \subseteq \mathbf{Q}^t(x)$ and $\mathbf{Q}^t(x) \backslash \mathbf{Q}^{(t-1)'}(x)$ contains only packets injected in round t. We induct on the buffers in the network from right to left, showing that for all $i \geq 0$, the claim holds for $x = n - i$. Note that we call a packet p **new** if $t_p = t$.

<u>Base Case</u> $(i = 0)$: From Algorithm 1, we know that $\mathbf{Q}^{(t-1)'}(n) = \mathbf{C}^{(t-1)'}(n)$ and that $\mathbf{Q}^t(n) = \mathbf{C}^t(n)$. During injection for round t, buffer n cannot lose any packets, and only gains some new packets (or potentially none). Thus, $\mathbf{C}^{(t-1)'}(n) \subseteq \mathbf{C}^t(n)$ with the difference containing only new packets, and the claim holds for $x = n$.

<u>Inductive Step</u> $(i > 0)$: Let $x = n - i$, and assume that $\mathbf{Q}^{(t-1)'}(x + 1) \subseteq \mathbf{Q}^t(x + 1)$ with only new packets in the difference between the two sets. If $\left| \mathbf{Q}^{(t-1)'}(x + 1) \right| = 0$, the same argument from the base case applies to x.

Otherwise, by the inductive hypothesis, the oldest packet p in $\mathbf{Q}^{(t-1)'}(x+1)$ is the oldest packet in $\mathbf{Q}^t(x+1)$ as well, so p is dequeued from both. Letting $R = \mathbf{Q}^{(t-1)'}(x + 1) \setminus \{p\}$ and $R' = \mathbf{Q}^t(x + 1) \setminus \{p\}$ denote the states of the queue over $x + 1$ after dequeueing at timesteps $(t - 1)'$ and t, respectively, we see that $R \subseteq R'$ as well with only new packets in the difference, and from Algorithm 1, $\mathbf{Q}^{(t-1)'}(x) = R \cup \mathbf{C}^{(t-1)'}(x)$ and $\mathbf{Q}^t(x) = R' \cup \mathbf{C}^t(x)$, where $R \cap \mathbf{C}^{(t-1)'}(x) = R' \cap \mathbf{C}^t(x) = \emptyset$. Thus, since x only gains new packets during injection, $\mathbf{Q}^{(t-1)'}(x) \subseteq \mathbf{Q}^t(x)$ with the difference containing only new packets, we get that the claim holds for buffer x.

This claim implies the lemma since, for each $x \in (-\infty, n]$ with $\mathbf{SL}^{(t-1)'}(x) > 0$, $p = s^{(t-1)'}(x) \in \mathbf{Q}^t(x)$ and since all of the packets in $\mathbf{Q}^t(x) \setminus \mathbf{Q}^{(t-1)'}(x)$ are younger than p, we have $p = s^t(x)$. Thus, the lemma holds.

Lemma 6 is analogous to a special case of Lemma 3.4 in [7] (with $j = 0$). While the proof in [7] is stated only for OED forwarding, the proof for $j = 1$ only relies upon the properties shared by all solid protocols.

Appendix: Total Load Analysis

Proof (Lemma 7). For the $\Longrightarrow$ direction, suppose $\mathbf{L}^{t+1} = \mathbf{L}^t + 1$. This implies that the total load increased after forwarding and injection. Since one packet was injected in round $t + 1$, buffer n must not have forwarded, hence $\mathbf{L}^t(n) = 0$. Therefore, buffer n must be empty in $\mathbf{S}^t$ as well, since the smoothing process never moves packets forward.

Conversely, suppose $\mathbf{S}^t$ has a gap in buffer n. Then $\mathbf{L}^t(n) = 0$, hence no packet is absorbed during the forwarding step: $\mathbf{L}^{t'} = \mathbf{L}^t$. Since a packet is injected in the next injection step, we have $\mathbf{L}^{t+1} = \mathbf{L}^{t'} + 1 = \mathbf{L}^t + 1$.

Proof (Lemma 9). We argue the contrapositive. Suppose that at least $n - \mathbf{L}^t + 1$ injections occur ahead of g. Observe that after the first forwarding step, there are at most $n - \mathbf{L}^t$ gaps ahead of g. By Lemma 3, the number of gaps ahead of g is non-increasing during each forwarding step. Further, by Lemma 5, the number of gaps ahead of g decreases each time a packet is injected ahead of g (unless the injection destroys g). Therefore, after (at most) $n - \mathbf{L}^t$ injections ahead of g, either no gaps remain ahead of g or g was destroyed. Any subsequent injection ahead of g will destroy g (if it wasn't already destroyed), and the desired result follows.

Proof (Corollary 3). We argue that when $\mathbf{L}^t \geq (\frac{1}{2} + \delta)n$, the probability that the gap created in round t survives is exponentially small in n. To this end, for each t and $i \in [n]$, we define the random variable X_i^t to be 1 if the injection in round $t + i$ occurs ahead of the gap formed in round t and 0 otherwise. For each round t, let j_t denote the buffer into which the adversary injects in round t. Then

$$X_i^t = \begin{cases} 1 & j_{t+i} \geq t + i - 1 \\ 0 & \text{otherwise.} \end{cases} \tag{1}$$

Thus, the sum $X^t = X_1^t + X_2^t + \cdots + X_n^t$ is the total number of packets injected ahead of the gap created in round t. Observe that for a fixed t, the variables $X_1^t, X_2^t, \ldots, X_n^t$ are independent, and that

$$\Pr(X_i^t = 1) = \frac{n - i + 1}{n}. \tag{2}$$

Therefore, linearity of expectation gives

$$\mu = E(X^t) = \sum_{i=1}^{n} \frac{n - i + 1}{n} = \frac{1}{2}(n + 1).$$

If $\mathbf{L}^t > (\frac{1}{2} + \delta)n$, then in order for the gap formed in round t to survive, at most $(\frac{1}{2} - \delta)n < (1 - \delta)\mu$ packets have been injected ahead of g. The corollary follows by applying the following Chernoff bound to bound the probability of this event.

Lemma 11 (Cf. Theorem 4.5 in [6]). *Suppose $Y_1, Y_2, \ldots, Y_n$ are independent Poisson trials with $\Pr(Y_i) = p_i$, let $Y = \sum_{i=1}^{n} Y_i$, and $\mu = E(Y)$. Then for $0 < \delta < 1$, we have*

$$\Pr(Y \leq (1 - \delta)\mu) \leq e^{-\mu \delta^2 / 2}.$$

Proof (Lemma 10). Let $\mathbf{S}$ be a smoothed configuration with total load $\mathbf{L}^t = (\frac{1}{2} + \delta)n$. Let $g_1, g_2, \ldots, g_k$ be the gaps in the configuration at time t, where $k = (\frac{1}{2} - \delta)n$. We bound $\mathbf{L}^{t+n}$ by giving an upper bound on the number of the gaps $g_1, \ldots, g_k$ that survive. Specifically, we will show that a constant fraction of the gaps are destroyed with very high probability. To this end, assume that g_1 is the rightmost gap, g_2 the next gap to the left, and so on. Thus, each g_j is in buffer $i < n - j$. Let $s = (\frac{1}{2} - \delta)n/4$. Observe that in rounds $t + 1, t + 2, \ldots, t + s$, at most s gaps could survive. Further, in these rounds, any injection into a buffer

i satisfying $i > n - 2s$ is guaranteed to destroy one of the gaps $g_1, g_2, \ldots, g_k$ (and not a gap that was created in some round $t' \geq t$). We call an injection into a buffer $i > n - 2s$ a **good** injection. Observe that if the are $\gamma \leq s$ good injections in rounds $t + 1, t + 2, \ldots, t + 4s$, then in round $t + n$, the total load satisfies $\mathbf{L}^{t+n} \leq n - \gamma$.

The probability that any injection is good is $\frac{2s}{n} = (\frac{1}{2} - \delta)/2$. Therefore, the expected number of good injections in rounds $t + 1, \ldots, t + s$ is $\mu = s(\frac{1}{2} - \delta)/2 = (\frac{1}{2} - \delta)^2/8 = 2\alpha$. Applying Lemma 11, with $\delta = \frac{1}{2}$ gives the desired result.

References

1. Adler, M., Rosén, A.: Tight bounds for the performance of longest-in-system on DAGs. In: Alt, H., Ferreira, A. (eds.) STACS 2002. LNCS, vol. 2285, pp. 88–99. Springer, Heidelberg (2002). https://doi.org/10.1007/3-540-45841-7_6, https://dx.doi.org/10.1007/3-540-45841-7_6
2. Borodin, A., Kleinberg, J., Raghavan, P., Sudan, M., Williamson, D.P.: Adversarial queuing theory. J. ACM **48**(1), 13–38 (2001)
3. Dobrev, S., Lafond, M., Narayanan, L., Opatrny, J.: Optimal local buffer management for information gathering with adversarial traffic. In: Scheideler, C., Hajiaghayi, M.T. (eds.) Proceedings of the 29th ACM Symposium on Parallelism in Algorithms and Architectures, SPAA 2017, Washington DC, USA, 24–26 July 2017, pp. 265–274. ACM (2017). https://doi.org/10.1145/3087556.3087577
4. Miller, A., Patt-Shamir, B.: Buffer size for routing limited-rate adversarial traffic. In: Gavoille, C., Ilcinkas, D. (eds.) DISC 2016. LNCS, vol. 9888, pp. 328–341. Springer, Heidelberg (2016). https://doi.org/10.1007/978-3-662-53426-7_24, http://dx.doi.org/10.1007/978-3-662-53426-7_24
5. Miller, A., Patt-Shamir, B., Rosenbaum, W.: With great speed come small buffers: space-bandwidth tradeoffs for routing. In: Robinson, P., Ellen, F. (eds.) Proceedings of the 2019 ACM Symposium on Principles of Distributed Computing, PODC 2019, Toronto, ON, Canada, July 29–2 August 2019, pp. 117–126. ACM (2019). https://doi.org/10.1145/3293611.3331614
6. Mitzenmacher, M., Upfal, E.: Probability and Computing, 2nd edn. Cambridge University Press, Cambridge (2017)
7. Patt-Shamir, B., Rosenbaum, W.: The space requirement of local forwarding on acyclic networks. In: Schiller, E.M., Schwarzmann, A.A. (eds.) PODC 2017: Proceedings of the ACM Symposium on Principles of Distributed Computing, Washington, DC, USA, 25–27 July 2017, pp. 13–22. ACM (2017). https://doi.org/10.1145/3087801.3087803
8. Patt-Shamir, B., Rosenbaum, W.: Space-optimal packet routing on trees. In: 2019 IEEE Conference on Computer Communications, INFOCOM 2019, Paris, France, April 29– 2 May 2019, pp. 1036–1044. IEEE (2019). https://doi.org/10.1109/INFOCOM.2019.8737596
9. Rosén, A., Scalosub, G.: Rate vs. buffer size-greedy information gathering on the line. ACM Trans. Algorithms **7**(3), 32:1–32:22 (2011). https://doi.org/10.1145/1978782.1978787
10. Rosenbaum, W.: Packet forwarding with a locally Bursty adversary. In: Scheideler, C. (ed.) 36th International Symposium on Distributed Computing, DISC 2022, 25–27 October 2022, Augusta, Georgia, USA. LIPIcs, vol. 246, pp. 34:1–34:18. Schloss Dagstuhl - Leibniz-Zentrum für Informatik (2022). https://doi.org/10.4230/LIPIcs.DISC.2022.34

Exact Distributed Sampling

Sriram V. Pemmaraju[ID] and Joshua Z. Sobel[(✉)][ID]

Department of Computer Science, University of Iowa, Iowa City, IA, USA
{sriram-pemmaraju,joshua-sobel}@uiowa.edu

Abstract. Fast distributed algorithms that output a feasible solution for constraint satisfaction problems, such as maximal independent sets, have been heavily studied. There has been much less research on distributed *sampling* problems, where one wants to sample from a distribution over all feasible solutions (e.g., uniformly sampling a feasible solution). Recent work (Feng, Sun, Yin PODC 2017; Fischer and Ghaffari DISC 2018; Feng, Hayes, and Yin arXiv 2018) has shown that for some constraint satisfaction problems there are distributed Markov chains that mix in $O(\log n)$ rounds in the classical LOCAL model of distributed computation. However, these methods return samples from a distribution close to the desired distribution, but with some small amount of error. In this paper, we focus on the problem of *exact* distributed sampling. Our main contribution is to show that these distributed Markov chains in tandem with techniques from the sequential setting, namely *coupling from the past* and *bounding chains*, can be used to design $O(\log n)$-round LOCAL model exact sampling algorithms for a class of weighted local constraint satisfaction problems. This general result leads to $O(\log n)$-round exact sampling algorithms that use small messages (i.e., run in the CONGEST model) and polynomial-time local computation for some important special cases, such as sampling weighted independent sets (aka the *hardcore* model) and weighted dominating sets.

Keywords: Distributed Sampling · Bounding Chains · Perfect Sampling · Coupling from the Past

1 Introduction

There is a vast body of literature on the distributed complexity of solving local constraint satisfaction problems (CSPs) on graphs [2,3,5,21,22,24,28]. Here "local" refers to the fact that the constraints span vertices with a constant diameter in the underlying graph. These local CSPs include classic "symmetry breaking" problems such as maximal independent sets, $(\Delta+1)$-colorings, and maximal matchings [1,3,12,26]. A distributed algorithm solving one of these local CSPs is required to construct some feasible solution of the local CSP. In contrast, this paper focuses on the problem of *sampling* a feasible solution of a local CSP. In the

The authors were supported, in part, by NSF grant IIS-1955939.

sampling problem the algorithm is required to output a solution sampled from the set of all feasible solutions of the CSP according to some desired probability distribution. Clearly, the sampling problem is at least as hard as the construction problem because solving the sampling problem requires the construction of a feasible solution. More precisely, we are interested in sampling solutions of local *weighted* CSPs. Here "weighted" refers to an assignment of a weight to each feasible solution of the CSP with the stipulation that the solutions be sampled with probabilities proportional to the weights. When the weights are identical, the sampling distribution is uniform.

Traditionally, in the sequential setting, sampling from a weighted local CSP on a graph involved running an ergodic Markov chain with a stationary distribution matching the desired distribution. After running for a long enough time, the distribution of the current state of the chain becomes arbitrarily close to (within any ϵ in terms of *total variation distance*) the desired distribution. For a given ϵ, the time required for this is known as the *mixing time*. After the mixing time is reached the current state of the chain is returned. The two most simple examples of Markov chains that are used to sample from weighted local CSPs are the *Metropolis-Hastings algorithm* and the *Glauber dynamics* [23]. For some weighted local CSPs (e.g., proper colorings) with certain parameters, these chains have mixing times of $O(n \log \frac{n}{\epsilon})$, where n is the number of nodes in the graph.

A *distributed* Markov Chain for sampling from a weighted local CSP, the *Local Metropolis* chain, was introduced by Feng, Sun, and Yin [8]. This chain allows every vertex to simultaneously propose a new label, rather than a single selected vertex. This chain is easily implemented in the LOCAL model of distributed computing, taking a constant number of rounds for each step of the chain. For certain weighted local CSPs, the chain can also be implemented in the CONGEST model, taking a constant number of rounds for each step of the chain. Both Fischer and Ghaffari [10] and Feng, Hayes, and Yin [7] showed that this chain could be improved, at least in the case of colorings[1], by only updating a small fraction of marked nodes in each step instead of attempting to update every node at every step. For colorings, the Local Metropolis chain [8] has a mixing time of $O(\log \frac{n}{\epsilon})$ when the palette has at least $\alpha \Delta$ colors for any constant $\alpha > 2 + \sqrt{2}$. The improvement in [7,10] only requires $\alpha > 2$, while achieving the same mixing time. Here, n is the number of vertices and Δ is the maximum degree of the graph. The point to note about these algorithms is that they return a state drawn from a distribution that approximates the desired distribution within a total variation distance of ϵ. Furthermore, the bound on the mixing time grows as ϵ becomes smaller. The current paper focuses on exact distributed sampling, i.e., the setting where $\epsilon = 0$.

---

[1] Fischer and Ghffari [10] claim that their approach, where not every vertex is marked, has an $O(\log \frac{n}{\epsilon})$-round mixing time for a more general class of weighted CSPs than colorings, though the proof of this does not appear in the paper.

In the sequential setting, one elegant and well known method for sampling exactly from the stationary distribution of a Markov chain is *coupling from the past (CFTP)* [29], introduced by Propp and Wilson. In its original form, CFTP took exponential time in general, so its use was limited to Markov chains that had state spaces with special properties (e.g., the *monotonicity* property). Subsequently, Nelander and Häggström [16] and Huber [17] showed that in cases where the original CFTP algorithm may not be tractable, augmenting the algorithm with *bounding chains* may still allow the algorithm to be used. The main contribution of this paper is showing that it is possible to use CFTP and the bounding chain technique in the LOCAL model. We use CFTP and the bounding chain technique in conjunction with the Markov chains described in the previous paragraph, to sample *exactly* from weighted local CSPs. Our algorithms are fast and in some cases run in the CONGEST model. Our results are described in more detail in the next section.

Comparable Results. While the papers [7,8,10] focus on approximate distributed sampling, there are two recent papers on exact sampling of certain weighted local CSPs. Feng and Yin [9] use a seminal result of Jerrum, Valiant, and Vazirani [19], showing that for certain problems exact sampling reduces to approximate counting by using a rejection sampling procedure. They present a distributed implementation of that approach. Guo, Jerrum, and Liu [14] present a sampling version of the Lovász Local Lemma that can be applied in the distributed setting to exactly sample from some weighted local CSPs.

Our approach using CFTP and bounding chains differs significantly from both of these techniques. All three techniques also differ in terms of the weighted local CSPs they are able to sample from. However, our techniques lead to results for the hardcore model that improve upon the results of [9,14]. A precise comparison appears further below.

1.1   Main Results

The main results of our paper can be summarized as follows.

1. We present a distributed Markov chain for sampling from weighted local CSPs and prove that it has the correct stationary distribution. This chain is based on the *Local Metropolis* chain from [8] and the coloring chains from [7,10]. We believe that this chain may be the generalization briefly alluded to by Fischer and Ghaffari in [10].

2. We apply the CFTP with bounding chains approach to the above-mentioned distributed Markov chain. We then present a condition that guarantees termination of this algorithm in $O(\log n)$ rounds with high probability in the LOCAL model. Thus, under a fairly general condition, we obtain an $O(\log n)$-round algorithm in the LOCAL model for exact sampling from weighted local CSPs. This result is an improvement by a factor of n over the sequential setting running time of $O(n \log n)$, for a slightly different condition, given by [16].

3. We finally show that the general algorithm described above leads to $O(\log n)$-round, small-message (i.e., CONGEST model), sampling algorithms for the weighted independent sets (aka hardcore model) problem and the weighted dominating sets problem. For the hardcore model, we are able to sample within a constant multiple of the hardness threshold for this problem (see the end of this section). Our algorithms do not abuse the power of the CONGEST model and ensure that every local computation runs in polynomial time.

The hardcore model is governed by a parameter $\lambda > 0$, called the *fugacity* of the model. Each independent set of size x is assigned a weight of λ^x; therefore, the desired probability distribution assigns the same probability to all independent sets of the same size. Furthermore, when $\lambda < 1$, small independent sets are more likely than large independent sets. Our algorithm[2] for the hardcore model requires $\lambda \leq \frac{\alpha}{\Delta}$, for any constant $\alpha < 1$. This almost matches the condition $\lambda \leq \frac{\alpha}{\Delta-1}$ that [16] gives for $O(n \log n)$ time in the sequential setting. We also derive a similar condition for weighted dominating sets.

Feng and Yin [9] also present results for exactly sampling from the hardcore model. Their algorithm is much slower than ours, taking $O(\log^3 n)$ rounds, and it also uses large messages and exponential-time local computations. However, their result holds for a wider range of the fugacity parameter, specifically when $\lambda < \frac{(\Delta-1)^{\Delta-1}}{(\Delta-2)^\Delta}$. Note that from [8], sampling from the hardcore model is hard in the LOCAL model for $\lambda > \frac{(\Delta-1)^{\Delta-1}}{(\Delta-2)^\Delta}$. In the sequential setting, the same threshold is a barrier between polynomial and non-polynomial sampling, unless RP=NP [11,31,32,34], given the connection between approximate counting and sampling [19,30]. Like us, Guo, Jerrum, and Liu [14] provide an $O(\log n)$-round w.h.p. CONGEST algorithm for exact sampling from the hardcore model using polynomial-time local computation; however, they require a smaller range of λ than us, specifically $\lambda \leq \frac{1}{2\sqrt{e}\Delta-1}$.

Our algorithm improves over both [9,14], as every vertex always outputs its label in an exact sample. The algorithm from [9] succeeds with high probability and returns an exact sample conditioned on success; however, it may fail and failures cannot be detected by every vertex locally. On the other hand, the algorithm from [14] always succeeds in a random amount of time like ours; however, a vertex cannot locally determine when its portion of the output is finalized.

2   Technical Preliminaries

2.1   Sampling Weighted Local CSPs

A weighted CSP on a graph $G = (V, E)$ consists of a set L of vertex labels and a collection $\mathcal{S} \subseteq 2^V$ of *constraint sets*, in addition to a *constraint* C_R for each

---

[2] Results on the hardcore often assume that algorithms run on graphs of constant degree. In this sense, we require $\lambda < \frac{1}{\Delta}$. Furthermore, on graphs of bounded degree, our results can likely be slightly improved by using the LubyGlauber chain from [8].

constraint set R. A labeling assigns an element in L to each vertex in V; thus L^V is the set of all labelings of G. For a labeling $\ell \in L^V$, $v \in V$, and $R \subseteq V$, we use $\ell(v)$ to denote the label that ℓ assigns to vertex v and $\ell \restriction_R$ to denote the restriction of ℓ to R. Each constraint C_R maps the set of all restricted labelings $\ell \restriction_R$ to the non-negative reals. The *weight* of labeling ℓ is $\prod_{R \in \mathcal{S}} C_R(\ell \restriction_R)$. We call a labeling *valid* if it has weight greater than zero. A *CSP* is typically defined as the problem of finding an arbitrary valid labeling. The *weighted CSP* is to sample valid labelings, where the probability of choosing a labeling is proportional to its weight. This probability distribution of labelings will be referred to as π.

For notational convenience, we will assume that $\mathcal{S}$ does not contain any singleton sets. Instead, we will assume that for each vertex v there is a separate unary constraint $b_v : L \to \mathbb{R}^{\geq 0}$, that maps a label assigned to v to a non-negative real number. Note that this is without loss of generality because for any $v \in V$, we can set $b_v(x) = 1$ for all $x \in L$. With this additional notation, the weight of a labeling ℓ can be written as

$$\prod_v b_v(\ell(v)) \cdot \prod_{R \in \mathcal{S}} C_R(\ell \restriction_R). \tag{1}$$

when there is a constant k such that every constraint set has a diameter in the graph G bounded by k, then the weighted CSP is called a *weighted local CSP*. Note that the diameter here refers to the distance in G, not the distance in the subgraph of G induced by R.

We consider three examples of weighted local CSPs in this paper: weighted independent sets (the hardcore model), weighted dominating sets, and (briefly) the Ising model.

– *Weighted independent sets* are a weighted local CSP taking a parameter $\lambda > 0$ commonly called the *fugacity*. Here, the set of labels $L = \{0, 1\}$, the vertices labeled 1 form the independent set. Each unary constraint b_v is the function $b_v(0) = 1, b_v(1) = \lambda$. The collection $\mathcal{S}$ of constraint sets is E, the set of all edges in the graph. For every edge $e = \{u, v\}$, the constraint $C_e(\ell(u), \ell(v)) = 0$ if $\ell(u) = \ell(v) = 1$; otherwise, $C_e(\ell(u), \ell(v)) = 1$. C_e is simply ensuring that the valid labelings are independent sets of G. Here an independent set of size x has weight λ^x. Thus all independent sets of the same size have uniform probability; however, for $\lambda < 1$ small independent sets have a higher probability than large independent sets.
– *Weighted dominating sets* are a weighted local CSP very similar to weighted independent sets. This CSP also takes a parameter $\lambda > 0$ and has the same set of labels and unary constraints. The collection $\mathcal{S}$ of constraint sets consists of inclusive neighborhoods $N_v = Nbr^+(v)$ for each vertex v. The constraint C_{n_v} maps to 1 if at least one of the vertices in the inclusive neighborhood has the label 1 and 0 otherwise.
– The *Ising model*, in a simple form as given by [8], is another weighted local CSP. Here there are two possible labels, $\{-1, 1\}$, for each vertex, and $b_v(-1) = b_v(1) = 1$. $\beta > 0$ is provided with the model. Each pair of adjacent vertices has a constraint that maps to β if the vertices have the same label and 1

otherwise. The weight given to a labeling is β^a, where a is the number of edges that have both of their vertices assigned the same label.

2.2 Related Work on Lower Bounds

The discussion thus far has been on upper bounds. However, there are interesting lower bounds for sampling from the hardcore model in the LOCAL model [8, 14]. Feng, Sun, and Yin show an $\Omega(Diam)$ round lower bound on n-vertex graphs with diameter $Diam = \Omega(n^{1/11})$, when $\lambda > \frac{(\Delta-1)^{\Delta-1}}{(\Delta-2)^\Delta}$. Guo, Jerrum, and Liu [14] show a more general $\Omega(\log n)$-round lower bound. Similar bounds exist or can be derived for other weighted local CSPs, see the two cited papers.

3 Distributed Markov Chain

In this section, we present a distributed Markov chain for sampling from weighted local CSPs (see Algorithm 1). In subsequent sections we show that it is possible to use this chain to exactly sample solutions of weighted local CSPs efficiently in the LOCAL model (and in the CONGEST model in some cases). Our Markov chain is a simple modification of the LOCALMETROPOLIS Markov chain, given in [8]. The LOCALMETROPOLIS chain contains a *propose* step in which each vertex v independently proposes a label σ_v in L with probability proportional to $b_v(\sigma_v)$. This is followed by a probabilistic *local filter* that is applied to each constraint set. We describe this in more detail in the next paragraph. If all the constraint sets containing a vertex v pass the local filter, then v adopts σ_v; otherwise it retains its old label. We modify this Markov chain by first marking each vertex independently with a fixed probability p and then allowing only the marked vertices to be active in each step. This idea – of randomly sampling vertices which will be active – is a standard idea in randomized distributed algorithms (for example, Luby's algorithm [1, 26]), but more to the point it was used in [7, 10] to speed up their distributed Markov chain for coloring. In particular, [7, 10] both present the chain resulting from augmenting the LOCALMETROPOLIS chain for colorings to have a set of marked vertices. Furthermore, we infer that the Markov chain we present is the generalization mentioned by [10].

We now describe the local filtering step. During each step of the chain, if a vertex v is marked then it has a current label X_v and a proposed label σ_v; otherwise, it only has a current label X_v. For each constraint set R, we now consider a collection $\mathcal{L}(R)$ of labelings of R. In particular, we call $(\ell(v_1), \ell(v_2), \ldots, \ell(v_{|R|}))$ a *potential* labeling of R, if each label $\ell(v_i)$ is chosen from either X_{v_i} or σ_{v_i}, and as long as at least one of the $|R|$ choices was made from σ. We now let $\mathcal{L}(R)$ be the collection of all potential labelings, with the note that it can contain the same element multiple times if there are multiple valid ways of choosing it. To be technically correct, each potential labeling should be represented as a binary vector, however we bend notation and treat a potential labeling as a labeling of the constraint set. Each constraint set R passes the local filter with probability

$$\prod_{\ell \in \mathcal{L}(R)} \frac{C_R(\ell)}{C_R^*}. \tag{2}$$

Here, C_R^* refers to the maximum value that the constraint can take over its entire domain L^R. C_R^* can be assumed to be nonzero; otherwise, every labeling would be invalid.

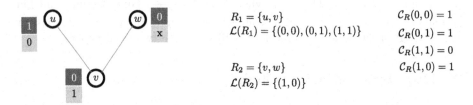

Fig. 1. Example illustrating the computation of local filter probabilities for the weighted independent set problem.

Figure 1 shows an example of how to apply the local filter to the weighted independent set problem. The figure shows a subgraph with 3 vertices and alongside each vertex we show its current label (top) and proposed label (bottom). Vertex w has "x" as its proposed label because it is not marked active in this iteration. Vertex v participates in two constraint sets R_1 and R_2 corresponding to the two edges incident on it. The potential labelings $\mathcal{L}(R_1)$ and $\mathcal{L}(R_2)$ of each constraint set are shown. Each tuple in $\mathcal{L}(R_1)$ (respectively, $\mathcal{L}(R_2)$) shows u's label (respectively, v's label) followed by $v's$ label (respectively, w's label). Note that tuple $(1, 0)$ does not appear in $\mathcal{L}(R_1)$ because that would correspond to both labels being chosen from the current labels. Finally, the values of the constraint function are shown on the right. In this case we use the same constraint function for every constraint set. Note that the expression in (2) evaluates to 0 for R_1 indicating that R_1 does not pass the local filter. For R_2 the local filter evaluates to 1, indicating that R_2 does pass the local filter. Since not all constraint sets of v pass the local filter, v does not update its label.

We next prove that the Markov chain given by Algorithm 1 is ergodic (aperiodic and irreducible) with π as its stationary distribution under a mild condition; for any two valid states X and Y, we have a sequence of valid states $X = Z_1, ..., Z_n = Y$, where adjacent states differ at only a single vertex. Recall that π is the distribution over labelings of G in which each labeling ℓ has probability proportional to its weight, expression (1). This proof is based on and an extension of the proof from [8]. The chain is clearly aperiodic since every state can transition to itself. Furthermore, the condition mentioned above ensures irreducibility. It remains to show that π is the stationary distribution. A standard way of showing that a distribution π is stationary for a Markov chain with transition matrix M is to prove the *detailed balance equations*,

Algorithm 1

Require: each vertex v initially has label X_v

each $v \in V$ is marked active with probability p

for each active vertex $v \in V$ **do**   ▷ Propose step

$\quad v$ chooses $\sigma_v \in L$ with probability proportional to $b_v(\sigma_v)$

end for

for each constraint set $R \in \mathcal{S}$ **do**   ▷ Local Filter step

$\quad R$ passes the local filter with probability $\prod_{\ell \in \mathcal{L}(R)} \frac{C_R(\ell \mid R)}{C_R^*}$

end for

for each active vertex $v \in V$ **do**   ▷ Finalize labels

$\quad$ **if** all constraint sets in $\mathcal{S}$ containing v pass their checks **then**

$\quad\quad X_v = \sigma_v$

$\quad$ **end if**

end for

$$\pi(X) \cdot M[X, Y] = \pi(Y) \cdot M[Y, X]$$

for all states of the chain X, Y. As noted in [8], a slightly stronger condition is needed for the chain to have the correct limit distribution if it starts in a state with weight 0. This point will not concern us, as we will view the chain as only being defined over valid states.

Theorem 1. *Let M be the transition matrix of the Markov chain defined by Algorithm 1. For all states X, Y we have $\pi(X) \cdot M[X, Y] = \pi(Y) \cdot M[Y, X]$.*

Proof. Let $X \neq Y$ be two valid labelings, thus X and Y are also states in the Markov chain. Let X_v be the current label of vertex v and σ_v be the proposed label if the vertex is marked, otherwise let $\sigma_v = \sqcup$. We also define the binary vector $\mathbf{I}$, where for each constraint set $T \in \mathcal{S}$, $\mathbf{I}_T = 1$ if the constraint set T passes its check and $\mathbf{I}_T = 0$ otherwise. If a vertex v is contained in any constraint set T such that $\mathbf{I}_T = 0$, we call v *restricted*. Note that $(\sigma, \mathbf{I})$ defines a function from one state to another.

Given a pair of states, X and Y, there could be many tuples $(\sigma, \mathbf{I})$ that map X to Y and similarly many tuples $(\sigma', \mathbf{I}')$ that map Y to X. We now show a bijection between tuples mapping X to Y and tuples mapping Y to X. Let $(\sigma, \mathbf{I})$ be a tuple mapping X to Y. From $(\sigma, \mathbf{I})$, we construct $(\sigma', \mathbf{I}')$, a function mapping Y to X as follows. For every vertex where $Y_v \neq X_v$, let $\sigma'_v = X_v$. For all other vertices, let $\sigma'_v = \sigma_v$. Finally, let $\mathbf{I} = \mathbf{I}'$. To see that $(\sigma', \mathbf{I}')$ maps Y to X, first note that for any vertex v where $X_v \neq Y_v$, we must have $\sigma_v = Y_v$ and v must be unrestricted in $\mathbf{I}$. This means that v is unrestricted in $\mathbf{I}'$ and since $\sigma'_v = X_v$, we see that $(\sigma', \mathbf{I}')$ maps Y_v to X_v. For a vertex v where $X_v = Y_v$, either (a) $\sigma_v = Y_v = X_v$, (b) $\sigma_v = \sqcup$, or (c) v is restricted in $\mathbf{I}$. In case (a), $\sigma'_v = Y_v = X_v$, in case (b) $\sigma'_v = \sqcup$, and in case (c) v is restricted in $\mathbf{I}'$. In all three cases, $(\sigma', \mathbf{I}')$ maps Y_v to X_v. It can be checked that this construction gives a bijection $(\sigma, \mathcal{C}) \leftrightarrow (\sigma', \mathcal{C}')$ between tuples that map from X to Y and those that map from Y to X.

It is now sufficient to show

$$\frac{P(\sigma)P(\mathbf{I}|\sigma, X))}{P(\sigma')P(\mathbf{I}'|\sigma', Y))} = \frac{\pi(Y)}{\pi(X)}. \tag{3}$$

This is because

$$\pi(X) \cdot M[X, Y] = \pi(X) \sum P(\sigma)P(\mathbf{I}|\sigma, X)) = \pi(Y) \sum P(\sigma')P(\mathbf{I}'|\sigma', X)) = \pi(Y)M[Y, X],$$

where the middle equality follows from (3).

To show (3), we first observe that

$$\frac{P(\sigma)}{P(\sigma')} = \prod_{v | X_v \neq Y_v} \frac{p \cdot b_v(Y_v)}{p \cdot b_v(X_v)} = \prod_v \frac{b_v(Y_v)}{b_v(X_v)}.$$

We now consider $\frac{P(\mathbf{I}|\sigma, X))}{P(\mathbf{I}'|\sigma', Y))}$. Since each constraint is passed or failed independently,

$$\frac{P(\mathbf{I}|\sigma, X))}{P(\mathbf{I}'|\sigma', Y))} = \prod_{T \in \mathcal{S}} \frac{P(\mathbf{I}_T|\sigma, X))}{P(\mathbf{I}'_T|\sigma', Y))}.$$

There are now two cases to consider.

Case $\mathbf{I}_T = 0$. In this case, every vertex in the constraint set T is restricted. For each of these vertices, we must have $X_v = Y_v$ and also $\sigma_v = \sigma'_v$. This means that the set of potential labelings $\mathcal{L}(T)$ used in the local filter probability (2) are identical for the chain in state X and the chain in state Y. Thus $P(\mathbf{I}_T = 0|\sigma, X) = P(\mathbf{I}'_T = 0|\sigma', Y)$. Therefore, we can can rewrite the ratio $P(\mathbf{I}_T = 0|\sigma, X)/P(\mathbf{I}'_T = 0|\sigma', Y)$ as

$$\frac{P(\mathbf{I}_T = 0|\sigma, X))}{P(\mathbf{I}'_T = 0|\sigma', Y))} = 1 = \frac{C_T(Y_{v_1}, \ldots, Y_{v_{|T|}})}{C_T(X_{v_1}, \ldots, X_{v_{|T|}})}.$$

Case $\mathbf{I}_T = 1$. We establish a mapping between potential labels in set $\mathcal{L}(T)$ for state X and potential labels in set $\mathcal{L}(T)$ for state Y that is almost a bijection. Let $\mathcal{X} = (\mathcal{X}_{v_1}, \mathcal{X}_{v_2}, \ldots, \mathcal{X}_{v_{|T|}})$ be a potential labeling for state X. From $\mathcal{X}$, we create a potential labeling $\mathcal{Y} = (\mathcal{Y}_{v_1}, \mathcal{Y}_{v_2} \ldots, \mathcal{Y}_{v_{|T|}})$ for state Y, as follows.

- If $\mathcal{X}_{v_i}$ was chosen from X_{v_i} and $X_{v_i} = Y_{v_i}$ we can let $\mathcal{Y}_{v_i}$ be chosen from Y_{v_i}.
- If $\mathcal{X}_{v_i}$ was chosen from X_{v_i} and $X_{v_i} \neq Y_{v_i}$ we can let $\mathcal{Y}_{v_i}$ be chosen from σ'_i. Note that in this case, $\sigma'_{v_i} = X_{v_i}$.
- If $\mathcal{X}_{v_i}$ was chosen from σ_{v_i} and $X_{v_i} = Y_{v_i}$ we can let $\mathcal{Y}_{v_i}$ be chosen from σ'_{v_i}.
- If $\mathcal{X}_{v_i}$ was chosen from σ_{v_i} and $X_{v_i} \neq Y_{v_i}$ we can let $\mathcal{Y}_{v_i}$ be chosen from Y_{v_i}.

Note that in all 4 cases, $\mathcal{X}_{v_i} = \mathcal{Y}_{v_i}$.

Note that $\mathcal{Y}$ is a potential labeling from $\mathcal{L}(T)$ as long as not every choice was from Y_{v_i}. There is exactly one choice for the labels in $\mathcal{X} = (\mathcal{X}_{v_1}, \mathcal{X}_{v_2}, \ldots, \mathcal{X}_{v_{|T|}})$ that is mapped to $\mathcal{Y} = (Y_{v_1}, Y_{v_2}, \ldots, Y_{v_{|T|}})$. In this choice, every $\mathcal{X}_{v_i}$ was chosen from σ_{v_i} when $X_{v_i} \neq Y_{v_i}$ and X_i when

$X_{v_i} = Y_{v_i}$. We denote this label $\mathcal{X}'$ and see that $\mathcal{X}' = (Y_{v_1}, Y_{v_2}, \ldots, Y_{v_{|T|}})$. Furthermore, since $X \neq Y$ at least one choice for $\mathcal{X}'$ was made from σ so $\mathcal{X}'$ is a valid labeling.

Now recall that $P(\mathbf{I}_T = 1|\sigma, X) = \prod_{\ell \in \mathcal{L}(T)} \frac{C_T(\ell|T)}{C_T^*}$ according to (2). Thus, in the ratio $P(\mathbf{I}_T = 1|\sigma, X)/P(\mathbf{I}_T' = 1|\sigma, Y)$, all terms in the numerator cancel out except for $C_T(Y_{v_1}, Y_{v_2}, \ldots, Y_{v_{|T|}})$. By a symmetric argument, there is a single term $C_T(X_{v_1}, X_{v_2}, \ldots, X_{v_{|T|}})$ left in the denominator.

Altogether, this shows

$$\frac{P(\mathbf{I}_T = 1|\sigma, X))}{P(\mathbf{I}_T' = 1|\sigma', Y))} = \frac{C_T(Y_{v_1}, \ldots, Y_{v_{|T|}})}{C_T(X_{v_1}, \ldots, X_{v_{|T|}})},$$

completing the proof.

4   Distributed Coupling from the Past

4.1   Coupling from the Past

Coupling from the past (CFTP), introduced in [29], is a technique for sampling *exactly* from the stationary distribution of an ergodic Markov chain. A chapter covering coupling from the past can be found in [23]. Suppose the Markov chain is defined over a set of states Ω and has a transition matrix M. Following the notation in [33], let $f : \Omega \times \{0, 1\}^* \to \Omega$ be a function such that $P(f(X, r) = Y) = M[X, Y]$, when r is a string chosen uniformly at random from $\{0, 1\}^*$. The function f is called a *random mapping* representation of the transition matrix M and is known to always exist (see Proposition 1.5 in [23]).

The function f allows us to use a common source of randomness and evolve a chain beginning from each state $\sigma \in \Omega$ in a "coupled" manner. More precisely, for integer t, let $r_t \in \{0, 1\}^*$ be chosen uniformly at random. Let $f_t : \Omega \to \Omega$ be the function $f_t(X) = f(X, r_t)$. Now define a function $F_{t_1}^{t_2} : \Omega \to \Omega$ for integers $t_2 \geq t_1$ as

$$F_{t_1}^{t_2}(X) = (f_{t_2-1} \circ f_{t_2-2} \circ \cdots \circ f_{t_1})(X) = f_{t_2-1}(f_{t_2-2}(\ldots f_{t_1}(X))).$$

The function $F_{t_1}^{t_2}$ defines a coupled evolution of states from time t_1 to time t_2 with the property that $P(F_{t_1}^{t_2}(X) = Y) = M^{t_2-t_1}[X, Y]$. For each $\sigma \in \Omega$, $F_0^t(\sigma)$ defines the state at time t of a Markov chain beginning at σ and with transition matrix M. The entire collection of Markov chains that evolve in this manner, one starting in each state of Ω and using common random strings, is called a *grand coupling* [23].

Now consider the function

$$F_{-T}^0(X) = (f_{-1} \circ f_{-2} \circ \cdots \circ f_{-T})(X) = f_{-1}(f_{-2}(\ldots f_{-T}(X))).$$

The insight of the CFTP technique is that if we can show that with probability 1 there exists a T such that F_{-T}^0 is a constant function (i.e., every state in Ω

is mapped to the same state by F^0_{-T}), then the unique element in this image is drawn exactly from the stationary distribution π of the Markov chain. We now provide some intuition for this possibly surprising claim. First note that if $F^0_{-T_1}$ maps all elements in Ω to $\omega \in \Omega$, then for any $T_2 > T_1$, the function $F^0_{-T_2}$ also maps all elements in Ω to ω. Note that this is not true for the symmetric and invalid sampling technique, "coupling to the future", where we use F^T_0 instead of F^0_{-T}. The critical difference here is that even if F^T_0 becomes constant for a sufficiently large T, we may not have $F^{T+1}_0 = F^T_0$. We can therefore intuitively think of CFTP as computing $F^0_{-\infty}$. Since the stationary distribution is the limit distribution of an ergodic Markov chain, the unique element in the image of $F^0_{-\infty}$ is drawn exactly from the stationary distribution.

This insight suggests a natural algorithm for exact sampling. Starting with $T' = 1$, compute $F^0_{-T'}(\Omega)$ and check if $|F^0_{-T'}(\Omega)| = 1$. If it is, we output the unique element (state) in $F^0_{-T'}(\Omega)$, otherwise we double the value of T' and repeat. Propp and Wilson point out that by choosing to double T' at each iteration we only overshoot the smallest value of T where $|F^0_{-T}(\Omega)| = 1$ by a constant multiple [29]. To avoid biasing the samples it is critical that the same choice of functions $f_{-1}, f_{-2}, \ldots$ is used to compute $F^0_{-T'}$ each time T' is increased and that the process is not stopped even if T' grows large without $F^0_{-T'}$ becoming constant.

Given that $|\Omega|$ can be exponentially large relative to the size of the input, a problem with this algorithm is efficiently checking if $F^0_{-T'}$ is a constant function. Another issue is that in general the final value of T' needed may be large.

4.2   Bounding Chains

It is easy to check if F^0_{-T} is a constant function in the special case where the grand coupling has a property called *monotonicity* [29]; however, we want to sample from weighted local CSPs where monotonicity may not exist. Häggström and Nelander [16] and Huber [17] describe the *bounding chain* technique for determining when coupling from the past gives a constant function. The idea, is that for each vertex we compute a superset of the labels it could be assigned by F^0_{-T} for an unknown valid input labeling. If this label set is a singleton for every vertex, F^0_{-T} is a constant function. Note that the converse does not necessarily hold. When the label set of every vertex is a singleton, we call F^0_{-T} singular (acknowledging the slight abuse of notation, since singularity also depends on the method of computing the label sets). We can implement CFTP by checking whether F^0_{-T} is singular, instead of constant. Note that a trivial choice for each label set is L, however, this will lead to an algorithm that never terminates.

To give a concrete practical example, consider sampling from the hardcore model in the sequential setting using bounding chains [16]. One step of the Markov chain is defined as follows. A uniformly random vertex v is selected and a coin is flipped with probability of heads $\frac{\lambda}{1+\lambda}$. If the coin flip is tails then v is removed from the independent set. If the coin flip is heads and if no neighbors of v are in the current independent set, then v joins the independent set. The random mapping representation of this Markov chain, f, can be defined in a natural way with the random string r encoding a vertex and a coin flip, chosen with the correct distribution from all (vertex, coin flip) pairs.

We now compute a superset of labels each vertex can be assigned by F^0_{-T}. We use a recursive approach. In the trivial base case, $T = 0$, every vertex receives the label set $\{0, 1\}$, indicating that in some states the vertex is outside the independent set and in some states the vertex is inside the independent set. For $T \geq 1$, we first compute the label set for every vertex for F^0_{-T+1}. If the operation of f_{-T} is the removal of vertex v from the independent set, we can be sure that regardless of the input state, v has the label 0. Thus v's set of possible labels is set to $\{0\}$. On the other hand, when f_{-T} represents v attempting to join the independent set, there are a few cases. If every neighboring vertex of v has the label set $\{0\}$ from the recursive call, then v joins the independent set regardless of the input state, and is assigned the label set $\{1\}$. If any neighboring vertex of v has the label set $\{1\}$ from the recursive call, the join will fail regardless of the input state, and v will get the label set $\{0\}$. In the remaining case, whether the join succeeds may depend on the input state. Thus we assign v the label set $\{0, 1\}$. Every other vertex simply keeps its set from the recursive call. It is shown in [16], that singularity occurs for this process for $T = O(n \log n)$ in expectation, as long as the fugacity $\lambda \leq \frac{\alpha}{\Delta - 1}$, for any constant $\alpha < 1$. This also holds with high probability.

4.3 Distributed Bounding Chains

This section contains our main result in which we show how to apply CFTP and the bounding chain technique to the distributed Markov chain defined in Algorithm 1. This yields Theorem 2 which shows that if the weighted local CSP satisfies a general condition we can sample exactly in $O(\log n)$ rounds in the LOCAL model. With additional conditions we get an $O(\log n)$-round CONGEST algorithm. Note that on some weighted local CSPs, such as colorings, this algorithm will run forever. Furthermore, as stated earlier, terminating long running instances of the algorithm before a sample is returned may bias the results.

We first note that there is a clear choice of a random mapping representation of the chain described by Algorithm 1. In Algorithm 1, three random choices are made: (i) each $v \in V$ is marked active with probability p, (ii) each active vertex v picks a label σ_v with probability proportional to $b_v(\sigma_v)$, and (iii) each constraint set R passes the local filter with a certain probability (defined in Eq. (2)). These random choices can be collectively specified by string $r \in \{0, 1\}^*$ chosen uniformly at random. Since we want to execute this algorithm in the LOCAL model, we note that r_t can be generated in a distributed fashion. Each vertex v can locally pick a random bit specifying whether it is marked active and choose σ_v at random from L proportional to its b_v values. To determine if the constraint set R passes the local filter, we could have one vertex in R (e.g., the vertex with the highest ID in R) pick a number uniformly at random from $[0, 1]$. To avoid worrying about precision, we actually let this vertex generate a binary table (with arbitrary precision) specifying whether the constraint will pass or not given every possible combination of current labels and proposals.

The algorithm starts with $T' = 1$ and every vertex having label set $\{l \in L : b_v(l) > 0\}$. The algorithm then proceeds in *stages* $1, 2, \ldots$. After each stage, T' is doubled. At the start of a stage, some vertices have already output a label; we call these vertices *coalesced* and the remaining vertices *uncoalesced*. Each stage is initiated by uncoalesced vertices. The coalesced vertices are in a "stand by" mode for the stage and will only become active if and when prompted by uncoalesced vertices. At the end of a stage, any uncoalesced vertices that now have singleton label sets output the unique label in their label sets. The algorithm is complete after every vertex outputs a label.

Each stage consists of two phases, a *preprocessing* phase and a *main* phase. We now describe both of these phases separately. Recall that k is the largest diameter for any constraint.

Preprocessing Phase. First, each uncoalesced vertex v notifies every vertex w in its (inclusive) kT' hop neighberhood that w will be active in the stage. Each active vertex also learns its shortest path length from an uncoalesced vertex. Next, each active vertex generates its portion of the random strings $r_{-T'}, \ldots, r_{-(T'/2)-1}$. These correspond to functions $f_{-(T'/2+1)}, f_{-(T'/2+2)}, \ldots, f_{-T'}$. Note that the random strings r_i, $i > -(T'/2) - 1$, are retained from the previous stage.

Main Phase. Every active vertex begins the phase by resetting its label set to $\{l \in L : b_v(l) > 0\}$. The main phase is composed of T' steps $\{0, \ldots, T' - 1\}$.

At step i, every active vertex v with distance at most $k(T' - 1 - i)$ from an uncoalesced vertex collects the part of $r_{-T'+i}$ as well as the current label set from every vertex in its k hop neighborhood. Now v has enough information to compute the label $f_{-T'+i}$ assigns to v for every labeling of its k hop neighborhood. v can now update its label set to be the union of the label it is assigned by $f_{-T'+i}$ for every *possible* labeling of its k hop neighborhood. A labeling is only possible if every vertex is given a label from its label set.

At the end of the phase, any vertices that have singleton label sets output the single label in their label set.

4.4 Analysis

At the end of the main phase, every vertex that was uncoalesced at the beginning of the stage has computed an upper bound on the set of labels that it could possibly be assigned by $F^0_{-T'}$. Now consider a vertex v that has a singleton label set at the end of the main phase. This means that regardless of the input labeling, $F^0_{-T'}$ assigns v a single label l. Since functions are composed in a "backwards" order, we also know that $F^0_{-T''}$ assigns v the single label l for all $T'' > T'$. Once every vertex has output a label, we see that $F^0_{-T'}$ is singular. Furthermore, the labeling output by the vertices is the unique labeling in the image of $F^0_{-T'}$. Therefore the output labeling is exactly drawn from π, the desired sampling distribution, since π is also the stationary distribution of the Markov chain by Theorem 1.

The algorithm runs in $O(k \cdot T^*)$ rounds in the LOCAL model, where T^* is the smallest value such that $F^0_{-T^*}$ is singular. When $k = 1$ and the set of labels for the vertices has constant size, the algorithm runs in $O(T^*)$ rounds in the CONGEST model. In theory, our algorithm requires exponential work per machine; however, for some practical examples such as the hardcore model and weighted dominating sets only polynomial work is required per machines. This is because we can determine all of the possible labels of a vertex with a simple rule.

We now prove a theorem that shows that in some cases $T^* = O(\log n)$ with high probability. This theorem statement is similar to Theorem 2 from [16]; however, our method of proof is slightly different. Following the lead from that paper (they credit some ideas to Murdoch and Green [27]), we choose γ to be a lower bound, over every vertex v, on the probability that f_i assigns v a single label $l \in L$ regardless of the input labeling, conditioned on v being marked active. We also choose β to be an upper bound, over every vertex v, on the probability that for any two labelings l_1, l_2 with $l_1(v) = l_2(v)$ we have $(f_i(l_1))(v) \neq (f_i(l_2))(v)$, conditioned on v being marked active. Intuitively, γ and β describe the likelihood of a vertex moving towards or respectively away from a singleton label.

Theorem 2. *The distributed bounding chain has* $T^* = O(\frac{1}{p\gamma - \Delta_k p\beta} \log n)$ *with high probability if* $\gamma > \Delta_k \beta$, *where* Δ_i *is the number of nodes in the largest (exclusive) i hop neighborhood of the graph and k is the maximum diameter of any constraint.*

Proof. We will use the standard trick of considering the value of T needed for F^T_0 to be singular, instead of directly considering T^*. While 'Coupling to the Future' is not a valid sampling technique, the distributions of time needed for singularity are the same for F^T_0 and F^0_{-T}. Let Y^v_t be the indicator random variable for whether vertex v is always given a singleton label set by F^t_0. Let Y_t be the sum of all the Y^v_t. Assuming that the set of possible labels for each vertex contains at least two elements, we have $Y_0 = n$. The only way for a vertex with a singleton label set to grow in size, is if it shares a constraint with a vertex with a non-singleton label set. We can now see

$$E[Y_{t+1}] \leq Y_t + Y_t \Delta_k p\beta - Y_t p\gamma = Y_t(1 + \Delta_k p\beta - p\gamma) = Y_t(1 - (p\gamma - \Delta_k p\beta)).$$

We want to have $p\gamma - \Delta_k p\beta > 0$, which is equivalent to the condition $\gamma > \Delta_k \beta$. Setting $\alpha = (p\gamma - \Delta_k p\beta)$, we now have, for $0 < \alpha < 1$,

$$E[Y_{t+1}] = \sum_{i=0}^{n} E[Y_{t+1}|Y_t = i]P(Y_t = i) \leq \sum_{i=0}^{n}(1 - \alpha)iP(Y_t = i) = (1 - \alpha)E[Y_t].$$

By induction, it follows that $E[Y_t] \leq (1 - \alpha)^t n$. By Markov's inequality, for $T = O(\frac{1}{\alpha} \log n)$, it follows that $P(Y_T \geq 1) \leq \frac{1}{n}$, which completes the proof.

Corollary 1. *When* $\gamma - \Delta_k \beta > r$ *for some constant $r > 0$, the algorithm runs in $O(k \log n)$ rounds in the LOCAL model. Furthermore, if $k = 1$ and L has finite size, the algorithm runs in $O(\log n)$ rounds in the CONGEST model.*

5   Results for Specific CSPs

Theorem 3. *An $O(\log n)$ round* CONGEST *algorithm exists for sampling from the hardcore model when the fugacity $\lambda \leq \frac{\alpha}{\Delta}$ for any constant $\alpha < 1$.*

Proof. Note that a choice of γ is $\frac{1}{1+\lambda}(1 - \Delta p \frac{\lambda}{1+\lambda})$, since a vertex will always accept a proposal to leave the independent set if no neighboring vertex is marked and proposing to join the independent set. A choice of β is $\frac{\lambda}{1+\lambda}$, since the label set of a vertex can only grow if the vertex is proposing to join the independent set. Finally, $\Delta_k = \Delta$. Thus we need $\gamma - \Delta\beta > r > 0$, for some constant r.

Note that

$$\frac{1}{1+\lambda}\Big(1 - \Delta p \frac{\lambda}{1+\lambda}\Big) - \Delta\frac{\lambda}{1+\lambda} \geq \frac{1}{1+\lambda}\Big(1 - p\frac{\alpha}{1+\lambda}\Big) - \frac{\alpha}{1+\lambda} =$$

$$\frac{1-\alpha}{1+\lambda} - \frac{p\alpha}{(1+\lambda)^2} \geq \frac{1 - \alpha - p\alpha}{1+\lambda} \geq \frac{1 - \alpha(1+p)}{1+\alpha}$$

Choosing p small enough we are done.

Theorem 4. *An $O(\log n)$* CONGEST *algorithm exists for sampling weighted dominating sets when $\lambda \geq \alpha\Delta^2$ for any constant $\alpha > 1$.*

Proof. While there are now constraints that are not unary or binary, we are still able to run the algorithm in the CONGEST model. This is because necessary information can be aggregated at the vertex at the center of each constraint and then distributed to the other vertices of the constraint.

Similar to the hardcore model we can choose $\gamma = \frac{\lambda}{1+\lambda}(1 - \Delta_k p \frac{1}{1+\lambda})$ and $\beta = \frac{1}{1+\lambda}$. Note that here $\Delta_k = \Delta + \Delta(\Delta - 1) = \Delta^2$. Again it is sufficient for $\gamma - \Delta_k\beta > r > 0$, for some constant r.

$$\frac{\lambda}{1+\lambda}\Big(1 - \Delta_k p \frac{1}{1+\lambda}\Big) - \Delta_k\frac{1}{1+\lambda} \geq \frac{\alpha\Delta_k}{1+\alpha\Delta_k}\Big(1 - \frac{\Delta_k p}{1+\alpha\Delta_k}\Big) - \frac{\Delta_k}{1+\alpha\Delta_k} =$$

$$\frac{\Delta_k}{1+\alpha\Delta_k}\Big(\alpha\big(\frac{1 + \Delta_k(\alpha - p)}{1+\alpha\Delta_k}\big) - 1\Big) \geq \frac{1}{1+\alpha}\Big(\alpha\big(\frac{\alpha - p}{\alpha}\big) - 1\Big) = \frac{a - p - 1}{1+a}$$

We are done as long as p is sufficiently small.

Remark 1. The simplified Ising model with parameter $\beta > 1$ remains monotone in the distributed setting; see [29] for a more detailed explanation in the sequential setting. Our algorithm can be modified to give samples from the Ising model. Instead of keeping track of potential labels for each vertex, each vertex keeps track of its label when the input of F_{-T} is $\top$ and $\bot$. When they are equal, the vertex outputs this label. We don't have a bound on the runtime of this approach; however, combining the analysis of monotone CFTP [29] and the remark of [10][3] may imply an efficient, $O(\log^2 n)$ round, CONGEST algorithm for some range of β.

---

[3] They mention that the Dobrushin condition [6] is enough for fast mixing of their chain; however, we cannot guarantee that we are using the generalization they mention and also they don't provide a proof of their claim.

6   Conclusion

We have shown that for certain weighted local CSPs, CFTP combined with the bounding chain technique allows efficient exact sampling in the LOCAL and sometimes even the CONGEST model. Previous work [9,14] that achieved results for exact distributed sampling used very different techniques, so a conceptual contribution of our paper is showing that CFTP and bounding chains should also be added to our toolkit for exact distributed sampling. We wish to highlight two open questions suggested by this work.

Does there exist a logarithmic-round CONGEST or LOCAL algorithm that exactly samples uniform colorings from a palette of size $O(\Delta)$? In the sequential setting, one of the original papers describing bounding chains showed that there is a polynomial-time exact coloring algorithm when the number of colors is $\Theta(\Delta^2)$ [17]. Very recent work on bounding chains in the sequential setting has shown that uniform exact sampling of colorings is possible in polynomial time using only $\Theta(\Delta)$ colors [4,18].

Compared to the CONGEST model, many problems can be solved much faster in "all-to-all" communication models such as the CONGESTEDCLIQUE model [13,15,20,25]. A question suggested by recent work on distributed sampling is whether much faster distributed sampling algorithms – exact or approximate – can be designed for the CONGESTEDCLIQUE model.

References

1. Alon, N., Babai, L., Itai, A.: A fast and simple randomized parallel algorithm for the maximal independent set problem. J. Algorithms **7**(4), 567–583 (1986). https://doi.org/10.1016/0196-6774(86)90019-2
2. Awerbuch, B., Luby, M., Goldberg, A.V., Plotkin, S.A.: Network decomposition and locality in distributed computation. In: Proceedings of the 30th Annual Symposium on Foundations of Computer Science (SFCS 1989), pp. 364–369, IEEE Computer Society, USA (1989). https://doi.org/10.1109/SFCS.1989.63504
3. Barenboim, L., Elkin, M., Pettie, S., Schneider, J.: The locality of distributed symmetry breaking. J. ACM **63**(3), 1–45 (2016). https://doi.org/10.1145/2903137
4. Bhandari, S., Chakraborty, S.: Improved bounds for perfect sampling of k-colorings in graphs. In: Proceedings of the 52nd Annual ACM SIGACT Symposium on Theory of Computing (STOC 2020), pp. 631–642, Association for Computing Machinery, New York, NY, USA (2020). https://doi.org/10.1145/3357713.3384244
5. Chang, Y.J., Kopelowitz, T., Pettie, S.: An exponential separation between randomized and deterministic complexity in the local model. SIAM J. Comput. **48**(1), 122–143 (2019). https://doi.org/10.1137/17M1117537
6. Dobruschin, P.L.: The description of a random field by means of conditional probabilities and conditions of its regularity. Theory Probab. Appl. **13**(2), 197–224 (1968)
7. Feng, W., Hayes, T.P., Yin, Y.: Distributed symmetry breaking in sampling (optimal distributed randomly coloring with fewer colors) (2018). https://doi.org/10.48550/ARXIV.1802.06953
8. Feng, W., Sun, Y., Yin, Y.: What can be sampled locally? Distrib. Comput. **33**(3), 227–253 (2020)

9. Feng, W., Yin, Y.: On local distributed sampling and counting. In: Proceedings of the 2018 ACM Symposium on Principles of Distributed Computing (PODC 2018), pp. 189–198. Association for Computing Machinery, New York, NY, USA (2018). https://doi.org/10.1145/3212734.3212757

10. Fischer, M., Ghaffari, M.: A simple parallel and distributed sampling technique: local glauber dynamics. In: Schmid, U., Widder, J. (eds.) 32nd International Symposium on Distributed Computing (DISC 2018). Leibniz International Proceedings in Informatics (LIPIcs), vol. 121, pp. 26:1–26:11. Schloss Dagstuhl-Leibniz-Zentrum fuer Informatik, Dagstuhl, Germany (2018). https://doi.org/10.4230/LIPIcs.DISC.2018.26

11. Galanis, A., Štefankovič, D., Vigoda, E.: Inapproximability of the partition function for the antiferromagnetic Ising and hard-core models. Comb. Probab. Comput. **25**(4), 500–559 (2016). https://doi.org/10.1017/S0963548315000401

12. Ghaffari, M.: An improved distributed algorithm for maximal independent set. In: Proceedings of the Twenty-Seventh Annual ACM-SIAM Symposium on Discrete Algorithms (SODA 2016), pp. 270–277. Society for Industrial and Applied Mathematics, USA (2016)

13. Ghaffari, M., Parter, M.: MST in log-star rounds of congested clique. In: Proceedings of the 2016 ACM Symposium on Principles of Distributed Computing (PODC 2016), Chicago, IL, USA, 25–28 July 2016, pp. 19–28 (2016). https://doi.org/10.1145/2933057.2933103

14. Guo, H., Jerrum, M., Liu, J.: Uniform sampling through the Lovasz local lemma. In: Proceedings of the 49th Annual ACM SIGACT Symposium on Theory of Computing (STOC 2017), pp. 342–355, Association for Computing Machinery, New York, NY, USA (2017). https://doi.org/10.1145/3055399.3055410

15. Hegeman, J.W., Pandurangan, G., Pemmaraju, S.V., Sardeshmukh, V.B., Scquizzato, M.: Toward optimal bounds in the congested clique: graph connectivity and MST. In: Proceedings of the 2015 ACM Symposium on Principles of Distributed Computing (PODC 2015), pp. 91–100. ACM, New York, NY, USA (2015). https://doi.org/10.1145/2767386.2767434

16. Häggström, O., Nelander, K.: On exact simulation of Markov random fields using coupling from the past. Scand. J. Stat. **26**(3), 395–411 (1999). https://www.jstor.org/stable/4616564

17. Huber, M.: Exact sampling and approximate counting techniques. In: Proceedings of the Thirtieth Annual ACM Symposium on Theory of Computing (STOC 1998), pp. 31–40. Association for Computing Machinery, New York, NY, USA (1998). https://doi.org/10.1145/276698.276709

18. Jain, V., Sah, A., Sawhney, M.: Perfectly sampling $k \geq (8/3 + o(1))\Delta$-colorings in graphs. In: Proceedings of the 53rd Annual ACM SIGACT Symposium on Theory of Computing (STOC 2021), pp. 1589–1600. Association for Computing Machinery, New York, NY, USA (2021). https://doi.org/10.1145/3406325.3451012

19. Jerrum, M.R., Valiant, L.G., Vazirani, V.V.: Random generation of combinatorial structures from a uniform distribution. Theoret. Comput. Sci. **43**, 169–188 (1986)

20. Jurdziński, T., Nowicki, K.: MST in O(1) rounds of congested clique. In: Proceedings of the Twenty-Ninth Annual ACM-SIAM Symposium on Discrete Algorithms (SODA 2018), pp. 2620–2632. Society for Industrial and Applied Mathematics, Philadelphia, PA, USA (2018). https://dl.acm.org/citation.cfm?id=3174304.3175472

21. Kuhn, F., Moscibroda, T., Wattenhofer, R.: Local computation: lower and upper bounds. J. ACM **63**(2), 1–44 (2016). https://doi.org/10.1145/2742012

22. Kuhn, F., Moscibroda, T., Wattenhofer, R.: What cannot be computed locally! In: Proceedings of the Twenty-Third Annual ACM Symposium on Principles of Distributed Computing (PODC 2004), pp. 300–309. Association for Computing Machinery, New York, NY, USA (2004). https://doi.org/10.1145/1011767.1011811

23. Levin, D.A., with, Y.P.: Markov chains and mixing times. American Mathematical Society (2017). https://doi.org/10.1090/mbk/107

24. Linial, N.: Locality in distributed graph algorithms. SIAM J. Comput. **21**(1), 193–201 (1992). https://doi.org/10.1137/0221015

25. Lotker, Z., Patt-Shamir, B., Pavlov, E., Peleg, D.: Minimum-weight spanning tree construction in $O(\log \log n)$ communication rounds. SIAM J. Comput. **35**(1), 120–131 (2005)

26. Luby, M.: A simple parallel algorithm for the maximal independent set problem. In: Proceedings of the Seventeenth Annual ACM Symposium on Theory of Computing (STOC 1985), pp. 1–10. Association for Computing Machinery, New York, NY, USA (1985). https://doi.org/10.1145/22145.22146

27. Murdoch, D.J., Green, P.J.: Exact sampling from a continuous state space. Scand. J. Stat. **25**(3), 483–502 (1998). https://www.jstor.org/stable/4616516

28. Naor, M., Stockmeyer, L.: What can be computed locally? In: Proceedings of the Twenty-Fifth Annual ACM Symposium on Theory of Computing (STOC 1993), pp. 184–193. Association for Computing Machinery, New York, NY, USA (1993). https://doi.org/10.1145/167088.167149

29. Propp, J.G., Wilson, D.B.: Exact sampling with coupled Markov chains and applications to statistical mechanics. Random Struct. Algorithms **9**(1–2), 223–252 (1996)

30. Sinclair, A., Jerrum, M.: Approximate counting, uniform generation and rapidly mixing Markov chains. Inf. Comput. **82**(1), 93–133 (1989)

31. Sly, A.: Computational transition at the uniqueness threshold. In: Proceedings of the 2010 IEEE 51st Annual Symposium on Foundations of Computer Science (FOCS 2010), pp. 287–296. IEEE Computer Society, USA (2010). https://doi.org/10.1109/FOCS.2010.34

32. Sly, A., Sun, N.: Counting in two-spin models on d-regular graphs. Ann. Probab. **42**(6), 2383–2416 (2014). https://www.jstor.org/stable/24519110

33. Vigoda, E.: Lecture notes in Markov chain Monte Carlo methods, Georgia tech (2006). https://faculty.cc.gatech.edu/~vigoda/MCMC_Course/CouplingFromPast.pdf

34. Weitz, D.: Counting independent sets up to the tree threshold. In: Proceedings of the Thirty-Eighth Annual ACM Symposium on Theory of Computing (STOC 2006), pp. 140–149. Association for Computing Machinery, New York, NY, USA (2006). https://doi.org/10.1145/1132516.1132538

Weighted Packet Selection for Rechargeable Links in Cryptocurrency Networks: Complexity and Approximation

Stefan Schmid[1]($\boxtimes$) (iD), Jakub Svoboda[2]($\boxtimes$) (iD), and Michelle Yeo[2]($\boxtimes$) (iD)

[1] Technische Universität Berlin, Berlin, Germany
stefan.schmid@tu-berlin.de
[2] Institute of Science and Technology, Klosterneuburg, Austria
{jsvoboda,myeo}@ist.ac.at

Abstract. We consider a natural problem dealing with weighted packet selection across a rechargeable link, which e.g., finds applications in cryptocurrency networks. The capacity of a link (u, v) is determined by how much nodes u and v allocate for this link. Specifically, the input is a finite ordered sequence of packets that arrive in both directions along a link. Given (u, v) and a packet of weight x going from u to v, node u can either accept or reject the packet. If u accepts the packet, the capacity on link (u, v) decreases by x. Correspondingly, v's capacity on (u, v) increases by x. If a node rejects the packet, this will entail a cost affinely linear in the weight of the packet. A link is "rechargeable" in the sense that the total capacity of the link has to remain constant, but the allocation of capacity at the ends of the link can depend arbitrarily on the nodes' decisions. The goal is to minimise the sum of the capacity injected into the link and the cost of rejecting packets. We show that the problem is NP-hard, but can be approximated efficiently with a ratio of $(1 + \varepsilon) \cdot (1 + \sqrt{3})$ for some arbitrary $\varepsilon > 0$.

Keywords: network algorithms · approximation algorithms · complexity · cryptocurrencies · payment channel networks

1 Introduction

This paper considers a novel and natural throughput optimization problem where the goal is to maximise the number of packets routed through a network. The problem variant comes with a twist: link capacities are "rechargeable", which is primarily motivated by payment-channel networks routing cryptocurrencies.

We confine ourselves to a single capacitated network link and consider a finite ordered sequence of packet arrivals in both directions along the link. This can be modelled by a graph that consists of a single edge between two vertices u and v, where b_u and b_v represent the capacity u and v inject into the edge respectively. Each packet in the sequence has a weight (or value) and a direction

© The Author(s), under exclusive license to Springer Nature Switzerland AG 2023
S. Rajsbaum et al. (Eds.): SIROCCO 2023, LNCS 13892, pp. 576–594, 2023.
https://doi.org/10.1007/978-3-031-32733-9_26

(either going from u to v, or from v to u). When u forwards a packet going in the direction u to v, u's capacity b_u decreases by the packet weight and v's capacity b_v correspondingly increases by the packet weight (see Fig. 1 for an example). Node u can also reject to forward a packet, incurring a cost linear in the weight of the packet. The links we consider are rechargeable in the sense that the total capacity $b_u + b_v$ of the link can be arbitrarily distributed on both ends, but the total capacity of the link cannot be altered throughout the lifetime of the link. Given a packet sequence, our goal is to minimise the sum of the cost of rejecting packets and the amount of capacity allocated to a link.

The primary motivating example of our model is payment channel networks [7,8] supporting cryptocurrencies [1,12]. These networks are used to route payments of some amount (i.e. weighted packets in our model) in a multi-hop fashion between any two users of the network. In this way, users can directly transact with other users off-chain, and in so doing avoid the hefty transaction fees as well as long delays they would incur when transacting on the blockchain. Any two users in a payment channel network can create a channel (i.e. rechargeable link in our model) between themselves and deposit some funds only to be used in this channel (i.e. the initial capacity injected at each endpoint in our model). We note that users can always retrieve their funds in the payment channel at any time, but this would involve closing the channel and taking out the funds. For users that transact frequently and hence use payment channel networks, frequently closing channels and withdrawing their funds would defeat the purpose of them using payment channels as they would now need go back to transacting on the blockchain which is costly. Thus, the amount of funds injected into the payment channel can be seen as a "cost" for keeping the payment channel open to avoid using the blockchain. The total amount of funds deposited in the channel is its total capacity and remains invariant for the lifetime of the channel. Each payment moving across the channel simply updates the current balances (i.e. capacity at each end point of the link) of the two users in the channel, while maintaining that the total amount of funds in the channel remains the same.

Routing payments in payment channel networks comes with a profit: intermediate nodes on a payment route typically charge a fee for forwarding payments that is linear in the payment amount. Hence, if users reject to forward a payment, they would lose out on profiting from this fee and thereby incur the fee amount as opportunity cost. However, a depleted channel (i.e. a link with capacity 0 at one end) due to indiscriminate forwarding of payments can also impact transaction throughput. In particular, a depleted channel cannot forward any further payments unless the channel is closed and reopened with larger capacity, which also incurs corresponding cost. Hence the choice of how much capacity to inject into a channel and which transactions to forward and which to reject is crucial to maintain the lifetime of a payment channel [4,10]. Channels in payment channel networks are also rechargeable for security reasons, see [12] for more details.

Here we stress a crucial difference between our problem and problems on optimising flows and throughput in typical capacitated communication networks [5,13]. In traditional communication networks, the capacity is usually independent in the two directions of the link [9]. In our case, however, the amount of packets u sends to v in a link (u, v) directly affects v's capability to send packets, as each packet u send to v increases v's capacity on (u, v).

We start with a description of rechargeable links, then explain the actions nodes can take and corresponding costs. Finally, we state our main results.

Rechargeable Links. One unique aspect of our problem is that the links we consider are rechargeable. Rechargeable links are links that satisfy the following properties:

1. Given a link (u, v) with total capacity M, the capacity can be arbitrarily split between both ends based on the number and weight of packets processed by u and v. That is, b_u and b_v can be arbitrary as long as $b_u + b_v = M$ and $b_u, b_v \geq 0$. See Fig. 1 for an example of how b_u and b_v can vary in the course of processing packets.
2. The total capacity of a link is invariant throughout the lifetime of the link. That is, it is impossible for nodes to add to or remove any part of the capacity in the link. In particular, if a node is incident to more than one link in the network, the node cannot transfer part of their capacity in one link to "top up" the capacity in the other one.

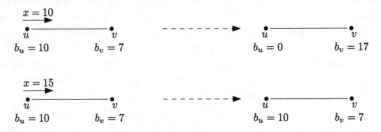

Fig. 1. The diagram on the top shows the outcome of u successfully processing a packet x of weight 10 along the link (u, v). The subsequent capacities of u and v are 0 and 17 respectively. The diagram on the bottom shows the outcome where, even though the total capacity of the (u, v) link is 17, u's capacity of 10 on (u, v) is insufficient to forward a packet x of weight 15. As such, the subsequent capacities of u and v on the link (u, v) remain the same.

Node Actions and Costs. First, we note that creating a link incurs an initial cost of the amount the node allocates in the link. That is, if node u allocates b_u in link (u, v), the cost of creating the link (u, v) for u would be b_u. Consider a link (u, v) in the network and a packet going from u to v along the edge. Node u can choose to do the following to the packet:

- **Accept packet.** Node u can accept to forward the packet if their capacity in (u, v) is at least the weight of the packet. The result of doing so decreases their capacity by the packet weight and increases the capacity of v by the packet weight. Note that apart from gradually depleting a node's capacity, accepting the packet does not incur any cost.
- **Reject packet.** Node u can also reject the packet. This could happen if u's capacity is insufficient, or if accepting the packet would incur a larger cost in the future. For a packet of weight x, the cost of rejecting the packet is $f \cdot x + m$ where $f, m \in \mathbb{R}^+$.

We note that node u does not need to take any action for packets going in the opposite direction (i.e. from v to u) as these packets do not affect u's cost. See Sect. 2 for more detail regarding packets.

Our Contributions. We introduce the natural weighted packet selection problem and show that it is NP-hard by a reduction from subset sum. Our main contribution is an efficient constant-factor approximation algorithm. We further initiate the discussion of how our approach can be generalised from a single link to a more complex network.

Organisation. Section 2 introduces the requisite notations and definitions we use in our paper, and also a formal statement of the weighted packet selection for a link problem. Section 3 provides the necessary algorithmic building blocks we use to construct our main algorithm. In Sect. 4, we present our main approximation algorithm and prove that it achieves an approximation ratio of $(1+\varepsilon)(1+\sqrt{3})$ for weighted packet selection in Theorem 1. We show that weighted packet selection for a link is NP-hard in Sect. 5. Finally, we discuss some possible generalisations of our algorithm from a single link to a larger network in Sect. 6. We conclude our work by discussing future directions in Sect. 7.

2   Notation and Definitions

Packet Sequence. Let (u, v) be a link. We denote an ordered sequence of packets by $X_t = (x_1, \ldots, x_t)$. Each packet $x_i \in X_t$ has a weight and a direction. We simply use $x_i \in \mathbb{R}^+$ to denote the weight of the packet x_i. We say a packet x_i goes in the left to right direction (resp. right to left) if it goes from u to v (resp. from v to u). Let $X_\rightarrow$ denote the subsequence of X_t that consists of packets going from left to right and $X_\leftarrow$ the subsequence of X_t that consists of packets going from right to left. For an integer $t \geq 1$, we use $[t]$ to denote $\{1, \ldots, t\}$.

Problem Definition. We now formally define weighted packet selection for a link . The input to our problem is a rechargeable link (u, v) and a sequence of packets X_t arriving on that link. We adopt the optimisation problem perspective over the *entire link*, instead of individual nodes. That is, we suppose nodes u and v collaborate and act as a coalition regardless of how they decide to initially split the capacity on both ends. The problem therefore is to compute the initial capacity and distribution (how it should be split on both ends) on the link as well as to decide on whether to accept or reject each packet in X_t such that the overall solution minimises the sum of the rejection cost as well as the cost of the capacity locked in the link.

Optimal Algorithm and Costs. Let $x_{\min}$ be the weight of the packet with the smallest weight in X_t and $M_{\max}$ be the total capacity in the link needed to accept all packets. $M_{\max}$ for X_t is easy to compute in time $\mathcal{O}(t)$ and is upper bounded by the sum of the weight of all packets in X_t. Similarly, given any sequence of *decisions*, we can compute the minimal cost of the capacity locked in the link and optimal initial distribution of capacity by greedy simulation. Let OPT be the cost of the optimal algorithm and OPT_M be the cost of the optimal algorithm using a capacity of M in the link. Additionally, we use OPT^R to denote the cost of the optimal algorithm for rejecting packets and OPT^C to denote the corresponding capacity cost (i.e., amount of capacity injected in the link). Similarly, we use OPT_M^R to denote the cost for rejecting packets of the optimal algorithm using a capacity of M in the link (note that $OPT_M^C = M$, $OPT = OPT^C + OPT^R$, and $OPT_M^R \leq OPT^R$).

3   Preliminary Insights and Algorithmic Building Blocks

We start our investigation of the weighted packet selection problem by describing a procedure to approximate the optimal capacity in a link using binary search and use this approximation to derive a lower bound on the cost of the optimal algorithm. We then describe a linear program that fractionally accepts packets (i.e. part of a packet can be accepted) given a fixed link capacity M and show that the solution of the linear program given M is a lower bound on the cost of the optimal algorithm given M. These results are used as building blocks for our main algorithm and theorem in Sect. 4. Nevertheless, we also present a simpler example of how to use the solution of the linear program that also comes with some guarantees in Appendix A which may be of independent interest.

3.1   Approximating the Optimal Capacity

We present a lemma that allows us to fix the capacity of the link to some value $M \in \mathbb{R}^+$ for a small trade-off in the approximation ratio. Recall that $x_{\min}$ is the weight of the smallest packet in X_t and $M_{\max}$ is the capacity needed to accept all packets. Observe that if the optimal capacity is not 0, it has to lie in the interval $[x_{\min}, M_{\max}]$. We thus fix some $\varepsilon > 0$ and perform a search for M over all $k \in \mathbb{N}$ such that $x_{\min}(1 + \varepsilon)^k \leq M_{\max}$. Let us denote by LB_M any lower bound on OPT_M^R, the optimal rejection cost using at most capacity M.

Lemma 1. *For any $\varepsilon > 0$, let $\mathcal{M} = \{x_{\min}(1+\varepsilon)^k | k \in \mathbb{N}$ and $x_{\min}(1+\varepsilon)^k \leq M_{\max}\} \cup \{0\}$. Then, the following inequality holds:*

$$\min_{M \in \mathcal{M}} \left(LB_M + \frac{M}{1+\varepsilon} \right) \leq OPT$$

Proof. We first analyse the case where the optimal algorithm rejects all packets. In this case, we know that since $\mathcal{M}$ contains 0, $LB_0 \leq OPT_0^R = OPT_0 = OPT$, so the inequality holds.

Now, suppose that the optimal algorithm accepts at least one packet. This means $OPT^C \geq x_{\min}$. So there exists a $k \in \mathbb{N}$ such that $x_{\min}(1+\varepsilon)^{k-1} \leq OPT^C \leq x_{\min}(1+\varepsilon)^k$. Set $M = x_{\min}(1+\varepsilon)^k$. We need to show that that $LB_M + \frac{M}{1+\varepsilon} \leq OPT = OPT^R + OPT^C$. From the way we choose M, we know that $\frac{M}{1+\varepsilon} \leq OPT^C$.

Now we just need to show $LB_M \leq OPT^R$. Observe that the optimal rejection cost for any link with larger capacity is always at most the rejection cost for any link with smaller capacity, as in the worst case the algorithm in the former setting accepts the same set of packets that the algorithm in the latter setting accepts. Thus, for any $M' \geq M$, $OPT_{M'}^R \leq OPT_M^R$. And since we chose M as an upper bound on OPT^C, it means $OPT^R = OPT_{OPT^C}^R \geq OPT_M^R \geq LB_M$. $\square$

Looking ahead, we describe an algorithm that is a $(1 + \sqrt{3})$-approximation of LB_M in Sect. 4. Thus, together with Lemma 1, we can use this algorithm to approximate weighted packet selection with a ratio of $(1+\varepsilon)(1+\sqrt{3})$ by running the algorithm at most $\frac{1}{\varepsilon} \log \frac{M_{\max}}{x_{\min}}$ times. We note that choosing a smaller value of ε yields a better approximation, but increases the running time.

3.2 Linear Program Formulation

Here, we describe a linear program that computes a lower bound for OPT_M^R. We first observe that due to the capacity constraints, the optimal algorithm with capacity M cannot accept packets with weight larger than M. Hence, for the rest of the analysis, we assume that all packets in X_t have weight less than M.

In the linear program, we allow accepting a fractional amount of a packet. That is, we create a variable $0 \leq y_i \leq x_i$ for every packet $x_i \in X_t$ that represents the extent to which the packet is accepted. For instance, $y_i = \frac{x_i}{2}$ means that half of x_i is accepted. We introduce variables $S_{L,i}$ and $S_{R,i}$ denoting the capacity on the left and right ends of the link after processing first i packets from X_t. We reiterate that due to the rechargeable property of the link, $S_{L,i} + S_{R,i} = M$, and $0 \leq S_{L,i}, S_{R,i} \leq M$.

We can now formulate the linear program in Eq. (1):

$$\text{minimise} \quad \sum_i f(x_i - y_i) + m\frac{x_i - y_i}{x_i} \tag{1}$$

$$\text{subject to} \quad \forall i : y_i, S_{L,i}, S_{R,i} \geq 0$$

$$\forall i : y_i \leq x_i$$

$$\forall i : S_{L,i} + S_{R,i} = M$$

$$\forall x_i \in X_\rightarrow : S_{L,i} = S_{L,i-1} - y_i$$

$$\forall x_i \in X_\rightarrow : S_{R,i} = S_{R,i-1} + y_i$$

$$\forall x_i \in X_\leftarrow : S_{L,i} = S_{L,i-1} + y_i$$

$$\forall x_i \in X_\leftarrow : S_{R,i} = S_{R,i-1} - y_i$$

Let LP_M be the solution of the linear program with capacity parameter M. The following lemma states that LP_M is a lower bound of the optimal cost of the weighted packet selection for a link problem with capacity M.

Lemma 2. $LP_M \leq OPT_M$ for all M.

Proof. OPT_M is an admissible solution to the linear program. If some other (fractional) solution is found, we know that it is at most OPT_M. ☐

The linear program can be solved in time $\mathcal{O}(n^\omega)$ where n is the number of variables in the linear program and ω the matrix multiplication exponent [6] (currently ω is around 2.37).

4   A Constant Approximation Algorithm

Based on the insights in the previous section, we now present a $(1 + \sqrt{3})$-approximation algorithm for the weighted packet selection for a link problem with fixed capacity M. We present the formal description of the algorithm in Algorithm 1 for packets $x_i \in X_\rightarrow$ and omit the procedure for $x_i \in X_\leftarrow$ as the decision process is symmetric.

In a nutshell, Algorithm 1 consists of three main ideas: first, it uses M capacity to follow the decisions made by the linear program solution as much as possible, using an additional $\sqrt{3}M$ as reserve capacity to fully accept some packets that were fractionally accepted by the linear program. Second, Algorithm 1 tries to maintain a balance in the distribution of capacity in both ends of the link. Intuitively, any packet that is accepted by the linear program can be accepted when the algorithm is in this balanced state.

Lastly, whenever the capacity is unbalanced: one side (wlog left side) has too little capacity, the algorithm prioritises accepting packets that come from right to left as well as rejecting packets that go from left to right. This brings the capacity at both sides to the balanced state, and our analysis shows that the approximation ratio is maintained below $1 + \sqrt{3}$.

Input and Initial Capacity Distribution. Algorithm 1 takes as input X_t and the solution of linear program given a fixed capacity M. Recall that $S_{L,i}$ and $S_{R,i}$ for $i \in [t]$ are the capacity distributions from the linear program solution on the left and right end of the link respectively after processing the ith packet. The algorithm uses the initial distribution $S_{L,0}$ and $S_{R,0}$, and additionally creates 2 "reserve capacity buckets" R_L and R_R of size $\frac{\sqrt{3}}{2}M$ each on both ends. Thus, the initial capacity of the left node would be $S_{L,0} + R_L$ and the initial capacity of the right node would be $S_{R,0} + R_R$. Intuitively, one can think of the additional capacity in R_L and R_R as a reserve source of capacity that is used to help Algorithm 1 *fully* accept packets that are *fractionally* accepted in the linear program solution.

Algorithm 1 accepts packets in the following way: for a packet of size x_i wlog in $X_\rightarrow$, assuming there is sufficient capacity in R_L, the packet is accepted using $(x_i - y_i)$ capacity from R_L and y_i capacity from $S_{L,i}$. The capacity of $S_{L,i}$ decreases by y_i and the capacity of $S_{R,i}$ increases by y_i, and the capacity in R_L decreases by $(x_i - y_i)$ while the capacity in R_R increases by the same amount. If the algorithm rejects x_i, the algorithm takes y_i from $R_{R,i}$ and adds it to $S_{R,i+1}$, and takes y_i from $S_{L,i}$ and adds it to $R_{L,i+1}$. We stress that in doing so, the algorithm always ensures that the updates to $S_{L,i}$ and $S_{R,i}$ at each step are exactly the same as the solution to the linear program. We also note that Algorithm 1 always maintains the invariant that $S_{L,i} + S_{R,i} = M$ and $R_L + R_R = \sqrt{3}M$ for all i.

We distinguish between three phases of Algorithm 1. We say the algorithm is in the *balanced phase* if both $R_L \geq \frac{\sqrt{3}-1}{2}$ and $R_R \geq \frac{\sqrt{3}-1}{2}$. If $R_L < \frac{\sqrt{3}-1}{2}$, we say the algorithm is in the *left phase*, and if $R_R < \frac{\sqrt{3}-1}{2}$, we say the algorithm is in the *right phase*. We also distinguish between 2 types of packets: little-accepted and almost-accepted packets. We say a packet is *little-accepted* if $\frac{y_i}{x_i} < \frac{\sqrt{3}}{1+\sqrt{3}}$, and *almost-accepted* if $\frac{y_i}{x_i} \geq \frac{\sqrt{3}}{1+\sqrt{3}}$.

Balanced Phase. In the balanced phase, Algorithm 1 accepts all packets that are almost-accepted in the linear program solution. It also accepts little-accepted packets that allow it to remain in the balanced phase. That is, for a little-accepted packet x_i wlog in $X_\rightarrow$, it first checks if the left reserve R_L is sufficient to forward the packet, and that doing so keeps the algorithm in the balanced phase (Line 4). If R_L does not have sufficient capacity, the algorithm rejects x_i (Line 8).

We first show in the following lemma that rejecting any little-accepted packet is safe in the sense that doing so will not push the approximation ratio of the algorithm above $1 + \sqrt{3}$.

Lemma 3. *All little-accepted packets can be rejected while keeping the approximation ratio below $1 + \sqrt{3}$.*

Algorithm 1. $(1 + \sqrt{3})$-approximation algorithm

Input: packet sequence X_t, capacity M, solution of LP_M:$S_{L,i}, S_{R,i}, y_i$.
Output: decisions to accept or reject
1: initialise $R_L = \frac{\sqrt{3}}{2}M$, $R_R = \frac{\sqrt{3}}{2}M$
2: **for** $i \in [t]$ **do**
3:     **if** $x_i \in X_\rightarrow$ **then**
4:         **if** $R_L - (x_i - y_i) \geq \frac{\sqrt{3}-1}{2}M$ **then**
5:             **Accept**
6:             $R_L = R_L - (x_i - y_i)$
7:             $R_R = R_R + (x_i - y_i)$
8:         **else if** x_i is little-accepted **then**
9:             **Reject**
10:             $R_L = R_L + y_i$
11:             $R_R = R_R - y_i$
12:         **else**
13:             $\phi_A, \phi_R, U, R'_L, j \leftarrow$ DIVIDER_L, LP_M, X_t, i
14:             $U_R \leftarrow \{\}$
15:             **if** $R'_L < 0$ **then**
16:                 $U_R, R'_L \leftarrow$ REJECTBIGX_t, U, R'_L
17:             **Accept** all $x_i \in \phi_A \cup (U \setminus U_R)$
18:             **Reject** all $x_i \in \phi_R \cup U_R$.
19:             $R_L = R'_L$
20:             $R_R = \sqrt{3}M - R_L$
21:             $i = j$

Proof. Recall that rejecting a packet x_i incurs a cost of $fx_i + m$. From Eq. (1), the cost of a little-accepted packet x_i for the linear program is $f \cdot (x_i - y_i) + m \frac{x_i - y_i}{x_i} \geq \frac{fx_i}{1+\sqrt{3}} + \frac{m}{1+\sqrt{3}} = \frac{1}{1+\sqrt{3}}(fx_i + m)$. From Lemma 2, we know that the solution of the linear program for a fixed capacity M is a lower bound on the optimal solution with capacity M, hence rejecting little-accepted packets will not increase the approximation ratio above $1 + \sqrt{3}$. □

In the next lemma (with proof in Appendix B.1), we show that processing little-accepted packets does not affect whether the algorithm stays in the balanced phase or not. This simplifies the decision making process of Algorithm 1 as it can just focus on the decision problem for almost-accepted packets.

Lemma 4. *Algorithm 1 never leaves the balanced phase after processing a little-accepted packet.*

Left Phase. Since Algorithm 1 accepts all almost-accepted packets in the balanced phase, it would sometimes have to enter the left or right phase. Here we describe the procedure for what happens in the left phase (the right phase is analogous).

Suppose Algorithm 1 enters the left phase after processing packet x_{i-1}. The objective of Algorithm 1 in this phase is to accept all almost-accepted packets among the unprocessed packets (i.e. packets $x_i, \ldots, x_t$). If this is not possible, the algorithm rejects some of them such that both of the following conditions hold: first, the approximation ratio remains $1 + \sqrt{3}$, and second, the algorithm returns to a balanced phase.

To do so, Algorithm 1 calls a subroutine DIVIDE (Line 13 in Algorithm 1) to sort all unprocessed packets into three sets: ϕ_A, ϕ_R, U. Set ϕ_A contains all packets from $X_{\leftarrow}$. These will be accepted as they will increase the left capacity reserve R_L and help to bring Algorithm 1 back into the balanced phase. Set ϕ_R contains little-accepted packets from $X_{\rightarrow}$. These will be rejected and from Lemma 3 we know that doing so does not increase the approximation ratio. Set U contains almost-accepted packets from $X_{\rightarrow}$. Some of these packets will be accepted and some rejected in a way that maintains the approximation ratio.

DIVIDE (described in Algorithm 2) takes as input the packet sequence X_t, the solution of the linear program as well as the current capacity in the left reserve R_L. DIVIDE creates the sets ϕ_A, ϕ_R, U incrementally by processing each unprocessed packet and accepting packets from $\phi_A \cup U$ and rejecting packets from ϕ_R until one of the following stopping conditions occurs:

1. $R_L < 0$ which would mean the left capacity reserves are depleted
2. $R_L > \frac{\sqrt{3}-1}{2}$
3. all packets are processed

If the first stopping condition is reached (Line 15 in Algorithm 1), the procedure REJECTBIG is called. REJECTBIG (described in Algorithm 3) takes as input the set U and outputs another set $U_R \subset U$. This set U_R is created by greedily selecting the biggest sized packets in U (Line 3 in Algorithm 3) and adding them to U_R. These packets will be rejected and the left capacity reserves will be accordingly updated after each rejected packet (Line 6 in Algorithm 3). The procedure REJECTBIG terminates when the left capacity reserves $R_L \geq \frac{\sqrt{3}-1}{2}$.

Now we show that if DIVIDE terminates on either the second and third stopping condition, Algorithm 1 will either be in the balanced phase (second stopping condition) or all packets will be processed and Algorithm 1 terminates (third stopping condition).

Lemma 5. *If* DIVIDE *returns* $R'_L \geq 0$ *and* j, *all almost-accepted packets between* i *and* j *are accepted by Algorithm 1 and either all packets are processed or* $R_{R,j} \geq \frac{\sqrt{3}-1}{2}$ *and* $R_{L,j} \geq \frac{\sqrt{3}-1}{2}$.

Proof. There are two reasons why DIVIDE returned $R'_L \geq 0$: either $R'_L \geq \frac{\sqrt{3}-1}{2}$ or $j = t$.

In both cases, we note that DIVIDE simulated accepting all packets from ϕ_A and U and rejecting all packets from ϕ_R, and at no time R'_L went below 0. That means that Algorithm 1 just repeats decisions of DIVIDE.

Finally, since $R_L + R_R = \sqrt{3}$ and all packets are smaller than M, this means that Algorithm 1 after emerging from left-phase cannot plunge to a right-phase right away. □

Algorithm 2. Function DIVIDE to create sets ϕ_A, ϕ_R, and U.

Input: packet sequence X_t, solution of LP : $S_{L,i}, S_{R,i}, y_i$, value R_L, capacity M,
Output: sets ϕ_A, ϕ_R, U, resulting R_L
1: $R_L = R_L - (x_i - y_i)$
2: $\phi_A, \phi_R, U \leftarrow \{\}, \{\}, \{x_i\}$
3: $j = i$
4: **while** $R_L \geq 0$ and $R_L < \frac{\sqrt{3}-1}{2}$ and $j < t$ **do**
5:    $j = j + 1$
6:    **if** $x_j \in X_\rightarrow$ and x_j is almost-accepted **then**
7:       $R_L = R_L - (x_j - y_j)$
8:       $U \leftarrow U \cup x_j$
9:    **else if** $x_j \in X_\rightarrow$ and x_j is little-accepted **then**
10:       $R_L = R_L + y_j$
11:       $\phi_R \leftarrow \phi_R \cup x_j$
12:    **else**
13:       $R_L = R_L + (x_j - y_j)$
14:       $\phi_A \leftarrow \phi_A \cup x_j$
15: **return** $\phi_A, \phi_R, U, R_L, j$

Algorithm 3. Function REJECTBIG to prune out packets from U.

Input: packet sequence X_t, set U, value R'_L
Output: set U_R, value R'_L
1: $U_R \leftarrow \{\}$
2: **while** $R'_L < \frac{\sqrt{3}-1}{2}$ **do**
3:    $x_k \leftarrow$ biggest packet from U
4:    $U \leftarrow U \setminus x_k$
5:    $U_R \leftarrow U_R \cup \{x_k\}$
6:    $R'_L = R'_L + x_k$
7: **return** U_R, R'_L

As the penultimate step in our analysis, we show in the next lemma (with proof in Appendix B.2) that REJECTBIG does not bring the approximation ratio of Algorithm 1 over $1 + \sqrt{3}$. We do this by computing the rejection cost incurred by Algorithm 1 on packets from U_R and showing that it is always lower than $(1 + \sqrt{3})$ times the cost of the linear program on U. As the solution of the linear program is a lower bound on the cost of the optimal algorithm, this shows that Algorithm 1 maintains the approximation ratio even when rejecting packets from U_R.

Lemma 6. *In Algorithm 1, for sets U and U_R the following inequality holds:*

$$(1 + \sqrt{3}) \sum_{x_i \in U} f \cdot (x_i - y_i) + m \frac{x_i - y_i}{x_i} \geq \sum_{x_i \in U_R} f x_i + m$$

We now have all the necessary ingredients to state and prove our main theorem, which is that weighted packet selection can be approximated with an approximation ratio of $(1 + \varepsilon)(1 + \sqrt{3})$.

Theorem 1. *The weighted packet selection for a link problem can be approximated with a ratio $(1 + \varepsilon)(1 + \sqrt{3})$ in time $\mathcal{O}(n^\omega \cdot \frac{1}{\varepsilon} \cdot \log \frac{M_{\max}}{x_{\min}})$, where ω is the exponent of n in matrix multiplication.*

Proof. We perform a search for the capacity of the link according to Lemma 1 and for every capacity M searched we solve the linear program (as stated in Eq. (1)) and run Algorithm 1. The solution is the output of Algorithm 1 with the smallest cost.

We know that $x_{\min}(1 + \varepsilon)^{\frac{1}{\varepsilon} \cdot \log \frac{M_{\max}}{x_{\min}}} \geq M_{\max}$. That means we need to solve the linear program and run Algorithm 1 at most $\frac{1}{\varepsilon} \cdot \log \frac{M_{\max}}{x_{\min}}$ times.

From Lemma 2 we know that the solution of the linear program with parameter M is a lower bound for OPT_M^R.

From Lemma 7, we know that Algorithm 1 accepts all fully-accepted packets. The algorithm can reject any little-accepted packets by Lemma 3. We also know from Lemma 4 that in the balanced phase Algorithm 1 accepts all almost-accepted packets and never leaves the phase after processing little-accepted packets. Finally, Lemma 6 shows that even in a left (or right) phase the approximation ratio of Algorithm 1 on almost-accepted packets is $1 + \sqrt{3}$. This means Algorithm 1 is $(1 + \sqrt{3})$-approximation algorithm for the solution of the linear program. Moreover, the algorithm uses $(1 + \sqrt{3})$ times more capacity that the linear program.

Using Lemma 1, we find that the selected solution is a $(1 + \varepsilon)(1 + \sqrt{3})$-approximation of the weighted packet selection for a link problem. $\qquad \square$

5   Hardness

In this section, we show that weighted packet selection for a link is generally NP-hard.

Theorem 2. *Weighted packet selection for a link is NP-hard.*

Proof. We show a reduction from the subset sum problem, which is known to be NP-hard [2]. In the subset sum problem, we are given a multiset of integers $\mathcal{I} := \{i_1, i_2, \ldots, i_n\}$ and a target integer S. The goal is to find a subset of $\mathcal{I}$ with a sum of S.

Consider the following question in the weighted packet selection for a link problem: "is the cost below a given value?" We show this question is NP-hard.

We set the constants to $m = 0$ and $f = \frac{3}{4}$. We create a packet sequence consisting of $i_1, i_2, \ldots, i_n$ where the jth packet in the sequence has weight i_j which is the jth element in $\mathcal{I}$. These packets all go from left to right. Then we add a packet of weight S going from right to left.

Suppose that there exists $\mathcal{I}' \subseteq \mathcal{I}$, such that $\sum_{j \in \mathcal{I}'} i_j = S$. Then we show that the cost is at most $\frac{1}{4}S + \frac{3}{4}\sum_{j \in \mathcal{I}} i_j$.

The solution reaching that cost is as follows: nodes start with capacity S on the right and accept all packets from $\mathcal{I}'$ and then accept the last packet of weight S. The cost is then $S + \frac{3}{4}\sum_{j \in \mathcal{I} \setminus \mathcal{I}'} i_j$. Since $\sum_{j \in \mathcal{I}'} i_j = S$, the bound holds.

Now, suppose that there is no subset of $\mathcal{I}$ summing to S. Let $\mathcal{A} \subseteq \mathcal{I}$ be any set with sum A. We follow the same procedure as described above by starting with A capacity on the right and accept all packets from $\mathcal{A}$. The cost for the packets going from left to right is $A + \frac{3}{4} \sum_{j \in \mathcal{I} \backslash \mathcal{A}} i_j = \frac{1}{4}A + \frac{3}{4}\sum_{j \in \mathcal{I}} i_j$. Depending on whether $A < S$ or $A > S$, the last packet of size S can be either accepted or rejected. Thus, we need to add $\min(\max(S - A, 0), \frac{3}{4}S)$ to the overall cost, which represents either the additional capacity cost of "topping up" the initial capacity of A on the right side by $S - A$ such that we can accept the last packet, or the additional cost of rejecting the last packet S, whichever is smaller. Since $A \neq S$, we know that

$$\frac{1}{4}A + \frac{3}{4}\sum_{j \in \mathcal{I}} i_j + \min(\max(S - A, 0), \frac{3}{4}S) > \frac{1}{4}S + \frac{3}{4}\sum_{j \in \mathcal{I}} i_j$$

Rearranging, we get $\min(\max(S - A, 0), \frac{3}{4}S) > \frac{1}{4}(S - A)$, which means that weighted packet selection for a link is NP-hard. $\qquad\square$

6   Extensions

We highlight two natural and interesting directions to generalise our approach from a link to a network.

6.1   Cyclic Redistribution of Capacity to Reduce Cost

Suppose node u on link (u, v) is incident to ≥ 2 links (let us call one of the incident links (u, w)). From our definition of rechargeable links (see Sect. 1), we know it is not possible for u to increase the capacity on the (u, v) link by transferring excess capacity from (u, w). However, if (u, v) and (u, w) are part of a larger cycle in the network, u can send excess capacity from link to link in a cyclic fashion starting from the (u, w) link and ending at (u, v) while maintaining the invariant that the total capacity on each link as well as the sum of all the capacities of a node on their incident links remains the same. This can be done at any point in time without the need to transfer packets. We call this cyclic redistribution (note that this is possible on payment channel networks [4, 10, 11] and is known as rebalancing) and illustrate it with an example in Fig. 2. In some situations, especially if the cost of closing and recreating a link is extremely large, the possibility of cheaply shifting capacities in cycles can reduce the overall cost to nodes in the network.

Let us denote the cost of decreasing capacities by x on the right and increasing it by x on the left using cyclic redistribution by $C(fx + m)$ for some $C \geq 1$ (one can view C as a function of the length of the cycle one sends the capacities along).

Here, we sketch an approximation algorithm that solves the weighted packet selection for a link problem with the possibility of cyclic redistribution. Note that our sketch is not precise, we simply modify Algorithm 1 where we assume the constants are already optimised for the basic problem.

We modify the linear program by adding variables $o_i, i \in [t]$ with constraints $0 \leq o_i \leq M$. The variable o_i denotes the capacity that was shifted from one side to the other before the algorithm processes packet x_i. We also modify the capacity constraints in the following way (for the case where $x_{i-1} \in X_{\leftarrow}$ and $x_i \in X_{\leftarrow}$): $S_{L,i} = S_{L,i-1} - o_i + y_{i-1}$ and $S_{R,i} = S_{R,i-1} + o_i - y_{i-1}$. We change the signs of variables for the other cases. Finally, we add $\sum_i C(fo_i + \frac{o_i}{M}m)$ to the objective in Eq. (1).

We divide the algorithm into epochs. We sum all o_i in the current epoch. If the sum is above $\frac{1}{1+\sqrt{3}}M$ we perform cyclic distribution if needed and start a new epoch. Note that in the current epoch, the optimal algorithm already paid at least $Cf\frac{M+m}{1+\sqrt{3}}$, so, we can move M capacity, incurring a cost at most $1 + \sqrt{3}$ times bigger than the optimal algorithm for cyclic redistribution.

To deal with capacity changes inside each epoch, we increase R_L and R_R. We initialise them in a way that they absorb changes of capacity in the first epoch of our algorithm. After an epoch, we reset them by cyclic redistribution such that they absorb changes of capacity in the next. The increase in R_L and R_R is at most $\frac{1}{1+\sqrt{3}}M$. These changes increase the approximation ratio of our algorithm from $1 + \sqrt{3}$ to $1 + \sqrt{3} + \frac{1}{1+\sqrt{3}}$, which is $\frac{1+3\sqrt{3}}{2}$.

Fig. 2. The graph on the left depicts three nodes u, v, w connected in a cycle. The red numbers by each link represent the capacity of a node in a certain link. u can increase their capacity by 10 on the (u, v) link by first sending the excess capacity of 10 to w along the (u, w) link. Then w sends the excess capacity of 10 to v along (w, v). Finally, v sends capacity of 10 back to u along (v, u). The graph on the right depicts the updated capacities of each node on each link after cyclic redistribution. (Color figure online)

6.2   Going from a Link to a General Network

Here we show how to extend the weighted packet selection problem from a single link to a general network. We begin by describing the problem for a general network.

Weighted Packet Selection for General Graph. The input to the problem is a general graph $G = (V, E)$ where each link in the graph is rechargeable, and an ordered sequence of packet requests $X_t = ((x_1, p_1), \ldots (x_t, p_t))$. $x_i \in \mathbb{R}^+$ denotes the weight of the ith packet and p_i represents the directed path through the graph that the ith packet needs to be routed though. The actions and costs per node are the same as described in Sect. 1 for the weighted packet selection for a link problem. As in the case of the problem confined to a single link, the goal in this setting is also to optimise over the entire graph. That is, the goal is for involved nodes to collaboratively decide on the initial capacity and distribution for all involved links as well as which packets to accept or to reject, so as to minimise the *total* rejection and capacity costs.

Solution Where there are Few Long Paths. We now present a simple solution in the case where the input only contains a few packets that have to be routed over multiple links. We first observe that if a packet has to be routed through a path with length > 1, the packet has to be be accepted or rejected by *all* the links in its routing path. Let us call a packet *long* if it needs to pass through more than one link.

Thus far, we showed an approximation algorithm that solves the problem if all packets only go through a single link. Suppose we are given the situation where we can bound the number of long packets, say by ℓ. Given a network and packet sequence for which we know only ℓ are long, we can approximate the extended problem with approximation ratio $(1+\varepsilon)(1+\sqrt{3})$ in time 2^ℓ times the time needed for the problem confined to a single link by simply trying to accept all subsets of paths of long packets. If the packet is accepted or rejected, we can reflect it in the linear program by requiring $y_i = x_i$. Then Algorithm 1 surely accepts this packet and the condition that the packet needs to be accepted by all links it passes through is satisfied.

Heuristic for General Graphs. We now describe a heuristic for the general case where the input sequence may contain many packets that have to be routed over multiple links. In particular, some links can be used in more than 1 routing path.

The idea of the algorithm for the general case is as follows:

– We create and solve a linear program similar to the one defined in Sect. 3.2. Compared to the linear program on a single link, this new linear program returns how much a packet should be accepted (the fractional solution, output of the linear program) on the whole path.
– Having the linear program solution for the whole graph, we look only at decisions inside a single (arbitrary) link ℓ and use Algorithm 1. This gives us decisions for packets going through ℓ.
– Respecting the decisions on ℓ, we solve the problem recursively (link ℓ is removed).

Our approach does not give any theoretical guarantees, only that the decisions on a link ℓ are a $(1+\sqrt{3})$- approximation while respecting the previous decisions. We believe this opens an exciting avenue for future work.

7   Conclusion

We initiated the study of weighted packet selection over a rechargable capacitated link, a natural algorithmic problem e.g., describing the routing of financial transactions in cryptocurrency networks. We showed that this problem is NP-hard and provided a constant factor approximation algorithm.

We understand our work is a first step, and believe that it opens several interesting avenues for future research. In particular, it remains to find a matching lower bound for the achievable approximation ratio, and to study the performance of our algorithm in practice. More generally, it would be interesting to study the online version of the weighted packet selection problem, and explore competitive algorithms. This version of the problem, when extended to a network, can be seen as a novel version of the classic online call admission problem [3].

Acknowledgments. We thank Mahsa Bastankhah and Mohammad Ali Maddah-Ali for fruitful discussions about different variants of the problem. This work is supported by the European Research Council (ERC) Consolidator Project 864228 (AdjustNet), 2020-2025, the ERC CoG 863818 (ForM-SMArt), and the German Research Foundation (DFG) grant 470029389 (FlexNets), 2021–2024.

A   Example Algorithm

We present an example of how to use the solution of the linear program in Sect. 3.2 for some fixed capacity M to create an algorithm that uses twice as much capacity as the linear program but guarantees that all packets that are fully accepted by the linear program (i.e. $x_i = y_i$) will also be fully accepted by the algorithm.

Algorithm 4 describes the decision making process only for packets coming from left to right on the link, i.e. $X_\rightarrow$. As the decision process for $X_\leftarrow$ is symmetric, we omit it to avoid repetition. The algorithm takes as input the solution to the linear program and the packet sequence X_t. Recall that $S_{L,i}$ and $S_{R,i}$ for $i \in [t]$ are the capacity distributions from the linear program solution on the left and right end of the link respectively after processing the ith packet. The algorithm uses the initial distribution $S_{L,0}$ and $S_{R,0}$, and additionally splits the extra M capacity into 2 "reserve buckets" R_L and R_R of size $\frac{M}{2}$ each on both ends. Thus, the initial capacity of the left node would be $S_{L,0} + R_L$ and the initial capacity of the right node would be $S_{R,0} + R_R$. Intuitively, one can think of the additional capacity in R_L and R_R as a reserve source of capacity that is used to help Algorithm 4 *fully* accept packets that are *fractionally* accepted in the linear program solution. We stress that Algorithm 4 always maintains the invariant that $S_{L,i} + S_{R,i} = M$ and $R_L + R_R = M$ for all i.

When processing each packet, say packet i which is wlog in $X_\rightarrow$, the algorithm first checks if there is sufficient excess capacity in R_L to accept the remaining fraction of packet i (Line 4 in Algorithm 4). If so, the packet is accepted using $(x_i - y_i)$ capacity from R_L and y_i capacity from $S_{L,i}$. The capacity of $S_{L,i}$ decreases by y_i and the capacity of $S_{R,i}$ increases by y_i, and the capacity in R_L decreases by $(x_i - y_i)$ while the capacity in R_R increases by the same amount. If there is insufficient capacity in $R_{L,i}$, i.e. $R_{L,i} < x_i - y_i$, the algorithm takes y_i from $R_{R,i}$ and adds it to $S_{R,i+1}$, and takes y_i from $S_{L,i}$ and adds it to $R_{L,i+1}$

Algorithm 4. Algorithm accepting all fully accepted packets

Input: packet sequence X_t, capacity M, solution of LP: $S_{L,i}, S_{R,i}, y_i$.

Output: decisions to accept or reject

1: initialise $R_L = \frac{M}{2}$, $R_R = \frac{M}{2}$
2: **for** $i \in [t]$ **do**
3:      **if** $x_i \in X_\rightarrow$ **then**
4:          **if** $R_L \geq x_i - y_i$ **then**
5:             **Accept**
6:             $R_L = R_L - (x_i - y_i)$
7:             $R_R = R_R + (x_i - y_i)$
8:             $S_{L,i} = S_{L,i} - y_i$
9:             $S_{R,i} = S_{R,i} + y_i$
10:          **else**
11:             **Reject**
12:             $R_L = R_L + y_i$
13:             $R_R = R_R - y_i$
14:             $S_{L,i} = S_{L,i} - y_i$
15:             $S_{R,i} = S_{R,i} + y_i$

(see Lines 12 to 15 in Algorithm 4). Note that the updates to $S_{L,i}$ and $S_{R,i}$ at each step are exactly as the solution to the linear program (Lines 8 and 9 and Lines 14 and 15).

Lemma 7. *Given the solution of the linear program, a sequence of packets X_t, and link capacity M, Algorithm 4 incurs a link capacity cost of $2M$ and accepts all packets fully accepted in the linear program.*

Proof. Wlog, let the ith packet in the sequence belong to $X_\rightarrow$. After processing the ith packet x_i, we denote R_L (resp. R_R) at that step as $R_{L,i}$ (resp. $R_{R,i}$). We show that the link, at time i, has capacity at least $S_{L,i}$ on the left and at least $S_{R,i}$ on the right.

When $R_{L,i}$ is large enough to accept packet x_i, we use y_i capacity from $S_{L,i}$ and $x_i - y_i$ capacity from $R_{L,i}$. The capacity y_i from the accepted packet goes to $S_{R,i+1}$ and the rest $(x_i - y_i)$ of the capacity goes to $R_{R,i+1}$.

If the packet is forced to be rejected, we know that $R_{L,i} < x_i - y_i$. Since $R_{R,i} = M - R_{L,i}$, we know that $R_{R,i} > M - x_i + y_i$, and because all packets have weight smaller than M, $R_{R,i} > y_i$ follows. This means we can take y_i from $R_{R,i}$ and add it to $S_{R,i+1}$ and remove y_i from $S_{L,i}$ (because the capacity disappeared from there) and add it to $R_{L,i+1}$.

If the packet is fully accepted, then $x_i - y_i = 0$. This means that the condition $R_{L,i} \geq x_i - y_i$ is satisfied and the algorithm accepts it. □

We conclude this example with two remarks.

Remark 1. Lemma 7 holds for any initial distribution of R_L and R_R so long as $R_L + R_R = M$.

Remark 2. Algorithm 4 is greedy and accepts all packets as long as $R_L \geq x_i - y_i$. This could be suboptimal as it might not have enough capacity in R_L to accept important packets later in the sequence. However, to maintain the condition $R_L, R_R \geq 0$ in line 4 of Algorithm 4, we can substitute the conditional check $R_L \geq x_i - y_i$ with $R_R < y_i$ at any point. Then, the proof of Lemma 7 still holds. We note that one could use this as a heuristic to develop a better approximation as it allows more fine-grained control over the greediness of the algorithm.

B   Omitted Proofs

B.1   Proof of Lemma 4

Proof. Each little-accepted packet moves at most $\frac{1}{1+\sqrt{3}}M$ from the left side to the right side of a link, and at most $\frac{\sqrt{3}}{1+\sqrt{3}}M$ from the right side to the left side of a link.

Because $R_L + R_R = \sqrt{3}M$ and any packet has a weight at most M. If $R_L - \frac{1}{1+\sqrt{3}}M < \frac{\sqrt{3}-1}{2}M$, then $R_R - \frac{\sqrt{3}}{1+\sqrt{3}}M \geq \frac{\sqrt{3}-1}{2}M$.

That means that rejecting a little-accepted packet from $X_\rightarrow$ does not create a situation where $R_R < \frac{\sqrt{3}-1}{2}M$. □

B.2   Proof of Lemma 6

Proof. If $R'_L \geq 0$, we know that all almost-accepted packets are accepted from Lemma 5. For $R'_L < 0$ we prove that $(1 + \sqrt{3}) \sum_{x_i \in U} (x_i - y_i) \geq \sum_{x_i \in U_R} x_i$, then we argue that the whole theorem holds.

Let $D = R_{L,i-1} - R'_L$ where R'_L is the value returned by DIVIDE in Algorithm 1 on Line 13. We know that $D \geq \frac{\sqrt{3}-1}{2}M$.

By following the changes of R'_L in DIVIDE, we get

$$\sum_{x_i \in U} x_i - y_i = D + \sum_{x_i \in \phi_R} y_i + \sum_{x_i \in \phi_A} x_i - y_i$$

By this we know that $\sum_{x_i \in U} x_i - y_i \geq D$.

Algorithm REJECTBIG removes packets from U until $\sum_{x_i \in U_R} x_i \geq D$. If the condition is satisfied, we know that REJECTBIG returns U_R, because $R'_L \geq \frac{\sqrt{3}-1}{2}$.

If $|U_R| = 1$, we know that $\sum_{x_i \in U_R} x_i \leq M$, because every $x_i \leq M$. So in that case $\sum_{x_i \in U_R} x_i \leq M \leq (1 + \sqrt{3}) \frac{\sqrt{3}-1}{2} M \leq (1 + \sqrt{3})D$.

If $|U \setminus U_R| > 1$, we know that rejecting just one packet is not enough. This means the biggest packet has weight at most D, so $\sum_{x_i \in U_R} x_i \leq 2D \leq (1+\sqrt{3})D$.

Now, we know that Algorithm 1 rejects less weight than the linear program times $(1 + \sqrt{3})$. It implies that $(1 + \sqrt{3}) \sum_{x_i \in U} f \cdot (x_i - y_i) \geq \sum_{x_i \in U_R} f x_i$ and leaves us to prove $(1 + \sqrt{3}) \sum_{x_i \in U} \frac{x_i - y_i}{x_i} \geq \sum_{x_i \in U_R} m$.

But we know that the packets are moved to U_R from the biggest. For every $x_k \in U_R$ and $x_l \in U$ holds $\frac{x_l - y_l}{x_l} \geq \frac{x_l - y_l}{x_k}$. That means rejecting smaller packets incurrs on average bigger cost than rejecting bigger packets, so $(1 + \sqrt{3}) \sum_{x_i \in U} \frac{x_i - y_i}{x_i} \geq \sum_{x_i \in U_R} m$. □

References

1. Raiden network (2017). https://raiden.network/
2. Arora, S., Barak, B.: Computational Complexity: A Modern Approach, 1st edn. Cambridge University Press, USA (2009)
3. Aspnes, J., Azar, Y., Fiat, A., Plotkin, S., Waarts, O.: On-line routing of virtual circuits with applications to load balancing and machine scheduling. J. ACM (JACM) **44**(3), 486–504 (1997)
4. Avarikioti, Z., Pietrzak, K., Salem, I., Schmid, S., Tiwari, S., Yeo, M.: Hide and seek: privacy-preserving rebalancing on payment channel networks. In: Proceedings of the Financial Cryptography and Data Security (FC) (2022)
5. Chekuri, C., Khanna, S., Shepherd, F.B.: The all-or-nothing multicommodity flow problem. In: Proceedings of the 36th Annual ACM Symposium on Theory of Computing (STOC), pp. 156–165 (2004)
6. Cohen, M.B., Lee, Y.T., Song, Z.: Solving linear programs in the current matrix multiplication time. J. ACM **68**(1), 3:1–3:39 (2021). https://doi.org/10.1145/3424305
7. Decker, C., Wattenhofer, R.: A fast and scalable payment network with bitcoin duplex micropayment channels. In: Pelc, A., Schwarzmann, A. (eds.) Stabilization, Safety, and Security of Distributed Systems. SSS 2015. LNCS, vol. 9212, pp. 3–18. Springer, Cham (2015). https://doi.org/10.1007/978-3-319-21741-3_1
8. Dotan, M., Pignolet, Y.A., Schmid, S., Tochner, S., Zohar., A.: Survey on blockchain networking: context, state-of-the-art, challenges. In: Proceedings of the ACM Computing Surveys (CSUR) (2021)
9. Gupta, P.K., Kumar, P.R.: The capacity of wireless networks. IEEE Trans. Inf. Theory **46**, 388–404 (2000)
10. Khalil, R., Gervais, A.: Revive: rebalancing off-blockchain payment networks. In: Thuraisingham, B.M., Evans, D., Malkin, T., Xu, D. (eds.) Proceedings of the 2017 ACM SIGSAC Conference on Computer and Communications Security, CCS 2017, Dallas, TX, USA, October 30–03 November 2017, pp. 439–453. ACM (2017). https://doi.org/10.1145/3133956.3134033
11. Pickhardt, R., Nowostawski, M.: Imbalance measure and proactive channel rebalancing algorithm for the lightning network. In: IEEE International Conference on Blockchain and Cryptocurrency, ICBC 2020, Toronto, ON, Canada, 2–6 May 2020, pp. 1–5. IEEE (2020). https://doi.org/10.1109/ICBC48266.2020.9169456
12. Poon, J., Dryja, T.: The bitcoin lightning network: scalable off-chain instant payments (2015). https://lightning.network/lightning-network-paper.pdf
13. Raghavan, P., Thompson, C.D.: Provably good routing in graphs: regular arrays. In: Proceedings of the Seventeenth Annual ACM Symposium on Theory of Computing, pp. 79–87 (1985)

Author Index

S. Rajsbaum et al. (Eds.): SIROCCO 2023, LNCS 13892, pp. 595–596, 2023.
https://doi.org/10.1007/978-3-031-32733-9

Printed in the United States
by Baker & Taylor Publisher Services